Handbook of Experimental Pharmacology

Volume 88

Drugs for the Treatment of Parkinson's Disease

Contributors

R. D. G. Blair, D. B. Calne, S. G. Diamond, S. Fahn, A. Fine, V. Glover, M. Goldstein, G. Gopinathan, O. Hornykiewicz, R. Horowski, J. Irwin, J. Jankovic, K. Jellinger, H. Kaufmann, H. L. Klawans, A. E. Lang, J. W. Langston, P. A. LeWitt, A. N. Lieberman, C. H. Markham, W. R. W. Martin, E. G. McGeer, P. L. McGeer, J. A. Obeso, J. D. Parkes, J. B. Penney, R. F. Peppard, H. A. Robertson, M. Sandler, W. Schultz, I. Shoulson, S. M. Stahl, I. Suchy, C. M. Tanner, H. Teräväinen, J. K. Tsui, M. D. Yahr, A. B. Young

Editor

Donald B. Calne

Springer-Verlag Berlin Heidelberg New York
London Paris Tokyo Hong Kong

DONALD B. CALNE, Professor D.M., F.R.C.P., F.R.C.P.C.

Head, Division of Neurology
UBC Health Sciences Centre Hospital
2211 Wesbrook Mall
Vancouver, British Columbia,
Canada V6T 1W5

With 62 Figures

ISBN-13:978-3-642-73901-9 e-ISBN-13:978-3-642-73899-9
DOI: 10.1007/978-3-642-73899-9

Library of Congress Cataloging-in-Publication Data
Drugs for the treatment of Parkinson's disease. (Handbook of experimental pharmacology; v.88) Includes bibliographies and index. 1. Parkinsonism—Chemotherapy. 2. Antiparkinsonian agents. I. Blair, R.D.G. (R.D. Gordon) II. Calne, Donald B. (Donald Brian) III. Series. [DNLM: 1. Antiparkinson Agents—therapeutic use. 2. Parkinson Disease—drug therapy. W1 HA51L v.88 / WL 359 D794] QP905.H3 vol.88 [RC382] 615'.1 s 88-29468 ISBN-13:978-3-642-73901-9 (U.S.) [616.8'33061]

Softcover reprint of the hardcover 1st edition 1989

2122/3020 – 543210 – Printed on acid-free paper

List of Contributors

R. D. G. Blair, The Movement Disorders Clinic, Toronto Western Hospital, 399 Bathurst Street, Toronto, Ontario, Canada M5T 2S8

D. B. Calne, Division of Neurology, U.B.C. Health Sciences Centre Hospital, 2211 Wesbrook Mall, Vancouver, British Columbia, Canada V6T 1W5

S. G. Diamond, University of California Los Angeles, Department of Neurology, Reed Neurological Research Center, UCLA School of Medicine, Center for the Health Sciences, Los Angeles, CA 90024, USA

S. Fahn, Department of Neurology, Columbia University College of Physicians and Surgeons and Neurological Institute of New York Presbyterian Hospital, 710 West 168th Street, New York, NY 10032, USA

A. Fine, Department of Physiology and Biophysics, Faculty of Medicine, Dalhousie University, Halifax, Nova Scotia, Canada B3H 4H7

V. Glover, Bernhard Baron Memorial Research Laboratories, Department of Chemical Pathology, Queen Charlotte's Maternity Hospital, Goldhawk Road, London W6 OXG, Great Britain

M. Goldstein, New York Infirmary, Beekman Downtown Medical Center, New York, NY, USA

G. Gopinathan, New York University Medical Center, 650 First Avenue, New York, NY 10016, USA

O. Hornykiewicz, Institute of Biochemical Pharmacology, Faculty of Medicine, University of Vienna, Borschkegasse 8a, A-1090 Vienna

R. Horowski, Clinical Research Division of Schering AG, Main Department, Postfach 650311, D-1000 Berlin 65

J. Irwin, Institute for Medical Research and the Santa Clara Valley Medical Center, Parkinson's Research and Clinical Programs, 2250 Clove Drive, San José, CA 95128, USA

J. Jankovic, Department of Neurology, Baylor College of Medicine, Texas Medical Center, Houston, TX 77030, USA

K. Jellinger, L.B.I. Klinische Neurobiologie, Wolkersbergenstr. 1, A-1130 Vienna

H. Kaufmann, Clinical Center for Research in Parkinson's Disease and Allied Disorders, Department of Neurology, Mount Sinai School of Medicine of the City University of New York, 14th Floor Annenberg Building, 1 Gustave Levy Place, New York, NY 10029, USA

H. L. Klawans, Department of Neurological Sciences, Rush-Presbyterian-St. Lukes Medical Center, 1725 West Harrison, Chicago, IL 60612, USA

A. E. Lang, The Movement Disorders Clinic, Division of Neurology, Toronto Western Hospital, 339 Bathurst Street, Toronto, Ontario, Canada M5T 2S8

J. W. Langston, The Institute for Medical Research and the Santa Clara Valley Medical Center, Parkinson's Research and Clinical Programs, 2260 Clove Drive, San José, CA 95128, USA

P. A. Le Witt, Department of Neurology, Lafayette Clinic, 951 East Lafayette Ave., Detroit, MI 48207, USA

A. N. Lieberman, New York University School of Medicine, 4th Floor, Neurology, 650 First Avenue, New York, NY 10016, USA

C. H. Markham, University of California Los Angeles, Department of Neurology, Reed Neurological Research Center, UCLA School of Medicine, Center for the Health Sciences, Los Angeles, CA 90024, USA

W. R. W. Martin, Division of Neurology, Department of Medicine, University of British Columbia, Health Sciences Centre Hospital, 2211 Wesbrook Mall, Vancouver, British Columbia, Canada V6T 1W5

E. G. McGeer, The University of British Columbia, Department of Psychiatry, Kinsmen Laboratory of Neurological Research, 2255 Wesbrook Mall, Vancouver, British Columbia, Canada V6T 1W5

P. L. McGeer, The University of British Columbia, Department of Psychiatry, Kinsmen Laboratory of Neurological Research, 2255 Wesbrook Mall, Vancouver, British Columbia, Canada V6T 1W5

J. A. Obeso, Movement Disorders Unit, Department of Neurology, University Clinic Medical School, University of Navarra, Apartado 192, E-31080 Pamplona

J. D. Parkes, The Bethlehem Royal Hospital and The Mandsley Hospital, Department of Neurology, Institute of Psychiatry, De Crespigny Park, Denmark Hill, London SE5, 8AF, Great Britain

J. B. Penney, Department of Neurology, The University of Michigan, Neuroscience Laboratory Building, 1103 E. Huron, Ann Arbor, MI 48105, USA

R. F. Peppard, Division of Neurology, U.B.C. Health Sciences Centre Hospital, 2211 Wesbrook Mall, Vancouver, British Columbia, Canada V6T 1W5

H. A. Robertson, Department of Pharmacology, Faculty of Medicine, Dalhousie University, Halifax, Nova Scotia, Canada B3H 4H7

M. Sandler, Royal Postgraduate Medical School, Institute of Obstetrics and Gynaecology, Department of Chemical Pathology, Queen Charlotte's and Chelsea Hospital, Goldhawk Road, London W6 OXG, Great Britain

W. Schultz, Institut de Physiologie, Université de Fribourg, Rue du Musée 5, CH-1700 Fribourg

I. Shoulson, Department of Neurology and Pharmacology, University of Rochester, Medical Center, 601 Elmwood Avenue, Box 673, Rochester, NY 14642, USA

S.M. Stahl, Neuroscience Research Centre, Merck Sharp & Dohme Research Laboratories, Terlings Park, Eastwick Road, Harlow, Essex CM20, Great Britain

I. Suchy, Clinical Research of Schering AG, Postfach 650311, D-1000 Berlin 65

C. M. Tanner, Department of Neurological Sciences, Rush-Presbyterian-St. Lukes Medical Center, 1725 W. Harrison Street, Chicago, IL 60612, USA

H. Teräväinen, Department of Neurology, University of Helsinki, Haartmaninkatu 4, 00290 Helsinki, Finland

J. K. Tsui, Department of Medicine, Division of Neurology, U.B.C. Health Sciences Centre Hospital, 2211 Wesbrook Mall, Vancouver, British Columbia, Canada V6T 1W5

M. D. Yahr, Clinical Center for Research in Parkinson's Disease and Allied Disorders, Department of Neurology, Mount Sinai School of Medicine of the City University of New York, 14th Floor, Annenberg Building, 1 Gustave Levy Place, New York, NY 10029, USA

A. B. Young, Department of Neurology, The University of Michigan, Neuroscience Laboratory Building, 1103 E. Huron, Ann Arbor, MI 48105, USA

Preface

Over the last 25 years, few topics in medicine, and none in neurology, surpass Parkinson's disease from the viewpoint of progress in understanding mechanisms and treating symptoms. Our entire concept of anatomy (the very existence of a nigrostriatal pathway) and physiology (dopaminergic transmission) has undergrone a revolution as the result of studies on Parkinson's disease leading to (a) the recognition of dopamine depletion as a crucial biochemical feature, and (b) the ability to alleviate symptoms by replenishing dopamine with levodopa. From this background has emerged a subclassification of dopamine receptors into D1 and D2 types, together with the development and therapeutic application of synthetic molecules that function as agonists at dopamine receptors. The pharmacological interrelationship between parkinsonism (inadequate dopamine) and chorea (excessive dopamine) has been elucidated because dopaminomimetic agents were found to alleviate parkinsonism and induce chorea, while dopamine blocking drugs induced parkinsonism and alleviated chorea. Pharmacokinetic manipulation of levodopa achieved by adding extracerebral decarboxylase inhibitors (carbidopa, benserazide) decreased certain side effects and resulted in efficacy being attained with lower dosage. Extracerebral dopamine receptor blockers have proved invaluable in decreasing the emesis of dopaminomimetics, because the dopaminoceptive chemoreceptor trigger zone is located outside the blood-brain barrier. Recently, novel routes of administration of antiparkinson drugs, such as subcutaneous infusion, have been explored in an attempt to achieve more evenly sustained blood concentrations of therapeutic agents.

There have been speculations concerning the etiology of loss of neurons in the substantia nigra in Parkinson's disease. These have led to new studies directed at delaying the inexorable progress of the pathology of Parkinson's disease, by long-term administration of antioxidants. Finally, after decades of maintaining a low profile, neurosurgeons have suddenly become active in the experimental therapy of Parkinson's disease; this development stemmed from recognition that some pathways, such as the nigrostriatal projection, may operate by the diffuse release of neurotransmitters, in a way reminiscent of endocrine function rather than classical neurophysiology. With this knowledge, neurosurgeons are pursuing the transplantation of dopaminergic cells into the striatum of parkinsonian patients.

The treatment of Parkinson's disease has undergone remarkable and sustained progress as the result of substantial steps forward every 4–5 years. Each step has solved some old problems, but has generated new ones. There are many unanswered questions that need to be formulated with greater precision in order to apply new technologies. MPTP has brought us a better animal model of the disease than was previously available. Current imaging methodology has allowed us to analyze the anatomy and biochemistry of the brain so that we now have access to information in vivo which hitherto only became available postmortem.

This book brings together authoritative accounts of the history, current status, and future directions of all these aspects of the pharmacology and therapeutics of Parkinson's disease.

Vancouver, January 1989 DONALD B. CALNE

Contents

CHAPTER 7

Pyridine Toxins

CHAPTER 8

The Relationship Between Parkinson's Disease and Other Movement Disorders

CHAPTER 1

Neurophysiology of Basal Ganglia

W. SCHULTZ

A. Introduction

Consider the case of a parkinsonian patient suffering from hypokinesia at an advanced stage. Knowing that the disease is predominantly due to neuronal deficits in the basal ganglia would suggest at first sight a role of this brain structure in motor control. However, an absence of movement can have many causes. When discussing movement-related neuronal processes it is helpful to distinguish between the initiation and the execution of movements (ALLEN and TSUKAHARA 1974). The slowing of movements in parkinsonism is compatible with deficits in movement execution, and in fact highly reproducible increases in movement time in controlled reaction time situations have been noted (EVARTS et al. 1981). Although deficits in movement execution emphasize the motor functions of the basal ganglia, they do not represent the full extent of parkinsonian akinesia. The fact that movements are simply not initiated, or occur with greatly reduced frequency and speed, argue for defects in neuronal processes that need to be operational before the motor act begins. These are less easy to define and may comprise mechanisms such as volition, incentive, motivation, memory, arousal, intention, attention, reactivity, sensory discrimination, response selection, planning, preparation, and preprogramming. Most of these can be grouped as motivational, cognitive, or motor set. With this reasoning, the deficits underlying parkinsonian hypokinesia relate to a broad spectrum of neuronal mechanisms involved in the control of behavioral acts that include but go far beyond the addressing and control of muscle activity.

The approximate role of the basal ganglia can be outlined by considering the phylogenetic development of their connections to other brain structures of known function. The mammalian basal ganglia receive the bulk of afferents from cerebral cortex, midline thalamic nuclei, the limbic system, and the dorsal raphe. Their efferents are primarily directed back to the frontal cortex while only a minority of them influences descending midbrain structures. Through their close, reciprocal contacts with brain structures dominating telencephalic brain functions, particularly the cerebral cortex, they are in a position to participate significantly in cognitive, emotional, and sensorimotor regulation. In nonmammalian vertebrates, such as amphibians, reptiles, and birds, homologs of striatum, pallidum, and substantia nigra (SN) can be defined from similarities in neurotransmitters, in particular from the presence of cholinesterase, catecholamines, and peptides (KARTEN and DUBBELDAM 1973; BRAUTH and KITT 1980; BRAUTH et al. 1983). Striatal homolog regions are recipients of afferents from homologs of the cerebral cortex (KARTEN and DUBBELDAM 1973; WEBSTER 1979; BRAUTH et al. 1978; BRAUTH and KITT 1980). However, in contrast to mammals, the basal ganglia in these species do not feed back information to the cerebral cortex, but instead influence tectal areas that control orienting and motor behavior. This occurs by way of pallidal output to the nucleus spiriformis lateralis in birds (KARTEN and DUBBELDAM 1973), the pretectum in amphibians (WILCZYNSKI and NORTHCUTT 1983) and the dorsal nucleus of the posterior commissure in reptiles (BRAUTH and KITT 1980). These tectal areas give rise to descending projections to reticulospinal neurons of brain stem motor centers (REINER and KARTEN 1982; REINER et al. 1982). Main catecholamine afferents to the striatum arise from the midbrain nucleus tegmenti pedunculopontinus that should be considered as the homolog of the mammalian SN (LOHMAN and VAN WOERDEN-VERKLEY 1978; WEBSTER 1979; BRAUTH et al. 1978; BRAUTH and KITT 1980; PARENT 1984). The phylogenetic development of basal ganglia output away from brain stem motor centers in nonmammalian vertebrates towards the cerebral cortex in mammals reflects the increased use of the cerebral cortex for the control of behavior. These comparisons suggest that the vertebrate basal ganglia participate as higher control centers in the association and integration of behavior-related neuronal information.

B. Electrophysiology of Connections

I. Peripheral and Cerebral Input to Basal Ganglia

The major afferent structure of the basal ganglia is the striatum (putamen and caudate nucleus) and its ventral extension, the nucleus accumbens. Additional input structures are the nucleus subthalamicus and the SN. These structures receive information from: (a) all areas of the cerebral cortex; (b) midline thalamic nuclei; (c) the limbic system at the level of the amygdala; and (d) the raphe.

1. Striatum

a. Peripheral Input

Neurons of the caudate nucleus in chloralose-anesthetized or locally anesthetized and immobilized cats respond with sequences of excitatory and inhibitory postsynaptic potentials (EPSPs and IPSPs) and impulse discharges to electrical stimulation of the contralateral and ipsilateral spinal and trigeminal nerves with latencies of 30–50 ms (SEDGWICK and WILLIAMS 1967; WILSON et al. 1983; LIDSKY et al. 1978). Most of these neurons also respond to auditory stimuli and light flashes. Natural somatosensory stimuli of the face, such as skin indentation and hair movement, are equally effective in driving neurons in the caudate and pallidum of conscious cats without encoding stimulus magnitude (SCHNEIDER and LIDSKY 1981; SCHNEIDER et al. 1982). Striatum neurons of urethane-anesthetized rats respond to noxious input in a moderate rostrocaudal topographic fashion (RICHARDS and TAYLOR 1982). Visual responses with large receptive fields and without orientation specificity and movement sensitivity are seen in the caudate of chloralose-anesthetized cats (POUDEROUX and FRETON 1979). Thus, neurons in the striatum receive sensory information from all three major modalities. Sensory responses in behaving animals will be further reviewed later, in conjunction with motor control and higher functions.

b. Cerebral Input

Electrical stimulation of several areas of the cerebral cortex in rats and cats elicits EPSPs and impulse discharges in the ipsilateral striatum at latencies of less than 10 ms (BUCHWALD et al. 1973; LILES 1974; KITAI et al. 1976; SCHULTZ and UNGERSTEDT 1978; HERRLING and HULL 1980; VAN der MAELEN and KITAI 1980). Neurons receive convergent excitatory inputs from the contralateral and ipsilateral cortex (WILSON 1986). A pronounced inhibition, often longer than 200 ms, follows the initial excitations. The cortically evoked excitation is due to an excitatory amino acid, probably glutamate (SPENCER 1976; DIVAC et al. 1977; KIM et al. 1977; MCGEER et al. 1977), which activates a quisqualate- or kainate-type receptor (HERRLING 1985). At least the early part of the following inhibition is mediated by γ-aminobutyric acid (GABA) released from striatal neurons (BERNARDI et al. 1976; HERRLING 1984).

The intralaminar thalamic nuclei centre médian and parafascicularis give rise to an ipsilateral monosynaptic excitatory input to the striatum. Electrical stimulation in these nuclei elicits EPSPs in the caudate of cats at latencies of 2–20 ms, followed by IPSPs of durations longer than 150 ms (PURPURA and MALLIANI 1967; BUCHWALD et al. 1973; KOCSIS et al. 1976). Thalamic neurons projecting monosynaptically to the caudate are excited by electrical stimulation of the forepaw (KUNZE et al. 1979). This suggests a second pathway by which somatosensory information could reach striatal neurons, in addition to the input from the somatosensory cortex to the putamen (KÜNZLE 1977).

Although a matter of earlier debate, the serotonergic input from the ipsilateral dorsal raphe nucleus is predominantly excitatory. Electrical stimulation of this nucleus elicits EPSPs in the striatum at latencies of 2–10 ms that are followed by long-lasting IPSPs (VAN der MAELEN et al. 1979). EPSPs are strongly reduced after depletion of cerebral serotonin levels by *p*-chlorophenylalanine, which suggests the serotonergic nature of the observed excitation (PARK et al. 1982b).

The nucleus accumbens, as the ventral part of the striatum, receives its major input from the limbic system. Recent studies report short latency orthodromic activation of accumbens neurons after stimulation of basal nuclei of the amygdala in cats and rats (ITO et al. 1974; YIM and MOGENSON 1982). The monosynaptic nature of the projection from the amygdala to the nucleus accumbens is suggested by antidromic responses of amygdala neurons at latencies of 6–30 ms after accumbens stimulation (ITO et al. 1974).

In summary, the striatum receives excitatory influences from all extrabasal ganglia afferents, while inhibitions are largely generated by intrinsic collaterals of GABAergic output neurons. An important convergence of excitatory inputs onto striatum neurons occurs in cats between cortical and thalamic sites (KOCSIS et al. 1976) and in rats between cortex and dorsal raphe (VAN der MAELEN et al. 1979).

2. Nucleus Subthalamicus

Neurons in the lateral part of this structure are activated by electrical or phasic innocuous mechanical stimulation of predominantly contralateral vibrissae in rats (HAMMOND et al. 1978b). Responses often show unilateral and bilateral convergence between trigeminal and spinal inputs. A major role of the sensorimotor cortex in mediating these responses is suggested by their suppression following cortical cooling or chronic ablation (HAMMOND et al. 1978b).

Direct electrical stimulation of several ipsilateral and contralateral areas of the cerebral cortex in rats elicits large and long-lasting EPSPs of 1-5 ms latency (KITAI and DENIAU 1981) and discharges at 8-14 ms latency in the majority of subthalamic neurons (ROUZAIRE-DUBOIS and SCARNATI 1985). Lowest effective stimulation currents for these responses are found in cortical layers V and VI (ROUZAIRE-DUBOIS and SCARNATI 1985). As with corticostriatal afferents, the transmitter of the corticosubthalamic excitation is probably glutamate acting on a quisqualate- or kainate-type receptor (ROUZAIRE-DUBOIS and SCARNATI 1987).

3. Substantia Nigra

Electrical stimulation of the median raphe nucleus leads to a reduction of neuronal activity in both parts of the SN at mean latencies of 4 ms. These responses could be mediated by serotonergic transmission since similar depressions of activity are obtained by iontophoretic application of serotonin (DRAY et al. 1976). Other extrabasal ganglia afferents to the SN include a probably glutamatergic cortical input (KÜNZLE 1978; KERKERIAN et al. 1983), as well as fibers arising from the dorsal raphe, the central amygdala and the hypothalamus (for review see GRAYBIEL and RAGSDALE 1979), which have not been the subjects of electrophysiological studies.

II. Internal Basal Ganglia Connections

The striatal dopamine (DA) innervation originates from cell bodies of the pars compacta of the SN (area A9) and dorsolateral neighboring neurons of the ventral reticular formation (area A8). The nucleus accumbens receives DAergic afferents from neurons of the medial ventral tegmental region (area A10). The striatum projects to both segments of the globus pallidus and both parts of the SN. The external segment of the globus pallidus projects of the nucleus subthalamicus and probably to the pars compacta of the SN. The Nucleus subthalamicus projects to both pallidal segments and to the pars reticulata of the SN, and has reciprocal connections with the nucleus tegmenti pedunculopontinus. The pedunculopontinus also has reciprocal connections with the pallidum, particularly its internal segment, receives afferents from the pars reticulata of the SN, and projects to the pars compacta of the SN.

1. Nigrostriatal Dopamine System

a. Substantia Nigra

DA-containing cells of the SN and adjoining areas of the ventral midbrain have electrophysiologic and pharmacologic properties that distinguish them from all other neurons in this area of the brain. They discharge polyphasic impulses of relatively long duration (1.5-5.5 ms) at rates of 0.1-8.0 Hz. This was first observed in extracellular recordings in rats by means of fluorescence microscopy for detection of their catecholamine content (Bunney et al. 1973b), and subsequently substantiated by intracellular labeling (Grace and Bunney 1983a). DA neurons have similar distinctive characteristics in other species, such as mice (Sanghera et al. 1984), cats (Steinfels et al. 1981), and monkeys (Aebischer and Schultz 1984). The relatively long impulse is composed of an initial short axon hillock spike and a subsequent longer somatodendritic component which is dependent on Ca^{2+} conductance (Grace and Bunney 1983b; Llinas et al. 1984; Kita et al. 1986).

DA neurons possess non-adenylate cyclase-coupled receptors for DA (for review see Kebabian and Calne 1979) which are responsible for membrane hyperpolarization and depression of impulse activity following systemic or iontophoretic application of DA agonists (Bunney et al. 1973a; Aghajanian and Bunney 1977; Grace and Bunney 1983a). Blockade of DA autoreceptors by systemically administered neuroleptics increases the impulse rate moderately (Bunney et al. 1973b). Although direct electrophysiologic evidence is lacking, autoreceptors on cell bodies can be stimulated and neuronal activity depressed by DA that is released from dendrites of DA neurons (Groves et al. 1975; Cheramy et al. 1981) independent of impulses (Llinas et al. 1984). Dendritically released DA may also reach receptors located on non-DA neurons of the pars reticulata of the SN. After iontophoretic DA application, neurons projecting to the thalamus increase their discharge rate (Ruffieux and Schultz 1980) through interaction with an adenylate cyclase-coupled DA receptor (Matthews and German 1986) while the depression of impulse activity following iontophoretic application of GABA is reduced (Waszszak and Walters 1983). In this way, dendritically released DA is able to exert a direct influence on neurons projecting outside the basal ganglia without involving the striatum.

b. Striatum

Iontophoretic application of DA in the striatum predominantly leads to a reduction in impulse activity (Bloom et al. 1965). In intracellular recordings, DA elicits slow depolarizations with concomitant reductions of impulse activity (Bernardi et al. 1978; Herrling and Hull 1980), or, probably more specifically, true membrane hyperpolarizations (Herrling and Hull 1980). Results obtained with electrical stimulation of the nigrostriatal tract are difficult to evaluate in view of concomitant antidromic activation of closely passing axons of striatonigral neurons, which possess widely branching collaterals in the striatum (for more detailed discussion see Schultz 1982).

In view of this uncertain action of DA on striatal neurons, a more functional investigation of DAergic influences on striatal input and output is necessary. By increasing striatal DA receptor stimulation with iontophoretically applied DA, systemically administered amphetamine, or electrical stimulation of the SN, a reduction of responses to stimulation of the cerebral cortex or peripheral nerves is noted while spontaneous activity remains largely unchanged (Hirata et al. 1984; Abercrombie and Jacobs 1985; Mercuri et al. 1985). Iontophoretic application of sulpiride, an antagonist of the non-adenylate cyclase-coupled DA receptor, increases responses to cortical stimulation without influencing spontaneous activity (Brown and Arbuthnott 1983). This is compatible with a reduction of the normal inhibitory effects of DA on glutamate release from corticostriatal terminals (Mitchell and Dogett 1980). These data suggest that DAergic neurotransmission reduces inputs to the striatum. Signs of a damping effect of DAergic neurotransmission are also found when investigating the responses of pallidum neurons after cortical stimulation while modifying striatal DAergic transmission. Responses increase with DA receptor blockade, and decrease after electrical stimulation of the SN (Hirata and Mogenson 1984; Toan and Schultz 1985). Combined evidence from these data suggests that DAergic transmission exerts a focusing effect on neuronal mechanisms in the striatum whereby weak striatal inputs are suppressed and only the strongest and most salient information is able to pass. In this way, DA would restrict and control the functional use of the anatomically established striatal connections which show considerable divergence and convergence (Chang et al. 1981; Percheron et al. 1984).

2. Striatopallidal

The major output of the striatum is directed ipsilaterally to both segments of the globus pallidus and the SN. These projections originate from medium spiny neurons which represent the large majority of striatal neurons (Preston et al. 1980). Neurons in the lateral and medial segments of the globus pallidus are regularly inhibited by electrical stimulation of the striatum or nucleus accumbens with conduction velocities of about 1 m/s in rats (Dray and Oakley 1978; Jones and Mogenson 1980; Pacitti et al. 1982; Park et al 1982a), cats (Yoshida et al. 1972), and monkeys (Ohye et al. 1976). Striatopallidal inhibitions are most probably due to GABAergic neurotransmission (Obata and Yoshida 1973). The participation of an opiate mechanism is suggested by the naloxone-induced reduction of pallidal responses to striatal stimulation (Napier et al. 1983). In contrast to these pure inhibitions, earlier studies also report striatally evoked EPSPs in a portion of pallidal neurons (Malliani and Purpura 1967; Levine et al. 1974). This may reflect excitations provided by a substance P-ergic striatopallidal connection (Ljungdahl et al. 1978), although it is unclear why recent studies have failed to detect them.

In agreement with an excitatory corticostriatal and a predominantly inhibitory striatopallidal connection, electrical stimulation of the cerebral cortex leads to depression of discharge rate in most pallidal neurons while about a quarter of them show a preceding activation (NODA et al. 1968; HIRATA and MOGENSON 1984; TOAN and SCHULTZ 1985). Considerable convergence occurs from even distant cortical sites onto single pallidum neurons in the rat (TOAN and SCHULTZ 1985), which may be mediated by the large dendritic trees of striatal neurons (PRESTON et al. 1980), by striatal interneurons or internal collaterals (MARCO et al. 1973; PARK et al. 1980), and by the striatopallidal connection (WILSON and PHELAN 1982). Similarly, pallidal neurons receiving afferents from the nucleus accumbens respond predominantly with reduction of activity to stimulation of the amygdala (YIM and MOGENSON 1983), thus providing evidence for a functional connection between the limbic system and a major basal ganglia output nucleus.

3. Striatonigral

Electrical stimulation of the striatum or nucleus accumbens regularly evokes monosynaptic, GABA-mediated IPSPs and depressions of discharge rate in neurons of the pars reticulata and pars compacta of rat and cat SN with conduction velocities of about 1 m/s (PRECHT and YOSHIDA 1971; YOSHIDA and PRECHT 1971; CROSSMAN et al. 1973; COLLINGRIDGE and DAVIES 1981; YOSHIDA et al. 1981; SCARNATI et al. 1983). Excitations preceding these inhibitions are occasionally seen in the SN after striatal stimulation in *encéphale isolé* cats (FRIGYESI and PURPURA 1967), conscious monkeys (FEGER and OHYE 1975), and anesthetized rats and cats (KANAZAWA and YOSHIDA 1980; COLLINGRIDGE and DAVIES 1981). The presence of the putative excitatory transmitter substance P in some striatonigral neurons may explain these excitations (DAVIES and DRAY 1976; WALKER et al. 1976).

4. Nucleus Subthalamicus

Besides the cortical input to this structure described already, the subthalamicus receives input from the external segment of the globus pallidus and nucleus tegmenti pedunculopontinus. Electrical stimulation of the external pallidal segment elicits monosynaptic IPSPs in subthalamic neurons in rats, with a conduction velocity of 3.8 m/s (KITA et al. 1983), and suppresses impulse activity in rats (ROUZAIRE-DUBOIS et al. 1980), cats (TSUBOKAWA and SUTIN 1972), and monkeys (OHYE et al. 1976). An involvement of GABAergic transmission in the inhibitory responses to pallidal stimulation is suggested by their blockade following iontophoretically applied picrotoxin and bicuculline (ROUZAIRE-DUBOIS et al. 1980). Monosynaptic input from the pedunculopontine nucleus is shown by orthodromic activation of subthalamic neurons after electrical stimulation of the pedunculopontinus, and by antidromic activation of pedunculopontine neurons from the subthalamicus in rats (HAMMOND et al. 1983a). Conduction velocity in this projection is 1.7 m/s.

Efferent projections of the nucleus subthalamicus are directed to both segments of the globus pallidus, pars reticulata of the SN, and nucleus tegmenti pedunculopontinus. These neurons have conduction velocities ranging from less than 1 to 5–10 m/s, and the majority branch to both the pallidum and SN (Deniau et al. 1978b; Hammond et al. 1983b). Electrical stimulation of the subthalamicus leads to reductions of impulse activity in neurons of the internal pallidal segment (entopeduncular nucleus) of rats and cats (Larsen and Sutin 1978; Perkins and Stone 1980; Hammond et al. 1983b) which are blocked by the GABA antagonists picrotoxin and bicuculline (Rouzaire-Dubois et al. 1983). In contrast to the depressant responses in the pallidum, SN neurons are clearly activated by subthalamic stimulation (Hammond et al. 1978a, 1983b). This is difficult to understand since the subthalamic nucleus sends divergent axons to the pallidum and SN. Neurons of the nucleus tegmenti pedunculopontinus are depressed by subthalamic stimulation (Hammond et al. 1983a).

5. Nucleus Tegmenti Pedunculopontinus

Besides its reciprocal links with the subthalamic nucleus reviewed already, this nucleus has reciprocal connections with the pallidum. The majority of neurons of the internal pallidal segment project to the pedunculopontinus (50% in cats: Filion and Harnois 1978; 87% in monkeys: Harnois and Filion 1982). These fibers have conduction velocities of about 6 m/s in monkey (Harnois and Filion 1982). Some pallidopedunculopontine fibers also originate in the external pallidal segment (Gonya-Magee and Anderson 1983). Electrical stimulation of the pedunculopontinus elicits monosynaptic IPSPs in both segments of the pallidum at latencies of 2.0–5.0 ms (Gonya-Magee and Anderson 1983).

The pedunculopontinus receives inhibitory afferents from the pars reticulata of the SN with mean conduction velocities of 1.7 m/s (Noda and Oka 1984; Scarnati et al. 1984). Neurons of both parts of the SN, predominantly the pars compacta, are monosynaptically excited from the pedunculopontinus, with mean latencies of 3–6 ms. This projection probably involves glutamatergic transmission acting on a quisqualate-type amino acid receptor (Scarnati et al. 1984, 1986).

III. Output of Basal Ganglia

The output of the basal ganglia is directed to: (a) the frontal cortex via pallidothalamic and nigrothalamic fibers; (b) the limbic system via pallidohabenular axons, and (c) subcortical motor centers via nigrocollicular fibers and projections from the pedunculopontinus to the reticular formation and spinal cord.

1. Pallidothalamic

The large majority of neurons of the internal pallidal segment of cat and monkey project to the ventroanterior (VA) and ventrolateral (VL) thalamic nuclei, and send collaterals to the pedunculopontinus (Harnois and Filion 1982). Pallidothalamic fibers have conduction velocities of 5–11 m/s and inhibit VA and VL neurons, as evidenced by monosynaptic IPSPs (Uno et al. 1978).

2. Nigrothalamic

Many neurons of the pars reticulata of the SN in rat and cat project to the ipsilateral ventromedial thalamus (VM) with conduction velocities of 0.8–10 m/s (Deniau et al. 1978a; Anderson and Yoshida 1980). Some of these neurons in the rat have branches to the contralateral VM. Electrical stimulation of the SN depresses the impulse rate (Deniau et al. 1978d) and elicits monosynaptic IPSPs in cat and rat VM neurons (Ueki et al. 1977; Chevalier and Deniau 1982). The involvement of GABA as inhibitory transmitter is suggested by the attenuating effects of systemically or iontophoretically applied GABA antagonists on SN-evoked inhibitions in VM neurons (Yoshida and Omata 1979; MacLeod et al. 1980).

Local injection of GABA into the SN leads to activation of VM neurons (MacLeod et al. 1980; Deniau and Chevalier 1985) which is probably caused by direct inhibition of nigrothalamic neurons and subsequent release of VM neurons from the tonic nigrothalamic inhibitory influence. When striatal neurons are excited by locally injected glutamate, pars reticulata neurons of the SN are inhibited via the striatonigral GABAergic projection. Subsequently, VM neurons are activated (Deniau and Chevalier 1985). These data are in line with the observation that striatonigral neurons are able to inhibit nigrothalamic cells (Deniau et al. 1976) and indicate that activations in striatal neurons may ultimately result in an increased basal ganglia drive on VM thalamocortical neurons via a double inhibitory transmission in SN and VM.

Considerable species differences exist in the convergence on thalamic neurons. In the rat, the majority of thalamic neurons receiving SN input are also influenced by the cerebellum. SN-evoked IPSPs are effective in preventing cerebellum-evoked impulse discharges (Chevalier and Deniau 1982). This convergence has not been observed electrophysiologically in the cat (Ueki et al. 1977), and its absence in the monkey is suggested by anatomical studies (Schell and Strick 1984; Ilinsky et al. 1985).

3. Pallidohabenular

The majority of neurons of the entopeduncular nucleus in the rat and a sizeable portion of neurons of the internal pallidal segment in cat and monkey project to the lateral habenula (Nauta 1974; Kim et al. 1976; Herkenham and Nauta 1977). Many entopeduncular neurons respond antidromically to electrical stimulation of the lateral habenula (52% in rat: Pacitti et al. 1982; 25% in cat: Filion and Harnois 1978). Orthodromic activating or depressing responses in lateral habenula neurons are evoked by electrical stimulation of the entopeduncular nucleus in rats at latencies of 2–29 ms (Garland and Mogenson 1983).

4. Nigrocollicular

Pars reticulata neurons of the SN exert an inhibitory influence on the superior colliculus. Thus, electrical stimulation in the SN results in reductions of impulse rate in tectal neurons (Deniau et al. 1978a) which are strongly diminished by the iontophoretically administered GABA antagonist bicuculline (CHEVALIER et al. 1981). Conduction velocities in nigrotectal neurons are 1.9–7.0 m/s in rats (DENIAU et al. 1978a) and up to 10 m/s in cats (ANDERSON and YOSHIDA 1980). Some nigrocollicular neurons are branches of nigrothalamic cells (DENIAU et al. 1978a). By analogy to the nigrothalamic projection, intranigral application of GABA, resulting in inhibition of reticulata cells, leads to an activation of tectal neurons. Excitation of striatal neurons by locally applied glutamate results in an activation of tectospinal neurons, again mediated by two consecutive steps of inhibitory transmission in the SN and superior colliculus (CHEVALIER et al. 1985).

5. Nucleus Tegmenti Pedunculopontinus

With electrophysiologic data presently lacking, anatomic studies reveal pedunculopontine projections to the pontine and medullary reticular formation, and to the spinal cord (JACKSON and CROSSMAN 1983; MOON-EDLEY and GRAYBIEL 1983).

C. Motor Control Functions

I. Anatomic Considerations

Neuronal activity related to motor commands and sensory feedback from ongoing motor acts reaches the basal ganglia primarily from the motor and somatosensory cortex. In monkeys, the motor cortex projects somatotopically to lateral parts of the ipsilateral and contralateral putamen (KÜNZLE 1975) and to the lateral part of the ipsilateral subthalamic nucleus (HARTMANN-VON MONAKOW et al. 1978). The putamen also receives somatotopically organized input from the ipsilateral somatosensory cortex (KÜNZLE 1977). In view of the patch-like distribution of terminals in the putamen, it is presently unclear whether inputs from motor and somatosensory cortex converge on single putamen neurons. Thus, the origin of cortical inputs attributes a major role in motor control to lateral parts of the putamen and subthalamicus.

The internal connections of the primate basal ganglia maintain a tendency for segregation of motor functions (ALEXANDER et al. 1986). The projections of the putamen are predominantly directed to the ventral two-thirds of both pallidal segments and to the lateral SN, while the caudate projects with little overlap to other parts of the pallidum, and predominantly to the SN (SMITH and PARENT 1986).

The majority of basal ganglia output of monkeys is directed back to the cerebral cortex Briefly, the pallidum projects via the anterior part of the VL thalamus to the supplementary motor area (SCHELL and STRICK 1984), whereas the pars reticulata influences the prefrontal cortex via VA and anterior parts of VM thalamic nuclei (ILINSKY et al. 1985). Thus, movement-related neuronal information from the putamen reaches the supplementary motor area, and is ultimately directed to the motor cortex through corticocortical connections (Jürgens 1984). Although the mammalian basal ganglia are traditionally considered parts of the extrapyramidal motor system, it should be emphasized that the majority of their output does not bypass the pyramidal system, but rather feeds back to the cortical level. A limited noncortical motor output is provided by the projection of the pars reticulata of the SN to tectospinal neurons (Chevalier et al. 1985) and by the pedunculopontine output to the reticular formation and spinal cord (Jackson and Crossman 1983).

II. Lesions, Cooling, and Local Drug Injections

1. Striatum

Lesions of the striatum in monkeys result in strikingly little gross motor disturbances, other than a decreased use and clumsiness of contralateral limbs (for detailed review see DeLong and Georgopoulos 1981). More refined analysis of motor deficits is done during performance of controlled behavioral tasks. Reversible cooling of the putamen and external pallidum in monkeys leads to flexion drift of the forearm and to slowing of elbow flexion and extension movements, probably caused by muscular cocontractions (Hore and Vilis 1980). For these movements in a reaction time task, onset times of EMG activity following a trigger stimulus remain unchanged during cooling of the putamen—external pallidum (Hore and Vilis 1980). Reaction times of monkeys performing an arm-reaching task are increased by 5%–14% during cooling of the contralateral hand area in the putamen, while the time required for execution of the movement is normal (Beaubaton et al. 1980). Reaction times in cats lifting the forepaw in response to a sound stimulus are increased by 10% during cooling of the contralateral caudate nucleus (Conde et al. 1979). Thus, reversible inactivation of striatal motor regions induced by cooling leads to reduced behavioral reactivity, except in the simplest task (elbow flexion and extension), while the speed of movements is inconsistently affected.

2. Globus Pallidus

Unilateral or bilateral lesions of either or both segments of the globus pallidus in monkeys lead to few gross neurologic deficits apart from a transient hypokinesia (see cf DeLong and Georgopoulos 1981). Specific impairments are seen in monkeys during performance of controlled motor tasks. Animals show forearm flexor deviation after kainic acid-induced lesions of the internal pallidal segment (DeLong and Coyle 1979) and during cooling of the pallidum—putamen complex (Hore and Vilis 1980). A severe breakdown of active flexion—extension movements is observed during cooling of the pallidum when sight of the moving arm is prevented (Hore et al. 1977). This suggests that treatment of proprioceptive information is impaired during pallidal dysfunction, which can be largely compensated by visual feedback. In an arm-reaching task, monkeys with lesions or cooling of the internal pallidal segment show a persistent 9%–18% decrease of reaction time when moving to a randomly changed target (AMATO et al. 1978). Normal reaction times, but considerable slowing of movements (about 40%-50% increased movement time), are seen during the first week after kainic acid-induced lesions of both pallidal segments in monkeys reaching for a fixed target (HORAK and ANDERSON 1984). Changes in reaction time are equally absent in elbow flexion—extension movements during cooling of the pallidum—putamen complex (HORE and VILIS 1980).

3. Substantia Nigra Pars Reticulata

In view of the close proximity between DA and non-DA neurons of the SN, extensive selective electrolytic lesions or cooling are difficult to perform. Carefully done local injections of GABA agonists into the pars reticulata of rats lead to contralateral head and body turning and stereotyped oral motor behavior that is independent of the existence of DA neurons (for review see PYCOCK 1980; SCHEEL-KRÜGER et al. 1980). Contralateral head turning is also seen after local injections of the GABA agonist muscimol in cats (BOUSSAOUD and JOSEPH 1985). Local muscimol application in trained cats and monkeys elicits irrepressible saccades of the eyes to the contralateral side when visual fixation is required, and leads to inappropriate performance of triggered saccades to contralateral targets (BOUSSAOUD and JOSEPH 1985; HIKOSAKA and WURTZ 1985b). The most likely explanation considers a direct inhibition of nigrocollicular neurons and a subsequent disinhibition of neurons in the superior colliculus. In line with this reasoning, local injection of the GABA antagonist bicuculline into intermediate layers of the superior colliculus of monkeys results in eye movements similar to those produced by muscimol injection into the pars reticulata of the SN (HIKOSAKA and WURTZ 1985a). This suggests that the output of the striatum operates via two inhibitory synapses (in the SN and superior colliculus), and leads to an activation of the target structure superior colliculus through disinhibition. The different behavioral results obtained in the various species may reflect a phylogenetic development in orienting behavior away from simple head turning in rats and toward increased eye motility in primates (HIKOSAKA and WURTZ 1985b).

4. Nucleus Subthalamicus

Besides lesions of the nigrostriatal DA system producing parkinsonism, destruction of the subthalamicus leads to the most pronounced movement disturbances of all nuclei in the basal ganglia. In agreement with the human neurologic disorder (BONHOEFFER 1897), electrolytic or kainic acid-induced destruction of this nucleus results in hemiballism, consisting of contralateral irrepressible involuntary movements of the proximal limb musculature (WHITTIER and METTLER 1949; HAMMOND et al. 1979). The importance of GABAergic afferents to the subthalamicus for control of movements is demonstrated by the fact that injections of GABA antagonists into this nucleus also lead to severe hemiballism in the monkey (CROSSMAN et al. 1980).

III. Neuronal Recordings

1. Striatum

During neurophysiological experiments in conscious, behaving monkeys, neuronal activity related to limb movements is found in a low percentage throughout the head of the caudate nucleus (BUSER et al. 1974; ROLLS et al. 1983) and with a high incidence in areas of the putamen receiving input from the primary motor and somatosensory cortex (DELONG 1973; LILES 1983). Neurons showing activity related to target-directed, but not spontaneous saccades of the eyes are located in the head of the caudate (HIKOSAKA and SAKAMOTO 1986).

Separate regions in monkey putamen are activated during movements of the face, arm, and leg, and overlap with areas receiving somatosensory input from corresponding body parts, thus indicating a somatotopic organization (CRUTCHER and DELONG 1984a; LILES 1985). Although neurons in the forelimb area of the putamen are preferentially related to proximal movements (CRUTCHER and DELONG 1984a) and show changes in relation to maintenance of posture (ANDERSON 1977), clear relations to distal movements of fingers and wrist are equally observed in these regions (LILES and UPDYCKE 1985). Somatosensory input to the putamen predominantly arises from afferents of joints and very little from cutaneous receptors. Between one-third and nearly one-half of movement-related neurons are driven by passive joint rotation (CRUTCHER and DELONG 1984a; LILES 1985). Neurons with the same topographic motor or somatosensory relationship tend to occur in multiple clusters of 100–500 μm diameter, which could represent a functional correlate to the patchy arrangement of corticostriatal terminals (KÜNZLE 1975; 1978). A similar somatotopic and clustered organization is found when muscle contractions are elicited by electrical microstimulation in the putamen (ALEXANDER and DELONG 1985b). Ibotenic acid-induced destructions of neurons in the putamen, while sparing axons, abolish the microstimulation effects, which suggests that output fibers and not axon collaterals of corticospinal neurons mediate the observed effects (ALEXANDER and DELONG 1985a).

Neuronal impulse frequency of movement-related neurons in monkey putamen encodes the level of active force (LILES 1985). Another parameter encoded in putamen neurons is the direction of movement. Muscle activity is different when movements in the same direction are performed together with or against an imposed load. Using this paradigm in an elbow flexion—extension task, CRUTCHER and DELONG (1984b) found about half of task-related putamen neurons to have preferential relations to the direction of movement, independent of the underlying pattern of muscle activity, while only 13% of putamen neurons showed EMG-related changes.

Comparisons of onset times between neuronal changes and EMG activity in quick behavioral reactions show that the majority of putamen neurons are activated after EMG onset (81%: CRUTCHER and DELONG 1983b; 91%: LILES 1985). In contrast, a high percentage of pyramidal tract neurons in the motor cortex of monkeys are activated before the earliest EMG activity (EVARTS

1966). Although input from the motor cortex probably causes the observed movement-related changes in putamen neurons, it remains to be shown that the earliest active motor cortex neurons actually project to this structure. In any case, neurons in the putamen are not actived before those of the motor cortex during the initiation of quick movements in reaction time situations.

2. Globus Pallidus

The earliest neurophysiological study on movement-related activity in the basal ganglia was performed in the globus pallidus of monkeys. About 20% of pallidal neurons increase or decrease activity in relation to rapidly alternating push—pull or flexion—extension movements, mostly of the contralateral arm or leg (DELONG 1971). With a broader spectrum of tests, between 10 % and 49 % of neurons in the external and internal segments of primate globus pallidus are related to movements of the arm, leg, or face of the contralateral, ipsilateral, or both sides (IANSEK and PORTER 1980; DELONG et al. 1985; NISHINO et al. 1985). Changes in activity are equally seen with postural adjustments to tilt of the monkey's chair (ANDERSON 1977). Relationships to distal movements of the arm are not uncommon and comprise up to 30% of neurons in both segments (IANSEK and PORTER 1980; DELONG et al. 1985). A somatotopy is suggested by the fact that neurons related to movements of different parts of the body are found in fairly separated areas of pallidum, while neurons with similar functional properties tend to occur in small clusters (IANSEK and PORTER 1980; DELONG et al. 1985). Many pallidal neurons are activated or depressed by specific movements at only one joint (IANSEK and PORTER 1980), and show differential changes with extension or flexion movements (DELONG 1971; ALDRIDGE et al. 1980a; GEORGOPOULOS et al. 1983; DELONG et al. 1985; NISHINO et al. 1985). Discharge rates show a linear relationship to amplitude or peak velocity of movement (GEORGOPOULOS et al. 1983). About one-third of movement-related pallidal neurons respond to passive rotation of joints, whereas input from muscles or skin is largely ineffective (DELONG et al. 1985).

As with the putamen, few neurons in the pallidum change activity in advance of the earliest muscle activity. In a flexion—extension arm-tracking task, 24% and 11% of neurons in the external and internal segment of pallidum, respectively, are activated before EMG onset (GEORGOPOULOS et al. 1983). In a reaction time task employing an arm-reaching movement, the earliest neuronal changes only began at the time of EMG onset (ANDERSON and HORAK 1985). In support of these data, electrical stimulation in the pallidum impairs execution of arm-reaching movements most severely when applied at the beginning of EMG activity, but not before (ANDERSON and HORAK 1985). These data show that pallidum neurons are closely related to the execution of movements by encoding important movement parameters, while not providing evidence for an active participation in initiating quick movement reactions.

3. Substantia Nigra Pars Reticulata

Neurons in this structure are activated or depressed during arm-reaching movements and mouth movements, and depressed in relation to saccades of the eyes. Nearly half of reticulata neurons in monkeys show changes with contralateral and often ipsilateral reaching movements of the arm toward a target (Schultz 1986a), although very few of them are related to distal flexion–extension or hand prehension movements (DeLong et al. 1983; Schultz 1986a). None of these changes resemble in their time course the activity of muscles. Reticulata neurons in cats also show rhythmic activity changes during locomotion on a treadmill (Schwarz et al. 1984). Whereas in cats the majority of reticulata neurons are influenced by mechanical stimulation of the skin (Schwarz et al. 1984), somatosensory input is not consistently found in conscious monkeys (DeLong et al. 1983; Schultz 1986a).

Neuronal activity related to mouth protrusion, licking, and chewing occurs in a substantial number of reticulata neurons of the monkey (Mora et al. 1977; DeLong et al. 1983; Schultz 1986a) and cat (Joseph et al. 1985). Some of the apparent mouth movement relations are caused by somatosensory input from the tongue and jaw in conscious monkeys (DeLong et al. 1983). Afferent input from skin and intraoral structures is also found in anesthetized cats (Harper et al. 1979).

Decreases of activity are regularly observed in relation to saccadic eye movements in monkeys (Hikosaka and Wurtz 1983a) and cats (Joseph and Boussaoud 1985). These changes are conditional on the presence of a visual target, and little or no modulations are seen with spontaneously occurring saccades. Changes in activity begin rapidly 100 ms or more before onset of the saccade. Many saccade-related reticulata neurons project to the superior colliculus, as shown by their antidromic responses after electrical stimulation of this structure (Hikosaka and Wurtz 1983d). Other reticulata neurons show contralateral sensory inputs of visual or auditory modality (Hikosaka and Wurtz 1983a; Joseph and Boussaoud 1985; Schultz 1986a), and visual responses in many neurons are enhanced when the stimulus is used as the target for a subsequent saccade. The conditional nature of the relationship to saccades and the enhancement of visual responses with saccades suggest a complex involvement in eye movements beyond pure oculomotor control.

4. Nucleus Subthalamicus

A sizeable portion of neurons in this structure of monkeys show increases or decreases of activity in relation to active movements of the arm (28%), leg (15%), face (18%), or eyes (DeLong et al. 1985). Although most of these neurons are related to proximal limb movements, about 20% of arm-related neurons show preferential changes with distal movements of wrist or fingers. Neurons with similar somatotopic relationships tend to occur in clusters. About 20% of neurons respond to passive rotation of joints, while inputs from muscles or skin are ineffective.

The activity of subthalamic neurons is significantly related to movement direction, as tested during arm flexion and extension movements (Georgo-

POULOS et al. 1983). They also show a linear relationship to the amplitude of movement or its peak velocity. Onset of movement-related changes occurs in about one-third of neurons before the earliest EMG activity. Median onset times of subthalamic neurons lead the movement by about 50 ms. Thus, changes in activity prior to movement occur slightly earlier than in the pallidum, but later than in the motor cortex (GEORGOPOULOS et al. 1983).

D. Higher Functions

I. Anatomy of Connections

1. Cortical Input

The first studies using modern anatomic tracer techniques in the monkey report heavy connections from the prefrontal, parietal, occipital, and temporal association cortex to the caudate and, to a much lesser extent, the putamen (GOLDMAN and NAUTA 1977; KÜNZLE 1978; YETERIAN and VAN HOESEN 1978; VAN HOESEN et al. 1981). Lateral and medial area 6 (premotor cortex and supplementary motor area, respectively) projects predominantly to the putamen (KÜNZLE 1978). Connections are mostly ipsilateral, but minor contralateral projections to the striatum are seen from virtually all of these cortical areas. Terminals in the striatum occur in clusters or patches which are distributed over wide areas and thus provide for an inhomogeneous corticostriatal projection. The patches of corticostriatal terminals match in several areas the inhomogeneities of acetylcholinesterase staining (RAGSDALE and GRAYBIEL 1981).

Recent studies addressing the topography of corticostriatal projections in the monkey emphasize the mediolaterally arranged longitudinal stripes of terminal fields in the striatum (SELEMON and GOLDMAN-RAKIC 1985; ARIKUNI and KUBOTA 1986), whereas earlier studies using degeneration methods report an anteroposterior correspondence between cortical origin and striatal terminals. Thus, the posterior parietal association cortex projects to the dorsolateral caudate, while the orbitofrontal, temporal, and cingulate cortex influence the ventromedial caudate. The dorsolateral prefrontal cortex sends fibers to a central stripe along the whole longitudinal extent of this structure. While the projections of some interconnected cortical areas overlap in the striatum (YETERIAN and VAN HOESEN 1978), this cannot be generally applied to all cortical afferents (SELEMON and GOLDMAN-RAKIC 1985). Rather, inputs from different cortical areas appear to interdigitate so that separated small clusters of afferents influence distinct subsets of striatal neurons.

2. Limbic Input

In the monkey, the major limbic afferents to the striatum and nucleus accumbens originate from the orbitofrontal cortex, the cingulate gyrus, and the amygdala. The orbitofrontal cortex projects to the whole longitudinal extent of the ventromedial caudate, ventromedial putamen, and possibly nucleus accumbens (SELEMON and GOLDMAN-RAKIC 1985). The anterior cingulate gyrus (area 24) projects heavily on the caudate, although, without obvious reasons, its terminals are found in varying regions in different studies. These comprise the dorsolateral caudate (YETERIAN and VAN HOESEN 1978), the dorsomedial caudate, dorsomedial putamen, and nucleus accumbens (BALEYDIER and MAUGUIERRE 1980), and the ventromedial caudate, ventromedial putamen, and possibly nucleus accumbens (SELEMON and GOLDMAN-RAKIC 1985). The posterior cingulate gyrus (area 23) projects to the extreme dorsomedial cap of the caudate (BALEYDIER and MAUGUIERRE 1980). Amygdala neurons of the basolateral nucleus and, to a smaller extent, the central nucleus project to the putamen of squirrel monkeys (PARENT et al. 1983). A similar projection mainly from the basolateral amygdala to the ventral putamen and nucleus accumbens exists in rats and cats (KRETTEK and PRICE 1978).

3. Output

Neuronal information from all areas of the association cortex arriving predominantly in the caudate is mostly passed to the SN (SMITH and PARENT 1986) and leaves the basal ganglia from the pars reticulata of the SN toward the superior colliculus and VA thalamus. This thalamic nucleus projects to the frontal association cortex (ILINSKY et al. 1985). Influences from area 6 directed to the putamen leave the basal ganglia, mostly via the internal pallidal segment to the anterior VL thalamus which projects to the supplementary motor area (SCHELL and STRICK 1984). Neurons of the internal pallidal segment also project to the lateral habenula, and thus influence the limbic system (KIM et al. 1976). Although these flow charts suggest the existence of segregated limbic, associational, premotor, and motor loops linking cortex and basal ganglia (ALEXANDER et al. 1986), strong anatomic convergence and divergence is found across these serially arranged synaptic links (CHANG et al. 1981; PERCHERON et al. 1984) which remain to be elucidated in functional respects. Nevertheless, influences of the basal ganglia derived from heterogeneous sensory, motor, associational, and limbic inputs converge onto a few areas of the frontal cortex, particularly the supplementary motor area and prefrontal cortex, and are also directed to the limbic system and midbrain.

II. Neuropsychologic Deficits After Lesions

Some of the typical cognitive deficits occurring after lesions of associational cortical areas are also found after destruction of even small parts of the caudate and putamen in monkeys. Thus, deficits in delayed alternation and delayed response are specific for lesions of the prefrontal cortex as compared with other cortical areas. Impairments in a delayed alternation task are found with small lesions of the head of the caudate (ROSVOLD and DELGADO 1956). However, even with larger caudate lesions, deficits in delayed alternation and delayed response learning are less severe as compared with prefrontal lesions in the same study (BÄTTIG et al. 1960). Dorsoanterior and particularly ventroposterior parts of the caudate are the most sensitive areas for delayed alternation deficits (DIVAC et al. 1967). Deficits in an object reversal task, in which reward contingency is alternated between two objects every few trials, occur with lesions of the orbitofrontal cortex and are found equally after destruction of the ventrolateral part of the head of the caudate (DIVAC et al. 1967). Impairments in visual discrimination learning, characteristic of lesions of the inferotemporal cortex, are also found with lesions in the tail of the caudate (DIVAC et al. 1967).

The general similarity of cognitive impairments between the striatum and areas of the association cortex suggest that these connected brain structures may function in parallel or are even two components of a single neuronal system (BÄTTIG et al. 1960). The observed differential deficits after lesions in distinct parts of the striatum appeared to fit well with results from earlier anatomic degeneration studies, suggesting an anteroposteriorly organized topographic correspondence between cortical and striatal areas (DIVAC et al. 1967). However, the recently described longitudinal bands of corticostriatal connections (SELEMON and GOLDMAN-RAKIC 1985) suggest that future studies should reevaluate and assess in more detail the effects of regional striatal lesions and compare them with deficits following lesions in cortical afferent areas.

III. Neuronal Recordings

1. Untrained Behavior

A classic early study describes the responses of caudate neurons in the monkey to visual and auditory stimuli that arouse the attention of the animal (BUSER et al. 1974). Data from a subsequent study support the conditional nature of sensory input to the head of the caudate in which only 3 %–10 % of neurons respond to visual stimuli independent of the behavioral context, while about 25 % of neurons are activated when food objects or cues indicating their availability are presented to the animal (ROLLS et al. 1983). The only areas of the striatum where visual responses are found are the tail of the caudate and the adjoining ventral putamen. Here, neurons respond to complex patterns of stimuli, but show rapid habituation to the same stimulus, a property that makes them rather unresponsive in a visual discrimination task where the same few stimuli are repeatedly presented (CAAN et al. 1984). These neurons appear to be particularly sensitive to changes in complex patterns of visual input, and may thus participate in the orientation and attention to the changing visual surroundings.

2. Context-Dependent Responses to Directly Triggering Stimuli

In trying to standardize the description of common components of the various behavioral tasks used in different independent studies, we will call those stimuli that lead to a direct behavioral reaction "triggers". Reactions in these studies normally consist of movements, which can either be a single movement, as in simple reaction time paradigms, or the choice of one of a small repertoire of movements which may be triggered by a particular stimulus (e.g., arm flexion or extension following exposure to a light on the left or the right side). For studying the initiation of triggered movements, a go—no go paradigm appears suitable in which two different stimuli call for a movement reaction or its suppression (e.g., pressing a key after a green light and remaining motionless after a red light). Many neurons are found in the basal ganglia which respond to such trigger stimuli, often specifically to a stimulus demanding a particular reaction, although they would not respond when the stimulus is given without a behavioral context.

Between one-third and three-quarters of neurons in the caudate nucleus and in both segments of the pallidum respond with mostly phasic activations (caudate and pallidum) or depressions (caudate) at latencies of 200 ms to visual stimuli that trigger arm flexion or extension movements in monkeys (ALDRIDGE et al. 1980b). Responses in some of these neurons are reduced when the stimulus demands suppression of movement, some responses are differentially related to stimuli calling for flexion or extension, and only very few responding neurons are activated by light flashes without a behavioral context. The majority of caudate neurons in cats respond with less than 100 ms latency to an auditory stimulus triggering key release while the stimulus is ineffective in some neurons when the animal fails to respond (AMALRIC et al. 1984). In monkeys executing or suppressing mouth movements in a visual discrimination task, about one-third of caudate neurons respond to the trigger stimulus, and one-quarter of responding neurons show responses only in go trials (ROLLS et al. 1983). The effective stimulus of the majority of these differentially responding neurons reversed when the reward contingency of the stimulus was reversed, signifying that the neuronal response was related to the behavioral reaction and not to the stimulus properties. Again, most of these neurons activated by visual trigger stimuli do not respond to unrelated visual input. Similar differential mouth movement-dependent responses to auditory stimuli are found in the monkey putamen in the form of increases and in the globus pallidus as decreases or increases (KIMURA et al. 1984). Responses in both structures disappear entirely when the animal does not react to an identical but unrewarded stimulus. When monkeys are presented with different food objects, caudate neurons respond to their sight according to the efforts animals undertake to obtain them (NISHINO et al. 1984). Thus, for a given series of food objects, caudate neurons respond strongly to the sight of beans or pieces of orange, moderately to bread or cookies, and not at all to carrots or screws. This gradient is paralleled by an earlier onset of bar pressing to obtain beans or orange as compared with cookies, and by an absence of bar pressing for carrots and screws. When the animal is satiated by a specific food and refrains from bar pressing, caudate neurons fail to respond to its sight, but respond to a more attractive food object for which the animal will work. These responses are not caused by muscular contractions since latencies for bar pressing can be dissociated from those of neuronal responses (NISHINO et al. 1984).

Recent studies investigating stimulus-triggered eye movements have shown that visual and auditory responses in different parts of the basal ganglia are enhanced when the stimulus is used as a target for a saccadic eye movement. A sizeable portion of neurons in the pars reticulata of the SN in monkeys and cats respond with a short depression of activity to visual stimuli while the animal fixates a spot of light (HIKOSAKA and WURTZ 1983a; JOSEPH and BOUSSAOUD 1985). These neurons show large, predominantly contralateral receptive fields, respond best to small spots of light, and are not particularly sensitive to moving or complex stimuli. The visual response becomes more prominent in about half of the neurons when the animal makes a saccade toward the stimulus. Detailed analysis reveals that neurons showing enhancement are also depressed in relation to saccades and that this depression is additive to the depressant response to the stimulus occurring at the same time (HIKOSAKA and WURTZ 1983a). A separate group of reticulata neurons in the SN shows reductions in activity in response to the onset of a visual stimulus to which the monkey makes a saccade without having fixated another light spot, and in response to offset of a stimulus that the animal has been fixating (HIKOSAKA and WURTZ 1983b). Some reticulata neurons in the SN of monkeys and cats respond to contralateral auditory stimuli, the responses again being enhanced when the animal uses the stimulus as the target for a saccade (HIKOSAKA and WURTZ 1983a; JOSEPH and BOUSSAOUD 1985). In the monkey caudate nucleus, discharges in response to a visual stimulus and response enhancement with a saccade toward it are seen in a small fraction of neurons (HIKOSAKA and SAKAMOTO 1986). These data on eye movements demonstrate that apparently purely sensory visual and auditory responses are modified when the stimuli causing them are used as targets for behavioral reactions. It would nevertheless be conceivable that even these pure sensory responses, apart from their enhancement with eye movements, are dependent on a behavioral context since they all occurred in situations in which correct performance was rewarded in each trial.

These results show that most responses in neurons of the basal ganglia to external stimuli are not primarily sensory in nature, but occur when the stimuli are used for direct behavioral reactions. Thus, these basal ganglia neurons participate in the initiation of rapid behavioral reactions in response to external stimuli.

3. Preparation to Act

A sensory signal indicating that a subject should react to a forthcoming trigger stimulus is called an instruction. The instruction may simply contain information indicating *that* a reaction to the trigger should occur, and the trigger determines how to react if a choice is involved. This is called a nondiscriminative instruction. If the instruction specifically determines *how* to react when a choice is possible, it will be called a discriminative instruction, and then the trigger should not contain information on how to react. A delay, during which the instruction signal is kept on, exists in both situations between instruction onset and trigger stimulus. However, to use a discriminative instruction in a classic delayed reaction paradigm, the instruction signal must be turned off several seconds before a trigger stimulus occurs. Thus, to perform correctly in a delayed response task, the subject needs to remember the particular instruction, and care must be taken to avoid uncontrolled internal and external cues during the delay period.

a. Nondiscriminative Instructions

Neurons in several structures of the monkey basal ganglia are tonically activated or depressed during the presentation of an instruction that precedes a trigger stimulus. Changes of activity normally subside after the trigger stimulus. A nondiscriminative tone or light cue preceding by 0.5 s the trigger for a licking response is accompanied by a tonically increased activity of 15% of caudate neurons (ROLLS et al. 1983). The activation disappears when the licking tube is withdrawn and a reward-related behavioral response is excluded, in which case the signal loses its instructive significance. When these neurons are examined outside a behavioral task they frequently show activity when the experimenter indicates that food is about to be presented, for example, by looking at or reaching out to a tray of food for the animal (ROLLS et al. 1983). Similar activations in caudate neurons and activations or depressions in pallidum neurons are seen when a delay period of 4 s after the presentation of a food object is introduced before the animal is allowed to press a bar to obtain the food (NISHINO et al. 1984, 1985). The presentation of a sound cue, which indicates that a discriminative instruction in a go—no go task will follow after 0.5 s, increases or decreases the activity of some neurons in the pars reticulata of the SN (SCHULTZ 1986a). In one of the earliest such investigations, SOLTYSIK et al. (1975) find gradually increasing or decreasing activity in caudate and pallidum neurons between a light signal and an arm-pulling movement. Interestingly, the monkeys in this task perform the movement between 6.5 and 10 s after instruction onset without a further trigger stimulus, i.e., the task employs essentially a loosely timed, self-paced movement following an instruction signal.

These data show that signals indicating an upcoming behavioral act are effective in changing the impulse activity in several structures of the basal ganglia. Since behavioral reactions in operantly conditioned animals are ultimately linked to reward, the relationship of each neuron could be difficult to discriminate between the preparation to respond and the expectancy of reward.

b. Discriminative Instructions

Instructions containing specific information on how to respond to a subsequent trigger stimulus are effective in influencing the impulse activity in several structures of primate basal ganglia. In a go—no go task, about 10% of pars reticulata neurons in the SN are tonically activated or depressed between instruction onset and the trigger stimulus about 2 s later, most of them only when the instruction indicates reaction with a movement to the trigger (go situation) and not when it indicates withholding the movement (no go) (SCHULTZ 1986a). In the caudate and putamen, 22% of 320 and 34% of 188 neurons, respectively, are activated in relation to instructions in a similar go—no go task (SCHULTZ and ROMO 1988). Responses occur either phasically in response to instruction onset, tonically during the total instruction period until the trigger occurs, or with a slow increase preceding the expected time of triggering. Prolongation of the regular 2 s interval between instruction and trigger would result either in a decrease or a further increase of tonic responses. In most neurons, instruction-related increases are specifically seen during the go situation, in which the animal is required to make a movement after the trigger to obtain a reward. However, about 5 % of caudate and putamen neurons are specifically activated in the unrewarded no go situation during the instruction which indicates withholding the movement after the trigger stimulus.

These data suggest that there are neurons in the basal ganglia that are actively involved in the preparation of specific behavioral responses that are only executed at a later time. The instruction-related increases during the no go situation demonstrate that some of these responses are certainly not related to expectancy of reward.

c. Delayed Response Tasks

In view of the deficits in delayed alternation and delayed response occurring after lesions in the head of the monkey caudate nucleus, the first neurophysiological study investigating higher functions in the basal ganglia employed a delayed alternation paradigm. NIKI et al. (1972) found that only 14 of thousands of neurons in the head and body of the monkey caudate showed impulse discharges during performance in this task. All of these neurons were only activated in relation to lever pressing after the delay, their activity preceding EMG activity in the moving arm by about 200 ms, while none of them showed increased activity during the delay period. In a recent study, ALEXANDER (1987) uses a delayed reaction task in which the instruction consists of an elbow movement performed in a specific direction by following a cursor on an oscilloscope. After the delay period, the monkey has to repeat the same movement in response to a trigger stimulus which does not contain information on the movement direction. A number of putamen neurons show tonically increased activity during the delay period. Most of these changes occur for a specific movement direction.

Using saccadic eye movements in a delayed response paradigm, HIKOSAKA and WURTZ (1983c) require monkeys to make a saccadic eye movement after the fixation point for the eyes is turned off. The target for the saccade is indicated 1-3 s before the trigger by a brief light stimulus. The delay in this task consists of the period between light stimulus (spatially specific instruction) and offset of fixation point (trigger stimulus). Neurons in the pars reticulata of the SN show three types of specific responses in this task: a phasic depressant response to the instruction light (but no response when the light is used as the target for an immediate saccade); a tonic and often gradually increasing depression during the whole delay period; and a decrease in relation to the saccade after the delay (but no change with an undelayed saccade and greater changes with longer delays). Some neurons in the caudate nucleus show a specific activation in relation to delayed saccades in the same delayed response task (HIKOSAKA and SAKAMOTO 1986).

These data suggest that neurons in the pars reticulata of the SN, caudate, and putamen are activated by specific instructions for impeding behavior. Since instructions in delayed response paradigms disappear before the behavioral reaction occurs, mechanisms of information storage, maintenance, and retrieval need to be operational. Therefore, these responses appear to be related to short-term memory, as explicitly proposed before (HIKOSAKA and WURTZ 1983c).

d. Self-Initiated Acts

The investigation of neuronal mechanisms leading to "spontaneous" movements is necessary for elucidating the initiation of movements. In trying to circumvent the considerable conceptual problems involved in terms such as "spontaneous", it is easier to define some of these behavioral acts in a more operational and descriptive way. Thus, the term "self-paced movements" denotes those that are made by a subject without an acutely triggering external stimulus, although they may be performed automatically, regularly, rhythmically, and at a trained pace. The term "self-initiated movements" can then be used for describing a situation that excludes these obvious constraints of self-paced movements which separate them from more "spontaneous" acts. Thus, self-initiated movements should be performed without rhythmicity and without external cues for timing, including temporary ones during training periods. To further distinguish them from self-paced movements, they may also include the purpose of action. In using self-initiated movements, one may investigate the planning, preparation, and initiation of behavioral acts that are not reactions to immediate external constraints.

Self-initiated movements were studied in cats rewarded for depressing a bar with their forepaw for at least 2 s before releasing it, without external trigger stimuli (NEAFSEY et al. 1978). In both pallidal segments, 23% of 141 neurons increase or decrease their activity by more than 500 ms and up to 1200 ms in advance of this forelimb flexion movement, while EMG activity precedes the movement by about 200 ms. In an other study, monkeys perform self-initiated arm-reaching movements from a resting key into a covered food-containing box ahead of them, once every 5-30 s (SCHULTZ and ROMO 1988). Increases in activity are seen in 23% of 167 neurons in the putamen and 18% of 215 neurons in the caudate, beginning 700-3000 ms before movement onset, while the earliest EMG activity precedes the movement by maximally 350–400 ms. These changes begin slowly and reach a maximum before or close to movement onset. Since most of these neurons have a very low background discharge rate, the premovement activity often provides an impressive indicator of the upcoming movement more than 1 s before EMG activity is observable on the oscilloscope. Many of these neurons are exclusively activated in advance of self-initiated movements and not during instruction signals preparing for a triggered movement in a second task.

Although few studies have so far employed self-initiated movements, it appears that neurons in some of the main structures of the basal ganglia are activated while movements with minimal external constraints are initiated. Together with the data from tasks employing external instruction signals, this presents evidence for an active participation of basal ganglia in early neuronal processes preceding behavioral acts.

E. Dopaminergic Functions

I. Animal Models of Parkinsonism

1. Monkey

Destruction of the nigrostriatal DA system by electrolytic lesions of cell bodies in the SN or local injections of the catecholamine neurotoxin 6-hydroxydopamine (6-OHDA) into the nigrostriatal axon bundle results in a profound and permanent hypokinesia (POIRIER 1960; GOLDSTEIN et al. 1973; CROSSMAN and SAMBROOK 1978). The striatal DA depletion needs to be higher than 75% for hypokinesia to occur, while less important lesions are without overt neurological deficits (POIRIER et al. 1966). Tremor and rigidity are less readily obtained by these lesion methods. They occur only when the parvocellular part of the red nucleus is also destroyed while leaving its magnocellular part and rubrospinal connections intact (PÉCHADRE et al. 1976). Systemic injections of DA precursors or receptor agonists alleviate the symptoms (PÉCHADRE et al. 1976; CROSSMAN and SAMBROOK 1978).

Replications of parkinsonian symptoms in monkeys through local lesions are compromised by the difficulty of sufficiently destroying the extended group of DA neurons without inflicting damage on neighboring structures, and by the necessity of also lesioning the red nucleus which is normally not affected in human idiopathic parkinsonism. A major advance in this field of experimentation occurred with the discovery that accidental systemic injection of 1-methyl-4-phenyl-1,2,3,6-tetrahydropyridine (MPTP) in humans leads to a neurologic syndrome that is virtually identical with idiopathic parkinsonism (DAVIS et al. 1979; LANGSTON et al. 1983). Subsequent experimental MPTP administrations to animals showed a severe and permanent parkinsonian syndrome in monkeys (BURNS et al. 1983), while in rodents only certain strains of mice are affected (HEIKKILA et al. 1984; HALLMAN et al. 1984). Monkeys develop severe hypokinesia with grossly reduced spontaneous and stimulus-triggered limb and eye movements, and, to varying degrees, rigidity and activation tremor, while a true resting tremor is virtually absent. In addition, they show periods of frozen posture, paradoxical kinesia following external stimuli, and motivational deficits such as adipsia and aphagia. Drug therapy with DA precursor or receptor agonists is effective in alleviating the symptoms unless the lesion is very complete, in which case on—off phenomena are frequent. Although MPTP causes a rather selective destruction of the nigrostriatal DA system (90% or more depletion of DA in the caudate and putamen), in some monkeys less complete lesions are seen to varying degrees in other DA, norepinephrine, and serotonin systems. After entering the brain, the nontoxic MPTP is transformed by monoamine oxidase (type B) after an intermediate step to the neurotoxin 1-methyl-4-phenylpyridine (MPP^+) (LANGSTON et al. 1984; COHEN et al. 1984). MPP^+ is concentrated in DA neurons by their reuptake mechanism which largely explains its selectivity (JAVITCH et al. 1985; SUNDSTRÖM and JONSSON 1985; SCHULTZ et al. 1986). Once inside the neuron, MPP^+ causes cell death by interfering with aerobic glycolysis in mitochondria (NICKLAS et al. 1985; POIRIER and BARBEAU 1985). For further details on MPTP-induced parkinsonism in monkeys see Chap. 7 and SCHULTZ (1988).

Neurophysiological studies in thalamic and cortical regions in parkinsonian patients and in monkeys with large ventral tegmental lesions reveal changes of neuronal activity in phase with tremor cycles (JASPER and BERTRAND 1966; LAMARRE and JOFFROY 1979). The rhythmic activity in monkeys persists after elimination of peripheral feedback by cutting the dorsal roots or after neuromuscular blockade, demonstrating an involvement in the tremorgenic pacing mechanism (LAMARRE and JOFFROY 1979). Hypokinesia after lesions of the DA system is quantitatively assessed in reaction time tasks, in which the initiation of arm movement in response to a directly triggering stimulus (reaction time) is separated from the subsequent execution of movement (movement time). Monkeys with electrolytic or 6-OHDA-induced lesions show increases in both measurements in a choice reaction pointing task with randomly varying targets (VIALLET et al. 1983; APICELLA et al. 1986). In monkeys performing thousand of movements in a simple reaction time task under computer control, the onset of EMG activity in prime mover muscles and the onset of movement following the trigger stimulus are delayed by 31%–129% after MPTP treatment, while the duration of the reaching movements is prolonged by 24%–100% (SCHULTZ et al. 1985). These data demonstrate a deficit in the initiation and execution of stimulus-triggered movements as signs of reduced behavioral reactivity. This occurs with a specific lesion of the nigrostriatal DA system while the cortical monoamine innervation is intact (SCHULTZ et al. 1985). Similar deficits are found with elbow flexion—extension movements in a choice reaction task in MPTP-treated monkeys with unknown lesion extent (DOUDET et al. 1985). Initiation deficits are also seen with triggered and spontaneous saccadic eye movements (BROOKS et al. 1986; SCHULTZ, ROMO, SCARNATI, STUDER, JONSSON and SUNDSTRÖM, in preparation).

2. Rodent

Rats are commonly used to assess the effects of 6-OHDA-induced lesions of the nigrostriatal DA system. After bilateral stereotactic injections, animals develop hypokinesia, adipsia, and aphagia (UNGERSTEDT 1971b). With unilateral lesions, rats turn in tight circles when DA receptor stimulation prevails strongly on the contralateral side (UNGERSTEDT 1971a). Rats with 6-OHDA-induced lesions show reduced reactions to sensory stimuli (LJUNGBERG and UNGERSTEDT 1976) which is possibly due to a deficit in initiating the behavioral response (CARLI et al. 1984). The adaptive neuronal mechanisms to striatal

DA depletions comprise increased metabolic activity of remaining DA neurons, reduced uptake of synaptically released DA by the smaller number of surviving DA terminals, and, with DA depletions above 90 %, postsynaptic supersensitivity of mainly the non-adenylate cyclase-coupled DA receptor (for further details see Schultz 1982). The neurotoxic mechanism of action of 6-OHDA is reviewed by Jonsson (1980).

II. Impulse Activity of Dopamine Neurons

1. Peripheral Input Under Anesthesia

DA neurons of the pars compacta of the SN and the neighboring ventral tegmental area identified histologically and by their electrophysiologic characteristics (see Sect. B.II.1.a), respond predominantly with reduced discharge activity to electrical stimulation of afferent input of various modalities, and to natural somatosensory stimulation of mostly high intensity. Olfactory input is suggested by increases and decreases of activity in two-thirds of rat DA neurons at latencies of 8.5–120 ms after electrical stimulation of the ipsilateral anterior olfactory nucleus (Tulloch and Arbuthnott 1979). Direct olfactory stimuli, as well as light flashes, are effective in decreasing or increasing the activity of DA neurons (Chiodo et al. 1980). Electrical stimulation of A- and C-fiber groups in the contralateral and ipsilateral sciatic nerve reduces the activity of virtually all SN DA cells in rats at mean latencies of 35–40 ms, while activating responses are rare (Hommer and Bunney 1980; Tsai et al. 1980).

With somatosensory stimulation in rats, decreases or increases of impulse activity are seen after mild or intense pinching of the tail, intense pinching of the paws, cervical probing, and thermal stimulation of the tail (Chiodo et al. 1980; Tsai et al. 1980; Maeda and Mogenson 1982). Intradermal injections of a substance producing pain in conscious subjects also depresses DA neurons in anesthetized rats (Tsai et al. 1980). In monkeys, half of the nigrostriatal DA neurons are depressed and a quarter are activated during high intensity pinch stimulation of face, hands, legs, dorsum, and tail, while intense but innocuous somatosensory stimuli are ineffective in altering their discharge rate (Schultz and Romo 1987). Convergence occurs regularly between trigeminal and spinal inputs, and from both sides. The DA receptor antagonist haloperidol strongly reduces responses to sciatic nerve stimulation in rats and to pinch stimulation in monkeys, demonstrating that DAergic neurotransmission is necessary for this peripheral input (Hommer and Bunney 1980; Tsai et al. 1980; Schultz and Romo 1987).

The potential behavioral implications of these responses under anesthesia are difficult to evaluate. While the noxious character of most stimuli is apparent, it should be emphasized that all effective stimuli elicit strong activating behavioral responses when applied to conscious subjects. When comparing the responses of DA neurons with those of the ascending pain system, it appears that DA neurons do not subserve a specific pain detection mechanism, but may rather be involved in behavioral reactions to harmful stimuli (for further discussion see SCHULTZ and ROMO 1987).

2. Relations to Behavior

The fact that impulses of DA cells display typical and distinctive electrophysiologic characteristics offers the unique possibility of investigating the discharge activity of neurons with a psychoactive neurotransmitter during behavioral acts.

a. Execution of Movements

In view of the hypokinetic movement disorder in parkinsonian patients and experimentally lesioned animals, attempts were made to find activity changes in DA neurons during movements. However, none of the studies reports any relationship to individual movements during walking in cats (STEINFELS et al. 1981; TRULSON et al. 1981), although a slight overall increase in impulse activity occurs during active as compared with restful waking (TRULSON et al. 1981). When testing DA neurons in monkeys during a visually guided wrist movement task typical of motor control studies, a conspicuous absence of phasic changes is noted (DELONG et al. 1983). Only large, individual forelimb reaching movements toward a target directly associated with food reward are accompanied by mild increases of activity in about one-half of DA neurons in the lateral two-thirds of the monkey SN (SCHULTZ et al. 1983). Changes are slow and not related to particular phases of the movement. These data suggest that the activity of SN DA neurons may vary together with the level of behavioral activation during motor acts, but is not engaged in encoding detailed movement parameters.

b. Responses to Stimuli

DA neurons fail to respond to changing visual signals which continuously inform a monkey about the position of its moving wrist in a visuomotor tracking task (DELONG et al. 1983). Conditioned instructive auditory or light signals preceding by several seconds the trigger stimulus for movement are ineffective in changing the impulse activity of SN DA neurons in a go—no go task (SCHULTZ et al. 1983). Quite different from these sensory signals, trigger stimuli for a direct licking or forelimb movement response in rats are effective in eliciting discharges of A9 and A10 DA neurons (MILLER et al. 1981). Responses to intense auditory clicks or light flashes are seen in untrained, conscious cats (STEINFELS et al. 1983). Although the animal's reactions to these stimuli are unclear, a certain behavioral involvement of responses is suggested by the fact that they are easily extinguished by distractions (STRECKER and JACOBS 1985).

In a reaction time task, the majority of DA neurons in areas A8, A9, and A10 of monkeys respond with a brief burst of impulses to the combined auditory and visual stimulus triggering an arm-reaching movement toward a box containing a reward (SCHULTZ 1986b). Responses do not habituate over consecutive trials, and occur at median latencies of 65 ms for onset and 95 ms for peak. They remain present when visual and auditory components of the trigger stimulus are separated. Neurons also respond when the animal refrains from moving in response to an identical trigger stimulus presented in the no go situation of a go—no go task (Schultz and Romo 1986). However, the same stimuli of either modality are entirely ineffective when presented outside the behavioral schedule.

DA neurons in monkeys performing self-initiated reaching movements toward a morsel of food hidden behind a cover respond with a brief burst of impulses when the hand touches the food (Romo and Schultz 1986). No response occur when the hand touches a wire instead of the food at the same position. The somatosensory response to touching the food is absent in the same neurons when the movement is triggered by an immediately preceding stimulus. Thus, the excitatory and context-dependent responses to mild somatosensory input in the behaving animal are entirely different from the predominant depressions seen after high intensity pinching under anesthesia (Schultz and Romo 1987).

These data suggest that DA neurons respond to sensory stimuli on the condition that they are of immediate consequence for the behavior of the animal. Effective stimuli are triggers for direct behavioral reactions in the form of execution of movement, suppression of movement, or obtaining vital objects such as food. Stimuli that contain specific information, but do not trigger direct reactions, are ineffective (e.g., lights indicating the wrist position, instruction signals preparing the animal for a later behavioral response, or a wire touched by the hand).

c. Preparation to Act

Although DA neurons in monkeys fail to respond to conditioned instructive signals, a brief burst of impulses is seen in response to these stimuli after they have not been used for several weeks (Schultz and Romo, unpublished). Responses occur only during the first few days after reintroduction of the instructions, and disappear entirely thereafter. Their transient character suggests that they are not related to the preparation for behavioral response, but rather reflects the capacity of the unfamiliar stimulus to change the level of behavioral activation.

DA neurons in monkeys were also tested during self-initiated movements in which the animal releases a touch-sensitive key from a relaxed position without any phasic external cues and reaches into a box to obtain a hidden food object (Romo and Schultz 1986). Only a few DA neurons are activated several hundred milliseconds in advance of these movements, and even these changes are very small and only detectable with statistical methods. The absence of major changes in DA neurons is in striking contrast to the strong activations of striatum neurons seen in the same experimental situation (see Sect. D.III.3.d).

3. Comparison with Deficits

With the data available, one may attempt to assess the function of nigrostriatal DA neurons in behavior by comparing their neuronal activity (positive image) with the behavioral deficits after their lesion (negative image). A striking lack of correspondence is observed between the severe deficit in initiating spontaneous movements in DA-deficient subjects and the minor activation of DA neurons in advance of self-initiated movements in unlesioned monkeys. This suggests that DA neurons are not actively implicated in the initiation process. A similar conclusion can be drawn for the execution of movements. However, the mild increases of DA neuron impulse rate during both initiation and execution of large reaching movements may reflect the behavioral activation occurring concomitantly with these acts. Strong and synchronous activations in the large majority of DA neurons occur in response to stimuli that are important for the animal and elicit direct reactions. These responses of DA neurons match the reduced behavioral reactivity after their lesion, as seen clinically and assessed quantitatively in reaction time tasks. Thus, there exist distinct classes of behavioral situations during which impulse activity corresponds to lesion-induced deficits, and others without an apparent match. The rate of change in behavioral activation appears to be common to situations which are effective in influencing impulse activity in DA neurons. Within the range of tests performed so far, this is maximal with a directly triggering stimulus and also high when the animal touches a vital object without a preceding trigger stimulus. Behavioral activation changes less rapidly during the subsequent execution of the movement and in relation to the investigated self-initiated acts, and only minor variations of impulse rate are seen in these situations.

The burst of impulses occurring synchronously in the majority of DA neurons following strong activating stimuli should lead to a rapid and short increase of DA release in the target areas. In particular, impulse-dependent striatal DA release is exponentially related to the frequency of discharge (Gonon and Buda 1985) which suggests a disproportionately higher DA release with the high frequency burst observed in conscious animals, as compared with spontaneous activity. In the striatum, DAergic transmission exerts a focusing effect through which the information transfer can be restricted to the necessary operations, whereas less prominent input would be suppressed (Toan and Schultz 1985). Through a short, burst-induced increase in DAergic influences, DA neurons would be able to focus neuronal processing in the target areas to the most essential operations necessary for quick reactions to salient stimuli.

Several behavioral acts are not accompanied by major increases of impulse rate, such as execution of movements and initiation of self-generated behavior. However, deficiencies of these acts in DA-depleted subjects demonstrate the crucial presence of DA. It is possible that the release of DA caused by spontaneous impulses and its mild impulse-dependent changes are sufficient to sustain this behavioral repertoire. In these situations, DAergic neurotransmission would exert its focusing effects with only limited phasic changes on the diverging and converging striatal connections. Glutamate, the probable transmitter of the corticostriatal pathway, exerts a facilitory influence on striatal DA release (GIORGUIEFF et al. 1977; ROMO et al. 1986; for review see CHESSELET 1984). Although impossible to measure at sufficiently high temporal resolution with present methods, fast, behavior-related changes in corticostriatal activity may lead to variations in DA release that are more important than those of impulse activity. In this way. DA release could be increased in those striatal areas where active cortical input is found, which according to the known electrophysiologic effects of DA, would lead to a reduction of the effects of cortical input while other areas in the striatum would remain uninfluenced. Another role for presynaptically mediated DA release may concern processes employing a slower time course. Indeed, short pain pinch stimulation in anesthetized monkeys produces a rapid and short change in DA neuron discharge rate and a slower variation of excitability of striatal DAergic axons over several minutes, which is suggestive of increased DA release (ROMO and SCHULTZ 1985). In this way, presynaptically induced DA release could participate in a slowly changing focusing of striatal neurotransmission that would be controlled by cortical input, while impulse-dependent DA release would be operational when rapid changes in behavioral activation occur.

F. Conclusions

I. Functional Connectivity

1. Lateralization of Function

Although corticostriatal projections are predominantly ipsilateral, considerable contralateral influences are also exerted from most cortical areas, one notable exception being the somatosensory cortex. Internal basal ganglia connections remain largely, but not exclusively ipsilateral. Basal ganglia outputs at the level of the pallidum and pars reticulata of the SN are also ipsilateral with a modest contralateral component. Thus, when judged from their connectivity, the basal ganglia are less lateralized in their functions than primary sensory and motor areas of the cerebral cortex. This may simply be a remnant of the phylogenetic development without much functional importance, but it may also suggest that the basal ganglia are engaged in control functions dealing with behavioral acts that involve coordination of both sides of the body rather than precise, topographically related input or output.

2. Disinhibition of Target Structures

Cerebral input to the basal ganglia is largely excitatory in nature. In contrast, major internal connections are inhibitory and use GABA as a fast amino acid transmitter. (A particular role is played by the nigrostriatal DA system, exerting a focusing effect on striatal neurotransmission.) Output of the basal ganglia, e.g., pallidothalamic, nigrothalamic, and nigrocollicular projections, are also inhibitory in nature. Through two consecutive inhibitory synapses, activation of striatal neurons would principally cause a disinhibition of the main target structures of the basal ganglia, which has in fact been demonstrated experimentally for output of the SN (DENIAU and CHEVALIER 1985; CHEVALIER et al. 1985). The high spontaneous impulse activity in the two main output stations, the pallidum and pars reticulata of the SN, would be ideally suited for fine tuning and integration of inhibitory influences arising from the striatum.

3. The "Extrapyramidal Motor System"

Whereas earlier anatomical studies suggested a predominantly noncortical, nonpyramidal output of the basal ganglia involved in motor functions, more recent data show that the majority of influences are directed to the cerebral cortex, particularly the supplementary motor area and the prefrontal cortex. This connectivity with the cortex is considered to be the result of increased differentiation of cortical areas with phylogenetic development, and so one would expect the noncortical output of mammalian basal ganglia from the SN and nucleus tegmenti pedunculopontinus to be engaged in more basic behavioral acts, such as orienting behavior. This is not to say that these forms of behavior have not undergone phylogenetic development, which, for example, is suggested in mammals by the increased use of eye movements in primates, as compared with head and body turning in rats (HIKOSAKA and WURTZ 1985b).

II. Dopamine System

1. Mismatch Between Negative and Positive Image

Major changes of impulse activity of DA neurons are only found in a subset of behavioral acts that are deficient after DAergic lesions. Whereas DA neurons respond well to behaviorally important stimuli they show only minor and slow changes during execution of movements and initiation of self-generated acts. This is a remarkable example of the difficulty of assessing the function of a neuronal system on the basis of deficits arising after its destruction. In the case of the DA system, this mismatch suggests that DA neurons are actively involved in reactions to external stimuli, and sustain a number of other functions which should be encoded in structures influenced by DAergic terminals. Thus, DA neurons are neither sensory nor motor in function, but are engaged in more basic processes of behavioral reactivity.

2. Paradoxical Kinesia

It is well known that DA-depleted, akinetic subjects are able to move for a short time when strong stimuli are presented. Although speculative, it is possible to imagine that these reactions are facilitated by a short release of DA from the few remaining terminals, caused by bursts of impulses, similar to the responses of DA neurons in intact animals. It can not be ruled out that other neuronal systems are engaged in this kind of emergency reaction as long as recordings from DA neurons in partially lesioned animals are lacking.

3. Neuronal Activity in Target Areas of DA Neurons

If DA neurons are engaged in behavioral reactivity, but do not encode details of sensory and motor information, some of the activity related to functions deficient after lesions should be sought in the striatum. In particular, neuronal activity in relation to initiation of behavioral acts as the key component of hypokinesia is very minor in DA neurons, but is frequently found in striatal neurons. This is found when animals are prepared for an upcoming movement by external signals (instructions) and also when movements are initiated at a self-determined rate without phasic external cues. Similarly, changes of neuronal activity during execution of movements are minor in DA neurons, but are reproducibly found in the lateral putamen.

III. Corticostriatal Activity

Outputs of the basal ganglia in non mammalian vertebrates are directed to the midbrain. In contrast, in mammals, the majority of influences of the basal ganglia are directed back to the cerebral cortex, giving rise to cortical—basal ganglia—cortical loops. The closer relationships to the phylogenetically differentiating cortex suggest a role of the basal ganglia as an important forebrain center for controlling a multitude of behavioral acts. It is then necessary to distinguish between the functions of the cortex and basal ganglia. This appears easier when considering functions for which only minor corticostriatal links exist, such as vision and audition. In respect to motor functions, onset of neuronal activity in direct relation to individual movements occurs earlier in the motor cortex than in its projection area in the putamen, suggesting that the motor cortex is "driving" these basal ganglia neurons. Putamen neurons encode the direction of movements independent of the underlying muscular pattern, whereas corticomotoneuronal cells have a very direct relation to muscle activity. Recent data on neuronal activity in relation to higher functions, such as preparation for behavioral acts, attention, and short-term memory, show certain similarities to those obtained from different areas of the association and premotor cortex. Although too early for detailed comparisons, it appears that many of these cortical functions are executed in close conjunction with basal ganglia neurons. Much less is known about limbic functions of the basal ganglia, the existence of which is suggested by the heavy afferent and efferent connections with several cortical and subcortical limbic structures. The fact that many neurons in the striatum are activated when the animal is given indications that a reward is about to be presented may partly reflect limbic influences and should be investigated more specifically.

Acknowledgments. The comments of Dr. E. Scarnati on an earlier version of the manuscript are gratefully acknowledged. The author's work is supported by the Swiss National Science Foundation.

References

Abercrombie ED, Jacobs BL (1985) Dopaminergic modulation of sensory responses of striatal neurons: single unit studies. Brain Res 358:27-33

Aebischer P, Schultz W (1984) The activity of pars compacta neurons of the monkey substantia nigra is depressed by apomorphine. Neurosci Lett 50:25-29

Aghajanian GK, Bunney BS (1977) Dopamine "autoreceptors": pharmacological characterization by microiontophoretic single cell recording studies. Naunyn Schmiedebergs Arch Pharmacol 297:1-7

Aldridge JW, Anderson RJ, Murphy JT (1980a) Sensory-motor processing in the caudate nucleus and globus pallidus: a single-unit study in behaving primates. Can J Physiol Pharmacol 58:1192-1201

Aldridge JW, Anderson RJ, Murphy JT (1980b) The role of the basal ganglia in controlling a movement initiated by a visually presented cue. Brain Res 192:3-16

Alexander GE, (1987) Selective neuronal discharge in monkey putamen reflects intended direction of planned limb movements. Exp Brain Res 67:623-634

Alexander GE, DeLong MR (1985a) Microstimulation of the primate neostriatum. I. Physiological properties of striatal microexcitable zones. J Neurophysiol 53:1401-1416

Alexander GE, DeLong MR (1985b) Microstimulation of the primate neostriatum. II. Somatotopic organization of striatal microexcitable zones and their relation to neuronal response properties. J Neurophysiol 53:1417-1430

Alexander GE, DeLong MR, Strick PL (1986) Parallel organization of functionally segregated circuits linking basal ganglia and cortex. Annu Rev Neurosci 9:357-381

Allen GI, Tsukahara N (1974) Cerebrocerebellar communication systems. Physiol Rev 54:957-1006

Amalric M, Condé H, Dormont JF, Farin D, Schmied A (1984) Activity of caudate neurons in cat performing a reaction time task. Neurosci Lett 49:253-258

Amato G, Trouche E, Beaubaton D, Grangetto A (1978) The role of internal pallidal segment in the initiation of a goal directed movement. Neurosci Lett 9:159-163

Anderson ME (1977) Discharge patterns of basal ganglia neurons during active maintenance of postural stability and adjustment to chair tilt. Brain Res 143:325-338

Anderson ME, Horak FB (1985) Influence of the globus pallidus on arm movements in monkeys. III. Timing of movement-related information. J Neurophysiol 54:433-448

Anderson ME, Yoshida M (1980) Axonal branching patterns and location of nigrothalamic and nigrocollicular neurons in the cat J Neurophysiol 43:883-895

Apicella P, Legallet E, Nieoullon A, Trouche E (1986) Differential time-course of reaction time recovery depending on variations in the amplitude of a goal-directed movement after nigrostriatal lesion in monkeys. Neurosci Lett 68:79-84

Arikuni T, Kubota K (1986) The organization of prefrontocaudate projections and their laminar origin in the macaque monkey: a retrograde study using HRP-gel. J Comp Neurol 244:492-510

Baleydier C, Mauguierre F (1980) The duality of the cingulate gyrus in monkey. Neuroanatomical study and functional hypotheses. Brain 103:525-554

Bättig K, Rosvold HE, Mishkin M (1960) Comparison of the effects of frontal and caudate lesions on delayed response and alternation in monkeys. J Comp Psychol 53:400-404

Beaubaton D, Amato G, Trouche E, Legallet E (1980) Effects of cooling on the latency, speed and accuracy of a pointing movement in the baboon. Brain Res 196:572-576

Bernardi G, Marciani MG, Morocutti C, Giacomini P (1976) The action of picrotoxin and bicuculline on rat caudate neurons inhibited by GABA. Brain Res 102:379-384

Bernardi G, Marciani MG, Morocutti C, Pavone F, Stanzione P (1978) The action of dopamine on rat caudate neurons intracellularly recorded. Neurosci Lett 8:235-240

Bloom FE, Costa E, Salmoiraghi GC (1965) Anesthesia and the responsiveness of individual neurons of the caudate nucleus of the rat to acetylcholine, norepinephrine and dopamine administered by microelectrophoresis. J Pharmacol Exp Ther 150:244-252

Bonhoeffer K (1897) Ein Beitrag zur Lokalisation der choreatischen Bewegungen. Monatsschr Psychiatr Neurol 1:6-41

Boussaoud D, Joseph JP (1985) Role of the cat substantia nigra pars reticulata in eye and head movements. II. Effects of local pharmacological injections. Exp Brain Res 57:297-304

Brauth SE, Kitt CA (1980) The paleostriatal system of *Caiman crocodilus.* J Comp Neurol 189:437-465

Brauth SE, Ferguson JL, Kitt CA (1978) Prosencephalic pathways related to the paleostriatum of the pigeon *(Columba livia).* Brain Res 147:205-221

Brauth SE, Reiner A, Kitt CA, Karten HJ (1983) The substance P-containing striatotegmental path in reptiles: an immunohistochemical study. J Comp Neurol 219:305-327

Brooks BA, Fuchs AF, Finocchio D (1986) Saccadic eye movement deficits in the MPTP monkey model of Parkinson's disease. Brain Res 383:402-407

Brown JR, Arbuthnott GW (1983) The electrophysiology of dopamine (D_2) receptors: a study of the actions of dopamine on corticostriatal transmission. Neuroscience 10:349-355

Buchwald NA, Price DD, Vernon L, Hull CD (1973) Caudate intracellular response to thalamic and cortical inputs. Exp Neurol 38:311-323

Bunney BS, Aghajanian GK, Roth RH (1973a) Comparison of effects of *L*-dopa, amphetamine and apomorphine on firing rate of rat dopaminergic neurons. Nature 245:123-125

Bunney BS, Walters JR, Roth RH, Aghajanian GK (1973b) Dopaminergic neurons: effects of antipsychotic drugs and amphetamine on single activity. J Pharmacol Exp Ther 185:560-571

Burns RS, Chiueh CC, Markey SP, Ebert MH, Jacobowitz DM, Kopin IJ (1983) A primate model of parkinsonism: selective destruction of dopaminergic neurons in the pars compacta of the substantia nigra by *N*-methyl-4-phenyl-1,2,3,6-tetrahydropyridine. Proc. Natl Acad Sci USA 80:4546-4550

Buser P, Pouderoux G, Mereaux J (1974) Single unit recording in the caudate nucleus during sessions with elaborate movements in the awake monkey. Brain Res 71:337-344

Caan W, Perrett DI, Rolls ET (1984) Responses of striatal neurons in the behaving monkey. 2. Visual processing in the caudal neostriatum. Brain Res 290:53-65

Carli M, Evenden JL, Robbins TW (1984) Depletion of unilateral striatal dopamine impairs initiation of contralateral actions and not sensory attention. Nature 313:679-682

Chang HT, Wilson CJ, Kitai ST (1981) Single neostriatal efferent axons in the globus pallidus: a light and electron microscopic study. Science 213:915-918

Chéramy A, Leviel V, Glowinski J (1981) Dendritic release of dopamine in the substantia nigra. Nature 289:537-542

Chesselet MF (1984) Presynaptic regulation of neurotransmitter release in the brain: facts and hypothesis. Neuroscience 12:347-375

Chevalier G, Deniau JM (1982) Inhibitory nigral influence on cerebellar evoked responses in the rat ventromedial thalamic nucleus. Exp Brain Res 48:369-376

Chevalier G, Thierry AM, Shibazaki T, Feger J (1981) Evidence for a GABAergic inhibitory nigrotectal pathway in the rat. Neurosci. Lett 21:67-70

Chevalier G, Vacher S, Deniau JM, Desban M (1985) Disinhibition as a basic process in the expression of striatal functions. I. The striato-nigral influence on tecto-spinal/tecto-diencephalic neurons. Brain Res 334:215-226

Chiodo LA, Antelman SM, Caggiula AR, Lineberry CG (1980) Sensory stimuli alter the discharge rate of dopamine (DA) neurons: evidence for two functional types of DA cells in the substantia nigra. Brain Res 189:544-549

Cohen G, Pasik P, Cohen B, Leist A, Mytilineou C, Yahr MD (1984) Pargyline and deprenyl prevent the neurotoxicity of 1-methyl-4-phenyl-1,2,3,6-tetrahydropyridine (MPTP) in monkeys. Eur J Pharmacol 106:209-210

Collingridge GL, Davies J (1981) The influence of striatal stimulation and putative neurotransmitters on identified neurones in the rat substantia nigra. Brain Res 212:345-359

Condé H, Benita M, Dormont JF, Schmied A, Cadoret A (1979) Control of reaction time performance involves the striatum. J Physiol (Paris) 77:97-105

Crossman AR, Sambrook MA (1978) Experimental torticollis in the monkey produced by unilateral 6-hydroxydopamine brain lesions. Brain Res 149:498-502

Crossman AR, Walker RJ, Woodruff GN (1973) Picrotoxin antagonism of gamma aminobutyric acid inhibitory responses and synaptic inhibition in the rat substantia nigra. Br J Pharmacol 49:696-698

Crossman AR, Sambrook MA, Jackson A (1980) Experimental hemiballism in the baboon produced by injection of a gamma-aminobutyric acid antagonist into the basal ganglia. Neurosci Lett 20:369-372

Crutcher MD, DeLong MR (1984a) Single cell studies of the primate putamen. I. Functional organization. Exp Brain Res 53:233-243

Crutcher MD, DeLong MR (1984b) Single cell studies of the primate putamen. II. Relations to direction of movement and pattern of muscular activity. Exp Brain Res 53:244-258

Davies J, Dray A (1976) Substance P in the substantia nigra. Brain Res 107:623-627

Davis GC, Williams AC, Markey SP, Ebert MH, Caine ED, Reichert CM, Kopin IJ (1979) Chronic parkinsonism secondary to intravenous injection of meperidine analogues. Psychiatry Res 1:249-254

DeLong MR (1971) Activity of pallidal neurons during movement. J Neurophysiol 34:414-427

DeLong MR (1973) Putamen: activity of single units during slow and rapid arm movements. Science 179:1240-1242

DeLong MR, Coyle JT (1979) Globus pallidus lesions in the monkey produced by kainic acid: histologic and behavioral effects. Appl Neurophysiol 42:95-97

DeLong MR, Georgopoulos AP (1981) Motor functions of the basal ganglia. In: Handbook of physiology. The nervous system, II. American Physiological Society, Bethesda, pp 1017-1061

DeLong MR, Crutcher MD, Georgopoulos AP (1983) Relations between movement and single cell discharge in the substantia nigra of the behaving monkey. J Neurosci 3:1599-1606

DeLong MR, Crutcher MD, Georgopoulos AP (1985) Primate globus pallidus and subthalamic nucleus: functional organization. J Neurophysiol 53:530-543

Deniau JM, Chevalier G (1985) Disinhibition as a basic process in the expression of striatal function. II. The striato-nigral influence on thalamocortical cells of the ventromedial thalamic nucleus. Brain Res 334:227-233

Deniau JM, Feger J, LeGuyader, C Striatal evoked inhibition of identified nigrothalamic neurons. Brain Res 104:152-156

Deniau JM, Chevalier GM, Feger J (1978a) Electrophysiological study of the nigrotectal pathway in the rat Neurosci Lett 10:215-220

Deniau JM, Hammond C, Chevalier G, Feger J (1978b) Evidence for branched subthalamic nucleus projections to substantia nigra, entopeduncular nucleus and globus pallidus. Neurosci Lett 9:117-121

Deniau JM, Hammond C, Riszk A, Feger J (1978c) Electrophysiological properties of identified output neurons of the rat substantia nigra (pars compacta and pars reticulata): evidence for the existence of branched neurons. Exp Brain Res 32:409-422

Deniau JM, Lackner D, Feger J (1978d) Effect of substantia nigra stimulation on identified neurons in the VL-VA thalamic complex: comparison between intact and chronically decorticated cats. Brain Res 145:27-35

Divac I, Rosvold HE, Szwarcbart MK (1967) Behavioral effects of selective ablation of the caudate nucleus. J Comp Psychol 63:184-190

Divac I, Fonnum F, Storm-Mathisen J (1977) High affinity uptake of glutamate in terminals of corticostriatal axons. Nature 26:377-378

Doudet D, Gross C, Lebrun-Grandie P, Bioulac B (1985) MPTP primate model of Parkinson's disease: a mechanographic and electromyographic study. Brain Res 335:194-199

Dray A, Oakley NR (1978) Projections from nucleus accumbens to globus pallidus and substantia nigra in the rat. Experientia 34:68-70

Dray A, Gonye TJ, Oakley NR, Tanner T (1976) Evidence for the existence of a raphe projection to the substantia nigra in rat. Brain Res 113:45-57

Evarts EV (1966) Pyramidal tract activity associated with a conditioned hand movement in the monkey. J Neurophysiol 29:1011-1027

Evarts EV, Teräväinen H, Calne DB (1981) Reaction time in Parkinson's disease. Brain 104:167-186

Feger J, Ohye C (1975) The unitary activity of the substantia nigra following stimulation of the striatum in the awake monkey. Brain Res 89:155-159

Filion M, Harnois C (1978) A comparison of projections of entopeduncular neurons to the thalamus, the midbrain and the habenula in the cat. J Comp Neurol 181:763-780

Frigyesi TL, Purpura DP (1967) Electrophysiological analysis of reciprocal caudato-nigral relations. Brain Res 6:440-456

Garland JC, Mogenson GJ (1983) An electrophysiological study of convergence of entopeduncular and lateral preoptic inputs on lateral habenular neurons projecting to the midbrain. Brain Res 263:33-41

Georgopoulos AP, DeLong MR, Crutcher MD (1983) Relations between parameters of step-tracking movements and single cell discharge in the globus pallidus and subthalamic nucleus of the behaving monkey. J Neurosci 3:1586-1598

Giorguieff MF, Kemel ML, Glowinski J (1977) Presynaptic effect of 1-glutamic acid on the release of dopamine in rat striatal slices. Neurosci Lett 6:73-77

Goldman PS, Nauta WJH (1977) An intricately patterned prefronto-caudate projection in the rhesus monkey. J Comp Neurol 171:369-386

Goldstein M, Battista AF, Ohmoto T, Anagnoste B, Fuxe K (1973) Tremor and involuntary movements in monkeys: effect of L-Dopa and of a dopamine receptor stimulating agent. Science 179:816-817

Gonon FG, Buda MJ (1985) Regulation of dopamine release by impulse flow and by autoreceptors as studied by in vivo voltammetry in the rat striatum. Neuroscience 14:765-774

Gonya-Magee T, Anderson M (1983) An electrophysiological characterization of projections from the pedunculopontine area to entopeduncular nucleus and globus pallidus in the cat. Exp Brain Res 49:269-279

Grace AA, Bunney BS (1983a) Intracellular and extracellular electrophysiology of nigral dopaminergic neurons—1. Identification and characterization. Neuroscience 10:301-315

Grace AA, Bunney BS (1983b) Intracellular and extracellular electrophysiology of nigral dopaminergic neurons—2. Action potential generating mechanisms and morphological correlates. Neuroscience 10:317-331

Graybiel AM, Ragsdale CW (1979) Fiber connections of the basal ganglia. Prog Brain Res 51:239-283

Groves PM, Wilson CJ, Young SJ, Rebec GV (1975) Self-inhibition by dopaminergic neurons. Science 190:522-529

Hallman H, Olson L, Jonsson G (1984) Neurotoxicity of the meperidine analog *N*-methyl-4-phenyl-1,2,3,6-tetrahydropyridine on brain catecholamine neurons in the mouse. Eur J Pharmacol 97:133-136

Hammond C, Deniau JM, Rizk A, Feger J (1978a) Electrophysiological demonstration of an excitatory subthalamonigral pathway in the rat. Brain Res 151:235-244

Hammond C, Deniau JM, Rouzaire-Dubois B, Feger J (1978b) Peripheral input to the rat subthalamic nucleus, an electrophysiological study. Neurosci Lett 9:171-176

Hammond C, Feger J, Bioulac B, Souteyrand JP (1979) Experimental hemiballism in the monkey produced by unilateral kainic acid lesion in corpus Luysii. Brain Res 171:577-580

Hammond C, Rouzaire-Dubois B, Feger J, Jackson A, Crossman AR (1983a) Anatomical and electrophysiological studies on the reciprocal projections between the subthalamic nucleus and nucleus tegmenti pedunculopontinus in the rat. Neuroscience 9:41-52

Hammond C, Shibazaki T, Rouzaire-Dubois B (1983b) Branched output neurons of the rat subthalamic nucleus: electrophysiological study of the synaptic effects on identified cells in the two main target nuclei, the entopeduncular nucleus and the substantia nigra. Neuroscience 9:511-520

Harnois C, Filion M (1982) Pallidofugal projections to thalamus and midbrain: a quantitative antidromic activation study in monkeys and cats. Exp Brain Res 47:277-285

Harper JA, Labuszewski T, Lidsky TI (1979) Substantia nigra unit responses to trigeminal sensory stimulation. Exp Neurol 65:462-470

Hartmann-von Monakow K, Akert K, Künzle H (1978) Projections from the precentral motor cortex and other cortical areas of the frontal lobe to the subthalamic nucleus in the monkey. Exp Brain Res 33:395-403

Heikkila RE, Hess A, Duvoisin RC (1984) Dopaminergic neurotoxicity of 1-methyl-4-phenyl-1,2,3,6-tetrahydropyridine in mice. Science 224:1451-1453

Herkenham M, Nauta WJH (1977) Afferent connections of the habenular nuclei in the rat. A horseradish peroxydase study, with a note on the fibre-of-passage problem. J Comp Neurol 173:123-146

Herrling PL (1984) Evidence for GABA as the transmitter for early cortically evoked inhibition of cat caudate neurons. Exp Brain Res 55:528-534

Herrling PL (1985) Pharmacology of the corticocaudate excitatory postsynaptic potential in the cat: evidence for its mediation by quisqualate- or kainate-receptors. Neuroscience 14:417-426

Herrling PL, Hull CD (1980) Iontophoretically applied dopamine depolarizes and hyperpolarizes the membrane of cat caudate neurons. Brain Res 192: 441-362

Hikosaka O, Sakamoto M (1986) Cell activity in monkey caudate nucleus preceding saccadic eye movements. Exp Brain Res 63:659-662

Hikosaka O and Wurtz RH (1983a) Visual and oculomotor functions of monkey substantia nigra pars reticulata. I. Relation of visual and auditory responses to saccades. J Neurophysiol 49:1230-1253

Hikosaka O, Wurtz RH (1983b) Visual and oculomotor functions of monkey substantia nigra pars reticulata. II. Visual responses related to fixation of gaze. J Neurophysiol 49:1254-1267

Hikosaka O, Wurtz RH (1983c) Visual and oculomotor functions of monkey substantia nigra pars reticulata. III. Memory-contingent visual and saccade reponses. J Neurophysiol 49:1268-1284

Hikosaka O, Wurtz RH (1983d) Visual and oculomotor functions of monkey substantia nigra pars reticulata. IV. Relation of substantia nigra to superior colliculus. J Neurophysiol 49:1285-1301

Hikosaka O, Wurtz RH (1985a) Modification of saccadic eye movements by GABA-related substances. I. Effect of muscimol and bicuculline in monkey superior colliculus. J Neurophysiol 53:266-291

Hikosaka O, Wurtz RH (1985b) Modification of saccadic eye movements by GABA-related substances. II. Effects of muscimol in monkey substantia nigra pars reticulata. J Neurophysiol 53:292-308

Hirata K, Mogenson GJ (1984) Inhibitory response of pallidal neurons to cortical stimulation and the influence of conditioning stimulation of substantia nigra. Brain Res 321:9-19

Hirata K, Yim CY, Mogenson GJ (1984) Excitatory input from sensory motor cortex to neostriatum and its modification by conditioning stimulation of the substantia nigra. Brain Res 321:1-8

Hommer DW, Bunney BS (1980) Effect of sensory stimuli on the activity of dopaminergic neurons: involvement of non-dopaminergic nigral neurons and striato-nigral pathways. Life Sci 27:377-386

Horak F, Anderson ME (1984) Influence of globus pallidus on arm movements in monkeys. I. Effects of kainic acid-induced lesions. J Neurophysiol 52:290-304

Hore J, Vilis T (1980) Arm movement performance during reversible basal ganglia lesions in the monkey. Exp Brain Res 39:217-228

Hore J, Meyer-Lohmann J, Brooks VB (1977) Basal ganglia cooling disables learned arm movements of monkeys in the absence of visual guidance. Science 195:584-586

Iansek R, Porter R (1980) The monkey globus pallidus: neuronal discharge properties in relation to movement. J Physiol (Lond) 301:439-455

Ilinsky IA, Jouandet ML, Goldman-Rakic PS (1985) Organization of the nigrothalamocortical system in the rhesus monkey. J Comp Neurol 236:315-330

Ito N, Ishida H, Miyakawa F, Naito H (1974) Microelectrode study of projections from the amygdaloid complex to the nucleus accumbens in the cat. Brain Res 67:338-341

Jackson A, Crossman AR (1983) Nucleus tegmenti pedunculopontinus: efferent connections with special reference to the basal ganglia, studied in the rat by anterograde and retrograde transport of horseradish peroxydase. Neuroscience 10:725-765

Jasper HH, Bertrand G (1966) Recording from microelectrodes in stereotactic surgery for Parkinson's disease. J Neurosurg 24:219-221

Javitch JA, D'Amato RJ, Strittmatter SM, Snyder SH (1985) Parkinsonism-inducing neurotoxin, *N*-methyl-4-phenyl-1,2,3,6-tetrahydropyridine: uptake of the metabolite *N*-methyl-4-phenylpyridine by dopamine neurons explains selective toxicity. Proc Natl Acad Sci USA 82:2173-2177

Jones DL, Mogenson GJ (1980) Nucleus accumbens to globus pallidus GABA projection: electrophysiological and iontophoretic investigations. Brain Res 188:93-105

Jonsson G (1980) Chemical neurotoxins as denervation tools in neurobiology. Annu Rev Neurosci 3:169-187

Joseph JP, Boussaoud D (1985) Role of the cat substantia nigra pars reticulata in eye and head movements. I Neural activity. Exp Brain Res 57:286-196

Joseph JP, Boussaoud D, Biguer B (1985) Activity of neurons in the cat substantia nigra pars reticulata during drinking. Exp. Brain Res 60:375-379

Jürgens U (1984) The efferent and afferent connections of the supplementary motor area. Brain Res 300:63-81

Kanazawa I, Yoshida M (1980) Electrophysiological evidence for the existence of excitatory fibers in the caudo-nigral pathway in the cat. Neurosci Lett 20:301-306

Karten HJ, Dubbeldam JL (1973) The organization and projections of the paleostriatal complex in the pigeon *(Columba livia)*. J Comp Neurol 148:61-89

Kebabian JW, Calne DB (1979) Multiple receptors for dopamine. Nature 277:93-96

Kerkerian L, Nieoullon A, Dusticier N (1983) Topographic changes in high-affinity uptake in the cat red nucleus, substantia nigra, thalamus, and caudate nucleus after lesions of sensorimotor cortical areas. Exp. Neurol. 81:598-612

Kim JS, Hassler R, Hang P, Paik KS (1977) Effect of frontal cortex ablation on striatal glutamic acid level in rat. Brain Res 132:370-374

Kim R, Nakano K, Jayaraman A, Carpenter MB (1976) Projections of the globus pallidus and adjacent structures: an autoradiographic study in the monkey. J Comp Neurol 169: 263-289

Kimura M, Rajkowski J, Evarts E (1984) Tonically discharging putamen neurons exhibit set-dependent responses. Proc Natl Acad Sci USA 81:4998-5001

Kita H, Chang HT, Kitai ST (1983) Pallidal inputs to subthalamus: intracellular analysis. Brain Res 264:255-265

Kitai ST, Deniau JM (1981) Cortical inputs to the subthalamus: intracellular analysis. Brain Res 214:411-415

Kitai ST, Kocsis JD, Preston RJ, Sugimori M (1976) Monosynaptic inputs to caudate neurons identified by intracellular injection of horseradish peroxydase. Brain Res 109:601-606

Kita T, Kita H, Kitai ST (1986) Electrical membrane properties of rat substantia nigra compacta neurons in an in vitro slice preparation. Brain Res 372:21-30

Kocsis JD, Sugimori M, Kitai ST (1976) Convergence of excitatory inputs to caudate spiny neurons. Brain Res 124:403-413

Krettek JE, Price JL (1978) Amygdaloid projections to subcortical structures within the basal forebrain and brainstem in the rat and cat. J Comp Neurol 178:225-254

Kunze W, McKenzie JS, Bendrups AP (1979) An electrophysiological study of thalamocaudate neurones in the cat. Exp Brain Res 36:233-244

Künzle H (1975) Bilateral projections from precentral motor cortex to the putamen and other parts of the basal ganglia. An autoradiographic study in *Macaca fascicularis*. Brain Res 88:195-209

Künzle H (1977) Projections from the primary somatosensory cortex to basal ganglia and thalamus in the monkey. Exp Brain Res 30:481-492

Künzle H (1978) An autoradiographic analysis of the efferent connections from premotor and adjacent prefrontal regions (areas 6 and 9) in *Macaca fascicularis*. Brain Behav Evol 15:185-234

Lamarre Y, Joffroy AJ (1979) Experimental tremor in monkey. Activity of thalamic and precentral cortical neurons in the absence of peripheral feedback. Adv Neurol 24:109-122

Langston JW, Ballard P, Tetrud JW, Irwin I (1983) Chronic parkinsonism in humans due to a product of meperidine-analog synthesis. Science 219:979-980

Langston JW, Irwin I, Langston EB (1984) Pargyline prevents MPTP-induced parkinsonism in primates. Science 225:1480-1482

Larsen KD, Sutin J (1978) Output organization of the feline entopeduncular and subthalamic nuclei. Brain Res 157:21-31

Levine MS, Hull CD, Buchwald NA (1974) Pallidal and entopeduncular intracellular responses to striatal, cortical, thalamic and sensory inputs. Exp Neurol 44:448-460

Lidsky TI, Labuszewski T, Avitable MJ, Robinson JH (1978) The effects of stimulation of trigeminal sensory afferents upon caudate units in cats. Brain Res Bull 4:9-14

Liles SL (1974) Single-unit responses of caudate neurons to stimulation of frontal cortex, substantia nigra and entopeduncular nucleus in cats. J Neurophysiol 37:254-265

Liles SL (1983) Activity of neurons in putamen associated with wrist movements in the monkey. Brain Res 263:156-161

Liles SL (1985) Activity of neurons in putamen during active and passive movements of the wrist. J Neurophysiol 53:217-236

Liles SL, Updyke BV (1985) Projection of the digit and wrist area of precentral gyrus to the putamen: relation between topography and physiological properties of neurons in the putamen. Brain Res 339:245-255

Ljungberg T, Ungerstedt U (1976) Sensory inattention produced by 6-hydroxydopamine-induced degeneration of ascending dopamine neurons in the brain. Exp Neurol 53:585-600

Ljungdahl A, Hökfelt T, Nilsson G (1978) Distribution of substance P-like immunoreactivity in the central nervous system of the rat—I. Cell bodies and nerve terminals. Neuroscience 3:861-943

Llinas R, Greenfield SA, Jahnsen H (1984) Electrophysiology of pars compacta cells in the in vitro substantia nigra—a possible mechanism for dendritic release. Brain Res 294:127-132

Lohman AHM, Van Woerden-Verkley I (1978) Ascending connections to the forebrain in the tegu lizard. J Comp Neurol 182:555-594

MacLeod NK, James TA, Kilpatrick IC, Starr MS (1980) Evidence for a GABAergic nigrothalamic pathway in the rat. II. Electrophysiological studies. Exp Brain Res 40:55-61

Maeda H, Mogenson GJ (1982) Effects of peripheral stimulation on the activity of neurons in the ventral tegmental area, substantia nigra and midbrain reticular formation. Brain Res Bull 8:7-14

Malliani A, Purpura DP (1967) Intracellular studies of the corpus striatum: II. Patterns of synaptic activities in lenticular and entopeduncular neurons. Brain Res 6:341-354

Marco LA, Copack P, Edelson AM (1973) Intrinsic connections of caudate neurons: locally evoked intracellular responses. Exp Neurol 40:683-698

Matthews RT, German DC (1986) Evidence for a functional role of dopamine type-1 (D-1) receptors in the substantia nigra of rats. Eur J Pharmacol 120:87-93

McGeer PL, McGeer EG, Scherer U, Singh K (1977) A glutamatergic corticostriatal path? Brain Res 128:369-373

Mercuri N, Bernardi G, Calabresi P, Cotugno A, Levi G, Stanzione P (1985) Dopamine decreases cell excitability in rat striatal neurons by pre- and postsynaptic mechanisms. Brain Res 358:110-121

Miller JD, Sanghera MK, German DC (1981) Mesencephalic dopaminergic unit activity in the behaviorally conditioned rat. Life Sci 29:1255-1263

Mitchell PR, Dogett NS (1980) Modulation of striatal ^{3}H-glutamic acid release by dopaminergic drugs. Life Sci 26: 2073-2081

Moon-Edley S, Graybiel AM (1983) The afferent and efferent connections of the feline nucleus tegmenti pedunculopontinus, pars compacta. J Comp Neurol 217:187-215

Mora F, Mogenson GJ, Rolls ET (1977) Activity of neurons in the region of the substantia nigra during feeding in the monkey. Brain Res 133:267-276

Napier TC, Pirch JH, Strahlendorf HK (1983) Naloxone antagonizes striatally-induced suppression of globus pallidus unit activity. Neuroscience 9:53-59

Nauta WJH (1974) Evidence of a pallidohabenular pathway in the cat. J Comp Neurol 156:19-28

Neafsey EJ, Hull CD, Buchwald NA (1978) Preparation for movement in the cat. II. Unit activity in the basal ganglia and thalamus. Electroencephalogr Clin Neurophysiol 44:714-723

Nicklas WJ, Vyas I, Heikkila RE (1985) Inhibition of NADH-linked oxidation in brain mitochondria by 1-methyl-phenylpyridine, a metabolite of the neurotoxin, 1-methyl-4-phenyl-1,2,3,6-tetrahydropyridine. Life Sci 36:2503-2508

Niki H, Sakai M, Kubota K (1972) Delayed alternation performance and unit activity of the caudate head and medial orbitofrontal gyrus in the monkey. Brain Res 38: 343-353

Nishino H, Ono T, Sasaki K, Fukuda M, Muramoto KI (1984) Caudate unit activity during operant feeding behavior in monkeys and modulation by cooling prefrontal cortex. Behav Brain Res 11:21-33

Nishino H, Ono T, Muramoto KI, Fukuda M, Sasaki K (1985) Movement and nonmovement related pallidal unit activity during bar press feeding behavior in the monkey. Behav Brain Res 15:27-42

Noda H, Manohar S, Adey WR (1968) Responses of cat pallidal neurons to cortical and subcortical stimuli. Exp. Neurol 20:585-610

Noda T, Oka H (1984) Nigral inputs to the pedunculopontine region: intracellular analysis. Brain Res 322:332-336

Obata K, Yoshida M (1973) Caudate evoked inhibition and actions of GABA and other substances on cat pallidal neurons. Brain Res 64:455-459

Ohye C, LeGuyader C, Feger J (1976) Responses of subthalamic and pallidal neurons to striatal stimulation: an extracellular study on awake monkeys. Brain Res 111:241-252

Pacitti C, Fiadone G, Gasbarri A, Civitelli D, Scarnati E (1982) Electrophysiological evidence for an inhibitory accumbens-entopeduncular pathway in the rat. Neurosci Lett 33:35-40

Parent A (1984) Functional anatomy and evolution of monoaminergic systems. Am Zool 24:783-790

Parent A, Mackey A, De Bellefeuille L (1983) The subcortical afferents to caudate nucleus and putamen in primate: a fluorescence retrograde double labeling study. Neuroscience 10:1137-1150

Park MR, Lighthall JW, Kitai ST (1980) Recurrent inhibition in the rat neostriatum. Brain Res 194: 359-369

Park MR, Falls WM, Kitai ST (1982a) An intracellular HRP study of the rat globus pallidus. I. Responses and light microscopic analysis. J Comp Neurol 211:284-294

Park MR, Gonzalez-Vegas JA, Kitai ST (1982b) Serotonergic excitation from dorsal raphé stimulation recorded intracellularly from rat caudate-putamen. Brain Res 243:49-58

Péchadre JC, Larochelle L, Poirier LJ (1976) Parkinsonian akinesia, rigidity and tremor in the monkey. J Neurol Sci 28:147-157
Percheron G, Yelnik J, François C (1984) A Golgi analysis of the primate globus pallidus. III. Spatial organization of the striopallidal complex. J Comp Neurol 227:214-227
Perkins MN, Stone TW (1980) Subthalamic projections to the globus pallidus: an electrophysiological study in the rat. Exp Neurol 68:500-511
Poirier LJ (1960) Experimental and histological study of midbrain dyskinesias. J Neurophysiol 23:534-551
Poirier J, Barbeau A (1985) 1-methyl-4-pyridinium-induced inhibition of nicotinamide adenosine dinucleotide cytochrome c reductase. Neurosci Lett 62:7-11
Poirier LJ, Sourkes TL, Bouvier G, Boucher R, Carabin S (1966) Striatal amines, experimental tremor and the effect of harmaline in the monkey. Brain 89:37-52
Pouderoux G, Freton E (1979) Patterns of responses to visual stimuli in the cat caudate nucleus under chloralose anesthesia. Neurosci Lett 11:53-58
Precht W, Yoshida M (1971) Blockade of caudate-evoked inhibition of neurons in the substantia nigra by picrotoxin. Brain Res 32:229-233
Preston RJ, Bishop GA, Kitai ST (1980) Medium spiny neuron projection from the rat striatum: an intracellular horseradish peroxydase study. Brain Res 183:253-263
Purpura DP, Malliani A (1967) Intracellular studies of the corpus striatum. 1. Synaptic potentials and discharge characteristics of caudate neurons activated by thalamic stimulation. Brain Res 6:325-340
Pycock C (1980) Turning behavior in animals. Neuroscience 5:461-514
Ragsdale CW, Graybiel AM (1981) The fronto-striatal projection in the cat and monkey and its relationship to inhomogeneities established by acetylcholinesterase histochemistry. Brain Res 208:259-266
Reiner A, Karten HJ (1982) Cells of origin of descending tectofugal pathways in the pigeon *(Columba livia)*. J Comp Neurol 204:165-187
Reiner A, Brecha NC, Karten HJ (1987) Basal ganglia pathways to the tectum: the afferent and efferent connections of the lateral spiriform nucleus of pigeon. J Comp Neurol 208:16-36
Richards CD, Taylor DCM (1982) Electrophysiological evidence for a somatotopic sensory projection to the striatum of the rat. Neurosci Lett 30:235-240
Rolls ET, Thorpe SJ, Maddison SP (1983) Responses of striatal neurons in the behaving monkey. 1. Head of the caudate nucleus. Behav Brain Res 7:179-210
Romo R, Schultz W (1985) Prolonged changes in dopaminergic terminal excitability and short changes in dopaminergic discharge rate after short peripheral stimulation in monkey. Neurosci Lett 62:335-340
Romo R, Schultz W (1986) Discharge activity of dopamine cells in monkey midbrain: comparison of changes related to triggered and spontaneous movements. Soc. Neurosci Abstr 12:207
Romo R, Chéramy A, Godeheu G, Glowinski J (1986) In vivo presynaptic control of dopamine release in the cat caudate nucleus—III. Further evidence for the implication of corticostriatal glutamatergic neurons. Neuroscience 19:1091-1099
Rosvold HE, Delgado JMR (1956) The effect on delayed-alternation test performance stimulating or destroying electrically structures within the frontal lobes of the monkey brain. J Comp Psychol 49:365-372
Rouzaire-Dubois B, Scarnati E (1985) Bilateral corticosubthalamic nucleus projections: an electrophysiological study in rats with chronic cerebral lesions. Neuroscience 15:69-79
Rouzaire-Dubois B, Scarnati E (1987) Pharmacological study of the cortical-induced

excitation of subthalamic nucleus neurons in the rat: evidence for amino acids as putative neurotransmitters. Neuroscience 21:429-440.

Rouzaire-Dubois B, Hammond C, Hamon B, Feger J (1980) Pharmacological blockade of the globus pallidus-induced inhibitory response of subthalamic cells in the rat. Brain Res 200:321-329

Rouzaire-Dubois B, Scarnati E, Hammond C, Crossman AR, Shibazaki T (1983) Microiontophoretic studies on the nature of the neurotransmitter in the subthalamo-entopeduncular pathway of the rat. Brain Res 271:11-20

Ruffieux A, Schultz W (1980) Dopaminergic activation of reticulata neurons in the substantia nigra. Nature 285:240-241

Sanghera MK, Trulson ME, German DC (1984) Electrophysiological properties of mouse dopamine neurons: in vivo and in vitro studies. Neuroscience 12:793-801

Scarnati E, Campana E, Pacitti C (1983) The functional role of the nucleus accumbens in the control of the substantia nigra: electrophysiological investigations in intact and striatum-globus pallidus lesioned rats. Brain Res 265:249-257

Scarnati E, Campana E, Pacitti C (1984) Pedunculopontine-evoked excitation of substantia nigra neurons in the rat. Brain Res 304:351-361

Scarnati E, Proia A, Campana E, Pacitti C (1986) A microiontophoretic study on the nature of the putative synaptic neurotransmitter involved in the pedunculopontine-substantia nigra pars compacta excitatory pathway of the rat. Exp Brain Res 62:470-478

Scheel-Krüger J, Arnt J, Magelund G, Olianas M, Przewlocka B, Christensen AV (1980) Behavioural functions of GABA in basal ganglia and limbic system. Brain Res Bull 5 (Suppl 2):261-267

Schell GR, Strick PL (1984) The origin of thalamic inputs to the arcuate premotor and supplementary motor areas. J Neurosci 2:539-560

Schneider JS, Lidsky TI (1981) Processing of somatosensory information in striatum of behaving cats. J Neurophysiol 45:841-851

Schneider JS, Morse JR, Lidsky TI (1982) Somatosensory properties of globus pallidus neurons in awake cats. Exp Brain Res 46:311-314

Schultz W (1982) Depletion of dopamine in the striatum as experimental model of parkinsonism: direct effects and adaptive mechanisms. Prog Neurobiol 18:121-166

Schultz W (1986a) Activity of pars reticulata neurons of monkey substantia nigra in relation to motor, sensory and complex events. J Neurophysiol 55:660-677

Schultz W (1986b) Responses of midbrain dopamine neurons to behavioral trigger stimuli in the monkey. J Neurophysiol 56:1439-1461

Schultz W (1988) MPTP-induced parkinsonism in monkeys: mechanism of action, selectivity and pathophysiology. Gen Pharmacol 19:153-161

Schultz W, Romo R (1986) Dopamine neurons of monkey midbrain discharge in response to sensory stimuli implicated in behavioral reactions. Soc Neurosci Abstr 12:207

Schultz W, Romo R (1987) Responses of nigrostriatal dopamine neurons to high intensity somatosensory stimulation in the anesthetized monkey. J Neurophysiol 57:201-217

Schultz W, Romo R (1988) Neuronal activity in the monkey striatum during the initiation of movements. Exp Brain Res 71:431-436

Schultz W, Ungerstedt U (1978) A method to detect and record from striatal cells of low spontaneous activity by stimulating the corticostriatal pathway, Brain Res 142:357-362

Schultz W, Ruffieux A, Aebischer P (1983) The activity of pars compacta neurons of the monkey substantia nigra in relation to motor activation. Exp Brain Res 51:377-38

Schultz W, Studer A, Jonsson G, Sundström E, Mefford I (1985) Deficits in behavioral initiation and execution processes in monkeys with 1-methyl-4-phenyl-1,2,3,6-tetrahydropyridine-induced parkinsonism. Neurosci Lett 59:225-232

Schultz W, Scarnati E, Sundström E, Tsutsumi T, Jonsson G (1986) The catecholamine uptake blocker nomifensine protects against MPTP-induced parkinsonism in monkeys. Exp Brain Res 63:216-220

Schwarz M, Sontag KH, Wand P (1984) Sensory-motor processing in substantia nigra pars reticulata in conscious cats. J Physiol (Lond) 347:129-147

Sedgwick EM, Williams TD (1967) The response of single units in the caudate nucleus to peripheral stimulation. J Physiol (Lond) 189:281-298

Selemon LD, Goldman-Rakic PS (1985) Longitudinal topography and interdigitation of corticostriatal projections in the rhesus monkey. J Neurosci 5:776-794

Smith Y, Parent A (1986) Differential connections of caudate nucleus and putamen in the squirrel monkey *(Saimiri sciureus).* Neuroscience 18:347-371

Soltysik S, Hull CD, Buchwald NA, Fekete T (1975) Single unit activity in basal ganglia of monkeys during performance of a delayed response task. Electroencephalogr Clin Neurophysiol 39:65-78

Spencer HJ (1976) Antagonism of cortical excitation of striatal neurons by glutamic acid diethylester: evidence for glutamic acid as an excitatory transmitter in the rat striatum. Brain Res 102:91-101

Steinfels GF, Heym J, Jacobs BL (1981) Single unit activity of dopaminergic neurons in freely moving animals. Life Sci 29:1435-1442

Steinfels GF, Heym J, Strecker RE, Jacobs BL (1983) Behavioral correlates of dopaminergic unit activity in freely moving cats. Brain Res 258:217-228

Strecker RE, Jacobs BL (1985) Substantia nigra dopaminergic unit activity in behaving cats: effect of arousal on spontaneous discharge and sensory evoked activity. Brain Res 361:339-350

Sundström E, Jonsson G (1985) Pharmacological interference with the neurotoxic action of 1-methyl-4-phenyl-1,2,3,6-tetrahydropyridine (MPTP) on central catecholamine neurons in the mouse. Eur J Pharmacol 110:293-299

Toan DL, Schultz W (1985) Responses of rat pallidum cells to cortex stimulation and effects of altered dopaminergic activity. Neuroscience 15:683-694

Trulson ME, Preussler DW, Howell GA (1981) Activity of substantia nigra units across the sleep-waking cycle in freely moving cats. Neurosci Lett 26:183-188

Tsai CT, Nakamura S, Iwama K (1980) Inhibition of neuronal activity of the substantia nigra by noxious stimuli and its modificatuion by the caudate nucleus. Brain Res 195:299-311

Tsubokawa T, Sutin J (1972) Pallidal and tegmental inhibition of oscillatory slow waves and unit activity in the subthalamic nucleus. Brain Res 41:101-118

Tulloch IF, Arbuthnott GW (1979) Electrophysiological evidence for an input from the anterior olfactory nucleus to substantia nigra. Exp Neurol 66:16-29

Ueki A, Uno M, Anderson M, Yoshida M (1977) Monosynaptic inhibition of thalamic neurons produced by stimulation of substantia nigra. Experientia 33:1480-1482

Ungerstedt U (1971a) Postsynaptic supersensitivity after 6-hydroxydopamine induced degeneration of the nigro-striatal dopamine system. Acta Physiol Scand [Suppl] 367:69-93

Ungerstedt U (1971b) Adipsia and aphagia after 6-hydroxydopamine induced degeneration of the nigro-striatal dopamine system. Acta Physiol Scand [Suppl] 367:95-117

Uno M, Ozawa N, Yoshida M (1978) The mode of pallido-thalamic transmission investigated with intracellular recording from cat thalamus. Exp Brain Res 33:493-507

Van der Maelen CP, Bonduki AC, Kitai ST (1979) Excitation of caudate-putamen neurons following stimulation of the dorsal raphé nucleus in the rat. Brain Res 175:356-361
Van der Maelen CP, Kitai ST (1980) Intracellular analysis of synaptic potentials in rat neostriatum following stimulation of the cerebral cortex, thalamus, and substantia nigra. Brain Res Bull 5:725-733
Van Hoesen GW, Yeterian EH, Lavizzo-Mourey R (1981) Widespread corticostriate projections from temporal cortex of the rhesus monkey. J Comp Neurol 199:205-219
Viallet F, Trouche E, Beaubaton D, Nieoullon A, Legallet E (1983) Motor impairment after unilateral electrolytic lesions of the substantia nigra in baboons: behavioral data with quantitative and kinematic analysis of a pointing movement. Brain Res 279:193-206
Walker RJ, Kemp JA, Yajima H, Kitagawa K, Woodruff GN (1976) The action of substance *P* on mesencephalic reticular and substantia nigra neurones of the rat. Experientia 32:214-215
Waszczak BL, Walters JR (1983) Dopamine modulation of the effects of γ-aminobutyric acid on substantia nigra pars reticulata neurons. Science 220:218-221
Webster KE (1979) Some aspects of the comparative study of the corpus striatum. In: Divac I, Oberg RGE (eds) The neostriatum. Pergamon, Oxford, pp 107-126
Whittier JR, Mettler FA (1949) Studies on the subthalamus of the monkey. II. Hyperkinesia and other physiologic effects of subthalamic lesions, with special reference to the subthalamic nucleus of Luys. J Comp Neurol 90:319-372
Wilczynski W, Northcutt RG (1983) Connections of the bullfrog striatum: efferent projections. J Comp Neurol 214:333-343
Wilson CJ (1986) Postsynaptic potentials evoked in spiny neostriatal projection neurons by stimulation of ipsilateral and contralateral neocortex. Brain Res 367:201-213
Wilson CJ, Phelan KD (1982) Dual topographic representation of neostriatum in the globus pallidus of rat. Brain Res 243:354-359
Wilson JS, Hull CD, Buchwald NA (1983) Intracellular studies of the convergence of sensory input on caudate neurons of cat. Brain Res 270:197-208
Yeterian EH, Van Hoesen GW (1978) Cortico-striate projections in the rhesus monkey: the organization of cerain cortico-caudate connections. Brain Res 139:43-63
Yim CY, Mogenson GJ (1982) Response of nucleus accumbens neurons to amygdala stimulation and its modification by dopamine. Brain Res 239:401-415
Yim CY, Mogenson GJ (1983) Response of ventral pallidal neurons to amygdala stimulation and its modulation by dopamine projections to nucleus accumbens. J Neurophysiol 50:148-161
Yoshida M and Omata S (1979) Blocking by picrotoxin of nigra-evoked inhibition of neurons of ventromedial nucleus of the thalamus. Experientia 35:794
Yoshida M, Precht W (1971) Monosynaptic inhibition of neurons of the substantia nigra by caudato-nigral fibers. Brain Res 32:225-228
Yoshida M, Rabin A, Anderson M (1972) Monosynaptic inhibition of pallidal neurons by axon collaterals of caudato-nigral fibers. Exp Brain Res 15:333-347
Yoshida M, Nakajima N, Nijima K (1981) Effect of stimulation of the putamen on the substantia nigra in the cat. Brain Res 217:169-174

CHAPTER 2

Pathology of Parkinson's Syndrome

K. JELLINGER

A. Introduction

Parkinson's syndrome or parkinsonism, involving the clinical symptoms first described by JAMES PARKINSON (1817) occurs in a variety of disorders of the central nervous system (CNS), basically characterized by dysfunction of the dopaminergic nigrostriatal system. It may or, rarely, may not be associated with distinct anatomic damage to the melanin-containing neurons of the substantia nigra, changes to the neuronal cytoskeleton, including the Lewy bodies, and pathology in other non-nigral neuronal systems, often as part of a more widespread process. The term Parkinson's disease (PD) is restricted to paralysis agitans, the idiopathic form of parkinsonism associated with formation of Lewy bodies and loss of neurons in the pars compacta of the substantia nigra, known since TRÉTIAKOFF (1919) as the principal at-risk system in this disorder. It can be accompanied by nonspecific or age-related brain pathology and a variety of other coincidental lesions elsewhere in the CNS. Unilateral parkinsonism is associated with predominant damage to the contralateral brain stem or substantia nigra (BLOCQ and MARINESCO 1893; LINDENBERG 1964; MARTINEZ and UTTERBACK 1973; OPPENHEIMER 1984), while only in exceptional cases of parkinsonism no morphological lesions of the substantia nigra can be found (DENNY-BROWN 1960; LEVERENZ and SUMI 1986; MORRIS et al. 1987). On the other hand, moderate or even considerable damage to the substantia nigra, e.g., in aged subjects or senile dementia of the Alzheimer type (SDAT), may not be necessarily associated with parkinsonian symptoms (HAKIM and MATHIESON 1979; JACOB 1983; HEILIG et al. 1985; LEVERENZ and SUMI 1986; JELLINGER 1987b; DITTER and MIRRA 1987). Since FOIX and NICOLESCO (1925) reported on the major pathological findings in PD, suggesting a "diffuse and localized degeneration," many morbid anatomical studies of this and related disorders have been published (HASSLER 1938; KLAUE 1940; GREENFIELD and BOSANQUET 1953; HALLERVORDEN 1957; RICHARDSON 1965; STADLAN et al. 1966; FORNO 1966; EARLE 1968; FORNO and ALVORD 1971; LEWIS 1971; ESCOUROLLE et al. 1971; ALVORD et al. 1974; JELLINGER 1974; 1986a, b, 1987a, b; PEARCE 1979; BLACKWOOD 1981; OPPENHEIMER 1984; SCHOENE 1985; GIBB 1986, 1988; GRAY 1988). This chapter will outline the cytoskeletal pathology occurring in the neuronal population in parkinsonism, the pathomorphology of the major types of parkinsonism, and stress recent data on the morphological correlates of pathobiochemical findings and dementia in PD.

Table 1. Neuronal cytoskeletal pathology in parkinsonism

Type of Lesion (location)	Ultrastructure	Immunocytochemistry							
		NFP	Tau protein	MAP	PHF	Ubiquitin	Actin	Tubulin	Thioflavin-S
Lewy body (cytoplasm)	8 to 10 nm filaments (intermediate type)	++	−	−(+)	+	++	−	+	+
a) Early/cortex	Random arrangement (mainly 10 nm filaments)	+ (diff)	−	−	+	++	−	−	+
b) Mature/brain stem	Central condensation (core)	+/++	−	−	+/++	++	−	+	+
	Peripheral radiation (halo)	(++)	−	+	+(++)	+	−	+	+
Hirano Body (juxtanuclear)	8 to 10 nm filaments (intermediate)	−	+	+	+	?	+	−	−
Eosinophilic Granules (cytoplasm)	Paracrystalline 8.5-nm filaments	−	−	−	−	−	+	−	−
Marinesco bodies (intranuclear)	Lattice-like arrangement of 10 nm Φ filaments	+	−	?	−	+	−	−	−
ANFT (cytoplasm)	10 nm PHFs twisted at 80-nm intervals (andoccasional 15-nm straight filaments)	++	++	+/++	++	++	−	−	+++
NFTs in PSP (cytoplasm)	15 to 20 nm straight filaments (andoccasional PHF/twisted filaments)	++	+	?	(+)	?	−	−	++
Granulovacuolar degeneration	Granules and membrane-bound vesicles (hippocampal neurons)	+	+	?	+	?	−	+	−
Axonal Spheroids	tubulovesicular, dense bodies, 10 nm filaments, lamellar SER structures	++	−	?	−	++	−	−	+
Neuritic plaque	Dystrophic neuritic terminals, filaments, abnormal synapses, glia, amyloid	++	+	+	++	+	−	−	++

NFP neurofilament proteins; MAP microtubule-associated proteins; PHF paired helical filaments; SER smooth endoplasmic reticulum

B. Cytoskeletal Pathology

In the parkinsonian brain, neurons exhibit several types of cytoskeletal abnormalities, including Lewy bodies and other cytoplasmic inclusions, paranucleolar Marinesco bodies, Hirano bodies, granulovacuolar degeneration, neurofibrillary tangles, grumelose degeneration, and dystrophic axons (Table 1). Each of these forms of structural pathology can be found in normal and aged brains without neurologic disorders and in various CNS diseases. However, in PD and other degenerative types of parkinsonism, some of these structural abnormalities are abundant, and their presence may serve as diagnostic criteria. Because cytoskeletal elements are important in the maintenance of neuronal size and shape, in intracellular transport processes, and in the organization of both cytoplasmic elements and membrane constituents, nerve cells exhibiting these abnormalities may not be able to carry out normal functions. In addition, demented and senile individuals with parkinsonism may have senile (neuritic) plaques (NP) and deposits of amyloid in the neural parenchyma.

I. Lewy Bodies

These cytoplasmic inclusions, first described by Lewy (1913) in the substantia innominata and the dorsal vagal nucleus in a patient with idiopathic PD, appear as single or multiple eosinophilic bodies surrounded by a less dense amorphous halo. They vary in size from 3 to 20 μm (mean 9-10 μm), and can be located in a nerve cell body (classical perikaryal), within nerve cell processes (intraneuronal), and extraneuronally free in the neuropil (Forno 1982, 1986). They are composed of proteins, free fatty acids, sphingomyelin, and polysaccharides, their core containing aromatic α-amino acids (Hartog-Jager 1969). Electron probe microanalysis demonstrated a high sulfur content, indicating breakdown products of proteins (Kimula et al. 1983). Ultrastructurally, Lewy bodies are composed of intermediate-type filaments of diameter 8-10 nm, admixed with vesicular and granular material (Duffy and Tennyson 1965; Roy and Wolman 1969). Immunocytochemically, they react with monoclonal antibodies that recognize all three of the neurofilament proteins (Forno et al. 1983; Goldman et al. 1983; Nakazato et al. 1984; Kahn et al. 1985) and that recognize both phosphorylated and nonphosphorylated neurofilament epitopes (Forno et al. 1986; Sima et al. 1986; Galloway et al. 1986), tubulin (Galloway et al. 1986) and occasionally microtubule-associated protein MAP 2, particularly in the periphery of some Lewy bodies (Dickson et al. 1987), but not actin and not antibodies to human brain microtubules reacting with both Alzheimer neurofibrillary tangles (NFTs) (Yen et al. 1986) and with filaments in Pick bodies (Rasool and Selkoe 1985), and microtubule-associated tau proteins (Galloway et al. 1986, 1988; Yen et al. 1986; Kuzuhara et al. 1987; Dickson et al. 1987; Bancher et al. 1989). Their core reacts strongly with monoclonal antibodies to paired helical filaments (PHFs) (Popovitch et al. 1987; Kuzuhara et al. 1987; Bancher et al. 1989), and with ubiquitin, a polypeptide required for the ATP-dependent nonlysosomal protein breakdown (Mori et al. 1987; Kuzuhara et al. 1988) which has recently been shown to be associated with PHF (MORI et al. 1987; G PERRY et al. 1987b). Lewy bodies show different ultrastructural arrangements:

1. A mature or *brain stem type*, consisting of a central core composed of densely packed filements and granular material, and peripheral radiation of filaments, showing immune reaction against neurofilament proteins, mainly in the core (PAPPOLA 1986), and against ubiquitin and PHFs, particularly in the periphery (KUZUHARA et al. 1988), while we have seen strong reaction with monoclonal antibodies against isolated PHFs only in the central core (BANCHER et al. 1989);
2. The early or *cortex type* which is rather homogeneous with random arrangement of the filaments without a marked dense core (IKEDA et al. 1978; KOSAKA 1978; KOSAKA et al. 1984; DICKSON et al. 1987; SIMA et al. 1986), and shows diffuse reaction with antibodies to neurofilament proteins and PHFs (SIMA et al. 1986; GALLOWAY et al. 1986; POPOVITCH et al. 1987; DICKSON et al. 1987; BANCHER et al. 1989).

Similar homogeneous, poorly staining intraneuronal hyaline (colloid) inclusions or "pale bodies" (HASSLER 1938), composed of randomly arranged collections of 10 nm filaments (ROY and WOLMAN 1969) are found in many neurons in PD and are suggested to be related inclusions from which the Lewy bodies may form as a result of concentration of filaments (GIBB and LEES 1987). The results of recent ultrastructural and immunohistochemical studies suggest ectopic phosphorlyation of neurofilament proteins associated with the formation of Lewy bodies occurring as a cause or a consequence of impaired assembly and transport of neurofilament proteins (CLARK et al. 1986), the different forms of Lewy bodies probably representing different stages of impaired assembly or transport of neurofilament units. Monoclonal antibodies against Lewy bodies have been raised from parkinsonian brain (HIRSCH et al. 1985), but the molecular basis and principal mechanisms for their development are unknown. They have been suggested to represent accumulation of neurofilaments undergoing progressive structural breakdown with formation of the body's core (GIBB and LEES 1987), or sequestrated neurofilaments after neuronal regeneration (PRICE et al. 1986), or retrograde degeneration (APPEL 1981). Lewy bodies in catecholaminergic neurons show immunoreactivity with tyrosine hydroxylase (TH) antisera, suggesting that the TH enzyme plays a role in their production (NAKASHIMA and IKUTA 1984b). Although the Lewy bodies frequently affect aminergic neurons (OHAMA and IKUTA 1976) and in the cerebral cortex show a distribution partly corresponding to that of dopaminergic axon terminals (YOSHIMURA 1983), they appear not to represent a cytoskeletal abnormality specific for any chemically defined neuronal system (see Table 3). Lewy bodies occur in most cases of idiopathic PD with a wide distribution (see Table 4), but also in pigmented brain stem nuclei in 5% of normal controls over the age of 30 years, and in 8%-15% of normal brain over the age of 65 years (FORNO 1969, 1982; FORNO and ALVORD 1971; GIBB 1988), but in 45%-60% of all cases of Alzheimer's disease (AD) and senile dementia of the Alzheimer type (SDAT) (JELLINGER 1986a, 1987a; DITTER and MIRRA 1987; see also Table 4). In addition, Lewy bodies in the substantia nigra and brain stem have been found in several cases of progressive supranuclear palsy

(PSP) (Yahr et al. 1972; Mori et al. 1986), and in other CNS disorders, e.g., ataxia-telangiectasia, Hallervorden-Spatz disease, spino-olivopontocerebellar atrophy, and chronic panencephalitis (Forno 1982, 1986; Gibb 1986, 1988; Mori et al. 1986; Jellinger 1986a, 1987a). Hence, they are not pathognomonic for parkinsonism, but are considered probably to be more characteristic of this disorder than the NFTs for AD/SDAT (Forno 1982, 1986).

II. Hirano Bodies

These rod-shaped eosinophilic intracytoplasmic inclusions, first described in the Ammon's horn of Guamanian patients with amyotrophic lateral sclerosis and parkinsonism-dementia complex (Hirano et al. 1966), consist of crystalloid arrays of interlacing 6 to 10 nm filaments of lattice-like or "herringbone" structure (Hirano 1981). They are composed of actin, a contractile protein (Yen et al. 1986; Galloway et al. 1987) and react with tau, but not with neurofilament proteins and tubulin (Peterson et al. 1988). Hirano bodies are found in the hippocampus and neocortex of normal individuals from the age of 20 years onward, increasing with age, occasionally combined with granulovacuolar degeneration, and increased numbers are seen in Parkinson's, Alzheimer's, and Pick's disease (Gibson 1978; Hirano 1981).

III. Intracytoplasmic Eosinophilie Granules

The melanin-containing neurons of the substantia nigra and locus ceruleus in normal aged human brain, and various disorders may contain small intracytoplasmic eosinophilie granules composed of aggregates of parallel banded or twisted 8.5-nm filaments (Schochet et al. 1970); others regard them as altered mitochondria (Hirano 1981).

IV. Marinesco Bodies

These intranuclear eosinophilic bodies, described by Marinesco (1912) in the pigmented neurons of the substantia nigra and locus ceruleus, consist of aggregates of moderately osmiophilic granular material with a lattice-like arrangement of filaments (Leestma and Andrews 1969) that stain with a monoclonal antibody to PHF which also recongizes determinants of ubiquitin (Bancher et al. 1989). They are found in increasing number with advancing age (Yuen and Baxter 1963).

V. Neurofibrillary Tangles

These flame-shaped or globose intracytoplasmic lesions in neurons appear in two major forms.

1. Alzheimer's Neurofibrillary Tangles

(ANFTs) are composed of close-packed 10 nm PHFs twisted at 80 nm intervals (Kidd 1964) which can be admixed with 10 nm neurofilaments and 12 to

15 nm straight filaments (YAGISHITA et al. 1981; G. PERRY et al. 1987 a). Each of the 10-nm filaments consists of four protofilaments composed of eight longitudinally connected globules with short side arms (WISNIEWSKI and WEN 1985). They are composed of bonds of cross-connected proteins containing antigen determinants of various abnormal neurofilament and non-neurofilament proteins and microtubule epitopes reacting with anti-NFT, monoclonal anti-PHF, microtubule-associated anti-MAP_1 and anti-MAP_2, anti-tau protein (PRICE et al. 1986; MILLER et al. 1986; YEN et al. 1986), and with ubiquitin (MORI et al. 1987; KUZUHARA et al. 1988; G. PERRY et al. 1987b), indicating some chemical cytoskeletal abnormalities with deposition in the neuronal cytoplasma. ANFTs are found in increasing density in brains of normal aged subjects, and are particularly abundant in AD/SDAT, postencephalitic parkinsonism, Guam ALS-Parkinson-dementia complex, and boxer's dementia (TOMLINSON and CORSELLIS 1984; HIRANO 1981; YEN et al. 1986; JELLINGER 1989). NFTs involve neurons of varying transmitter specificities, including cholinergic neurons in the basal forebrain (RASOOL et al. 1986). They may also occur in a variety of other conditions, and have been suggested to indicate premature (local) aging processes (WISNIEWSKI et al. 1979, 1985).

2. Tangles in Progressive Supranuclear Palsy

The NFTs in PSP differ from ANFTs by their topography (JELLINGER et al. 1980) and ultrastructure, since they are composed of 15 to 20 nm straight filaments that are sometimes mixed with typical PHFs (TAKAUCHI et al. 1983; TELLEZ-NAGEL and WISNIEWSKI 1973; YAGISHITA et al. 1979). They are recognized by anti-neurofilament, anti-tau and anti-ANFT antibodies (DICKSON et al. 1985; YEN et al. 1986), while isolated monoclonal anti-PHF recognized only few PSP tangles owing to admixed PHF (BANCHER et al. 1987). Both PHF and straight filaments were recognized by all antibodies after extraction with an ionic detergent (G. PERRY et al. 1987a). These findings suggest that the straight filaments contain most, if not all of the antigens known to be present in PHF, and result from alternative pathways of organization of the same components as the PHF, rather than being PHF precursors. The pathways for formation of NFTs with different ultrastructural morphology are unknown.

VI. Granulovacuolar Degeneration

These cytoplasmic inclusions first described by SIMCHOWICZ (1911) are present in pyramidal neurons of the hippocampus and consist of electron-dense granules surrounded by a membrane-bound vacuole (HIRANO 1981). The contain tubulin, phosphorylated neurofilament epitopes and tau (DICKSON et al 1987b). These bodies are common in elderly people, increasing in number with age, and particularly in AD/SDAT, and correlations have been reported between their frequency and the loss of pyramidal cells and numbers of ANFTs (BALL 1977).

VII. Axonal Dystrophy and Grumelous Degeneration

"Grumelous degeneration" in the reticular zone of the SN (TRÉTIAKOFF 1919) represents spheroid-like or ovoid swellings of axons, considered as a special form of axonal degeneration referred to as axonal dystrophy. The spheroids, which include accumulations of organelles, tubulovesicular structures, disarranged 10 nm neurofilaments, dense bodies, abnormal mitochondria, and lamellar membranes derived from the smooth endoplasmic reticulum, involve presynaptic axon terminals and have been related to disorders of axonal transport. The lesion affects primarily the nucleus gracilis and reticular zone of the substantia nigra in elderly subjects, with close correlation to age, but is also seen in a variety of disorders referred to as "neuroaxonal dystrophies" (SEITELBERGER 1986). In PD the incidence of axonal spheroids in the reticulata nigrae is significantly increased after the age of 50 years, and may be related to the depletion of pigmented neurons in the compacta nigrae with loss of contact to the postsynaptic neuron (JELLINGER 1973). Grumose degeneration of dentate neurons in PSP also consists of clustered terminal axons (MIZUSAWA et al. 1989).

VIII. Neuritic Plaques and Amyloid

Senile or neuritic plaques (NPs) are homogeneous globular, spherical dense regions in the neuropil, consisting of swollen presynaptic and postsynaptic degenerating, regenerating, or dystrophic neuritic terminals, abnormal synapses, phagocytes, glia, filamentous proteins, and extracellular amyloid. Some axon terminals in the plaques contain serotonin, somatostatin, substance P, and acetylcholinesterase (TERRY 1985). Formation of the NPs has been related to degeneration of neuronal processes without any relationship to specific transmitter systems (STRUBLE et al. 1985). The origin of plaque amyloid and its relation to amyloid angiopathy in the senile brain are hitherto unknown (TERRY 1985; PRICE et al. 1986; JELLINGER 1989). Neuritic plaques are seen in the hippocampus and neocortex, but also in subcortical nuclei and brain stem, in aged subjects (with a close correlation to age), and in increased density and distribution in AD/SDAT (JAMADA and MEHRAEIN 1968; TOMLINSON and CORSELLIS 1984; MANN 1985; TERRY 1985; CRYSTAL et al. 1987; JELLINGER 1989).

C. Major Types of Parkinsonism

In large autopsy series, including a personal sample of 580 cases, degenerative forms of parkinsonism account for 75%-95%, including PD (60-75%) and multisystem degenerations (about 15%), while the incidence of postencephalitic forms is decreasing. In two personal autopsy series, it was 12% between 1957 and 1970, and 2.6% in recent years (JELLINGER 1986a, 1987a); 5%-7% of parkinsonism is associated with cerebrovascular disease, about 6% with Alzheimer's disease or SDAT, while confirmed cases due to trauma, tumors, intoxication, and drugs are rare, some of the latter presenting anatomic changes characteristic of idiopathic PD (RAJPUT et al. 1982). A small number of cases remain unclassified. A survey of the different types of parkinsonism, their average incidence in autopsy materials, and the major types and distribution of CNS lesions is given in Table 2.

I. Parkinson's Disease

Representing one of the commonest types of adult onset, chronic degenerative disorders of the CNS, and accounting for 75%–95% of all confirmed cases of parkinsonism, paralysis agitans is characterized by anatomic lesions mainly confined to the brain stem, with progressive degeneration of the nigrostriatal and mesocorticolimbic dopaminergic systems, and less frequently the hypothalamic dopaminergic neurons associated with progressive loss of melanin-containing mesencephalic neurons. Other non-nigral neuronal systems are affected as well, while neurons in the basal ganglia and cerebral cortex are comparatively spared. Dysfunction of at-risk neurons is associated with cytoskeletal abnormalities, the principal type being the Lewy body.

Grosslys, the brain is generally unremarkable, except in old and demented patients who may show diffuse cortical atrophy and additional Alzheimer changes. The brain stem shows pallor of pigmented nuclei (substantia nigra and locus ceruleus) with variable unilateral and, more often, bilateral symmetric loss of melanin-containing neurons and gliosis. Particularly involved are the ventral (central) and caudal parts of the zona compacta nigrae projecting to the putamen, the ventral tegmental area (VTA), locus ceruleus, and dorsal vagal nucleus. In addition, there is frequent involvement of the nucleus basalis of Meynert (NBM), the dorsal raphe nucleus, and tegmental pedunculopontine nucleus, pars compacta (see Tables 5–9). The nigrostriatal and mesolimbic parts of the mesencephalic melanin cell groups are affected with different intensity Although both parts lose about two-thirds of their neurons, the volume of the remaining cells in the medial group seems to be better preserved. In the most medial nucleus niger suboculomotorius, the neuronal loss (about 40%) is least severe (BOGERTS et al. 1983), while the hypothalamic dopaminergic neurons appear to be intact (MATZUK and SAPER 1985). Ultrastructurally, melanin granules in nigral or locus ceruleus neurons often cannot be distinguished from those in normal brain, but there can be a decrease in the very dense component of the melanin granules within neurons of parkinsonian cases. The substructure of the altered melanin granules, including the presence of linear arrays, resembles that of lipofuscin (DUFFY and TENNYSON 1965), but the relationship between these ultrastructural changes of melanin and the disintegration of melanin-containing cells remains obscure. Neuronal loss has also been reported in the striatopallidum and other brain stem nuclei, while studies revealing neuronal loss in the cerebral cortex are inconsistent (ALVORD et al. 1974). In the third layer of the isocortex, an increased number of small to medium-sized pyramidal cells with lipofuscin-filled extensions of their proximal axons have been observed (STOCKHAUSEN and BRAAK 1984), but the significance of these age-related cortical changes is unknown.

Table 2. Classification, incidence, and histopathology of different types of parkinsonism (P)

Disease	Incidence (%)		Age of onset	Duration	Neuropathological findings				
	Literature[a] (mean)	JELLINGER 1957-86[b]	(years)		Substantia nigra	LB (%)	NFT (%)	Extranigral lesions: Brain stem	Extranigral lesions: Other CNS regions
Paralysis agitans (idiopathic P)	44-94 (75)	68,5	50-79	9-12	Focal damage; neuron + melanin loss + gliosis	99	18	Neuron loss + gliosis: LC, dorsal vagus nucleus, nucleus basalis, nucleus dorsalis raphe	nonspecific or age-related
Other degenerative P	0-30 (7)	14.1							
Senile P (PD/SDAT)	?	5.7	~80	2-3	Mild neuron + melanin loss	45	60	Neuron loss LC, NFT + SP brain stem	Severe senile lesions, SDAT
Diffuse Lewy body disease	?	0.5	38-75	4-13	Focal neuron + melanin loss + gliosis	100	80	Widespread LB + neuron loss	Widespread LB and frequent NFT + SP
Striatonigral degeneration	?	2.8	45-60	2-7	Focal neuron loss + gliosis	90	0	Multiple system degenerations, no NFT	Atrophy + hyperpigmented putamen, often associated OPCA

Table 2 (continued)

Disease	Incidence (%)		Age of onset	Duration	Neuropathological findings				
	Literature[a] (mean)	JELLINGER 1957–86[b]	(years)		Substantia nigra	LB	NFT	Extranigral lesions	
						(%)		Brain stem	Other CNS regions
Multisystem atrophies	?	2.2	40–50	Many	Focal or diffuse neuron loss	Frequent	Rare	Olivopontocerebellar, spinocerebellar degenerations	Joseph's disease, nigrocerebellar degeneration
Progressive supranuclear palsy	?	2.2	20–60	4–20	Diffuse neuron loss + gliosis	0	100	Systemic neuron loss + gliosis + straight NFT	Little or no cortical lesions
Parkinson-dementia complex	? (Guam, Europe)	0.7	40–50	4	Neuron loss + gliosis	0	90	Neuron loss + gliosis + NFT tegmentum + nucleus basalis of Meynert	NFT hippocampus, thalamus, pallidum
Vascular P (multi-infarct)	0–15 (2)	6.9	70–80	3–4	Vascular lesions, lacunar state	13	10	Vascular lesions, lacunar state basal ganglia	Multi-infarct encephalopathy
Posttraumatic P	0–1 (0.2)	0.5	Any	Many	Traumatic or vascular focal necroses	0	0	Primary or secondary traumatic lesions in brain stem and/or basal ganglia	

Boxer's dementia	?	0	40–60	Many	Neuron loss + NFT	0	100	Multiple NFT brain stem	Multiple NFT, no senile plaques
Toxic (CO, CS, MPTP, drugs, manganese)	?	0.9	Any	Many	Diffuse or focal damage + gliosis	0	0	Damage to globus pallidus, nonspecific lesions	Nonspecific caudate nucleus (neuroleptic-induced P)
Symptomatic P	?	2.9	Any	Many	Local damage	Rare	0	Various lesions: tumors, multiple sclerosis, Alzheimer's disease, Hallervorden–Spatz disease, neurolipidoses, SSPE, Creutzfeldt–Jakob disease, etc.	
Unclassified	0–19 (2.6)	1.9	Any	Many	Focal damage	Occasional	?	Various nonspecific lesions	

[a] About 1400 autopsy cases: HASSLER 1938, KLAUE 1970, BEHEIM-SCHWARZBACH 1952, GREENFIELD and BOSANQUET 1952, HALLERVORDEN 1957, FORNO 1966, EARLE 1968, RICHARDSON 1965, STADLAN et al. 1966, FORNO and ALVORD 1971, ESCOUROLLE et al. 1975, YAHR et al. 1972, ALVORD et al. 1974, JACOB 1983, GRILIA et al. 1987

[b] 580 autopsy cases

[c] 12% between 1957 and 1970, but 2.7% between 1971 and 1986

LB Lewy bodies; LC locus ceruleus; NFT neurofibrillary tangles; OPCA olivopontocerebellar atrophy; SP senile plaques; SSPE subacute sclerosing encephalitis; ST straight filament tangles

Significant differences in the intensity and pattern of pathologic lesions between the various clinical subtypes of PD, i.e, one with predominant rigidity and bradykinesia/akinesia and one with prominent tremor (MORTIMER et al. 1982; ZETUSKY et al. 1985), or the akinetic-rigid, tremor-dominant, and mixed types (POEWE and GERSTENBRAND 1986; BARBEAU 1986) appear not to have been observed up to the present, although rigidity and akinesia much more than tremor seem to be correlated with progressive dysfunction of the striatonigral system and nigral cell loss (BERNHEIMER et al. 1973; AGID et al. 1987; JELLINGER 1987a).

The principal cytoskeletal pathology of PD is the Lewy body which, in 85%-100% of cases, occurs in many aminergic and other subcortical nuclei,

Table 3. Distribution of Lewy bodies in Parkinson's Disease (HARTOG-JAGER and BETHLEM 1960; ISHII 1966; OHAMA and IKUTA 1976; HUNTER 1985; JELLINGER 1987a)

Affected Region[a]	Major (putative) neuromediators[b]	Frequent	Rare
Cerebral cortex	Multiple		*
Substantia innominata	Acetylcholine	**	*
Hypothalamus, lateral, posterior	Norepinephrine		*
Hypothalamus, periventricular	Multiple	*	
Subthalamic nucleus	Dopamine		*
Periaqueductal gray	Multiple		*
Substantia nigra	Dopamine	**	
Nucleus parabrachialis pigmentosus	Dopamine	**	
Nucleus paranigralis	Dopamine	**	
Westphal-Edinger nucleus	Acetylcholine	*	
Darkschewitsch nucleus	Acetylcholine		*
Supratrochlear nucleus	Serotonin	*	
Nucleus tegmenti pedunculopontinus	Acetylcholine	*	
Central pontine gray	multiple		*
Locus ceruleus	Norepinephrine	**	
Nucleus subceruleus	Norepinephrine	*	
Nucleus pontis centralis oralis	Serotonin	*	
Central superior nucleus of raphe	Serotonin		*
Processus griseum pontis supralemniscalis	?	*	
Dorsal motor nucleus of vagus	Norepinephrine	**	
Nucleus of Roller	?		*
Nucleus gigantocellularis	Serotonin		*
Nucleus paragigantocellularis lateralis	?		*
Nucleus medullae oblongatae centralis	Serotonin		*
Spinal cord, intermediolateral column	Multiple		*
Spinal cord, intermediomedial column	Acetylcholine		*
Spinal cord, anterior horn			*
Autonomic (sympathetic) ganglia		**	
Enteric (parasympath.) nerve plexuses[c]	Catecholamines		*

[a] Nomenclature of the brain stem nuclei according to OLSZEWSKI and BAXTER 1982
[b] NIEUWENHUYS 1985; [c] WAKABAYASHI et al. 1988
* Mild; ** Severe

spinal cord, sympathetic ganglia and, less frequently, in the cortex, parasympathetic myenteric plexuses and adrenal medulla (Table 3). However, they are not found in all cases of PD (EARLE 1968; FORNO 1982), and some regions, e.g., the periventricular hypothalamus, are relatively spared (LANGSTON and FORNO 1978). Degeneration of the dopaminergic nigrostriatal system, being the hallmark of PD, can be variably associated with pathology in non-nigral aminergic pathways and other neuronal systems, and other cytoskeletal pathology, including Hirano bodies, granulovacuolar degeneration, dystrophic axons, and Alzheimer pathology, particularly in old and demented patients (ALVORD et al. 1974; BOLLER et al. 1979; HAKIM and MATHIESON 1979; BOLLER 1985; JELLINGER 1987a). In autopsy cohorts of PD individuals, the incidence of additional AD/SDAT pathology ranges from 5 % to 47 % (HAKIM and MATHIESON 1979; GASPAR and GRAY 1984; BOLLER 1985; E. K. PERRY et al. 1985, 1987; SUDHAKER et al. 1987; GRILIA et al. 1987; JELLINGER 1987a, b). Additional cerebrovascular lesions, i.e., about 30 % with lacunar state of the basal ganglia and 6 %-12 % of old cerebral infarcts, show about the same incidence as in age-related controls (JELLINGER 1986a, 1987a; DUBINSKY and JANKOVIC 1987). Concurrence of PD with other degenerative multiple system degenerations and aging processes may occur and is associated with Parkinson-plus syndromes (FISHER 1984; BARBEAU 1986).

II. Parkinson's Disease and Alzheimer's Disease

The relationship and communality between the two age-related neurodegenerative disorders has received increasing attention (BOLLER 1985; LEVERENZ and SUMI 1986; DITTER and MIRRA 1987), and has been the subject of some speculation (QUINN et al. 1986). Postmortem demonstration of severe brain atrophy, with significantly reduced brain weight (HAKIM and MATHIESON 1979) and a six- to ninefold increase of AD pathology over that in age-matched controls (BOLLER et al. 1979) have suggested an increased incidence of AD in patients with PD and some links between these two disorders. The neuropathologic criteria for the diagnosis of AD/SDAT have been summarized by KHATCHATURIAN (1985) and WISNIEWSKI and MERZ (1985); the comparative neuropathology of AD and PD by JELLINGER (1987b, c).

The prevalence of extrapyramidal signs in AD patients ranges from 33 % to 90 %, and particularly includes rigidity and bradykinesia (PEARCE 1974; SULKOVA et al. 1983; MOLSA et al. 1984; LEVERENZ and SUMI 1986; CHUI et al. 1986; MORRIS et al. 1987) and that of PD symptoms in AD subjects was found to be 2.4-3.6 times greater than in aged-matched control populations (SULKOVA et al. 1983; LEVERENZ and SUMI 1986), while MORRIS et al. (1987), in a small cohort of patients with SDAT, reported development of symptoms in 34 %, which was ten times higher than in age-matched controls. PD signs and symptoms in AD are often, but not invariably, associated with damage to the substantia nigra and locus ceruleus, with the presence of Lewy bodies and with or without ANFTs (SUGIMURA et al. 1977; ROSENBLUM and GHATAK 1979; JACOB 1983; DITTER and MIRRA 1987). In autopsy cohorts of AD, additional PD pathology has been reported in 4 % (XUEREB et al. 1987) up to

66 % of cases (Perl et al. 1984; Leverenz and Sumi 1986; Jellinger 1987a; McGeer et al. 1988). In one series, 85 % of the AD subjects with clinical PD signs had typical pathology of PD (Leverenz and Sumi 1986), while in some AD patients with clinical PD symptoms, no nigral damage or other PD markers were found (Sulkova et al. 1983; Leverenz and Sumi 1986; Ditter and Mirra 1987; Morris et al. 1987). In general, the substantia nigra in AD

Table 4. Brain stem lesions in parkinsonism, Alzheimer's disease, senile dementia and controls (percentage of total cases)

Type of disorder	PD	Senile P (PD/SDAT)	Vascular P (MID)	PEP	AD	SDAT	Controls
	$N = 306$	$N = 33$	$N = 39$	$N = 18$	$N = 114$	$N = 325$	$N = 284$
Mean age (years ± SEM)	74.0 ± 0.9	83.6 ± 1.1	77.5 ± 1.2	63.6 ± 1.9	65 ± 0.5	81.6 ± 0.6	74 ± 0.6
Substantia nigra							
Neuron loss 3+	93.8	0	30.0	100.0	8.8	2.2	0
2+	6.2	9.1	70.0	0	6.1	5.9	1.1
1+	0	90.9	0	0	68.6	81.9	26.0
Lewy bodies	99.5	45.4	12.8	0	39.5	40.3	6.0
NFT	17.6	57.6	10.2	94.4	68.6	38.3	1.8
Locus ceruleus							
Cell loss	100.0	94.0	25.6	100.0	100.0	100.0	13.4
Lewy bodies	98.2	51.5	18.0	5.9	45.6	57.4	18.3
NFT	16.7	63.7	10.2	94.4	77.1	65.4	4.1
Dorsal vagus nucleus							
Cell loss	90.2	31.2	5.1	11.8	NE	NE	2.8
Lewy bodies	86.1	9.1	5.1	11.8	NE	NE	2.8
Nucleus basalis	($N = 94$)				($N = 36$)	($N = 171$)	($N = 43$)
Lewy bodies	97.9	NE	NE	NE	8.0	15.0	7.5
NFT	36.2	NE	NE	NE	100.0	97.0	7.5
Westphal-Edinger nucleus	($N = 30$)				($N = 10$)	($N = 48$)	
Lewy bodies	96.7	NE	NE	NE	10.0	10.4	NE
NFT	16.7	NE	NE	NE	100.0	85.5	NE
Pontine SP	5.6	45.5	16.3	0	65.8	77.5	1.8
Tegmentum NFT	21.3	57.8	23.1	98.0	77.7	78.1	1.9
Lacunar state	31.5	39.4	100.0	44.4	10.2	30.2	20.8
Old infarcts	8.2	12.1	86.3	5.9	9.0	11.4	4.6
Amyloid Angiopathy	12.1	30.3	5.1	0	97.7	86.8	0.5
Cortical Lewy bodies	7.0	0	0	0	0.9	1.8	0
PD/AD (< 70 years)	18.6				14.9/16.7[a]		
PD/SDAT (> 70 years)	35.0	100.0				17.0/8.1[a]	

[a] Clinical/anatomic
NE not examined; NFT neurofibrillary tangles; SP senile plaques

shows less severe damage than in PD, and Lewy bodies in the substantia nigra without clinical signs have been observed in 10%-25% of all AD brains, suggesting a preclinical stage of PD (LEVERENZ and SUMI 1986). Neuronal loss in the substantia nigra in AD/SDAT has been reported as 8.4%-26% (MANN et al. 1983, 1985; CHUI et al. 1986; GIBB 1988) up to an average of 28% (JELLINGER 1987b, c), and in AD up to 48%-54% with an average of 38%-48% (TABATON et al. 1985; JELLINGER 1987b, d), while the VTA shows a much higher neuronal loss of 40%-60% (MANN et al. 1987a). Among 114 autopsy cases of AD, 14% of which had clinical signs of PD, 69% showed mild and 15% moderate to severe nigral damage, but Lewy bodies in the substantia nigra and locus ceruleus were present in 40% and 45.6%, respectively (Table 4). Among 325 SDAT cases, 20% showed extrapyramidal signs, but moderate to severe nigral lesions were present in 8%, whereas Lewy bodies in the substantia nigra and locus ceruleus were seen in 40% and 57% of cases, respectively, but they were outnumbered by ANFTs in these nuclei. The incidence of Lewy bodies in AD and SDAT was significantly higher than in aged controls (Table 4). In another autopsy cohort of AD, the incidence of Lewy bodies in the substantia nigra was 35%, i.e., 3-3.8 times higher than in age-matched controls (LEVERENZ and SUMI 1986). While DITTER and MIRRA (1987) observed Lewy bodies in the substantia nigra and locus ceruleus in 55% of a cohort of AD cases, GIBB (1988) reported only an incidence of 14%.

A small group of demented patients clinically presenting moderate parkinsonian symptoms (akinesia, rigidity, gait disorders and, less often, tremor) with onset around the age of 80 years, short duration (average 2 years), and severe mental side effects in reaction to otherwise ineffective levodopa treatment, showed only mild damage to the substantia nigra and locus ceruleus, associated with Lewy bodies in 45%-50% and ANFTs in about 60% of all cases (Table 4). This group, representing less than 6% of a total of 580 autopsy cases of parkinsonism showing the clinical signs of both AD/SDAT and PD, and the pathologic features of SDAT with mild damage to pigmented brain stem nuclei, has been tentatively called "senile parkinsonism" (JELLINGER and RIEDERER 1984; JELLINGER 1986a, 1987b). Whether this group, otherwise referred to as PD/AD (PRICE et al. 1986), represents a subset of PD populations with late onset and only mild or incipient PD pathology, or a subset of AD/SDAT with mild but clinically relevant degeneration of dopaminergic brain stem systems, remains to be elucidated.

On the other hand, a considerable percentage of PD brains show widespread NFTs and NPs, that have the character and distribution of the lesions seen in AD/SDAT. In autopsy cohorts of PD individuals, the incidence of additional markers of AD/SDAT ranges from 7% to 47% (BOLLER et al. 1979; HAKIM and MATHIESON 1979; R. H. PERRY et al. 1983, 1987; PERL et al. 1984; TOMLINSON and CORSELLIS 1984; GASPAR and GRAY 1984; GRILIA et al. 1986; JELLINGER 1986a, 1987b, d; MCGEER et al. 1988) and additional AD pathology is seen in 5%-18.6% of demented PD patients dying before the age of 70 years, and in 35%-45% of the older PD subjects (see Fig. 5).

However, based on pathologie findings, a distinction between AD/SDAT with PD lesions and PD with severe cortical Alzheimer pathology appears difficult, if not impossible. One difference may be in a distinctive nigral pathology, where the majority of tangled neurons in AD are clustered within a small group in the most medial parts of the substantia nigra, and not distributed throughout the nigral cells (TOMLINSON and CORSELLIS 1984). These data need further confirmation, as do results from a personal study indicating that another distinct feature is the preservation of TH-immunoreactive fibers in the striatonigral system in AD/SDAT, but severe reduction in the classical PD (JELLINGER 1987a). Both clinical and pathologic data suggest that AD patients have an increased risk of PD, and vice versa, and that the association between the two disorders is more than would be expected by pure coincidence. However, there may also be overlapping pathologies; this association appears to be less frequent than previously suggested, and its basis remains unclear.

III. Diffuse Lewy Body Disease

This recently recognized disorder, clinically presenting as PD with or without dementia, Shy-Drager syndrome, or as progressive dementia with or without parkinsonian symptoms (YOSHIMURA 1983; KOSAKA et al. 1984) is characterized by widespread distribution of Lewy bodies and Lewy body-like cytoplasmic inclusions in the cortex and brain stem, combined with degenerative changes in the brain stem and with variable cortical Alzheimer pathology (FORNO et al. 1978; BURKHARDT et al. 1988; SIMA et al. 1986; BYRNE et al. 1987). Some of the severely demented cases show "plaque only" AD pathology, i.e., NPs without ANFTs (SIMA et al. 1986; DICKSON et al. 1987a). A juvenile sporadic type of PD with widespread Lewy bodies was recently observed in Japan (YOSHIMURA et al. 1987), and a young adult from of sporadic dementia with diffuse ANFTs and Lewy bodies without NPs was described by POPOVITCH et al. (1987). Three types of diffuse Lewy body disease (DLBD) have been described: a "diffuse" type, with generalized occurrence of Lewy bodies and severe Alzheimer lesions suggests a combination with AD; a "transitional" type shows numerous Lewy bodies in the brain stem and diencephalon, but less frequent in the basal ganglia and cortex; while the "brain stem" type, with many Lewy bodies limited to the brain stem appears identical with "classical" PD (KOSAKA et al. 1984; YOSHIMURA et al. 1987). A case of PSP with large numbers of Lewy bodies in the brain stem and cerebral cortex has been reported by MORI et al. (1986). DLBD appears not to be an entity, since there are transitions to classical PD with cortical Lewy bodies (IKEDA et al. 1978; YOSHIMURA 1983; YOSHIMURA et al. 1987), while the diffuse type with generalized Lewy bodies and severe AD pathology shows marked cell loss in the VTA and NBM which correlates with a profound decrease of neocortical choline acetyltransferase (CAT) activities (CLARK and LEHMAN 1983) and the distribution of cortical Lewy bodies in the proposed mesocortical dopaminergic projections. While these lesions suggest a defect in both dopaminergic and cholinergic innervation of the cortex (EGGERSTON and SIMA 1986; SIMA et al. 1986), recent demonstration of decreased somatostatin-like immunoreactivity in the neocortex suggests an additional degeneration of intrinsic cortical neurons, similar to that found in AD; thus, DLBD often presents both morphological and biochemical similarities with AD/SDAT (DICKSON et al. 1987a).

IV. Multisystem Degenerations

A variety of disorders frequently associated with parkinsonism is morphologically characterized by multiple system degenerations, including the dopaminergic nigrostriatal pathways. In *striatonigral degeneration* (ADAMS et al. 1964), often clinically misdiagnosed as PD, focal damage to the substantia nigra and locus ceruleus with or without Lewy bodies is associated with severe atrophy of the hyperpigmented putamen, owing to deposition of hematin, neuromelanin, and lipofuscin in the astroglia (BORIT et al. 1975; KOEPPEN et al. 1971), moderate damage to the caudate nucleus, and multiple brain stem lesions (TAKEI and MIRRA 1973). It can be associated with olivopontocerebellar atrophy (ADAMS et al. 1964) which also shows frequent damage to the substantia nigra and other brain stem systems in both autosomal and sporadic cases, clinically presenting with or without parkinsonism (BERCIANO 1982) which may result from the same mutant gene (DUVOISIN 1987). Orthostatic hypotension or *Shy-Drager syndrome* shows multisystem degeneration, one subtype associated with widespread occurrence of Lewy bodies, the other with striatonigral and olivopontocerebellar atrophy, while intermediolateral column cell loss is the only damage common to all cases (OPPENHEIMER 1984). *Familial juvenile parkinsonism* represents a heterogeneous group, the pathologic features of which may or may not differ from those in idiopathic PD (NARABAYASHI et al. 1986; YOSHIMURA et al. 1987). The neuronal loss extends to several structures, with only a few Lewy bodies, if any (YOKOCHI et al. 1984) and with pallidopyramidal degeneration (DAVISON 1954) or pallidonigroluysiodentate atrophy and dorsal column degeneration (MAYER et al. 1986). Akinetic-rigid parkinsonism also occurs in pallidoluysian, pallidonigroluysian atrophy (KOSAKA et al. 1981), and pallidonigroluysiothalamic degeneration (TAKAHASHI et al. 1977; IIZUKA and HIRAYAMA 1986) and related multisystem diseases (MAYER et al. 1986; JELLINGER 1986c). *Joseph disease*, an autosomal dominant ataxic multisystem disorder, probably resulting from the same mutant gene as the autosomal dominant olivopontocerebellar atrophy (OPCA) of the Schut-Swier type (ROSENBERG 1984), shows variable combinations of nigrosubthalamopallidal atrophy with degeneration of the cerebellofugal and cerebellopetal system (MIZUTANI et al. 1983; ROSENBERG 1984), while some hereditary combined spinocerebellar and extrapyramidal degenerations are associated with taurine deficiency (PURDY et al. 1979).

V. Parkinson-Dementia Complex

This combination of parkinsonism with dementia (PDC) and frequent amyotrophic lateral sclerosis, occurring endemically in the Chamorro population of Guam, is characterized by diffuse cortical atrophy with abundant NFTs and Hirano bodies associated with multisystem neuronal loss and gliosis, rare Lewy bodies in the locus ceruleus, but absence of senile plaques (HIRANO et al. 1981; HIRANO and LLENA 1986; CHEN and CHASE 1986). The PD-ALS complex was thought to be of genetic origin for many years, because frequently several family members were found to be affected (HUDSON 1981). Now it is known that the disease is caused by environmental factors (CHEN and CHASE 1986), probably due to the local cycad seed containing *N*-methylamine-D-alamine (SPENCER et al. 1986). Similar syndromes have been reported in Japan, the United States, and the Federal Republic of Germany (MATA et al. 1983; SCHMITT et al. 1984), and a sporadic case of ALS-PD with diffuse Lewy bodies was recently reported by DELISLE et al. (1987). Concurrence of parkinsonism with motor neuron disease has been reported infrequently (HUDSON 1981), although considerable damage to the substantia nigra and locus ceruleus is often seen in the latter disorder (ESCOUROLLE et al. 1971; JELLINGER 1974; JACOB 1983).

VI. Progressive Supranuclear Palsy

This progressive sporadic PD-like disorder with ophthalmoplegia, axial dystonica, rigid akinesia, pseudobulbar palsy, and dementia (STEEL et al. 1964) shows widespread NFTs and multisystem neuronal loss and gliosis, the hippocampus and neocortex being preserved. The distribution of lesions and the ultrastructure of the tangles, consisting of 15 nm wide straight filaments, are different from those in postencephalitic parkinsonism (PEP) and Guam PDC (TELLEZ-NAGEL and WISNIEWSKI 1973; JELLINGER et al. 1980), but concurrence of straight and paired helical filaments in the same patient, or even in the same diseased nerve cell have been observed (TAKAUCHI et al. 1983; YAGISHITA et al. 1979). NFTs are also present in the spinal cord, most frequently in the posterior horn (KATO et al. 1986). The clinical syndrome of PSP can also be produced by multi-infarct encephalopathy (DUBINSKY and JANKOVIC 1987).

VII. Postencephalitic Parkinsonism

Residual deficits of encephalitis lethargica and other viral infections of the CNS show extensive diffuse loss of pigmented neurons and gliosis in the substantia nigra (range 91.6%-95%; Table 5) and locus ceruleus, with nearly complete disappearance of all melanin cell groups, and damage to other parts of the upper brain stem, with widerspread occurrence of ANFTs, but little or no involvement of the neocortex (HASSLER 1938; KLAUE 1940; GREENFIELD and BOSANQUET 1953; RICHARDSON 1965; JELLINGER 1974; ISHII and NAKAMURA 1981; RAIL et al. 1981; GIBB and LEES 1987). Their ultrastructure is that of paired helical filaments, as seen in AD/SDAT and Guam PDC, but concurrence with straight filament tangles has been observed (ISHII and NAKAMURA 1981). The tangles show positive immunoreactivity with neurofilament proteins and human brain microtubular fraction (YEN et al. 1983). The distribution of neuronal loss and of the fibrillary tangles in the most severely affected sites, and the demonstration of antibodies to influenza B by the immunofluorescence technique (GAMBOA et al. 1974) suggest pathogenetic relations with a viral or slow viral agent. The shrunken volume of both parts of the pigmented

Table 5. Quantitative changes in the substantia nigra in parkinsonism and in Alzheimer's disease

Reference	Disease	*N*	Mean age (years)	Fresh volume (% control)	Neuronal loss (% control)	Nuclear[1] nucleolar[2] volume (% control)	Volume perikarya (% control)	Melanin content (% control)
Pakkenberg and Brody 1965	PD	10	68		59[c] (total)			
					66[d] (pigm)			
Javoy-Agid et al. 1984	PD	2	?		77.3			
Mann and Yates 1983 a[1]	PD	8	68		80[d]	15[b]		15[c]
	PD untreated	4	63		76.3[d]	11.4		17.4[c]
	PD treated	4	73		85.4[d]	11.9		12.7[b]
Bogerts et al. 1983[2]	PD	5	·72	28[c] (l)	66[d] (l)	60[b] (l)	39[b] (l)	
				24[a] (m)	63[c] (m)	17[a] (m)	24 (m)	
	PEP	6	54	70[d] (l)	94[d] (l)	39[b] (l)	52[c] (l)	
				62[d] (m)	95[d] (m)	24[b] (m)	36[b] (m)	
Chui et al. 1986	PD	3	65		63.3[d] (47–87)			
Gibb and Lees 1987	PD	12	57		72.0[d]			
	PEP	12	60		91.6[d]			
Hirsch et al. 1988	PD	4	?		84 (pigm.)			
					77 (TH-IR)			
Takeda et al. 1982	PDC	?	?		80[d] (total)			
					96[d] (pigm)			
					42[b] (unpigm)			
	OPCA	?	?		61.8[d] (total)			
					83.3[d] (pigm)			
					23 (unpigm)			
Mann et al. 1983[2]	SDAT	19	84.7		6.3 ± 3.6[a]	8.4 ± 3.9[a]		
	AD	12	66.0		11.8 ± 3.7[a]	23.8 ± 4.5[b]		
Mann et al. 1984[2]	AD + SDAT	22	74.5		7.5[a]	13.0[a]		
Tabaton et al. 1985	AD	4	63.0		38.1[c] (18–54)			
Chui et al. 1986	AD + SDAT	5	75.6		16.0[b] (3–72)			
Mann et al. 1987[2]	AD	15	73.3		13.6[b]	18.5[b]		
Jellinger 1987b	SDAT	16	82.6		28.1[c] (22–48)			
	AD	1	70.0		48.5[c]			
Gibb 1988	SDAT	11	74.5		26.0			

[a] Not significant; [b] $p < 0.05$; [c] $p < 0.01$; [d] $p < 0.001$

AD Alzheimer's disease; PD Parkinson's disease; PDC Parkinson-dementia complex; PEP postencephalitic parkinsonism; OPCA olivopontocerebellar atrophy; SDAT senile dementia of Alzheimer type; TH-IR tyrosin hydroxylase immunoreactive

1 lateral; m medial; pigm pigmented; unpigm unpigmented

nigra leads to increased glial densities with formation of glial scars. However, the absolute number of glial cells is 40%–50% lower than in normal controls. This loss of glial cells indicates that the damage affects both neurons and glia (BOGERTS et al. 1983).

VIII. Vascular or Multi-infarct Parkinsonism

Cerebrovascular lesions occur in 6%–10% of typical PD brains (ALVORD et al. 1974; EARLE 1968; ESCOUROLLE et al. 1971; JACOB 1983; JELLINGER 1986a, 1987a; DUBINSKY and JANKOVIC 1987), while vascular of "arteriosclerotic" parkinsonism has been clinically well characterized (CRITCHLEY 1929; EADIE and SUTHERLAND 1964; SCHWAB and ENGLAND 1968; PARKES et al. 1974; TOLOSA and SANTAMARIA 1984; FRIEDMAN et al. 1986), but has been recognized by some as pathologically unproved (LEWIS 1971). However, in autopsy series of parkinsonism, 1.8%–7% of cases have been classified as being of vascular origin (EARLE 1968; BERNHEIMER et al. 1973; TOGHI 1977; JELLINGER 1974), while in clinical surveys 5.7%–9% of PD patients showed evidence of stroke (see DUBINSKY and JANKOVIC 1987). In a personal autopsy series of 580 parkinsonian patients, 6.9% revealed a multi-infarct state, hypertensive encephalopathy of Binswanger-type, with a lacunar state or small infarcts in the basal ganglia and brain stem, but with little or no degenerative damage to the substantia nigra and locus ceruleus. The presence of Lewy bodies in about 10% of these cases was twice as high as in age-matched controls (see Table 4), suggesting a combination of idiopathic PD with cerebrovascular disorders or a multi-infarct state, thus reducing the incidence of acceptable vascular or multi-infarct parkinsonism to 5%–6% of all anatomically confirmed cases of parkinsonism.

IX. Toxic Parkinsonism

Pathology of parkinsonism due to *carbon monoxide, carbon disulfide*, and *cyanide intoxication* or postnarcotic encephalopathy shows anoxic lesions of the globus pallidus and/or substantia nigra (LAPRESLE and FARDEAU 1967; ESCOUROLLE et al. 1971; UITTI et al. 1985; JELLINGER 1986b). *Manganese encephalopathy* is characterized by widespread neuronal loss and gliosis, mainly in the pallidum and caudoputamen with little or nigral damage with or without Lewy bodies (PARNITZKE and PEIFFER 1956; BERNHEIMER et al. 1973; YAMADA et al. 1986). Since no increase of manganese content has been found in postmortem brains (PARNITZKE and PEIFFER 1956; YAMADA et al. 1986), the continuance of neurologic disorders is obviously not linked to elevated manganese in the brain. Contrary to previously suggested similarities (BARBEAU 1984), chronic manganese encephalopathy is different from PD in neuropathology and manganese distribution in the brain. Akinetic-rigid parkinsonian symptoms in chronic *lead* intoxication are attributed to lesions of the substantia nigra (VAN BOGAERT 1956). The pathology of *neuroleptic* and *drug-induced* parkinsonism is poorly understood, since the minor changes in the reticulata nigrae and midbrain (CHRISTENSEN et al. 1970) are considered as normal age-related findings (JELLINGER 1977), while neuronal loss with Lewy bodies in the substantia nigra and locus ceruleus, with reduced homovanillic acid levels in the striatum, indicate subclinical PD in some cases (RAJPUT et al. 1982).

Neuropathology of parkinsonism induced by *MPTP* and other piperindines in one human (DAVIS et al. 1979) and in experimental animals shows neuronal loss and gliosis, particularly involving the zona compacta nigrae without typical Lewy bodies, and selective damage to the dopamine-containing neurons of the A9 area with sparing of most other neuronal systems (CHIUEH et al. 1985; WILKENING et al. 1986). FORNO et al. (1986) described locus ceruleus lesions and eosinophilic neuronal inclusions resembling Lewy bodies in MPTP-treated aged monkeys, suggesting similarities with the pathology of human PD (RICHARDSON 1986), but the ultrastructure of these inclusions differs from that of typical Lewy bodies (FORNO et al. 1988).

X. Symptomatic Parkinsonism

Parkinsonism has been observed in a wide variety of disorders involving the brain stem and/or substantia nigra and its dopaminergic projections, e.g., following *head injury* involving destruction of the substantia nigra by bullet injury (DE MORSIER 1960), direct traumatic impact or herniation contusion of the upper brain stem, or secondary damage to the midbrain and basal ganglia resulting from vascular compression in raised intracranial pressure (LINDENBERG 1964; HUHN and JACOB 1971; JELLINGER 1974). Concurrence of posttraumatic brain damage with idiopathic PD may occur. *Boxer's dementia*, associated with clinical signs of parkinsonism, shows diffuse cortical atrophy, severe neuronal loss from the locus ceruleus and substantia nigra with numerous NFTs, but no senile plaques, throughout the CNS (CORSELLIS et al. 1973). Parkinsonism has also been observed in rare cases of tuberculoma (BLOCQ and MARINESCO 1893) or tumor of the brain stem (GHERARDI et al. 1985), solid tumors causing brain stem compression (GARCIA de YEBENES et al. 1982), calcification of the basal ganglia (KLAWANS et al. 1976), syringomyelia, viral encephalitis, subacute panencephalitis, multiple sclerosis, neurolipidoses, and late onset Hallervorden-Spatz disease (see FAHN 1977; JANKOVIC et al. 1985; JELLINGER 1986a, 1987a; GIBB 1988).

D. Morphological Correlates of Pathobiochemistry

The morphological basis of the major neuromediator changes in PD and related disorders is now well established, although the structural and molecular background of many biochemical findings, their mutual relations, and causative mechanisms remain to be elucidated.

I. Dopaminergic System

Degeneration of the central dopaminergic systems is the hallmark of idiopathic PD and most other parkinsonian syndromes, where both the ascending (nigrostriatal, mesocorticolimbic, and hypothalamic) and descending (to the spinal cord) pathways arising from the pigmented neurons of the zona compacta nigrae (area A9 in mammalian brain) and the ventral tegmental area (area A10 of mammalian brain) are lesioned to different degrees.

1. Substantia Nigra and Ventral Tegmental Area

Damage to the dopaminergic nigral neurons has been confirmed by morphometric studies in various types of parkinsonism (Table 5). In PD, the substantia nigra shows a decrease of fresh volume of about 25 %, with loss of pigmented neurons ranging from 60 % to 85 %, considerable reduction in nuclear, nucleolar, and perikaryal volume, and loss of melanin up to 88 % of controls.

The nigrostriatal parts of the lateral substantia nigra are much more affected than the mesolimbic parts arising from the medial groups of the nigral neurons, and in the most medial suboculomotor part the neuronal loss (40 %) is less severe than in the lateral parts (66 %). Between 1 % and 25 % of the remaining neurons contain Lewy bodies (GIBB and LEES 1987). Comparing young- and old-onset PD cases, GIBB and LEES (1988) observed 25 % greater nigral cell loss in the young-onset cases having much longer duration, and greater cell loss was also seen in advanced stages of PD (MANN et al. 1984). The number of glial cells and the volume of glial cell nuclei in the substantia nigra in PD are unchanged (BOGERTS et al. 1983). Lesions in the VTA, the source of the mesolimbic-mesocortical dopaminergic system which projects to the amygdaloid nucleus and cerebral cortex, has been reported in PD (GREENFIELD and BOSANQUET 1953) and in PEP (FOIX and NICOLESCO 1925). Morphometric studies of the VTA contralateral to therapeutic lesions placed in the basal ganglia or thalamus in PD showed an average neuronal loss of 64 % in the anterior VTA (nucleus paranigralis) and 45 % in the posterior VTA (area of ventral supratrochlear) (UHL et al. 1985). In the hypothalamus, the melanin-pigmented neurons of the arcuate and periventricular nuclei are preserved (MATZUK and SAPER 1985). In OPCA, a similar loss of over 60 % of nigral neurons is seen with more severe depletion of pigmented cells (83 %), while Guam PDC shows a total nigral cell loss of 80 % with almost complete disappearance of pigmented cells. In PEP, there is also nearly complete depletion of all mesencephalic melanin cell groups, with loss of 92 %–96 % of nigral cells (Table 5) and increased glial density with formation of glial scars, but reduction in the total number of glial cells by 40 %–50 % (BOGERTS et al. 1983). Progressive cell loss preferentially involves the neuromelanin-pigmented and TH-IR cells, suggesting a selective vulnerability of dopaminergic mesencephalit neurons in PD (HIRSCH et al. 1988). The degree of pigmented cell loss in the substantia nigra correlates significantly with the decrease of dopamine and homovanillic acid in the striatum and ventral tegmentum, that is more severe in the putamen than in the caudate nucleus (FAHN et al. 1971; BERNHEIMER et al. 1973), and with loss of activity of TH, the rate-limiting enzyme of catecholamine synthesis, in these regions (P. L. MCGEER and E. G. MCGEER 1976; JAVOY-AGID et al. 1981; NAGATSU et al. 1981; GASPAR and GRAY 1984), while the extent of loss of pigmented neurons in the VTA is comparable to the

50 % decrease of TH activity in the nucleus accumbens (JAVOY-AGID and AGID 1980). Dopamine loss is more severe in the caudal putamen and oral caudate nucleus because cell loss in the substantia nigra is more severe in the ventral and caudal parts projecting to the putamen (KISH et al. 1988). In the hypothalamus, the pigmented neurons in the arcuate and periventricular nuclei are preserved, but there is a 60 % decrease in dopamine content (AGID et al. 1987). TH is an iron-dependent enzyme, but while its basal activity in the PD caudate nucleus is decreased by 49 % of controls, it does not show any changes of its rate of stimulation by 1 m*M* Fe^{2+} (RIEDERER et al. 1985). Increased levels of iron have been reported in the PD brain (LHERMITTE et al. 1927; EARLE 1968; ROJAS et al. 1965), and recent significant studies revealed iron in total iron III and ferritin of different areas of the PD brain (RIEDERER et al. 1988). In patients with Parkinson-plus syndromes due to multisystem atrophies and PSP, increased concentrations of iron in the putamen, and less prominent increases in the caudate nucleus, globus pallidus, and lateral pars compacta nigrae have been found by high field magnetic resonance imaging (DRAYER et al. 1986; RUTLEDGE et al. 1987). The reduced activity of TH might result from depletion of its cofactor iron (II) (RIEDERER et al. 1988).

Immunocytochemistry shows a substantial loss of TH-immunoreactive (TH-IR) neurons in the substantia nigra, nucleus paranigralis, and ventral tegmentum (PEARSON 1983; NAKASHIMA et al. 1983; JAVOY-AGID et al. 1984), and a marked diminution of TH-IR processes (axons and dendrites) in the striatonigral pathways and VTA (Fig. 1). While Lewy bodies in catecholaminergic neurons show positive reaction with TH antisera, suggesting that the TH enzyme may play a role in their formation (NAKASHIMA and IKUTA 1984b), positive TH immunoreactivity of neurons in the substantia nigra and locus ceruleus affected by NFTs in PD, PSP, PDC, and AD indicate that these neurons still contain TH enzyme proteins and that formation of the tangles, except in the final stages when the perikarya are entirely replaced by them, develops independently of TH protein synthesis (NAKASHIMA and IKUTA 1984a). Most of the remaining melanin-laden neurons of the brain stem show negative im-

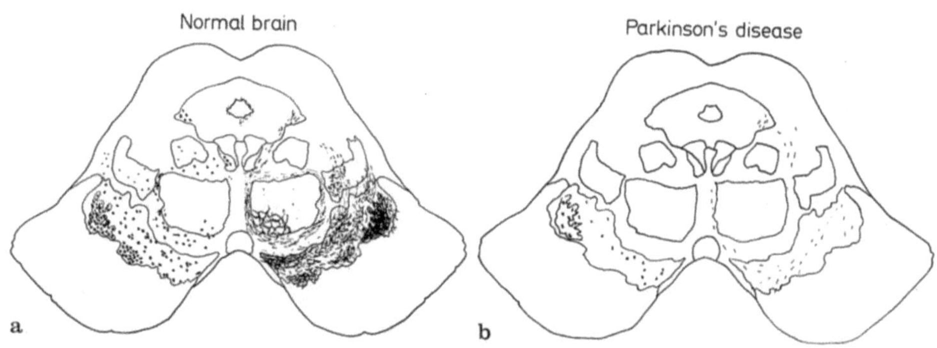

Fig. 1 a, b. Schematic distribution of tyrosine hydroxylase immunoreactive neurons (left) and fibers or terminals (right) in midbrain of normal human (**a**) and Parkinson's disease brain (**b**). Camera lucida drawing

Table 6. Quantitative changes of the striatopallidum in parkinsonism

Reference	Disease	N	Mean age (years)	Volume, weight			Cell count			Cell density
				Caudate	Putamen	Pallidum	Caudate	Putamen	Pallidum, external	Pallidum, external
				(% control)						
SABUNCU 1969	PD	8	62						100	
BÖTTCHER 1975	PD	9	75					100	100	
WUKETICH et al. 1980	PD	11	76	96.2 (W)	96 (W)	99 (W)				
BUGIANI et al. 1980	PD	14	63				50 (L)	65 (S)		
LANGE and BOGERTS 1982	PD	5	72	93 (V)	100 (V)	100 (V)	89 (S)	100	85	
	PEP	6	54	85 (V)	100 (V)	93 (V)	90 (S)	93 (L)	71	
ARENDT et al. 1983	PD	14	59						92	105
	PEP	7	59						100	98

L large cells; PD Parkinson's disease; PEP postencephalitic parkinsonism; S small cells; V volume; W weight (fresh)

munoreaction with TH, indicating loss of TH enzyme activity in dying neurons (HIRSCH et al. 1988). TH immunoreactivity is related to the quantity of this rate-limiting enzyme of catecholamine synthesis in the tissue (BENNO et al. 1982) and, therefore, is a good marker for catecholaminergic pathways in the human postmortem CNS by the peroxidase-antiperoxidase (PAP) method (PEARSON 1983; DIETL 1985). Loss of TH immunoreactivity indicates severe demage to the catecholaminergic neurons and fibers. It is generally concluded that parkinsonian symptoms do not appear until 70 %-80 % of the nigral dopamine neurons and an equivalent percentage of striatal dopamine have been lost (RIEDERER and WUKETICH 1976; AGID et al. 1987). This critical level correlates well with the morphometric findings in various types of parkinsonism, while the progressive degeneration of the nigrostriatal pathway is documented by the greater neuronal loss in the substantia nigra in advanced stages of PD (MANN et al. 1983). However, no definite morphological distinction of the various clinical subtypes of PD (tremor-dominant, akinesia-dominant, or combination type) has been found so far.

2. Striatopallidum, and Other CNS Areas

Although deprivation of the normal input may cause degeneration of the dopaminergic terminals in the striatum (MANN and YATES 1983a), there is only little change to *striatal* morphology. In PD, neither the putamen nor the caudate nucleus show any considerable reduction of their fresh volume and wet weight, nor definite loss of striatal neurons, the shrinkage of the caudate nucleus being only 7 %, with loss of 11 % of its small neurons (Table 6). No definite ultrastructural abnormalities have been found in the synaptic organization of the striatum in PD (FORNO and NORVILLE 1984). The immunoelectrophoretic demonstration of normal concentrations of synaptic marker antigens related to rat brain D_2 and D_3 proteins in putamen extracts of PD patients (JØRGENSEN et al. 1982) also suggests the absence of major neuronal degeneration in the striatum. In PEP, there is no atrophy of the striatum, except for the caudate nucleus which loses 15 % in volume, 10 % of its small neurons and only 7 % of its large neurons (Table 6).

The *globus pallidus*, in contrast to previously suggested severe neuronal loss (MARTIN 1965) shows only a 7 % reduction of its volume in PEP, but not in PD, and about 30 % neuronal loss in the external segment without changes in cell density in both PD and PEP (Table 6). Following experimentally induced damage to the substantia nigra, loss of TH-IR fibers and boutons in the striatum suggests denervation with disappearance of dopaminergic terminals (PEARSON 1983; ARLUISON et al. 1984). However, demonstration of TH immunoreactivity in the human postmortem brain is difficult (JAVOY-AGID et al. 1984; DIETL 1985), but PET studies indicate a decrease in 6-fluorodopamine uptake in the striatum in PD (MARTIN et al. 1986; GUTTMAN and CALNE 1988).

On the other hand, there is considerable reduction of TH-IR cell bodies and fibers in the mesolimbic system and hypothalamus (JAVOY-AGID et al. 1984; GASPAR and GRAY 1984). In the *spinal cord* of parkinsonian subjects, reduced density of TH-IR cell bodies, fibers, and varicosities have been demonstrated with almost complete disappearance of TH-IR fibers in layers I and X, suggesting involvement of the spinal catecholaminergic system (JELLINGER 1986b) which has been identified in the rat spinal cord by immunohistochemistry of three specific catecholamine-synthesizing enzymes (MOUCHET et al. 1986). Although dopamine levels in dorsal, ventral, and lateral parts of the spinal cord were found not to be different from controls, suggesting that the descending dopaminergic pathways are preserved (SCATTON et al. 1986), together with reduced TH activity and significant decrease of dopamin in the adrenal medulla (RIEDERER et al. 1979) and the occurrence of Lewy-like bodies in the adrenal medulla (CARMICHAEL et al. 1988), these data suggest *generalized* demage to the catecholaminergic system in PD. On the other hand, preservation of hypothalamic dopaminergic neurons indicates that the disorder does not involve the indiscriminate degeneration of all catecholaminergic cell groups in the brain (MATZUK and SAPER 1985).

The *red nucleus* in PD loses about 15% of its neurons without shrinkage in volume, while in PEP this nucleus loses 14% of its volume, 28% of its neurons, and 19% of its glial cells (LANGE and BOGERTS 1982).

II. Noradrenergic System

1. Locus Ceruleus

The locus ceruleus is the main source of widespread noradrenergic innervation of most CNS regions, with the dorsal vagal nucleus and the supraoptic and paraventricular nuclei of the hypothalamus forming its major projection fields (MANN et al. 1983; NIEUWENHUYS 1985). In PD, PSP, PDC, and OPCA, this nucleus shows a depletion of pigmented neurons ranging from 50% to 80% with 75%–80% loss of melanin, but the capacity for protein synthesis in the few remaining cells, expressed by their nucleolar volumes, seems to be maintained, even in the late stages of PD (Table 7). Neuronal losses from the locus ceruleus are less in aged controls where they amount to 35%–40% without reduction of nucleolar volume (MANN et al. 1984), while in AD, young SDAT patients, and boxer's dementia, up to 85% cell loss approaches that in PD (Table 6). In PD, gross cell depletion of the locus ceruleus is associated with Lewy bodies in almost 99% of cases, which in one PD patient aged 74 involved 32% of its neurons, while 0.8% showed neurofibrillary degeneration, as compared with Lewy bodies in 0.07% and ANFTs in 0.5%–6.6% of the locus ceruleus neurons in aged controls (TOMONAGA 1983). Rare coexistence of Lewy bodies and ANFTs in the same neuron has been reported (TOMONAGA 1981). In PSP, boxer's dementia, and AD/SDAT, many of the remaining locus ceruleus neurons are atrophied and contain ANFTs (CORSELLIS et al. 1973; MANN et al. 1983; TOMONAGA 1983). Topographic arrangement of locus ceruleus cell loss with highest intensity in the anterior and central parts projecting to the temporal cortex and hippocampus (LOCKHART et al. 1984, INGRAM et al. 1987; ZWEIG et al. 1988) and significant correlation with the density of senile plaques in the cortex observed in AD/SDAT (MARCYNIUK et al. 1986), suggesting a retrograde degeneration of subcortical nuclei owing to primary damage

Table 7. Quantitative changes in locus ceruleus in parkinsonism and senile dementia

Reference	Disease	*N*	Mean age (years)	Cell loss	Nucleolar volume	Melanin content
				(% age-matched controls)		
TOMONAGA 1983	PD	1	74.0	90.0[a]		
MANN et al. 1983	PD	6	65.4	78.5[a]	101	
MANN and YATES 1983a	PD	8	68.2	78.8[a]	100	24.6[a]
	PD untreated	4	63.0	72.2[a]	104.8	24.0
	PD treated	4	73.2	85.3[a]	97.3	20.0[a]
MANN et al. 1983	PSP	4	66.0	50.7[a]		
TAKEDA et al. 1982	PDC	?	?	49.6[a]		
	OPCA	?	?	63.9[a]		
TOMONAGA 1983	OPCA	2	?	85-90[a]		
MANN et al. 1983b	SDAT	19	84.7	54.4[a]	82.3[a]	
MANN et al. 1984	AD + SDAT	22	74.5	70.3[a]	68.5[a]	
MANN et al. 1983c	BD	4	55.0	65.8[a]	74.4[a]	
VIJAYASHANKAR-BRODY 1979	SDAT	24	75-87	40.0[a]		
TOMLINSON et al. 1981	SDAT	15	81.0	52-56[a]		
BONDAREFF et al. 1982	SDAT	20	78±7	80.0[a]		
TOMONAGA 1983	AD	1	67	~79.0[a]		
CHUI et al. 1986	AD + SDAT	5	75.6	41.2[a] (30-86)		
ICHIMIJA et al. 1986	AD + SDAT	7	67.2	68.6[a]		
JELLINGER 1987a	SDAT	16	81.6	41.5[a] (36-59)		
	AD	1	69.0	60.3[a]		
INGRAM et al. 1987	AD	1	66.0	68.0[a]		
ZWEIG et al. 1988	AD	25	72.5	67-81[a]		

AD Alzheimer's disease; BD Boxer's dementia; OPCA olivopontocerebellar atrophy; PD Parkinson's disease; PDC Parkinson-dementia complex; PSP progressive supranuclear palsy; SDAT senile dementia of Alzheimer type

[a] $p < 0,001$

Table 8. Quantitative changes in some subcortical nuclei in Parkinson's and in Alzheimer's disease

Nucleus	Neuronal loss	Nucleolar volume	RNA content	Lewy bodies	NFTs	Disease	N	Mean age (years)	Reference
	(% control)			(% neurons)					
Supraoptic	0	72.6[a]	76.4[b]			PD	8	68.2	MANN et al. 1983
nucleus	65.1[b]					SDAT	19	84.7	MANN et al. 1985
Paraventricular	0	68.8[a]	71.3[b]			PD	8	68.2	MANN and YATES 1983
nucleus		48.1[b]				SDAT	19	84.7	MANN et al. 1985
Dorsal raphe	0	91.3	83.7[b]			PD	8	68.2	MANN and YATES 1983
nucleus	44.5[b]			7.0	4.5	PD	28	78.5	JELLINGER 1987b
	17.4 ± 6[c]	34.3[b]				AD	12	66.0	MANN et al. 1984
	6.1 ± 1.6	22.9[b]				SDAT	19	84.7	
	14.6 ± 4[a]	(Total)			2.25	SDAT	7	87.7	CURCIO and KEMPER 1984
	27.3 ± 5[a]	(Large)							
	76.9 ± 15[a]				90.0	AD+SDAT	5	74.2	YAMAMOTO and HIRANO 1985
	36.8 (23–48)[a]				14.6	AD	4	63.0	TABATON et al. 1985
	21.0[a]					AD+SDAT	7	67.2	ICHIMIJA et al. 1986
	55.8 ± 9.0[a]					AD+SDAT	22	75.0	JELLINGER 1987b
	58.4 ± 6[a]	(Large)			33.7 ± 9.0	AD	9	67.2	JELLINGER unpublished data
	42.7 ± 7[a]	(Large)			27.4 ± 6.0	SDAT	14	81.9	
	36[c]	(caudal)				AD	25	72.5	ZWEIG et al. 1988
	10	(oral)							
Westphal-Edinger nucleus	54.0[a]			3.0	2.0	PD	2	76.5	HUNTER 1985
Dorsal motor	0	83.4[c]	87.2[c]			PD	8	68.2	MANN et al. 1983
vagus nucleus	29.9 ± 8.2[a]	36.2[a]				SDAT	19	84.7	MANN et al. 1984
	43.9 ± 12.2[a]	43.9[a]				AD	12	66.0	

[a] $P < 0.01$; [b] $P < 0.001$; [c] $P < 0.05$

of cortical projection areas (MANN et al. 1985), have not been reported so far in PD. Here, depletion of the locus ceruleus is combined with Lewy bodies in 98 % of all PD cases and ANFTs in 16.7 % as compared with neuronal depletion and Lewy bodies in only 6 % of postencephalitic parkinsonism and in 18 % of aged controls, while in AD/SDAT, Lewy bodies and AFNTs are present in 45 %-58 % and 65 %-77 %, respectively (see Table 4). Synaptic morphology of the locus ceruleus in both PD and AD shows little change except for an accumulation of large dense core vesicles in axonal terminals and nerve cell processes, which may be due to accumulation of biogenic amines, particularly norepinephrine, in the affected terminals deprived of their postsynaptic components by degeneration of nerve cells and their dendrites (FORNO and NORVILLE 1981). In PD, Lewy bodies and ANFTs are widely scattered throughout the hypothalamus, while its periventricular region is relatively spared (LANGSTON and FORNO 1978).

Despite the unaffected protein synthesis capacity of the remaining locus ceruleus cells in normal aged individuals and in PD and PSP patients (MANN et al. 1983, 1985), the 25 %-32 % reduction of nucleolar volume and cytoplasmic RNA within the neurons which are not reduced in number in the related hypothalamic paraventricular and supraoptic nuclei, (Table 8), indicate that the loss of locus ceruleus neurons leads to a decreased function of the noradrenergic system with reduction in cerebral eginephrine, MHPG, and related enzymes in PD and AD (JAVOY-AGID et al. 1986; BIRKMAYER and RIEDERER 1985; AGID et al. 1987). Degeneration of the noradrenergic locus ceruleus neurons has been shown to be less severe in nondemented than in demented PD subjects and in AD/SDAT (CASH et al. 1987). The central noradrenergic deficiency due to degeneration of cerulocortical neurons has been related to the intellectual impairment in PD (AGID et al. 1987).

2. Dorsal Vagal Nucleus

The noradrenergic *dorsal vagal nucleus* has been reported to show little or no neuronal loss in aged controls, PD, and PSP (MANN et al. 1983), with frequent occurrence of Lewy bodies and/or ANFTs, while in AD/SDAT, cell loss and reduction of nucleolar volume is seen in 33 %-44 % of cases (Table 8). In an autopsy series of 306 PD cases, we observed neuronal loss with Lewy bodies in 85 %-95 % as compared with 12 % in postencephalitic parkinsonism and 2.8 % in normal aged controls (see Table 4). In the periphery, reduced TH activity in the adrenal medulla and occurrence of Lewy bodies in sympathetic ganglia and similar inclusions in the adrenal medulla (RIEDERER et al. 1977) indicate damage to the sympathetic noradrenergic system which may be related to vegetative clinical symptoms in PD (BIRKMAYER and RIEDERER 1985; AGID et al. 1987).

III. Serotonergic System

Damage to the dorsal raphe nucleus (DRN) or nucleus supratrochlearis (OLSZEWSKI and BAXTER 1982), giving rise to the ascending serotonergic pathways

(BRODAL 1981; DESCARRIES et al. 1982; NIEUWENHUYS 1985), has been reported in both PD and PEP (LEWIS 1971; ESCOUROLLE et al. 1971), with reduction of the RNA content and nucleolar volume by 9%–19% (Table 8). Some DRN neurons contain Lewy bodies, while ANFTs in PD are rare, in contrast to AD/SDAT, where the DRN has the highest ANFT incidence in the brain stem (ISHII 1966; JELLINGER et al. 1980; YAMADA and MEHRAEIN 1977). While we observed an average loss of large DRN neurons of 44.5 % in PD, of 43.7 % in SDAT, and 58.4 % in AD (Fig. 2). ZWEIG et al. (1988) in AD/SDAT reported only 10 to 36 % cell loss. In PD, 4.5 % of the DRN neurons contain Lewy bodies and 6.5 % have NFTs (JELLINGER 1987a). In both Guam PDC and AD/SDAT, similar losses of large DRN neurons ranging on average from 36.8 % to 80 %, with reduction of the nucleolar volumes by 33.7 % ± 9 % (YA-

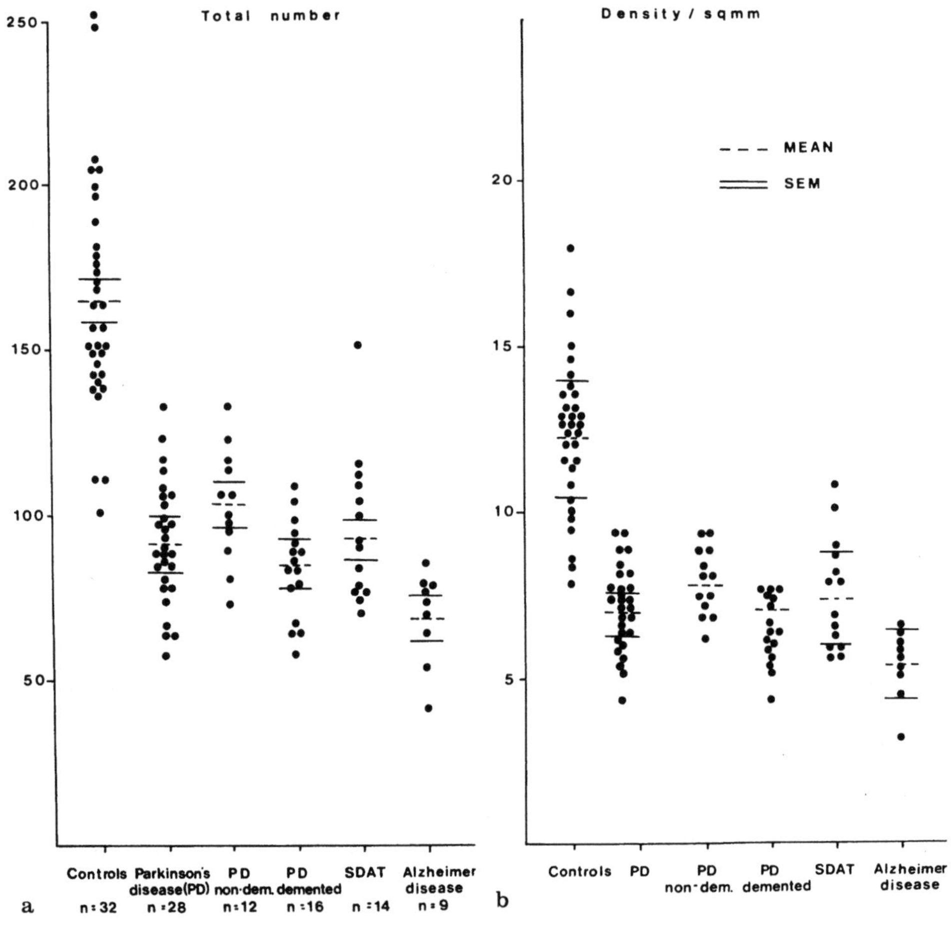

Fig. 2 a, b. Total number (**a**) and mean density (**b**) of large neurons in human nucleus dorsalis raphe in Parkinson's disease, Alzheimer's disease, and senile dementia of Alzheimer type (SDAT). Mean values and SEM are indicated by *broken* and *full horizontal lines*

MAMOTO and HIRANO 1985a; TABATON et al. 1985), and affection of 2.25%-91% of the neurons by ANFTs (CURCIO and KEMPER 1984; YAMAMOTO and HIRANO 1985a) have been reported. These data indicate considerable functional damage to the serotonergic system in PD and related disorders as well as in AD/SDAT, correlated with the considerable reduction of serotonin and its metabolites in many brain areas in these disorders (JAVOY-AGID et al. 1984, 1986; BIRKMAYER and RIEDERER 1985; ICHIMIYA et al. 1986; AGID et al. 1987). Alterations of serotonin metabolism have been tentatively related to depression in PD (MAYEUX et al. 1984).

IV. Cholinergic Systems

1. Nucleus Basalis of Meynert

The NBM in the substantia innominata of the basal forebrain, providing the major cholinergic input to the neocortex, amygdala, and hippocampus (HEDREEN et al. 1984; MESULAM and MUFSON 1984; EZRIN-WATERS and RESCH 1986), was one of the nuclei in which Lewy bodies were first observed in PD (LEWY 1913; FOIX and NICOLESCO 1925). The extent of neuronal loss which was found not to correlate with the severity of the extrapyramidal syndrome in PD (HASSLER 1938) was suggested as the anatomic substrate of bradyphrenia commonly observed in PD (HASSLER 1965). Normal aged subjects (80-90 years) show an average neuronal reduction in NBM of 30%-40% with predominant involvement of the anterior part projecting to the hippocampus (MCGEER et al. 1984; DOUCETTE et al. 1986), and decrease of the nucleolar volume and RNA content by 21% and 18%, respectively (MANN et al. 1984; JACOBS and BUTCHER 1986), while a 70% cell loss in the basal forebrain of a subject aged 95 years was associated with age-related decrease in the cortical CAT activity (MCGEER et al. 1984). In PD, NBM cell depletion ranges from 32% to 77% with a mean of 50%-60% (Table 9); it is 44% in PSP and over 60% in OPCA (TAGLIAVINI and PILLERI 1984), while PEP shows little or no damage to the NBM (Table 9), except for frequent occurrence of ANFTs in the neurons (GREENFIELD and BOSANQUET 1953; BUTTLAR-BRENTANO 1955). NBM cell loss is much higher in demented PD cases irrespective of cortical pathology (range 52%-77%), approaching the values seen in AD (57%-90%; Table 9), than in nondemented PD subjects (32%-60%), where it is similar to the losses in SDAT (33%-60%; Table 9). Cell loss affects particularly the large cholinergic neurons in the intermediate and posterior parts of the NBM projecting to the neocortex, and to a smaller degree the anterior portion, with comparative preservation of the nucleus of Broca's diagonal bundle innervating the hippocampus (MCGEER et al. 1984; ROGERS et al. 1985), where significant neuronal loss is rarely seen in PD (ARENDT et al. 1983). The extent of NBM neuronal loss shows no correlation with age or the duration of the illness (TAGLIAVINI et al. 1984a). Severe depletion of the NBM is seen in Guam PDC, showing an average 93% loss of the large cells (Table 9) and in most cases of DLBD (SIMA et al. 1986; POPOVITCH et al. 1987; BURGHARDT et al. 1988) with neuronal losses as high as 75%-80% (DICKSON et al. 1987a). In these conditions, usually associated with dementia, as well as in severely demented PD

Table 9. Neuronal loss in the nucleus basalis of Meynert in parkinsonism and dementias (mean values in % of controls)

Disease	*N*	Mean age (years)	Nerve cell loss	Reduction in maximum density	Reduction in mean density	Reference
Parkinson's disease (PD)	14	58.5	77.0	70.0		Arendt et al. 1983
	32	73.8	47.0	46.0		Gaspar and Gray 1984
	11	74.5	60.0		52.0	Nakano and Hirano 1984
	6	71.1	45.8			Tagliavini and Pilleri 1984
	50	76.7	59.9	53.5	57.5	Jellinger 1986a
PD nondemented	14	71.8	32.0	33.2		Gaspar and Gray 1984
	4	64.0	33.3	+6.2		Whitehouse et al. 1983
	1	?	60.0			Pendlebury and Perl 1984
	1	73.0	40.0			Rogers et al. 1985
	18	72.9	49.3	42.7	48.3	Jellinger 1986a
PD demented	18	75.3	60.0	58.0		Gaspar and Gray 1984
	3	70.0	77.1	53.1		Whitehouse et al. 1983
	3	73.3	51.7			Tagliavini and Pilleri 1984
	3	62.0	75.0			Rogers et al. 1985
	33	80.4	69.5	64.2	66.6	Jellinger 1986a
Parkinson–dementia complex	2	61.1	93.0		85.0	Nakano and Hirano 1983
Progressive supranuclear palsy	6	69.3	44.0			Tagliavini et al. 1984b
Postencephalitic Parkinsonism	7	59.0	0	26.0		Arendt et al. 1983
	7	?	0			Candy et al. 1983
	1	50.0	15.7	0		Whitehouse et al. 1983
	1	67.0	38.2	18.3		Jellinger 1986a
Alzheimer's disease	34	64,2	55.8–90.0		54.0	See Ezrin-Waters and Resch 1986; Ichimiya et al. 1986; Jellinger 1986a, b; Jacobs and Butcher 1986
	15	63.9	69.1	64.8	68.0	Jellinger 1987b, c
	4	66.0	71–91	(Ch4 > Ch2)		Etienne et al. 1986
SDAT	56	79.7	33.0–59.9			See Mann et al. 1984, 1986; Doucette et al. 1986; Jellinger 1987b
	30	80.9	56.6	52.5	59.3	Jellinger 1987c
	6	86.8	49.95	(Ch4 > Ch2)		Etienne et al. 1986

and in AD/SDAT, all subdivisions of the NBM are severely involved (Doucette et al. 1986; Sima et al. 1986; Dickson et al. 1987; Jellinger 1986a). There is similar reduction of the maximum and mean neuronal densities in various types of parkinsonism ranging from 33% to 85% of age-matched controls (Table 9). The nucleolar volumes of the remaining NBM neurons in PD remain either unchanged (Tagliavini and Pilleri 1984e) or show considerable decrease in protein synthesis capacity, with reduction of nucleolar volume and RNA content by 22%-24% as compared with 30%-35% in AD (Mann et al. 1983; see Table 8). The incidence of Lewy bodies in the NBM ranges from 84% to 97.3%, and of ANFTs from 29% to 65% of all PD cases (Gaspar and Gray 1984; Jellinger 1986a, 1987a) as compared with a 96%-100% prevalence of ANFTs in AD/SDAT, with only 8%-15% Lewy bodies (see Table 4). The ultrastrucuture of the Lewy bodies in NBM shows the same characteristics as those in the substantia nigra and locus ceruleus (Morimura et al. 1985). While senile plaques are frequent in the basal forebrain and NBM in AD/SDAT (Rudelli et al. 1984, Tagliavini and Pilleri 1984; Saper et al. 1985; Jacobs and Butcher 1986), they are extremely rare in cases of PD (Gaspar and Gray 1984; Nakano and Hirano 1984; Jellinger 1986a, 1987b,).

While in AD/SDAT the severity and topographic arrangement of cell loss in NBM is related to the density of ANFTs or senile plaques in the temporal cortical target areas (Mann et al. 1985; Arendt et al. 1985; Jacobs and Butcher 1986; Etienne et al. 1986), suggesting either retrograde or, less probably, anterograde degeneration of the cholinergic NBM neurons (Pearson and Powell 1987), to the best of our knowledge, no such corticosubcortical correlations have been observed in PD. In this condition, severe depletion of the NBM or other subcortical nuclei may occur without impressive cortical neuronal loss or AD pathology (Gaspar and Gray 1984; Nakano and Hirano 1984; Pendlebury and Perl 1984; Jellinger 1986a, 1987a). These data suggest independent or concomitant degenerations of the cortical and subcortical neuronal systems in PD similar to that which has been prosposed for AD (Saper et al. 1985; Morrison et al. 1986; Etienne et al. 1986).

Neuronal loss in the NBM is associated with decreased activities of the cholinergic marker enzymes CAT and acetylcholinesterase (AChE) in the neocortex, hippocampus, and NBM in PD (Candy et al. 1983, Perry et al. 1983a, 1985; Gaspar and Gray 1984; Dubois et al. 1985; Javoy-Agid et al. 1984, 1986; Agid et al. 1987) as well as in AD/SDAT (Coyle et al. 1983; Davies 1983; Perry et al. 1983; Rossor et al. 1984; Fishman et al. 1986; Etienne et al. 1986), supporting the contention that the innominatocortical and septohippocampal cholinergic systems are involved in these disorders. The greater cholinergic deficits in the neocortex and NBM are seen in demented PD subjects (Perry et al. 1983, 1985; Dubois et al. 1985; Agid et al. 1987), suggesting that lesion of the innominatocortical cholinergic pathway is related to dementia in PD. However, in PD patients with no intellectual deterioration, a cortical cholinergic deficiency (Dubois et al. 1985, Perry et al. 1985), and neuronal loss in the NBM (Nakano and Hirano 1984) indicate that the cholinergic system is already in the process of degeneration.

2. Nucleus Tegmenti Pedunculopontinus

The parabrachial pedunculopontine nucleus, pars compacta (PPNc), lying in the dorsolateral part of the tegmental reticular formation, is recognized as an important "loop nucleus" (GRAYBIEL 1984). It receives fibers from the internal pallidal segment and zona reticulata nigrae, while it projects to the zona compacta nigrae, subthalamic nucleus, globus pallidus, and striatum (GRAYBIEL 1984; SCARNATI et al. 1987), with portions of the neuraxis that modulate somatic motor activities at a cortical level (CARPENTER 1986). This putative cholinergic nucleus in the lateral mesencephalon, projecting to the neocortex, lies in a region containing many cholinergic neurons, and may represent an extrastriatal mechanism for affecting the balance between cho-

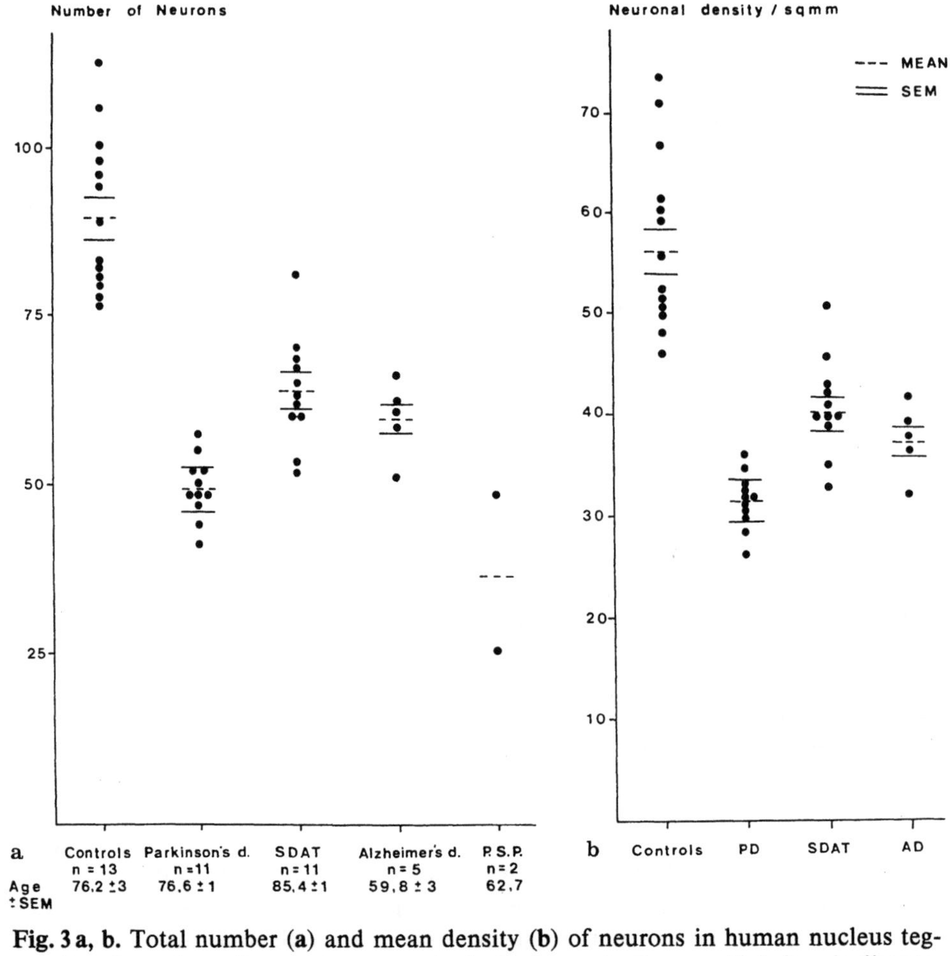

Fig. 3a, b. Total number (**a**) and mean density (**b**) of neurons in human nucleus tegmenti pedunculopontinus, pars compacta, in Parkinson's disease, Alzheimer's disease, senile dementia of Alzheimer type (SDAT), progressive supranuclear palsy (PSP), and controls

linergic and dopaminergic functions of the basal ganglia (GRAYBIEL 1984, 1986). Significant neuronal loss in the PPNc has been demonstrated in PSP, PD and PD/AD, with frequent occurrence of Lewy bodies and NFTs (ZWEIG et al. 1987; HIRSCH et al. 1987; JELLINGER 1988). PPNc neurons are reduced by an average of 43 to 51 % in PD and by 60 to 80 % in PSP, but only by an average of 25.7 % in SDAT, and 33.8 % in AD, as compared with age-matched controls (Fig. 3). Lewy bodies are abundant in all cases of PD, and NFTs are common in AD/SDAT and PSP, but both types of lesion are infrequent in aged controls. These data suggest that putative cholinergic neurons of the PPNc are more selectively affected in PD and PSP, and may be related to certain disorders of motor performance and coordination in these disorders. The PPNC is also involved in experimental MPTP parkinsonism in monkeys (CROSSMAN et al. 1988).

3. Westphal-Edinger Nucleus

The Westphal-Edinger nucleus, a visceral subdivision of the oculomotor complex, giving rise to cholinergic fibers to the ciliary ganglion regulating pupilloconstriction (BENDER 1980), has been shown to suffer a 54% neuronal loss in PD, and 2 %-3 % of its cells are affected by Lewy bodies and NFTs (HUNTER 1985). Among 30 PD brains, we saw Lewy bodies and NFTs in 97 % and 17 %, respectively, while in AD/SDAT their prevalence was 12 % and 90 %, respectively. The frequent involvement of this nucleus may explain ophthalmoplegia and other neuro-ophthalmic disorders in PD and AD (CORIN et al. 1972; GUILOFF et al. 1980).

V. Peptidergic Systems

Mild reduction in PD of cholecystokinin (CCK-8), substance P, Met-enkephalin, and Leu-enkephalin in the compacta nigrae and globus pallidus, and of somatostatin in the frontal cortex and hippocampus, particularly in demented subjects (JAVOY-AGID et al. 1986; DUBOIS et al. 1985; AGID et al. 1987), are not fully explained by immunocytochemistry. The severe reduction of substance P and Met-enkephalin in the globus pallidus and substantia nigra in Huntington's chorea is not seen in PD and PEP, where intense immunoreactivity for both substances (GRAFE et al. 1983; ZECH and BOGERTS 1985) or even increased immunoreactivity for substance P has been observed in these nuclei (CONSTANTINIDIS et al. 1983; TABAN et al. 1984). On the other hand, biochemical studies have shown a reduction of substance P (TENOVUO et al. 1984) and of Met-enkephalin in the substantia nigra and caudate nucleus (LLORENS-CORTES et al. 1984), and both these peptides are known to occur in fibers constituting the striatonigral pathways (NIEUWENHUYS 1985). Immunoreactivity for CCK in the substantia nigra is comparatively preserved in PD (DIETL 1985), although it coexists with dopamine in the nigrostriatal pathway (UHL et al. 1986). On the other hand, somatostatin, in contrast to normal or even increased immunoreactivity in the striatum in Huntington's chorea (MARTIN 1984), in both PD and AD shows considerable reduction or even absence of immunoreactive cells and fibers in the basal ganglia (FORNO et al. 1985a; FERRANTE et al. 1985). These data suggest that degeneration of peptidergic neuronal systems affects limited areas in the PD brain (JAVOY-AGID et al. 1986; AGID et al. 1987).

E. Morphological Effects of Levodopa Treatment

No correlation can be drawn between the therapeutic effects of levodopa substitution therapy and the anatomic type and degree of damage to the substantia nigra (YAHR et al. 1972; BERNHEIMER et al. 1973). The brain pathology findings in PD patients after long-term levodopa treatment show no change compared with untreated subjects (YAHR et al. 1972; JELLINGER 1974). There is no indication for remelaninization or increased neuromelanin content in the nigral or other pigmented neurons, and no ultrastructural differences have been observed (DUFFY 1972). Long-term administration of Levodopa in mice did not induce damage to dopaminergic nigrostriatal neurons, suggesting that prolonged Levodopa therapy may not accelerate the degeneration of human nigrostriatal neurons or the rate of progression of PD (HEFTI et al. 1981). On the other hand, progressive degeneration of the nigrostriatal system with signs of progressive cell destruction in the substantia nigra is known to occur in PD (GIBB and LEES 1987), and is documented by an increased loss of neurons and melanin in the substantia nigra in advanced stages of PD treated with levodopa (MANN et al. 1983; GIBB 1988). The morphological effects of autologous transplantation of adrenal medullary tissue to the caudate nucleus (GAGE et al. 1986) or of fetal nigral tissue to the brain of PD patients are currently being studied in several laboratories (JANKOVIC 1988).

F. Pathology of Dementia in Parkinson's Disease

Intellectual impairment, estimated to occur in 15%–93% of PD patients with an average of about 30% (LIEBERMAN et al. 1979; MAYEUX and STERN 1983; BALL 1984; BOLLER 1985; ELIZAN et al. 1986; HUBER et al. 1986; MAYEUX et al. 1988), has been attributed by some to cerebral atrophy and extranigral brain lesions (ALVORD et al. 1974; JACOB 1983; YOSHIMURA 1983), to concomitant cortical Alzheimer pathology (BOLLER et al. 1979; HAKIM and MATHIESON 1979; BOLLER 1985), and involvement of various subcortical and subcortico-cortical neuromediator-specific systems (MANN and YATES 1983a; PERRY et al. 1983b, 1985; WHITEHOUSE et al. 1983; HORNYKIEWICZ and KISH 1984; GASPAR and GRAY 1984; JAVOY-AGID et al. 1984, 1986, HEILIG et al. 1985; DUBOIS et al. 1985; CHUI et al. 1986; WHITEHOUSE 1986; GROWDON and CORKIN 1986; AGID et al. 1987; CASH et al. 1987). Computerized tomography studies showed a more severe cerebral atrophy in demented PD patients than in nondemented ones and age-matched controls (SCHNEIDER et al. 1979), only cortical atrophy being related to both age and duration of PD (SROKA et al. 1981). Postmortem demonstration of brain atrophy with significantly reduced brain weight (HA-

KIM and MATHIESON 1979), more pronounced cortical atrophy (ALVORD et al. 1974), and increased cortical Alzheimer pathology in an appreciable percentage of PD brains (BOLLER et al. 1979; GASPAR and GRAY 1984) suggested an increased incidence of AD in patients with PD, and some links between the (as then unknown) disturbances underlying both PD and AD (HAKIM and MATHIESON 1979). However, while in aged subjects without PD and in AD/SDAT a positive correlation was found between the degree of dementia and the number density of NFTs (and to a smaller degree senile plaques) in the neocortex and hippocampus (WILCOCK and ESIRI 1982; GIBSON 1983; JELLINGER and RIEDERER 1984; KATZMAN 1986; NEARY et al. 1986; CRYSTAL et al. 1987), in small autopsy cohorts of PD no significant differences in the density of cortical or hippocampal Alzheimer lesions were found between demented and nondemented patients (HESTON 1981; BALL 1984; MANN et al. 1984; PERRY et al. 1985; CHUI et al. 1986).

In AD/SDAT, the average fresh brain weight is significantly reduced by 4%–10% compared with age-matched controls (Fig. 4; TERRY 1985), while in a cohort of PD subjects, 54% of whom had some degree of dementia, there was only a nonsignificant 2% decrease in brain weight (after opening of the ventricles) compared with age-matched controls, the latter being within the range of comparable samples (DEKABAN 1978; HO et al. 1980). By contrast, PEP showed a significant reduction of brain weight by 7.8% (Fig. 4). Independent evaluation of brain weights for males and females showed similar results, only males

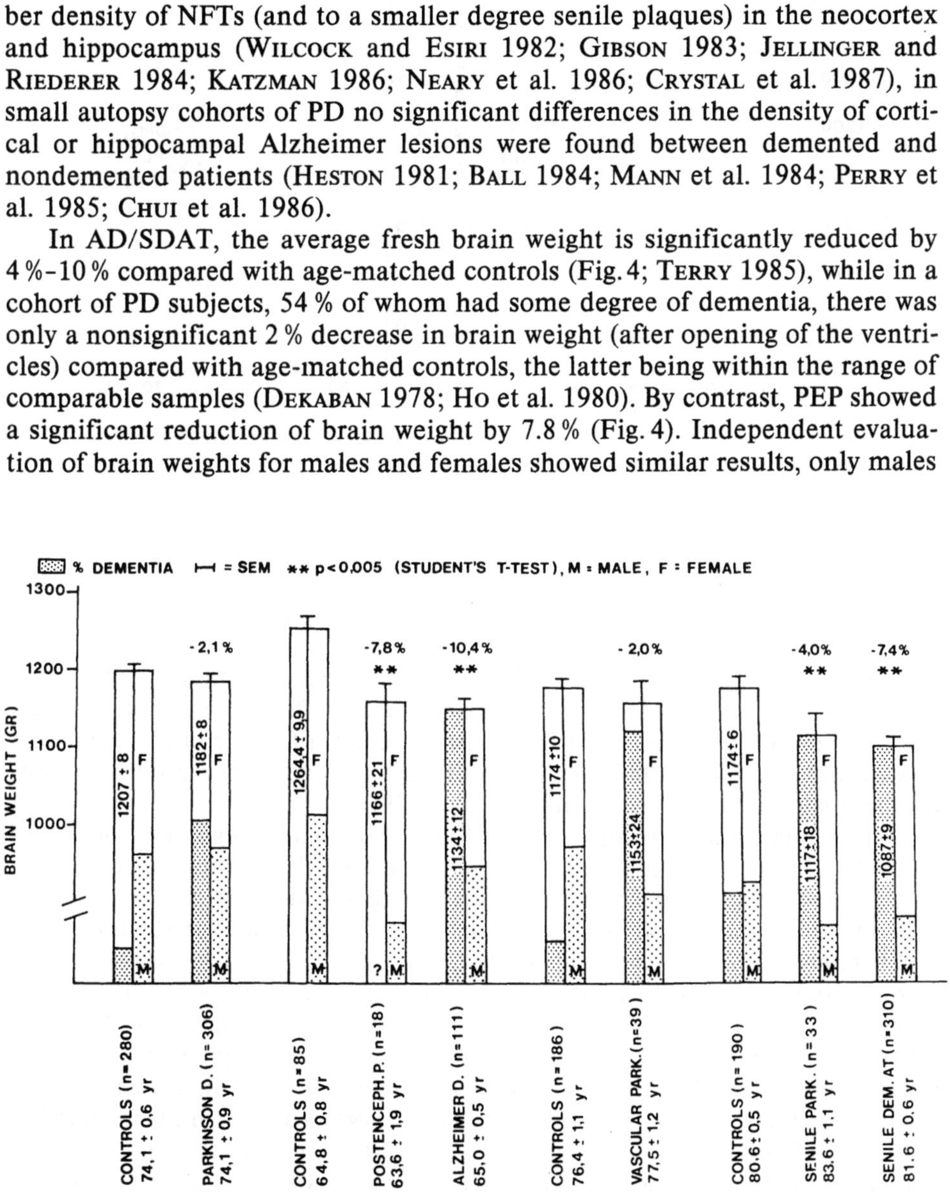

Fig. 4. Comparison of fresh brain weights in different types of parkinsonism, Alzheimer's disease, senile dementia of Alzheimer type (SDAT), and age-matched controls

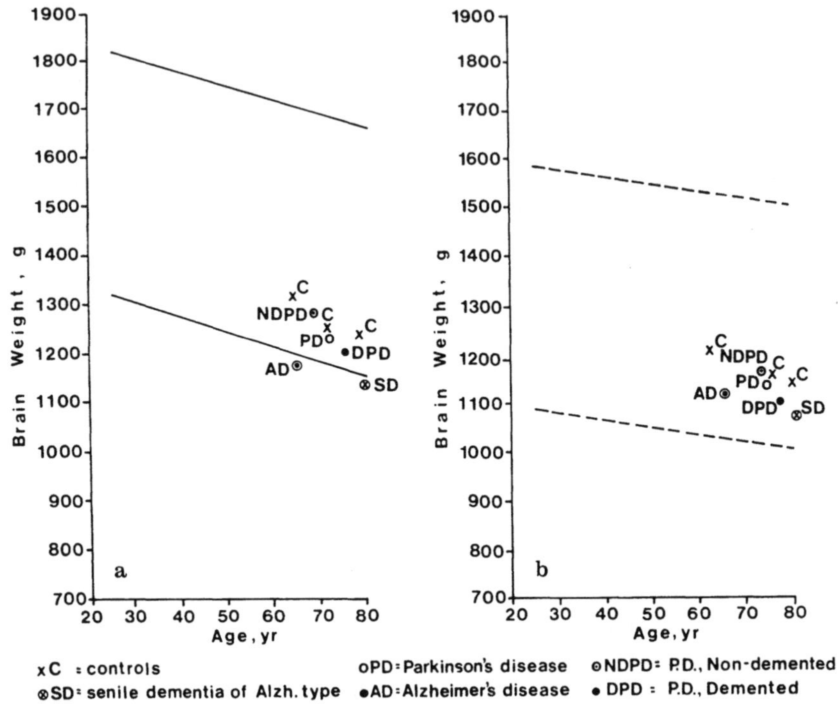

Fig. 5 a, b. The 95% confidence range of brain weight for white men (**a**) and women (**b**) aged 25–80 years. (Ho et al. 1980)

with AD and SDAT were below the 95 % confidence range for white humans (Fig. 5). Regression analysis showed a positive correlation between the loss of brain weight in PD — but not in AD — to both the age and duration of illness, although reduction in brain weight was accelerated after the onset of PD. This was more pronounced in PD patients with onset after the age of 60 years (Jellinger and Riederer 1984), which confirms time course computerized tomographie studies in PD (Schneider et al. 1979).

While quantitative studies in AD/SDAT show significant neuronal loss (40 %-60 %) in both the hippocampus and neocortex (Hubbard and Anderson 1985; Terry 1985) that correlate strongly with the density of NFTs and/or NPs and the intensity of dementia (Gibson 1983; Katzman 1986; Jellinger and Riederer 1984; Neary et al. 1986; Crystal et al. 1987), we are not aware of such studies in PD. Evaluation of histologic changes, on a four-degree intensity scale, showed that neuronal loss and severe AD lesions (ANFTs and NPs) in the frontal cortex and hippocampus were more frequent in PD brains than in age-matched controls, but these differences are not significant (Mann et al. 1983; Jellinger and Riederer 1984; Jellinger 1986a, b). Considerable reduction of large cortical neurons, as seen in SDAT (Terry 1985), associated with a 45 % reduction and shrinkage of large somatostatin neurons (Joynt and McNeill 1984; Beal et al. 1986), is suggested by reduced somatostatin content in PD brain (Javoy-Agid et al. 1984, 1986; Epelbaum et al. 1986; Agid et

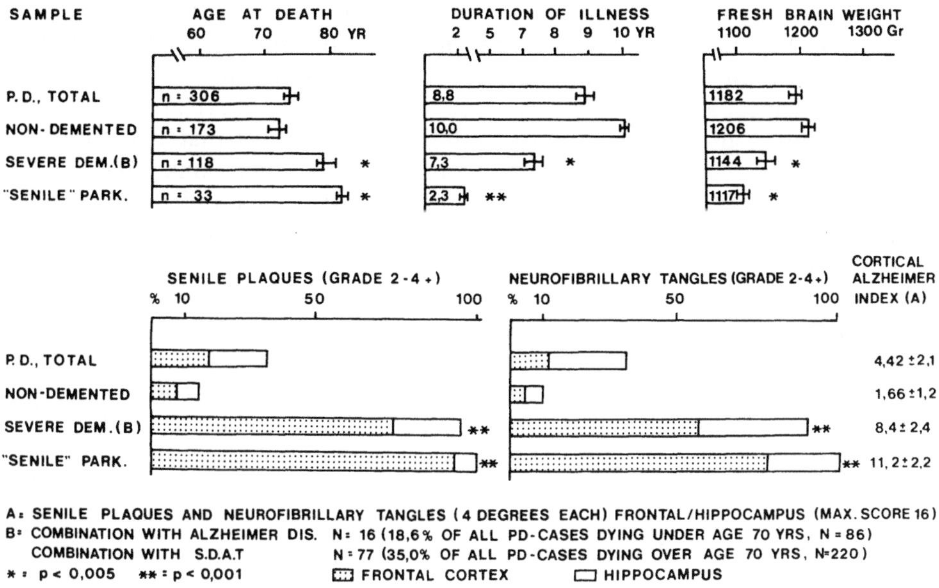

Fig. 6. Comparison of age, duration of illness, brain weight, and cortical Alzheimer scores in Parkinson's disease patients of both sexes with and without severe dementia

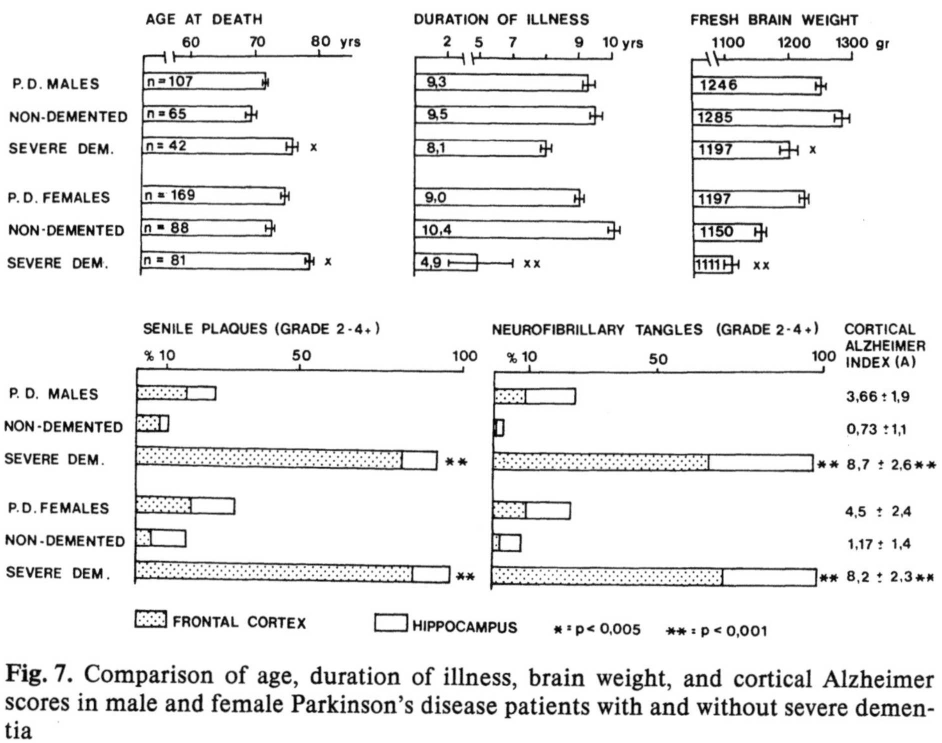

Fig. 7. Comparison of age, duration of illness, brain weight, and cortical Alzheimer scores in male and female Parkinson's disease patients with and without severe dementia

al. 1987), but has not been quantitatively substantiated. On the other hand, cortical AD lesions were found to be significantly more numerous in demented PD patients (GASPAR and GRAY 1984; JELLINGER 1987a), and a comparison of a semiquantitative three-degree intensity scale of NFTs and NPs in the frontal cortex and hippocampus of 240 PD cases and age-matched controls revealed good correlation with the degree of dementia (JELLINGER 1986a). PD patients with severe dementia, in general have a later onset and shorter duration of illness (LIEBERMAN et al. 1979; SROKA et al. 1981; DESMEDT et al. 1982; JELLINGER 1986b), while no such difference was seen by GASPAR and GRAY (1984). Demented PD subjects are significantly older at death, have much lower brain weight, and significantly more AD lesions in the frontal cortex and hippocampus than PD patients without dementia (Figs. 6 and 7). The severity of cortical AD lesions, expressed as a total Alzheimer score on a four-degree intensity scale for both NFTs and NPs in the frontal cortex and hippocampus (maximal score 16), in PD cases with little or no dementia, was similar to age-matched controls, and showed a positive correlation with age, while no such age correlation has been observed in demented PD subjects with additional AD or SDAT (Fig. 8). This negative age relation of AD pathology is

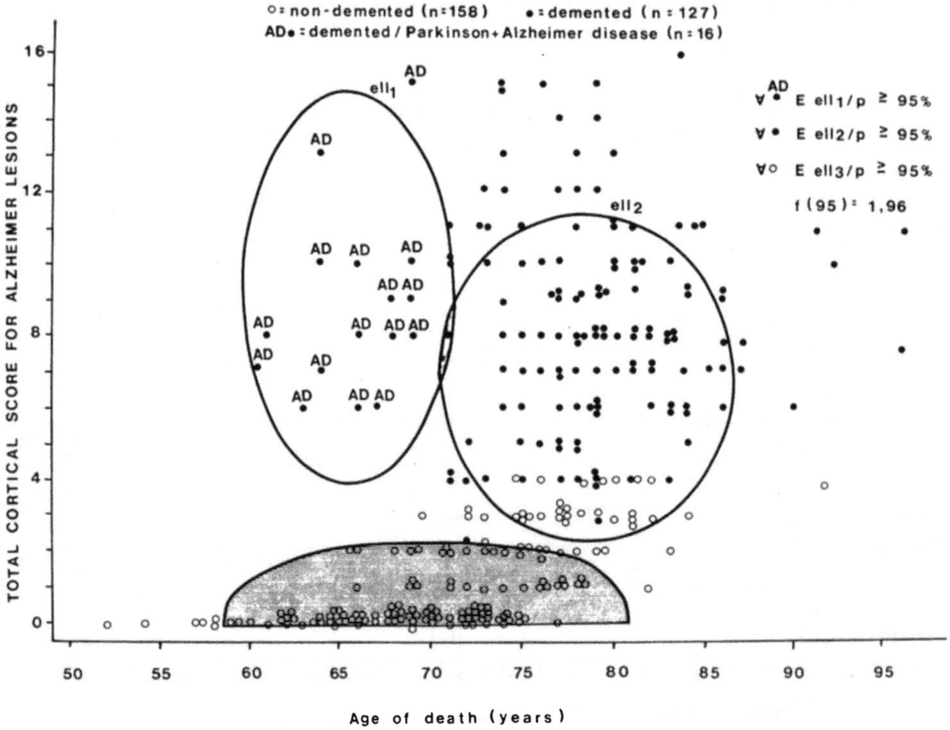

Fig. 8. Relation of cortical Alzheimer score (senile plaques and neurofibrillary tangles in frontal cortex and hippocampus) to age in 301 patients with Parkinson's disease; cluster analysis

well established for AD and younger SDAT subjects (GIBSON 1983; ALAFUZOFF et al. 1987), while about 40% of demented PD cases show associated SDAT pathology (see Fig. 6).

In general, Alzheimer pathology in the neocortex, hippocampus, and brain stem is such more severe in AD than in very old SDAT subjects (JAMADA and MEHRAIN 1968; MOUNTJOY et al. 1984; MANN et al. 1985; MANN 1985; JACOBS and BUTCHER 1986), separating premature or "malignant" from physiologic aging of the brain (JELLINGER 1989). In small autopsy cohorts of PD individuals, the incidence of additional pathologic markers of AD/SDAT ranges from 27% (PERL et al. 1984) to 33% or even 47% (BOLLER et al. 1979; HAKIM and MATHIESON 1979), while among 306 autopsy cases of PD, we observed additional AD pathology in 18.6% of the subjects dying before the age of 70 years, whereas 35% of the older subjects dying over the age of 70 had additional cortical AD/SDAT lesions (see Fig. 6). Most of these patients have been demented. These data are in accordance with those of other autopsy cohorts of parkinsonism, where the incidence of coexisting pathology of PD and AD ranges from 5% or 7% (GRILIA et al. 1987; XUEREB et al. 1987) up to 16.7% (PERRY et al. 1983a; DITTA and MIRRA 1987), while additional cortical AD pathology is seen in 35%-40% of the older PD subjects (PERRY et al. 1985; 1987; JELLINGER 1987a, b). Although most PD cases having additional cortical Alzheimer pathology show mental impairment, a number of severely demented cases with nigral damage and Lewy bodies, but little or no cortical AD lesions have been reported (HESTON 1981; GASPAR and GRAY 1984; PERRY et al. 1985, 1987; HEILIG et al. 1985; CHUI et al. 1986; OYANAGI et al. 1986). Hence, dementia in PD does not necessarily imply the coexistence of cortical Alzheimer lesions, but other mechanisms of dementia have been suggested.

In many such cases, a subcortical explanation for mental impairment is likely (CUMMINGS and BENSON 1984; ALBERT 1984; WHITEHOUSE 1986), e.g., involvement of the mesocorticolimbic dopaminergic system with neuronal loss in the VTA (JAVOY-AGID et al. 1986; AGID et al. 1987), the noradrenergic locus ceruleus (MANN et al. 1986), the reticular formation (FUJIMURA and UMBACH 1987), and the cholinergic forebrain systems (WHITEHOUSE et al. 1983; TAGLIAVINI and PILLERI 1984; EZRIN-WATERS and RESCH 1986; WHITEHOUSE 1986; AGID et al. 1987). While TAGLIAVINI et al. (1984a) found no correlation between cell loss in the NBM and mental status in six PD subjects, in many demented PD individuals NBM cell loss has been reported to be significantly greater than in nondemented ones (GASPAR and GRAY 1984; JELLINGER 1986a, b, 1987a; CHUI et al. 1986; see also Table 9). Almost total deprivation of the NBM has been observed in Guam PDC (NAKANO and HIRANO 1983) and in DLBD (SIMA et al. 1986; BURGHARDT et al. 1988; DICKSON et al. 1987); severe dementia and reduced CAT levels are present in both disorders. Comparative studies in 50 PD subjects showed that, in the nondemented ones, cell loss in the NBM ranging from 15% to 62% was associated with little or no AD lesions in the frontal cortex and hippocampus, while in severely demented patients, NBM cell loss ranged from 64% to 90% and was often, but inconsistently, accompanied by severe cortical AD pathology (Fig. 9).

These data and the demonstration of reduced levels of the marker en-

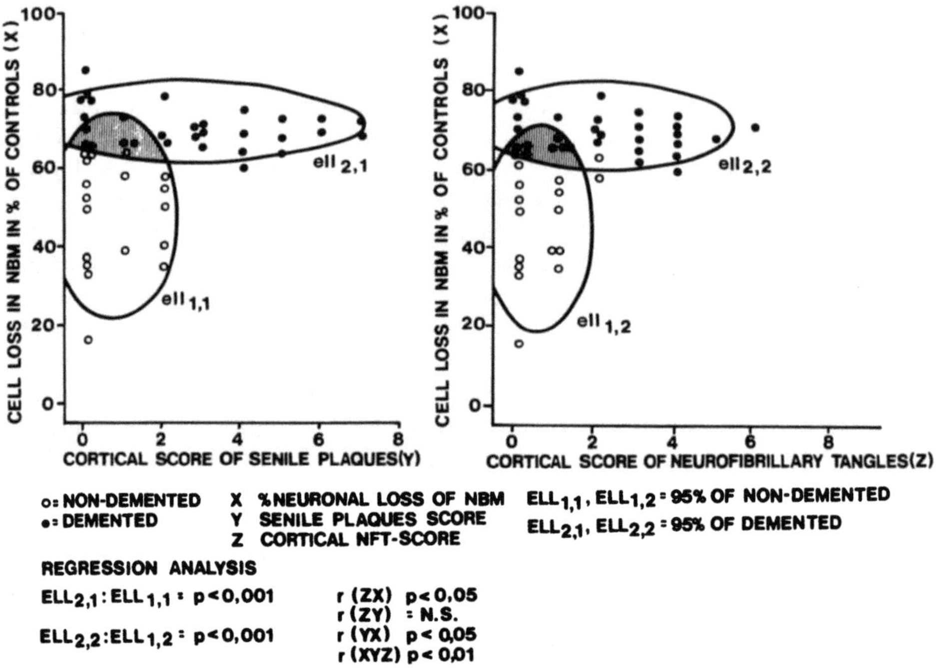

Fig. 9. Relation between neuronal loss in nucleus basalis of Meynert (NBM) and Alzheimer pathology in frontal cortex and hippocampus (4 degrees of intensity for each region) in 50 autopsy cases of Parkinson's disease with and without dementia

zymes CAT and AChE in the neocortex and NBM of demented compared with nondemented PD subjects (CANDY et al. 1986; E. K. PERRY et al. 1983, 1985, 1987; GASPAR and GRAY 1984; DUBOIS et al. 1985; JAVOY-AGID et al. 1986; AGID et al. 1987) reinforce the postulated role of the cholinergic forebrain system in the development of cognitive impairment in PD. Reduction of AChE activity in the frontal cortex of nondemented PD patients (DUBOIS et al. 1985) suggests that degeneration of the cholinergic system may precede the appearance of intellectual deterioration. Some evidence exists suggesting that there may be a critical threshold for the cholinergic deficit before dementia becomes apparent (CANDY et al. 1983; WHITEHOUSE et al. 1983; HORNYKIEWICZ and KISH 1984; AGID et al. 1987). It can be concluded that this threshold level — similar to that in the striatonigral dopaminergic system producing motor PD signs (BERNHEIMER et al. 1973; BIRKMAYER and RIEDERER 1985; HORNYKIEWICZ and KISH 1986; AGID et al. 1987) — lies around 65 %–80 % cell loss in the NBM and equivalent reduction of cortical cholinergic markers, which is due to the large compensatory capacity of the remaining neurons in these neuromediator-specific systems (HORNYKIEWICZ and KISH 1984; AGID et al. 1987). Although in most severely demented PD patients, NBM cell loss is comparable to the attrition seen in AD, the unevenness of cholinergic cell loss in the various subdivisions of the NBM (ARENDT et al. 1983) and its variable relationship with CAT reduction in both the neocortex and NBM (R.H. Perry et al. 1983, 1985; Gaspar and Gray 1984; Dubois et al. 1985) emphasizes that the relationship between the NBM and intellectual impairment and the basic mechanisms for the dysfunction of the cholinergic forebrain remain unresolved.

In AD/SDAT, the neuroanatomic distribution of the pathologic changes suggests an early and severe involvement of the amygdala/hippocampus (Jamada and Mehraein 1968; Ball 1977; Hyman et al. 1984; Pearson et al. 1985; Mann et al. 1987a) and a progression of the disease process from the medial temporal cortex and amygdala into the parietotemporal association area (Pearson et al. 1985), and neurochemical evidence also points to the early and severe involvement of the amygdala (Palmer et al. 1986; R.H. Perry et al. 1987). The close reciprocal connection of the amygdala with the NBM may underlie the degeneration of the latter in AD as a secondary phenomenon through both anterograde and retrograde mechanisms (Pearson and Powell 1987). In AD/SDAT, all severely demented cases have dramatic cell loss in the NBM, with marked reduction of cortical levels of CAT and AChE activity and nicotinic receptors (Davies 1983; R.H. Perry et al. 1983, 1987; Rossor et al. 1984; Fishman et al. 1986), but only half of them have reduced CAT levels in the NBM (Younkin et al. 1986; Etienne et al. 1986). These data and the lack of strict correlation between NBM cell loss and decreased CAT activities in both the NBM and neocortex in PD (R.H. Perry et al. 1983e; Gaspar and Gray 1984; Dubois et al. 1985), and in some cases of AD (Etienne et al. 1986; Younkin et al. 1986), suggest both a reduction of CAT synthesis and disorders of axonal CAT transport, with accumulation of the enzyme in NBM neurons (Younkin et al. 1986; Etienne et al. 1986). Immunocytochemical studies of CAT and AChE in AD revealed a reduction of neuronal diameters in the NBM by 13.6% (Pearson et al. 1985) and a decreased AChE staining of the neuropil in the NBM (Etienne et al. 1986), suggesting a reduced enzyme synthesis. Assuming that cortical CAT is mostly synthesized in the NBM, the more severe reduction of the cholinergic marker in the cortex than in the NBM, which shows an increased CAT: neuronal cell ratio, and the paucity of AChE neuritic staining with preserved cytoplasmic reaction in the NBM, suggest axonal transport defects of cholinergic enzymes in AD/SDAT (Younkin et al. 1986; Etienne et al. 1986). It is hypothesized that the defect more selectively involves the magnocellular population of the NBM, since early functional metabolic abnormalities of large neurons tend to be predominantly manifest in distal intracortical preterminal branches as a "dying back" process. On the other hand, the failure to demonstrate a consistent relationship between NBM cell loss and the density of cortical NPs in PD (Candy et al. 1983; Gaspar and Gray 1984; Nakano and Hirano 1984; E.K. Perry et al. 1985; Chui et al. 1986; Jellinger 1986a, b, 1987a), in Guam PDC (Nakano and Hirano 1983), and boxer's dementia (Uhl et al. 1982), and the fact that NPs and plaque-like amyloid-containing structures are frequently observed within the NBM (Rudelli et al. 1984; Saper et al. 1985; Jacobs et al. 1985), do *not* support the dying back hypothesis of cholinergic terminals as the initiator of NP formation (Price et al. 1983). Others suggest retrograde or secondary transneuronal degeneration of the NBM and other subcortical nuclei, owing to primary cortical lesions (PEARSON et al. 1985, 1986; MANN et al. 1986, MARCYNIUK et al. 1986; JACOBS and BUTCHER 1986; PEARSON and POWELL 1987). However, the frequent inconsistencies of neuronal cell and plaque counts in the cortex, and other subcortical nuclei in PD, the occurrence of NP in both the neocortex and basal forebrain, and the neuronal pathology of the NBM in both PD and AD, suggest a parallel or concomitant degeneration of both cortical and subcortical systems in both PD and AD (SAPER et al. 1985; MORRISON et al. 1986; ETIENNE et al. 1986; JELLINGER 1987a).

Regardless of whether or not the basal forebrain is a primary or secondary structural locus for the pathologic processes leading to mental impairment in AD/SDAT, PD, and other non-Alzheimer dementias, it seems clear that a full understanding of dementia will not be achieved until their various causal and predisposing factors are understood and their place in the overall fabric of these disorders is established. From the distribution pattern of the degenerative lesions in basal ganglia and brain stem, and from both the distribution pattern of senile plaques and the demonstration of many neuromediator-specific markers in abnormal neurites and terminals in and around senile plaques (TERRY 1985), it can be concluded that pathology in both PD and AD/SDAT is not restricted to a single specific neuronal system, but involves *multiple* mediator systems (STRUBLE et al. 1985; JELLINGER 1987a). Both the causative or predisposing factors in these degenerative disorders and the pathogenetic connections of degeneration in various systems, e.g., the dopaminergic nigrostriatal and the cholinergic forebrain systems in PD, are unknown, although the demonstration of TH immunoreactivity of part of the cholinergic neurons in the magnocellular part of the NBM (HENDERSON 1987) and of TH-TR fibers and terminals in the NBM (JELLINGER 1987a), as well as the biochemical demonstration of dopamine and homovanillie acid in human NBM (RIEDERER et al. 1985) suggest some dopaminergic innervation of the basal forebrain, while a cholinergic innervation of the substantia nigra has recently been suggested (HENDERSON and GREENFIELD 1987). In AD, the NBM shows a significant decrease in both spiperone binding and in the levels of dopamine and serotonin, indicating a loss of both dopaminergic and serotonergic synapses (SPARKS et al. 1986). Decreased somatostatin levels and reduced numbers of somatostatin-containing large neurons and immunoreactivity in the neocortex have also been recognized as possible factors contributing to dementia in both AD and PD (JAVOY-AGID et al. 1986; BEAL et al. 1986; MORRISON et al. 1986; AGID et al. 1987) and in DLBD (DICKSON et al. 1987a).

Other parkinsonian syndromes with dementia are distinguishable from PD by several differences: in PSP and boxer's dementia, both characterized by the abundance of NFTs in the absence of neocortical lesions and NPs, the cholinergic subcortical systems are at most only mildly involved, and the somatostatin levels in the anatomically preserved neocortex are intact (AGID et al. 1986, 1987). In these and other disorders, the basic mechanisms for dementia remain unclear. Mental impairment in PD cannot be related to the pathology of a single neuronal or mediator-specific system, but, based on the principal pathology, several subtypes can be distinguished:

— *"Cortical dementia."* Neuronal and dendritic loss, and Alzheimer pathology in the neocortex, with reduced activities of specific neuromediators and their synthesizing enzymes, e.g., somatostatin and acetylcholine, which, as in other aged people, show good correlation with the degree of dementia. The most severe changes may cause a breakdown of the functional and anatomic integrity of the corticocortical association systems which may extend to the "global cortical disconnection syndrome" observed in AD (MORRISON et al. 1986).

— *"Limbic or hippocampal dementia."* Similar severe cell-specific Alzheimer pathology may cause a functional and anatomic isolation of the hippocampus and other nontelencephalic cortical structures (amygdaloid nucleus) as another substrate of cognitive and behavioral disorders (HYMAN et al. 1984; BALL 1984).

— *"Subcortical dementia"* due to dysfunction of ascending subcortical systems (CUMMINGS 1986; ALBERT 1984; WHITEHOUSE 1986): (a) dysfunction of the cholinergic forebrain system related to degeneration of the NBM, with disorders of synthesis and transport of specific enzymes to the neocortex and hippocampus; and (b) dysfunction of other neuromediator systems owing to degeneration or dysfunction in their projecting nuclei, e.g., in the ascending noradrenergic system, owing to damage to the locus ceruleus, the mesocorticolimbic system, owing to damage to the ventral tegmental area (TORACK and MORRIS 1988), the serotonergic system, related to damage to the dorsal raphe nucleus, and some peptidergic systems, etc., with both disorders of synthesis and transport of specific neuromediators and synthesizing enzymes. It should be emphasized that degeneration of subcortical neuronal systems related to dementia is not necessarily associated with cortical Alzheimer pathology.

Combined cortical and subcortical dementia may be due to variable association of degenerativ edamage to subcortical systems with cortical and/or subcortical Alzheimer pathology.

G. Concluding Remarks

The pathomorphology of PD and related parkinsonism is characterized by a variety of degenerative changes in several subcortical neuronal systems, with or without cortical or other nervous system lesions. Damage to the striatonigral dopaminergic pathways, being the principal at-risk system in these disorders, is associated with lesions in most of the main ascending pathways: dopaminergic mesocorticolimbic, noradrenergic cerulocortical, serotonergic raphe, cholinergic innominatocortical and septohippocampal, part of the cholinergic oculomotor and pedunculopontine systems, and some peptidergic pathways, while other systems, e.g., GABAergic and some intracortical pathways, appear to remain preserved in "classical" PD (AGID et al. 1987). These lesions, which are often associated with various cytoskeletal neuronal abnormalities, have been shown to cause a multimediator dysfunction, owing to neuronal dysfunction and cell loss, reduction of synthesis capacity of the remaining neurons, and disorders of transport mechanisms, that result in the characteristic clinical syndromes. Additional superimposed age-related or coincidental lesions of the CNS will influence the clinical course, signs, and symptoms of the disorders. The causative mechanisms and predisposing factors underlying the degenerative changes in the at-risk and other affected neuronal systems are unknown. Whereas recently demonstrated correlations between neuronal depletion in some subcortical nuclei and cell loss and/or plaque density in cortical target areas suggest retrograde or anterograde degeneration as a consequence of primary cortical damage (JACOBS and BUTCHER 1985; MANN et al. 1986; MARCYNIUK et al. 1986; ARENDT et al. 1985; PEARSON et al. 1985; PEARSON and POWELL 1987), no such relations have been found in PD and related multisystem degenerations. It remains unclear whether degenerative neuronal changes in at-risk neuronal systems represent a "downregulation" of neuromediator production, enhanced aminergic activity (WILLIS 1988) loss of specific enzyme synthesis, or neurotrophic protein production (APPEL 1981) as the primary sequelae of neuronal dysfunction also expressed by cytoskeletal abnormalities, or whether they are caused by transsynaptic or retrograde transport disorders, interaction with other systems, or other endogenous or exogenous mechanism of hitherto unknown origin.

Based on morbid anatomy and cytoskeletal pathology, several subtypes of parkinsonian syndromes, with possible differences in etiology and pathogenesis, can be distinguished, while an anatomic distinction of the various clinical subtypes of PD appears impossible from the data available. According to the principal cytoskeletal changes, two major types of parkinsonism have been distinguished (ALVORD et al. 1974):

1. The predominant "Lewy body type", including idiopathic PD which is relatively stereotyped and shows good correlation of the degree of clinical parkinsonian symptoms with that of both nigral cell loss and striatal dopamine and TH deficiency (BERNHEIMER et al. 1973; HORNYKIEWICZ 1982; AGID et al. 1987), and DLBD that includes a variety of disorders with progressive dementia transitions to idiopathic PD and AD/SDAT;
2. The "Alzheimer type", characterized by widespread NFTs, which includes a wide variety of disorders, ranging from AD with concurrent PD (AD/PD), AD extending to the brain stem, PSP, Guam PDC, PEP, and boxer's dementia (SUDHAKER et al. 1987).

In addition to idiopathic PD, a wide range of "atypical" PD (SROKA et al. 1981) or Parkinson-plus syndromes have been reported (FISCHER 1984), the relation of which are to be further elucidated. More recent data allow subdivision of degenerative parkinsonian syndromes into several clinicopathologic groups:

— *"Typical" idiopathic PD* of the Lewy body type without dementia in which pathology is mainly restricted to pigmented brain stem systems without additional CNS lesions (increased cerebral atrophy, extranigral lesions, Alzheimer pathology), although cortical AD changes and cerebrovascular lesions can be superimposed by normal brain aging. Although the duration of PD is not predictive of either the mental status or the severity of nigral changes and cortical AD pathology, reduction of brain weight is related to both age and duration of PD, the age dependency of both brain weight (and cerebral atrophy as seen in the cranial CT) and cortical AD pathology being more pronounced. These data indiate that "normal" brain aging is superimposed on the basic degenerative process causing PD.

- *"Atypical" PC* (clinically referred to as Parkinson-plus syndromes), often associated with dementia and other signs or symptoms. They often show a significantly higher degree of brain atrophy and cortical AD pathology than nondemented PD and age-matched control populations which, at least in some cases, appears due to the more advanced age, while others are associated with AD or other multisystem degenerations. They can be subdivided into several groups;
- *Concurrence of PD and AD* (PD/AD), showing a true combination of both PD and AD pathology suggesting an association of both diseases which is seen in 5%–20% of all demented PD subjects dying under the age of 65 years.
- *PD with superimposed AD/SDAT*, occurring in 35%–45% of aged demented PD subjects showing typical PD pathology of Lewy body type with additional AD pathology in old age.
- *AD/SDAT extending to the brain stem*, showing severe cortical Alzheimer pathology and Alzheimer lesions in the brain stem with various degrees of degenerative damage to the pigmented nuclei with or without Lewy bodies or other PD markers. This "senile parkinsonism" or AD/SDAT/PD is seen in 5%–6% of proven PD cases and in 8%–45% of proven AD/SDAT subjects. They may present either a distinct subset of the aged PD population with early or mild stages of nigral degeneration and superimposed severe SDAT, or a variant of AD/SDAT with mild or clinically relevant nigral damage surpassing the age-related level.
- Demented PD subjects without cortical Alzheimer pathology showing degeneration or dysfunction of cholinergic or other mediator-specific subcortical (noradrenergic, dopaminergic, serotonergic, peptidergic) or intracortical (somatostatin) neuronal systems. This group often shows signs of "subcortical dementia" (Whitehouse 1986) and includes Guam PDC, DLBD, and other multisystem degenerations.
- Parkinsonism in other degenerative processes involving the pigmented brain stem systems with or without cortical AD pathology, e.g., in multisystem degenerations, PSP, etc, where the anatomic and biochemical basis of dementia is often unknown.
- Parkinsonism associated with or superimposed on multifocal cerebrovascular lesions or other brain damage, including toxic and drug-induced lesions (symptomatic or secondary parkinsonism).

In most of the parkinsonian syndromes, the variable topography and morphology of lesions suggests involvement of multiple neuromediator-specific subcortical and, occasionally, cortical neuronal systems inducing a multitransmitter dysfunction of variable extent and intensity which is related to motor, vegetative, and intellectual impairment in these syndromes. While motor impairment is particularly related to dysfunctions of the dopaminergic systems, and vegetative disorders to noradrenergic deficits, the intellectual and psychic impairment has been largely related to dysfunctions of the cholinergic, serotonergic, dopaminergic mesocorticolimbic, and somatostatin intracortical systems (see Agid et al. 1987). Although some cytoskeletal changes in PD, e.g., the Lewy bodies, may serve as diagnostic or even pathogenic markers, most of them are nonspecific and probably give no insight into the basic mechanisms of degenerative pathobiology of PD and related disorders. Further studies are needed to clucidate: (a) the clinicopathologic correlationsa of the different clinical subtypes of idiopathic PD and related parkinsonian syndromes; (b) the basic mechanisms underlying the degenerative changes in the at-risk and other affected neuronal systems and their reciprocal correlations and relative contributions to the clinical syndromes; (c) the relations of these disorders to aging processes of the brain: (d) the reliability of presently available experimental neurotoxic models of parkinsonism for human disease; and (e) the significance of the pathobiology of PD and related disorders for future therapeutic strategies.

References

Adams RD, Bogaert L van, Van der Eecken H (1964) Striato-nigral degeneration. Neuropathol Exp Neurol 23:584-608

Agid Y, Javoy-Agid F, Ruberg M, Pillon B, Bubous B (1986) Progressive supranuclear palsy: anatomoclinical and biochemical considerations. Adv Neurol 45:101-206

Agid Y, Javoy-Agid F, Ruberg M (1987) Biochemistry of neurotransmitters in Parkinson's disease. In: Marsden CD, Fahn ST (eds) Movement disorders 2. Butterworths, London, pp 166-230

Alafuzoff I, Iqbal K, Friden H, Adolfsson R, Winblad B (1987) Histopathological criteria for progressive dementia disorders: clinico-pathological correlation and classification by multivariate data analysis. Acta Neuropathol (Berl) 74:204-225

Albert ML (1984) The controversy of subcortical dementia. Neurol Neurosurg Update Series 5:2-8

Alvord ED, Forno L, Kusske JA, Kaufmann BJ, Rhodes JS, Goetowski CR (1974) The pathology of parkinsonism. A comparison of degeneration in cerebral cortex and brainstem. Adv Neurol 5:175-193

Appel SH (1981) A unifying hypothesis for the cause of amyotrophic lateral sclerosis, parkinsonism and Alzheimer disease. Ann Neurol 10:499-505

Arendt T, Bigl V, Arendt A, Tennstedt A (1983) Loss of neurons in the nucleus basalis of Meynert in Alzheimer's disease, paralysis agitans and Korsakoff's disease. Acta Neuropathol (Berl) 61:101-108

Arendt A, Bigl V, Tennstedt A (1985) Neuronal loss in different parts of the nucleus basalis is related to neuritic plaque formation in cortical areas in Alzheimer's disease. Neuroscience 14:1-14

Arluison M, Dietl M, Thibault J (1984) Ultrastructural morphology of dopaminergic nerve terminals and synapses in the striatum of the rat using tyrosine hydroxylase immunocytochemistry. A topographical study. Brain Res Bull 13:269-285

Ball MJ (1977) Neuronal loss, neurofibrillary tangles and granulovacuolar degeneration in the hippocampus with aging and dementia. Acta Neuropathol (Berl) 37:111-118

Ball MJ (1984) The morphological basis of dementia in Parkinson's disease. Can J Neurol Sci 11:180-184

Bancher C, Lassmann H, Budka H, Grundke-Iqbal I, Iqbal K, Wiche G, Seitelberger F, Wisniewski HM (1987) Neurofibrillary tangles in Alzheimer's disease and progressive supranuclear palsy: antigenic similarities and differences. Acta Neuropathol (Berl) 74:39-46

Bancher C , Lassmann H, Budka H, Jellinger K, Grundke-Iqubal I, Iqbal K, Wiche G, Seitelberger F, Wisniewski HM (1989) Antigenic profile of Lewy bodies: immunocytochemical evidence for protein phosphorylation and ubiquitation. J Neuropathol Exp Neurol 48: (in press)

Barbeau A (1984) Manganese and extrapyramidal disorders. Neurotoxicology (Park Forest Il) 5:13-36

Barbeau A (1986) Parkinson's disease: clinical features and etiopathology. In: Vinken PJ, Bruyn GW, Klawans HL (eds) Handbook of clinical neurology, vol 5 (49). Elsevier, Amsterdam, pp 87-152

Beal MF, Mazwak MF, Sevendsen CN et al. (1986) Widespread reduction of somatostatin-like immunoreactivity in cerebral cortex in Alzheimer's disease. Ann Neurol 20:489-495

Beheim-Schwarzbach D (1952) Über Zelleib-Veränderungen im Nucleus coeruleus bei Parkinson-Symptomen. J Nerv Ment Dis 116:619-627

Bender MB (1980) Brain control of conjugate horizontal and vertical eye movements. Brain 103:25-69

Benno RH, Tucker LW, Joh TH, Reis DJ (1982) Quantitative immunocytochemistry of tyrosine hydroxylase in rat brain. Brain Res 246:225-236

Berciano J (1982) Olivopontocerebellar atrophy. A review of 117 cases. J Neurol Sci 53:253-272

Bernheimer H, Birkmayer W, Hornykiewicz O, Jellinger K, Seitelberger F (1973) Brain dopamine and the syndromes of Parkinson and Huntington. J Neurol Sci 20:415-455

Birkmayer W, Riederer P (1985) Die Parkinson-Krankheit, 2nd edn. Springer, Vienna, New York

Blackwood W (1981) Morbid anatomy. In: Rose FC, Capildeo R (eds) Research progress in Parkinson's disease. Pitman, London, pp 25-31

Blocq P, Marinesco G (1893) Sur un cas de tremblement parkinsonien hémiplégique symptomatique d'une tumeur du pédoncule cérébral. C R Soc Biol (Paris) 45:105

Bogerts B, Häntsch J, Herzer M (1983) A morphometric study of the dopamin-containing cell groups in the mesencephalon of normals, Parkinson patients, and schizophrenics. Biol Psychiatry 18:951-969

Boller F (1985) Parkinson's disease and Alzheimer's disease: are they associated? In: Senile dementia of the Alzheimer type. Liss, New York, pp 119-129

Boller F, Mizutani T, Roessmann U, Gambetti P (1979) Parkinson's disease, dementia and Alzheimer disease: clinicopathological correlations. Ann Neurol 7:329-335

Bondareff N, Mountjoy CQ, Roth M (1982) Loss of neurons of origin of the adrenergic projections to the cerebral cortex (nucleus locus coeruleus) in senile dementia. Neurology (NY) 32:165-168

Borit A, Rubinstein LJ, Urich H (1975) The striatonigral degeneration—putaminal pigments and nosology. Brain 98:101-112

Bøttcher J (1975) Morphology of the basal ganglia in Parkinson's disease. Acta Neurol Scand 52 (Suppl 62):1-160

Brodal A (1981) Neurological anatomy in relation to clinical medicine, 3rd edn. Oxford University Press, New York, pp 411-416

Bugiani O, Perdelli F, Salvarini S, Leonardi A, Mancardi GL (1980) Loss of striatal neurons in Parkinson's disease: a cytometric study. Eur Neurol 19:339-344

Burkhardt C, Filley CM, Kleinschmidt-De Masters BK, de la Monte S, Norenberg MD, Schneck SA (1988) Diffuse Lewy body disease and progressive dementia. Neurology 38:1520-1528

Buttlar-Brentano K von (1955) Das Parkinsonsyndrom im Lichte der lebensgeschichtlichen Veränderungen des Nucleus basalis. J Hirnforsch 2:55-76

Byrne EJ, Lowe J, Godwin-Austin RB, Aire T, Jones R (1987) Dementia and Parkinson's disease associated with diffuse cortical Lewy bodies. Lancet I:501

Candy JM, Perry RH, Perry EK, Irving D, Blessed G, Fairbairn AF, Tomlinson BE (1983) Pathological changes in the nucleus of Meynert in Alzheimer's and Parkinson's disease. J Neurol Sci 54:277-289

Candy JM, Perry EK, Perry RH, Court JA, Oakley AE, Edwardson JA (1986) The current status of the cortical cholinergic system in Alzheimer's disease and Parkinson's disease. In: Swaab DF, Fliers E, Mirmirian M, Van Gool WA, Van Hagren F (eds) Aging of the brain and Alzheimer's disease. Progress in brain research, vol 70. Elsevier Scientific, Amsterdam, pp 105-1322

Carmichael SW, Wilson RJ, Brimijoin WS et al. (1988) Decreased catecholamine levels in the adrenal medulla of patients with Parkinson's disease. New Engl J Med 319:254-256

Carpenter MB (1986) Anatomy of the basal ganglia. In: Vinken PJ, Bruyn GW, Klawans HL (eds). Handbook of clinical neurology, vol5 (49): Elsevier Scientific, Amsterdam—New York: pp 1-18

Cash R, Dennis R, L'Heureux R, Raisman R, Javoy-Agid F, Scatton B (1987) Parkinson's disease and dementia: norepinephrine and dopamine in locus ceruleus. Neurology 37:42-46

Chen K-M, Chase TN (1986) Parkinson-dementia. In: Vinken PJ, Bruyn GW, Klawans HL (1986) Handbook of clinical neurology, vol 5 (49). Elsevier Scientific, Amsterdam—New York pp 167-183

Chiueh CC, Burns RS, Markey SP, Jacokobowitz DM, Kopin IJ (1985) Primate model of parkinsonism: selective lesion of nigrostriatal neurons by 1-methyl-4-phenyl-1,2,3,6-tetrahydropyridine produces an extrapyramidal syndrome in rhesus monkeys. Life Sci 36:213-218

Christensen E, Möller JE, Faurbye A (1970) Neuropathological investigation of 28 brains from patients with dyskinesias. Acta Psychiatr Scand 46:14-23

Chui HC, Mortimer JA, Slager U, Barrow C, Bondareff W, Webster DD (1986) Pathological correlates of dementia in Parkinson's disease. Arch Neurol 43:991-995

Clark AW, Lehmann J (1983) Dementia with widespread Lewy bodies. Studies of the neocortical cholinergic system (abstr). Canadian Association of Neuropathology, 23rd Annu Meeting, Sept 29-30, 1983

Clark AW, Sternberger NH, Parhad IM, Sima AAF, Sternberger LA (1986) Immunoreactivity of Lewy bodies with monoclonal antibodies to neurofilament epitopes (abstr). J Neuropathol Exp Neurol 45:333

Constantinidis J, Bouras C, Richard J (1983) Putative neurotransmitters in human neuropathology; a review of topography and clinical implications. Clin Neuropathol 2:47-54

Corin MS, Elizan TS, Bender MB (1972) Oculomotor function in patients with Parkinson's disease. J Neurol Sci 15:251-265

Corsellis JAN, Bruton CJ, Freeman-Browne D (1973)The aftermath of boxing. Psychol Med 3:270-303

Coyle JT, Price D, De Long MR (1983) Alzheimer's disease: a disorder of cortical cholinergic innervation. Science 219:1184-1190

Critchley M (1929) Arteriosclerotic parkinsonism. Brain 52:23-83

Crossman AR, Clarke CE, Boyce S et al. (1988). MPTP-incuded parkinsonism in the monkey. Can J Neurol Sci 14:428-435

Crystal HA, Dickson D, Fuld P, Maseu J, Grober E, Aronson M, Wolfson L, Goldman J (1987) Association between senile plaque count and cognitive status in dementia and normal aging (abstr). Neurology 37(Suppl 1):226

Cummings JL (1986) Subcortical dementia: neuropsychology, neuropsychiatry, and pathophysiology. Brit J Psychiatr 149:682-697

Curcio CA, Kemper T (1984) Nucleus raphe dorsalis in dementia of the Alzheimer type: neurofibrillary changes and neuronal packing density. J Neuropathol Exp Neurol 43:359-368

Davies P (1983) Neurotransmitters and neuropeptides in Alzheimer's disease. In: Katzman R (ed) Biological aspects of Alzheimer's disease, Banbury report vol 15. Cold Spring Harbor, New York, pp 255-261

Davis GC, Williams AC, Markey SP, Ebert MH, Caine ED, Reichert DM, Kopin IJH (1979) Chronic parkinsonism secondary to intravenous injection of meperidine analogues. Psychiat Res 1:249-254

Davison CH (1954) Pallido-pyramidal disease. J Neuropathol Exp Neurol 13:50-59

Dekaban AS (1978) Changes in brain weights during the span of human life. Ann Neurol 4:345-356

Delisle MN, Gorce P, Hirsch E, Hauw JJ, Boissou H, Rascol A (1987) Motor neuron disease, parkinsonism and dementia. Report of a case with diffuse Lewy body-like intracytoplasmic inclusions. Acta Neuropathol (Berl) 75:104-108

De Morsier G (1960) Parkinsonism consécutif à une lésion traumatique du noyau rouge et du locus niger. Psychiatr Neurol (Basel) 139:60-84

Denny-Brown D (1960) Diseases of the basal ganglia; their relations to disorders of movement. Lancet II:1099-1101

Descarries L, Watkins EC, Garcia S, Beaudet A (1982) The serotonin neurons in nucleus raphe dorsalis of adult rats: a light and electron microscopic radioautographic study. J Comp Neurol 297:239-254

De Smedt Y, Ruberg M, Serdarn M, Dubois B, Agid J (1982) Confusion, dementia and anticholinergics in Parkinson's disease. J Neurol Neurosurg Psychiatry 45:1161-1164

Dickson DW, Kress Y, Crowe A, Yen SH (1985) Monoclonal antibodies to Alzheimer neurofibrillary tangles (ANT). 2. Demonstration of a common antigenic determinant between ANT and neurofibrillary degenerative supranuclear palsy. Am J Pathol 120:292-303

Dickson DW, Davies P, Mayeux R, Crystal H, Horoupian DS, Thompson A, Goldman JE (1987a) Diffuse Lewy body disease: neuropathological and biochemical studies of six patients. Acta Neuropathol (Berl) 75:1-7

Dickson DW, Kszeizak-Reding H, Davies P, Yen SH (1987b) A monoclonal antibody that recognizes a phosphorylated epitope in Alzheimer neurofibrillary tangles, neurofilaments and tau proteins immunostains granulovacuolar degeneration. Acta Neuropathol (Berl) 73:254-258

Dietl M (1985) Etude immuncytochimique de la cholecystokinine et la tyrosine hydroxylase dans la moelle épinière chez le rat et chez l'homme. Thesis Univ Paris.

Ditter SM, Mirra SS (1987) Neuropathologic and clinical features of Parkinson's disease in Alzheimer disease patients. Neurology 37:754-760

Doucette R, Fisman M, Hachinski VC, Mersky H (1986) Cell loss from the nucleus basalis of Meynert in Alzheimer disease. Can J Neurol Sci 13:435-440

Drayer BP, Olanow W, Burger P, Johnson GA, Herfkens R, Riederer P (1986) Parkinson plus syndrome: diagnosis using high field MR imaging of brain iron. Radiology 159:493-498

Dubinsky RM, Jankovic J (1987) Progressive supranuclear palsy and a multi-infarct state. Neurology 37:570-576

Dubois B, Hauw JJ, Ruberg M, Serdaru M, Javoy-Agid F, Agid Y (1985) Dementia and Parkinson's disease: biochemical and clinico-pathological correlations. Rev Neurol (Paris) 141:184-193
Duffy P (1972) Discussion. Neurology (Minneap) 22 (Suppl II):66-67
Duffy P, Tennyson VM (1965) Phase and electron microscopic observations of Lewy bodies and melanin granules in the substantia nigra and locus coeruleus in Parkinson's disease. J Neuropathol Exp Neurol 24:398-414
Duvoisin RC (1987) The olivopontocerebellar atrophies. In: Marsden CD, Fahn ST (eds) Movement disorders 2. Butterworths, London, pp 249-269
Eadie MJ, Sutherland JM (1964) Arteriosclerosis in parkinsonism. J Neurol Neurosurg Psychiatry 27:237-240
Earle KM (1968) Studies on Parkinson's disease including X-ray fluorescent spectroscopy of formalia fixed brain tissue. J Neuropathol Exp Neurol 27:1-14
Eggerstron DE, Sima AAF (1986) Dementia with cerebral Lewy bodies. A mesocortical dopaminergic deficit? Arch Neurol 43:524-527
Elizan TS, Sroka M, Maker H, Smith M, Yahr MD (1986) Dementia in idiopathic Parkinson's disease. Variables associated with its occurrence in 203 patients. J Neural Transm 65:285-302
Epelbaum J, Ruberg M, Moyse E, Javoy-Agid F, Dubois B, Agid Y (1986) Somatostatin in Parkinson's disease. Brain Res 278:376-397
Escourolle R, Recodo J de, Gray F (1971) Etude anatomo-pathologique des syndromes parkinsonism. In: Ajuriaguerra J, Gauthier G (eds) Monoamines et noyaux gris centraux. Masson, Geneve, pp 173-229
Etienne P, Robitaille Y, Wood P, Gauthier S, Nair NPV, Quirion R (1986) Nucleus basalis neuronal loss, neuritic plaques and choline acetyl-transferase activity in advances Alzheimer's disease. Neuroscience 19:1279-1291
Ezrin-Waters C, Resch L (1986) The nucleus basalis of Meynert. Can J Neurol Sci 13:8-14
Fahn ST (1977) Secondary parkinsonism. In: Goldensohn ES, Appel SH (eds) Scientific approaches to clinical neurology. Lee and Febiger, Philadelphia, pp 1159-1189
Fahn S, Libsch LR, Cutler RMA Monoamines in the human neostriatum: topographic distribution in normals und in Parkinson's disease and their role in akinesia, rigiditgy, chorea, and tremor. J Neurol Sci 14:427-455.
Ferrante RJ, Kowall WW, Marint JB, Richardson EP (1985) Characteristics of a selectively spared subset of neurons in Huntington's chorea (abstr). J Neuropathol Exp Neurol 44:325
Fischer PA (ed) (1984) Parkinson plus. Zerebrale Polypathie beim Parkinson-Syndrom. Springer, Berlin Heidelberg New York Tokyo
Fishman EB, Siek GC, MacCallum RD, Bird ED (1986) Distribution of the molecular forms of acetylcholinesterase in human brain: alterations in dementia. Ann Neurol 19:246-252
Foix C, Nicolesco J (1925): Les noyaux gris centraux et la région mésencéphalo-sous-optique. Masson, Paris
Forno LS (1966) Pathology of parkinsonism. J Neurosurg Suppl II:266-271.
Forno LS (1969) Concentric hyaline intraneuronal inclusions of Lewy type in the brains of elderly persons (50 incidental cases); relationship in parkinsonism. J Am Geriatr Soc 17:557-575
Forno LS (1982) Pathology of Parkinson's disease. In: Marsden CD, Gahn S (eds) Movement disorders. Butterworths, London, pp 25-30
Forno LS (1986) The Lewy body in Parkinson's disease. Adv Neurol 45:35-43
Forno LS, Alvord EC (1971) The pathology of parkinsonism. In: Markham CH (eds) Recent advances in Parkinson's disease. Davis, Philadelphia, pp 120-130

Forno LS, Norville RL (1981) Synaptic morphology in the human locus coeruleus. Acta Neuropathol (Berl) 53:7-14

Forno LS, Barbour PJ, Norville RL (1978) Presenile dementia with Lewy bodies and neurofibrillary tangles. Arch Neurol 35:816-822

Forno LS, Sternberger NR, Eng LF (1983) Immunocytochemical staining of neurofibrillary tangles and of the periphery of Lewy bodies with a monoclonal antibody to neurofilaments (abstr). J Neuropathol Exp Neurol 42:342

Forno LS, Gardiner RE, Eng LF (1985a) Somatostatin-like immunoreactivity in the human basal ganglia (abstr). J Neuropathol Exp Neurol 44:326

Forno LS, Sternberger LA, Sternberger NH, Strefling AM et al. (1985b) Reaction of Lewy bodies with antibodies to phosphorylated and non-phosphorylated neurofilaments. Neurosci Lett 64:253-258

Forno LS, Langston JW, DeLanney LE, Irwin I, Ricaurte GA (1986) Locus ceruleus lesions and eosinophilic inclusions in MPTP-treated monkeys. Ann Neurol 20:449-455

Forno LS, Langston JW, DeLanney LE, Irwin I (1988) An electron microscopic study of MPTP-induced inclusion bodies in an old monkey. Brain Res 448:150-157

Friedman A, Kang UJ, Tatemichi TK, Burke RE (1986) A case of parkinsonism following striatal lacunar infarction. J Neurol Neurosurg Psychiatry 49:1087-1088

Fujimura M, Umbach J (1987) Pathological considerations between dementia, Parkinson's disease and lesions of the reticular formation. Rev Neurol (Paris) 143:108-114

Gage FH, Björklund A, Isacson O, Brundin P (1986) Uses of neuronal transplantation in models of neurodegenerative disease. In: Das GD, Wallace RB (eds) Neural transplantation and regeneration proceedings in life sciences. Springer, Berlin Heidelberg New York Tokyo, pp 103-124

Galloway P, Perry G, Grundke-Iqbal I, Antili-Gambetti L, Gambetti P (1986) Lewy bodies share epitopes with neurofilaments, tubulin, and neurofibrillary tangles (abstr) Proc 10th Int Congr Neuropath, Stockholm, p 298

Galloway PG, Perry G, Gambetti P (1987) Hirano body filaments contain actin and actin-associated proteins. J Neuropathol Exp Neurol 46:185-199

Galloway PG, Grundke-Iqbal I, Iqbal K, Perry G (1988) Lewy bodies contain epitopes both shared and distinct from Alzheimer neurofibrillary tangles. J Neuropathol Exp Neurol 47:654-663

Gamboa ET, Wolf A, Yahr MD et al. (1974) Influenza virus antigen in postencephalitic parkinsonism brain. Detection by immunofluorescence. Arch Neurol 31:228-232

Garcia de Yebenes J, Gervas JJ, Iglesias J, Menaj et al. (1982) Biochemical finding in a case of parkinsonism secondary to brain tumor. Ann Neurol 11:313-316

Gaspar P, Gray F (1984) Dementia in idiopathic Parkinson's disease. A neuropathological study of 32 cases. Acta Neuropathol (Berl) 64:43-52

Gherardi R, Roudes B, Fleury J, Probst C et al. (1985) Parkinsonian syndrome and central nervous system lymphoma involving the substantia nigra. Acta Neuropathol (Berl) 65:338-343

Gibb WRG (1986) Idiopathic Parkinson's disease and the Lewy body disorders. Neuropathol Appl Neurobiol 12:223-234

Gibb WRG (1988) The neuropathology of parkinsonian disorders. In: Jankovic J, Tolowa E (eds): Parkinson's disease and movement disorders. Baltimore—Munich: Urban & Schwarzenberg, pp 205-223

Gibb WRG, Lees AJ (1987) The progression of idiopathic Parkinson's disease is not explained by age-related changes. Clinical and pathological comparisons with post-encephalitic parkinsonian syndrome. Acta Neuropathol (Berl) 73:195-201

Gibb WRG, Lees AJ (1988) A comparison of clinical and pathological features of young- and old-onset Parkinson's disease. Neurology 38:1402-1406
Gibson PH (1978) Light and electron microscopic observations on the relationship between Hirano bodies, neuron and glial perikarya in the human hippocampus. Acta Neuropathol (Berl) 42:165-171
Gibson PH (1983) Form and distribution of senile plaque in silver impregnation sections in the brains of intellectually normal elderly people and people with Alzheimer-type dementia. Neuropathol Appl Neurobiol 9:379-389
Goldman JE, Yon SH, Chiu FC, Peress NS (1983) Lewy bodies of Parkinson's disease contain neurofilament antigens. Science 221:1082-1084
Grafe MR, Forno LS, Eng LF (1983) Substance P and met-enkephalin immunoreactivity in Parkinson's Huntington's and Alzheimer's disease (abstr). J Neuropathol Exp Neurol 42:345
Gray F (1988) Neuropathologie des syndromes parkinsoniens. Rev Neurol (Paris) 144:229-248
Graybiel AM (1984) Neurochemically specified subsystems in the basal ganglia. In: Functions of the basal ganglia. Ciba Found Symp 107:114-144
Graybiel AM (1986) Neuropeptides in the basal ganglia. In: Martin JB, Barchas J (eds). Neuropeptides in neurologic and psychiatric disease. Raven, New York, pp 135-161
Greenfield JG, Bosanquet FD (1953) The brainstem lesions in parkinsonism. J Neurol Neurosurg Psychiatry 16:213-226
Grilia M, Hénin D, Sazdovitch Y, Hauw JJ (1987) Clinicopathological correlation in 222 cases of parkinsonism syndrome (abstr). Neuropathol Appl Neurobiol 13:614
Growdon JH, Corkin S (1986) Cognitive impairment in Parkinson's disease. Adv Neurol 45:383-392
Guiloff RJ, George RJ, Marsden CD (1980) Reversible supranuclear ophthalmoplegia associated with parkinsonism. J Neurol Neurosurg Psychiatry 43:352-354
Guttman M, Calne DB (1988) In vivo characterization of cerebral dopamine systems in human parkinsonism. In: Jankovic J, Tolosa E (eds) Parkinsons' disease and movement disorders. Urban & Schwarzenberg: Baltimore—Munich, pp 49-58
Hakim AM, Mathieson G (1979) Dementia in Parkinson disease: a neuropathologic study. Neurology (NY) 20:1209-1214
Hallervorden J (1957) Paralysis agitans. In: Scholz W (ed) Handbuch der speziellen pathologischen Anatomie und Histologie, vol XIII/1A. Springer, Berlin Göttingen Heidelberg, pp 900-924
Hartog-Jager WA (1969) Sphingomyelin in Lewy inclusion bodies in Parkinson's disease. Arch Neurol 21:615-619
Hartog-Jager WA, Bethlem J (1960) The distribution of Lewy bodies in the central and autonomic systems in idiopathy paralysis agitans. J Neurol Neurosurg Psychiatry 23:283-290
Hassler R (1938) Zur Pathologie der Paralyse agitans und des postenzephalitischen Parkinsonism. J Psychol Neurol (Lpz) 48:387-476
Hassler R (1965) Extrapyramidal control of the speed of behaviour and its change by primary age processes. In: Welford AT, Birren JK (eds). Behavior, aging and the nervous System. Thomas, Springfield, pp 284-306
Hedreen JC, Struble RG, Whitehouse PJ, Price DL (1984) Topography of the magnocellular basal forebrain system in human brain. J Neuropathol Exp Neurol 43:1-21
Hefti F, Melamed E, Bhawan J, Wurtman RJ (1981) Long-term administration of L-dopa does not damage dopaminergic neurons in the mouse. Neurology (NY) 31:1194-1195
Heilig CW, Knopman DS, Mastri AR, Frey W (1985) Dementia without Alzheimer pathology. Neurology 35:762-765

Henderson Z (1987) A small proportion of cholinergic neurones in the nucleus basalis magnocellularis of ferret appear to stain positively for tyrosine hydroxylase. Brain Res 412:363-369
Henderson Z, Greenfield SA (1987) Does the substantia nigra have cholinergic innervation? Neurosci Lett 73:109-113
Heston LL (1981) Genetic studies of dementia. In: Mortimer A, Schulman LM (eds) Epidemiology of dementia. Oxford University Press, New York, pp 101-114
Hirano A (1981) A guide to neuropathology. Igaku-Shoin. Tokyo New York
Hirano A, Malamud N, Elisan TS, Kurland LT (1966) Amyotrophic lateral sclerosis and parkinsonism-dementia complex of Guam. Further pathologic studies. Arch Neurol 15:35-51
Hirano A, Zimmerman HM (1962) Alzheimer's neurofibrillary changes: a topographic study. Arch Neurol 7:227-242
Hirano A, Malamud N, Kurland LT (1981) Parkinsonism-dementia complex, an endemic disease on the island of Guam. II.—Pathological features. Brain 84:662-679
Hirano A, Llena J (1986) Neuropathological features of Parkinsonism dementia complex of Guam: reapraisal and comparative study with Alzheimer's disease and Parkinson's disease. Prog Neuropathol 6:17-31
Hirsch E, Ruberg M, Dardenne M, Portier MM, Javoy-Agid F, Bach JF, Agic Y (1985) Monoclonal antibodies raised against Lewy bodies in brains from subjects with Parkinson's disease. Brain Res 345:374-378
Hirsch EC, Graybiel AM, Duyckaerts C, Javoy-Agid F (1987) Neuronal loss in the pedunculopontine tegmental nucleus in Parkinson's disease and progressive supranuclear palsy. Proc Nat Acad Sci USA 84:5976-5981
Hirsch E, Graybiel AM, Agid YA (1988) Melanized dopaminergic neurons are differentially susceptible to degeneration in Parkinson's disease. Nature 334:345-348
Ho KC, Roessmann U, Straumfjord JV, Monroe G (1980) Analysis of brain weight. Pathol Lab Med 104:635-639
Hornykiewicz O (1982) Brain neurotransmitter changes in Parkinson's disease. In: Marsden CD, Fahn S (eds) Movement disorders. Butterworth, London, pp 41-58
Hornykiewicz O, Kish SJ (1984) Neurochemical basis of dementia in Parkinson's disease. Can J Neurol Sci 11:185-190
Hornykiewicz O, Kish SJ (1986) Biochemical pathophysiology of Parkinson's disease. Adv Neurol 45:19-34
Hubbard BM, Anderson JM (1985) Age related variations in the neurone content of the cerebral cortex in senile dementia of the Alzheimer type. Neuropathol Appl Neurobiol 11:369-382
Huber SJ, Shuttleworth EC, Paulson GW (1986) Dementia in Parkinson's disease. Arch Neurol 43:987-990
Hudson AJ (1981) Amyotrophic lateral sclerosis and its association with dementia, parkinsonism, and other neurological disorders. A review. Brain 104:217-247
Huhn B, Jakob H (1971) Traumatische Hirnstammläsionen mit vieljähriger Überlebensdauer. Beitrag zur Pathologie der Substantia nigra und der oralen Brückenhaube. Nervenarzt 41:326-334
Hunter S (1985) The rostral mesencephalon in Parkinson's and Alzheimer's disease. Acta Neuropathol (Berl) 68:53-58
Hymann BT, vanHoesen GW, Damasio AR, Barned CL (1984) Alzheimer's disease: cell specific pathology isolates the hippocampal formation. Science 225:1168-1170
Ichimija Y, Arai H, Kosaka K, Iizuka R (1986) Morphological and biochemical changes in the cholinergic and monoaminergic system in Alzheimer-type dementia. Acta Neuropathol (Berl) 70:112-116
Iizuka R, Hirayama K (1986) Dentato-rubro-pallido-luysian atrophy. In: Vinken PJ,

Bruyn GW, Klawans HL (eds) Handbook of Clinical Neurology, vol. 5 (49) Elsevier Scientific, Amsterdam New York: pp 437-443

Ikeda K, Ikeda S, Yoshimura T, Kato H, Namba M (1978) Idiopathic parkinsonism with Lewy-type inclusions in cerebral cortex. A case report. Acta Neuropathol (Berl) 41:165-168

Ingram VM, Koenig JH, Miller CH, Moore HE, Blanchard B, Perry DE (1987) The locus coeruleus: computer assisted 3-dimensional analysis of degeneration in Alzheimer's and Down's disease. In: Wurtman RJ, Corkin SH, Growden JH (eds) Alzheimer's disease: advances in basic research and therapy. Center for Brain Science and Metabolism Charitable Trust, Cambridge, pp 435-440.

Ishii T (1966) Distribution of Alzheimer's neurofibrillary tangles in the brainstem and hypothalamus in senile dementia. Acta Neuropathol (Berl) 6:181-187

Ishii T, Nakamura Y (1981) Distribution and ultrastructure of Alzheimer's neurofibrillary tangles in postencephalitic parkinsonism of Economo type. Acta Neuropathol (Berl) 55:59-62

Jacob H (1983) Klinische Neuropathologie des Parkinsonism. In: Gänshirt H, Berlit P, Haack G (eds) Pathophysiologie, Klinik und Therapie des Parkinsonism. Roche, Basel, pp 5-18

Jacobs RW, Butcher LL (1986) Pathology of the basal forebrain in Alzheimer's disease and other dementias. In: Scheibel AB, Wechsler AF (eds) The biological substrates of Alzheimer's disease. Academic, Orlando, pp 87-100

Jamada M, Mehraein P (1968) Verteilungsmuster der senilen Veränderungen im Gehirn. Arch Psychiatr Nervenkr 211:308-324

Jankovic J (1988) Neural transplants in the treatment of Parkinson's disease and other neurodegenerative disorders. In: Jankovic J, Tolosa E (eds) Parkinson's disease and movement disorders. Urban & Schwarzenberg: Baltimore—Munich, pp 471-480

Jankovic J, Kirkpatrick JB, Blomquist KA, Langlais PJ, Bird ED (1985) Late-onset Hallervorden-Spatz disease presenting as familial parkinsonism. Neurology (NY) 35:227-234

Javoy-Agid F, Agid Y (1980) Is the mesocortical dopaminergic system involved in Parkinson's disease? Neurology (NY) 30:1326-1330

Javoy-Agid F, Ploska A, Agid Y (1981) Microtopography of tyrosine hydroxylase, glutamic acid decarboxylase, and choline acetyltransferase in the sustantia nigra and ventral tegmental area of control and parkinsonian brain. J Neurochem 37:1218-1227

Javoy-Agid F, Ruberg M, Taquer H, Bobobza B, Agid Y (1984) Biochemical neuropathology of Parkinson's disease. Adv Neurol 40:189-197

Javoy-Agid F, Ruberg M, Hirsch E, Cash R, Raisman R, Taquet H, Epelbaum J, Scatton B, Duyckaerts C, Agid Y (1986) Recent progress in the neurochemistry of Parkinson's disease. In: Fahn S, Marsden CD, Jenner P, Teychenne P (eds) Recent developments in Parkinson's disease. Raven, New York, pp 67-83

Jellinger K (1973) Neuroaxonal dystrophy: its natural history and related disorders. Prog Neuropathol 2:129-180

Jellinger K (1974) Pathomorphologie des Parkinson-Syndroms. Aktuelle Neurol 1:83-98

Jellinger K (1977) Neuropathologic findings after neuroleptic long-term therapy. In: Shiraki H, Grcevic N (eds) Neurotoxicology. Raven, New York, pp 25-42

Jellinger K (1986a) Pathology of parkinsonism. In: Fahn S, Marsden CD, Jenner P, Teychenne R (eds) Recent developments in parkinsonism. Raven, New York, pp 33-66

Jellinger K (1986b) Overview of morphological changes in Parkinson's disease. Adv Neurol 45:1-18

Jellinger K (1986c) Pallidal, pallidonigral, pallidoluysionigral degenerations. In: Vincken P, Bruyn GW, Klawans W (eds) Handbook of clinical neurology, vol 49. Elsevier Scientific Amsterdam, New York pp 445-463
Jellinger K (1987a) The pathology of parkinsonism. In Marsden CD, Fahn S (eds) Movements disorders 2. Butterworths, London pp 124-165
Jellinger K (1987b) Neuropathological substrates of Alzheimer's disease and Parkinson's disease. J Neural Transm [Suppl]. 24:109-129
Jellinger K (1987c) Quantitative changes in some subcortical nuclei in aging, Alzheimer's disease and Parkinson's disease. Neurobiol Aging 8:556-561
Jellinger K (1987d) Pathologic correlates of dementia in Parkinson's disease Arch Neurol 44:692
Jellinger K (1988) The pedunculopontine nucleus in Parkinson's disease, supranuclear palsy and Alzheimer's disease. J Neurol Neurosurg. Psychiat 51:540-544
Jellinger K (1989) Morphologie des alternden Gehirns und der (prä)senilen Demenzen. In: Platt D, Oesterreich K (eds) Handbuch der Gerontologie, vol 5. Fischer, Stuttgart
Jellinger K, Riederer P (1984) Dementia in Parkinson's disease and (pre)senile dementia of Alzheimer type: morphological aspects and changes in the intracerebral MAO activity. Adv Neurol 40:199-210
Jellinger K, Riederer P, Tomonaga M (1980) Progressive supranuclear palsy: clinicopathological and biochemical studies. J Neural Transm 16:111-128
Jørgensen OS, Reynolds GP, Riederer P, Jellinger K (1982) Parkinson's disease putamen: normal concentration of synaptic membrane marker antigens. J Neural Transm 54:171-179
Joynt RJ, McNeill TH (1984) Neuropeptides in aging and dementia. Peptides (Fayetteville) 5 (Suppl 1):269-274
Kahn J, Anderton BH, Gibb WRG, Lees FR, Wells CD, Marsden D (1985) Neuronal filaments in Alzheimer's Pick's and Parkinson's diseases. N Engl J Med 313:520-521
Kato T, Hirano A, Winberg MN, Jacobs AK (1986) Spinal cord lesions in progressive supranuclear palsy: some new observations. Acta Neuropathol (Berl) 71:11-14
Katzman R (1986) Alzheimer's disease. N Engl J Med 324:964-972
Khachaturian ZS (1975) Diagnosis of Alzheimer's disease. Arch Neurol 42:1097-1105
Kidd M (1964) Alzheimer's disease. An electron microscopic study. Brain 87:307-320
Kimula Y, Utsuyama M, Yoshimura M, Tomonaga M (1983) Element analysis of Lewy and adrenal bodies in Parkinson's disease by electron probe microanalysis. Acta Neuropathol (Berl) 59:233-236
Kish SJ, Shannak K, Hornykiewicz O (1988) Uneven patterns of dopamin loss in the striatum of patients with idiopathic Parkinsons' disease. New Engl J Med 318:876-878
Klaue R (1940) Parkinsonsche Krankheit (Paralysis agitans) und postenzephalitischer Parkinsonismus. Versuch einer klinisch-anatomischen Differentialdiagnose. Arch Psychiatr Nervenkr 111:251-321
Klawans HL, Lupton M, Simon L (1976) Calcification of the basal ganglia as a cause of levodopa resistant parkinsonism. Neurology (NY) 26:221-225
Koeppen AH, Barron KD, Cox GL (1971) Striato-nigral degeneration. Acta Neuropathol (Berl) ;9:10-19
Kosaka K (1978) Lewy bodies in cerebral cortex. Report of three cases. Acta Neuropathol (Berl) 42:127-134
Kosaka K, Oyanagi S, Matsushita M, Hori A, Iwase S (1977) Multiple system degeneration involving thalamus, reticular formation, pallido-nigral, pallido-luysian, and dentatorubral system. Acta Neuropathol (Berl) 39:89-95

Kosaka K, Matsushita M, Oyanagi S, Uchiyama S, Iwase S (1981) Pallido-nigral-luysial atrophy with massive appearance of corpora amylacea. Acta Neuropathol (Berl) 53:169-171

Kosaka K, Yoshimura M, Ikeda K, Budka H (1984) Diffuse type of Lewy body disease Clin Neuropathol 3:185-192

Kuzuhara S, Mori H, Izumiyama N, Yoshimura M, Ihara Y (1988) Lewy bodies are ubiquinated. A light and electron microscopic immunocytochemical study. Acta Neuropathol (Berl) 75:345-353

Lange HN, Bogerts B (1982) Postencephalitic and idiopathic parkinsonism. Quantitative change of tel-, di-, mesencephalon and basal ganglia (abstr). Neuroscience (Suppl 17):127

Langston JW, Forno LS (1978) The hypothalamus in Parkinson's disease. Ann Neurol 3:129-133

Lapresle J, Fardeau M (1967) The central nervous system and carbon monoxide poisoning. II. Anatomical study of brain lesions following intoxication with carbon monoxide (22 cases). Prog Brain Res 24:31-74

Leestma JE, Andrews JM (1969) The fine structure of the Marinesco body. Arch Pathol 88:431-436

Leverenz J, Sumi SM (1986) Parkinson's disease in patients with Alzheimer's disease. Arch Neurol 43:662-664

Lewis PD (1971) Parkinsonism—neuropathology. Br Med J [Clin Res] 3:690-697

Lewy FH (1913) Zur pathologischen Anatomie der Paralyse agitans. Dtsch Z Nervenheilk 50:50-55

Lhermitte J, Kraus WM, McAlpine D (1927) Etude des produits de désintégration et des depots du globus pallidus dans un cas de syndrome parkinsonien. Rev Neurol (Paris) 1:326-361

Lieberman AM, Dziatolowski M, Kupersmith M, Serby B, Boodgeld A, Korbin J, Goldstein MA (1979) Dementia in Parkinson's disease. Ann Neurol 6:355-359

Lindenberg R (1964) Die Schädigungsmechanismen der Substantia nigra bei Hirntraumen und das Problem des posttraumatischen Parkinsonismus. Dtsch Z Nervenheilk 185:637-663

Llorens-Cortex C, Javoy-Agid F, Agid Y, Taquet H, Schwartz JC (1984) Enkephalinergic markers in substantia nigra and caudate nucleus from parkinsonian subjects. J Neurochem 43:874-877

Lockhart MP, Gibson CJ, Ball MJ (1984) Topographical loss of locus coeruleus cells in Alzheimer's disease. Soc Neurosci Abstr 10:995

Mann DMA, Yates PO (1983 a) Pathological basis for neurotransmitter changes in Parkinson's disease. Neuropathol Appl Neurobiol 9:3-19

Mann DMA, Yates PO (1983 b) Possible role of neuromelanin in the pathogenesis of Parkinson's disease. Mech Ageing Dev 21:193-203

Mann DMA, Yates PO, Hawkes J (1983) The pathology of the human locus coeruleus. Clin Neuropathol 2:1-7

Mann DMA, Yates PO, Marcyniuk B (1984) Changes in nerve cells of the nucleus basalis of Meynert in Alzheimer's disease and their relationship to aging and to the accumulation of lipofuscin pigment. Mech Ageing Dev 25:189-204

Mann DMA, Yates PO, Marcyniuk B (1985) Correlation between senile plaque and neurofibrillary tangle counts in the cerebral cortex and neuronal counts in cortex and subcortical structures in Alzheimer's disease. Neurosci Lett 56:51-55

Mann DMA, Tucken CM, Yates PO (1987 a) The topographic distribution of senile plaques and neurofibrillary tangles in the brains of non-demented persons of different ages. Neuropathol Appl Neurobiol 13:123-139

Mann DMA, Yates PO, Marcyniuk B (1987 b) Dopaminergic neurotransmitter systems in Alzheimer's disease and Down's syndrome at middle age. J Neurol Neurosurg Psychiatry 50:341-344

Marcyniuk B, Mann DMA, Yates PO (1986) Loss of nerve cells from locus coeruleus in Alzheimer's disease is topographically arranged. Neurosci Lett 64:247-252

Martin JP (1965) The globus pallidus in post-encephalitic parkinsonism. J Neurol Sci 2:344-365

Martin WRW, Adam MJ, Bergstrom M, Ammann W, Harrop R et al. (1986) In vivo study of DOPA metabolism in parkinson's disease. In: Fahn S, Marsden CD, Jenner P, Teychenne P (eds) Recent development in Parkinson's disease. Raven, New York, pp 97-102

Martinez AJ, Utterback RA (1973) Unilateral Parkinson's disease, clinical and neuropathologic findings. Neurology (Minneap) 23:164-170

Mata M, Dorovini-Zis K, Wilson M, Young AB (1983) A new form of familial Parkinson-dementia syndrome: clinical and pathologic findings. Neurology (CI) 33:1439-1443

Matzuk MM, Saper CB (1985) Preservation of hypothalamic dopaminergic neurons in Parkinson's disease. Ann Neurol 18:552-555

Mayer JM, Mikol J, Haguenau M, Dellanave J, Bépin B (1986) Familial juvenile parkinsonism with multiple systems degenerations. A clinicopathological study. J Neurol Sci 72:91-101

Mayeux R, Stern Y (1983) Intellectual dysfunction and dementia in Parkinson disease. In: Mayeux R, Rosen WG (eds): The dementias. Raven New York

Mayeux R, Stern V, Cote L, Williams JBW (1984) Altered serotonin metabolism in depressed patients with Parkinson's disease. Neurology (CI) 34:642-646

Mayeux R, Stern Y, Rosenstein Y et al. (1988) An estimate of the prevalence of dementia in idiopathic Parkinson's disease. Arch Neurol 45:260-262

McGeer Pl, McGeer EG (1976) Enzymes associated with the metabolism of catecholamines, acetylcholine and GABA in human controls and patients with Parkinson's disease and Huntington's chorea. J Neurochem 26:65-76

McGeer PL, McGeer EG, Suzuki J (1984) Aging, Alzheimer's disease, and the cholinergic system of the basal forebrain.Neurology (CI) 34:741-745

McGeer PL, Itagaki S, Boyes BE, McGeer EG (1988) Reactive microglia are positive for HLA-DR in the substantia nigra of Parkinson's and Alzheimer's disease brains. Neurology 38:1285-1291

Mesulam MM, Mufson EJ (1984) Neuronal inputs into the nucleus basalis of the substantia innominata (Ch 4) in the rhesus monkey. Brain 107:253-274

Miller C, Haugh M, Kahn J, Anderson B (1986) The cytoskeleton and neurofibrillary tangles in Alzheimer's disease. Trends Neurosci 9:76-81

Mizusawa H, Yen SH, Hirano A (1989) Grumose degeneration of the dentate nucleus in progressive supranuclear palsy Acta Neuropathol 77 (in press)

Mizutani T, Oda M, Abe H, Fukuda S, Oikawa H, Kosaka K (1983) Hereditary multisystemic degeneration with unusual combination of cerebellopetal, dentato-rubral, and nigro-subthalamo-pallidal degeneration. Clin Neuropathol 2:147-153

Molsa PK, Martilla RJ, Rinne UK (1984) Extrapyramidal signs in Alzheimer's disease. Neurology (CI) 34:1114-1116

Mori H, Yoshimura M, Tomonaga H, Yamanouchi H (1986) Progressive supranuclear palsy with Lewy bodies. Acta Neuropathol (Berl) 71:344-346

Mori H, Kondo J, Ihara Y (1987) Ubiquitin is a component of paired helical filaments in Alzheimer's disease. Science 235:1641-1644

Morimura Y, Hirano Y, Llena JF (1985) Electron microscopic observations of the nucleus basalis Meynert in human autopsy cases. Acta Neuropathol (Berl) 68:130-137

Morris JC, Drazner M, Fulling K, Berg L (1987) Parkinsonísm in senile dementia of the Alzheimer type. In: Wurtman RJ, Corkin SJ, Growden JH (eds) Alzheimer's

disease: advances in basic research and therapies. Center for Brain Science and Metabolism Charitable Trust, Cambridge, pp 499-504
Morrison JH, Scherr S, Lewis DA, Campbell MJ, Bloom FE (1986) The laminar and regional distribution of somatostatin and neuritic disconnection syndrome. In Scheibel AB, Wechsler AF (eds) The biological substrates of Alzheimer's disease. Academic, Orlando, pp 115-131
Mortimer JA, Pirozzlo FF Hansch EC, Webster DD (1982) Relationship of motor symptoms to intellectual deficits in Parkinson's disease. Neurology (NY) 32:133-137
Mouchet P, Manier M, Dietl M, Feuerstein C, Berod A, Arluison A, Denorys L, Thibaut J (1986) Immunohistochemical study of catecholaminergic cell bodies in the rat spinal cord. Brain Res Bull 16:341-353
Mountjoy CA, Rossor MN, Iversew LI, Roth M (1984): Correlation of cortical cholinergic and GABA deficits with quantitative neuropathological findings in senile dementia. Brain 107:507-518
Nagatsu T, Oka K, Yamamoto T, Matusui H et al. (1981) Catecholaminergic enzymes in Parkinson's disease and related extrapyramidal diseases. In: Riederer P, Usdin E (eds) Transmitter biochemistry of human tissue. Macmillan, London, pp 291-302
Nakano I, Hirano A (1983) Neuron loss in the nucleus basalis of Meynert in Parkinson-dementia complex of Guam. Ann Neurol 13:87-91
Nakano I, Hirano A (1984) Parkinson's disease: neuron loss in the nucleus basalis without concomitant Alzheimer's disease. Ann Neurol 15:415-418
Nakashima S, Ikuta F (1984a) Catecholamine neurons with Alzheimer's neurofibrillary changes and alteration of tyrosine hydroxylase: immunohistochemical investigation of tyrosine hydroxylase. Acta Neuropathol (Berl) 64:273-280
Nakashima S, Ikuta F (1984b) Tyrosin hydroxylase proteins in Lewy bodies of parkinsonism and senile brain. J Neurol Sci 66:91-96
Nakashima S, Kumanishi T, Ikuta F (1983) Immunohistochemistry on tyrosine hydroxylase in the substantia nigra of human autopsied cases. Brain Nerve 35:1023-1029
Nakazato X, Sasaki A, Hirato J, Ishida Y (1984) Immunohistochemical localization of neurofilament protein in neuronal degenerations. Acta Neuropathol (Berl) 64:30-36
Narabayashi H, Yokochi M, Iizuka R, Nagatsu T (1986) Juvenile parkinsonism. In: Vinken PJ, Bruyn GW, Klawans HL (eds) Handbook of clinical neurology, vol 49. Elsevier Scientific, Amsterdam New York, pp 153-165
Neary D, Snoweden JS, Mann DMA et al. (1986) Alzheimer's disease; a correlative study. J Neurol Neurosurg Psychiatry 49:129-137
Nieuwenhuys R (1985) Chemoarchitecture of the brain. Springer, Berlin Heidelberg New York, Tokyo
Ohama E, Ikuta F (1976) Parkinson's disease. Distribution of Lewy bodies and monoamine neuron system. Acta Neuropathol (Berl) 34:311-319
Olszewski J, Baxter D (1982) Cytoarchitecture of the human brain stem, 2nd edn. Karger, Basle New York
Oppenheimer N (1984) Diseases of the basal ganglia, cerebellum and motor neurons. In: Adams JH, Corsellis JAN, Duchen LW (eds) Greenfield's neuropathology, 4th ed. Arnold, London, pp 698-747
Oyanagi K, Nakashima S, Ikuta F, Honma Y (1986) An autopsy case of dementia and parkinsonism with severe degeneration exclusively in the substantia nigra. Acta Neuropathol (Berl) 71:
Pakkenberg H, Brody H (1965) The number of nerve cells in the substantia nigra in paralysis agitans. Acta Neuropathol (Berl) 5:320-324

Palmer AM, Procter AW, Stratman GC, Bowen DM (1986) Excitatory amino acidreleasing and cholinergic neurones in Alzheimer's disease. Neurosci Lett 66:199-204
Pappola MA (1986) Lewy bodies of Parkinson's disease. Immune electron microscopic demonstration of neurofilament antigens in constituent filaments Arch Pathol Lab Med 110:1160-1163
Parkes JSD, Marsden CD, Rees JE (1974) Parkinson's disease, cerebral arteriosclerosis and senile dementia: clinical features and response to levodopa. J Med 4:93:49-61
Parkinson J (1817) An essay on the shaking palsy. Willingham and Rowland London
Parnitzke KH, Peiffer J (1956) Zur Klinik der pathologischen Anatomie der chronischen Braunsteinvergiftung. Arch Psychiat Nervenkr 192:405-427
Pearce GW (1979) The neuropathology of parkinsonism. In: Smith WT, Canavagh JB (eds) Recent advances in neuropathology, vol I. Churchill-Livingstone, London, pp 299-320
Pearce J (1974) The extrapyramidal disorder of Alzheimer's disease. Eur Neurol 12:94-103
Pearson J (1983) Neurotransmitter immunocytochemistry in the study of human development, anatomy and pathology. Prog Neuropathol 5:41-97
Pearson RCA, Esiri MM, Hiorns RW, Woldock GK, Powell TPS (1985) Anatomical correlates of the distribution of the pathological changes in the neocortex in Alzheimer's disease. Proc Natl Acad Sci USA 82:4531-4534
Pearson RCA, Powell TPS (1987) Anterograde vs retrograde degeneration of the nucleus basilis medialis in Alzheimer's disease. J Neurol Transm Suppl 24:139-146
Pearson RCA, Sofroniew MV, Cuello AC et al. (1983) Persistence of cholinergic neurons in the basal nucleus in a brain with senile dementia of the Alzheimer's type demonstrated by immunohistochemical staining for choline acetyltransferase. Brain Res 289:375-379
Pearson RCA, Esiri MM, Hions RW, Waldeck GK, Powell TPS (1985) Anatomical correlates of the distribution of the pathological changes in the neocortex in Alzheimer's disease. Proc Natl Acad Sci USA 82:4531-4534
Pendlebury WV, Perl DL (1984) Nucleus basalis of Meynert: severe cell loss in Parkinson's disease without dementia (abstr). Ann Neurol 16:124
Perl DP, Pendlebury WV, Bird FD (1984) Detailed neuropathologic evaluation of banked brain specimens submitted with a clinical diagnosis of Alzheimer's disease. In: Wurtman RJ, Corkin SH, Growden JH (eds) Proc, 3rd meeting of the Int study group on the treatment of memory disorders associated with aging. p 463
Perry EK, Curtis M, Dick DJ, Candy JM, Atack JH, Bloxham CA, Blessed G, Fairburn A, Tomlinson BE, Perry RH (1985) Cholinergic correlates of cognitive impairment in Parkinson's disease: comparison with Alzheimer's disease. J Neurol Neurosurg Psychiatry 48:413-421
Perry EK, Perry RH, Smith CJ, Dick DJ, Candy JM, Edwardson JA, Fairbairn A, Blessed JG (1987) Nicotinic receptor abnormalities in Alzheimer's and Parkinson's diseases. J Neurol Neurosurg Psychiatry 50:806-809
Perry G, Manetto V, Autilio-Gambetti L, Gambetti P (1987a) Straight filaments of Alzheimer neurofibrillary tangles (abstr). J Neuropathol Exp Neurol 46:334
Perry G, Friedman R, Shaw G, Chau V (1987b) Ubiquitinis detected in neurofibrillary tanges and senile plaque neurites of Alzheimer disease brains. Proc Natl Acad Sci USA 84:3033-3036
Perry RH, Candy H, Perry EK, Irving D, Blessed C et al. (1983a) Extensive loss of choline acetyltransferase activity is not reflected by neuronal loss in the nucleus basalis of Meynert in Alzheimer's disease. Neurosci Lett 33:311-315
Perry RH, Tomlinson BE, Candy JM, Blessed G et al. (1983b) Cortical cholinergic deficit in mentally impaired Parkinsonian patients. Lancet 2:789-790

Perry RH, Perry EK, Smith CJ, Xuereb JH, Irving D, Whitford CA, Candy JM, Cross AJ (1987) Cortical neuropathological and neurochemical substrates of Alzheimer's and Parkinson's disease. J Neurol Transm [Suppl] 24:131-136
Perry TL, Bratty PLA, Hansen S et al. (1975) Hereditary mental depression and parkinsonism with taurine deficiency. Arch Neurol 32:108-113
Peterson C, Kress V, Vallee R, Goldman JE (1988) High molecular weight microtubule-associated proteins bind to actin lattices (Hirano bodies) Acta Neuropathol (Berl) 77
Poewe W, Gerstenbrand F (1986) Klinische Klassifikation des Parkinson-Syndroms: Subtypen und Übergänge zu Multi-System-Atrophien. In: Schnaberth G, Auff E (eds) Das Parkinson-Syndrom. Roche, Vienna; pp 39-46
Popovitch ER, Wisniewski HM, Kaufman MA, Grundke-Iqbal I, Yen GY (1987) Young adult form of dementia with neurofibrillary changes and Lewy bodies. Acta Neuropathol (Berl) 47-104
Price DL, Whitehouse PJ, Struble RG et al. (1983) Basal forebrain cholinergic neurons and neuritic plaques in primate brains. In: Katzman R (ed) Biological aspects of Alzheimer's Disease. Cold Spring Harbor, Banbury report, vol 15. New York, pp 65-77
Price DL, Whitehouse PJ, Struble RG (1986) Cellular pathology in Alzheimer's and Parkinson's disease. Trends Neurosci 9:29-33
Purdy A, Hahn A, Barnett HJM, Bratty P (1979) Familial fatal parkinsonism with alveolar hypoventilation and mental depression. Ann Neurol 6:523-531
Quinn NP, Rossor MN, Marsden CD (1986) Dementia and Parkinson's disease. Pathological and neurochemical considerations. Br Med Bull 42:86-96
Rail D, Scholtz C, Swash M (1981) Post-encephalitic parkinsonism: current experience. J Neurol Neurosurg Psychiatry 44:670-676
Rajput AH, Rozdilsky B, Hornykiewicz O, Shannak K et al. (1982) Reversible drug-induced Parkinsonism. Clinicopathologic study of two cases. Arch Neurol 39:644-646
Rasool CS, Selkoe DJ (1985) Recognition of Pick bodies by antibodies to neurofibrillary tangles in Alzheimer's disease. N Engl J Med 310:700-705
Rasool CS, Svendsen CN, Selkoe DJ (1986) Neurofibrillary degeneration of cholinergic and noncholinergic neurons of the basal forebrain in Alzheimer's disease. Ann Neurol 20:482-488
Richardson EP (1965) Remarks on the pathology of Parkinson's disease. In: Barbeau A, Doshay LJ, Spiegel EA (eds) Parkinson's disease: trends in research and treatment. Grune and Stratton, New York, pp 63-68
Richardson EP (1986) Editorial. Parkinson's disease. Ann Neurol 20:447-448
Riederer P, Wuketich S (1976) Time course of striatonigral degeneration in Parkinson's disease. J Neural Transm 38:277-301
Riederer P, Birkmayer W, Seeman D, Wuketich S (1977) Brain noradrenalin and 3-methoxy-4-hydroxy-phenylglycol in Parkinson's syndrome. J Neural Transm 41:241-251
Riederer P, Sofic E, Rausch WD, Kruzik P, Youdim MBH (1985) Dopaminforschung heute und morgen—L-Dopa in der Zukunft. In: Riederer P, Umek H (eds) L-Dopa-Substitution der Parkinson-Krankheit. Geschichte — Gegenwart — Zukunft. Springer, Vienna New York, pp 127-144
Riederer P, Rausch WD, Schmidt B et al. (1988) Biochemical fundamentals of Parkinson's disease. Mount Sinai J Med 55:21-28
Rogers JD, Brogan D, Mirra SS (1985) The nucleus basalis of Meynert in neurological disease: a quantitative morphological study. Ann Neurol 17:163-170
Rojas G, Asenhjo A, Chiorino R, Aranda L, Rocamora R, Donosa P (1965) Cellular

and subcellular structure of the ventrolateral nucleus of thalamus in Parkinson's disease: deposits of iron. Confin Neurol 26:362–376
Rosenberg N (1984) Joseph disease, an autosomal motor system degeneration. Adv Neurol 41:179–194
Rosenblum WI, Ghatak NR (1979) Lewy bodies in the presence of Alzheimer's disease. Arch Neurol 36:170–171
Rossor MN, Svedsen C, Hunt SP, Mountjoy CG, Roth M, Iversen LL (1984) The substantia innominata in Alzheimer's disease: a histochemical and biochemical study of cholinergic marker enzymes. Neurosci Lett 28:217–222
Roy S, Wolman L (1969) Ultrastructural observations in parkinsonism. J Pathol 99:39–44
Rudelli RD, Ambler MW, Wisniewski HM (1984) The morphology and distribution of Alzheimer neuritic (senile) and amyloid plaques in striatum and diencephalon. Acta Neuropathol (Berl) 64:273–281
Rutledge JN, Hilal SK, Silver AJ, Defendini R, Fahn S (1987) Study of movement disorders and brain iron by MR. Am J Roentgenol 149:365–379
Sabuncu N (1969) Quantitative Untersuchungen am Pallidum beim Parkinsonsyndrom. Dtsch Z Nervenheilk 196:40–48
Saper CB, German DC, White CL (1985) Neuronal pathology in the nucleus basalis and associated cell groups in senile dementia of the Alzheimer's type. Possible role in cell loss. Neurology (NY) 35:1089–1095
Scarnati E, Gasbarri A, Campana E, Pacitti C (1987) The organization of nucleus tegmenti pedunculopontine neurons projecting to basal ganglia and thalamus. Neurosci Lett 79:11–16
Scatton B, Dennis T, Lheureux R, Montofort J, Duyckaerts C, Javoy K, Agid F (1986) Degeneration of noradrenergic and serotonergic but not dopaminergic neurons in the lumbar spinal cord of parkinsonian patients. Brain Res 380:181–185
Schmitt HP, Emser W, Heimes C (1984) Familial occurrence of amyotrophic lateral sclerosis, parkinsonism and dementia. Ann Neurol 16:642–648
Schneider E, Becker H, Fischer PA et al. (1979) The course of brain atrophy in Parkinson's disease. Arch Psychiat Nervenkrankh 227:89–95
Schochet SS, Wyatt S, McCormick WF (1970) Intracytoplasmic acidophiliac granules in the substantia nigra. Arch Neurol 22:550–555
Schoene WC (1985) Diseases of the mesencephalon involving the substantia nigra and related systems of neurons. In: Davis RL, Robertson DH (eds) Textbook of neuropathology. Williams and Wilkins, Baltimore, pp 798–804
Schwab R, England AD (1968) Parkinson syndromes due to various specific causes. In: Vinken P, Bruyn GW (eds) Handbook of clinical neurology, vol 6. North-Holland, Amsterdam, pp 227–247
Seitelberger F (1986) Neuroaxonal dystrophy: its relation to aging and neurological diseases. In: Vinken PJ, Bruyn GW, Klawans HL (eds) Handbook of clinical neurology, vol 5 (49) Elsevier Scientific, Amsterdam, New York: pp 391–415
Sima AAF, Clark AW, Sternberger NA, Sternberger LA (1986) Lewy body dementia without Alzheimer changes. Can J Neurol Sci 13:490–497
Simchowicz T (1911) Histologische Studien über die senile Demenz. Histol Histopath Arb 4:267–444
Sparks DL, Markesberry WR, Slevin JT (1986) Alzheimer's disease: monoamines and spiperone binding reduced in nucleus basalis. Ann Neurol 19:602–604
Spencer PS, Nunn PB, Hugon JS (1986) Motoneuron disease of Guam: possible role of a food neurotoxin. Lancet I:965
Sroka H, Elizan TS, Yahr MD, Binger A, Mendazu MR (1981) Organic mental syndrome and confusional states in Parkinson's disease. Arch Neurol 38:339–342
Stadlan EM, Duvoisin R, Yahr MD (1966) The pathology of parkinsonism. Proc, 5th Int Congr Neuropathol, ICS 100. Excerpta Medica, Amsterdam, pp 569–571

Steele JC, Richardson JC, Olszewski J (1964) Progressive supranuclear palsy. Arch Neurol 10:333-359

Stockhausen P, Braak H (1984) Morphological changes of the isocortex in morbus Parkinson. Clin Neuropathol 3:206-209

Struble RG, Cork LC, Whitehouse PJ et al. (1982) Cholinergic innervation in neuritic plaques. Science 216:413-415

Struble RG, Powers RE, Casanova MF, Kitt CA, O'Connor DT, Price DL (1985) Multiple transmitter-specific markers in senile plaque in Alzheimer's disease (abstr) J Neuropathol Exp Neurol 44:325

Sudhaker S, Rajput AH, Rozdilsky B, Uitti RJ (1987) A clinicopathologic study of neurofibrillary tangles in Parkinsonism. Neurology 37 (Suppl 1):278-279

Sugimura K, Yamasaki Y, Ando K (1977) Parkinson's disease accompanied by dementia. A case of concurrent Parkinson's and Alzheimer's diseases. Rinsho Shin Keigaku 12:513-519

Sulkova R, Haltia M, Paltau A et al. (1983) Accuracy of clinical diagnosis in primary degenerative dementia. Correlation with neuropathological findings. J Neurol Neurosurg Psychiatry 46:9-13

Taban CH, Bouras C, Constantinidis J (1984) Substance P like immunoreactivity in the substantia nigra in one case of Huntington's chorea and in one case of Parkinson's disease compared to normal brain. In: Tagliavini LF, Pilleri G (eds) Brain pathology, vol 1. pp 219-228

Tabaton M, Schenone A, Romagnoli P, Mancardi GL (1985) A quantitative and ultrastructural study of substantia nigra and nucleus centralis superior in Alzheimer's disease. Acta Neuropathol (Berl) 68:218-223

Tagliavini F, Pilleri G (1983) Basal nucleus of Meynert. A neuropathological study in Alzheimer's disease, simple senile dementia, Pick's disease and Huntington's chorea. J Neurol Sci 62:243-260

Tagliavini F, Pilleri G (1984) The basal nucleus of Meynert in cerebral aging and degenerative dementias. In: Tagliavini F, Pilleri G (eds) Brain pathology, vol 1. Bern pp 181-218

Tagliavini F, Pilleri G, Bouras C, Constantinidis J (1984a) The nucleus basal of Meynert in idiopathic Parkinson's disease. Acta Neurol Scand 69:20-28

Tagliavini F, Pilleri G, Bouras C, Constantinidis J (1984b) The basal nucleus of Meynert in patients with progressive supranuclear palsy. Neurosci Lett 44:37-42

Takahashi K, Nakashima R, Takao T, Nakamura H (1977) Pallido-nigro-luysian atrophy associated with degeneration of the centrum medianum. Acta Neuropathol (Berl) 37:81-85

Takauchi S, Mizumara T, Miyoshi K (1983) Unusual paired helical filaments in progressive supranuclear palsy. Acta Neuropathol (Berl) 59:225-228

Takeda S, Ohama E, Izumo S et al. (1982) Substantia nigra and locus coeruleus in Parkinson-dementia cemplex of Guam and OPCA. In: Seitelberger F, Lassmann H, Jellinger K (eds) Abstracts of the IXth int congr neuropathol, Vienna, Wiener Medizinische Akademie, pp 115

Takei Y, Mirra SS (1973) Striatonigral degeneration, a form of multisystem atrophy with clinical parkinsonism. Prog Neuropathol 2:217-251

Tellez-Nagel I, Wisniewski HM (1973) Ultrastructure of neurofibrillary tangles in Steele-Richardson-Olszewski syndrome. Arch Neurol 29:324-327

Tenovuo O, Rinne UK, Viljanen MK (1984) Substance P immunoreactivity in the post-mortem Parkinsonian brain. Brain Res 303:113-116

Terry RD (1985) Alzheimer's disease. In: Davis RD, Robertson DM (eds) Textbook of neuropathology. Williams and Wilkins, Baltimore, pp 824-841

Tohgi H (1977) Arteriosclerotic parkinsonism: a reassessment. 11th world congr neurol, ICS 427 Excerpta Medica, Amsterdam, p 200

Tolosa ES, Santamaria J (1984) Parkinsonism and basal ganglia infarcts Neurology (CI) 34:1516-1518
Tomlinson BE, Corsellis JAN (1984) Ageing and the dementias. In: Adams JH, Corsellis JAN, Duchen LW (eds) Greenfield's neuropathology, 4th edn. Arnold, London, pp 951-1025
Tomlinson BE, Irving D, Blessed G (1981) Cell loss in the locus coeruleus in senile dementia of the Alzheimer type. J Neurol Sci 49:419-428
Tomonaga M (1981) Neurofibrillary tangles and Lewy bodies in the locus coeruleus neurons of the aged brain. Acta Neuropathol (Berl) 53:165-168
Tomonaga M (1983) Neuropathology of the locus coeruleus: a semiquantitative study. J Neurol 230:231-240
Torack RM, Morris JC (1988) The association of ventral tegmental area histopathology with adult dementia. Arch Neurol 45:497-501
Trétiakoff C (1919) Contribution à l'étude de l'anatomie pathologique du locus niger. Thesis, University of Paris
Uhl GR, McKinney M, Hedreen JC, White CL, Coyle JT, Whitehouse PJ, Price DL (1982) Dementia pugilistica: loss of basal forebrain cholinergic neurons and cortical cholinergic markers. Ann Neurol 12:99
Uhl GR, Hedreen JC, Price DL (1985) Parkinson's disease: loss of neurons from the ventral tegmental area contralateral to therapeutic surgical lesions. Neurology 35:1215-1218
Uhl GR, Hackney GO, Torchia M, Stranov V, Tourtellotte WW, Whitehouse PJ, Vinh Tran, Strittmacher S (1986) Parkinson's disease. Nigral receptor changes support peptidergic role in nigrostriatal modulation. Ann Neurol 20:194-203
Uitti RJ, Rajput AH, Ashenhust EM, Rozdilsky B (1985) Cyanide-induced parkinsonism: a clinico pathologic study. Neurology (NY) 35:921-925
Van Bogaert L (1956) Sur le parkinsonism saturnien avec paralysie de movements oculaires associées. Etude anatomique. Mschr Psychiatr Neurol 131:73-88
Wakabayashi K, Takahashi H, Takeda S et al. (1988) Parkinson's disease: the presence of Lewy bodies in Auerbach's and Meissner's plexuses. Acta Neuropathol (Berl.) 76:217-221
Whitehouse PJ (1986) The concept of subcortical and cortical dementia: another look. Ann Neurol 19:1-6
Whitehouse PJ, Hendreen JC, White CL, Price DL (1983) Basal forebrain neurons in the dementia of Parkinson's disease. Ann Neurol 13:243-248
Wilcock GK, Esiri MM (1982) Plaques and dementia: a quantitative study. J Neurol Sci 56:343-356
Wilkening G, Vernier VG, Arthaud LE, Treacy G, Kenney JP, Nickolson VJ, Clark R, Smith DH, Boswell G (1986) A parkinson-like neurologic deficit in primates is caused by a novel 4-substituted piperidine. Brain Res 368:239-246
Willis GL (1987) Amine accumulation in Parkinson's disease and other disorders. Neurosci Biobehavior Rev 11:97-105
Wisniewski HM, Merz GS (1985) Neuropathology of the aging brain and dementia of Alzheimer type. In: Haitz CH, Samorajski T (eds) Aging 2000; our health care destiny, Vol 1. Biomedical issue. Springer, Berlin Heidelberg New York, pp 231-243
Wisniewski HM, Wen GY (1985) Substructures of paired helical filaments from Alzheimer neurofibrillary tangles. Acta Neuropathol (Berl) 66:173-176
Wisniewski HM, Terry RD, Hirano A (1970) Neurofibrillary pathology. J Neuropathol Exp Neurol 29:163-176
Wisniewski K, Jervis GA, Moretz RC, Wisniewski HM (1979) Alzheimer neurofibrillary tangles in diseases other than senile and presenile dementia. Ann Neurol 5:288-294

Wisniewski K, Wisniewski HM, Wen GY (1985) Occurrence of neuropathological changes and dementia of Alzheimer's disease in Down's syndrome. Ann Neurol 17:278-282

Wuketich ST, Riederer P, Jellinger K, Ambrozi L (1980) Quantitative dissection of human brain areas. J Neural Transm [Suppl] 16:53-67

Xuereb JHM, Perry RH, Irving D (1987) The dementia in Parkinson's disease (abstr). Neuropathol Appl Neurobiol 13:231.

Yagishita S, Itoh Y, Amano N, Nakano T (1979) Ultrastructure of neurofibrillary tangles in progressive supranuclear palsy. Acta Neuropathol (Berl) 48:27-30

Yagishita S, Oto Y, Wang N, Amano N (1981) Reapprasial of the fine structure of Alzheimer's neurofibrillary tangles. Acta Neuropathol (Berl) 54:239-245

Yahr MD, Wolf A, Antunes JL, Miyoshi K, Duffy P (1972) Autopsy findings in parkinsonism following treatment with levodopa. Neurology (Minneap) 22 (Suppl II):56-65

Yamada M, Mehraein P (1977) Verteilungsmuster der senilen Veränderungen in den Hirnstammkernen. Folia Psychiatr Neurol Jpn 31:219

Yamada M, Ohno S, Okayasu I, Okeda R, Hatakeyama S, Watanabe H, Ushio K, Tsukagoshi H (1986) Chronic manganese poisoning: a neuropathological study with determination of manganese distribution. Acta Neuropathol (Berl) 70:273-277

Yamamoto T, Hirano A (1985 a) Nucleus raphe dorsalis in Alzheimer's disease; neurofibrillary tangles and loss of neurons. Ann Neurol 17:573-577

Yamamoto T, Hirano A (1985 b) Nucleus raphe dorsalis in Parkinson's dementia complex of Guam. Acta Neuropathol (Berl) 67:296-299

Yen SH, Horoupian DS, Terry RD (1983) Immunocytochemical comparison of neurofibrillary palsy, and postencephalitic parkinsonism. Ann Neurol 13:172-175

Yen SH, Dickson DW, Peterson C, Goldman JE (1986) Cytoskeletal abnormalities in neuropathology. Prog Neuropathol 6:63-90

Yokochi M, Narabayashi H, Ilzuka R, Nagatsu T (1984) Juvenile parkinsonism—some clinical, pharmacological and neuropathological aspects. Adv Neurol 40:407-413

Yoshimura M (1983) Cortical changes in the parkinsonian brain: a contribution to the delineation of "diffuse Lewy body disease". J Neurol 229:17-32

Yoshimura N, Yoshimura I, Kudo H, Asada M, Hayashi S, Fukushima Y, Sato T (1989) Juvenile Parkinson's disease with widespread Lewy bodies in the cerebral cortex. Acta Neuropathol (Berl) 77:

Younkin SG, Goodridge B, Lockett G, Usiak M, Younkin LH (1986) Morphometry of cholinergic neurons in Alzheimer's disease (abstr). J Neuropathol Exp Neurol 45:341

Yuen P, Baxter DW (1963) The morphology of Marinesco bodies (paranucleolar corpuscles) in the melanin-pigmented nuclei of the brain stem. J Neurol Neurosurg Psychiatry 26:178-183

Zech M, Bogerts B (1985) Methionine, enkephalin and substance P in the basal ganglia of schizophrenics; a quantitative immunohistochemical comparison with Huntington and Parkinson patients. Acta Neuropathol (Berl) 68:32-38

Zetusky WJ, Jankovic J, Pirozzolo FJ (1985) The heterogeneity of Parkinson disease: clinical and prognostic implications. Neurology (NY) 35:522-526

Zweig RM, Whitehouse PJ, Casanova MF, Walker LC, Jankel WR, Pride CL (1987) Loss of pedunculopontine neurons in progressive supranuclear palsy. Ann Neurol 22:18-25

Zweig RM, Rose CA, Hedren IC et al. (1988) The neuropathology of aminergic nuclei in Alzheimer's disease. Ann Neurol 24:233-242

CHAPTER 3

Biochemical Neuroanatomy of the Basal Ganglia

E. G. McGeer and P. L. McGeer

A. Introduction

The biochemical neuroanatomy of the basal ganglia has been under intensive investigation for at least a quarter of a century. It has been a favorite topic since the early days of biochemical neuroanatomy because many of the classical neurotransmitters are found in very high concentration in extrapyramidal nuclei, these nuclei are relatively distinct and easily dissected, and classical pathological studies had demonstrated a clear association between pathology in these nuclei and various movement disorders. In the past decade, the subject has become increasingly complex with the identification of many new neurotransmitters, the discovery through axonal transport tracing techniques of innumerable previously unsuspected connections, and the demonstration, both through such tracing techniques and behavioral studies, that many parts of the limbic system are involved in movement control. We have reviewed the literature on the afferents and efferents of the caudate/putamen (CP), globus pallidus (GP), and substantia nigra (SN), and on the neurotransmitters occurring in the various tracts (McGeer et al. 1984). We present updates of those diagrams in the Appendix to this chapter, as well as new diagrams on the afferents and efferents of the ventral pallidum and the nucleus accumbens which forms the major part of the ventral striatum. The concepts of the ventral striatum and ventral pallidum are becoming ever more important with increasing evidence that they are concerned in movement control as well as in functions more generally associated with the limbic system. The ventral striatum has been defined by Heimer and colleagues to include the substriatal gray, striatal cell bridges, parts of the olfactory tubercle, and the nucleus accumbens (N.Ac); the definition is made on morphological grounds (see for example Zahm and Heimer 1985), but is also supported by similarities in connectional patterns (cf. Carlsen and Heimer 1986). The ventral pallidum (VnGP) is defined by the unique pattern of so-called woolly fibers using enkephalin (ENK), dynorphin (DYN), or substance P (SP) (Haber and Nauta 1983); it comprises a subcommissural projection of the pallidum which reaches beyond the substantia innominata into the deep, polymorph layer of the olfactory tubercle (cf. Meyer and Wahle 1986; Wahle and Meyer 1986). The morphology is similar to that in the GP, but the area embraces, even in humans (Haber and Watson 1985), parts of the basal forebrain with its large corticopetal cholinergic neurons.

In the text of this chapter, however, we would like to discuss a general classification of neurotransmitters with implications for basal ganglial as well as other brain functions, to mention developing, but as yet little understood, aspects of the biochemical neuroanatomy, such as the subnuclear organization and interconnections between various pathways, and, finally, to put forth some very hypothetical diagrams illustrating how some aspects of the system may work and be affected by disease. A complete review of the literature on these various aspects is not attempted and most references given in our earlier review (McGEER et al. 1984) are not repeated here.

B. Neurotransmitters in the Extrapyramidal System

A bewildering variety of neurotransmitters has been shown to exist in the extrapyramidal system, and many of them are present in higher concentration in nuclei of that system than in any other brain region of comparable size. Table 1 lists the neurotransmitters so far reported in the basal ganglia and gives the abbreviation for each which is used in this chapter. The neurotransmitters in brain can be classed into three general types (McGEER et al. 1984, 1987), and it may help to consider the general characteristics of each of these types of transmitters when attempting to hypothesize their possible roles in extrapyramidal function.

I. Type I: Amino Acid Neurotransmitters

The first transmitter type embraces the amino acids (glutamate, aspartate, glycine, and GABA) which occur in a concentration range of micromoles per gram of brain and have at almost all receptors an ionotropic action. This means that their receptors function to open ionic gates on the postsynaptic membranes, and thus to induce large conductance changes and consequent rapid excitatory or inhibitory actions. The majority of corticofugal paths appear to use glutamate or aspartate (Glu/Asp) as a major transmitter, and most or all nuclei associated with the extrapyramidal system receive such pathways (cf. McGEER et al. 1984, 1987). Of particular importance are the very large Glu/Asp tracts from the cortex to the striatum (including the ventral striatum) and the thalamus. Glutamate has also recently been shown to be the major or sole transmitter in the projections from the subthalamus to the GP and the SN, and there is considerable evidence that the projection from the pedunculopontine nucleus to the SN also contains an excitatory amino acid transmitter.

GABA is also clearly an extremely important neurotransmitter in the control of movement by the extrapyramidal system. The SN and VnGP have the highest levels of GABA and its synthetic enzyme, glutamic acid decarboxylase (GAD), found anywhere in brain. Besides the GABA interneurons, which are certainly found in the CP and N.Ac, and probably occur in most nuclei associated with the extrapyramidal system, a number of long axon GABA tracts are known. The most important of these are probably the projections of the

Table 1. Classification of neurotransmitters or putative neurotransmitters of the extrapyramidal system with their approximate levels in rat brain and the abbreviations used in this chapter

Type I: Amino acids (2-14 μmol/g)	Type II: Amines (1-25 nmol/g)
Glutamate (Glu)	Acetylcholine (ACh)
Aspartate (Asp)	Dopamine (DA)
γ-Aminobutyric acid (GABA)	Noradrenaline (NA)
Glycine (Gly)	Adrenaline (ADR)
	Serotonin (5-HT)
	Histamine (HIS)
	Adenosine (ADO)
	Tryptamine
Type III: Peptides[a] (<1-500 pmol/g)	
Cholecystokinin (CCK)[b]	Prodynorphins (DYN)
Somatostatin (SOM)[b]	Dynorphin A (1-8), A (1-17)
Neurotensin (NT)	Dynorphin B (1-13)
Neuropeptide Y (NPY)[c]	α-Neoendorphin
Vasoactive intestinal peptide (VIP)	Proenkephalins (ENK)[g]
Thyrotropin-releasing hormone (TRH)	Met^{5}-enkephalin
Angiotensin II (AII)	Met^{5}-enkephalin-Arg^{6}-Phe^{7}
Bombesin (BO)	Met^{5}-enkephalin-Arg^{6}-Phe^{7}-Leu^{8}
Cyclo(His-Pro)	Leu^{5}-enkephalin
Calcitonin gene-related peptide (CGRP)[d]	Tachykinins (SP/K)
Galanin (GAL)[e]	Substance P (SP)
P7 of IB236[f]	Substance K (SK)[h]
Oxytocin (OXY)	Corticotropin-releasing factor (CRF)
Vasopressin (AVP)	

[a] Does not include some, such as bradykinin, FMRFamine, or members of the proopiomelanocortin family which are present only in very low concentrations in the extrapyramidal system (cf. GRAYBIEL 1986).

[b] Probably occurs in various forms.

[c] Early reports of avian pancreatic polypeptide probably were of NPY.

[d] SKOFITSCH and JACOBOWITZ (1985a).

[e] SKOFITSCH and JACOBOWITZ (1985b).

[f] This peptide, whose existence was predicted from the amino acid sequence of the precursor, is found in highest concentrations in the human GP and SN (SUZUKI et al. 1986).

[g] The proenkephalin mRNA is said to be expressed in >50% of neurons in rat CP (SHIVERS et al. 1986).

[h] Neurokinin A.

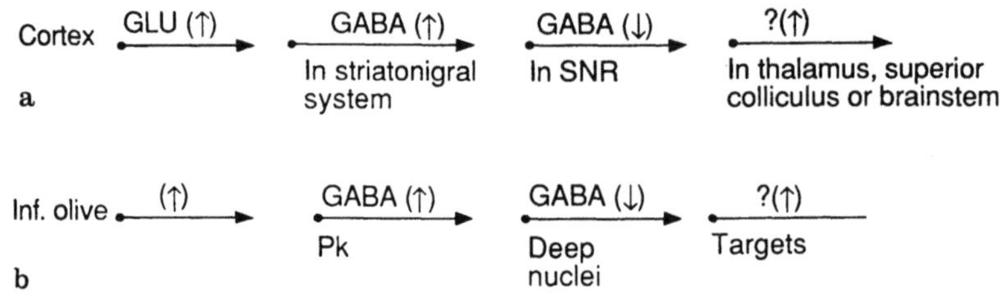

Fig. 1 a, b. Diagrams of how initial excitation of (**a**) the striatum or (**b**) cerebellar Purkinje cells (Pk) may lead through two GABA neurons to disinhibition of neurons in the target nuclei. The effect of stimulation of the excitatory neuron on other neurons in the chain is indicated by (upward arrows) increased or (downward arrows) decreased activity. There may be other excitatory neurons interspersed and more than one system linking cortical excitation to SNR output neurons. For example, there is some evidence for: Glu → GABA → SP → SNR GABA output. (Melis and Gale 1984)

pallidum and the substantia nigra pars reticulata (SNR) which comprise the major output of the basal ganglia.

There is far less evidence for glycine as an important neurotransmitter in the extrapyramidal system, but some turning experiments in rats suggest that glycine innervation of the SN and A10 area (Hartgraves and Kelly 1985) may play a role in movement control.

The fast ionotropic action of amino acid transmitters (type I), as compared with the metabotropic action of the amines (type II) and neuropeptides (type III), which is both slower in onset and extends over a longer time, suggests that the amino acids should serve the major input command and output systems of the basal ganglia, and this is in accord with the known biochemical neuroanatomy. It is of interest that the circuitry of these amino acid systems between the cortex and the eventual targets (thalamus, superior colliculus, pedunculopontine and other brain stem nuclei) appears to involve a double inhibitory chain, as illustrated in Fig. 1a, so that activation of the corticostriatal fibers results in disinhibition of the target neurons. This pattern is reminiscent of the circuitry of the cerebellum (Fig. 1b). There is probably provision also for rapid feedback from the target neurons to the cortex; stimulation of SNR neurons, for example, has been shown to increase Glu release in the ipsilateral striatum by a thalamic-dependent mechanism (Girault et al. 1986).

II. Type II: Amine Neurotransmitters

The amine neurotransmitters occur in brain in a concentration range of nanomoles per gram of tissue, roughly 1000-fold lower than type I transmitters. As previously mentioned, their action at most receptors is typically slow in onset and relatively long-lasting, and almost certainly involves second messenger systems which may be varied, but have a common mechanism of changing the degree of phosphorylation of membrane proteins (McGeer et al. 1987). The physiologic actions are diverse and are frequently considered modulatory; the overall effect may be inhibitory or excitatory, depending on the type of receptor and other inputs into the target neuron. All of the known amine neurotransmitters appear to be present in the extrapyramidal system. The nigrostriatal DA tract is probably one of the most studied projections in the central nervous system, and its importance to normal movement is clearly demonstrated in the parkinsonian symptoms which seem linked to major deficiencies in this tract. It seems worthwhile, however, pointing out that the DA deficiency must be very great (approximately 90%) before overt symptoms of parkinsonism appear. The cholinergic interneurons of the CP and ventral striatum are well known. So too now are the corticopetal cholinergic neurons of the VnGP which, in some recent investigations, also appear to play a role in movement control. Amine projections to extrapyramidal nuclei from the raphe (5-HT), hypothalamus (histamine, HIS), and locus ceruleus/subceruleus (NA/ADR) are also known, although their exact roles in extrapyramidal function are not. It is of interest that the 5-HT levels of the SN and GP are almost twice those found in the CP and subthalamus (Pasik et al. 1984), suggesting that 5-HT may be principally employed in modulating the output of the basal ganglia.

Athough no projection using ADR as the transmitter has been definitively established, either in the extrapyramidal system or elsewhere in brain, there is considerable evidence that ADR can act on specific receptors in the CP and other extrapyramidal nuclei to modulate the release of neurotransmitters, in particular DA and Glu.

Another candidate amine transmitter found in the extrapyramidal system is tryptamine. Relatively high levels of tryptamine in the caudate and of specific tryptamine binding sites in the N. Ac (McCormack et al. 1986) have been reported, and, on the basis of lesion experiments, this indole has been proposed as a transmitter in a small percentage of the nigrostriatal afferents in the rat (Jurio and Greenshaw 1986). Its identification as a neurotransmitter, however, remains highly controversial (for review see Jones 1982).

There is a little evidence that the amines may show some interesting species differences. Colocalization of DA and CCK seems, for example, almost entirely limited to A10 neurons in the rat, whereas virtually all the mesencephalic DA neurons in the cat, including those of the substantia nigra, pars compacta (SNC), seem to contain CCK (Markstein and Hokfelt 1984). The modulatory effects of CCK on DA release are seen in both the accumbens and caudate of the cat, but only in the former in the rat (see Sect. B.III).

The pattern of DA metabolites in the striatum varies considerably among species. 3,4-Dihydroxyphenylacetic acid (DOPAC) is the major metabolite in the rat; homovanillic acid (HVA) and DOPAC are formed in about equal amounts in gerbils (Upchurch 1985); and HVA seems to be the major metabolite in humans (McGeer et al. 1985). Whether such differences reflect differences in release or merely in the levels of the metabolizing enzymes is not known.

ADR is found in the striatum in some species such as the cat, and ADR neurons occur in the subceruleus in many species. The caudate of pigs is said to contain 2-3 times as much ADR as NA (Draper et al. 1984). On the other hand, the guinea pig, and possibly other species, lack ADR neurons (Cum-

MING et al. 1986). The HIS neurons of the hypothalamus, which innervate the striatum as well as many other brain regions, seem to increase in number disproportionately as one moves from rodents to cats to primates (P. REINER 1986, personal communication). The 5-TH innervation of basal ganglia nuclei in primates has been said to be generally rather similar to that in subprimates, although some differences in nuclear organization, fiber myelination (AZMITIA and GANNON 1986), and fiber distribution (MORI et al. 1985) are reported. Most work in biochemical neuroanatomy is done in rodents, but it is clearly essential to check the results as fully as possible in primates.

III. Type III: Peptide Neurotransmitters

This is by far the largest category since, at this stage, it includes at least two dozen peptides. Those shown in Table 1 have been reported to be prominent in the extrapyramidal system. The peptides exist in brain in much lower concentrations than transmitters of types I and II, usually in the range of picomoles per gram of tissue. Their functions and modes of actions are as yet largely unknown, but, apparently, relatively few molecules of peptide need to be released to affect their receptors. They probably have special metabotropic actions and, in some cases, have been shown to modulate the action of amine or amino acid transmitters. Some of these peptides have been shown to exist in interneurons in the striatum (SOM, NPY, ENK, CCK), in projection neurons from the striatum (SP/K, ENK, DYN), or in afferents to the striatum and accumbens (CCK). ENK neurons have also been reported in the SNR and entopeduncular nucleus (BECKSTEAD and KERSEY 1985), SOM neurons in the SN compacta and lateralis (TABER-PIERCE et al. 1985), CRF neurons in the N. Ac (JOSEPH et al. 1985), and NT neurons in the A10 area (HOKFELT et al. 1984). In some cases, high concentrations of the peptide (NT, GAL, CGRP, P7), or of peptide binding sites (NT, CGRP), or motor effects of locally injected peptides (e.g., NT, KALIVAS and TAYLOR 1985; SK, INNIS et al. 1985) have been demonstrated in extrypyramidal nuclei and thus, although the exact projections have not always been shown, the peptides are thought to play important roles.

An aspect of neurotransmitters which has been largely demonstrated by work in the peptide field is the colocalization of two or more transmitters within a single neuron. It is now known that this can involve colocalization of two classical neurotransmitters as well as colocalization of a peptide with a classical neurotransmitter or with another peptide, although the last seems by far the most frequent. Examples of colocalization reported in the extrapyramidal system are given in Table 2. The extent of colocalization in some cases may be species dependent (see Sect. B.II). Normally, one of the cotransmitters (the minor one) is released in much smaller amounts than the other, and may be released only under conditions of high stimulation rates. The physiologic need for such cotransmission is not yet clear, but it appears that the cotransmitter may sometimes modulate the release or postsynaptic action of the other. Thus, CCK, which is a cotransmitter in A10 but not nigral dopaminergic neurons in the rat, has been reported to modulate the release and/or ef-

Table 2. Reported occurrences of two neurotransmitters in single neurons of the extrapyramidal system[a]

		Region	*Reference*
Type I and type II transmitters			
GABA	5-HT	Ascending raphe neurons to striatum	BELIN et al. 1983; McGEER et al. unpublished work 1984
Type I and type III transmitters			
GABA	ENK	Rat caudate	OERTEL et al. 1983
		Rat and cat medium-size aspiny neurons	ARONIN et al. 1984 PENNY et al. 1986
		Terminals in ventral GP	ZAHM et al. 1985
GABA	β-END	Rat neostriatum	OERTEL et al. 1983
GABA	SP	Rat and cat neostriatum	PENNY et al. 1986
Type II and type III transmitters			
DA	CCK	Some rat A10 projections (partly crossed)	HOKFELT et al. 1980 FALLON et al. 1983
		Cat nigrostriatal A10	MARKSTEIN and HOKFELT 1984
DA	NT[c]	Nigrostriatal	GRAYBIEL (1986)
		A10 neurons	HOKFELT et al. 1984
Two or more type III transmitters			
SOM	NPY	Rat striatal and nucleus accumbens interneurons	BEAL et al. 1986; VINCENT and JOHANSSON 1983; SMITH and PARENT 1986a
SOM[b]	ENK[b]	Rat and cat CP neurons	PENNY et al. 1986
SP	DYN	CP neurons	BESSON et al. 1986

[a] This table does not include many instances of reported colocalizations in neurons which may or may not project to extrapyramidal nuclei; for example, we have not included the reports of 5-HT coexisting in raphe neurons with SP, TRH, and/or ENK, or of NA or ADR coexisting in locus ceruleus/subceruleus neurons with ENK, SOM, NPY, NT, CCK, or AVP, or of ACh coexisting with SP and/or NPY in the pedunculopontine and similar brain stem nuclei in neurons which may well project to the SN and VnGP (McGEER et al. 1987).

[b] About 50% of the medium cell population containing each of these substances contains the other (ARONIN et al. 1984).

[c] NT increases both basal and K^+-evoked release of DA in rat striatal slices (BATTAINI et al. 1986).

fects of DA in the A10 target (the N.Ac), but not in the nigral target (the striatum) (cf. Katsuura et al. 1985; Voigt et al. 1985; Crawley et al. 1984). The interaction is complex and highly concentration dependent. It has even been reported that, within the rat accumbens, CCK potentiates the effect of DA on D_1 receptors in the caudal area innervated by mixed DA/CCK fibers, but reduces the effect in the anterior portion which is innervated by distinct and separate DA and CCK fibers (Studler et al. 1985). Clearly, in this, as in many other aspects of neurotransmitter action, much more information is required.

Many reported examples of the colocalization of two peptides in a single neuron have turned out to depend on the fact that peptide neurotransmitters are made from large preprohormones which may encode the amino acid sequence of more than one active peptide. Thus, for example, SP and SK are both tachykinins which are encoded in the same preprohormone. The findings that SK exists with SP in striatonigral afferents (Lee et al. 1986; Brodin et al. 1986), and that the SN has a high concentration of both SK (Innis et al. 1985) and SK binding sites (Quirion and Dam 1985; Mantyh and Hunt 1986) probably explain the oft-cited anomaly that, despite the apparently large SP striatonigral projection, there is a relatively low density of SP binding sites in the SN. The two peptides have different pharmacologic profiles and therefore presumably different, though possibly related, physiologic roles (Nawa et al. 1984). It may be that the relative amounts of two such cotransmitters formed from a single precursor may be constant in a given projection or may vary in different neurons of the projection, or even in a given neuron, depending on the excitation conditions (cf. Bannon et al. 1986a). This is another question requiring further exploration.

Species differences may be expected in peptide as well as amine systems. For example, Inagaki and Parent (1984) have indicated that, in the SN, the distribution of the large number of SP fibers follows a similar pattern in rat, cat, and monkey, while the number of ENK fibers and the complexity of their organization increases strikingly from rodent to primate. More comparative studies are clearly needed.

IV. Receptor Subtypes

An aspect of neurotransmitter action which has been touched on before but deserves further emphasis is that it depends as much on the particular receptor as on the particular neurotransmitter. All neurotransmitters so far described seem to have multiple types of receptors which are differentiated most easily by pharmacologic specificity, but also frequently by their action on various second messengers and/or neuronal excitability. The D_1 and D_2 receptors for dopamine are the best known example in the extrapyramidal system, but many other examples are known in that system as well as in the rest of brain.

C. Heterogeneities in Extrapyramidal Nuclei

I. Mosaics or Gradients

The mammalian striatum has been found to show marked heterogeneity by a variety of neuroanatomic, histochemical, and biochemical markers. One type of heterogeneity is the patches or striosomes which appear particularly in the

Table 3. Reported differences in the distribution of various neuronal-related indices in patches vs matrix in the caudate[a]

Indices reported to be highly concentrated in: Patches (striosomes, islands)	Matrix
Very well defined	
μ Opiate binding sites[b]	AChE staining[b]
Inputs from thalamus[c]	Inputs from sensory and motor cortex[d]
Inputs from prelimbic cortex[d]	Neurons projecting to SNR and pallidum[d]
Neurons projecting to SNC[d, e]	Inputs from A8 (rubrorostral) and A10 (VTA)[e]
	Calcium binding protein-positive neurons[k]
Not as well defined	
Opiate immunoreactivity[b, f]	SOM-positive neurons[l]
SP-positive neurons[g]	SOM/NPY/NADPH diaphorase-positive neuropil[l]
GAD immunoreactivity[h]	NT binding sites[f]
NT immunoreactivity[f]	
Benzodiazepine binding sites[j]	
DYN-positive neurons	

[a] For review see GRAYBIEL (1983).
[b] In caudates of carnivores and primates, including humans (GRAYBIEL and RAGSDALE 1978; HERKENHAM and PERT 1981); although AChE is in matrix, QNB binding is evenly distributed (BRAND 1980).
[c] HERKENHAM and PERT (1981); BECKSTEAD (1985).
[d] GERFEN (1984).
[e] JIMINEZ-CASTELLANOS and GRAYBIEL (1985, 1986); projection from patches is to medial SN and this may well be SNC.
[f] See GOEDERT et al. (1984) and references therein; DYN seems more concentrated than ENK (GRAYBIEL 1986).
[g] In rat and cat caudate (PENNY et al. 1986), but GRAYBIEL et al. (1981) say that SP-positive patches were aligned with both AChE-rich and AChE-poor zones in cat caudate. See BEACH and MCGEER (1984) for patchy staining in baboon and humans.
[h] GRAYBIEL et al. (1983).
[j] In human caudate; in the ventral striatum, higher densities were found, but were aligned with both AChE-rich and AChE-poor regions (FAULL and VILLIGER 1986).
[k] In the striatum of rats and monkeys (GERFEN 1985).
[l] In cat caudate; neurons are mainly in matrix, but close to patch borders and may serve as interfaces between patches and matrix (CHESSELET and GRAYBIEL 1986; SANDELL et al. 1986).

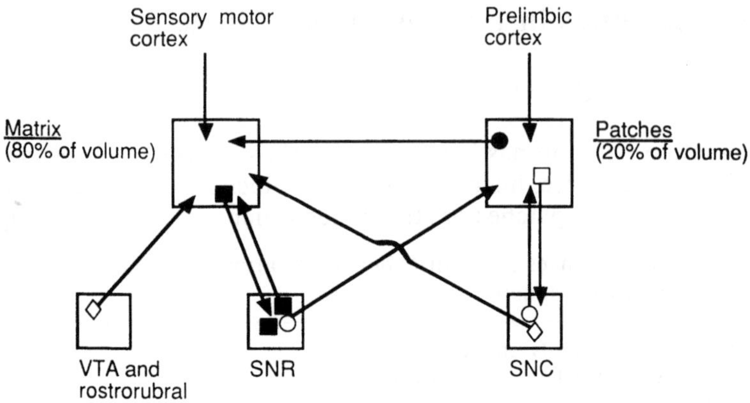

Fig. 2. Diagram of some of the hypothesized differences in connectivities between the matrix and patch (striosome) compartments. This is based largely on the report of Gerfen (1986). Inputs from cortex probably use Glu. Full circles SOM; full squares GABA neurons; open squares probably SP/K/ENK/DYN. Lozenges type 1 and open circles type 2 DA neurons differ in calcium binding protein and sensitivity to neurotoxins; the report of Fukui et al. (1986) that patches of DA staining appear in adult rat striatum after administration of reserpine or methamphetamine also suggests different sensitivities to these catecholamine depleters

caudate of carnivores and primates. Since the original report of Graybiel and Ragsdale (1978) of acetylcholinesterase (AChE)-poor striosomes with a darker AChE-rich matrix, a number of other biochemical indices, as well as connectivities, of the patches and matrix compartments have been found to be different (Table 3 and Fig. 2). The patchy distribution of staining is usually much less evident in the putamen than in the caudate and, while it may occur in the ventral striatum or nucleus accumbens, the correspondence between various neurochemical indices may not be the same as in the caudate.

Other indices within the caudate may occur in patches and yet not coincide with the patches defined by opiate binding, AChE-staining, or similar index, as indicated in Table 3. Examples are the clusters of caudate neurons in the rat, which project to either the pallidum or the SN, but not to both (Loopuijt and Van der Kooy 1985), or the patches of high D_2 binding site density in monkey caudate (Kohler and Radesater 1986).

Patchy distribution of some indices such as DA terminals and muscarinic binding sites may occur in neonatal rat CP, but is largely lost in adult rodents (Nastuk and Graybiel 1985; see also Fig. 2). Striatal heterogeneities in the form of gradients have, however, been reported. This has been particularly true of the DA and 5-HT systems. The striatal innervation from midbrain DA cells is said to be greater rostrally than caudally, while the 5-HT innervation terminates principally ventrally and caudally (Widmann and Sperk 1986; Mori et al. 1985; Beal and Martin 1985). Analogous gradient differences have been reported in the densities of D_2 (Dubois and Scatton 1985) and 5-HT_2 (Altar et al. 1986) binding sites. D_2 binding sites are also reported to show a lateral > medial gradient in rat CP, not consistent with gradients in

DA levels (JOYCE et al. 1985), but D_1 sites are concentrated more in the ventrolateral and medial as compared with the ventromedial and dorsolateral aspects (SAVASTA et al. 1986). With increasing use of quantitative autoradiographic techniques for binding site studies, more such reports can be expected.

Concentration gradients have also been reported for cholinergic indices (lateral > medial, MARSHALL et al. 1983; REA and SIMON 1981), for NPY-positive cells in both monkey and rat (caudal > rostral, CHRISTIE et al. 1986; SMITH and PARENT 1986a), and for NPY and SOM staining in the rat (ventromedial > dorsolateral, BEAL et al. 1986).

Inhomogeneities most probably also occur in other nuclei in or associated with the extrapyramidal system. Relatively little work has been done except in the superior colliculus where the complex internal organization involves not only distinct layers, but patches within the layers which are again distinguishable by staining for AChE or ENK, and by the convergence of afferents from the frontal cortex and SN (ILLING and GRAYBIEL 1985, 1986; ROSS and GODFREY 1985; cf. WIENER 1986).

II. Differences Between Caudate, Putamen, and Nucleus Accumbens

Overall chemical differences between the three major nuclei of the striatum and ventral striatum, i.e., the caudate, putamen, and N. Ac, are also of interest since all three are clearly concerned in motor control, but probably in different fashions. It is of interest in this regard that parkinsonism and dystonia may involve largely putamen abnormalities (BURTON et al. 1984). Until recently, the caudate and putamen were usually considered as part of a single homogeneous entity. Detailed studies of the striatal connections in monkeys by retrograde tracing techniques have suggested that the two structures are differently organized. For example, the sensory motor cortex projects extensively to the putamen where somatotropic representation of the leg, arm, and face occurs in the form of obliquely arranged strips. On the other hand, associative areas of the prefrontal, temporal, parietal, and anterior cingulate cortex project in the form of patches, almost exclusively to the caudate (cf. SMITH and PARENT 1986b). In addition, PARENT et al. (1984) suggest that the striatopallidal and striatonigral projections are largely (90%) distinct, with the neurons projecting to the pallidum occurring mainly in the putamen, while those projecting to the nigra occur mainly in the caudate. In this study, surprisingly few striatal neurons were found to project to both pallidum and SN. The projections from the caudate to the pallidum which do exist also terminate in different pallidal areas from the putamenal pallidal projections. A different, but largely complementary pattern is seen in the organization of thalamic projections to the caudate and putamen in the monkey; the caudate projections come largely from the central superolateral and parafascicular nuclei, while the putamenal projections come mainly from the central median nucleus (SMITH and PARENT 1986b).

In other work, Smith and Parent (1986a) have reported that the numerical density of NPY-positive cells is greater in the caudate than in the putamen in both cat and monkey, and Graveland and DiFiglia (1985) suggest from morphological studies that aspiny interneurons with indented nuclei may play a greater role in monkey caudate than putamen.

Biochemical distinctness is more widely recognized in relation to the nucleus accumbens or ventral striatum. A well-known difference is the presence of the peptide CCK in the dopaminergic innervation of the N. Ac and its relatively general absence from most of the dopaminergic innervation of the CP in the rat (but see Sect. B.II). Measurements of DA, 5-HT, and metabolites have suggested marked differences in turnover rate of the amines in the ventral striatum as compared with the CP, but reported data are so far not completely consistent (Widmann and Sperk 1986; Beal and Martin 1985). Freed and Yamamoto (1985) have suggested that DA metabolism in the N. Ac is more strongly linked to speed and direction of movement in rats, while in the caudate it is more affected by posture and direction. Beal et al. (1986) found NPY and SOM concentrations in rats higher in the N. Ac and ventromedial striatum than in the dorsolateral striatum. The accumbens also seems to differ from the CP in that it contains some CRF cells, as well as much higher levels of VIP and TRH (Graybiel 1986).

These are only examples of the type of information accumulating on the patterned organization within the extrapyramidal system, but it is clear that the subnuclear organization pattern will eventually have to be taken into account in any hypotheses relating the biochemical neuroanatomy to function, or biochemical pathology to disease processes.

D. Interconnections in the Extrapyramidal System

Identifying the projections and the transmitters which they use is only part of the task of the biochemical neuroanatomist. The complete "wiring diagram" will not be known until the interconnections of the various neurons are determined. The only definitive way to obtain such information is by electron microscopic studies, involving some kind of double-labeling technique. There have so far been only a few such studies. The first in the extrapyramidal system were those of Hattori et al. (1975, 1976) who showed that DA nerve endings (marked by 6-hydroxydopamine (6-OHDA)-induced degeneration) were in frequent contact with cholinergic dendrites (marked by choline acetyltransferase immunohistochemistry) in rat CP, and that projections (marked by radioactive axonal transport) to dopaminergic dendrites in the SN (marked by degeneration) came preferentially from the GP in the rat, while projections from the head of the caudate landed preferentially on nondopaminergic elements of the SN. Similar electron microscopic work in the rat caudate has provided evidence for: (a) innervation of common recipient neurons by cortical and DA efferents, and suggested a possible direct axonal interrelation between these two primary inputs (Bouyer et al. 1984a), as well as between corticostriatal afferents and striatal ENK neurons (Bouyer et al. 1984b); (b) DA innervation of both ENK and SP neurons (KUBOTA et al. 1986a, b); and (c) innervation of striatonigral neurons by DA afferents (FREUND et al. 1984), as well as GABA and ENK terminals (ARONIN et al. 1986), and SP-containing terminals in contact with cholinergic dendrites and cells (BOLAM et al. 1986).

In the GP, large pallidonigral neurons have been shown to be innervated by striatal afferents through symmetric contacts (TOTTERDELL et al. 1984). Electron microscopic studies of GAD-positive elements have revealed GABA innervation of GABA neurons in the GP and entopeduncular nucleus (EP) (OERTEL et al. 1984), of cholinergic neurons in the VnGP (INGHAM et al. 1986), and of both GABAergic and non-GABAergic neurons in the SN (HOLSTEIN et al. 1986); the non-GABAergic neurons include DA cells, as has been repeatedly demonstrated at the electron microscopic level (e.g., VAN DEN POL et al. 1985). In the rat SN, WILLIAMS and FAULL (1985) have demonstrated that striatonigral neurons synapse with nigrotectal fibers.

Light microscopic studies are frequently used to suggest interconnections, although they cannot be considered definitive. Such studies have, for example, been said to indicate that the medium spiny GABA projection neurons in the CP receive considerable cholinergic innervation (PHELPS et al. 1985), that DA neurons of the human SN and A10 are innervated by ENK terminals (GASPAR et al. 1983), and that the cholinergic neurons of the VnGP (labeled by AChE staining) are heavily innervated by SP-positive puncta (BEACH et al. 1987; also shown by ultrastructural investigations, BOLAM et al. 1986). These SP terminals may arise from the CP since both light microscopic (GROVE et al. 1986) and physiological studies (BUCHWALD et al. 1986) suggest a CP-VnGP projection.

Various other types of studies may suggest connections, although they also cannot be considered definitive. These include physiological studies and the attempted localization of high affinity binding sites, usually by lesion studies (Table 4), as well as studies of the effect of one transmitter on the release or turnover of another in a particular brain region. Good examples of the latter type of study are the innumerable ones that have shown that DA agonists inhibit ACh release and increase ACh levels in the striatum, while DA antagonists have the opposite effects. Such studies have been taken to support the direct innervation of cholinergic neurons of the CP by nigral DA afferents. In recent work, it has been suggested that one striatal D_2 receptor regulates the metabolism of at least 400 molecules of ACh (KORF et al. 1985), and that the D_2 agonist effect in inhibiting striatal ACh release is suppressed by both D_1 agonists and antagonists (GORREL et al. 1986; GORELL and CZARNECKI 1986) and is highly dependent on DA concentrations (PARKER and CUBEDDU 1986). The innervation of striatal ACh neurons by cortical, as well as DA, afferents is supported by the depression of ACh synthesis seen in decorticated rats (CONSOLO et al. 1986).

The interaction between DA and ACh in the CP appears to be a mutual one with ACh agonists and antagonists acting to affect DA release; a synaptic dialog between DA nerve endings and ACh dendrites has frequently been suggested, with some evidence for ACh receptors on DA systems being derived from both lesion (McGEER et al. 1979) and electron microscopic studies (HAT-

Table 4. Some possible localizations of binding sites or receptors in the extrapyramidal system indicated by binding (B) or physiological studies (P)

Type of binding site (receptor)	Nucleus	Localization	Type of study	Reference
Glu/Asp	N. Ac	GABA neurons	P	Yang and Mogenson 1985
Glu	CP	Intrinsic neurons and DA terminals	B	Butcher et al. 1986
GABA	CP	Non-5-HT systems	P	Mennini et al. 1986
GABA	SN	Non-DA cells	B	Cross and Waddington 1980
Muscarinic	SN	DA neurons	B	Cross and Waddington 1980
D_2 [a]	Rat CP	Intrinsic neurons	B	Neve et al. 1984; Trugman et al. 1986
D_1 [a]	SN	Striatonigral afferent terminals	B	Porceddu et al. 1986
NA	N. Ac	Neurons receiving hippocampal input	P	Unemoto et al. 1985
ADO	CP	Intrinsic neurons	B	Geiger 1986
ENK	SN and CP	Non-DA systems	P	Foote and Maurer 1986
NT	SNC	DA neurons	P, B	Pinnock 1985; Singer 1984
NT	VTA	DA neurons	B, P	Glimsher et al. 1984; Singer 1984; Palacios and Kuhar 1981
NT	CP	DA afferents	B	Quirion et al. 1985
NT	N. Ac	DA and intrinsic neurons	B	Quirion et al. 1985

[a] Many articles have appeared on the localization of D_2 binding sites and DA-stimulated adenylate cyclase (index of the D_1 site) in rat CP as indicated by studies in lesioned animals. Most agree that the D_1 site is almost entirely on intrinsic neurons while the D_2 is on intrinsic neurons, and may or may not be on corticostriatal (Glu) and nigrostriatal (DA) afferents; this is discussed in the text where recent reviews are cited. Autoradiographic studies indicate that D_2 sites in the SN are largely in the SNC, while D_1 sites are largely in the SNR (Kohler and Radesater 1986).

tori et al. 1979). A notable difference between the N. Ac and the CP is that DA agonists decrease the release of GABA and ACh in CP slices, but only of GABA in slices of the accumbens (de Belleroche and Gardiner 1983).

Since DA agonists inhibit K^+-induced, Ca^{2+}-dependent Glu release in striatal slices (Mitchell and Doggett 1980) and Glu facilitates DA release (Butcher et al. 1986), the existence of DA receptors on Glu terminals and of Glu receptors on DA terminals has been suggested; both are supported to some extent by lesion studies (Table 4; J. Kornhuber and M. Kornhuber

1986), but these are controversial. The controversy with regard to the possible location of DA receptors on corticostriatal afferents (cf. PALACIOS 1986; TRUGMAN et al. 1986) emphasizes the need for definitive ultrastructural studies before interconnections are accepted.

Many such interactions between DA and various peptidergic systems of the CP have also been reported. For example, the content of SP/K in striatonigral neurons is increased by DA agonists (RITTER et al. 1984, 1985), and decreased by DA antagonists (OBLIN et al. 1984; BANNON et al. 1986b; LINDEFORS et al. 1986); the reduction in nigral SP induced by chronic impairment of DA transmission is reversed by the GABA analog, progabide (ZIVKOVIC et al. 1985). DA appears to have an opposite effect on NPY cells of the CP since there is a marked increase in the NPY staining in 6-OHDA-lesioned animals (KERKERIAN et al. 1986). Whether these changes are due to changes in the rate of peptide synthesis or metabolism is unknown, although the report that the mRNAs for striatal preprotachykinin are decreased by DA antagonists suggests an effect on synthesis (BANNON et al. 1986b). This effect occurs surprisingly rapidly; an analogous decrease in the mRNAs for striatal preproenkephalin is seen only after relatively long activation of GABA systems (8 days and not at 1-4 days) (SIVAM and HONG 1986).

These are only a few examples of the many reports of transmitter interactions in the extrapyramidal system. Such transmitter interactions may not involve direct connections and are often difficult to interpret in terms of other anatomic evidence. Figure 3 gives a summary diagram of some of the proven and more probable interconnections in the striatum and SN.

It should be pointed out that suggested interconnections are often of a axono-axonal or even dendroaxonal type. In the SN, there is considerable evidence that dendritically released DA can influence GABAergic activity, including GABAergic inhibition of both SNR (WASZCZAK and WALTERS 1986)

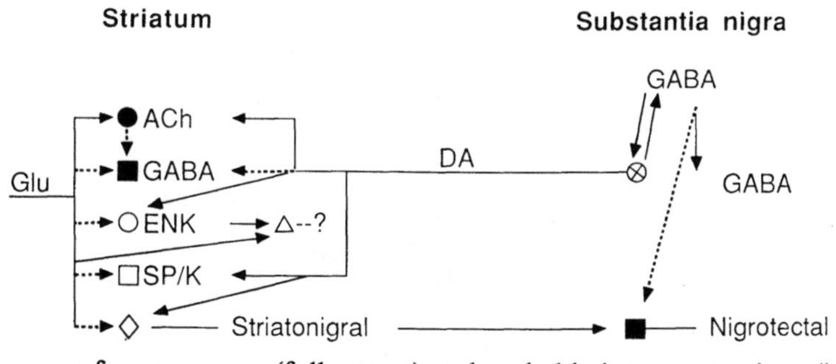

Fig. 3. Diagram of some proven (full arrows) and probable interconnections (broken arrows) in the striatum and substantia nigra. This is diagrammatic and not meant, for example, to imply that the same GABA neurons innervate DA and nigrotectal neurons in the SN, or that the ACh neurons innervating the GABA neurons in the CP are necessarily the same as the ACh neurons receiving DA innervation. The striatonigral neurons innervating nigrotectal neurons may be SP/K, DYN (PORCEDDU et al. 1983), ENK, or even GABA. The recently reported innervation of striatal ACh neurons by SP terminals is not included

and DA neurons, the latter possibly by a synaptic dialog such as that discussed for the DA-ACh interconnection in the CP (McGeer et al. 1979; Hattori et al. 1979).

E. Some Hypotheses as to the Role of Various Tracts in the Extrapyramidal System

In Sect. B.IA (Fig. 1), we have already proposed that the main messages in the extrapyramidal system are carried by Glu and GABA. In an overall hypothetical scheme (Fig. 4), which borrows heavily from Nauta (1979), action in the basal ganglia is initiated when a "command" for movement comes from the cortex via excitatory Glu pathways. The DA input to the CP/GP is a "hormonal" one which acts as a closed feedback loop, providing a damping of "static" so that the precise topography of the excitatory command can be appreciated. Parkinson's disease interferes with this loop. The inputs from the thalamus and many other regions are seen as being primarily a means of keeping the CP continually informed on the status of all the external systems involved in movement. Feedback loops are again involved, with the GP again playing a "central substation" role. The CP has an integrating role and feeds its results via both excitatory and inhibitory pathways to the SNR/EP where

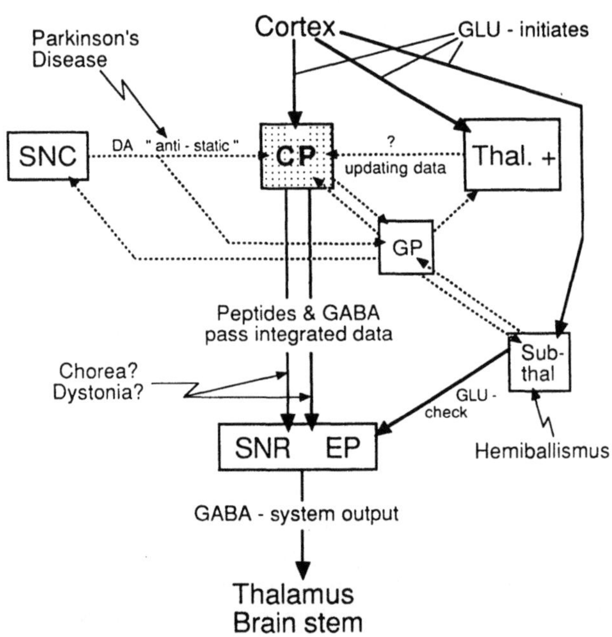

Fig. 4. A hypothetical diagram of information flow in the basal ganglia. Thal.$^+$ stands for all regions projecting to the CP except the cortex and SNC. *Broken arrows* indicate feedback loops

they are translated into excitation or inhibition of GABA pathways which carry the detailed orders contributing to the desired movement.

One feedback loop of particular interest may be that involving the subthalamus. The recent demonstration that the projection back to the GP uses Glu and has collaterals going to the SNR suggests that the subthalamus, like the CP, may serve as a station for passing integrated information to the SNR. Moreover, as ROUZAIRE-DUBOIS and SCARNATI (1985) point out, the subthalamus also receives an extensive excitatory input from the cortex, and hence is in an excellent position to compare direct cortical information with cortical input processed at the striatopallidal complex level; the subthalamo—SNR projection may therefore provide a check on the striatonigral input.

In this overall scheme, the connections of the SNR/EP become of paramount importance. In our previous review (MCGEER et al. 1984), we mentioned some of the pharmacological, physiological, and anatomical studies which support such importance, particularly of the nigrotectal and nigrotegmental pathways, and relevant literature has continued to appear. We will not belabor this point in this chapter, but rather consider what possible anatomic connections might be involved in the "antistatic" effect of the DA system, with particular relation to its failure in Parkinson's disease.

Figure 5 outlines a hypothetical scheme whereby activation of particular DA neurons linked to particular Glu inputs might result in excitation of topographically selected CP outputs. In this very crude scheme, the DA neuron (I) acts through D_1 receptors to inhibit the GABA neuron (I), and the ACh neuron (I), but through a D_2 receptor to activate the GABA neuron (II). The

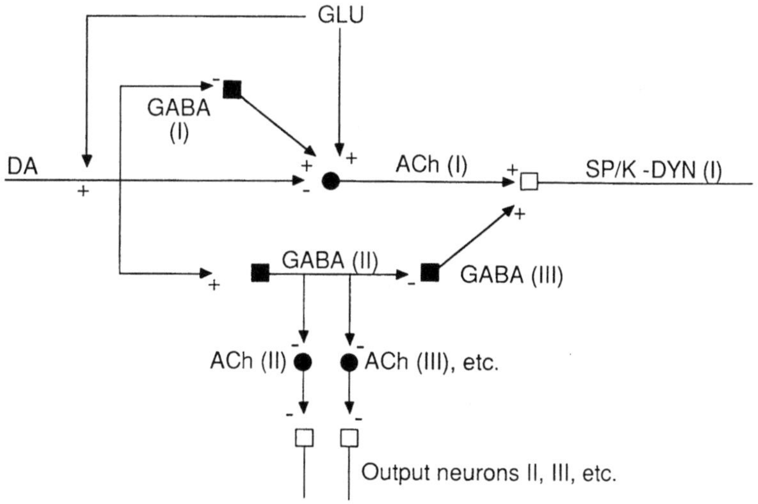

Fig. 5. A hypothetical scheme of connections by which dopamine neurons might exert an antistatic effect. The + and − signs indicate the supposed effect (+ being excitation or disinhibition and − being inhibition or a decrease in excitation) of stimulation of the Glu tract. ACh neurons II, III, etc., would, in turn, each be presumed to have a similar system of activating DA/$GABA_B$-Glu inputs. See text for details

GABA neuron (I) is supposed to exert a tonic inhibitory effect (through a $GABA_B$ receptor) on Glu release. Inhibition of the GABA neuron (I) therefore results in the release of enough Glu to overcome the minor inhibitory effect of the DA on the ACh neuron (I); excitation of this neuron leads to excitation of the SP/K-DYN output (I). Meanwhile, the excitatory action of DA on GABA neuron (II) through a D_2 site results in inhibition of other ACh neurons (II, III, etc.), resulting in decreased excitation of output neurons II, III, etc., and of the GABA neuron (III). This latter action reinforces the effect on output I since it is disinhibitory. Such an arrangement could lead to activation of only a selected output (I) and inhibition of others (II, III, etc.). DA agonists acting at both D_1 and D_2 sites would inhibit ACh release, but the effects on GABA release would be slight or mixed, as would the effects on SP/K-DYN (see Sect. D). This scheme is in accord with the report that the striatonigral SP system has opposite responses to D_1 and D_2 agonists (SONSALLA et al. 1984).

If the DA system failed, as in parkinsonism, movement commands would be lost unless "strong" enough to overcome the chronic $GABA_B$ inhibition. Giving a D_1 agonist would result in some excitation of the output neuron (I), but not as much as in the intact system, and would not result in the antistatic damping of other outputs. D_2 agonists would be more beneficial because the antistatic effect would be in operation and any Glu-induced excitation of output neuron (I) would be reinforced by disinhibition. This hypothetical scheme does not attempt to embrace all the neurotransmitters or intricacies, but is rather advanced as representing the type of complex diagrams that may have to be formulated as a basis for future work aimed at understanding the functioning of the extrapyramidal system.

F. Conclusions

Much progress has been made in identifying tracts and neurotransmitters of the extrapyramidal system, but these advances have opened new vistas of complexity, not only with regard to the number of transmitters and interconnected nuclei, but also with regard to factors such as subnuclear organization and species differences. Present indications are that the extrapyramidal system is involved in other functions, as well as in movement control, but the mechanisms of its role even in that one important, and long-recognized, function remain unclear.

G. Appendix: Afferents and Efferents of Extrapyramidal Nuclei

The diagrams in this section (Fig. 6-12) are largely updated versions of those given in McGEER et al. (1984), and only additional references will be cited here. In these diagrams, the topographical patterns and possible species differences are ignored and some of the very minor tracts are omitted. Abbreviations for transmitters are given in Table 1. CP caudate-putamen; EP entopeduncular nucleus (internal pallidum); GP globus pallidus (external pallidum); LC locus ceruleus and subceruleus; N. Ac nucleus accumbens; SN substantia nigra; RRA retrorubral area; VnGP ventral pallidum; VTA ventral tegmental (A10) area; X unknown transmitter. X is only used where there is definite evidence of an unknown transmitter in the projection. Where more than one transmitter is indicated in a projection, the minor ones are enclosed in parentheses. Small superscript numbers refer to the notes given after the figures. In these notes, the abbreviations for the neurotransmitters are also used as adjectives, e.g., DA dopaminergic; ACh cholinergic.

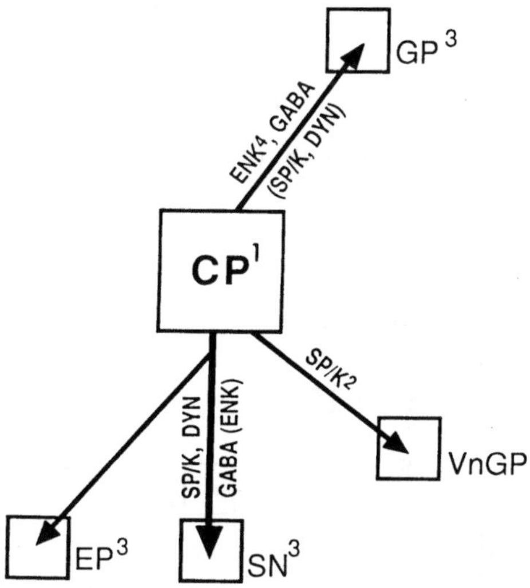

Fig. 6. Efferents of the caudate-putamen. (Notes see p. 134)

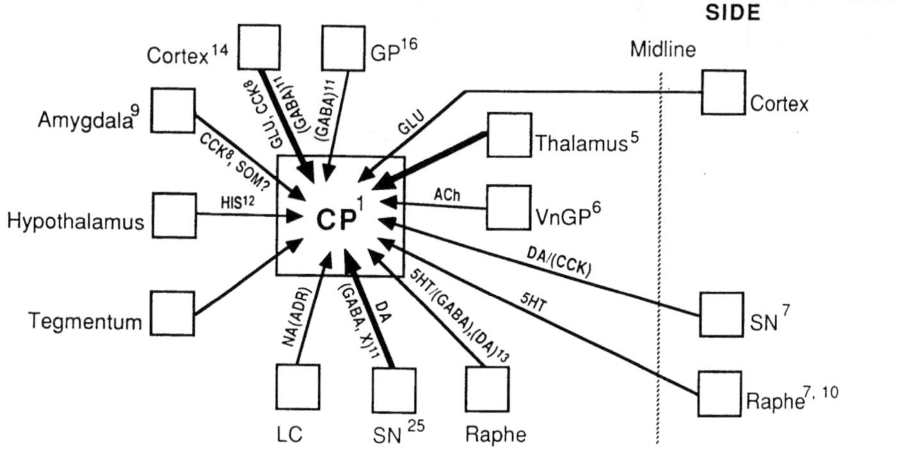

Fig. 7. Afferents to the caudate-putamen. (Notes see p. 134)

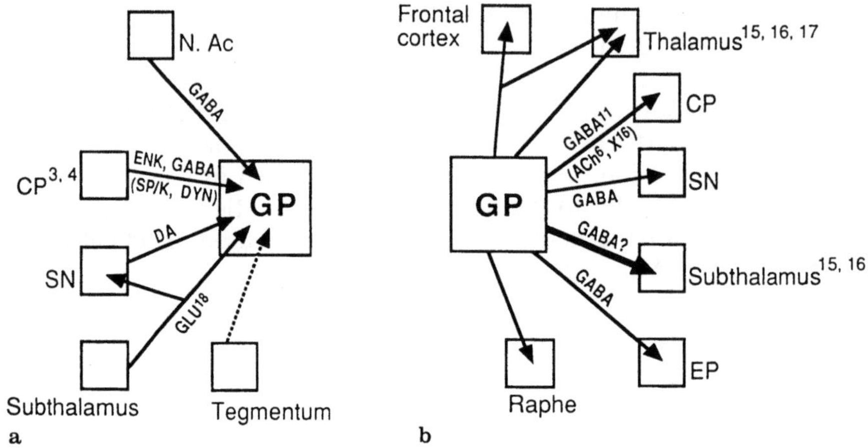

Fig. 8 a, b. Afferents (a) and efferents (b) of the globus pallidus. (Notes see p. 134)

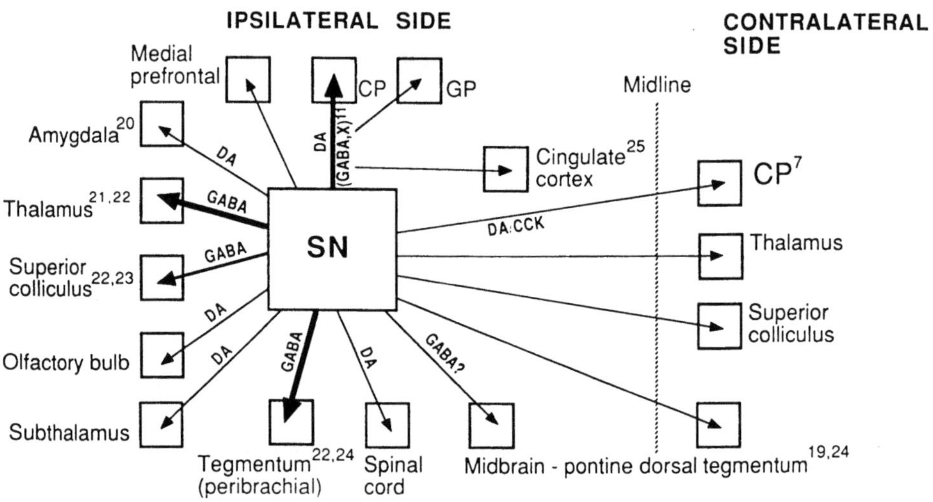

Fig. 9. Efferents of the substantia nigra[33]. (Notes see p. 134)

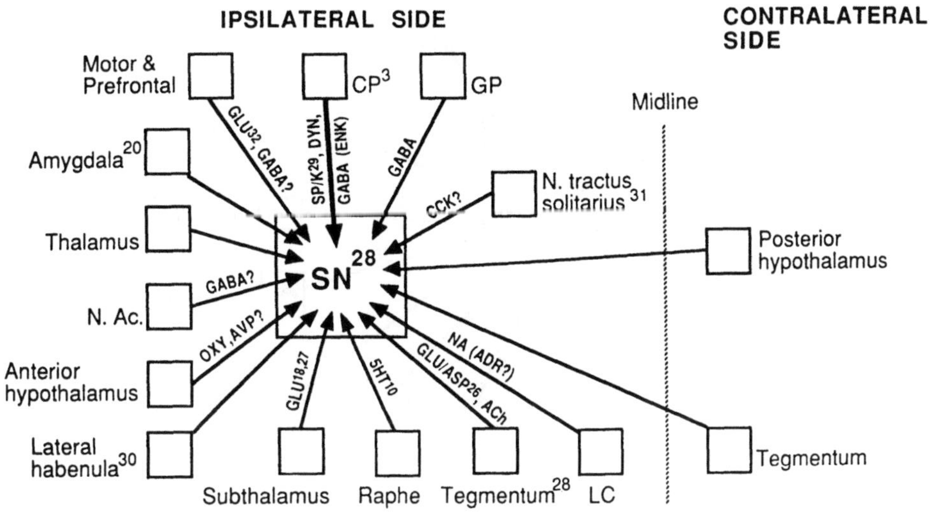

Fig. 10. Afferents to the substantia nigra[33]. (Notes see p. 134)

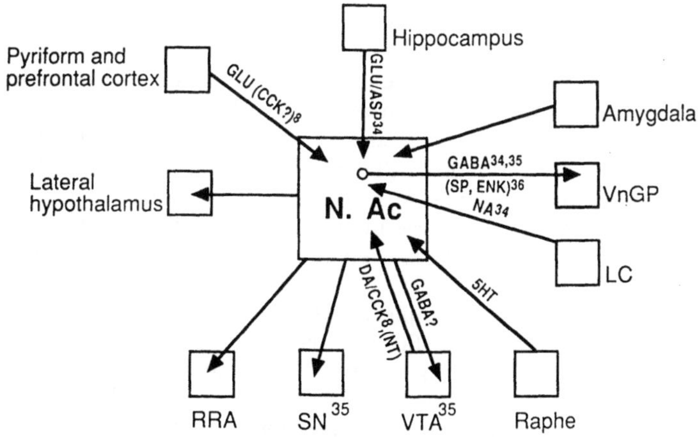

Fig. 11. Afferents and efferents of the nucleus accumbens. (Notes see p. 134)

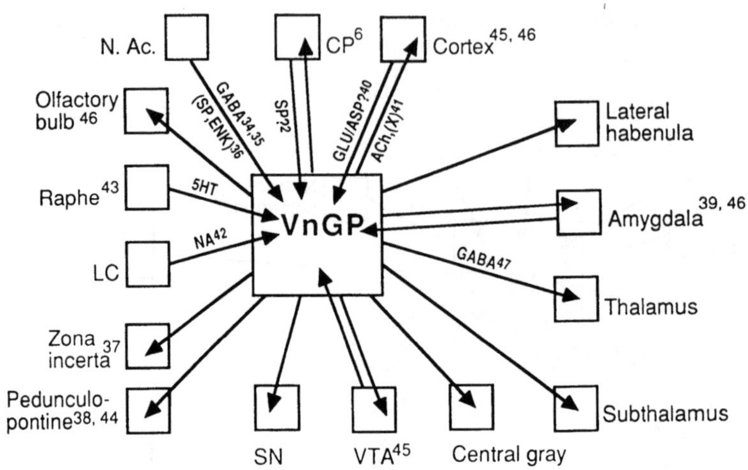

Fig. 12. Afferents and efferents of the ventral pallidum. (Notes see p. 134)

Notes on the Figures

[1] Interneurons in the CP have been found to use ACh, GABA, and NPY/SOM (e.g., see Chesselet and Graybiel 1986; Adrian et al. 1983). IB236-positive neurons (Bloom et al. 1985) and some CCK-positive aspiny neurons (Takagi et al. 1984) have been reported in rat CP, and these may be interneurons or projection neurons.

[2] A projection from caudate to VnGP has been shown (Grove et al. 1986; Buchwald et al. 1986) and SP-positive axons seem to innervate ACh neurons of the VnGP/basal forebrain in humans (Beach et al. 1987).

[3] Previously reported (in primates) differential distribution of ENK (mainly to GP) and SP/K-DYN (mainly to EP/SN) shown in the cat (Beckstead and Kersey 1985). The extent of collateralization is in question. For example, Loopuijt and van der Kooy (1985) report that about 40 % of rat CP neurons projecting either to the GP or SN project to both, while Parent et al. (1984) suggest that the striatopallidal and striatonigral projections are largely distinct in the monkey. Loss of SP and ENK in the SN in three cases of striatopallidal infarction is evidence of these tracts in humans (Pioro et al. 1984).

[4] ENK neurons in the CP make multiple contacts, usually symmetric, on dendrites and perikarya of CP and GP neurons (Somogyi et al. 1982).

[5] The major sources of thalamic afferents to the CP are the various subdivisions of the intralaminar nuclei, but recent studies have shown that the CP also receives projections from other thalamic nuclei. The medial geniculate projects to portions of the posterior CP and amygdala (LeDoux et al. 1986); the suprageniculate nucleus of the posterior thalamus projects selectively to the medial and intermediate regions of the caudate nucleus in cats (Hu and Jayaraman 1986); and the projection from the medial subdivision of the posterior group of the thalamus (POM) goes mainly to the lateral half of the CP at midrostral—caudal levels (Beckstead 1984). Retrograde tracing experiments indicate that single intralaminar thalamic neurons project to the cerebral cortex, CP, and nucleus reticularis thalami (NRT) in the rat. Some neurons were also doubly labeled for collaterals to the cortex-CP, CP-NRT, and cortex-NRT (Cesaro et al. 1985).

[6] Many neurons located in the GP and in the adjacent basal nucleus of Meynert (i.e., the VnGP) have branched axonal collaterals projecting to the neocortex and to the head of the caudate nucleus (Fisher et al. 1985). Retrograde tracing indicates a projec-

tion to the head of the caudate from the substantia innominata in monkeys (Arikuni and Kubota 1984), and a topographical projection of GP cells to the CP in cats with the morphology of some of the neurons resembling the ACh neurons of the nucleus basalis (Jayaraman 1983).

[7] Relatively small numbers of crossed projections to the CP come from cells which do not project ipsilaterally, but also occur in the SN and raphe (Fisher et al. 1984). Pritzel et al. (1983) say that the crossed nigrostriatal projection comes from a small percentage of neurons which bifurcate in the general region of the thalamus. In the rat, 93% of the minor (1%-2%) contralateral mesostriatal pathway is reportedly DA (cf. Consolazione et al. 1985), and 50% of the neurons providing this crossed pathway also contain CCK (Fallon et al. 1983).

[8] Midbrain CCK-containing cells supply the posteromedial parts of the N. Ac and olfactory tubercle, as well as the subcommissural part of the CP. They also reach more rostral parts of the ventral striatum, including the rostrolateral olfactory tubercle. A second major CCK pathway, with possible origin in the piriform and medial prefrontal cortices and/or the amygdala, projects to the subcommissural CP, the olfactory tubercle, the lateral part of the N. Ac, and the dorsal part of the bed nucleus of stria terminalis. Finally, the rostral part of the dorsal CP receives a substantial CCK innervation from the basal lateral amygdala, and possibly from the neocortex. According to radioimmunologic data, the descending telencephalic CCK system accounts for about 60% of all CCK in the rostral forebrain (Zaborszky et al. 1985). Lesion studies suggest that CCK/DA neurons of the ipsilateral VTA project to the frontal pole, piriform cortex, and N. Ac, but not to the CP. The CCK in the CP appears to arise from telencephalic structures (Giles et al. 1983).

[9] Retrograde tracing experiments indicate that the amygdala CP pathway is more extensive than previously believed, suggesting that the basolateral amygdala may serve as an important link between the limbic and extrapyramidal system. The basolateral amygdala projects to the dorsal lateral precommissural CP, characterized recently as a "nonlimbic" region of the neostriatum (Fass et al. 1984).

[10] Studies on the effects of habenular stimulation or application of various agents to the raphe suggest that the lateral habenula controls 5-HT transmission in the basal ganglia, and that this regulation may be different for those 5-HT neurons innervating the CP compared with those projecting to the SN (Reisine et al. 1982).

[11] Some lesion experiments suggest that tryptamine may be associated specifically with nigrostriatal neurons (Juorio and Greenshaw 1986). A small GABA projection from the SN to the CP has also been suggested (see Fig. 2), and Fisher et al. (1986) have used retrograde tracing plus GAD immunohistochemistry to indicate GABA inputs to the CP from the neocortex, SNR, and GP.

[12] Both HIS and non-HIS projections go from the posterior hypothalamus and premamillary region to the CP. About 20%-25% of the HIS neurons in the nuclei caudalis magnocellularis and caudalis magnocellularis postmamillaris innervate the CP in the rat (Steinbusch et al. 1986).

[13] Most, if not all, non-5-HT neurons projecting from the dorsal raphe to the CP appear to be DA (Descarries et al. 1986).

[14] The corticocaudate projections in the monkey are similar in laminar organization to the corticocortical connections (Arikuni and Kubota 1986).

[15] Van der Kooy and Kolb (1985) report the existence of a substantial population of small GP cells projecting to the cortex which are separate from the medium-size GP cells projecting to the subthalamus, and also distinct from the large ACh cells (VnGP) projecting to the cortex. They believe that the small cells are not ACh and some give off axonal collaterals to the paraventricular thalamic nucleus.

[16] Takada et al. (1986) report that a population of GP neurons, distinct from the more predominant group projecting to the subthalamus, sends collaterals to the CP and to the paraventricular nucleus of the thalamus; this subpopulation may also send afferents to the prefrontal/frontal cortex.
[17] Direct projections from the GP to the paraventricular nucleus of the thalamus in the rat have asymmetric axodendritic terminations (Sugimoto and Hattori 1984).
[18] Kitai and Kita (1986).
[19] Electrical stimulation of the rat SN has both excitatory and inhibitory effects on neurons in the medullary reticular formation (Duggal and Barasi 1983).
[20] Electrophysiological evidence for reciprocal neuronal connections between the SN and the amygdala in the rat (Barasi and Pay 1980).
[21] In the cat, neurons projecting to the posterior thalamic nuclei are restricted to the pars lateralis of the SN (Takada et al. 1984).
[22] The degree of collateralization of SNR output neurons was examined by retrograde transport studies using different dyes in the thalamus, the superior colliculus, and the reticular formation in both monkey and cat (Beckstead 1983; Parent et al. 1983).
[23] Electron microscopic studies in lesioned cats indicate that the termination of SNR afferents to the superior colliculus is predominantly in the stratum griseum intermediale with a few in the stratum griseum profundum (Warton et al. 1983).
[24] Physiological evidence that the nigrotegmental projection exerts an inhibitory influence on neurons located in a wide variety of nuclei of the midbrain—pontine dorsal tegmentum, including the pedunculopontine tegmental nucleus, the cuneiform nucleus, the central gray substance, the parabrachial nucleus, and the tegmentum between the inferior colliculus and the central gray (Noda and Oka 1984, 1986).
[25] Using a fluorescent retrograde double-labeling technique, Takada and Hattori (1986) found the existence of collateral SNC projections to the cingulate cortex and the CP in the rat. No labeled neurons were found in the VTA after tracer injections into the posterior cingulate cortex.
[26] Physiological evidence that the pedunculopontine—SNC path in the rat is excitatory (Scarnati et al. 1984) and probably uses Glu/Asp (Scarnati et al. 1986).
[27] The subthalamic projection to the SN forms asymmetric synapses on dendrites (Chang et al. 1984).
[28] Parent and deBellefeuille (1983) report tracing studies on efferents from the internal pallidum (EP) and afferents to the SN in the monkey; the EP does not project to the SN.
[29] This SP/K projection has been hypothesized as of primary importance in Parkinson's disease (Barker 1986), and a role in epilepsy has been suggested since antagonists in the SN are anticonvulsants (Garant et al. 1986).
[30] Stimulation of the lateral habenula inhibits DA neurons in both SN and VTA of the rat via a fasciculus retroflexus projection (Christoph et al. 1986).
[31] Hommer et al. (1985) suggest a tract from the nucleus tractus solitari to the SN, possibly involving CCK; such a projection might serve to modulate rhythmic (day/night) differences in motor activity (cf. McGeer et al. 1987).
[32] Further lesion evidence (Lindgren and Anden 1985).
[33] Afferents and efferents of the EP are similar to those of the SNR; projections which are probably GABAergic go to the tegmentum (Parent and De Bellefeuille 1983) and thalamus (Parent and De Bellefeuille 1983; Yamamoto et al. 1985), and a mixed GABA and GABA/SOM projection (Vincent and Brown 1986) goes to the habenula.
[34] Innervation of N. Ac to VnGP efferents by excitatory hippocampal afferents (Yang and Mogenson 1985) and of same neurons in N. Ac by hippocampal and LC afferents (Unemoto et al. 1985).

[35] Neurons projecting from the N.Ac to VTA-SN or VnGP are medium spiny neurons similar to those in the CP (CHANG and KITAI 1985). NT is found in some of the ascending projections (KALIVAS and MILLER 1984).
[36] Lesion studies suggest that SP and ENK may occur in the N.Ac-VnGP tract (HABER and NAUTA 1983).
[37] MOGENSON et al. (1985).
[38] SWANSON et al. (1984) provide physiological evidence for both excitatory and inhibitory inputs.
[39] ZABORSZKY et al. (1984).
[40] LEMANN and SAPER (1985) suggest an excitatory tract from cortex to VnGP.
[41] RYE et al. (1984).
[42] JACOBOWITZ and PALKOVITS (1974).
[43] PARENT et al. (1981).
[44] YANG and MOGENSON (1986).
[45] HARING and WANG (1986).
[46] WOOLF et al. (1986).
[47] YOUNG et al. (1984).

References

Adrian TE, Allen JM, Bloom SR, et al. (1983) Neuropeptide Y distribution in human brain. Nature 306:584-586

Altar CA, Boyar WC, Marien MR (1986) ^{125}I-LSD-Autoradiography confirms the preferential localization of caudate-putamen S_2 receptors to the caudal (peripallidal) region. Brain Res 372:130-136

Arikuni T, Kubota K (1984) Substantia innominata projection to caudate nucleus in macaque monkeys. Brain Res 302:184-189

Arikuni T, Kubota K (1986) The organization of prefrontocaudate projections and their laminar origin in the macaque monkey: a retrograde study using HRP. J. Comp Neurol 244:492-510

Aronin N, Difiglia M, Graveland GA, Schwartz WJ, Wu J-Y (1984) Localization of immunoreactive enkephalin in GABA synthesizing neurons of the rat neostriatum. Brain Res 300:376-380

Aronin N, Chase K, Difiglia M (1986) Glutamic acid decarboxylase and enkephalin immunoreactive axon terminals in the rat neostriatum synapse with striatonigral neurons. Brain Res 365:151:158

Azmitia E, Gannon P (1986) The primate serotonergic system: a review of human and animal studies and a report on *Macaca fascicularis*. Adv Neurol 43:407-468

Bannon MJ, Deutch AY, Tam S-Y et al. (1986a) Mild footshock stress dissociates substance P from substance K and dynorphin from Met- and Leu-enkephalin. Brain Res 381:393-396

Bannon MJ, Lee JM, Giraud P, Young A, Affolter HU, Bonner TI (1986b) Dopamine antagonist haloperidol decreases substance P, substance K and preprotachykinin mRNAs in rat striatonigral neurons. J Biol Chem 261:6640-6642

Barasi S, Pay S (1980) Electrophysiological investigation of the connection between the substantia nigra and the amygdala in rat. Neurosci Lett 17:265-269

Barker R (1986) Substance P and Parkinson's disease: a causal relationship? J. Theor Biol 120:353-362

Battaini F, Govoni S, Di Giovine S et al. (1986) Neurotensin effect on dopamine release and calcium transport in rat striatum: interactions with diphenylalkylamine calcium antagonists. Naunyn Schmiedebergs Arch Pharmacol 332:267-270

Beach TG, McGeer EG (1984) The distribution of substance P in the primate basal ganglia: an immunohistochemical study of baboon and human brain. Neuroscience 13:29-52

Beach TG, Tago H, McGeer EG (1987) Light microscopic evidence for a substance P-containing innervation of the human nucleus basalis of Meynert. Brain Res 408:251-257

Beal MF, Martin JB (1985) Regional catecholamine distribution and turnover in rat striatum. Brain Res 358:10-15

Beal MF, Chattha GK, Martin JB (1986) A comparison of regional somatostatin and neuropeptide Y distribution in rat striatum and brain. Brain Res 377:240-245

Beckstead RM (1983) Long collateral branches of substantia nigra pars reticulata axons to thalamus, superior colliculus and reticular formation in monkey and cat. Multiple retrograde neuronal labeling with fluorescent dyes. Neuroscience 10:767-779

Beckstead RM (1984) A projection to the striatum from the medial subdivision of the posterior group of the thalamus in the cat. Brain Res 300:351-356

Beckstead RM (1985) Complementary mosaic distributions of thalamic and nigral axons in the caudate nucleus of the cat: double anterograde labeling combining autoradiography and wheat germ-HRP histochemistry. Brain Res 335:153-159

Beckstead RM, Kersey KS (1985) Immunohistochemical demonstration of differential substance P-, met-enkephalin-, and glutamic-acid-decarboxylase-containing cell body and axon distributions in the corpus striatum of the cat. J Comp Neurol 232:481-498

Belin MF, Nanopoulos D, Didier M et al. (1983) Immunohistochemical evidence for the presence of γ-aminobutyric acid and serotonin in one nerve cell. A study on the raphe nuclei of the rat using antibodies to glutamate decarboxylase and serotonin. Brain Res 275:329-339

Besson M-J, Graybiel AM, Quinn B (1986) Coexistence of dynorphin B-like and substance P-like immunoreactivity in striatal neurons in the cat. Soc Neurosci Abstr 12:876

Bloom F, Battenberg E, Milner R, Sutcliffe J (1985) Immunocytochemical mapping of IB236, a brain-specific neuronal polypeptide deduced from the sequence of a cloned mRNA. J Neurosci 5:1781-1802

Bolam JP, Ingham CA, Izzo PN et al. (1986) Substance P-containing terminals in synaptic contact with cholinergic neurons in the neostriatum and basal forebrain; a double immunocytochemical study in the rat. Brain Res 397:179-289

Bouyer JJ, Park DH, Joh TH, Pickel VM (1984a) Chemical and structural analysis of the relationship between cortical inputs and tyrosine hydroxylase-containing terminals in rat neostriatum. Brain Res 302:267-275

Bouyer JJ, Miller RJ, Pickel VM (1984b) Ultrastructural relation between cortical efferents and terminals containing enkephalin-like immunoreactivity in rat neostriatum. Regul Pept. 8:105-115

Brand S (1980) A comparison of the distribution of acetylcholinesterase and muscarinic cholinergic receptors in the feline neostriatum. Neurosci Lett 17:113-117

Brodin E, Lindefors N, Dalsgaard CJ, Theodorsson-Norheim E, Rosell S (1986) Tachykinin multiplicity in rat central nervous system as studied using antisera raised against substance P and neurokinin A. Regul Pept 13:253:272

Buchwald NA, Fisher RS, Ceped C, Hull CD, Levine MS (1986) Neurophysiological and morphological studies of nucleus basalis of Meynert neurons in the cat. XXX congress IUPS, Vancouver, Abstr 10:511.16

Burton K, Farrell K, Li D, Calne DB (1984) Lesions of the putamen and dystonia: CT and magnetic resonance imaging. Neurology (NY) 34:962-965

Butcher SP, Roberts PJ, Collins J (1986) The distribution and function of DL-[^{3}H] 2-amino-4-phosphonobutyrate binding sites in the rat striatum. Brain Res 381:305-313

Carlson J, Heimer L (1986) The projection for the parataenial thalamic nucleus, as demonstrated by the *Phaseolus vulgaris*-leucoagglutinin (PHA-L) method, identifies a subterritorial organization of the ventral striatum. Brain Res 374:375-379

Cesaro P, Nguyen-Legros J, Pollin B, LaPlante S (1985) Single intralaminar thalamic neurons project to cerebral cortex, striatum and nucleus reticularis thalami. A retrograde anatomical tracing study in the rat. Brain Res 325:29-37

Chang HT, Kitai ST (1985) Projection neurons of the nucleus accumbens: an intracellular labeling study. Brain Res 347:112-116

Chang HT, Kita J, Kitai ST (1984) The ultrastructural morphology of the subthalamic-nigral axon terminals intracellularly labeled with horseradish peroxidase. Brain Res 299:182-185

Chesselet MF, Graybiel AM (1986) Striatal neurons expressing somatostatin-like immunoreactivity: evidence for a peptidergic interneuronal system in the cat. Neuroscience 17:547-571

Christie MJ, Beart PM, Jarrott B, Maccarrone C (1986) Distribution of neuropeptide Y in the rat basal ganglia; effect of excitotoxin lesions to caudate-putamen. Neurosci Lett 63:305-309

Christoph G, Leonzio R, Wilcox K (1986) Stimulation of the lateral habenula inhibits dopamine-containing neurons in the substantia nigra and ventral tegmental area of the rat. J Neurosci 6:613-619

Consolazione A, Bentivoglio M, Goldstein M, Toffano G (1985) Evidence for crossed catecholaminergic nigrostriatal projections by combining wheat germ agglutinin-horseradish peroxidase retrograde transport and tyrosine hydroxylase immunocytochemistry. Brain Res 338:140-143

Consolo S., Sieklucka M, Florentini F, Forloni G, Ladinsky H (1986) Frontal decortication and adaptive changes in striatal cholinergic neurons in the rat. Brain Res 363:128-134

Crawley JN, Hommer DW, Skirboll LR (1984) Behavioral and neurophysiological evidence for a facilatory interaction between co-existing transmitters: cholecystokinin and dopamine. Neurochem. Int 6:755-760

Cross AJ, Waddington JL (1980) [^{3}H] Quinuclidinyl benzylate and [^{3}H] GABA receptor binding in rat substantia nigra after 6-hydroxydopamine lesions. Neurosci Lett 17:271-275

Cumming P, von Krosigk M, Reiner PB, McGeer EG, Vincent SR (1986) Absence of adrenaline neurons in the guinea pig brain: a combined immunohistochemical and high-performance liquid chromatographic study. Neurosci Lett 63:125-130

de Belleroche J, Gardiner IM (1983) Action of apomorphine, bromocriptine and lergotrile on γ-aminobutyric acid and acetylcholine release in nucleus accumbens and corpus striatum. J Neural Transm 58:153-168

Descarries L, Berthelet F, Garica S et al. (1986) Dopaminergic projection from nucleus raphe dorsalis to neostriatum in the rat. J Comp Neurol 249:511-520

Draper DD, Rothschild MF, Beitz DC, Christian LL (1984) Age- and genotype-dependent differences in catcholamine concentrations in the porcine caudate nucleus. Exp Gerontol 19:377-381

Dubois A, Scatton B (1985) Heterogeneous distribution of dopamine D_2 receptors within the rat striatum as revealed by autoradiography of [^{3}H] *N-n*-propylnorapomorphine binding sites. Neurosci Lett 57:7-12
Duggal KN, Barasi S (1983) Investigation of the connection between the substantia nigra and the medullary reticular formation in the rat. Neurosci Lett 36:237-242
Fallon JF, Wang C, Kim Y, Canepa N, Loughlin J, Seroogy K (1983) Dopamine- and cholecystokinin-containing neurons of the crossed mesostriatal projection. Neurosci Lett 40:233-238
Fass B, Talbot K, Butcher LL (1984) Evidence that efferents from the basolateral amygdala innervate the dorsolateral neostriatum in rats. Neurosci Lett 44:71-75
Faull RLM, Villinger JW (1986) Heterogeneous distribution of benzodiazepine receptors in the human striatum: a quantitative autoradiographic study comparing the pattern of receptor labeling with the distribution of acetylcholinesterase staining. Brain Res 381:153-158
Fisher RS, Shiota C, Levine MS, Hull CF, Buchwald NA (1984) Subcortical crossed axonal projections to the caudate nucleus of the cat: a double-labelling study. Neurosci Lett 51:25-30
Fisher RS, Boylan MK, Hull CD, Buchwald NA, Levine MS (1985) Branched projections of pallidal and peripallidal neurons to neocortex and neostriatum: a double-labeling study in the cat. Brain Res 326:156-159
Fisher RS, Levine MS, Hull CD, Buchwald NA (1986) Extrinsic origin of GABAergic inputs to the neostriatum. XXX congress IUPS, Vancouver, Abstr no. 511.15
Foote RW, Maurer R (1986) Opioid binding sites in the nigrostriatal system of the guinea pig are not located on nigral dopaminergic neurons. Neurosci Lett 65:341-345
Freed CR, Yamamoto BK (1985) Regional brain dopaminergic metabolism: a marker for the speed, direction, and posture of moving animals. Science 229:62-65
Freund T, Powell J, Smith A (1984) Tyrosine hydroxylase-immunoreactive boutons in synaptic contact with identified striatonigral neurons, with particular reference to dendritic spines. Neuroscience 13:1189-1215
Fukui K, Kariyama H, Kashiba A, Kato N, Kimura H (1986) Further confirmation of heterogeneity of the rat striatum: different mosaic patterns of dopamine fibers after administration of methamphetamine or reserpine. Brain Res 382:81-86
Garant D, Iadarola M, Gale K (1986) Substance P antagonists in substantia nigra are anticonvulsant. Brain Res 382:372-378
Gaspar P, Berger B, Gay M et al. (1983) Tyrosine hydroxylase and methionine-enkephalin in the human mesencephalon. Immunocytochemical localization and relationships. J Neurol Sci 58:247-267
Geiger JD (1986) Localization of [^{3}H] cyclohexyladenosine and [^{3}H] nitrobenzylthioinosine binding sites in rat striatum and superior colliculus. Brain Res 363:404-408
Gerfen CR (1984) The neostriatal mosaic: compartmentalization of corticostriatal input and striatonigral output systems. Nature 311:461-463
Gerfen CR (1985) The neostriatal mosaic: compartmental distribution of calcium-binding protein and parvalbumin in the basal ganglia of the rat and monkey. Proc Natl Acad Sci USA 82:8780-8784
Gerfen CR (1986) Compartmentation in the basal ganglia. 2nd Triennial symposium, International Basal Ganglia Society, Victoria, BC
Giles C, Lotstra F, Vanderhaeghen J-J (1983) CCK nerve terminals in the rat striatal and limbic areas originate partly in the brain stem and partly in telencephalic structures. Life Sci 32:1683-1690

Girault JA, Spampinato U, Desban M et al. (1986) Enhancement of glutamate release in the rat striatum following electrical stimulation of the nigrothalamic pathway. Brain Res 374:362-366

Glimsher PW, Margolin DH, Giovino AA, Hoebel BG (1984) Neurotensin: a new "reward peptide". Brain Res 291:119-124

Goedert M, Mantyh PW, Emson PC, Hunt SP (1984) Inverse relationship between neurotensin receptors and neurotensin-like immunoreactivity in cat striatum. Nature 307:543-546

Gorell JM, Czarnecki B (1986) Pharmacologic evidence for direct dopaminergic regulation of striatal acetylcholine release. Life Sci 38:2239-2246

Gorell JM, Czarnecki B, Hubbel S (1986) Functional antagonism of D_1 and D_2 dopaminergic mechanisms affecting striatal acetylcholine release. Life Sci 38:2247-2254

Graveland GA, Difiglia M (1985) The frequency and distribution of medium-sized neurons with indented nuclei in the primate and rodent neostriatum. Brain Res 327:307-311

Graybiel AM (1983) Compartmental organization of the mammalian striatum. Prog Brain Res 58:247-256

Graybiel AM (1986) Neuropeptides in the basal ganglia. Res Publ Assoc Res Nerv Ment Dis 64:135-161

Graybiel AM, Ragsdale CW (1978) Histochemically distinct compartments in the striatum of human, monkey, and cat demonstrated by acetylthiocholinesterase staining. Proc Natl Acad Sci USA 75:5723-5726

Graybiel AM, Ragsdale CW, Yoneoka ES, Elde RP (1981) An immunohistochemical study on enkephalins and other neuropeptides in the striatum of the cat with evidence that the opiate peptides are arranged to form mosaic patterns in register with the striosomal compartments visible by acetylcholinesterase staining. Neuroscience 6:377-397

Graybiel AM, Chesselet M-F, Wu J-Y, Eckenstein F, Joh TE (1983) The relation of striosomes in the caudate nucleus of the cat to the organization of early-developing dopaminergic fibers, GAD-positive neurons and CAT-positive neurons. Soc Neurosci Abstr 9:14

Grove EA, Domesick VB, Nauta WJH (1986) Light microscopic evidence of striatal input to intrapallidal neurons of cholinergic cell group Ch4 in the rat: a study employing the anterograde tracer *Phaseolus vulgaris* leucoagglutinin (PHA-L). Brain Res 367:379-384

Haber SN, Nauta WJ (1983) Ramifications of the globus pallidus in the rat as indicated by patterns of immunohistochemistry. Neuroscience 9:245-260

Haber SN, Watson SJ (1985) The comparative distribution of enkephalin, dynorphin and substance P in the human globus pallidus and basal forebrain. Neuroscience 14:1011-1024

Haring JH, Wang RY (1986) The identification of some sources of afferent input to the rat nucleus basalis magnocellularis by retrograde transport of horseradish peroxidase. Brain Res 366:152-158

Hartgraves SL, Kelly PH (1985) Role of mesencephalic glycine in locomotor activity. Neurosci Lett 62:175-180

Hattori T, Fibiger HC, McGeer PL (1975) Demonstration of a pallidonigral projection innervating dopaminergic neurons. J Comp Neurol 162:487-504

Hattori T, Singh VK, McGeer PL (1976) Immunohistochemical localization of choline acetyltransferase-containing neostriatal neurons and their relationship with dopaminergic synapses. Brain Res 102:164-173

Hattori T, McGeer PL, McGeer EG (1979) Dendroaxonic transmission. II. Morphological sites for the synthesis, binding and release of neurotransmitters in dopaminergic dendrites in the substantia nigra and cholinergic dendrites in the neostriatum. Brain Res 170:71-83

Herkenham M, Pert CB (1981) Mosaic distribution of opiate receptors, parafascicular projections and acetylcholinesterase in rat striatum. Nature 291:415-417

Hokfelt T, Skirboll L, Rehfeld JF, Goldstein M, Markey K, Dann O (1980) A subpopulation of mesencephalic dopamine neurons projecting to limbic areas contains a cholecystokinin like peptide: evidence from immunohistochemistry combined with retrograde tracing. Neuroscience 5:2093-2124

Hokfelt T, Everitt BJ, Theodorsson-Norheim E, Goldstein M (1984) Occurrence of neurotensinlike immunoreactivity in subpopulations of hypothalamic, mesencephalic, and meduallary catecholamine neurons. J Comp Neurol 222:543-559

Holstein GR, Pasik P, Hamori J (1986) Synapses between GABA-immunoreactive axonal and dendritic elements in monkey substantia nigra. Neurosci Lett 66:316-322

Hommer DW, Palkovits M, Crawley JN, Paul SM, Skirboll LR (1985) Cholecystokinin-induced excitation in the substantia nigra; evidence for peripheral and central components. J Neurosci 5:1387-1392

Hu H, Jayaraman A (1986) The projection pattern of the suprageniculate nucleus to the caudate nucleus in cats. Brain Res 368:201-203

Illing RB, Graybiel AM (1985) Convergence of afferents from cortex and substantia nigra onto acetylcholinesterase-rich patches in superior colliculus. Neuroscience 14:455-482

Illing RB, Graybiel AM (1986) Complementary and non-matching afferent compartments in the cat's superior colliculus: innervation of the acetylcholinesterase-poor domain of the intermediate gray layer. Neuroscience 18:373-394

Inagaki S, Parent A (1984) Distribution of substance P and enkephalin-like immunoreactivity in the substantia nigra of rat, cat and monkey. Brain Res Bull 13:319-329

Ingham CA, Bolam JP, Smith AT (1986) Glutamate decarboxylate immunoreactive terminals in synaptic contact with basal forebrain neurons that project to neocortex. Neurosci Lett [Suppl] 24:S9

Innis RB, Andrade R, Aghajanian GK (1985) Substance K excites dopaminergic and non-dopaminergic neurons in rat substantia nigra. Brain Res 335:381-383

Jacobowitz DM, Palkovits M (1974) Topographic atlas of catecholamine and acetylcholinesterase-containing neurons in the rat brain. J Comp Neurol 157:12-28

Jayaraman A (1983) Topographic organization and morphology of peripallidal and pallidal cells projecting to the striatum in cats. Brain Res 275:279-286

Jiminez-Castellanos J, Graybiel AM (1985) The dopamine-containing innervation of striosomes; nigral subsystems and their striatal correspondents. Soc Neurosci Abstr 11:1249

Jiminez-Castellanos J, Graybiel AM (1986) Innervation of striosomes and extrastriosomal matrix by different subdivisions of the midbrain A8-A9-A10 dopamine-containing cell complex Soc Neurosci Abstr 12:1327

Jones RS (1982) Tryptamine: a neuromodulator or neurotransmitter in mammalian brain? Prog Neurobiol 19:117-139

Joseph SA, Pilcher WH, Knogge KM (1985) Anatomy of the corticotropin-releasing factor and opiomelanocortin systems of the brain. Fed Proc 44:100-107

Joyce JN, Loeschen SK, Marshall JF (1985) Dopamine D_2 receptors in rat caudate-putamen: the lateral to medial gradient does not correspond to dopaminergic innervation. Brain Res 338:209-218

Juorio AV, Greenshaw AJ (1986) Tryptamine depletion in the rat striatum following electrolytic lesions of the substantia nigra. Brain Res 371:385-389

Kalivas PW, Miller JS (1984) Neurotensin neurons in the ventral tegmental area project to the medial nucleus accumbens. Brain Res 300:157-160

Kalivas PW, Taylor S (1985) Behavioral and neurochemical effects of daily injections of neurotensin in the ventral tegmental area. Brain Res 358:70-76

Katsuura G, Itoh S, Hsiao S (1985) Specificity of nucleus accumbens to activities related to cholecystokinins in rats. Peptides (Fayetteville) 6:91-96

Kerkerian L, Bosler O, Pelletier G, Nieoullon A (1986) Striatal neuropeptide Y neurons are under the influence of the nigrostriatal dopaminergic pathway: immunohistochemical evidence. Neurosci Lett 66:106-112

Kitai ST, Kita H (1986) Subthalamic neurons and their afferents to the globus pallidus and the substantia nigra in the rat; anatomical and electrophysiological study. 2nd Triennial symp, International Basal Ganglia Society, Victoria, BC

Kohler C, Radesater AC (1986) Autoradiographic visualization of dopamine D_2 receptors in the monkey brain using the selective benzamide drug [^{3}H] raclopride. Neurosci Lett 66:85-90

Korf J, Sebens JB, Flentge F, van der Werf JF (1985) Occupation of dopamine receptors by *N-n*-propylnorapomorphine or spiperone and acetylcholine levels in the rat striatum. J Neurochem 44:314-318

Kornhuber J, Kornhuber M (1986) Presynaptic dopaminergic modulation of cortical input to the striatum. Life Sci 39:669-674

Kubota Y, Inagaki S, Kito S (1986a) Innervation of substance P neurons by catecholaminergic terminals in the neostriatum. Brain Res 375:163-167

Kubota Y, Inagaki S, Kito S, Takagi H, Smith AD (1986b) Ultrastructural evidence of dopaminergic input to enkephalinergic neurons in rat neostriatum. Brain Res 367:374-378

LeDoux JE, Sakaguchi A, Iwata J, Reis DJ (1986) Interruption of projections from the medial geniculate body to an archi-neostriatal field disrupts the classical conditioning of emotional responses to acoustic stimuli. Neuroscience 17:615-627

Lee JM, McLean S, Maggio JE et al. (1986) The localization and characterization of substance P and substance K in striatonigral neurons. Brain Res 371:152-154

Lemann W, Saper CB (1985) Evidence for a cortical projection to the magnocellular basal nucleus in the rat: an electron microscopic axonal transport study. Brain Res 334:339-343

Lindefors N, Brodin E, Ungerstedt U (1986) Neuroleptic treatment induces region-specific changes in levels of neurokinin A and substance P in rat brain. Neuropeptides 7:265-280

Lindgren S, Anden N-E (1985) Effect of the normal nerve impulse flow on the synthesis and utilization of GABA in the rat substantia nigra. J Neural Transm 61:21-34

Loopuijt LD, van der Kooy D (1985) Organization of the striatum; collateralization of its efferent axons. Brain Res 348:86-90

Mantyh PW, Hunt SP (1986) Changes in [^{3}H] substance P receptor binding in the rat brain after kainic acid lesion of the corpus striatum. J Neurosci 6:1537-1544

Markstein R, Hokfelt T (1984) Effect of cholecystokinin-octapeptide on dopamine release from slices of cat caudate nucleus. J Neurosci 4:570-575

Marshall JF, Van Oordt K, Kozlowski MR (1983) Acetylcholinesterase associated with dopaminergic innervation of the neostriatum: histochemical observations of a heterogeneous distribution. Brain Res 274:283-289

McCormack JK, Beitz AJ, Larson AA (1986) Autoradiographic localization of tryptamine binding sites in the rat and dog central nervous system. J Neurosci 6:94-101
McGeer EG, Staines WA, McGeer PL (1984) Neurotransmitters in the basal ganglia Can J Neurol Sci 11:89-99
McGeer EG, Norman M, Boyes B, O'Kusky J, Suzuki J, McGeer PL (1985) Acetylcholine and aromatic amine systems in postmortem brain of an infant with Down's syndrome. Exp Neurol 87:557-570
McGeer PL, McGeer EG, Innanen VT (1979) Dendroaxonic transmission. I. Evidence from receptor binding of dopaminergic and cholinergic agents. Brain Res 169:433-441
McGeer PL, Eccles JC, McGeer EG (1987) Molecular Neurobiology of the Mammalian Brain, 2nd ed. Plenum, New York
Melis MR, Gale K (1984) Evidence that nigral substance P controls the activity of the nigrotectal GABAergic pathway. Brain Res 295:389-393
Mennini T, Gobbi M, Romndini S (1986) Localization of GABA-A and GABA-B receptor subtypes on serotonergic neurons. Brain Res 371:372-375
Meyer G, Wahle P (1986) The olfactory tubercle of the cat. I. Morphological components. Exp Brain Res 62:515-527
Mitchell PR, Doggett NS (1980) Modulation of striatal [^{3}H]glutamic acid release by dopaminergic drugs. Life Sci 26:2073-2081
Mogenson GJ, Swanson LW, Wu M (1985) Evidence that projections from substantia innominata to zona incerta and mesencephalic locomotor region contribute to locomotor activity. Brain Res 334:65-76
Mori S, Ueda S, Yamada H, Takino T, Sano Y (1985) Immunohistochemical demonstration of serotonin nerve fibers in the corpus striatum of the rat, cat and monkey. Anat Embryol (Berl) 173:1-5
Nastuk M, Graybiel M (1985) Patterns of muscarinic cholinergic binding in the striatum and their relation to dopamine islands and striosomes. J Comp Neurol 237:176-194
Nauta HJW (1979) A proposed conceptual reorganization of the basal ganglia and telencephalon. Neuroscience 4:1875-1881
Nawa H, Doteuchi M, Igano K, Inouye K, Nakanmishi S (1984) Substance K: a novel mammalian tachykinin that differs from substance P in its pharmacological profile. Life Sci 34:1153-1160
Neve KA, Altar CA, Wong CA, Marshall JF (1984) Quantitative analysis of [^{3}H]spiroperidol binding to rat forebrain sections: plasticity of neostriatal dopamine receptors after nigrostriatal injury. Brain Res 302:9-18
Noda T, Oka H (1984) Nigral inputs to the pedunculopontine region: intracellular analysis. Brain Res 322:332-336
Noda T, Oka H (1986) Distribution and morphology of tegmental neurons receiving nigral inhibitory inputs in the cat: an intracellular HRP study. J Comp Neurol 244:254-266
Oblin A, Zivkovic B, Bartholini G (1984) Involvement of the D_2 dopamine receptor in the neuroleptic-induced decrease in nigral substance P. Eur J Pharmacol 105:175-177
Oertel WH, Riethmuller G, Mugnaini E et al. (1983) Opioid peptide-like immunoreactivity localized in gabaergic neurons of rat neostriatum and central amygdaloid nucleus. Life Sci 33(Suppl 1):73-76
Oertel WH, Nitsch C, Mugnaini E (1984) Immunocytochemical demonstration of the GABA-ergic neurons in rat globus pallidus and nucleus entopeduncularis and their GABA-ergic innervation. Adv Neurol 40:91-98.

Palacios JM (1986) Dopamine receptor disputes. Nature 323:205
Palacios JM, Kuhar MJ (1981) Neurotensin receptors are located on dopamine containing neurons in rat midbrain. Nature 294:587-589
Parent A, DeBellefeuille L (1983) The pallidointralaminar and pallidonigral projections in primate as studied by retrograde double-labeling method. Brain Res 278:11-27
Parent A, Descarries L, Beaudet A (1981) Organization of ascending serotonin systems in the adult rat brain. A radiographic study after intraventricular administration of [^{3}H] 5-hydroxytryptamine. Neuroscience 6:115-138
Parent A, Mackey A, Smith Y, Boucher R (1983) The output organization of the substantia nigra in primate as revealed by a retrograde double labeling method. Brain Res Bull 10:529-537
Parent A, Bouchard C, Smith Y (1984) The striatopallidal and striatonigral projections: two distinct fiber systems in primate. Brain Res 303:385-390
Parker E, Cubeddu L (1986) Effects of *d*-amphetamine and dopamine synthesis inhibitors on dopamine and acetylcholine neurotransmission in the striatum. J Pharmacol Exp Ther 237:179-203
Pasik P, Pasik T, Pecci-Saavedra J, Holstein G, Yahr M (1984) Serotonin in pallidal neuronal circuits: an immunocytochemical study in monkeys. Adv Neurol 40:63-76
Penny GR, Afsharpour S, Kitai ST (1986) The glutamate decarboxylase-, leucine enkephalin-, methionine enkephalin- and substance P-immunoreactive neurons in the neostriatum of the rat and cat: evidence for partial population overlap. Neuroscience 17:1011-1045
Phelps PE, Houser CR, Vaughn JE (1985) Immunocytochemical localization of choline acetyltransferase within the rat neostriatum: a correlated light and electron microscopic study of cholinergic neurons and synapses. Neurology (NY) 238:286-307
Pioro EPJ, Hughes JT, Cuello AC (1984) Loss of substance P and enkephalin immunoreactivity in the human substantia nigra after striato-pallidal infarction. Brain Res 292:339-347
Pinnock RD (1985) Neurotensin depolarizes substantia nigra dopamine neurones. Brain Res 338:151-154
Porceddu ML, Imperato A, Melis MR, Di Chiara G (1983) Role of ventral mesencephalic reticular formation and related noradrenergic and serotonergic bundles in turning behavior as investigated by means of kainate, 6-hydroxydopamine and 5,7-dihydroxytryptamine lesions. Brain Res 262:187-200
Porceddu ML, Giogi O, Ongini E et al. (1986) [^{3}H] SCH 23 390 binding sites in the rat substantia nigra: evidence for a presynaptic localization and innervation by dopamine. Life Sci 39:321-328
Pritzel M, Sarter M, Morgan S, Huston JP (1983) Interhemispheric nigrostriatal projections in the rat: bifurcating nigral projections and loci of crossing in the diencephalon. Brain Res Bull 10:385-390
Quirion R, Dam TV (1985) Multiple tachykinin receptors in guinea pig brain. High densities of substance K (neurokinin A) binding sites in the substantia nigra. Neuropeptides 6:191-204
Quirion R, Chiueh CC, Everist HD, Pert A (1985) Comparative localization of neurotensin receptors on nigrostriatal and mesolimbic dopaminergic terminals. Brain Res 327:385-389
Rea MA, Simon JR (1981) Regional distribution of cholinergic parameters within the rat striatum. Brain Res 219:317-326

Reisine TD, Soubrié P, Artaud F, Glowinski J (1982) Involvement of lateral habenula-dorsal raphe neurons in the differential regulation of striatal and nigral serotonergic transmission in cats. J Neurosci 2:1062-1071

Ritter JK, Schmidt CJ, Gibb JW, Hanson GR (1984) Increases of substance P-like immunoreactivity within striatal-nigral structures after subacute methamphetamine treatment. J Pharmacol Exp Ther 229:487-492

Ritter JK, Schmidt CJ, Gibb JW, Hanson GR (1985) Dopamine-mediated increases in nigral substance P-like immunoreactivity. Biochem Pharmacol 34:3161-3166

Ross CD, Godfrey DA (1985) Distributions of choline acetyltransferase and acetylcholinesterase activities in layers of rat superior colliculus. J Histochem Cytochem 33:631-641

Rouzaire-Dubois B, Scarnati E (1985) Bilateral corticosubthalamic projections: an electrophysiological study in rats with chronic cerebral lesions. Neuroscience 15:69-79

Rye DB, Wainer BH, Mesulam M-M, Mufson EJ, Saper CB (1984) Cortical projections arising from the basal forebrain: a study of cholinergic and noncholinergic components employing combined retrograde tracing and immunohistochemical localization of choline acetyltransferase. Neuroscience 13:627-643

Sandell JH, Graybiel AM, Chesselet M-F (1986) A new enzyme marker for striatal compartmentalization: NADPH diaphorase activity in the caudate nucleus and putamen of the cat. Neurology (NY) 243:326-334

Savasta M, Dubois A, Scatton B (1986) Autoradiographic localization of D_1 dopamine receptors in the rat brain with [^{3}H] SCH 23 390. Brain Res 375:291-301

Scarnati E, Campana E, Pacitti C (1984) Pedunculopontine-evoked excitation of substantia nigra neurons in the rat. Brain Res 304:351-361

Scarnati E, Proia A, Campana E, Pacitti C (1986) A microiontophoretic study on the nature of the putative synaptic neurotransmitter involved in the pedunculopontine-substantia nigra pars compacta excitatory pathway of the rat. Exp Brain Res 62:470-478

Shivers B, Harlan R, Romano G, Howells R, Pfaff D (1986) Cellular localization of proenkephalin mRNA in rat brain: gene expression in the caudate-putamen and cerebellar cortex. Proc Natl Acad Sci USA 83:6221-6225

Singer E (1984) Zur Wechselwirkung der Neuropeptide Substanz P und Neurotensin mit zentralen Monoamintransmittersystemen. Wien Klin Wochenschr 96:809-816

Sivam S, Hong J-S (1986) GABAergic regulation of enkephalin in rat striatum: alterations in Met^5-enkephalin level, precursor content and preproenkephalin messenger RNA abundance. J Pharmacol Exp Ther 37:326-331

Skofitsch G, Jacobowitz DM (1985a) Autoradiographic distribution of ^{125}I calcitonin gene-related peptide binding sites in the rat central nervous system. Peptides (Fayetteville) 4:975-986

Skofitsch G, Jacobowitz DM (1985b) Immunohistochemical mapping of galanin-like neurons in the rat central nervous system. Peptides (Fayetteville) 6:509-546

Smith Y, Parent A (1986a) Neuropeptide Y-immunoreactive neurons in the striatum of cat and monkey; morphological characteristics, intrinsic organization and co-localization with somatostatin. Brain Res 372:241-252

Smith Y, Parent A (1986b) Differential connections of caudate nucleus and putamen in the squirrel monkey *(Saimiri sciureus)*. Neuroscience 18:347-371

Somogyi P, Prestley JV, Cuello AC, Smith AD, Takagi H (1982) Synaptic connections of enkephalin-immunoreactive nerve terminals in the neostriatum: a correlated light and electron microscopic study. J Neurocytol 11:779-807

Sonsalla P, Gibb J, Hanson G (1984) Opposite responses in the striato-nigral substance P system to D_1 and D_2 receptor activation. Eur J Pharmacol 105:185–187

Steinbusch HWM, Sauren Y, Groenewegen H, Watanabe T, Mulder AH (1986) Histaminergic projections from the premammillary and posterior hypothalamic region to the caudate-putamen complex in the rat. Brain Res 368:389–393

Studler JM, Reibaud M, Tramu G, Blanc G, Glowinski J, Tassin JP (1985) Distinct properties of cholecystokinin-8 and mixed dopamine-cholecystokinin-8 neurons innervating the nucleus accumbens. Ann NY Acad Sci 457:306–314

Sugimoto T, Hattori T (1984) Direct projections from the globus pallidus to the paraventricular nucleus of the thalamus in the rat. Brain Res 323:188–192

Suzuki H, Ghatei MA, Reynolds GP, Anand P, Bloom SR (1986) Regional distribution of a novel peptide (P7 of IB236) immunoreactivity in the human central nervous system. Neurosci Lett 67:58–62

Swanson LW, Mogenson GJ, Gerfen CR, Robinson P (1984) Evidence for a projection from the lateral preoptic area and substantia innominata to the "Mesencephalic locomotor region" in the rat. Brain Res 295:161–178

Taber-Pierce E, Lichtenstein E, Feldman SC (1985) The somatostatin systems of the guinea-pig brainstem. Neuroscience 15:215–235

Takada M, Hattori T (1986) Collateral projections from the substantia nigra to the cingulate cortex and striatum in the rat. Brain Res 380:331–335

Takada M, Itoh K, Yasui Y, Sugimoto T, Mizuno N (1984) Direct projections from the substantia nigra to the posterior thalamic regions in the cat. Brain Res 309:143–146

Takada M, Ng G, Hattori T (1986) Single pallidal neurons project both to the striatum and thalamus in the rat. Neurosci Lett 69:217–220

Takagi H, Mizuta H, Matsuda T, Inagaki S, Tateishi K, Hamaoka T (1984) The occurrence of cholecystokinin-like immunoreactive neurons in the rat neostriatum: light and electron microscopic analysis. Brain Res 309:346–349

Totterdell S, Bolam J, Smith A (1984) Characterization of pallidonigral neurons in the rat by a combination of Golgi impregnation and retrograde transport of horseradish peroxidase: their monosynaptic input from the neostriatum. J Neurocytol 13:593–616

Trugman J, Geary WII, Wooten G (1986) Localization of D_2 dopamine receptors to intrinsic striatal neurones by quantitative autoradiography. Nature 323:267–269

Unemoto H, Sasa M, Takaori S (1985) A noradrenaline-elicited inhibition from locus coeruleus of nucleus accumbens neuron receiving input from hippocampus. Jpn J Pharmacol 39:233–239

Upchurch M (1985) Evidence for species differences between rats and gerbils in striatal dopamine content and dopamine metabolism. Neurosci Lett 59:159–163

Van den Pol AN, Smith AD, Powell JF (1985) GABA axons in synaptic contact with dopamine neurons in the substantia nigra: double immunocytochemistry with biotin-peroxidase and protein A-colloidal gold. Brain Res 348:146–154

Van der Kooy D, Kolb B (1985) Non-cholinergic globus pallidus cells that project to the cortex but not to the subthalamic nucleus in the rat. Neurosci Lett 57:113–118

Vincent SR, Brown JC (1986) Somatostatin immunoreactivity in the entopeduncular projection to the lateral habenula in the rat. Neurosci Lett 68:160–164

Vincent SR, Johansson O (1983) Striatal neurons containing both somatostatin and avian pancreatic polypeptide (APP)-like immunoreactivities and NADPH-diaphorase activity: a light and electron microscopic study. J Comp Neurol 217:264–270

Voigt MM, Wang RY, Westfall TC (1985) The effects of cholecystokinin on the in vivo release of newly synthesized [^{3}H] dopamine from the nucleus accumbens of the rat. J Neurosci 5:2744-2749
Wahle P, Meyer G (1986) The olfactory tubercle of the cat. II. Immunohistochemical compartmentation. Exp Brain Res 62:528-540
Warton S, Jones DG, Ilinsky IA, Kultas-Ilinsky K (1983) Nigra and cerebellar synaptic terminals in the intermediate and deep layers of the cat superior colliculus revealed by lesioning studies. Neuroscience 10:789-800
Waszczak BL, Walters JR (1986) Endogenous dopamine can modulate inhibition of substantia nigra pars reticulata neurons elicited by GABA iontophoresis or striatal stimulation. J Neurosci 6:120-126
Widmann R, Sperk G (1986) Topographical distribution of amines and major amine metabolites in the rat striatum. Brain Res 367:244-249
Wiener SL (1986) Laminar distribution and patchiness of cytochrome oxidase in mouse superior colliculus. J Comp Neurol 244:137-148
Williams MN, Faull RLM (1985) The striatonigral projection and nigrotectal neurons in the rat, a correlated light and electron microscopic study demonstrating a monosynaptic striatal input to identified nigrotectal neurons using a combined degeneration and horseradish peroxidase procedure. Neuroscience 14:991-1010
Woolf NJ, Hernit MC, Butcher LL (1986) Cholinergic and non-cholinergic projections from the rat basal forebrain revealed by combined choline acetyltransferase and *Phaseolus vulgaris* leucoagglutinin immunohistochemistry. Neurosci Lett 66:281-286
Yamamoto T, Samejima A, Oka H (1985) An intracellular analysis of the entopeduncular inputs on the centrum medianum-parafascicular nuclear complex in cats. Brain Res 348:343-347
Yang CR, Mogenson GJ (1985) An electrophysiological study of the neural projections from the hippocampus to the ventral pallidum and the subpallidal areas by way of the nucleus accumbens. Neuroscience 15:1015-1024
Yang CR, Mogenson GJ (1986) Hippocampal signal transmission to the mesencephalic motor region (MRL). XXX congress IUPS, Vancouver, Abstr no. 511.04
Young WS III, Alheid GF, Heimer L (1984) The ventral pallidal projection to the mediodorsal thalamus: a study with fluorescent retrograde tracers and immunohistofluorescence. J Neurosci 4:1626-1638
Zaborszky L, Leranth CS, Heimer L (1984) Ultrastructural evidence on amygdalofugal axons terminating on cholinergic cells of the rostral forebrain. Neurosci Lett 52:219-225
Zaborszky L, Alheid B, Beinfeld MC, Eiden LE, Heimer L, Palkovits M (1985) Cholecystokinin innervation of the ventral striatum: a morphological and radioimmunological study. Neuroscience 14:427-453
Zahm DS, Heimer L (1985) Synaptic contacts of ventral striatal cells in the olfactory tubercle of the rat: correlated light and electron microscopy of anterogradely transported *Phaseolus vulgaris*-leucoagglutinin. Neurosci Lett 60:169-175
Zahm DS, Zaborszky L, Alones VE, Heimer L (1985) Evidence for the coexistence of glutamic decarboxylase and Met-enkephalin immunoreactivities in axon terminals of rat ventral pallidum. Brain Res 325:317-321
Zivkovic B, Oblin A, Bartholini G (1985) Progabide reverses the nigral substance P reduction induced by chronic impairment of dopaminergic transmission. Eur J Pharmacol 112:253-255

CHAPTER 4

Receptors in the Basal Ganglia

A. B. YOUNG and J. B. PENNEY

A. Introduction

The study of both presynaptic and postsynaptic markers of neurotransmitter systems in the basal ganglia has afforded a great deal of information about the function of specific neuronal pathways in health and disease. In Parkinson's disease, the emphasis has been on the study of dopamine presynaptic and postsynaptic elements. Other neurotransmitter abnormalities, however, have been observed in this and other basal ganglia diseases. This chapter will focus on the normal distribution of neurotransmitter receptors in human basal ganglia and the receptor abnormalities observed in Parkinson's disease.

B. Neurotransmitter Receptor Properties

A receptor is defined as the macromolecule in the neuronal membrane with which a hormone or neurotransmitter substance interacts to initiate the transduction of either a biochemical or electophysiologic signal in the receiving cell. The interaction of a hormone or neurotransmitter with its receptor does not result in any degradation or modification of the signaling agent, as would occur with the substrate for an enzyme reaction. Basically, receptor—ligand interactions can be measured in two different ways. One way is direct measurement of the biologic response (electrical, behavioral, or biochemical) to a receptor agonist or antagonist ligand. The other way is direct measurement of the binding of the ligand to a biologic preparation. By either method a ligand—receptor interaction should be defined by its saturability, pharmacologic and physiologic specificity, and its reversibility (YOUNG et al. 1985). Since the receptor molecules mediate the effects of the neurohormone or neurotransmitter, it is important to study changes in these receptor molecules in various disease states in order to understand the pathophysiology of the disorders.

Traditional biochemical measures of neurotransmitter receptors have examined the binding of high specific activity radiolabeled drugs or ligands to membrane homogenates from various regions of the brain or other tissues (YOUNG et al. 1985). Membrane fractions are incubated in various buffer solutions in the presence of low concentrations of radioactive compounds. After the binding of the ligand with the receptor has equilibrated, the membranes are separated from the rest of the incubation medium, either by rapid filtration, rapid centrifugation, or by chromatography. In all binding assays, the ionic conditions in the buffer medium, the temperature of the assays, the presence of various enzyme inhibitors, and the actual membrane concentration can greatly affect the magnitude and extent of binding. In assessing studies of receptors in animal or human tissues, it is necessary to compare the various measurement techniques before coming to a conclusion. In addition, there are very few pharmacologic agents which interact solely with one type of neurotransmitter receptor. It is therefore necessary to carry out detailed pharmacological studies to demonstrate that the measured binding in a particular assay does in fact represent a selective group of binding sites.

Autoradiographic techniques have been developed to look at both the biochemistry and pharmacology of receptor binding, as well as the anatomy and localization of the binding sites (YOUNG and KUHAR 1979; HERKENHAM and PERT 1981; PENNEY et al. 1981). These methods have allowed measurements of receptors in very small regions of the brain that could not be measured accurately by homogenate techniques.

C. Dopamine Receptors

I. Properties and Subtypes (Table 1)

The interaction of dopamine with its receptors has been measured by behavioral, electrophysiological, and biochemical techniques. Behavioral studies have assessed the effect of drugs on rotation after unilateral nigrostriatal lesions, or on locomotor behaviors and stereotypies induced by drugs (UNGERSTEDT 1971; PIJNENBERG et al. 1975; QUINTON and HALLIWELL 1963; RANDUP et al. 1963; UNGERSTEDT 1979). These behavioral models allow the screening of large numbers of agonists and antagonists with relative ease.

Table 1. Dopamine receptor properties

Property	D_1 receptor	D_2 receptor
Effect on adenylate cyclase	Stimulates	Inhibits
Number of affinity states	2	2
Affinity for high affinity state		
Dopamine	n*M*	n*M*
Ergot agonists	Inhibit cyclase	n*M*
Phenothiazines	n*M*	p*M*
Thioxanthenes	n*M*	n*M*
Benzamides	>10µ*M*	n*M*
Butyrophenones	µ*M*	p*M*
SKF-38393	n*M*	µ*M*
SCH-23390	n*M*	µ*M*
Effect of guanine nucleotides	Decreases DA affinity	Decreases DA affinity

The first biochemical measures of the dopamine receptor examined the effects of dopaminergic drugs on cyclic AMP formation in striatal slices (KEBABIAN et al. 1972). Dopamine was found to increase cyclic AMP by stimulating adenylate cyclase. Although this stimulation was blocked by dopamine antagonists, the pharmacology of dopamine's effects in this assay differed substantially from the pharmacology of the dopaminergic drugs in rotation studies. Furthermore, the clinical potency of various antischizophrenic drugs was found to correlate well with their potency in the rotation assay and their affinity in membrane binding assays, but not as well with their potency in blocking activation of adenylate cyclase (SEEMAN et al. 1975, 1976; CREESE et al. 1976, 1977).

Subsequently, membrane binding studies using radioactively labeled dopamine agonists and antagonists have demonstrated two types of binding sites in striatal membranes (GRIGORIADIS and SEEMAN 1984; LEFF and CREESE 1983; STOOF and KEBABIAN 1984). One site has a pharmacology very similar to that of the dopamine-stimulated adenylate cyclase, and is called the D_1 receptor subtype (KEBABIAN and CALNE 1979). The other binding site has a pharmacology resembling the pharmacology of neuroleptic potency in the treatment of schizophrenia and in modifying rotational behaviors in animals. This latter site is called the D_2 receptor subtype (GRIGORIADIS and SEEMAN 1984; LEFF and CREESE 1983; STOOF and KEBABIAN 1984). Recent studies suggest that agonist stimulation of the D_2 receptor actually results in the inhibition of adenylate cyclase in striatal and pituitary membranes (GRIGORIADIS and SEEMAN 1984; LEFF and CREESE 1983; ONALI et al. 1985). Some have thought that the two receptors occur together in the same membrane, one to stimulate and the other to inhibit adenylate cyclase. Lesion studies, however, have suggested that at least some D_1 receptors are located on presynaptic terminals whereas D_2 receptors appear to exist postsynaptically in the striatum, but also on the somata and dendrites of substantia nigra dopamine neurons (DUBOIS et al. 1986; DAWSON et al. 1986; FILLOUX et al. 1986; HERRERA-MARSCHITZ and UNGERSTED 1984; SPANO et al. 1977). For several years, investigators believed there were two additional binding sites, which were termed the D_3 and D_4 binding sites. However, it is currently believed that these two binding sites merely represented high affinity states of the D_1 and D_2 receptors (LEFF and CREESE 1983; GRIGORIADIS and SEEMAN 1984).

The behavioral consequences of D_1 receptor-induced activation of adenylate cyclase are poorly defined. However, they appear to have many similarities to those of D_2 activation (although important differences also exist) (BARONE et al. 1986; CHRISTENSEN et al. 1984). With the recent development of selective D_1 agonists and antagonists, specific D_1-induced behaviors can now be defined. The low affinity state of the D_1 receptor has micromolar affinity for dopamine and the high affinity state has nanomolar affinity for dopamine (LEFF and CREESE 1983; GRIGORIADIS and SEEMAN 1984). Thioxanthenes (such as *cis*-flupenthixol) have nanomolar affinity for blocking the D_1 receptor, whereas butyrophenones (such as haloperidol and spiroperidol) have micromolar affinity for blocking the D_1 receptor, while benzamides (such as sulpiride) are nearly inactive (GRIGORIADIS and SEEMAN 1984; STOOF and KEBABIAN

1984). Specific antagonists at the D_1 site include SCH-23 390 and SKF-83 566 (Hyttel et al. 1983; Stoof and Kebabian 1984). Selective D_1 agonists include dihydroxynomifensine and SKF-38 393. Labeled SCH-23 390 appears to be the most specific D_1 receptor ligand currently available (Anderson et al. 1985; Hyttel 1983).

D_2 receptors have very high affinity (nanomolar range) for all neuroleptics (butyrophenones, benzamides, and phenothiazines) (Leff and Creese 1983; Grigoriadis and Seeman 1984). The high affinity state of the D_2 receptor has nanomolar affinity for dopamine, and the low affinity state, micromolar affinity (Grigoriadis and Seeman 1984). The D_2 receptor inhibits adenylate cyclase in the pituitary and striatum (Leff and Creese 1983; Grigoriadis and Seeman 1984). Drugs that act at the D_2 site result in marked behavioral effects in animals (rotation, locomotion, antiparkinsonian action, psychotomimetic action, emesis, and stereotypy) (Creese et al. 1983). In humans, antagonists at the D_2 site are potent antipsychotic and antiemetic agents and produce signs of parkinsonism. The most potent D_2 antagonists include the butyrophenones (haloperidol, spiroperidol) and the benzamides (sulpiride, tiapride) (Stoof and Kebabian 1984). The most potent agonists at the D_2 site are the ergot derivatives, pergolide, lisuride, and LY171 555 (Stoof and Kebabian 1984; Titus et al. 1983). In binding studies, [^{3}H] spiroperidol is commonly used to measure D_2 receptors.

II. Distribution

In rodents, D_1 receptors are highest in the striatum and the nucleus accumbens (Dawson et al. 1985; Schulz et al. 1985; Richfield et al. 1987b). Receptors are also relatively high in the entopeduncular nucleus (the murine equivalent of the medial segment of the globus pallidus), ventral pallidum, and substantia nigra (particularly in the pars reticulata). In the cerebral cortex, particularly frontal regions, D_1 receptors are quite rich in the inner layers (V, VI). Moderate densities of D_1 receptors are seen in the amygdala and the hypothalamus.

D_2 receptors are fewer by about 66% than D_1 receptors in all neostriatal regions (caudate/putamen, nucleus accumbens, olfactory tubercle) (Richfield et al. 1987a, b). Nevertheless, they are higher in the neostriatum than in other areas. In the substantia nigra, D_2 receptors are higher in the pars compacta than in the pars reticulata. Very few D_2 receptors are present in pallidal regions (most of them in the lateral globus pallidus) or in the cerebral cortex.

In humans and primates, D_1 receptors are distributed in a fashion similar to that in rodents (Fig. 1; Richfield et al. 1987b; J. Palacios 1986, personal communication). D_1 receptors are very high in the caudate, putamen, nucleus accumbens, and olfactory tubercle. They are also high in the substantia nigra pars reticulata and the medial globus pallidus. The lateral globus pallidus and substantia nigra pars compacta have very few D_1 receptors. In the cerebral cortex, they predominate in both the outer and inner layers of various regions, particularly the prefrontal areas.

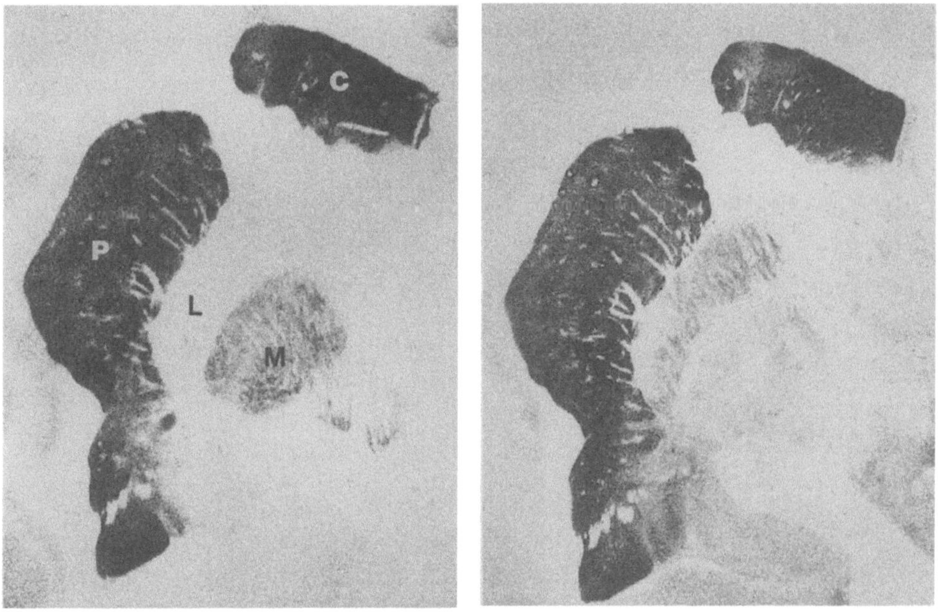

Fig. 1 a, b. Distribution of dopamine D_1 (**a**) and D_2 (**b**) receptors in basal ganglia of the monkey, as revealed by receptor autoradiography. D_1 receptors were imaged with 0.57 n*M* [^{3}H]SCH-23390. D_2 receptors were imaged with 0.25 n*M* [^{3}H]spiroperidol in the presence of 100 n*M* mianserin to block binding to serotonin receptors. Both receptor subtypes are highest in the caudate nucleus (C) and putamen (P) where there are twice as many D_1 as D_2 receptors. The density of both subtypes varies by about 15% within these structures. In addition, D_1 receptors are present in the medial globus pallidus (M) while D_2 receptors are present in the lateral globus pallidus (L)

D_2 receptors in humans and primates are present in levels about one-third those of D_1 receptors in the neostriatum (caudate, putamen, nucleus accumbens, and olfactory tubercle) (Fig. 1 and 2). The D_2 receptors are present in moderate density in the substantia nigra pars compacta where D_1 receptors are very low. In primates and humans, D_2 receptors are observed in mild to moderate amounts in the lateral globus pallidus, but not in the medial globus pallidus. Depending on the ligand used, there have been variable numbers of D_2 receptors observed in the cerebral cortex.

Inhomogeneities in the striatal distribution of both D_1 and D_2 receptors have been observed in cat, primate, and human brains (Richfield et al. 1987a, b; J. Palacios 1986, personal communication). However, the areas of densest binding seen with D_2 receptors do not correspond one for one with the areas of increased D_1 receptor binding. In adults, neither D_1 nor D_2 receptor inhomogeneities appear to correspond to striosomes as defined by acetylcholinesterase staining.

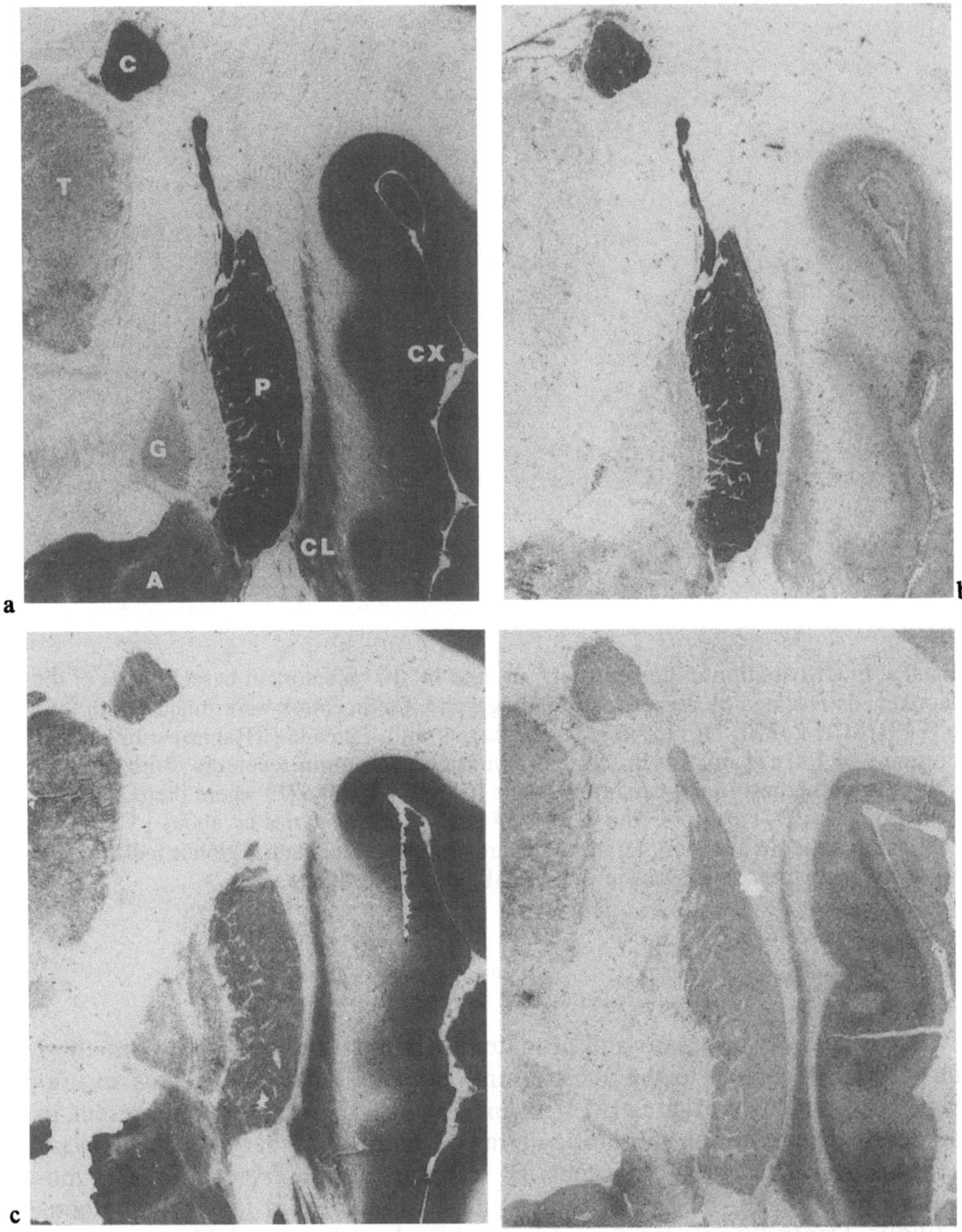

Fig. 2a–d. Distribution of neurotransmitter receptors in human basal ganglia as revealed by receptor autoradiography

a muscarinic cholinergic receptors imaged with 1 n*M* [^{3}H]quinuclidynyl benzilate. These receptors are very dense in the caudate (C), putamen (P), outer layers of the cortex (CX), claustrum (CL), and amygdala (A) and low in the globus pallidus (G) and thalamus (T),

b dopamine D_2 receptors imaged with 0.25 n*M* [^{3}H]spiroperidol in the presence of 100 n*M* mianserin. These receptors are densest in the caudate and putamen with very

III. Dopamine Receptors in Parkinson's Disease

Numerous homogenate studies of dopamine receptors in Parkinson's disease brains have been reported (GUTTMAN and SEEMAN 1985; GUTTMAN et al. 1985; LEE et al. 1978; PIMOULE et al. 1985; QUIKO et al. 1979; REISINE et al. 1977; RINNE 1982; RINNE et al. 1981, 1985). However, at times the studies have been conflicting, presumably because of the differences in assay technique, as well as differences in antemortem therapy. Although the issue of postmortem delay is important in measurements of all types of receptors, the D_1 and D_2 receptors do not appear to be particularly influenced by this factor.

In the initial homogenate studies by LEE et al (1978), D_2 receptors appeared to be substantially increased in untreated Parkinson's disease patients as compared with controls. In contrast, in treated patients, these receptors actually appeared to be reduced compared with controls. These D_2 receptor changes were present both in the caudate nucleus and in the putamen. D_1 receptors as measured with apomorphine, a somewhat nonselective ligand, appeared to be decreased in the Parkinson's disease caudate and putamen (LEE et al. 1978).

More recent studies of dopamine receptors in Parkinson's disease have confirmed the increases of D_2 receptors in untreated patients (GUTTMAN and SEEMAN 1985). In treated patients, D_2 receptors are normal (GUTTMAN et al. 1985; GUTTMAN and SEEMAN 1985). Unfortunately, most of the D_1 ligands that have been used interact with both D_1 and D_2 receptors (LEE et al. 1978; RINNE et al. 1985). Measurements of dopamine-stimulated adenylate cyclase in Parkinson's disease brain have also been conflicting (NAGATSU et al. 1978; SHIBUYA 1979).

The newly developed autoradiographic studies of dopamine receptors in Parkinson's disease patients have only been carried out in brains from treated patients. Studies using SCH-23 390 have indicated no apparent D_1 receptor changes in treated Parkinson's disease patients (PIMOULE et al. 1985; J. PALACIOS 1986, personal communication). No differences in D_2 receptors have been observed in these treated patients.

---

few receptors present elsewhere. Inhomogeneities of 15% less binding are present in the caudate and putamen,

c benzodiazepine receptors imaged with 8 nM [^{3}H] flunitrazepam. These receptors are densest in the middle layers of the cortex. Density is high in the cortex, putamen, claustrum and thalamus. Receptors are present at a lower density in the globus pallidus,

d μ opiate receptors imaged with 8 nM [^{3}H] naloxone. These receptors are dense in the amygdala, claustrum, and layers IV and VI of the cortex. Density is low in the caudate, putamen, and thalamus, and very low in the globus pallidus

IV. In Vivo Studies

Dopamine receptors have been imaged in humans by positron emission tomography (PET) (Wagner et al. 1983). A variety of different ligands have been used for these studies, but most data is available from studies with [^{11}C] spiroperidol. Presently, the PET receptor methodology is plagued by difficulties in modeling the pharmacokinetics of receptor binding (Perlmutter et al. 1986; Perlmutter and Raichle 1986; Frey et al. 1986). It has been very difficult to quantify receptor numbers by this technique, and only recently have adequate methods been developed for analyzing receptors in humans (Frey et al. 1985; Perlmutter et al. 1986; Wong et al. 1986a, b). In animal models of Parkinson's disease, in vivo administration of dopamine antagonists, such as [^{3}H] spiroperidol, has not shown a change in dopamine receptors after nigrostriatal lesions (Bennett and Wooten 1986). These in vivo studies contrast with the observed findings of increases in receptors as measured by in vitro homogenate studies of lesioned animals (Creese et al. 1977). Such studies make it even more difficult to interpret the meaning of the dopamine receptor PET studies in humans. Hopefully, in the future, more accurate pharmacokinetic models of in vivo receptor binding will be developed.

D. GABA Receptors

The levels of a number of other neurotransmitters and neuropeptides are abnormal in Parkinson's disease brains (Hornykiewicz 1983). The receptors for these agents have been measured in only a limited series of cases, however. GABA is the neurotransmitter used by many neurons of the striatum (Fonnum et al. 1978; Nagy et al. 1978). GABA receptors are closely linked to the receptors for the benzodiazepine class of drugs. These receptors are relatively dense in the striatum, i.e. caudate—putamen, nucleus accumbens, and olfactory tubercle (Fig. 2). A significant, but lesser number of these receptors are present in the lateral and medial globus pallidus and in the substantia nigra, both pars reticulata and pars compacta (Penney and Young 1982; Walker et al. 1984; Penney and Pan 1986; Uhl et al. 1986). In animal models of Parkinson's disease, the numbers of GABA and benzodiazepine receptors in the lateral globus pallidus decrease, while those in the medial globus pallidus and substantia nigra pars reticulata increase (Pan et al. 1985). Similar measurements have not been made from postmortmen Parkinson's disease brains. The few measurements that have been made indicate little change in the striatum and a decrease in the number of pars compacta receptors, consistent with the loss of the intrinsic dopamine neurons (Rinne 1982).

E. Acetylcholine Receptors

Muscarinic cholinergic receptors have been studied in normal human and Parkinson's disease brains. Muscarinic receptors are very dense in the caudate, the putamen, the nucleus accumbens, and the olfactory tubercle (Fig. 2; PENNEY and YOUNG 1982). In contrast, the receptors in the lateral and medial globus pallidus and substantia nigra are very low. Since the cholinergic neurons in the neostriatum are large aspiny interneurons, this localized high density of receptors in the striatum is not surprising. There are no known cholinergic afferents to or efferents from the neostriatum.

REISINE et al. (1977) found an increase in muscarinic receptors in the put amen of Parkinson's disease brains, but no change in caudate nucleus or globus pallidus. RINNE (1982) found those Parkinson's disease brains which had low numbers of dopamine receptors also had decreases in muscarinic cholinergic receptors. In contrast, patients with increased striatal dopamine receptors also had increased levels of striatal muscarinic cholinergic receptors (RINNE 1982). These data suggest that some Parkinson's disease patients may develop intrinsic striatal neuron damage to account for the receptor decreases.

F. Opiate Receptors

The striatum and its projection areas contain very high concentrations of Leu-enkephalin and dynorphin (WATSON et al. 1982; YOUNG and PENNEY 1984). These peptides are colocalized with GABA in medium-spiny neurons (OERTEL et al. 1983; AFSHARPOUR et al. 1984; ARONIN et al. 1984). Dynorphin/GABA-containing cells project primarily to the medial globus pallidus and substantia nigra pars reticulata, whereas enkephalin/GABA neurons have their major projection to the lateral globus pallidus (HABER and WATSON 1985).

Opiate receptors are differentially located in the basal ganglia. In rats, μ opiate receptors are present in dense patches in the striatum, and are localized both on presynaptic dopamine terminals and postsynaptically on striatal neurons (YOUNG and PENNEY 1984). Both δ and $\varkappa$ opiate receptors appear predominantly on striatal somata and terminals. After striatal kainic acid lesions in rats, δ and μ opiate receptors decrease in the globus pallidus and substantia nigra pars reticulata, suggesting that the pallidal and nigral receptors are on striatal afferent terminals (ABOU-KHALIL et al. 1984).

A striking property of rodent and primate opiate receptors is that they are present in very low concentrations in the globus pallidus, despite the very high levels of peptides in the same area. There is, thus, a distinct mismatch between peptide levels and receptors. The functional significance of this mismatch is unclear (KUHAR 1985).

In Parkinson's disease, only a few studies of opiate receptor binding have been carried out (RINNE 1982; PENNEY et al. 1983). Most of the studies were performed when only relatively nonselective ligands were available. Met-enkephalin and Leu-enkephalin binding have been reported to be increased in the putamen and limbic cortex of treated and untreated patients. In one study, naloxone binding was found to be decreased in the caudate nucleus, but not the putamen (REISINE et al. 1979). In the future, it may be possible to define these changes more carefully by selective μ, δ, and $\varkappa$ ligands.

G. Other Receptors

Uhl et al. (1984, 1986) have studied receptors in the substantia nigra in Parkinson's disease brain, and found a marked loss of somatostatin, neurotensin, μ, and κ opiate receptors in the pars compacta. They also found losses of D_2 receptors and benzodiazepine type 1 receptors in the pars compacta, with no change in angiotensin-converting enzyme, serotonin, or benzodiazepine type 2 receptors (Uhl et al. 1986).

Binding of [^{3}H] cocaine, which is a marker of dopamine uptake sites on the presynaptic dopamine terminals, has been found to be markedly decreased in the striatum and in the substantia nigra of Parkinson's disease brains, consistent with the loss of dopamine neurons (Pimoule et al. 1983). Labeled imipramine and paroxetine have been used to mark the presynaptic serotonin uptake sites (Raisman et al. 1986). These sites were found to be decreased in Parkinson's disease brain, and the amount of decrease in depressed Parkinson's disease patients is greater than that in nondepressed Parkinson's disease patients.

Adrenergic receptors have been measured in several regions of Parkinson's disease brain (Cash et al. 1984a, b). In the frontal cortex, α_1 and β_1 receptors are increased, while α_2 receptors are decreased and β_2 receptors are unchanged. In addition, α_1 receptor binding to cerebral microvessels has been measured (Cash et al. 1985). This binding represents a tiny fraction of the total binding, but is decreased in the putamen of Parkinson's disease brains.

Binding of 1-methyl-4-phenyl-1,2,3,6-tetrahydropyridine (MPTP) to monoamine oxidase has been measured in both normal and Parkinson's disease brains (Uhl et al. 1985). In Parkinson's disease brain, there is no loss of MPTP binding in the substantia nigra. This finding indicates that monoamine oxidase type B, to which MPTP binds (Parsons and Rainbow 1984) is not preferentially located within the dopamine neurons of the substantia nigra pars compacta, but instead is present in some other tissue compartment, such as glia or other neurons.

H. Summary

In general, very modest receptor changes at most have been observed in postmortem Parkinson's disease brains. Our understanding of receptor changes has been confounded by the use of multiple different receptor ligands for individual studies, and by the antemortem treatment of Parkinson's disease patients with numerous drugs, including dopamine agonists and antagonists. Investigations have concentrated on studies of dopamine receptors, in particular, the D_2 dopamine receptor. It does appear that most studies have observed an increase in D_2 receptors in the caudate and putamen of untreated Parkinson's patients. These receptors then decrease back to control levels with treatment. Very few studies of other receptor types have been done in the brains of untreated Parkinson's patients. The few studies that have measured receptors in the substantia nigra find changes that correlate with the documented cell loss in the substantia nigra pars compacta. In the future, studies of dopamine and other neurotransmitter receptors in the frontal cortex of untreated Parkinson's disease patients need to be carried out. Hopefully, the development of PET ligands that can be accurately quantitated in vivo will give even more information concerning the neurotransmitter and receptor interactions in this disease.

Acknowledgments. We would like to thank Dr. Eric Richfield for his photographs of dopamine receptors in primate brain, Darrell L. Debowey and Zane Hollingsworth for technical assistance, and Genell Fries for help in preparation of the manuscript. Supported by USPHS grants NS 19613 and 15655, the Arbogast Foundation, and the Fraternal Order of Eagles.

References

Abou-Khalil B, Young AB, Penney JB (1984) Evidence for the presynaptic localization of opiate binding sites on striatal efferent fibers. Brain Res 323:21-29

Afsharpour S, Penney GR, Kitai ST (1984) Glutamic acid decarboxylase, leucine-enkephalin and substance-P immunoreactive neurons in the neostriatum of the rat and cat (abstract). Soc Neurosci 14:702

Anderson PH, Gronvald FC, Jansen JA (1985) A comparison between dopamine-stimulated adenylate cyclase and [^{3}H] SCH 23390 binding in rat striatum. Life Sci 37:1971-1983

Aronin N, Di Figlia M, Graveland GA, Schwartz WJ, Wu JY (1984) Localization of immunoreactive enkephalins in GABA synthesizing neuron of the rat striatum. Brain Res 300:376-80

Barone P, Davis TA, Braun AR, Chase TN (1986) Dopaminergic mechanisms and motor function: characterizations of D_1 and D_2 dopamine receptor interactions. Eur J Pharmacol 123:109-114

Bennett JP, Wooten GF (1986) Dopamine denervation does not alter in vivo ^{3}H-spiperone binding in rat striatum: implications for external imaging of dopamine receptors in Parkinson's disease. Ann Neurol 19:378-383

Cash R, Raisman R, Ruberg M, Agid Y (1984a) Adrenergic receptors in frontal cortex in human brain. Eur J Pharmacol 108:225-232

Cash R, Ruberg M, Raisman R, Agid Y (1984b) Adrenergic receptors in Parkinson's disease. Brain Res 322:269-275

Cash R, Lasbennes F, Sercombe R, Seylaz J, Agid Y (1985) Adrenergic receptors on cerebral microvessels in control and parkinsonian subjects. Life Sci 37:531-536

Christensen AV, Arnt J, Hyttel J, Larsen JJ, Svendsen O (1984) Pharmacological effects of a specific dopamine D_1 agonist SCH 23390 in comparison with neuroleptics. Life Sci 34:1529-1540

Creese I, Burt D, Snyder SH (1976) Dopamine receptor binding predicts clinical and pharmacological potencies of anti-schizophrenic drugs. Science 192:481-483

Creese I, Burt D, Snyder SH (1977) Dopamine receptor binding enhancement accompanies lesion-induced behavioral supersensitivity. Science 197:596-598

Creese I, Hamblin MW, Leff SE, Sibley D (1983) The classification of dopamine receptors: relationship to radioligand binding. Annu Rev Neurosci 6:43-57

Dawson TM, Gehlert D, Yamamura HI, Barnett A, Wamsley JK (1985) D_1 dopamine receptors in rat brain: autoradiographic localization using [^{3}H] SCH 23 390. Eur J Pharmacol 108:323-325
Dawson TM, Gehlert DR, Filloux FM, Wamsley JK (1986) A quantitative autoradiographic comparison of the density and localization of dopamine D_1 and D_2 receptor in rat brain: effects of neurotoxins (abstract). Soc Neurosci 12:481
Dubois A, Savasta M, Curet O, Scatton B (1986) Autoradiographic distribution of the D_1 agonist [^{3}H] SKF 38 393, in the rat brain and spinal cord. Comparison with the distribution of D_2 dopamine receptors. Neuroscience 19:125-137
Filloux FM, Dawson TM, Gehlert DR, Wamsley JK (1986) A quantitative autoradiographic comparison of the effects of unilateral striatal and nigral neurotoxin lesion in the rat brain on [^{3}H] SCH 23 390 and [^{3}H]-Forskolin binding sites (abstract). Soc Neurosci 12:481
Fonnum F, Gottesfeld A, Grofova I (1978) Distribution of glutamate decarboxylase, choline acetyltransferase and aromatic amino acid decarboxylase in the basal ganglia of normal and operated rats. Evidence for striatopallidal, striatoentopeduncular and striatonigral GABAergic fibres. Brain Res 143:125-38
Frey KA, Hichwa RD, Ehrenkaufer RLE, Agranoff BW (1985) Quantitative in vivo receptor binding III: tracer kinetic modeling of muscarinic cholinergic receptor binding. Proc Natl Acad Sci USA 82:6711-6715
Frey KA, Agranoff BW, Young AB, Hichwa RD, Ehrenkaufer, RLE (1986) Human brain receptor distribution. Science 232:1269-1271
Grigoriadis D, Seeman P (1984) The dopamine/neuroleptic receptor. Can J Neurol Sci 11:108-113
Guttman M, Seeman P (1985) L-Dopa reverses the elevated density of D_2 dopamine receptors in Parkinson's diseased striatum. J Neural Transm 64:93-103
Guttman M, Seeman P, Reynolds GP, Riederer P, Jellinger K, Tourtellotte WW (1985) Dopamine D_2 receptor density remains constant in treated Parkinson's disease. Ann Neurol 19:487-492
Haber SN, Watson SJ (1985) The comparative distribution of enkephalin, dynorphin and substance P in the human globus and basal forebrain. Neuroscience 14:1011-24
Herkenham M, Pert CB (1981) In vitro autoradiography of opiate receptors in rat brain suggest loci of "opiatergic" pathways. Proc Natl Acad Sci USA 77:5532-5536
Herrera-Marschitz M, Ungerstedt U (1984) Evidence that striatal efferents relate to different dopamine receptors. Brain Res 323:269-278
Hornykiewicz O (1983) Brain neurotransmitter changes in Parkinson's disease. In: Marsden CD, Fahn S (eds) Movement disorders. Butterworth, London, pp 41-58
Hyttel J (1983) SCH 23 390—The first selective dopamine D_1 antagonist. Eur J Pharmacol 91:153-154
Kebabian JW, Calne DB (1979) Multiple receptors for dopamine. Nature 277:93-96
Kebabian JW, Petzold GL, Greengard P (1972) Dopamine sensitive adenylcyclase in caudate nucleus of rat brain and its similarity to the "dopamine receptor". Proc Natl Acad Sci USA 69:2145-2149
Kuhar MJ (1985) The mismatch problem in receptor mapping studies. Trends Neurosci 8:190-191
Lee T, Seeman P, Rajput A, Farlye IJ, Hornykiewicz O (1978) Receptor basis for dopaminergic supersensitivity in Parkinson's disease. Nature 273:59-61
Leff SE, Creese I (1983) Dopamine receptors re-explained. Trends Pharmacol Sci 463-467

Nagatsu T, Kanamori T, Kato T, Iizuka R, Narabayashi H (1978) Dopamine-stimulated adenylate cyclase activity in the human brain: changes in parkinsonism. Biochem Med 19:360-365

Nagy JI, Carter DA, Fibiger HC (1978) Anterior striatal projections to the globus pallidus, entopeduncular nucleus and substantia nigra in the rat: the GABA connection. Brain Res 158:15-29

Oertel WH, Riethmuler G, Mugnaini E et al. (1983) Opioid peptide-like immunoreactivity localized in GABAergic neurons of rat neostriatum and central amygdaloid nucleus. Life Sci 33 (Suppl 1):73-76.

Onali P, Olianas MC, Gessa GL (1985) Characterization of dopamine receptors mediating inhibition of adenylate cyclase activity in rat striatum. Mol Pharmacol 28:138-145

Pan HS, Penney JB, Young AB (1985) GABA and benzodiazepine receptor changes induced by unilateral 6-hydroxydopamine lesions of the medial forebrain bundle. J Neurochem 45:1396-1404

Parsons B, Rainbow TC (1984) High-affinity binding sites for [^{3}H]MPTP may correspond to monoamine oxidase. Eur J Pharmacol 102:375-377

Penney JB, Pan HS (1986) Quantitative autoradiography of GABA and benzodiazepine binding in studies of mammalian and human basal ganglia function. In: Boast C, Snowhill EW, Altar CA (eds) Quantitative receptor autoradiography. Liss, New York, pp 29-52

Penney JB, Young AB (1982) Quantitative autoradiography of neurotransmitter receptors in Huntington's disease. Neurology (NY) 32:1391-1395

Penney JB, Pan HS, Young AB, Frey KA, Dauth GW (1981) Quantitative autoradiography of [^{3}H]muscimol binding in rat brain. Science 214:1036-1038

Penney JB, Young AB, Walker FO, Shoulson I (1983) Quantitative autoradiography of opiate receptors in Huntington's and Parkinson's disease. Neurology (NY) 34 (Suppl 1):153

Perlmutter JS, Raichle ME (1986) In vitro and in vivo receptor binding: where does the truth lie? Ann Neurol 19:384-385

Perlmutter JS, Larson KB, Raichle ME, Markeham J, Mintun MA, Kilbourn MR, Welch MF (1986) Strategies for in vivo measurement of receptor binding using positron emission tomography. J Cereb Blood Flow Metab 6:154-169

Pijnenberg AJJ, Honig WMM, Van Rossum JM (1975) Inhibition of *d*-amphetamine-induced locomotor activity by injection of haloperidol into the nucleus accumbens of the rat. Psychopharmacology (Berlin) 41:87-95

Pimoule C, Schoemaker H, Javory-Agid F, Scatton B, Agid Y, Langer SZ (1983) Decrease in [^{3}H]cocaine binding to the dopamine transporter in Parkinson's disease. Eur J Pharmacol 95:145-146

Pimoule C, Schoemaker H, Reynolds GP, Langer SZ (1985) [^{3}H]SCH 23 390 labeled D_1 dopamine receptors are unchanged in schizophrenia and Parkinson's disease. Eur J Pharmacol 114:235-237

Quiko M, Spokes EG, Mackay AVP, Bannister R (1979) Alterations in ^{3}H-spiroperidol binding in human caudate nucleus, substantia nigra and frontal cortex in the Shy-Drager syndrome and Parkinson's disease. J Neurol Sci 43:429-437

Quinton RM, Halliwell G (1963) Effects of alpha-methyldopa and dopa on amphetamine excitatory response in reserpinized rats. Nature 200:178-179

Raisman R, Cash R, Agid Y (1986) Parkinson's disease: decreased density of ^{3}H-imipramine and ^{3}H-paroxetine binding sites in putamen. Neurology (NY) 36:556-560

Randup A, Munkvad I, Udsen P (1963) Adrenergic mechanisms and amphetamine induced abnormal behavior. Acta Pharmacol Toxicol 20:145-157

Reisine TD, Fields JZ, Yamamura HI, Bird ED, Spokes E, Schreiner PS, Enna SJ (1977) Neurotransmitter receptor alterations in Parkinson's disease. Life Sci 21:335-344

Reisine TD, Rossor M, Spokes E, Iverson LL, Yamamura HI (1979) Alterations in brain opiate receptors in Parkinson's disease. Brain Res 173:378-382

Richfield EK, Debowey DL, Penney JB, Young AB (1987a) Basal ganglia and cerebral cortical distribution of dopamine D_1 and D_2 receptors in neonatal and adult cat brain. Neurosci Lett 73:203-208

Richfield EK, Young AB, Penney JB (1987b) Comparative distribution of dopamine D_1 and D_2 receptors in the basal ganglia of turtle, pigeon, rat, cat and monkey. J Comp Neurol 262:446-463

Rinne UK (1982) Brain neurotransmitter receptors in Parkinson's disease. In: Marsden CD, Fahn S (eds) Movement disorders. Butterworth, London, pp 59-74

Rinne UK, Lonnberg P, Koskinen V (1981) Dopamine receptors in the parkinsonian brain. J Neural Transm 51:97-106

Rinne JO, Rinne JK, Laakso K, Lonnberg P, Rinne UK (1985) Dopamine D_1 receptors in the parkinsonian brain. Brain Res 359:306-310

Schulz DW, Stanford EJ, Wyrick SW, Mailman RB (1985) Binding of [^{3}H] SCH 23 390 in rat brain: regional distribution of effects of assay conditions and GTP suggest interactions at a D_1-like dopamine receptor. J Neurochem 45:1601-1611

Seeman P, Chau-Wong M, Tedesco J, Wong K (1975) Brain receptors for antipsyhotic-drugs and dopamine direct binding assays. Proc Natl Acad Sci USA 72:4376-4380

Seeman P, Lee T, Chau-Wong M, Wong K (1976) Antipsychotic drug doses and neuroleptic-dopamine receptors. Nature 26:717-719

Shibuya M (1979) Dopamine-sensitive adenylate cyclase activity in the striatum in Parkinson's disease. J Neural Transm 44:287-295

Spano PF, Trabucchi M, Di Chiara G (1977) Localization of nigral dopamine-sensitive adenylate cyclase on neurons originating in the corpus striatum. Science 196:1343-1345

Stoof JC, Kebabian JW (1984) Two dopamine receptors: biochemistry, physiology and pharmacology. Life Sci 35:2281-2296

Titus RD, Kornfeld EC, Jones ND, Clemens JA, Smalstig EB, Fuller RW, Hahn RA, Hynes MD, Mason NR, Wong DT, Foreman MM (1983) Resolution and absolute configuration of an ergoline-related dopamine agonist, *trans*-4,4a,5,6,7,8,8a,9-octohydro-5-propyl-1*H* (or 2*H*)-pyrazolo [3,4-g]quinoline. J Chem 26:1112-1116

Uhl GR, Whitehouse PJ, Price DL, Tourtelotte WW, Kuhar MJ (1984) Parkinson's disease: depletion of substantia nigra neurotensin receptors. Brain Res 308:186-190

Uhl GR, Javitch JA, Snyder SH (1985) Normal MPTP binding in parkinsonian substantia nigra: evidence for extraneuronal toxin conversion in human brain. Lancet 1:956-958

Uhl GR, Hackney GO, Torchia M, Stranov V, Tourtellotte WW, Whitehouse PJ, Tran V, Strittmatter S (1986) Parkinson's disease: nigral receptor changes support peptidergic role in nigrostriatal modulation. Ann Neurol 20:194-203

Ungerstedt U (1971) Postsynaptic supersensitivity after 6-hydroxydopamine induced degeneration of the nigrostriatal dopamine system. Acta Physiol Scand [Suppl] 367:69-93

Ungerstedt U (1979) Central dopamine mechanisms and unconditioned behavior. In: Horn AS, Kork J, Westerink BHC (eds) The Neurobiology of Dopamine. Academic, London, pp 577-596

Wagner HN Jr, Burns HD, Dannals RF, Wong DF, Langstrom B, Kuhar MJ (1983) Imaging dopamine receptors in human brain by positron tomography. Science 221:1264-1266

Walker FO, Young AB, Penney JB, Dorovini-Zis, Shoulson I (1984) Benzodiazepine receptors in early Huntington's disease. Neurology (NY) 34:1237-1240
Watson SJ, Khachaturian H, Akil H et al. (1982) Comparison of the distribution of dynorphin systems and enkephalin systems in brain. Science 218:1134-36
Wong DF, Gjedde A, Wagner HN (1986a) Quantification of neuroreceptors in the living human brain. I. Irreversible binding of ligands. J Cereb Blood Flow Metab 6:137-146
Wong DF, Gjedde A, Wagner HN, Dannals RF, Douglas KH, Links JM, Kuhar MJ (1986b) Quantification of neuroreceptors in the living human brain. II. Inhibition studies of receptor density and affinity. J Cereb Blood Flow Metab 6:147-153
Young AB, Penney JB (1984) Neurochemical anatomy of movement disorders. Neurol Clin 2(3):417-433
Young AB, Frey KA, Agranoff BW (1985) Receptor assays: in vivo and in vitro. In: Phelps M, Mazziotta JC, Schelbert H (eds) Tracer kinetic studies of cerebral and myocardial function: positron emission tomography and autoradiography. Raven, New York, pp 73-111
Young WS III, Kuhar MJ (1979) A new method for receptor autoradiography [^{3}H] opioid receptor labelling in mounted tissue sections. Brain Res 179:255-270

CHAPTER 5

Imaging the Basal Ganglia

W. R. W. MARTIN

A. Introduction

The rapid implementation of new techniques for imaging the central nervous system is one of the most impressive developments to have taken place in clinical neuroscience in the last decade. Not only have these techniques revolutionized the clinical management of patients with neurological disorders, but in many situations they have advanced our understanding of the pathophysiology of diseases which affect the brain.

Traditional concepts of imaging have focussed on the depiction of the anatomical detail of the brain. The development and widespread implementation in the 1970s of X-ray computed tomography (CT) as a structural imaging modality was the initial milestone in the establishment of the new science of "neuroimaging". More recently, the development of magnetic resonance imaging (MRI) has provided a further advance in our ability to depict anatomical detail.

By applying tomographic principles to radionuclide imaging methods, it has become possible to study cerebral physiology in human subjects with techniques such as positron emission tomography (PET). Although techniques for functional imaging have not yet seen widespread use in clinical patient management, they have provided new information concerning brain function in health and disease and show promise for providing yet further information in the future.

This chapter is divided into two major sections. The first part deals with structural imaging techniques and reviews the abnormalities which have been demonstrated in various movement disorders with these methods. The second section deals with functional imaging and discusses briefly the principles of methodology which are required to understand the associated potentials and difficulties. This section also discusses the application of these methods to the study of disorders of the basal ganglia.

B. Structural Imaging Techniques

I. Methodology

1. X-Ray Computed Tomography (CT)

This technique has seen widespread application in clinical neurosciences in the last decade and the results are well known. Current CT technology permits high quality images with a spatial resolution approaching 0.5 mm to be obtained and reconstructed in about 20 seconds per slice. With CT, it is usually possible to obtain reasonable distinction between the grey matter of the basal ganglia and the adjacent white matter and cerebrospinal fluid. Images correspond primarily to a map of electron density of the tissue being studied. The main limitations are the predilection for artifacts in tissues adjacent to bone, and the relative difficulty in discriminating minor differences in tissue density.

2. Magnetic Resonance Imaging

This technology has developed rapidly in recent years. Magnetic resonance imaging (MRI) is superior to CT in the depiction of structural detail, in particular with regard to the differentiation of grey matter from white matter. Other relative advantages include the fact that MRI does not involve ionizing radiation, and that direct multiplanar imaging, e.g. transverse, coronal, or sagittal, may be performed independently of patient positioning.

The technique itself is based on the magnetic properties of hydrogen nuclei (protons), their response to externally applied magnetic fields, and on the excitation of protons within such a field by a radiofrequency (RF) pulse. When the RF pulse is terminated, the protons lose energy by emitting an RF signal, the intensity and temporal decay pattern of which is related to the chemical milieu of the protons. This signal is recorded by the imaging device and is dependent upon the imaging parameters as well as tissue properties such as mobile proton density and T_1 and T_2 relaxation times. The theory underlying the MRI technique and details concerning the methodology itself have been summarized in recent reviews (Norman and Brant-Zawadzki 1985; Oldendorf 1985).

II. Parkinson's Disease

The major pathological abnormality in Parkinson's disease is the loss of dopaminergic neurons from the substantia nigra. An early claim (Bydder et al. 1982) that the substantia nigra could be visualized and appeared to be atrophied has not been confirmed. High field strength (1.5 T) MRI studies have been reported to show reduced signal intensity in the putamen and lateral substantia nigra on T_2 weighted images (Drayer et al. 1986), and this has been suggested to represent increased iron deposition in these regions.

III. Huntington's Disease

The cardinal structural feature of Huntington's disease is the striatal atrophy affecting most severely the caudate nucleus and anterior putamen (LANGE et al. 1976). In addition, there is a variable degree of cortical atrophy. An index for the estimation of the degree of caudate atrophy has been derived from CT measurements of the distance between the heads of the caudate nuclei, i.e. the bicaudate diameter. Measurements which show significantly increased bicaudate diameter in Huntington's disease as compared with normal individuals have been reported (BARR et al. 1978; STOBER et al. 1984; OEPEN and OSTERTAG 1981). Attempts to derive ratios between the bicaudate diameter and other brain or skull measurements have failed to increase the capacity of the bicaudate diameter to discriminate between patients with Huntington's disease and normal subjects (STOBER et al. 1984).

Proton MRI studies have also shown basal ganglia abnormalities in Huntington's disease. SAX et al. (1985) have suggested that when viewed in an axial plane parallel to Reid's base line, the ratio of the area of the caudate head to the frontal horn of the lateral ventricle decreases within 5 years of the diagnosis of Huntington's disease. These studies also demonstrate the complexity of MRI in that there is a difference in this area ratio on T_1 weighted images as compared with T_2 weighted images. SAX and BUONANNO (1986) reported that in the rigid form of Huntington's disease spin echo images show increased signal intensity in the putamina and suggest that this may reflect increased gliosis. Studies performed with a high field strength magnet (1.5 T) have shown decreased signal intensity in the globus pallidus on T_2 weighted images (KOZACHUK et al. 1986), possibly secondary to increased iron content. These studies illustrate that although the anatomical localization of the changes seen on MRI may be clear, the interpretation of the observed abnormalities is not without difficulty. Further studies are required to determine what role MRI has to play in the diagnosis and management of Huntington's disease.

IV. Dystonia

There is no convincing evidence that any structural changes can be detected with either CT or MRI in idiopathic or hereditary dystonia. This is not surprising since neither macroscopic nor microscopic morphological changes are evident in these disorders. In contrast, a large series of CT, and more recently MRI, reports have implicated lesions of the basal ganglia in dystonia secondary to various focal lesions of the brain. Although abnormalities have been reported in the putamen, pallidum, and caudate, the most consistent change is with pathology involving the putamen. Focal putamenal lesions secondary to infarction (BURTON et al. 1984) or to neoplasms (NARBONA et al. 1984) have been reported to produce dystonia. In children, structural lesions of the putamen associated with dystonia may be present in sporadic and familial forms of infantile bilateral striatal necrosis (GOUTIERES and AICARDI 1982; NOVOTNY et al. 1986).

Disorders characterized by metal deposition in the basal ganglia may also produce dystonia. Subcortical abnormalities have been reported in Wilson's disease with MRI (LAWLER et al. 1983; STAROSTA-RUBINSTEIN et al. 1985) and with CT (WILLIAMS et al. 1981). Dystonia was found to correlate best with lesions of the putamen, caudate and substantia nigra (STAROSTA-RUBINSTEIN et al. 1985). In two patients with a clinical diagnosis of Hallervorden—Spatz disease, MRI demonstrated marked atrophy affecting the brain stem and cerebellum (LITTRUP and GEBARSKI 1985). In addition, areas of decreased signal intensity, thought to be related to either disordered myelination or to abnormal iron storage, were evident in the lentiform nuclei and perilentiform white matter in these patients.

V. Other Movement Disorders

Focal lesions of the subthalamic nucleus have been associated with contralateral hemiballism because of well-established clinicopathological correlations. In vivo observations have been made with CT and these have confirmed that thalamic and striatal (LODDER and BAARD 1981; GLASS et al. 1984) lesions may also produce ballism as shown in pathological studies by MARTIN (1957). It has been suggested that the involuntary movements result from disinhibition of the globus pallidus.

C. Functional Imaging Techniques

I. Methodology

1. Positron Emission Tomography (PET)

PET is an imaging technique whereby the concentration of positron-emitting radioisotopes may be measured in the living brain. By combining this technique with appropriately labelled radiopharmaceuticals and the principles of tracer kinetic modelling, it is possible to quantify regionally selective biochemical and physiological processes in human subjects. The application of PET to basal ganglia disorders has involved two general types of studies, namely those of regional cerebral metabolism and blood flow, and those of presynaptic dopa metabolism and postsynaptic dopamine receptor function.

a. Cerebral Blood Flow and Metabolism

The regional distribution of cerebral glucose and oxygen metabolism, and of cerebral blood flow, is thought to be related to neuronal and synaptic function. This is because the major energy expenditure in the brain is for the ion transport mechanisms responsible for pumping sodium and potassium across neuronal membranes (MATA et al. 1980). This action is necessary to maintain the electrochemical gradients required for the generation of action potentials. Because aerobic glucose metabolism is responsible for most cerebral energy production, the mapping of regional cerebral metabolic rate for oxygen ($rCMRO_2$) or regional cerebral metabolic rate for glucose (rCMRG) provides a functional image of brain activity. Similarly, because regional cerebral blood flow (rCBF) is thought to be closely coupled to tissue metabolic demands, mapping of rCBF also provides an image of brain function. The degree of coupling between rCBF and $rCMRO_2$ has recently been questioned as a result of PET studies in which both of these aspects of brain function have been measured in response to tactile stimulation (Fox and Raichle 1986). Nevertheless, the general principle still applies that PET techniques are able to perform physiological measurements which relate to ongoing brain function.

The approach to measurement of rCMRG with PET is based on the [^{14}C] 2-deoxyglucose (DG) tissue autoradiographic method of Sokoloff et al. (1977). This method is based in turn on the unique biochemical properties of DG as compared with those of glucose. The bidirectional transport of DG across the blood—brain barrier and its hexokinase-catalysed phosphorylation is similar to that of glucose. In contrast to glucose, however, the phosphorylated product, DG-6-phosphate, is not metabolized further and is dephosphorylated very slowly. DG-6-phosphate is thus trapped within tissue and its accumulation is directly related to the rate of DG phosphorylation. An operational equation has been derived which expresses glucose utilization per unit mass of tissue in terms of measurable variables (Sokoloff et al. 1977).

The extension of the DG method to human studies was made possible by the development of PET and the tracer [^{18}F] 2-fluoro-2-deoxyglucose (FDG) which behaves in a similar fashion to DG (Reivich et al. 1979). The original model was modified to compensate for the slow dephosphorylation of FDG which was thought to be significant at the later scanning times employed with PET measurements (Phelps et al. 1979; Huang et al. 1980). This version has become the standard model employed for rCMRG measurement with PET in humans.

The measurement of $rCMRO_2$ makes use of inhaled ^{15}O-labelled compounds. Cerebral oxygen extraction is measured first, by either a steady state approach (Frackowiak et al. 1980) or a single-breath method (Mintun et al. 1984b). Laboratories which use the steady state method usually measure rCBF with inhaled [^{15}O] CO_2 (Frackowiak et al. 1980) whereas the single-breath method is usually accompanied by rCBF calculation from a bolus injection of [^{15}O] H_2O (Herscovitch et al. 1983; Raichle et al. 1983). With either method, oxygen metabolism is calculated as the product of regional oxygen extraction, rCBF and arterial oxygen content.

b. Presynaptic Dopaminergic Function

The development of the L-dopa analog 6-fluoro-L-dopa (6-FD), labelled with positron-emitting ^{18}F, has made it possible to label presynaptic dopaminergic nerve endings in vivo (Garnett et al. 1983a). Fluorodopa is a substrate for dopa decarboxylase (Firnau et al. 1975) and is thought to be decarboxylated by this enzyme to fluorodopamine within terminals of nigrostriatal neurons. Some of the accumulated striatal activity is discharged following the administration of reserpine (Garnett et al. 1983b), showing that at least some of this fluorodopamine is trapped within intraneuronal vesicles. In animals to which

6-FD has been administered, we have shown that fluorodopamine, as well as its metabolic products, is present in the striatum (CUMMING et al. 1987).

Radioactive imaging techniques such as PET are able to detect the presence of the radioactive label only. The compound (or compounds) to which the label is attached cannot be determined directly from these studies alone. The interpretation of 6-FD uptake data is complicated by the presence of multiple labelled compounds. There is significant "background" activity in both the striatum and nonstriatal areas. In carbidopa-pretreated animals this activity is in the form of 6-FD and 3-*O*-methyl-6-FD formed in the periphery by catechol *O*-methyltransferase (COMT); these are the only labelled compounds present in nondopaminergic brain structures such as the cerebellum (CUMMING et al. 1987). Both of these compounds are transported across the blood—brain barrier by the neutral amino acid transport system and competitive inhibition of uptake by concomitant infusion of amino acids has been shown (LEENDERS et al. 1986a). A normal 6-FD study is illustrated in Fig. 1.

The interpretation of 6-FD studies has thus far been semiquantitative and has not provided values which correspond directly to the kinetics of dopa metabolism itself. Application of tracer kinetic modelling techniques to the data obtained from 6-FD studies should permit more meaningful quantitative analysis. Initial attempts to develop such techniques have been reported (GARNETT et al. 1980; MARTIN et al. 1985). We have applied a model (MARTIN et al. 1985) based on a previously developed method for the analysis of unidirec-

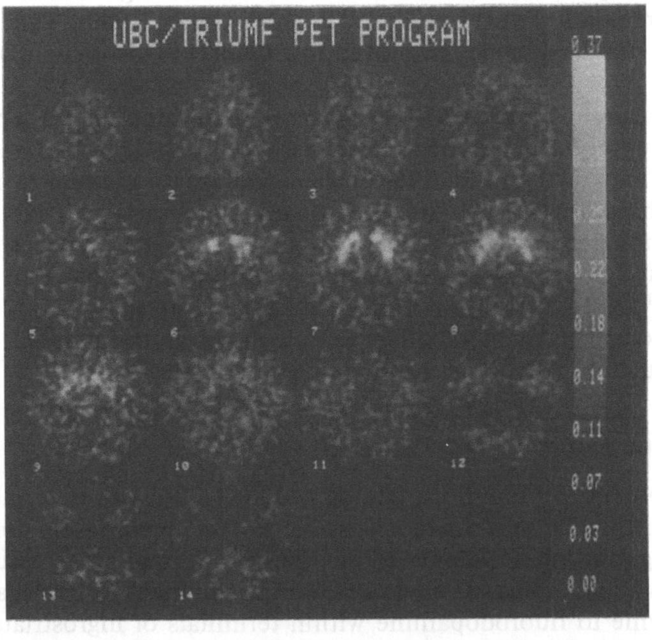

Fig. 1. A 6-fluorodopa PET scan obtained in a normal subject. Sequential transverse slices are placed at 7-mm intervals with the highest slice at the top left. The colour scale is calibrated in arbitrary units of regional radioactivity

tional tracer uptake data (PATLAK et al. 1983). This technique is of sufficient sensitivity to show an age-related decrease in striatal 6-FD-derived radioactivity (W.R.W. MARTIN 1987, unpublished work) which is similar in degree to the known age-related decrease in substantia nigra cell counts (MCGEER et al. 1977). Much work, however, remains to enable full application of tracer kinetic modelling techniques to 6-FD uptake data.

c. Dopamine Receptors

Radiolabelled ligands which bind to specific receptor sites have been utilized both in vivo and in vitro (YOUNG et al. 1986a). Initially studies with PET in human subjects involved the neuroleptic 3-*N*-methylspiperone (NMSP) which binds to both D_2 dopamine and S_2 serotonin receptors. [^{11}C] labelled NMSP in human subjects has been shown to accumulate maximally in the basal ganglia (WAGNER et al. 1983); this is considered to represent D_2 receptor binding. The lesser degree of uptake in the cerebral cortex was thought to reflect S_2 receptor binding (WONG et al. 1984). Numerous other positron-emitting dopamine receptor ligands have been synthesized and will undoubtedly undergo further investigation (ARNETT et al. 1984, 1985a, b; BARON et al. 1985; MAZIERE et al. 1984; MOERLEIN et al. 1986).

The goal of receptor studies with PET is to measure the receptor number B_{max} and, if possible, the dissociation constant for the ligand used K_d. The application of appropriate tracer kinetic methodology should make this possible. One such approach (MINTUN et al. 1984a; PERLMUTTER et al. 1986) utilizes sequential PET scans following tracer administration in order to characterize the dynamic interplay between ligand and receptor. These data are combined with measurement of local CBF and blood volume, also obtained with PET, to determine the blood—brain barrier permeability for the tracer and the "binding potential" of the tissue.

Other approaches to the quantitation of neuroreceptors with PET have been suggested. A method has been reported for use with NMSP based on its presumptive irreversible binding to receptors (WONG et al. 1986a, b). This method estimates the binding rate of a high affinity compound, permitting calculation of B_{max}. Values reported are similar to spiperone receptor density and haloperidol inhibitory potency in brain homogenates from human autopsy material (SEEMAN et al. 1985). FARDE et al. (1986) have reported the sequential administration of varying specific activities of [^{11}C] raclopride (which has very low nonspecific binding) to permit saturation analysis, and have calculated both B_{max} and K_d from these data.

The importance of using such techniques for the analysis of neuroreceptor data, and the inadequacy of techniques based on the measurement of activity ratios between areas of specific ligand binding (such as the striatum) and areas of nonspecific binding (such as the cerebellum), has been discussed elsewhere (WONG et al. 1986b).

2. Single-Photon Emission Computed Tomography (SPECT)

This nuclear medicine technique permits tomographic reconstruction of transverse images corresponding to the distribution of standard radionuclides. The technique has been reviewed by BUDINGER (1985). Positron-emitting radionuclides are not required and hence SPECT is not dependendent upon access to a cyclotron. There are other limiting factors, however. BUDINGER (1985) predicts a resolution as good as 9 mm for brain studies, but this theoretical figure has not yet been reached routinely in commercial SPECT systems. In contrast, current commercial PET systems have an in-plane resolution of about 5 mm. A major problem with SPECT in the past has been related to the difficulty in precise quantitation of regional radioactivity measurements. These problems may be resolved as improved methods to deal with attenuation correction and radiation scatter are implemented. The all-important tracer kinetic methodology which forms the basis for quantitative PET measurements cannot be applied to SPECT until it is possible to make accurate measurements of the regional concentration of the labelled tracer compound in use.

At present, the main application of SPECT in neurological studies is in the measurement of rCBF. Compounds such as $[^{123}I]$ *N*-isopropyl-*p*-iodoamphetamine which act as "metabolic microspheres" have been utilized for these measurements (HILL et al. 1982; KUHL et al. 1982a) which have not been applied extensively to studies of basal ganglia function. Regional cerebral metabolism has not been studied because of the unavailability of suitable single-photon emitting metabolic tracer compounds. The ability of SPECT to image cholinergic receptors has been reported (ECKELMAN et al. 1984). The potential of imaging with receptor-based radiopharmaceuticals with both PET and SPECT has been reviewed by KILBOURN and ZALUTSKY (1985).

II. Parkinson's Disease

Several investigators have reported measurements of cerebral metabolism and blood flow in patients with predominantly unilateral Parkinson's disease. In these patients one would anticipate that the major changes would occur contralateral to the most severely affected limbs. This expectation has been confirmed for local $CMRO_2$ (RAICHLE et al. 1984; WOLFSON et al. 1985), CMRG (MARTIN et al. 1984) and CBF (WOLFSON et al. 1985). In all these studies, increased metabolism or blood flow was described in the basal ganglia on the appropriate side. These findings are consistent with animal studies in which unilateral substantia nigra lesions led to increased glucose utilization in the ipsilateral pallidum (WOOTEN and COLLINS 1981). PERLMUTTER and RAICHLE (1985) have reported results obtained with a stereotactic method for anatomical localization (FOX et al. 1985). They found that pallidal blood flow (and hence, by implication, pallidal neuronal activity) was not necessarily increased although flows were significantly less tightly coupled in these patients than in control subjects.

These studies suggest that abnormalities in pallidal function are indeed present in Parkinson's disease, but that the situation is more complex than would be expected from the work of Wooten and Collins (1981). As suggested by Perlmutter and Raichle (1985), it may be that different clinical symptoms reflect different functional abnormalities of specific basal ganglia components. Further studies are required to investigate this possibility.

Cortical abnormalities in the form of depressed $CMRO_2$ and CBF have been reported in the frontal cortex contralateral to affected limbs in asymmetrically affected patients (Wolfson et al. 1985; Perlmutter and Raichle 1985). The depression in CBF has been localized to "mesocortex" receiving dopaminergic input from the ventral tegmental area (Perlmutter and Raichle 1985).

In patients with bilateral parkinsonism, a global decrease in CBF unaccompanied by changes in $CMRO_2$ has been observed (Wolfson et al. 1985). This was suggested as being secondary to vasoconstriction due to loss of dopaminergic innervation of blood vessels. Studies of local CMRG in patients have shown either no consistent difference when compared with normal subjects (Rougemont et al. 1984), or a moderate generalized reduction (Kuhl et al. 1984).

In all these studies, nondemented PD patients were studied almost exclusively. One moderately demented patient was included in the group studied by Kuhl et al. (1984). In this patient, a significant biparietal decrease in glucose metabolism was found, similar to that seen in Alzheimer's disease.

The effects of L-dopa on regional metabolism and blood flow have also been studied. Leenders et al. (1985) reported a diffuse increase in local CBF unaccompanied by changes in $CMRO_2$ in response to L-dopa administration in both normal and parkinsonian individuals and suggested that the increase was related to direct vasodilation of cerebral vasculature. This is consistent with the vasodilation induced by dopamine in cats (Edvinsson et al. 1978) and with the increases in CBF seen in humans following the administration of dopamine agonists (Guell et al. 1982; Bes et al. 1983). However, Perlmutter and Raichle (1985) found no change in global CBF in response to L-dopa administration, possibly because a smaller dose was given.

Several groups have reported results from 6-FD studies of dopaminergic pathways in Parkinson's disease (Fig. 2). Normal caudate isotope uptake, but impaired putamen activity has been reported in patients with purely unilateral symptoms (Nahmias et al. 1985). We have observed in patients with slightly more advanced disease (bilateral involvement, but significant clinical asymmetry) that there is a symmetric mild decrease in caudate radioactivity, but more severely depressed putamen activity (Martin et al. 1987). Significant putamen asymmetry was evident in both of these studies with the depression being most marked on the expected side, i.e. on the side opposite the affected limbs. Leenders et al. (1986b) suggest a decreased ability to retain striatal fluorodopamine in patients with Parkinson's disease. They suggest in addition that patients with the "on—off" effect have a lower ability to retain tracer than patients with less severe disease.

These 6-FD studies have been based largely on the ratio of striatal to background activity as an index of presynaptic dopaminergic function. The background area used has been either the cerebral cortex, or cerebellum. Further

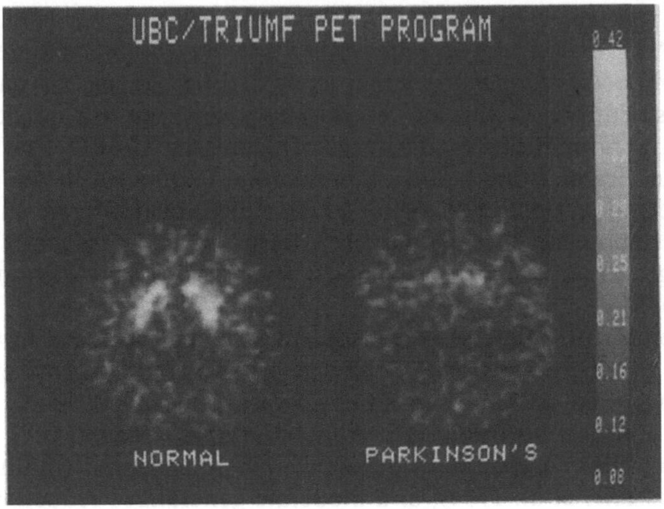

Fig. 2. A 6-fluorodopa PET scan obtained in a normal subject and a patient with Parkinson's disease. A comparable level through the basal ganglia is shown for each individual. The colour scale is calibrated in arbitrary units of regional radioactivity

refinement of the tracer kinetic methodology as it applies to 6-FD studies is required to obtain more precise information concerning dopa kinetics. With the appropriate techniques it should become possible to estimate the rate of 6-FD transport across the blood—brain barrier, the rate of 6-FD decarboxylation, and possibly the rate of fluorodopamine turnover. It should also be possible to study the importance of peripheral metabolism, especially the COMT-catalysed *O*-methylation. With the development of an appropriate "lumped constant" these values for 6-FD could be converted to corresponding values for dopa and dopamine. Should such a "lumped constant" be determined, it will be critical to the success of the method to determine how the constant itself is affected by disease processes which affect the blood—brain barrier and the processes involved in 6-FD metabolism.

III. Huntington's Disease

A characteristic decrease in striatal glucose utilization in Huntington's disease was first reported by KUHL et al. (1982b). Although some of their patients had CT evidence of ventricular dilation and caudate atrophy, it was suggested that the decreased striatal metabolism precedes bulk tissue loss. We have studied a series of patients with Huntington's disease selected on the basis of early clinical symptoms and minimal or no CT evidence of caudate atrophy and have confirmed that caudate hypometabolism occurs in the absence of significant tissue loss (HAYDEN et al. 1986). Similar observations were made by GARNETT et al. (1984). These observations support the hypothesis that the changes in caudate metabolism reflect true alterations in neuronal function rather than artifactual changes to caudate atrophy. In the appropriately se-

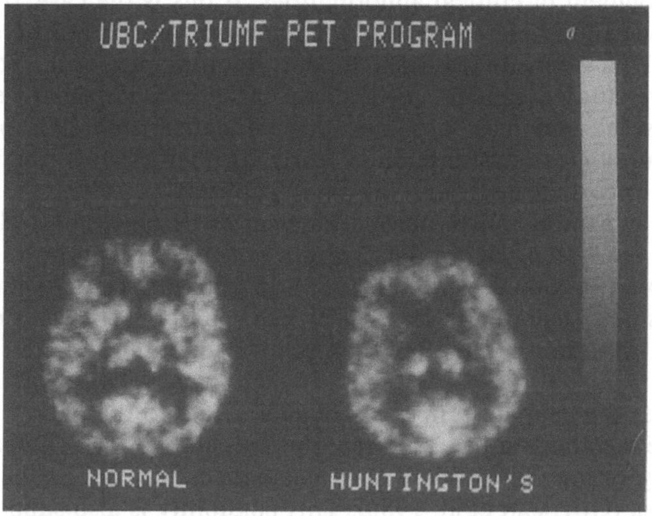

Fig. 3. A fluorodeoxyglucose PET scan obtained in a normal subject and a patient with Huntington's disease. A comparable level through the basal ganglia is shown for each individual. The colour scale is calibrated in units of regional cerebral glucose metabolism (mg per 100 g per minute)

lected patient, i.e. the patient with a positive family history of Huntington's disease who has early symptoms and a normal CT scan, impaired striatal glucose metabolism as demonstrated by PET provides useful supportive evidence for the diagnosis. A normal and a Huntington's disease scan are illustrated in Fig. 3.

Whether PET is useful in the diagnosis of Huntington's disease in at-risk individuals before the onset of clinical symptoms is a question which has generated intense interest from several groups. If caudate metabolism is depressed in patients before the onset of detectable structural changes, one might postulate the appearance of caudate hypometabolism preceding the onset of clinical symptoms. In their original publication, Kuhl et al. (1982) reported the presence of caudate hypometabolism in 6 of 15 asymptomatic subjects at risk for the disease. They found that 3 of the hypometabolic subjects developed clinical signs characteristic of the disease over the following 2 years (Kuhl et al. 1985). Mazziotta et al. (1986) found that 30% of at-risk individuals had abnormal caudate metabolism. We have studied a series of asymptomatic at-risk subjects by measuring caudate glucose metabolism with PET as well as performing DNA polymorphism studies in order to determine which subjects had a high probability of carrying the gene for Huntington's disease (Stoessl et al. 1986a). We found that all patients with abnormal caudate glucose metabolism had positive DNA studies, i.e. they were carrying the Huntington's disease gene. In addition, it was apparent that some patients with the gene did not have abnormal striatal metabolism. These findings suggest that, although the striatal metabolic abnormalities do appear prior to the onset of clinical symptoms, the hypometabolism occurs in close temporal relationship to the development of clinical abnormalities. If this is the case, then patients studied a long time before the onset of symptoms would not be expected to have striatal changes. In contrast to our findings, however, YOUNG et al. (1986b) have studied caudate metabolism in 29 individuals at risk for Huntington's disease and have observed that caudate metabolism parallels the signs of neurological dysfunction. Clearly, further studies are required to answer this important question with more certainty.

Caudate hypometabolism does not appear to be specific for Huntington's disease. We have reported similar changes in some patients with benign hereditary chorea (SUCHOWERSKY et al. 1986). In addition, caudate metabolic depression has been reported in Lesch—Nyhan syndrome (PALELLA et al. 1985). The clinical situation is therefore important in the interpretation of the PET findings.

How does the caudate functional change relate to the chorea itself? Normal caudate metabolism has been reported in patients with chorea secondary to lupus, suggesting that caudate hypometabolism may not be the PET correlate of chorea (GUTTMAN et al. 1986). This possibility is also supported by our studies in benign hereditary chorea; although abnormalities were present in some patients, this was not the case for all (SUCHOWERSKY et al. 1986). In contrast, we have not seen a patient with a clinical diagnosis of Huntington's disease who did not have caudate hypometabolism. Further studies are needed to answer this question concerning the functional substrate of chorea. One possibility is that chorea may be produced by different functional abnormalities acting on a final common pathway. Another possibility is that the caudate hypometabolism of Huntington's disease, although an important feature of the disease process, is not directly related to the chorea itself.

Although PET studies performed thus far in this disorder have concentrated on cerebral metabolism, some work with 6-FD has been reported. We have studied patients with typical adult onset choreic Huntington's disease as well as patients with the rigid juvenile onset variant (STOESSL et al. 1986b). In choreic patients 6-FD-derived activity did not differ significantly from age-matched controls. In contrast, rigid patients had markedly less striatal radioactivity than controls. This loss was not explainable by caudate atrophy. These findings suggest the presence of significant impairment of presynaptic dopaminergic function in the rigid variant and provide a theoretical basis for treatment of these patients with L-dopa or dopaminergic agonists. LEENDERS et al. (1986c) have reported a single individual with typical Huntington's disease in whom 6-FD uptake was normal, but D_2 receptor binding, assessed by measuring striatum: cortex ratios following [^{11}C]NMSP administration, was decreased. As more experience is gained with studies of dopaminergic and other receptors in Huntington's disease, significant further knowledge should accrue concerning the pathophysiological substrate of chorea and about the disease itself.

IV. Dystonia

Exploratory studies of cerebral glucose metabolism in both generalized and focal dystonia have been performed. Studies in idiopathic generalized dystonia have thus far failed to provide any major insights with regard to its pathophysiology. In focal dystonia (torticollis) we have reported that although there is no consistent abnormality in CMRG there is a breakdown in the normal relationships between thalamus and basal ganglia (Stoessl et al. 1986c). This suggests the presence of a functional abnormality in connections between these structures rather than pathology in a specific structure. Junck et al. (1986) found no definite abnormalities in three patients with torticollis, but in patients with more severe asymmetric dystonia, observed asymmetric metabolism in the sensorimotor cortex with lower metabolism in the hemisphere contralateral to the symptomatic side.

In all these studies it is important to determine whether the changes observed are secondary to the disease process which causes dystonia, or are secondary to the movement itself. In an attempt to answer this question, we have studied a group of normal subjects who were instructed to move the head in a fashion intended to simulate the involuntary movements of torticollis. The results were analysed with the same correlation technique used in our torticollis study (Stoessl et al. 1986c). We found that the two control groups (one group studied at rest and one studied during voluntary head movement) and the dystonic group were all quite different, suggesting that the findings in dystonia are significant.

Studies performed with 6-FD suggest that dystonia is a heterogeneous disorder. We have shown in patients with dystonia/parkinsonism that the calculated striatal influx constant in some patients is significantly lower than in age-matched controls (Martin et al. 1988). Similar findings have also been reported by Lang et al. (1988). In other patients with idiopathic dystonia unassociated with parkinsonian features we have observed a striatal influx constant which exceeds that found in age-matched controls (W. R. W. Martin 1986, unpublished work). A patient with progressive hemidystonia from childhood who subsequently developed parkinsonism on the same side and blepharospasm on the other side has been reported (Leenders et al. 1986d). A left upper brain stem lesion was evident on CT and MRI in this patient; the left striatum was structurally intact, but had decreased 6-FD uptake, because of nigrostriatal pathway involvement. On the right side, 6-FD uptake was normal. NMSP uptake was normal bilaterally, suggesting normal D_2 receptors (although analysis of receptor kinetics was not reported). Of note is the fact that impaired 6-FD uptake in all of these reports was associated with parkinsonism rather than with pure dystonia.

As described previously in this chapter, focal lesions of the putamen have been associated with dystonia. We have studied one such patient with infarction limited to the putamen by the 6-FD PET technique (Fross et al. 1986). There was evidence that dopaminergic input to the caudate was intact in this patient, providing additional support for the view that putamenal pathology is associated closely with dystonia.

The loss of dopaminergic input to the putamen with relative sparing of caudate input is similar to the situation seen in Parkinson's disease. While in Parkinson's disease this represents loss of dopaminergic input to intact neurons, in a patient with focal putamenal destruction there is loss not only of the dopaminergic input, but also of neurons intrinsic to the putamen. We have hypothesized that while parkinsonian motor manifestations derive from putamenal output released from nigral modulation, dystonia results from a loss of putamenal influence on other structures. This hypothesis is also consistent with our impression in torticollis that there is a functional abnormality in striatal—thalamic coupling.

D. Conclusions

Modern techniques of neuroimaging have permitted us to obtain much better information concerning the structural substrate of various disorders of the basal ganglia. In addition, with the development of PET, we have obtained an important new tool to study in vivo brain function in human subjects. This has resulted in improved understanding of striatal functioning both in health and in disease. These studies, however, are still in their infancy and a great deal of work remains to be done to elucidate the pathophysiology of these disease processes in more detail.

Studies performed with PET have thus far concentrated on the measurement of cerebral metabolism and blood flow. The potential of the technique is much greater, however. With the development of methods to study presynaptic and postsynaptic dopaminergic function, a very useful tool to assess neurotransmitter function in the basal ganglia is available. It is critical that methodological considerations receive close attention, however, and that these techniques are not applied indiscriminately. Studies of other receptors, although not addressed in this chapter, have also been proven possible. As the methodological aspects of these are developed, major advances are anticipated in understanding movement disorders and in the clinical management of affected patients.

References

Arnett CD, Fowler JS, Wolf AP, Logan J, MacGregor RR (1984) Mapping brain neuroleptic receptors in the live baboon. Biol Psychiatry 19:1365–1375

Arnett CD, Fowler JS, Wolf AP, Shiue CY, McPherson DW (1985a) [^{18}F] *N*-Methylspiroperidol: the radioligand of choice for PETT studies of the dopamine receptor in human brain. Life Sci 36:1359–1366

Arnett CD, Shiue CY, Wolf AP, Fowler JS, Logan J, Watanabe M (1985b) Comparison of three [^{18}F] labelled butyrophenone neuroleptic drugs in the baboon using positron emission tomography. J Neurochem 44:835–844

Baron JC, Comar D, Zarifian E, Agid Y, Crouzel C, Loo H, Deniker P, Kellershohn C (1985) Dopaminergic receptor sites in human brain: positron emission tomography. Neurology (NY) 35:16–24

Barr AN, Heinze WJ, Dogger GD, Valvassor GE, Sugar O (1978) Bicaudate index in computerized tomography of Huntington's disease and cerebral atrophy. Neurology (NY) 28:1196-1200

Bes A, Guell A, Fabre N, Arne-Bes MC, Geraud G (1983) Effects of dopaminergic agonists (piribedil and bromocriptine) on cerebral blood flow and parkinsonism. J Cereb Blood Flow Metab 3 (Suppl 1):490-491

Budinger TF (1985) Quantitative single-photon emission tomography for cerebral flow and receptor distribution imaging. In: Reivich M, Alavi A (eds) Positron emission tomography. Liss, New York

Burton K, Farrell D, Li D, Calne DB (1984) Lesions of the putamen and dystonia: CT and magnetic resonance imaging. Neurology (NY) 34:962-965

Bydder GM, Steiner RE, Young IR, Hall AS, Thomas DJ, Marshall J et al. (1982) Clinical NMR imaging of the brain: 140 cases. Am J Radiol 139:215-236

Cumming P, Boyes BE, Martin WRW, Adam M, Grierson J, Ruth T, McGeer EG (1987) The metabolism of [^{18}F] 6-fluoro-L-3,4-dihydroxyphenylalanine in the hooded rat. J Neurochem 48:601-608

Drayer BP, Olanow CW, Burger P (1986) High field strength resonance imaging in patients with Parkinson's disease. Neurology (NY) 36 (Suppl 1):309

Eckelman WC, Reba RC, Rzeszotarski WJ, Gibson RE, Hill T, Holman BL, Budinger T, Conklin JJ, Eng R, Grissom MP (1984) External imaging of cerebral muscarinic acetylcholine receptors. Science 223:291-293

Edvinsson L, Harebo JE, McCulloch J, Owman C (1978) Vasomotor response of cerebral blood vessels to dopamine and dopaminergic agonists. Adv Neurol 20:85-96

Farde L, Hall H, Ehrin E, Sedvall G (1986) Quantitative analysis of D_2 dopamine receptor binding in the living human brain by PET. Science 231:258-261

Firnau G, Garnett E, Sourkes TL, Missala K (1975) [^{18}F] Fluorodopa; a unique gamma emitting substrate for dopa decarboxylase. Experientia 31:1254-1255

Fox PT, Raichle ME (1986) Focal physiological uncoupling of cerebral blood flow and oxidative metabolism during somatosensory stimulation in human subjects. Proc Natl Acad Sci USA 83:1140-1144

Fox PT, Perlmutter JS, Raichle ME (1985) A stereotactic method of anatomical localization for positron emission tomography. J Comput Assist Tomogr 9:141-153

Frackowiak RSJ, Lenzi GL, Jones T, Heather JD (1980) Quantitative measurement of regional cerebral blood flow and oxygen metabolism in man using ^{15}O and positron emission tomography: theory, procedure, and normal values. J Comput Assist Tomogr 4:727-736

Fross RD, Martin WRW, Stoessl AJ, Adam MJ, Ruth TJ, Pate BD, Calne DB (1986) The anatomic basis of dystonia. Neurology (NY) 36 (Suppl 1):119

Garnett ES, Firnau G, Nahmias C, Sood S, Belbeck L (1980) Blood-brain barrier transport and cerebral utilization of dopa in living monkeys. Am J Physiol 238:318-327

Garnett ES, Firnau G, Nahmias C (1983 a) Dopamine visualized in the basal ganglia of living man. Nature 305:137-138

Garnett ES, Firnau G, Nahmias C, Chirakal R (1983 b) Striatal dopamine metabolism in living monkeys examined by positron emission tomography. Brain Res 280:169-171

Garnett ES, Firnau G, Nahmias C, Carbotte R, Bartolucci G (1984) Reduced striatal glucose consumption and prolonged reaction time are early features in Huntington's disease. J Neurol Sci 65:231-237

Glass JP, Jankovic J, Borit A (1984) Hemiballism and metastatic brain tumor. Neurology (Cleveland) 34:204-207

Goutieres F, Aicardi J (1982) Acute neurological dysfunction associated with destructive lesions of the basal ganglia in children. Ann Neurol 12:328-332

Guell A, Geraud G, Jauzac P, Victor G, Arne-Bes MC (1982) Effects of a dopaminergic agonist (piribedil) on cerebral blood flow in man. J Cereb Blood Flow Metab 2:255-257

Guttman M, Lang AE, Garnett S, Tyndel F, Gordon A (1986) No consistent abnormality of FDG scanning in SLE chorea: further evidence that striatal hypometabolism is not the correlate of chorea. Neurology (NY) 36 (Suppl 1):309

Hayden MR, Martin WRW, Stoessl AJ, Clark C, Hollenberg S, Adam MJ, Ammann W, Harrop R, Rogers J, Ruth T, Sayre C, Pate BD (1986) Positron emission tomography in the early diagnosis of Huntington disease. Neurology (NY) 36:888-894

Herscovitch P, Markham J, Raichle ME (1983) Brain blood flow measured with intravenous $H_2\,^{15}O$. I. Theory and error analysis. J Nucl Med 24:782-789

Hill TC, Holman BL, Lovett R, O'Leary DH, Front D, Magistretti P, Zimmerman RE, Moore S, Clouse ME, Wu JL, Lin TH, Baldwin RM (1982) Initial experience with SPECT (single photon computerized tomography) of the brain using *N*-isopropyl I-123-*p*-iodoamphetamine. J Nucl Med 23:191-195

Huang SC, Phelps ME, Hoffman EJ, Sideris K, Selin CJ, Kuhl DE (1980) Noninvasive determination of local cerebral metabolic rate of glucose in man. Am J Physiol 128:E69-E82

Junck L, Gilman S, Hickwa RD, Young AB, Markel DS, Ehrenkaufer RLE (1986) PET studies of local cerebral glucose metabolism in idiopathic torsion dystonia. Neurology (NY) 36 (Suppl 1):182

Kilbourn MR, Zalutsky MR (1985) Research and clinical potential of receptor based radiopharmaceuticals. J Nucl Med 26:655-662

Kozachuk W, Salanga V, Conomy J, Smith A (1986) MRI (magnetic resonance imaging) in Huntington's disease. Neurology (NY) 36 (Suppl 1):310

Kuhl DE, Barrio JR, Huang SC, Selin C, Ackermann RF, Lear JL, Wu JL, Lin TH, Phelps ME (1982a) Quantifying local cerebral blood flow by *N*-isopropyl-*p*-[^{123}I]iodoamphetamine (IMP) tomography. J Nucl Med 23:196-203

Kuhl DE, Phelps ME, Markham CH, Metter EJ, Riege WH, Winter J (1982b) Cerebral metabolism and atrophy in Huntington's disease determined by ^{18}FDG and computed tomographic scan. Ann Neurol 12:425-434

Kuhl DE, Metter EJ, Reige WH (1984) Patterns of local cerebral glucose utilization determined in Parkinson's disease by the [^{18}F]fluorodeoxyglucose method. Ann Neurol 15:419-424

Kuhl DE, Markham CH, Metter EJ, Riege WH, Phelps ME, Mazziotta JC (1985) Local cerebral glucose utilization in symptomatic and presymptomatic Huntington's disease. In: Sokoloff L (ed) Brain imaging and brain function. Raven, New York, pp 199-209

Lang AE, Garnett ES, Firnau G, Nahmias C, Talalla A (1988) Positron tomography in dystonia. Adv Neurol 50. Dystonia 2. Eds. Fahn S, Marsden CD, Calne DB. 249-253

Lange H, Thorner G, Hopf A, Schroder KF (1976) Morphometric studies of the neuropathological changes in choreatic disease. J Neurol Sci 28:401-425

Lawler GA, Pennock JM, Steiner RE, Jenkins WJ, Sherlock S, Young IR (1983) Nuclear magnetic resonance (NMR) imaging in Wilson's disease. J Comput Assist Tomogr 7:1-8

Leenders KL, Wolfson L, Gibbs JM, Wise RJS, Causon R, Jones T, Legg NJ (1985) The effects of L-dopa on regional cerebral blood flow and oxygen metabolism in patients with Parkinson's disease. Brain 108:171-191

Leenders KL, Poewe WH, Palmer AJ, Brenton DP, Frackowiak RSJ (1986a) Inhibition of L-[^{18}F]fluorodopa uptake into human brain by amino acids demonstrated by positron emission tomography. Ann Neurol 20:258-261

Leenders KL, Palmer AJ, Quinn N, Clark JC, Firnau G, Garnett ES, Nahmias C, Jones, Marsden CD (1986b) Brain dopamine metabolism in patients with Parkinson's disease measured with positron emission tomography. J Neurol Neurosurg Psychiatry 49:853-860

Leenders KL, Frackowiak R, Quinn N, Marsden CD (1986c) Brain energy metabolism and dopaminergic function in Huntington's disease measured in vivo using positron emission tomography. Movement Disorders 1:69-78

Leenders KL, Frackowiak RSJ, Quinn N, Brooks D, Sumner D, Marsden CD (1986d) Ipsilateral blepharospasm and contralateral hemidystonia and parkinsonism in a patient: CT, MRI, and PET scanning. Movement Disorders 1:51-58

Littrup PJ, Gebarski SS (1985) MR imaging of Hallervorden-Spatz disease. J Comput Assist Tomogr. 9:491-493

Lodder J, Baard WC (1981) Paraballism caused by bilateral hemorrhagic infarctions in the basal ganglia. Neurology (NY) 31:484-486

Martin JP (1957) Hemichorea (hemiballism) without lesions in the corpus luysii. Brain 80:1-10

Martin WRW, Beckman JH, Calne DB, Adam MJ, Harrop R, Rogers JG, Ruth TJ, Sayre CI, Pate BD (1984) Cerebral glucose metabolism in Parkinson's disease. Can J Neurol Sci 11(Suppl):169-173

Martin WRW, Boyes BE, Leenders KL, Patlak CS (1985) Method for the quantitative analysis of 6-fluorodopa uptake data from positron emission tomography. J Cereb Blood Flow Metab +(Suppl 1):593-594

Martin WRW, Stoessl AJ, Adam MJ, Ammann W, Bergstrom M, Harrop R, Laihinen A, Rogers JG, Ruth TJ, Sayre CI, Pate BD, Calne DB (1987) Positron emission tomography in Parkinson's disease: glucose and dopa metabolism. Adv Neurol 45:95-98

Martin WRW, Stoessl AJ, Palmer M, Adam MJ, Ruth TJ, Grierson JR, Pate BD, Calne DB (1988) PET scanning in dystonia. Adv Neurol 50. Dystonia 2. Eds. Fahn S, Marsden CD, Calne DB. 223-229

Mata M, Fink DJ, Gainer H, Smith CB, Davidsen L, Savaki H, Schwartz WJ, Sokoloff L (1980) Activity-dependent energy metabolism in rat posterior pituitary primarily reflects sodium pump activity. J Neurochem 34:213-215

Maziere JB, Loc'h C, Hantraye P, Guillon R, Duquesnoy N, Soussaline F, Naquet R, Comar D, Maziere M (1984) ^{76}Br-bromospiroperidol: a new tool for quantitative in vivo imaging of neuroleptic receptors. Life Sci 35:1349-1356

Mazziotta JC, Phelps ME, Pahl J, Huang SC, Wapenski J, Baxter LR, Riege W, Kuhl DE, Selin C, Sumida R, Markham CH (1986) Caudate hypometabolism in asymptomatic subjects at risk for Huntington's disease. J Nucl Med 27:920

McGeer PL, McGeer EG, Suzuki JS (1977) Aging and extrapyramidal function. Arch Neurol 34:33-35

Mintun MA, Raichle ME, Kilbourne MR, Wooten GF, Welch MJ (1984a) A quantitative model for the in vivo assessment of drug binding sites with positron emission tomography. Ann Neurol 15:217-227

Mintun MA, Raichle ME, Martin WRW, Herscovitch P (1984b) Brain oxygen utilization measured with 0-15 radiotracers and positron emission tomography. J Nucl Med 25:177-187

Moerlein SM, Laufer P, Stocklin G, Pawlak, Wienhard K, Heiss WD (1986) Evaluation of ^{75}Br-labelled butyrophenone neuroleptics for imaging cerebral dopaminergic receptor areas using positron emission tomography. Eur J Nucl Med 12:211-216

Nahmias C, Garnett ES, Firnau G, Lang A (1985) Striatal dopamine distribution in parkinsonian patients during life. J Neurol Sci 69:223-230

Narbona J, Obeso JA, Tonon T, Martinez-Lage JM, Marsden CD (1984) Hemidystonia secondary to localized basal ganglia tumor. J Neurol Neurosurg Psychiatry 47:704-709

Norman D, Brant-Zawadzki M (1985) Magnetic resonance imaging of the central nervous system. In: Sokoloff L (ed) Brain imaging and brain function. Raven, New York, p 259-269

Novotny EJ, Singh G, Wallace DC, Dorfman LJ, Louis A, Sogg RL, Steinman (1986) Leber's disease and dystonia: a mitochondrial disease. Neurology (NY) 36:1053-1060

Oepen G, Ostertag C (1981) Diagnostic value of CT in patients with Huntington's chorea and their offspring. J Neurol 225:189-196

Oldendorf WH (1985) Principles of imaging structure by NMR. In: Sokoloff L (ed) Brain imaging and brain function. Raven, New York, pp 245-257

Palella TD, Hichwa RD, Ehrenkaufer RC, Rothley JM, McQuillan MA, Young AB, Kelley WN (1985) ^{18}F-fluorodeoxyglucose PET scanning in HPRT deficiency. Am J Hum Genet 37:A70

Patlak CS, Blasberg RG, Fenstermacher JD (1983) Graphical evaluation of blood-to-brain transfer constants from multiple-time uptake data. J Cereb Blood Flow Metab 3:1-7

Perlmutter JS, Raichle ME (1985) Regional blood flow in hemiparkinsonism. Neurology (NY) 35:1127-1134

Perlmutter JS, Larsen KB, Raichle ME, Markham J, Mintun MA, Kilbourn MR, Welch MJ (1986) Strategies for in vivo measurement of receptor binding using positron emission tomography. J Cereb Blood Flow Metab 6:154-169

Phelps ME, Huang SC, Hoffman EJ, Selin C, Sokoloff L, Kuhl DE (1979) Tomographic measurement of local cerebral glucose metabolic rate in humans with (F-18)2-fluoro-2-deoxy-D-glucose: validation of method. Ann Neurol 6:371-388

Raichle ME, Martin WRW, Herscovitch P, Mintun MA, Markham J (1983) Brain blood flow measured with intravenous H_2 ^{15}O. II. Implementation and validation. J Nucl Med 24:790-798

Raichle ME, Perlmutter JS, Fox PT (1984) Parkinson's disease: metabolic and pharmacological approaches with positron emission tomography. Ann Neurol 15(Suppl):131-134

Reivich M, Kuhl D, Wolf A, Greenberg J, Phelps M, Ido T, Casella V, Fowler J, Hoffman E, Alavi A, Som P, Sokoloff L (1979) The [^{18}F] fluorodeoxyglucose method for the measurement of local cerebral glucose utilization in man. Circ Res 44:127-137

Rougemont D, Baron JC, Collard P, Bustany P, Comar D, Agid Y (1984) Local cerebral glucose utilisation in treated and untreated patients with Parkinson's disease. J Neurol Neurosurg Psychiatry 47:824-830

Sax DS, Buonanno FS (1986) Putaminal changes in spin-echo magnetic resonance imaging signal in bradykinetic/rigid form of Huntington's disease. Neurology (NY) 36(Suppl 1):311

Sax DS, Buonanno FS, Kramer C, Miatto O, Kistler JP, Martin JB, Brady TJ (1985) Proton nuclear magnetic resonance imaging in Huntington's disease. Ann Neurol 18:142

Seeman P, Ulpian C, Bergeron JC, Riederer P, Jellinger K, Gabriel E, Reynolds GP, Tourtelotte WW (1985) Bimodal distribution of dopamine receptor densities in brain of schizophrenics. Science 225:728-731

Sokoloff L, Reivich M, Kennedy C, Des Rosiers MH, Patlak CS, Pettigrew KD, Sakurada O, Shinohara N (1977) The [^{14}C]deoxyglucose method for the measurement of local cerebral glucose utilization: theory, procedure, and normal values in the conscious and anesthetized albino rat. J Neurochem 28:897-916

Stober T, Wussow W, Schimrigk K (1984) Bicaudate diameter—the most specific and simple CT parameter in the diagnosis of Huntington's disease. Neuroradiology 26:25-28

Stoessl AJ, Hayden MR, Martin WRW, Clark C, Pate BD (1986a) Predictive studies in Huntington's disease. Neurology (NY) 36(Suppl 1):310

Stoessl AJ, Martin WRW, Hayden MR, Adam MJ, Ruth TJ, Rajput A, Pate BD, Calne DB (1986b) Dopamine in Huntington's disease: studies using positron emission tomography. Neurology (NY) 36 (Suppl 1):310

Stoessl AJ, Martin WRW, Clark C, Adam MJ, Ammann W, Beckman JH, Bergstrom M, Harrop R, Rogers JG, Sayre CI, Pate BD, Calne DB (1986c) PET studies of cerebral glucose metabolism in idiopathic torticollis. Neurology (NY) 36:653-657

Starosta-Rubinstein S, Young AB, Kluin K, Hill GM, Aisen AM, Gabrielsen T, Brewer GJ (1985) Quantitative clinical assessment of 25 Wilson's patients: correlation with structural changes on MRI. Neurology 35(Suppl 1):175

Suchowersky O, Hayden MR, Martin WRW, Stoessl AJ, Hildebrand AM, Pate BD, (1986) Cerebral metabolism of glucose in benign hereditary chorea. Movement Disorders 1:33-44

Wagner HN Jr. Burns HD, Dannals RF, Wong DF, Langstrom B, Duelfer T, Frost JJ, Ravert HT, Links JM, Rosenbloom SB, Lukas SE, Kramer AV, Kuhar MJ (1983) Imaging dopamine receptors in the human brain by positron emission tomography. Science 221:1264-1266

Williams F, John B, Walshe JM (1981) Wilson's disease. An analysis of the cranical computerized tomographic appearances found in 60 patients and the changes in response to treatment with chelating agents. Brain 104:735-752

Wolfson LI, Leenders KL, Brown LL, Jones T (1985) Alterations of regional cerebral blood flow and oxygen metabolism in Parkinson's disease. Neurology (NY) 35:1399-1405

Wong DF, Wagner HN Jr, Dannals RF, Links JM, Frost JJ, Ravert HT, Wilson AA, Rosenbaum AE, Gjedde A, Douglass KH, Petronis JD, Folstein JK, Toung JKT, Burns HD, Kuhar MJ (1984) Effects of age on dopamine and serotonin receptors measured by positron tomography in the living human brain. Science 226:1393-1396

Wong DF, Gjedde A, Wagner HN (1986a) Quantification of neuroreceptors in the living human brain. I. Irreversible binding of ligands. J Cereb Blood Flow Metab 6:137-146

Wong DF, Gjedde A, Wagner HN Jr, Dannals RF, Douglass KH, Links JM, Kuhar MJ (1986b) Quantification of neuroreceptors in the living human brain. II. Inhibition studies of receptor density and affinity. J Cereb Blood Flow Metab 6:147-153

Wooten GF, Collins RC (1981) Metabolic effects of unilateral lesions of the substantia nigra. J Neurosci 1:285-291

Young AB, Frey KA, Agranoff BW (1986a) Receptor assays: in vitro and in vivo. In: Phelps M, Mazziotta J, Schelbert H (eds) Positron emission tomography and autoradiography: principles and applications for the brain and heart. Raven, New York,

Young AB, Penney JB, Starosta-Rubinstein S, Markel DS, Berent S, Jewett D, Rothley J, Betley A, Hichwa R (1986b) Persons at-risk for Huntington's disease: brain metabolism determined with ^{18}F-FDG. J Nucl Med 27:920

CHAPTER 6

The Neurochemical Basis of the Pharmacology of Parkinson's Disease

O. HORNYKIEWICZ

A. Introduction

Until 1960, the pharmacology of Parkinson's disease (PD) was purely empirical. In the second half of the last century, Charcot introduced the use of belladonna extracts as mildly effective remedies in PD. Since that time, there has been hardly a drug that has not been tried in patients suffering from this disorder. The list even includes such compounds as ferric sulfate, barium chloride, strychnine, pentylenetetrazol, and diverse thyroid and pyrathyroid preparations (see HASSLER 1953). Among these chemically more or less defined, but ineffective, agents and whole, or partly purified, organ extracts, striaphorin, an extract of the bovine striatal nuclei, deserves special mention. Although ineffective as an antiparkinsonian agent, it very likely contained dopamine (DA). Striaphorin's failure as a remedy for PD might have been due to such trivial factors as too small amounts of DA contained in the marketed preparation, or the impenetrability of the blood-brain barrier to DA. From the very wide variety of drugs and (tissue) preparations that have been proposed as potential remedies for PD, it can be concluded that the results of this century-long empirical approach to pharmacotherapy of PD were highly unsatisfactory clinically. Nevertheless, it is interesting that during this period of pure pharmacologic empiricism, the prototypes of two drug groups had been found to be of some limited benefit in the treatment of PD: namely, anticholinergics (in the shape of Charcot's belladonna extracts) on one hand and amphetamine (SOLOMON et al. 1937) and apomorphine (SCHWAB et al. 1951) on the other hand. According to our present concepts, the effectiveness of these chemically diverse drugs can be related to the most basic neurochemical abnormality in the brain of PD patients—the marked loss of DA in the nigrostriatal neuron system.

B. The Basic Neurochemical Pathology of Parkinson's Disease

The crucial finding by EHRINGER and HORNYKIEWICZ (1960) of reduced DA concentrations in the basal ganglia in patients with PD furnished the basis for the concept of DA substitution as a rational approach to drug treatment of PD. This concept was introduced by HORNYKIEWICZ in 1960-1961, when he initiated trials (with the assistance of clinical neurologists) of the DA precursor L-dopa (levodopa) in patients suffering from PD (see Hornykiewicz 1986).

I. The Nigrostriatal Dopamine Neuron System

The loss of DA in all components of the basal ganglia is the most consistent and the most specific neurochemical alteration in the brain of patients with PD; it includes: (a) severe loss of DA and its metabolites homovanillic acid, 3,4-dihydroxyphenylacetic acid, and 3-methoxytyramine in the caudate nucleus, putamen (both called the striatum), globus pallidus, and the compact zone of the substantia nigra; (b) reduced concentration of the DA-synthesizing enzymes tyrosine hydroxylase and dopa decarboxylase; and (c) a reduced number of specific presynaptic DA uptake (transporter) sites (Fig. 1; Hornykiewicz and Kish 1986).

The loss of striatal DA is highly specific for parkinsonian syndrome regardless of its etiology. It has been found in:

1. Idiopathic PD;
2. Postencephalitic parkinsonism;
3. Senile arteriosclerotic brain disease with parkinsonian symptomatology (Bernheimer et al. 1973);

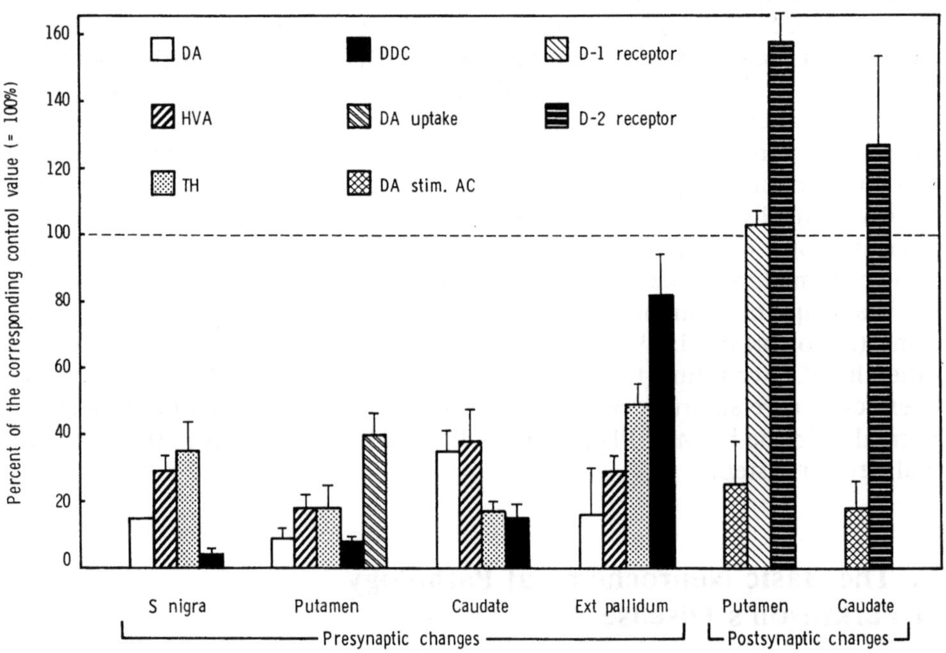

Fig. 1. Pre- and postsynaptic changes, expressed as percentage of controls (= 100%), in the nigrostriatal DA neuron system in PD. Abbreviations: DA stim. AC = DA-stimulated adenylate cyclase activity; DDC = dopa decarboxylase; HVA = homovanillic acid; TH = tyrosine hydroxylase. For references, see text

4. Other degenerative brain conditions presenting, inter alia, with parkinsonian features, such as: (a) progressive supranuclear palsy (KISH et al. 1985a; RUBERG et al. 1985), (b) striatonigral degeneration (SHARPE et al. 1973); and (c) neuronal intranuclear inclusion body disorder (KISH et al. 1985b);

5. Toxic parkinsonism, such as, (a) chronic manganese poisoning (BERNHEIMER et al. 1973); and (b) 1-methyl-4-phenyl-1,2,3,6-tetrahydropyridine (MPTP)-induced parkinsonian syndrome in humans and subhuman primates (LANGSTON et al. 1984; BURNS et al. 1985);

6. Reversible, drug-induced parkinsonism produced by reserpine or the antipsychotic neuroleptics of the phenothiazine and butyrophenone type; these drugs produce an actual or functional lack of DA (via DA depletion or receptor blockade) at the synaptic level (see HORNYKIEWICZ 1972).

II. Extrastriatal Dopamine Neurons

Moderate loss of DA (and homovanillic acid) has been observed in many subcortical limbic regions of the forebrain (nucleus accumbens, medial olfactory area, amygdaloid nucleus); in several limbic cortical and neocortical areas (entorhinal, cingulate, hippocampal, and frontal cortices) (FARLEY et al. 1977; PRICE et al. 1978; SCATTON et al. 1983, 1984); and in the mesencephalic ventral tegmental area (JAVOY-AGID et al. 1981), which is the area of origin of the DA

Table 1. Extrastriatal dopamine changes in Parkinson's disease: comparison with putamen

Brain region	Dopamine loss (%)
Subcortical regions	
Nucleus accumbens	58[a]
Medial olfactory area	22[b]
Lateral olfactory area (olfactory tubercle; substantia innominata; anterior perforate substance)	0.0[b]
Lateral hypothalamic area	90[a]
Amygdaloid nucleus	45[c]
Ventral tegmental area (VTA)	50[c]
Cortical regions	
Parolfactory cortex (Brodmann area 25)	90[b]
Frontal cortex	61[c]
Entorhinal cortex	68[c]
Cingulate cortex	52[c]
Hippocampus	68[c]
Striatum	
Putamen	91[a]

[a] PRICE et al. 1978
[b] FARLEY et al. 1977
[c] SCATTON et al. 1984

projections to subcortical limbic and cortical regions. The only extrastriatal forebrain regions with marked loss of DA were the lateral hypothalamus and the parolfactory gyrus (FARLEY et al. 1977; PRICE et al. 1978). A moderate reduction of DA in the spinal cord (SCATTON et al. 1984) was recently found to be statistically nonsignificant (Table 1; SCATTON et al. 1986).

III. Nondopamine Neuron Systems

Significant changes in several non-DA systems have also been observed in the parkinsonian brain. Reduced tissue concentrations in the striatum, substantia nigra, globus pallidus, nucleus accumbens, limbic and/or cortical regions, hippocampus, cerebellar cortex, and spinal cord have been reported for: (a) norepinephrine (EHRINGER and HORNYKIEWICZ 1960; BERNHEIMER et al. 1963; FARLEY and HORNYKIEWICZ 1976; KISH et al. 1984; RINNE and SONNINEN 1973; SCATTON et al. 1983, 1984); (b) serotonin (BERNHEIMER et al. 1961; LLOYD 1972; RINNE et al. 1974; SCATTON et al. 1983); (c) the neuropeptides Met- and Leu-enkephalin, cholecystokinin, substance P, somatostatin, neurotensin (BISSETTE et al. 1985; EPELBAUM et al. 1983; MAUBORGNE et al. 1983; STUDLER et al. 1982; TAQUET et al. 1983; TENOVUO et al. 1984); (d) glutamate decarboxylase (BERNHEIMER and HORNYKIEWICZ 1962; GASPAR et al. 1980; LAAKSONEN et al. 1976; LLOYD and HORNYKIEWICZ 1973; RINNE et al. 1974); and (e) choline acetyltransferase (LLOYD et al. 1975; PERRY RH et al. 1983; REISINE et al.

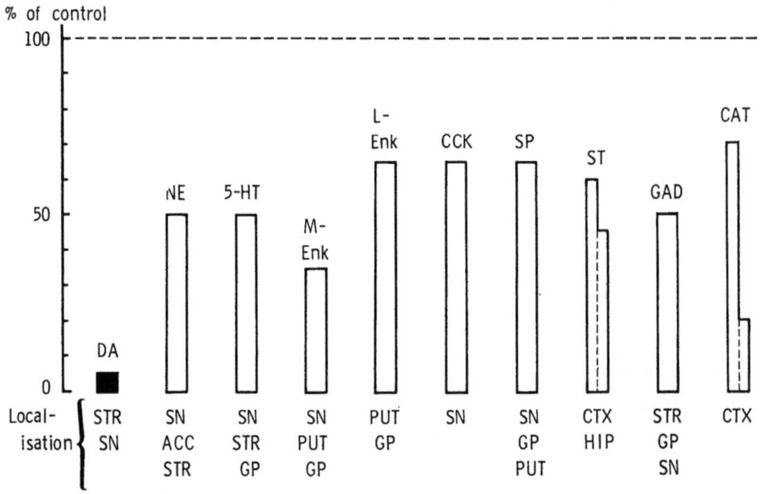

Fig. 2. Quantitative aspects of some non-DA changes in selected brain regions in PD; comparison with the nigrostriatal DA loss. Abbreviations: brain region — ACC = nucleus accumbens; CTX = cortex (cerebral); GP = globus pallidus; HIP = hippocampus; PUT = putamen; SN = substantia nigra; STR = striatum (putamen and caudate); neurotransmitter markers — CAT = choline acetyltransferase; CCK cholecystokinin; GAD = glutamate decarboxylase; 5-HT = serotonin; L-Enk = leucine-enkephalin; M-Enk = methionine-enkephalin; NE norepinephrine; SP = substance P; ST = somatostatin. For references, see text

1977; RUBERG et al. 1982). In contrast, the levels of γ-aminobutyric acid (GABA) in the parkinsonian striatum have been found above normal (PERRY TL et al. 1983; KISH et al. 1986); similarly, an increased cholinergic activity in the parkinsonian striatum, owing to loss of dopaminergic inhibition, can be deduced from observations in experimental animals which show that normally dopaminergic neurons strongly inhibit the activity of the striatal acetylcholine neurons (see BARTHOLINI 1980). It is noteworthy that the changes observed in non-DA systems are generally much less pronounced compared with the severe loss of nigrostriatal DA (Fig. 2).

C. Pathophysiologic and Pharmacologic Significance of the Biochemical Brain Abnormalities in Parkinson's Disease

I. Nigrostriatal Dopamine Loss

The loss of DA within the nigrostriatal neuron system of patients with PD is directly related to the main motor deficits of this disorder. The following evidence can be cited in support of this conclusion: (a) in fully developed PD the striatal DA loss is exceptionally severe in degree (Fig. 1 and 2); (b) striatal DA deficiency is present in degenerative brain disorders other than PD presenting with parkinsonian symptoms (see Sect. B.I), including progressive supranuclear palsy, striatonigral degeneration, and neuronal intranuclear inclusion body disorder; (c) as demonstrated by MPTP-induced parkinsonism, the loss of striatal DA is sufficient to produce all major symptoms of PD proper; (d) in patients with PD there exists a significant correlation between the degree of striatal DA loss and the severity of nigral cell loss (BERNHEIMER et al. 1973); (e) the severity of the parkinsonian symptoms, e.g., akinesia, is positively correlated with the degree of striatal DA loss (BERNHEIMER et al. 1973); this correlation also holds for patients with hemiparkinsonism, in whom a more

Table 2. Relationship between the loss of striatal DA and homovanillic acid (HVA) and the severity of parkinsonian symptoms

Diagnostic grouping	Caudate Loss (%)		Putamen Loss (%)	
	DA	HVA	DA	HVA
Hemiparkinsonism (right side) [a]				
Right striatum	62		80	
Left striatum	83		96.5	
PD (bilateral) [b]				
Mild akinesia	78	48	87	63
Marked akinesia	92	82	98.5	81

[a] BAROLIN et al. 1964
[b] BERNHEIMER et al. 1973

severe DA loss is found in the striatum contralateral to the affected side of the body (Table 2; BAROLIN et al. 1964); (f) the ultimate confirmation and most crucial evidence for the pivotal role played by the striatal DA loss for the symptomatology of PD is the therapeutic success of the DA substitution therapy, e.g., with L-dopa (BIRKMAYER and HORNYKIEWICZ 1961, 1962; BARBEAU et al. 1962; COTZIAS et al. 1967; see HORNYKIEWICZ 1986).

Together with striatal DA deficiency as the most characteristic neurochemical alteration in the PD brain, the high antiparkinsonian efficacy of dopaminergic drugs provides a sound basis for the concept of DA substitution therapy as a rational approach to the pharmacology of PD. The potentiation of L-dopa's clinical effectiveness by means of inhibitors of the extracerebral dopa decarboxylase (carbidopa; benserazide) or monoamine oxidase (MAO) (especially selective MAO-B inhibition with deprenyl) is based on, and further substantiates, the DA substitution concept. In addition, the fact that, in principle, DA substitution reverses all main motor symptoms of PD shows that DA is crucially involved in all major aspects of basal ganglia motor control (MARSDEN 1980). The role of brain DA for striatal (motor) function is substantially supported by a large body of experimental evidence obtained in laboratory animals, including MPTP-treated primates (see EVERED and O'CONNOR 1984; MARKEY et al. 1986).

II. Extrastriatal Dopamine Changes

There are no studies in the literature reporting possible correlations between the severity of the various motor deficits in patients with PD and the degree of DA changes in extrastriatal brain regions. From behavioral and pharmacologic studies in laboratory animals (PIJNENBURG et al. 1976; see also CHRONISTER and DE FRANCE 1981) it may be surmised that DA reduction in the nucleus accumbens may aggravate the motor deficits (especially akinesia) of PD, whereas DA loss in other limbic and/or cortical regions may contribute to disorders of affect (depression) or cognition (dementia) sometimes seen in patients with PD. It has been suggested that, in addition to their motor functions, the mesotelencephalic DA neurons may be involved in the control of nonmotor higher (cognitive) brain functions (see IVERSEN 1977). Despite these possibilities, it remains uncertain whether the comparatively mild degree of DA reduction found in the extrastriatal regions (see Table 1) is sufficient to produce overt (decompensated) clinical deficits. Therefore, it is unknown to what degree the correction, by means of DA substitution, of the possible DA-related dysfunction in the extrastriatal brain regions contributes to the beneficial antiparkinsonian effect of this therapy.

III. Nondopamine Changes

1. Changes in the Basal Ganglia

The importance, for the full clinical expression of PD, of the diverse changes in non-DA systems is at present difficult to assess because not enough is

Table 3. Actions of non-DA neurotransmitters on DA activity in the basal ganglia and their possible relation to PD symptomatology

Neurotrans-mitter	Physiologic mode and site of action and/or interaction with the brain DA system	Change in PD	Probable effect on PD motor symptoms due to DA deficiency
Norepinephrine	Acts synergistically with the mesotelencephalic DA system regarding motor activation	Decreased	Aggravation
Acetylcholine	Produces catalepsy and enhances striatal DA deficiency symptoms (especially rigidity and akinesia)	Increased (functio-nally; due to DA loss)	Aggravation
Substance P	Stimulates DA neurons in substantia nigra	Decreased	Aggravation
Cholecystokinin	Facilitates certain DA neurons (especially nucleus accumbens)	Decreased	Aggravation (?)
Serotonin	Inhibits DA neurons in substantia nigra	Decreased	Amelioration
GABA	Inhibits cholinergic activity in striatum	Increased (= deactiva-tion?)	Aggravation (?) (via choliner-gic activation)
Opioids/ neurotensin	Facilitate or inhibit DA neuron function, depending on site of action (cell bodies/terminals)	Decreased	?

For key references; see text

known about the role of these neurohumoral systems in normal brain function. Since most of these changes are small in comparison with the profound striatal DA loss, they may be fully compensated by the remaining neurons, and thus fail to produce any overt functional-clinical deficits. On the other hand, there exists a large body of experimental evidence obtained in laboratory animals suggesting the existence of functional interactions and a dynamic interplay between the brain DA neurons and several non-DA brain systems (HORNYKIEWICZ 1976; LLOYD 1977a; BARTHOLINI et al. 1981). The concept of a functional interaction implies that disturbance of one neuronal system will produce a disturbance in one or more of the other systems with which it is functionally connected. Of the many known interactions between brain DA and the non-DA systems, the following may be of special relevance to PD and its pharmacotherapy (Table 3): (a) the norepinephrine-DA interrelationship; (b) the acetylcholine-DA link; (c) the GABA-DA interplay; and (d) influences of serotonin and neuropeptides on nigrostriatal DA function.

a. Norepinephrine

Pharmacologic evidence obtained in laboratory animals suggests that, in respect to locomotor activity, brain norepinephrine enhances the effectiveness of the forebrain DA neurons (SVENSSON 1971; ANDÉN et al. 1973; HORNYKIEWICZ 1976; DONALDSON et al. 1978). The observed reduction of norepinephrine in the forebrain in PD (especially in the substantia nigra and nucleus accumbens) may, therefore, further aggravate the motor deficits primarily due to loss of striatal DA. Such an aggravating role of the norepinephrine changes would have obvious clinical and pharmacologic implications. It may also explain why L-dopa, which is converted in the body to DA as well as norepinephrine, is considerably more potent as an antiparkinsonian drug than the compounds acting directly on the brain DA receptors (but lacking a major norepinephrine agonist α_1 activity). This possibility should be kept in mind when trying to develop new DA substituting drugs for PD.

b. Acetylcholine

There is good evidence that cholinergic stimulation of the striatum produces catalepsy in laboratory animals (ZETLER 1968) and aggravates motor deficits, such as rigidity and akinesia, in patients with PD (DUVOISIN 1967). Since normally striatal DA activity exerts a tonic inhibitory influence on the activity of the cholinergic neurons (see HORNYKIEWICZ 1976; LEHMAN and LANGER 1983), the loss of this dopaminergic inhibitory influence in PD will result in cholinergic overactivity and a corresponding aggravation of the PD symptoms. In view of this possibility, the low efficacy of anticholinergic drugs in the pharmacotherapy of PD is somewhat surprising; the more so as anticholinergics are quite potent in reversing the Parkinson-like condition produced by brain DA receptor blocking drugs (neuroleptics). In this respect, it is noteworthy that the activity of choline acetyltransferase, the enzyme responsible for acetylcholine synthesis, has been reported to be reduced in PD striatum (LLOYD et al. 1975; REISINE et al. 1977). This finding has been interpreted by LLOYD (1977 b) as representing a downregulation of striatal acetylcholine synthesis so as to minimize the adverse consequences of the relative cholinergic overactivity. This interpretation would help to explain why the additional inhibition, by anticholinergic medication, of the already downregulated striatal cholinergic neuronal activity has only a modest therapeutic effect.

c. γ-Aminobutyric Acid

Although the precise role in PD symptomatology of the elevated GABA levels in PD striatum is at present unknown, these alterations deserve special attention in view of the importance of the DA-GABA interactions for the normal function of the basal ganglia (HORNYKIEWICZ 1976; LLOYD 1977 a; GALE and CASU 1981; SCHEEL-KRÜGER 1986). The finding that, in contrast to the 6-hydroxy-DA model of PD in the rat (VINCENT et al. 1978), the elevated GABA concentrations in PD striatum are not accompanied by a corresponding increase in the GABA-synthesizing enzyme glutamate decarboxylase, and, in addition, are inversely related to the severity of DA loss (KISH et al. 1986), suggests that there may be a DA-related deactivation (downregulation) of GABA neuron activity in PD striatum, with reduction of GABA release as its prominent feature. Since it has been suggested that GABA exerts an inhibitory influence on the striatal cholinergic activity (SCATTON and BARTHOLINI 1982), a deactivation of the striatal GABA neuron activity in response to the severe loss of DA would be expected to aggravate, via increase in cholinergic activity, the parkinsonian condition. This may be the reason why, in the 6-hydroxy-DA model of PD in the rat, GABAergic drugs potentiated the effect of DA agonists (BENNETT et al. 1987), and in patients with PD the GABAergic prodrug progabide potentiated the antiparkinsonian effect of the L-dopa treatment (BERGMAN et al. 1984). Clearly, elucidation of this aspect of the DA-GABA interaction may greatly contribute to the development of better drug regimes for PD.

d. Serotonin, Neuropeptides

Regarding the functional implications for PD of changes in systems other than the non-DA basal ganglia systems, pharmacologic evidence suggests that the serotonin system produces an overall inhibitory effect on the activity of the DA neurons in the substantia nigra (see DRAY 1981), whereas substance P stimulates these neurons (see GLOWINSKI et al. 1982; IVERSEN 1982); cholecystokinin also facilitates DA neurons, especially at the level of the nucleus accumbens (CRAWLEY et al. 1985; VANDERHAEGHEN and CRAWLEY 1985). Thus, it may be concluded that the reduction in PD of the basal ganglia substance P and cholecystokinin may aggravate, whereas the reduction in brain serotonin will tend to mitigate the motor symptoms due to striatal DA loss. No clinically applicable consequences have yet resulted from these possibilities.

2. Changes Outside the Basal Ganglia

One can only speculate about the clinical importance of the non-DA changes occurring in regions of the PD brain other than the striatum. In principle, some of these abnormalities are likely to contribute indirectly to the characteristic DA-related disturbances, whereas other changes may produce symptoms unrelated to PD proper. Thus, since norepinephrine has been found to facilitate the spinal flexor reflex activity (ANDÉN 1970), the (moderate) reduction of spinal norepinephrine may aggravate the parkinsonian movement deficits. Similarly, the loss of cerebellar norepinephrine in PD has been suggested to contribute to abnormalities of cerebellum-controlled motor performance as well as abnormalities of posture and equilibrium (KISH et al. 1984). In contrast, the cortical, limbic, and hypothalamic reduction of norepinephrine and serotonin may be involved in nonmotor symptoms such as autonomic and/or psychiatric disturbances sometimes found in PD. This may also apply to the loss of cholinergic forebrain neurons (especially nucleus basalis innervation) as well as cortical and hippocampal somatostatin found in many PD brains; these changes may be responsible for symptoms of dementia which may occur in up to 30% of patients with advanced PD (see HORNYKIEWICZ and KISH 1984, 1986). As in the case of the striatal non-DA changes, the changes of these systems outside the striatum have not yet resulted in any clinically significant pharmacotherapeutic advances.

D. Striatal Dopamine Deficiency and the Pharmacotherapy of Parkinson's Disease: Special Aspects of Dopamine Substitution

The profound brain DA loss in PD triggers several functional neurochemical changes in the affected basal ganglia areas. These changes are part of the general phenomenon of neuronal plasticity of the brain; they affect, in a typical manner, the function of the surviving presynaptic DA terminals as well as the postsynaptic DA elements. Since some of these changes play an important part in the effectiveness of the DA substitution therapy in PD, they will be discussed in the following subsections.

I. Compensatory Changes in the Nigrostriatal Dopamine Neurons

In PD, the nigrostriatal DA system responds to its partial damage with two major adaptive changes (see Hornykiewicz and Kish 1986); (a) presynaptically, there occurs an increase in DA metabolism in the remaining DA neurons, as judged by the shifting of the homovanillic acid: DA ratio in PD striatum in favor of the metabolite (Fig. 3; Bernheimer and Hornykiewicz 1965); and (b) postsynaptically, there develops an increased sensitivity of the affected striatum to DA, i.e., the phenomenon of "denervation supersensitivity," as evidenced by increase in the number of postsynaptic striatal D_2 DA binding sites (receptors) (Lee et al. 1978; Rinne 1982; Guttman and Seeman 1985).

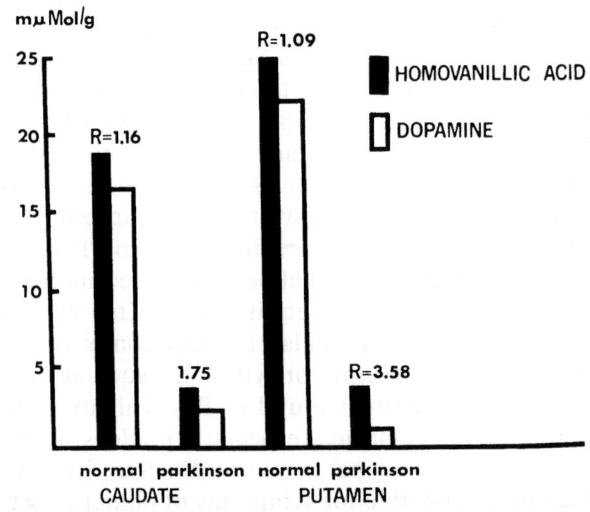

Fig. 3. Molar ratios (*R*) "homovanillic acid/DA" in caudate and putamen of patients with PD and control subjects. The greatly increased ratios in the PD striatum, especially the putamen, indicate that the remaining DA neurons are highly hyperactive, releasing (and synthesizing) considerably more DA than in (normal) subjects with intact nigrostriatal DA neurons. (Data from Bernheimer and Hornykiewicz 1965)

Both these changes are aimed at maximizing the physiologic effectiveness of the remaining DA neurons so as to maintain striatal function near normal levels despite severe DA neuron loss; they represent a powerful means of compensation for the partially lost brain DA function.

II. Compensated and Decompensated Stages of Parkinson's Disease and the Goal of Dopamine Substitution

The extent to which the partly damaged brain DA neuron system is capable of functionally compensating for the lost DA activity is clearly shown by the observation that PD symptoms do not become clinically manifest unless the striatum loses 80% or more of its DA (BERNHEIMER et al. 1973; HORNYKIEWICZ and KISH 1986).

On the basis of this unusually high capacity for compensation, PD as a progressive degenerative disorder of the basal ganglia DA neurons can be divided into two stages (HORNYKIEWICZ 1973; BERNHEIMER et al. 1973): (a) the compensated, clinically silent, stage during which the DA loss is less than 80%; and (b) the stage of functional decompensation which is characterized

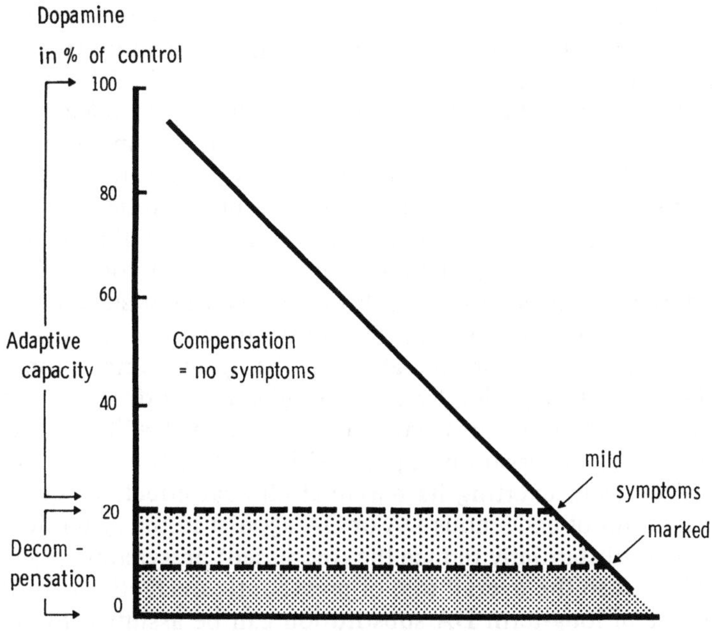

Fig. 4. Compensated and decompensated stages of PD as a function of critical loss of nigrostriatal DA. The large *(open)* triangular area represents the clinically asymptomatic (i.e., compensated) stage of striatal DA loss. Only when the DA loss exceeds the critical value of 80% *(stippleed areas)* do clinical symptoms become manifest (i.e., decompensation). The goal of DA substitution therapy is to achieve a functional recompensation of the decompensated stage of striatal DA deficiency by increasing dopaminergic activity above the critical mark of 80% reduction

by clinically manifest symptoms of striatal DA deficiency as a result of striatal DA loss exceeding the critical value of 80% (Fig. 4).

The concept of a critical level of striatal DA loss as a neurochemical correlate of the clinically compensated and decompensated stages of PD implies that the goal of DA substitution therapy will consist in reversing the clinically overt decompensated PD to the stage of functional compensation. In short, the goal of DA substitution therapy is the clinical "dopaminergic recompensation" of the striatal DA deficiency.

III. The Role of the Compensatory Changes for Dopamine Substitution

From the functional point of view, it is of special significance that, during the course of the progressive striatal DA loss, each of the two compensatory mechanisms evolves in a typical, time-related, manner and therefore plays its own specific role in the pathophysiology and pharmacotherapy of PD.

1. The Role of Presynaptic Overactivity

There is evidence indicating that the overactivity of the presynaptic DA terminals is switched on early in the course of the progressive loss of striatal DA: in patients with PD as well as in rats, a 30%–40% loss of nigral DA neurons is sufficient to increase the rate of DA metabolism significantly in the remaining neurons (HORNYKIEWICZ and KISH 1986; AGID et al. 1973; HEFTI et al. 1980). Thus, the presynaptic increase in DA synthesis and release (turnover) can be assumed to be a major physiologic mechanism of compensation for the progressive loss of striatal DA function, especially during the early stages of PD. During the stage of clinically overt PD, the presynaptic overactivity, although incapable of complete compensation, will undoubtedly contribute toward a reduction of the clinical consequences (i.e., symptoms) of the severe striatal DA deficiency present during this decompensated stage of PD.

With respect to DA substitution, the presynaptic increase in DA release can, pharmacodynamically, be thought of as being in competition with the DA or the DA agonist supplied by the DA substituting drug. The increased amounts of the endogenous, presynaptically released DA will also prevent the development of full denervation supersensitivity, and thus prevent the DA substituting drug from exerting its maximal clinical effect. However, by increasing the amount of exogenous DA or DA agonist in the synaptic area, DA substitution will substantially reduce the (increased) presynaptic release of endogenous DA, and so the overall role of presynaptic overactivity for the clinical effectiveness of long-term DA substitution can be assumed to be, on balance, only modest or negligible.

2. The Importance of Postsynaptic Supersensitivity for the Pharmacology of Parkinson's Disease

In contrast to the presynaptic overactivity of the remaining DA neurons, the postsynaptic supersensitivity requires a profound loss of nigral DA neurons in order to develop. In rats, the number of postsynaptic D_2 DA binding sites in the denervated striatum starts to rise significantly only when the loss of the nigrostriatal DA neurons exceeds 95% (CREESE and SNYDER 1979). This is in accord with the observation that behavioral supersensitivity (rotational behavior) to apomorphine and L-dopa in rats with unilateral lesions of the nigrostriatal path did not become apparent unless 90% or more of the nigrostriatal neurons had been destroyed (HEFTI et al. 1980). Consequently, this mechanism becomes functionally effective very late in the course of the striatal DA deficiency, namely when the striatal DA deficiency has already reached the stage of functional decompensation. This suggests that, compared with the presynaptic overactivity, the postsynaptic supersensitivity to DA may be of secondary importance as a mechanism involved in postponing the development of the decompensated stage of PD.

Unimportant as it may be as a compensatory mechanism, the postsynaptic supersensitivity of the PD striatum is of crucial importance for at least three aspects of the pharmacology of DA substitution in PD:

1. High efficacy of DA substitution therapy, both with dopaminergic drugs and DA-producing autografts of adrenal medullary tissue;
2. Relatively selective effect of DA substitution therapy on the most severly DA-depleted basal ganglia structures;
3. Effectiveness of the selective DA autoreceptor agonists on the denervated postsynaptic DA receptors—a potentially new class of very specific antiparkinsonian agents.

a. Special Sensitivity of the Parkinsonian Striatum to Dopamine Substitution

Clinical-pharmacologic experience indicates that PD patients are considerably more sensitive to the clinical (motor) effects of DA substitution than individuals with an intact basal ganglia DA neuron system (healthy volunteers). This special sensitivity to the pharmacologic effects of dopaminergic drugs is most likely due to the postsynaptic denervation supersensitivity of the PD striatum; this supersensitivity enables the clinician to keep the clinically effective doses of L-dopa or other drugs well below the doses that would be required to produce comparable motor effects in non-PD individuals with normosensitive striatal DA receptors.

b. Regional Selectivity of Dopamine Substitution Therapy

In clinically moderate doses, the DA substituting drugs can be expected to act relatively selectively on the brain regions most severely affected in PD. This follows from the fact that those basal ganglia regions with the most severe loss of DA will also be the regions developing the highest degree of denervation supersensitivity, i.e., they will be especially sensitive to the pharmacologic-therapeutic effects of DA and DA agonists. However, prolonged accumulation of high amounts of agonist molecules in the synaptic area is known to desensitize the respective receptor sites; therefore, long-term high dose DA substitution therapy will by necessity result in partial or total loss of the regional selectivity of this medication. In short, DA substitution therapy, especially in the higher dose range, is a self-limiting type of drug treatment (see HORNYKIEWICZ 1981).

c. New Approaches to Dopamine Substitution

α. Selective Dopamine Autoreceptor Agonists as a New Class of Specific Potential Antiparkinsonian Agents

As demonstrated in experimental models of PD, including MPTP-induced parkinsonism in the rhesus monkey, the denervated postsynaptic striatal (D_2) DA receptor becomes highly sensitive to a new class of dopaminergic drugs, the so-called selective DA autoreceptor agonists (see CARLSSON 1983; HINZEN et al 1986). In contrast to the "unselective" agonists, e.g., apomorphine and the ergoline derivatives, which in low doses stimulate the DA autoreceptors and, in addition, in higher doses stimulate the postsynaptic receptors, the "selective" DA autoreceptor agonists do not stimulate the normosensitive postsynaptic receptors; however, they are highly effective on the denervated striatal postsynaptic DA receptors. Therefore, in PD patients, the postsynaptic DA agonist effects of these drugs should be confined to those diseased basal ganglia regions with most severe DA loss (denervation), with little action (or an anti-DA effect) in other brain regions. By virtue of their dual DA agonist properties, i.e., their highly selective presynaptic site of action on intact DA systems and their strong postsynaptic action on the denervated DA receptor, the selective DA autoreceptor agonists can be expected to have more specific and more selective antiparkinsonian effects and, most probably, fewer side effects than the presently used (unselective) DA substituting drugs. In this respect, the "antidopaminergic" effect of the selective DA autoreceptor agonists on normosensitive DA systems may be of special advantage regarding the occurrence of psychotic reactions which are frequently elicited by the common DA substituting drugs. Of the DA autoreceptor agonists presently tested in laboratory animals, B-HT 920 (6-allyl-2-amino-5,6,7,8-tetrahydro-4*H*-thiazolo-[4,5-d]azepine) seems to be an especially selective and specific compound (HINZEN et al. 1986).

β. Autografting of Dopamine-Producing Cells into the Parkinsonian Striatum

As a result of exhaustive investigations in laboratory animals (see BJÖRKLUND and STENEVI 1985), autografting of DA-producing adrenal medullary (chromaffin) tissue to the striatum has been attempted in patients with PD (BACKLUND et al. 1985; MADRAZO et al. 1987). Although many basic research, medicoethical, and technical questions still have to be resolved, the favorable result of this procedure in two patients with advanced PD is highly encouraging (MADRAZO et al. 1987). Since grafts of DA-producing (fetal nigral) tissue only partially restore to normal the reduced DA concentrations in the denervated rat striatum (SCHMIDT et al. 1983), the clinical success of this novel approach to DA substitution in PD crucially depends on the presence of supersensitive postsynaptic DA receptor sites in the affected striatum.

E. Prospects of a Preventive Drug Treatment in Parkinson's Disease

The occurrence of MPTP-induced parkinsonism in humans and its duplication in subhuman primates has greatly stimulated research on the possible cause, or causes, of PD. To date, these studies have significantly contributed to our knowledge about the factors that may be involved in producing selective death of the DA-containing neurons in the substantia nigra (see MARKEY et al. 1986). The observations made in laboratory animals, showing that inhibition of brain MAO-B by pargyline or deprenyl (HEIKKILA et al. 1984; LANGSTON et al. 1984) as well as administration of DA uptake inhibitors (JAVITCH et al. 1985) and antioxidant agents (L-tocopherol, β-carotene, ascorbic acid, *N*-acetylcysteine) (PERRY et al. 1985) afforded complete or partial protection from the MPTP-induced effects on brain DA, has raised hopes that preventive drug treatment of the nigral cell death in PD may, in principle, be possible.

References

Agid Y, Javoy F, Glowinski J (1973) Hyperactivity of remaining dopaminergic neurons after partial destruction of the nigrostriatal dopaminergic system in the rat. Nature New Biol 245:150-152

Andén NE (1970) Effects of amphetamine and some other drugs on central catecholamine mechanisms. In: Costa E, Garattini S (eds) Amphetamines and related compounds, Raven, New York, p 447

Andén NE, Strömbom U, Svensson TH (1973) Dopamine and noradrenaline receptor stimulation: reversal of reserpine-induced suppression of motor activity. Psychopharmacology 29:289-298

Backlund EO, Granberg PO, Hamberger B, Sedvall G, Seiger Å, Olson L (1985) Transplantation of adrenal medullary tissue to striatum in parkinsonism. In: Björklund A, Stenevi U (eds) Neural grafting in the mammalian CNS. Elsevier, Amsterdam, p 551

Barbeau A, Sourkes TL, Murphy GF (1962) Les catécholamines dans la maladie de Parkinson. In: deAjuriaguerra J (ed) Monoamines et système nerveux centrale. Masson, Paris, p 247

Barolin GS, Bernheimer H, Hornykiewicz O (1964) Seitenverschiedenes Verhalten des Dopamins (3-Hydroxytyramin) im Gehirn eines Falles von Hemiparkinsonismus. Arch Neurol Neurochir Psychiatr 94:241-248

Bartholini G (1980) Interaction of striatal dopaminergic, cholinergic and GABA-ergic neurons: relation to extrapyramidal function. Trends Pharmacol Sci 1:138-141

Bartholini G, Scatton B, Worms P, Zivkovic B, Lloyd KG (1981) Interactions between GABA, dopamine, acetylcholine, and glutamate-containing neurons in the extrapyramidal and limbic systems. In: Di Chiara G, Gessa GL (eds) GABA and the basal ganglia. Raven, New York p 119

Bennett JP, Ferrari MB, Cruz CJ (1987) GABA-mimetic drugs enhance apomorphine-induced contralateral turning in rats with unilateral nigrostriatal dopamine denervation: implications for the therapy of Parkinson's disease. Ann Neurol 21:41-45

Bergmann KJ, Limongi JCP, Lowe YH, Mendoza MR, Yahr MD (1984) Potentiation of the "DOPA" effect in parkinsonism by a direct GABA receptor agonist. Lancet 1:559

Bernheimer H, Hornykiewicz O (1962) Das Verhalten einiger Enzyme im Gehirn normaler und Parkinson-kranker Menschen. Arch Exp Pathol Pharmakol 243:295-299

Bernheimer H, Hornykiewicz O (1965) Herabgesetzte Konzentration der Homovanillinsäure im Gehirn von Parkinson-kranken Menschen als Ausdruck der Störung des zentralen Dopaminstoffwechsels. Klin Wochenschr 43:711-715

Bernheimer H, Birkmayer W, Hornykiewicz O (1961) Verteilung des 5-Hydroxytryptamins (Serotonin) im Gehirn des Menschen und sein Verhalten bei Patienten mit Parkinson-Syndrom. Klin Wochenschr 39:1056-1059

Bernheimer H, Birkmayer W, Hornykiewicz O (1963) Zur Biochemie des Parkinsonsyndroms des Menschen. Klin Wochenschr 41:564-569

Bernheimer H, Birkmayer W, Hornykiewicz O, Jellinger K, Seitelberger F (1973) Brain dopamine and the syndromes of Parkinson and Huntington. J Neurol Sci 20:415-455

Birkmayer W, Hornykiewicz O (1961) Der L-3,4-Dioxyphenylalanin (=DOPA)-Effekt bei der Parkinson-Akinese. Wien Klin Wochenschr 73:787-788

Birkmayer W, Hornykiewicz O (1962) Der L-Dioxyphenylalanin (= L-DOPA)-Effekt beim Parkinson-Syndrom des Menschen: Zur Pathogenese und Behandlung der Parkinson-Akinese. Arch Psychiatr Nervenkr 203:560-574

Bissette G, Nemeroff CB, Decker MW, Kizer JS, Agid Y, Javoy-Agid F (1985) Alterations in regional brain concentrations of neurotensin and bombesin in Parkinson's disease. Ann Neurol 17:324-328

Björklund A, Stenevi (1985) Neural grafting in the mammalian CNS, Fernström Foundation Series vol 5. Elsevier, Amsterdam

Burns RS, LeWitt PA, Ebert MH, Pakkenberg H, Kopin IJ (1985) The clinical syndrome of striatal dopamine deficiency: parkinsonism induced by 1-methyl-4-phenyl-1,2,3,6-tetrahydropyridine (MPTP). N Engl J Med 312:1418-1421

Carlsson A (1983) Dopamine receptor agonists: intrinsic activity vs. state of receptor. J Neural Transm 57:309-315

Chronister RB, De France JF (eds) (1981) The neurobiology of the nucleus accumbens. Haer Inst Electrophysiol Research, Brunswick

Cotzias GC, Van Woert MH, Schiffer LM (1967) Aromatic amino acids and modification of parkinsonism. N Engl J Med 276:374-379

Crawley JN, Stivers JA, Blumstein LK, Paul SM (1985) Cholecystokinin potentiates dopamine-mediated behaviours: evidence for modulation specific to a site of coexistence. J Neurosci 5: 1972-1983

Creese I, Snyder SH (1979) Nigrostriatal lesions enhance striatal [^{3}H]apomorphine and [^{3}H]spiroperidol binding. Eur J Pharmacol 56:277-281

Donaldson IM, Dolphin AC, Jenner P, Pycock C, Marsden CD (1978) Rotational behavior produced in rats by unilateral electrolytic lesions of the ascending noradrenergic bundles. Brain Res 138:487-509

Dray A (1981) Serotonin in the basal ganglia: functions and interactions with other neuronal pathways. J Physiol (Paris) 77:393-403

Duvoisin RC (1967) Cholinergic-anticholinergic antagonism in parkinsonism. Arch Neurol 17:124-136

Ehringer H, Hornykiewicz O (1960) Verteilung von Noradrenalin und Dopamin (3-Hydroxytyramin) im Gehirn des Menschen und ihr Verhalten bei Erkrankungen des extrapyramidalen Systems. Klin Wochenschr 38:1236-1239

Epelbaum J, Ruberg M, Moyse E, Javoy-Agid F, Dubois B, Agid Y (1983) Somatostatin and dementia in Parkinson's disease. Brain Res 278:376-379

Evered D, O'Connor M (eds) (1984) Functions of the basal ganglia. Ciba Found Symp 107: p 281

Farley IJ, Hornykiewicz O (1976) Noradrenaline in subcortical brain regions of patients with Parkinson's disease and control subjects. In: Birkmayer W, Hornykiewicz O (eds) Advances in parkinsonism. Roche, Basel, p 178

Farley IJ, Price KS, Hornykiewicz O (1977) Dopamine in the limbic regions of the human brain: normal and abnormal. Adv Biochem Psychopharmacol 16:57-64

Gale K, Casu M (1981) Dynamic utilization of GABA in substantia nigra: regulation by dopamine and GABA in the striatum, and its clinical and behavioural implications. Mol Cell Biochem 39:369-405

Gaspar P, Javoy-Agid F, Ploska A, Agid Y (1980) Regional distribution of neurotransmitter synthesizing enzymes in the basal ganglia of human brain. J Neurochem 34:278-283

Glowinski J, Torrens Y, Beaujouan JC (1982) The striatonigral substance P pathway and dopaminergic mechanisms. In: Porter R, O'Connor M (eds) Substance P in the nervous system. Ciba Found Symp 91: 281

Guttman M, Seeman P (1985) L-Dopa reverses the elevated density of D_2 dopamine receptors in Parkinson's disease striatum. J Neural Transm 64:93-103

Hassler R (1953) Extrapyramidal-motorische Syndrome und Erkrankungen. In: Jung R Neurologie. Springer, Berlin Göttingen Heidelberg p 676 (Handbuch der Inneren Medizin. vol 5)

Hefti F, Melamed E, Wurtman RJ (1980) Partial lesions of the dopaminergic nigrostriatal system in rat brain: biochemical characterization. Brain Res 195:123-137

Heikkila RE, Manzino L, Cabbat FS, Duvoisin RC (1984) Protection against the dopaminergic neurotoxicity of 1-methyl-4-phenyl-1,2,5,6-tetrahydropyridine by monoamine oxidase inhibitors. Nature 311:467-469

Hinzen D, Hornykiewicz O, Kobinger W, Pichler L, Pifl C, Schingnitz G (1986) The dopamine autoreceptor agonist B-HT 920 stimulates denervated postsynaptic brain dopamine receptors in rodent and primate models of Parkinson's disease: a novel approach to treatment. Eur J Pharmacol 131:75-86

Hornykiewicz O (1972) Dopamine and extrapyramidal motor function and dysfunction. Res Publ Assoc Res Nerv Ment Dis 50:390

Hornykiewicz O (1973) Parkinson's disease: from brain homogenate to treatment. Fed Proc 32:183-190

Hornykiewicz O (1976) Neurohumoral interactions and basal ganglia function and dysfunction. In: Yahr MD (ed) The basal ganglia. Raven, New York, p 269

Hornykiewicz O (1981) Neurotransmitter substitution in neuropharmacology and psychiatry. In: Stjärne L, Hedquist P Lagercrantz H, Wennmalm A (eds) Chemical neurotransmission 75 Years. Academic, London, p 513

Hornykiewicz O (1986) A quarter century of brain dopamine research. In: Woodruff GN, Poat JA, Roberts PJ (eds) Dopaminergic systems and their regulation. Macmillan, London, p 3

Hornykiewicz O, Kish SJ (1984) Neurochemical basis of dementia in Parkinson's disease. Can J Neurol Sci 11:185-190

Hornykiewicz O, Kish SJ (1986) Biochemical pathophysiology of Parkinson's disease. Adv Neurol 45:19-34

Iversen SD (1977) Brain dopamine systems and behavior. In: Iversen LL, Iversen SD, Snyder SH (eds) Handbook of psychopharmacology, vol 8. Plenum, New York, p 333

Iversen SD (1982) Behavioural effects of substance P through dopaminergic pathways in the brain. Ciba Found Symp 91:307

Javitch JA, D'Amato RJ, Strittmatter SM, Snyder SH (1985) Parkinson-inducing neurotoxin, *N*-methyl-4-phenyl-1,2,3,6-tetrahydropyridine: uptake of the metabolite *N*-methyl-4-phenylpyridine by dopamine neurons explains selective toxicity. Proc Natl Acad Sci USA 82:2173-2177

Javoy-Agid F, Taquet H, Ploska A, Cherif-Zahar C, Ruberg M, Agid Y (1981) Distribution of catecholamines in the ventral mesencephalon of human brain with special reference to Parkinson's disease. J Neurochem 36:2101-2105

Kish SJ, Shannak K, Rajput AH, Gilbert JJ, Hornykiewicz O (1984) Cerebellar norepinephrine in patients with Parkinson's disease and control subjects. Arch Neurol 41:612-614

Kish SJ, Chang LJ, Mirchandani L, Shannak K, Hornykiewicz O (1985a) Progressive supranuclear palsy: relationship between extrapyramidal disturbances, dementia, and brain neurotransmitter markers. Ann Neurol 18:530-536

Kish SJ, Gilbert JJ, Chang LJ, Mirchandani L, Shannak K, Hornykiewicz O (1985b) Brain neurotransmitter abnormalities in neuronal intranuclear inclusion body disorder. Ann Neurol 17:405-407

Kish SJ, Rajput A, Gilbert J, Rozdilsky B, Chang LJ, Shannak K, Hornykiewicz O (1986) Elevated γ-aminobutyric acid level in striatal but not extrastriatal brain regions in Parkinson's disease: correlation with striatal dopamine loss. Ann Neurol 20:26-31

Laaksonen H, Riekkinen P, Rinne UK, Sonninen V (1976) Brain glutamic acid decarboxylase and γ-aminobutyric acid in Parkinson's disease. In: Birkmayer W, Hornykiewicz O (eds) Advances in parkinsonism: biochemistry, physiology, treatment. Roche, Basel, p 205

Langston JW, Langston EB, Irwin I (1984) MPTP-induced parkinsonism in human and non-human primates—clinical and experimental aspects. Acta Neurol Scand [Suppl] 70(100):49-54

Lee T, Seeman P, Tourtellotte WW, Farley IJ, Hornykiewicz O (1978) Binding of [^{3}H]neuroleptics and [^{3}H]apomorphine in schizophrenic brains. Nature 274:897-900

Lehmann J, Langer SZ (1983) The striatal cholinergic interneuron: synaptic target of dopaminergic terminals? Neurosci 10:1105-1120

Lloyd KG (1972) Biogenic amines and related enzymes in the human and animal brain. PhD thesis, Department of Pharmacology, University of Toronto

Lloyd KG (1977a) Neurotransmitter interactions related to central dopamine neurons. In: Youdim MBH, Lovenberg W, Sharman DE, Lagnado TR (eds) Essays in neurochemistry and neuropharmacology. Wiley, Chichester, p 131

Lloyd KG (1977b) CNS compensation to dopamine neuron loss in Parkinson's disease. Adv Exp Med Biol 90:255

Lloyd KG, Hornykiewicz O (1973) L-Glutamic acid decarboxylase in Parkinson's disease: effect of L-DOPA therapy. Nature 243:521-523

Lloyd KG, Möhler H, Hertz P, Bartholini G (1975) Distribution of choline acetyltransferase and glutamic acid decarboxylase within the substantia nigra and other brain regions from control and parkinsonian patients. J Neurochem 25:789-795

Madrazo I, Drucker-Colin R, Diaz V, Martinez-Mata J, Torres C, Becerril JJ (1987) Open microsurgical autograft of adrenal medulla to the right caudate nucleus in two patients with intractable Parkinson's disease. N Engl J Med 316:831-834

Markey SP, Castagnoli N, Trevor AJ, Kopin IJ (1986) MPTP: A neurotoxin producing a Parkinsonian syndrome. Academic, Orlando

Marsden CD (1980) The enigma of the basal ganglia and movement. Trends Neurosci 3:284-287

Mauborgne A, Javoy-Agid F, Legrand JC, Agid Y, Cesselin F (1983) Decrease of substance P-like immunoreactivity in the substantia nigra and pallidum of parkinsonian brains. Brain Res 268:167-170

Perry RH, Tomlinson BE, Candy JM, Blessed G, Foster JF, Bloxham CA, Perry EK (1983) Cortical cholinergic deficit in mentally impaired parkinsonian patients. Lancet 2:789-790

Perry TL, Javoy-Agid F, Agid Y, Fibiger HC (1983) Striatal GABAergic neuronal activity is not reduced in Parkinson's disease. J Neurochem 40:1120-1123

Perry TL, Yong V, Clavier RM, Jones K, Wright JM, Foulks JG, Wall RA (1985) Partial protection from the dopaminergic neurotoxin *N*-methyl-4-phenyl-1,2,3,6-tetrahydropyridine by four different antioxidants in the mouse. Neurosci Lett 60:109-114

Pijnenburg AJJ, Honig WMM, Van der Heyden JAM, Van Rossum JM (1976) Effects of chemical stimulation of the mesolimbic dopamine system upon locomotor activity. Eur J Pharmacol 35:45-58

Price KS, Farley IJ, Hornykiewicz O (1978) Neurochemistry of Parkinson's disease: relation between striatal and limbic dopamine. Adv Biochem Psychopharmacol 19:293-300

Reisine TD, Fields JZ, Yamamura HI, Bird E, Spokes E, Schreiner P, Enna SJ (1977) Neurotransmitter receptor alterations in Parkinson's disease. Life Sci 21:335-344

Rinne UK (1982) Brain neurotransmitter receptors in Parkinson's disease. In: Marsden CD, Fahn S (eds) Movement disorders. Butterworth, London, p 59

Rinne UK, Sonninen V (1973) Brain catecholamines and their metabolites in parkinsonian patients. Arch Neurol 28:107-110

Rinne UK, Sonninen V, Riekkinen P, Laaksonen H (1974) Dopaminergic nervous transmission in Parkinson's disease. Med Biol 52:208-217

Ruberg M, Ploska A, Javoy-Agid F, Agid Y (1982) Muscarinic binding and choline acetyltransferase activity in parkinsonian subjects with reference to dementia. Brain Res 232:129-139

Ruberg M, Javoy-Agid F, Hirsch E, Scatton B, L'Heureux R, Hauw JJ, Duyckaerts C, Gray F, Morel-Maroger A, Rascol A, Serdaru M, Agid Y (1985) Dopaminergic and cholinergic lesions in progressive supranuclear palsy. Ann Neurol 18:523-529

Scatton B, Bartholini G (1982) GABA receptor stimulation. IV: Effect of progabide (SL 76 002) and other GABAergic agents on acetylcholine turnover in rat brain areas. J Pharmacol Exp Ther 220:689-695

Scatton B, Javoy-Agid F, Rouquier L, Dubois B, Agid Y (1983) Reduction of cortical dopamine, noradrenaline, serotonin and their metabolites in Parkinson's disease. Brain Res 275:321-328

Scatton B, Javoy-Agid F, Montfort JC, Agid Y (1984) Neurochemistry of monoaminergic neurons in Parkinson's disease. In: Usdin E, Carlsson A, Dahlström A, Engel J (eds) Catecholamines: neuropharmacology and central nervous system-therapeutic aspects. Liss, New York p 43

Scatton B, Dennis T, L'Heureux R, Monfort JC, Duyckaerts C, Javoy-Agid F (1986) Degeneration of noradrenergic and serotonergic but not dopaminergic neurones in the lumbar spinal cord of parkinsonian patients. Brain Res 380:181-185

Scheel-Krüger J (1986) Dopamine-GABA interactions: evidence that GABA transmits, modulates and mediates dopaminergic functions in the basal ganglia and the limbic system. Acta Neurol Scand [Suppl] 73(107):1-54

Schmidt RH, Björklund A, Stenevi U, Dunnett SB, Gage FH (1983) Intracerebral grafting of neuronal cell suspensions. III. Activity of intrastriatal nigral suspension implants as assessed by measurements of dopamine synthesis and metabolism. Acta Physiol Scand[Suppl] 522:19-28

Schwab RS, Amador LV, Lettvin JY (1951) Apomorphine in Parkinson's disease. Trans Am Neurol Assoc 76:251

Sharpe JA, Rewcastle NB, Lloyd KG, Hornykiewicz O, Hill M, Tasker RR (1973) Striatonigral degeneration: response to levodopa therapy with pathological and neurochemical correlation. J Neurol Sci 19:275-286

Solomon P, Mitchell RS, Prinzmetal M (1937) The use of benzedrine sulfate in postencephalitic Parkinson's disease. Am Med Assoc J 108:1765-1770

Studler JM, Javoy-Agid F, Cesselin F, Legrand JC, Agid Y (1982) CCK-8 immunoreactivity distribution in human brain: selective decrease in the substantia nigra from parkinsonian patients. Brain Res 243:176-179

Svensson TH (1971) On the role of central noradrenaline in the regulation of motor activity and body temperature in the mouse. Arch Exp Pathol Pharmakol 271:111-120

Taquet H, Javoy-Agid F, Hamon M, Legrand JC, Agid Y, Cesselin F (1983) Parkinson's disease affects differently Met^5- and Leu^5-enkephalin in the human brain. Brain Res 280:379-382

Tenovuo O, Rinne UK, Viljanen K (1984) Substance P immunoreactivity in the postmortem parkinsonian brain. Brain Res 303:113-116

Vanderhaeghen JJ, Crawley JN (eds) (1985) Neuronal cholecystokinin. Ann NY Acad Sci 448: p 697

Vincent SR, Nagy JI, Fibiger HC (1978) Increased striatal glutamate decarboxylase after lesions of the nigrostriatal pathway. Brain Res 143:168-173

Zetler G (1968) Cataleptic state and hypothermia in mice, caused by central cholinergic stimulation and antagonized by anticholinergic and antidepressant drugs. Int J Neuropharmacol 7:325-335

CHAPTER 7

Pyridine Toxins

J. W. LANGSTON and I. IRWIN

A. Introduction

Just 4 years ago there were few if any hints of a connection between pyridines and Parkinson's disease. Yet a current review, particularly one focusing on experimental models, would be flawed without a discussion of this newly discovered relationship.This striking turn of events has been precipitated almost entirely by the discovery of a single compound, 1-methyl-4-phenyl-1,2,3,6-tetrahydropyridine, or, as it is now widely know, MPTP. In this chapter, we will explore the relevance of this compound and other pyridines to Parkinson's disease. As should become clear, investigations into the biologic effects of these "pyridine neurotoxins" have provided a surprising number of research avenues, ranging from the development of new animal models for the disease, to a serious reassessment of etiologic factors.

Before embarking on a journey into this new world of neurotoxicologic investigation, we will provide a brief review of pyridines, specifically addressing both what and where they are. These sections should set the stage for a comprehensive review of the potential importance of certain tetrahydropyridines to Parkinson's disease.

B. What are Pyridines?

I. Historical Background

The first hint of the existence of pyridines as a class of compounds was the foul odor of products obtained from the distillation of bones and other animal products by the early alchemists. In 1846, while investigating the components of coal tar, Anderson isolated the first pyridine derivative from pitch, which he named picoline, after the Latin *pix* (pitch) (BARNES 1960). Anderson later turned his attention to bone oil and, after much difficulty, isolated the parent compound of picoline. He named this new compound "pyridine" after the Greek name for fire *(pyr)* because he had to burn bones to obtain it.

II. Structure and Chemistry

Pyridines are all members of the broader class of heterocyclic compounds (i.e., organic ring structures containing one or more non-carbon atoms such as sulfur or nitrogen). Heterocyclic compounds are of great interest to chemists and biologists alike because of their reactivity and biochemistry. Among the heterocyclic compounds, pyridines have been considered the most important (BARNES 1960).

As a specific class of compounds, all pyridines consist of a six-membered ring which contains a single nitrogen atom. The parent compound (pyridine itself) contains three double bonds (Fig. 1a), and the ring is numbered clockwise, beginning with nitrogen as 1. A variety of substituents on the pyridine ring are possible, and the location of the substituents is indicated by a number. Hydrogen may be added to one or more double bonds, resulting in partially reduced or saturated pyridines. Thus, the reduction of a single double bond (which requires the addition of two hydrogen atoms, one to each of the two carbons adjacent to the double bond) results in a *dihydro*pyridine (Fig. 1b), and the saturation of two double bonds results in a *tetrahydro*pyridine (Fig. 1c). The locations of added hydrogens are indicated by their numerical position. By varying the location of the double bonds within the six-membered ring (remembering that two double bonds cannot be immediately adjacent), five different dihydropyridines and three tetrahydropyridines are possible. The full saturation of pyridine (i.e., when all of the carbons are fully hydrogenated and there are no double bonds remaining) results in piperidine.

Understanding this reduction scheme and its nomenclature is of more than just academic interest. Because the molecule can undergo the successive addition of hydrogen (which "reduces" the oxidation state) to form the various dihydropyridines and tetrahydropyridines, and this process can be reversed through the process of oxidation, a reactive chemical pathway is provided which has considerable biologic utility.

A second feature which is of importance to this discussion is the nitrogen in the pyridine ring. Nitrogen normally forms three covalent bonds, but still possesses an unshared electron pair—the "lone pair." Because of this configuration, the lone pair can form a *fourth* bond, and therefore a pyridine can act as an electron donor (such a compound is referred to as a "quaternary" compound or a pyridinium ion). The addition of the fourth substituent to the neutral nitrogen results in the formation of a positively charged species. This is because the nitrogen, in forming the fourth covalent bond, donates one of its electrons and in the process acquires a positive charge. Although quaternary pyridinium species can arise from a variety of reactions (e.g., nucleophilic displacement), a major route is via the oxidation of reduced pyridines which bear a substituent (e.g., a methyl group) on the nitrogen. As we will see, the ease with which pyridines undergo oxidation and reduction, and form quaternary pyridinium species, appears to be an important characteristic of the pyridine toxins to be discussed in this chapter.

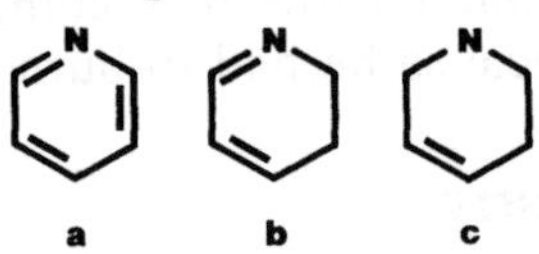

Fig. 1 a–c. Structures of: pyridine (**a**), 2,3-dihydropyridine (**b**), and 1,2,3,6-tetrahydropyridine (**c**)

III. Biologic Role

Pyridines are important in many biologic processes. For example, the 3-hydroxypyridine derivative (pyridoxine or vitamin B_6) and its oxidized phosphate ester act as enzyme cofactors in a number of biochemical transformations of amino acids. Other biologically active pyridines are important because of their ability to undergo oxidation-reduction and quaternarization. One of the examples of this process is the well-known cofactor nicotinamide adenine dinucleotide (NAD^+).

NAD^+ consists of the pyridine nicotinamide, anchored to adenine dinucleotide via a quaternary bond (hence the positive charge). One important function of NAD^+ in biologic systems is the oxidation of other compounds. In the process, NAD^+ is reduced to NADH (a 1,4-dihydropyridine). NADH can then be reoxidized to NAD^+ via various mitochondrial shuttle systems. NAD^+ and NADH (and their phosphorylated analogs $NADP^+$ and NADPH) are essential to many biologic processes, including cellular respiration, fatty acid synthesis, photosynthesis, and the processes of vision. The observation that over 40 biologic redox reactions are thought to depend on these two conjugated pyridines highlights the biologic utility of quaternarization and oxidation. As will become evident later in this chapter, this same principle, which is vital to many biologically important processes, may also be a key to the neurotoxic potential of certain pyridines.

IV. Distribution of Pyridines

Given the biologic importance of pyridines, it is not surprising that they are widely distributed in nature. In fact, a large variety of pyridines have been isolated from natural sources including animals, plants, fungi, and bacteria (Brody and Ruby 1960). The worldwide cultivation of tobacco alone (nearly 3 billion kilograms per year) represents between 30 and 160 million kilograms of pyridine compounds.

Industry represents another major source of pyridines in our environment. Most pyridines used in the chemical industry are derived from coal tar, which in turn is derived from the production of coke. This is no small matter, as about 20% of the coal produced in the United States in converted to coke. During the process of this conversion, about 30% of the coal is driven off as a gas which contains a large number of pyridines. In fact, for every 1000 kg of carbonized coal produced, approximately 0.1 kg of pyridine bases are made. From these, by way of the chemists' craft and the burgeoning chemical industry, a huge number of pyridines have been synthesized—in excess of 10,000 have been registered by the *Chemical Abstracts* service since 1965.

As might be expected, the pharmaceutical industry represents another major source of pyridines. The pyridine moiety is a key feature of many natural alkaloids, including cocaine, opiate, ergot, and quinine alkaloids. Pyridine is also well represented among synthetic drugs, including antidepressants, neuroleptics, antiseptics, antibiotics, and enzyme inhibitors. Nearly every class of drug has at least one pyridine member. Although a few of these (e.g., reserpine, haldol) have been associated with transient parkinsonian symptoms, none have produced degeneration of the neurons of the substantia nigra, or a permanent behavioral syndrome. As a result, none of these compounds could be used to produce an entirely satisfactory animal model of Parkinson's disease. However, as the result of the discovery of MPTP, not only is such an animal model available, but for the first time we can directly probe mechanisms underlying the process of neuronal degeneration in the nigrostriatal system, a phenomenon which lies at the very heart of Parkinsons's disease.

C. Pyridines as Toxins

The idea of pyridines as neurotoxins is not new. Zinc pyridinethione (ZPT) induces a central-peripheral axonal dystrophy in rodents (SAHENK and MENDELL 1980) and isoniazid (a derivative of isonicotinic acid) produces a peripheral neuropathy in humans and experimental animals (BLAKEMORE 1980). Another pyridine, 6-aminonicotinamide (6-AN), when administered systemically, is reported to cause selective depletions of striatal dopamine and produce a behavioral syndrome characterized by rigidity and spastic paralysis (HERKEN et al. 1976).This syndrome is responsive to therapy with L-dopa and dopamine agonists (LOOS et al. 1977). 6-AN is actually a protoxin which is converted to a toxic NADP analog in vivo, and it is the accumulation of this compound in specific brain regions that is thought to be responsible for the selectivity of 6-AN, via the inhibition of 6-phosphogluconate dehydrogenase, resulting in the accumulation of large amounts of 6-phosphogluconate. The excitotoxin quinolinic acid has been known for some time to be toxic to certain cell types in the neostriatum, and in fact has recently been shown to reproduce many of the neurochemical changes which are seen in Huntington's chorea (BEAL et al. 1986). However, the discovery of MPTP heralded an entirely new class of pyridine toxins which are proving to be highly relevant to the study of Parkinson's disease. We now turn to a detailed discussion of these "parkinsonogenic pyridines."

I. History

In the summer of 1982, the versatility of pyridines was once again exploited, but this time a clandestine chemist was at work. Taking a lead from the 1940s chemistry literature on synthetic opioid analgesics (ZIERING et al. 1947), a scheme intended to produce the potent meperidine analog MPPP (Fig. 2a) was attempted. However, through lack of care, skill, or quality control, MPTP (Fig. 2b), a by-product of the reaction scheme, was produced instead. For a brief period MPTP was inadvertently sold and consumed as heroin, but the effects on its users were devastating (LANGSTON et al. 1983). Some rapidly developed a syndrome which was virtually indistinguishable from severe Parkinson's disease (BALLARD et al. 1985). MPTP was quickly identified as the probable offending agent (LANGSTON et al. 1983), and its toxic effects were

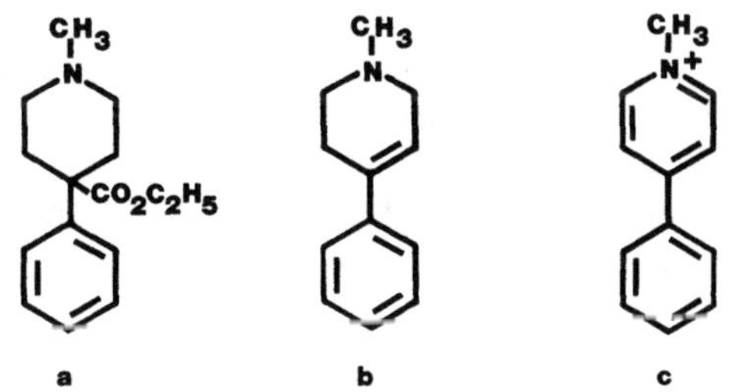

Fig. 2 a–c. Structures of: the meperidine analog, 1-methyl-4-phenyl-4-propionoxypiperidine, or MPPP (**a**); the neurotoxic by-product of MPPP synthesis, 1-methyl-4-phenyl-1,2,3,6-tetrahydropyridine, or MPTP (**b**); the putative toxic biotransformation product of MPTP, 1-methyl-4-phenylpyridinium ion, or MPP^+ (**c**)

confirmed in animals shortly thereafter (Burns et al. 1983; Langston et al. 1984 a).

The fact that MPTP so faithfully reproduced the clinical features of Parkinson's disease in humans (Langston et al. 1983; Ballard et al. 1985), including response to therapy, and even complications of therapy (Langston and Ballard 1984), gave rise to the hope that the compound would be effective in producing the long sought-after animal model of Parkinson's disease (Langston et al. 1983). This hope has been realized to a remarkable degree in the ensuing years (Lewin 1986).

II. Animal Models

The effects of MPTP have now been studied in a wide variety of animals, ranging from the medicinal leech to primates. In this section, we will summarize the effects of this compound in each of these species, ascending the phylogenetic scale. As might be expected, there is considerable interspecies (and even intraspecies) variability, but each of these models appears to have something to offer in regard to the study of MPTP and its usefulness as a tool to study Parkinson's disease.

1. Invertebrates

Perhaps the lowest animal on the evolutionary ladder to be studied to date is the medicinal leech *(Hirudo medicinalis)*. The study of MPTP in this animal makes more sense than might be immediately obvious because it has a population of large, grossly identifiable dopaminergic neurons; further, the animal has characteristic measurable behaviors, including feeding, swimming, and biting—even body tone can be studied. The injection of MPTP produces a variety of behavioral changes, including rigid body postures and the abolition of characteristic biting and swimming behaviors (Lent 1986). MPTP also produces a 70% depletion of DA in terminals, and an 85% depletion of DA in the roots derived from axons and somata. Interestingly, MPTP is even more potent at depleting 5-HT in this model, and species differences have been found even at this level. The American leech *(Macrobdella decora)*, a species which contains less pigment in the vasofibrous tissue, and possesses one-sixth of the MAO-B found in the European medicinal leech, requires more than tenfold greater doses of MPTP. Perhaps the most promising use for this model will be in direct studies of the effects of MPTP and its analogs and metabolites on the highly accessible dopaminergic neurons of this species.

2. Amphibia

MPTP also produces a range of behaviors reminiscent of Parkinson's disease in certain amphibia. Even before the discovery of MPTP, certain amphibia, because their skin pigment had spectroscopic and chemical similarities to neuromelanin, had found a place in research in Parkinson's disease (AMBANI et al. 1975). With the wave of interest in developing animal models for Parkinson's disease which followed the discovery of MPTP, amphibia once again received attention as a possible model for Parkinson's disease.

The salamander and frog have both been used as animal models (BARBEAU et al. 1986a,b). Slowness of movement, tremor, and rigidity were all reported to be features of MPTP-induced toxicity in frogs. Biochemical deficits of brain and adrenal dopamine (greater than 80%) 4 weeks after treatment were also reported in the frog. Once again, striking intraspecies variations have been observed. A 5 mg/kg dose was effective in *Rana clamitans*, while 40 mg/kg was required to achieve the same effect in *R. pipiens*. An admittedly intriguing observation in these animals was a striking dispersion of melanin pigments in the skin induced by MPTP, and thus the authors assertion that "the frog bears its neuromelanin on its skin." Amphibia will probably continue to represent an interesting, if not unusual, model to evaluate the biologic effects of MPTP.

3. Rodents

We now come to the first warm-blooded animals in our phylogenetic survey. The effects of MPTP have been studied in several rodent species. To date, although a great deal of evidence for neurochemical changes induced by MPTP has been forthcoming, no enduring behavioral syndrome has been convincingly demonstrated in any rodent species.

The rat was quickly found to be refractory to the effects of MPTP. Some investigators (CHIUEH et al. 1984b; BOYCE et al. 1984; SAHGAL et al. 1984) could find no effects on striatal dopamine after systemic injections of MPTP. A similar lack of effect on striatal dopamine has been reported even after infusion of MPTP into rat substantia nigra (CHIUEH et al. 1984b; BRADBURY et al. 1986). Others, however, have reported depletions of striatal dopamine in the rat after repeated subcutaneous injection (JARVIS and WAGNER 1985), after continuous infusion over 24 h (STERANKA et al. 1983), and after direct injections of very large doses of MPTP into the substantia nigra of rats (SAYRE et al. 1986). The only detailed neuropathologic study in the rat utilized a protocol that did not produce dopamine depletion (SAHGAL et al. 1984), and predictably little evidence of neuronal degeneration was found. Hence, whether or not the rat is as resistant to the effects of MPTP as is widely believed remains to be determined (LANGSTON and IRWIN 1986). For the moment, this species must be considered one of least attractive as a model for the study of MPTP-induced parkinsonism. As with the rat, the guinea pig first appeared to be insensitive to the neurotoxic effects of MPTP (CHIUEH et al. 1984a; PERRY et al. 1985). Recently however, CARVEY et al. (1986) have produced both dopamine depletion and nigral cell loss by directly administering MPTP into the substantia nigra.

In contrast, certain strains of mice appear to be much more sensitive to MPTP than the rat, and this species has been widely employed to study the effects of MPTP in vivo. HEIKKILA et al. (1984) were the first to report that MPTP produces nigral cell loss and dopamine depletion in the mouse following systemic administration. Although this original work was done in Swiss-Webster mice, the C57BL/6 may be one of the most sensitive to the effects of MPTP (HEIKKILA et al. 1984a). But, even in this species, conflicting results have been obtained. Depression of striatal dopamine for up to 1 month has been reported (HALLMAN et al. 1984; RICAURTE et al. 1986). Many studies, however, have shown that striatal dopamine begins to recover after 1 month (RICAURTE et al. 1986; HALLMAN et al. 1985; MELAMED et al. 1985a) and complete recovery occurs after 10 months (RICAURTE et al. 1986). Furthermore, the original observation of cell death in mice has not been confirmed by all investigators (RICAURTE et al. 1986; HALLMAN et al. 1985). Histologic studies using Fink-Haimer techniques reveal that, at least in young mice (6-8 weeks old), MPTP damages only the terminals of nigrostriatal neurons, sparing the cell bodies (RICAURTE et al. 1987). Many of these apparent discrepancies can probably be accounted for in terms of factors discussed in Sect. D. While transient rigidity and akinesia may be seen in the mouse, an enduring behavioral syndrome has yet to be reported.

4. Cat

The cat has also been used to study the effects of MPTP (SCHNEIDER et al. 1986; SCHNEIDER and MARKHAM 1986). Following administration of MPTP, the cat suffers an acute behavioral syndrome, characterized by akinesia, ataxia, and feeding difficulties. In addition, MPTP produces an 85% depletion of striatal dopamine and up to an 80% loss of nigral cell numbers in these animals. In spite of cell loss and a persistent dopamine depletion of around 75%, after 5 months a nearly complete behavioral recovery occurs.

5. Dog

Results in dogs are similar to those seen in cats. In beagles, MPTP produces extensive nigral cell death and changes in dopamine concentrations, but fails to produce an enduring behavioral syndrome (WILSON and WILSON 1986). In these studies both puppies and adult dogs were treated with MPTP. Interestingly, puppies were reported to be remarkably resistant to the effects of this compound (see Sect. D.V), but the age of adult animals was not mentioned.

6. Primates

Although numerous species have been used to study the effects of MPTP, nowhere are they more clearly demonstrated than in the primate. When given to primates, approximately one-tenth the amount of MPTP necessary to produce effects in other species produces an enduring syndrome which is remarkable in its resemblance to human parkinsonism (BURNS et al. 1983; LANGSTON et al. 1984a; COHEN et al. 1985; JENNER et al. 1984). The primate model is characterized by a clinical picture of bradykinesia, increased tone, postural freezing and instability which responds to dopamine replacement therapy (L-dopa) and the administration of dopamine agonists (bromocriptine). Recently, a resting tremor has been reported in the African green monkey (TETRUD et al. 1986), making the clinical similarity between the model and the disease complete.

The neuropathologic consequences of MPTP administration in the primate are strikingly similar to those seen in Parkinson's disease, and include a loss of neurons in the substantia nigra, and the presence of extraneuronal pigment, with gliosis and microglia proliferation (FORNO et al. 1986a; PARISI and BURNS 1986). The exquisite sensitivity of the primate's nigrostriatal neurons make this model the best to study the effects of this compound—a proposition which is further strengthened by the fact that they are phylogenetically closer to the human.

Subsequent work in the primate has, if anything, brought the animal model closer to the human disease. The original work in rhesus and squirrel monkeys reported a highly localized lesion in the substantia nigra (BURNS et al. 1983; LANGSTON et al. 1984a). Thus, although this model was (and is) very promising for the study of new forms of therapy, including both pharmacologic and surgical forms of intervention, the absence of two neuropathologic features of Parkinson's disease (i.e., locus ceruleus involvement and the presence of distinctive eosinophilic inclusions known as Lewy bodies) appeared to represent serious shortcomings. However, by manipulating variables such as route of administration and age, cell loss in the ceruleus and Lewy-body-like inclusions have recently been reported in primates after MPTP administration (FORNO et al. 1986b). The close resemblance between the neuropathology of Parkinson's disease and MPTP-induced parkinsonism suggests that the pathogenesis of the two disorders could be similar.

In summary, studies utilizing MPTP in a variety of animals suggest that primates are much more dependent on an intact nigrostriatal system than are other species. While this may have interesting evolutionary implications, the practical consequences cannot be ignored. For the moment, if a behavioral model is needed, the primate remains the only choice available.

D. Factors Affecting Toxicity

Each of the models described in the preceding section may serve a different experimental purpose or goal. For example, the primate appears to be by far the most useful for the testing of new therapeutic agents in view of the faithfully reproduced parkinsonian syndrome seen in these animals. However, there are many other investigational uses for MPTP, ranging from the study of the molecular and morphological aspects of neuronal degeneration to the search for the cause (or causes) of Parkinson's disease. Such studies require an understanding of the mechanism of action of MPTP (and related tetrahydropyridines) and the many factors which we now know to be important in regard to the final expression of neurotoxicity.

Obviously, should tetrahydropyridines ever become serious etiologic candidates for Parkinson's disease, these factors might also be highly relevant to understanding the process of exposure and individual susceptibility (i.e., why some people get the disease, and others do not). It should also be pointed out that it is only with the advent of a compound which selectively affects certain areas of the brain after systemic exposure that these questions become both relevant and possible to ask.

I. Mechanism of Action

Because there have been a number of recent detailed reviews regarding the mechanism of action of MPTP (LANGSTON 1985a, 1985b; SNYDER and D'AMATO 1986; IRWIN 1986; LANGSTON and IRWIN 1986), only a brief summary will be presented here. MPTP, as a lipophilic compound, readily crosses the blood-brain barrier (LANGSTON et al. 1984b; MARKEY et al. 1984), and is then rapidly oxidized via MAO-B (CHIBA et al. 1984) to its quaternary amine, 1-methyl-4-phenylpyridinium ion (MPP^+) (Fig. 2c) (LANGSTON et al. 1984b; MARKEY et al. 1984). This process probably takes place in glia (RANSOM et al. 1987), via a dihydropyridinium intermediate (CHIBA et al. 1985a). MPP^+ is then thought to enter dopaminergic neurons via the dopamine reuptake system (JAVITCH et al. 1985; CHIBA et al. 1985b). Exactly how MPP^+ kills dopaminergic neurons once it is intracellular is as yet uncertain, although there are at least five theories as to how this might occur (LANGSTON and IRWIN 1986). The hypothesis which is being most actively investigated at present relates to the inhibitory effects of MPP^+ on mitochondrial respiration (NICKLAS et al. 1987; TREVOR et al. 1987). Although most of these steps are fairly well accepted, they must still be regarded as only tentative since our knowledge regarding these events is growing on a daily basis, and many questions remain (LANGSTON 1985b).

II. Toxicokinetic and Toxicodynamic Effects

Before proceeding to a discussion of factors which affect toxicity, two concepts will be introduced which are exemplified by two terms from pharmacology—pharmacokinetic and pharmacodynamic. The former term refers to quantitative changes in the delivery of an agent to its site of action, and the latter to changes in response (i.e., an increase or decrease in sensitivity) to equivalent amounts of the agent. Paralleling these two pharmacologic concepts, we are suggesting the terms toxicokinetic and toxicodynamic.

The term toxicokinetic implies that, regardless of the ultimate mechanism by which a toxin injures or kills a cell, the pattern of biodisposition (i.e., how much of the toxin reaches its target) is what determines toxicity. In this scheme, cellular poisons selectively damage specific tissues simply because they reach higher concentrations in the target area. Toxicodynamic, on the other hand, implies that the sensitivity of the target cell changes, rather than the amount of the toxin reaching the target. Obviously, changes in one or both of these factors could dramatically alter the expression of toxicity.

Aminoglycoside antibiotics (AMGs) represent a simple model of a toxicokinetic effect. These agents produce irreversible ototoxicity. Animal studies have documented that these drugs selectively accumulate in the perilymph and endolymph of the inner ear (Huy et al. 1983). In addition, the half-life of AMGs is 5-6 times longer in the otic fluids than in the plasma. Although the biochemical events responsible for cell death in the case of AMGs are not known (they have a wide variety of effects, including the inhibition of various phospholipases, sphingolmyelinases, ATPases, and alteration of mitochondrial and ribosomal function), factors which alter the peak concentration and persistence of these drugs in target tissues are clearly determinants of toxicity. In the following sections, the importance of these concepts will become obvious in regard to various factors which affect MPTP toxicity, such as species differences and age.

III. Species Differences

Although it is as yet unclear why there are such differences in susceptibility to the toxic effects of MPTP between species, there is evidence that toxicokinetic factors may play an important role. In the mouse, although the doses of MPTP required for toxicity are approximately tenfold greater than in the primate, the regional concentrations of MPP^+ in striata and cortex are equivalent for both animals (Table 1). This suggests that the nigrostriatal nerve terminals of primates and rodents are equally sensitive to MPP^+, and the higher doses of MPTP required in the mouse are needed simply to deliver equivalent amounts of toxin to the terminal region. The fact that mouse striatal MPP^+ concentrations correlate with striatal dopamine depletion (Irwin et al. 1987) further

Table 1. Concentration of MPP^+ following administration of MPTP in the mouse and monkey

Species	Dosage	Tissue	[MPP^+] (μg per gram tissue)
Squirrel monkey	2 mg/kg/2 h × 4	Caudate	9.8
		Cortex	4.7
C57BL/6 mouse	20 mg/kg/h × 4	Caudate	12.7
		Cortex	4.0

supports this quantitative (toxicokinetic) hypothesis. KALARIA et al. (1986) have provided an interesting explanation for this phenomenon in that they have reported that the blood-brain barrier of rodents contains 30 times more MAO than that of primates.

Toxicokinetic factors may also help explain the fact that MPTP induces nigral cell loss in young primates, but not in young rodents. Some 72 h after MPTP administration in primates, the concentrations of MPP^+ in the central nervous system (CNS) are greatest in the substantia nigra (IRWIN and LANGSTON 1985), and actually appear to increase in this region. Thus, the concentration and retention of MPP^+ may be responsible for its selectivity. To date, MPP^+ concentrations in the mouse substantia nigra have not been measured; however, the data in primates suggest that, if conditions could be manipulated in such a way as to produce and retain greater amounts of MPP^+ in the mouse substantia nigra, then cell death similar to that seen in primates could be obtained. Recently, evidence suggests that diethyldithiocarbamate achieves precisely this effect in the rodent (IRWIN et al. 1987).

Until all of these factors have been clarified, however, generalizing from the rodent to the primate should be done only with great caution. Experience with dopamine uptake blockers provides an excellent example. In young mice, a variety of dopamine uptake blockers are effective in preventing the dopamine-depleting effects of MPTP (JAVITCH et al. 1985; RICAURTE et al. 1985; FULLER and HEMRICK-LUECKE 1985; MELAMED et al. 1985b; SUNDSTROM and JONSSON 1985), as might be expected based on the observation that MPP^+ gains entrance to dopaminergic neurons via the uptake system. However, to the best of our knowledge, it has not been possible to repeat these experiments in primates. Using at least two dopamine uptake blockers (mazindol and bupropion) in combination with MPTP, we have been unable to observe protection in any of seven monkeys (J. W. LANGSTON, L. E. DELANNEY and I. IRWIN 1987, unpublished work). Indeed, in two of these animals, uptake blockers actually appeared to exacerbate the toxicity of MPTP. While the explanation for these observations remains unclear, the lesson seems to be that caution should be exercised in generalizing from experiments based solely on data from rodents.

IV. Neuromelanin

The presence of neuromelanin in the cells of the human substantia nigra and locus ceruleus has long been of interest in regard to the process of neuronal degeneration in Parkinson's disease (MARSDEN 1961; MANN and YATES 1983; LANGSTON and IRWIN 1986; LANGSTON et al. 1987). The cellular content of neuromelanin increases with age, and of course, Parkinson's disease itself is known to have a predilection for the pigmented cells of the brainstem. Might neuromelanin play a role in MPTP neurotoxicity? There are a number of cogent arguments both for and against this possibility.

First, the increasing presence of this substance as one ascends the phylogenetic scale (MARSDEN 1961) seems an obvious explanation for species variability. Neuromelanin in present in primates, but has not been detected in rodents (MARSDEN 1961). Further, both MPTP (LYDEN et al. 1983) and MPP^+ (D'AMATO et al. 1986) have been found to bind to melanin, and the fact that MPP^+ accumulates in the primate substantia nigra (where neuromelanin concentrations are high) during the first 72 h after MPTP administration (IRWIN and LANGSTON 1985) suggests that MPP^+ may be binding to neuromelanin. Finally, it has recently been shown in primates that chloroquine (a substance which should block the binding of MPP^+ to neuromelanin) appears to produce at least a partial reduction in the dopamine-depleting effect of MPTP (D'AMATO et al. 1987).

Hence, it has been suggested that greater amounts of MPTP are required in the rodent because this species does not possess neuromelanin. One potential problem with this theory is the previously mentioned observation that, even though a tenfold increase in MPTP dosage is required in the mouse, the regional concentrations of MPP^+ in the rodent and in the primate CNS are approximately the same (Table 1). Thus, it appears that higher doses do not achieve greater levels in the brain, but are required to get an *equivalent* amount of metabolite to the target areas. Another problem with implicating neuromelanin as a key to toxicity relates to the fact that there are other pigmented dopaminergic regions in the brain, such as the ventral tegmental area, which do not appear to be particularly sensitive to the toxic effects of MPTP (JACOBOWITZ et al. 1984).

V. Age Differences

One of the most interesting results to come out of studies with MPTP in rodents is the relationship between MPTP and aging. For example, the dopamine-depleting effects of MPTP in the mouse striatum clearly increase with age (RICAURTE et al. 1987), and the compound induces cell loss in the substantia nigra of older, but not younger animals (RICAURTE et al. 1987). Although MPTP affects several markers of catecholaminergic cellular function in the substantia nigra of both young and very old mice, the aged animals exhibit changes in the locus ceruleus and ventral tegmental area as well (GUPTA et al. 1986).

To date, the explanation for these increased effects has yet to be fully elucidated, but recent data suggest they might arise, at least in part, from toxicokinetic effects. For example, after identical dosages of MPTP, MPP^+ concentrations are nearly twice as high in older animals (LANGSTON et al. 1987), and consequently this region of brain is exposed to greater amounts of toxin in older than in younger animals. In vitro data have shown that homogenates from the brains of older animals incubated with MPTP produce MPP^+ at a greater rate than those of younger animals (RICAURTE et al. 1987; LANGSTON et al. 1987), suggesting that the explanation for this effect appears, at least in part, to lie in changes in the CNS biotransformation of MPTP which occur with age.

As MAO activity is known to increase with age (ROBINSON et al. 1972; BENEDETTI and KEANE 1980; COTE and KREMZNER 1983), changes in the activity of this enzyme may explain the enhanced biotransformation of MPTP to

MPP$^+$. Because this enzyme can be easily inhibited by pharmacologic means, a role for MAO in this process could have therapeutic implications for Parkinson's disease. A second factor which may be responsible for the increased concentrations of MPP$^+$ in older animals may be related to a decline in the ability to eliminate this metabolite. In addition, there may be a shift in the burden of MPTP to the CNS, owing to reduced peripheral biotransformation. Certainly, reductions in hepatic metabolism and renal clearance of drugs with age are well documented.

It is interesting to speculate that one or more of these factors might explain the increasing incidence of Parkinson's disease with age. It may be that the brain is effectively protected from toxins by a variety of peripheral mechanisms in young animals. With aging, however, these mechanisms become less efficient or effective, resulting in an increasing exposure of the CNS to these putative toxins.

Finally, it is worth mentioning once again that eosinophilic inclusions have been described in very old primates exposed to MPTP (FORNO et al. 1986b). These intraneuronal inclusions have been seen in a variety of areas where Lewy bodies are known to occur in humans with Parkinson's disease. These findings also raise the question of a toxicodynamic effect in regard to aging, at least in the primate, as they suggest that the neurons in very old animals undergo a different form of degeneration. Of particular importance here is that derangement of the cytoarchitecture appears to be involved in a way that closely resembles the process which occurs in Parkinson's disease. These findings could provide a new avenue of investigation into the relationship between aging and the process of neuronal degeneration.

E. Pyridines as Protectors?

While it has been known for several years that a number of MAO inhibitors block the toxic effects of MPTP (MYTILINEOU and COHEN 1985; COHEN et al. 1985; HEIKKILA et al. 1984b; MARKEY et al. 1984; LANGSTON et al. 1984c), it has recently become apparent that pyridines similar to MPTP and MPP$^+$ may also have this effect. A compound which may be of particular interest in this regard is 4-phenylpyridine (4-PP), which is found in such common substances as peppermint, spearmint (SNYDER and D'AMATO 1986), and cigarette smoke (HECKMAN and BEST 1981). 4-PP is strikingly similar to both MPTP and MPP$^+$, and for this reason its activity both as a dopamine depletor and as an antagonist to the dopamine-depleting effects of MPTP were evaluated (IRWIN et al. 1987). Although it did not produce dopamine depletion, 4-PP provided partial protection against the dopamine-depleting effects of MPTP in the mouse (IRWIN et al. 1987).This effect is probably dependent on the ability of 4-PP to inhibit MAO-B reversibly (ARAI et al. 1986), and thus reduce the amount of MPP$^+$ formed.

These results raise two interesting questions. The first relates to Parkinson's disease and cigarette smoking. A repeated but somewhat puzzling observation has been the lower incidence of cigarette smoking among individuals who get the disease (Kessler and Diamond 1971; Ward et al. 1983; Baumann et al. 1980; Nefzger et al. 1968). The fact that 4-PP, which is reported to be present in cigarette smoke, blocks the effects of MPTP raises the intriguing possibility that there are compounds present in tobacco which might in some way protect against the agent (or agents) responsible for Parkinson's disease. Although there are certainly other possible explanations for this unusual relationship (Ward et al. 1983), the observations provide yet another avenue by which to explore cause and effect relationships in regard to Parkinson's disease and environmental agents.

The role of MAO in MPTP toxicity also points to the possibility that other factors in the environment may act to modulate the activity of this important enzyme. There is a substantial amount of data regarding platelet MAO activity which suggests that a number of factors may modulate its activity. The range for normal values is extremely wide and variable, even in the same individual at different times during the day. Because these changes occur over a relatively short time, it is likely that they are the result of changes in activity (not increased synthesis), suggesting that other modulating factors may be important. In fact, both inhibitory and activating factors have been found in human plasma and urine. One could speculate that exogenous factors might exist in the environment (4-PP-like compounds?) which could play a role as natural protectors against the disease. Should such phenomena ever be documented, it might help explain an age old question (which is very difficult to answer when it comes to the environmental hypothesis of Parkinson's disease) as to why everyone doesn't get the disease (at least if the toxin is widely distributed in the environment).

F. Toxic Tetrahydropyridines: A Growing Family

Factors such as age, route of administration, and species can be seen as variables which are likely to alter the biologic effects of MPTP. However, it is also possible to study the effects of MPTP by altering the molecule itself. The reasons for exploring the structure and activity of MPTP are twofold. The first and most obvious reason is to determine which features of the molecule are necessary for its action. Except for a single study utilizing primates, which established that the removal of the 4-5 double bond from MPTP produces a compound which is not neurotoxic in primates (Langston et al. 1984d), most studies have used rodents, or in vitro models of toxicity. With the caveat that these studies must be interpreted with caution, it is probably safe to say that the essential features of the molecule are the 4-5 double bond, a methyl group at the 1 position, and a substitution at the 4 position. In addition to saturation of the double bond, removal or substitution of the methyl group, opening of the pyridine ring, or addition of substituents to any other positions on the pyridine ring, all appear to reduce drastically or eliminate dopamine-depleting activity in rodent models (Youngster et al. 1986; Fries et al. 1986; Sayre et al. 1986; Zimmerman et al. 1986).

On the other hand, the nature of the substituent at the 4 position seems to

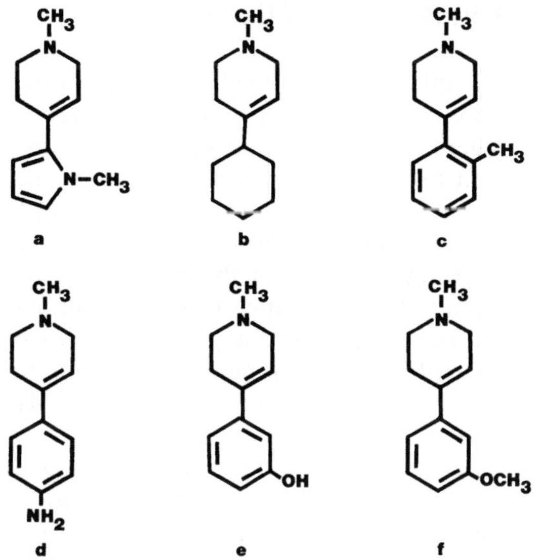

Fig. 3a–f. Structures of the six MPTP analogs known to deplete striatal dopamine in mice:
1,2,3,6-tetrahydro-1-methyl-4-(methylpyrrol-2-yl)pyridine (TMMP) (**a**);
1-methyl-4-cyclohexyl-1,2,3,6-tetrahydropyridine (MCTP) (**b**);
1-methyl-4-(2′-methylphenyl)-1,2,3,6-tetrahydropyridine (2′methyl-MPTP) (**c**);
1-methyl-4-(4′-aminophenyl)-1,2,3,6-tetrahydropyridine (4′amino-MPTP) (**d**);
1-methyl-4-(3′-hydroxyphenyl)-1,2,3,6-tetrahydropyridine (3′hydroxy-MPTP) (**e**);
and 1-methyl-4-(3′-methoxyphenyl)-1,2,3,6-tetrahydropyridine (3′methoxy-MPTP) (**f**)

have some flexibility. To date, six active analogs have been reported which bear a substituent other than a simple phenyl ring at the 4 position (Fig. 3). These are methylpyrrole (TMMP) (Fig. 3a; FINNEGAN et al. 1987), cyclohexyl (MCTP) (Fig. 3b; NICKLAS et al. 1987), 2′-methylphenyl (2′methyl-MPTP) (Fig. 3c; YOUNGSTER et al. 1986), 4′aminophenyl (4′amino-MPTP) (Fig. 3d; JOHANNESEN et al. 1987), 3′hydroxyphenyl (3′hydroxy-MPTP) (Fig. 3e; FULLER and HEMRICK-LUECKE 1986), and 3′methoxyphenyl (3′methoxy-MPTP) (Fig. 3f; ZIMMERMAN et al. 1986).

The study of these compounds is already yielding some interesting results. One of them (TMMP) is more potent in terms of striatal dopamine depletion in the rodent, though less effective in the primate (J. W. LANGSTON, L. E. DELANNEY and I. IRWIN 1986, unpublished work) than MPTP. The 2′-methyl variant of MPTP has been reported to be a substrate for both MAO-A and MAO-B. Treatment of rodents with MAO-B inhibitors alone does not provide protection against its effects (NICKLAS et al. 1987), but a combination of blockers to inactivate both the A and the B form (i.e., clorgyline + deprenyl) does provide protection. These data raise the interesting possibility that differences in species susceptibility could be the result of differences in MAO-A und MAO-B between species, and once again points out the potential hazards of generalizing from rodent to primate.

G. Pyridines and Parkinson's Disease

The very existence of a rapidly emerging list of toxic pyridines raises the question as to whether or not one or more of them might play a causative role in Parkinson's disease. There is no doubt that these tetrahydropyridines can induce parkinsonism from exogenous exposure (LANGSTON and BALLARD 1983; BARBEAU et al. 1985). Further, we have recently found that MPTP is effective via the oral route (DELANNEY et al. 1986), opening up an entire array of dietary compounds for investigation. Given the wide distribution of pyridines, their use in industry, and their importance in biologic processes, it is not too difficult to imagine that one or more of the pyridine toxins may eventually be found in the natural or artificial environment. Meanwhile, through manipulation of the various factors described in this chapter, the MPTP model continues to move closer to Parkinson's disease.

A second lesson which these studies may be teaching us is that changes in the kinetics and biodisposition of MPTP, which appear to vary with age, type of environmental exposure, and even evolution, may in turn be used to evaluate factors which could affect individual susceptibility and even the clinical expression of one or more neurodegenerative diseases of aging. The long-range hope would be that once these factors are fully understood, it might be possible to alter them favorably by using new therapeutic agents.

Finally, now that the MPTP model has been shown to exhibit so many of the features of Parkinson's disease (FORNO et al. 1986b), we might be bold enough to ask something entirely new of this model. Up to this point we have used Parkinson's disease as a standard against which to judge the validity of MPTP as a model. Each of the features of Parkinson's disease has been assiduously sought in the MPTP model and, by carefully varying different factors which affect toxicity, a surprising amount has been accomplished. Might it now be time to turn the search around? That is, to determine whether or not some of these modulators and effects of MPTP toxicity could also be variables which are important in Parkinson's disease? Actually, this process has already begun. UHL et al. (1985) have reevaluated variations in MAO activity in the brain in Parkinson's disease, and our own group is investigating the possibility that early MAO-B inhibitor therapy may slow or halt the progression of the disease. If these and other attempts to predict the features of idiopathic Parkinson's disease from the MPTP model are successful, they might provide valuable insights into the etiopathogenesis of Parkinson's disease itself.

References

Ambani LM, Van Woert MH, Murphy S (1975) Brain peroxidase and catalase in Parkinson disease. Arch Neurol 32:114–118

Arai Y, Kinemuchi H, Hamamichi N, Satoh N, Tadano T, Kisara K (1986) Inhibition of rat brain monoamine oxidase by some analogues of 1-methyl-4-phenyl-1,2,3,6-tetrahydropyridine and 1-methyl-4-phenylpyridinium ion. Neurosci Lett 66:43–48

Ballard PA, Tetrud JW, Langston JW (1985) Permanent human parkinsonism due to 1-methyl-4-phenyl-1,2,3,6-tetrahydropyridine (MPTP): seven cases. Neurology (NY) 35:949-956

Barbeau A, Roy M, Langston JW (1985) Neurological consequence of industrial exposure to 1-methyl-4-phenyl-1,2,3,6-tetrahydropyridine. Lancet 1:747

Barbeau A, Dallaire L, Buu NT, Veilleux F, Boyer H, DeLanney LE, Irwin I, Langston EB, Langston JW (1986a) New amphibian models for the study of 1-methyl-4-phenyl-1,2,3,6-tetrahydropyridine (MPTP). Life Sci 36:1125-1134

Barbeau A, Poirier J, Dallaire L, Rucinska E, Buu NT, Donaldson J (1986b) Studies on MPTP, MPP^+ and paraquat in frogs and in vitro. In: Markey SP, Castagnoli N Jr, Trevor AJ, Kopin IJ (eds) MPTP: a neurotoxin producing a parkinsonian syndrome. Academic, New York, p 85-103

Barnes RA (1960) Properties and reactions of pyridine and its hydrogenated derivatives. In: Klingsbert E (ed) Pyridine and derivatives, part 1. Interscience, New York, p 1-97

Baumann RJ, Jameson HD, McKean HE, Haack DG, Weisberg LM (1980) Cigarette smoking and Parkinson disease: 1. A comparison of cases with matched neighbors. Neurology (NY) 30:839-843

Beal MF, Kowall NW, Ellison DW, Mazurek MF, Swartz KJ, Martin JB (1986) Replication of the neurochemical characteristics of Huntington's disease by quinolinic acid. Nature 321:169-171

Benedetti MS, Keane PE (1980) Differential changes in monoamine oxidase A and B activity in the aging rat brain. J Neurochem 35:1026-1032

Blakemore WF (1980) Isoniazid. In: Spencer PS, Schaumburg HH (eds) Experimental and clinical neurotoxicology. Williams and Wilkins, Baltimore, pp 476-489

Boyce S, Kelly E, Reavill C, Jenner P, Marsden CD (1984) Repeated administration of *N*-methyl-4-phenyl-1,2,5,6-tetrahydropyridine to rats is not toxic to striatal dopamine neurons. Biochem Pharmacol 33:1747-1752

Bradbury AJ, Brossi A, Costall B, Domeney AM, Gessner W, Naylor RJ (1986) Biochemical changes caused by the infusion into the substantia nigra of the rat of MPTP and related compounds which antagonise dihydropteridine reductase. Neuropharmacology 25(6):583-586

Brody F, Ruby (1960) Synthetic and natural sources of the pyridine ring. In: Klingsbert E (ed) Pyridine and derivatives, part 1. Interscience, New York, p 99-589

Burns RS, Chiueh CC, Markey SP, Ebert MH, Jacobowitz DM, Kopin IJ (1983) A primate model of parkinsonism: selective destruction of dopaminergic neurons in the pars compacta of the substantia nigra by *N*-methyl-4-phenyl-1,2,3,6-tetrahydropyridine. Proc Natl Acad Sci USA 80:4546-4550

Carvey PM, Kao LC, Klawans HL (1986) Permanent postural effects of MPTP in the guinea pig. In: Markey SP, Castagnoli N Jr, Trevor AJ, Kopin IJ (eds) MPTP: a neurotoxin producing a parkinsonian syndrome. Academic, New York, p 407-411

Chiba K, Trevor A, Castagnoli N JR (1984) Metabolism of the neurotoxic tertiary amine, MPTP, by brain monoamine oxidase. Biochem Biophys Res Commun 120:574-578

Chiba K, Peterson LA, Castagnoli KP, Trevor AJ, Castagnoli N Jr (1985a) Studies on the molecular mechanism of bioactivation of the selective nigrostriatal toxin 1-methyl-4-phenyl-1,2,3,6-tetrahydropyridine. Drug Metab Dispos 13:342-347

Chiba K, Trevor AJ, Castagnoli N Jr (1985b) Active uptake of MPP^+, a metabolite of MPTP, by brain synaptosomes. Biochem Biophys Res Commun 128:1229-1232

Chiueh CC, Markey SP, Burns RS, Johannessen JN, Jacobowitz DM, Kopin IJ (1984a) Selective neurotoxic effects of *N*-methyl-4-phenyl-1,2,3,6-tetrahydropyridine

(MPTP) in subhuman primates and man: a new animal model of Parkinson's disease. Psychopharmacol Bull 20:548-553

Chiueh CC, Markey SP, Burns RS, Johannessen J, Pert A, Kopin IJ (1984b) Neurochemical and behavioral effects of systemic and intranigral administration of *N*-methyl-4-phenyl-1,2,3,6-tetrahydropyridine in the rat. Eur J Pharmacol 100:189-194

Cohen G, Pasik P, Cohen B, Leist A, Mytilineou C, Yahr MD (1985) Pargyline and deprenyl prevent the neurotoxicity of 1-methyl-4-phenyl-1,2,3,6-tetrahydropyridine (MPTP) in monkeys. Eur J Pharmacol 106:209-210

Cote LJ, Kremzner LT (1983) Biochemical changes in normal aging in human brain. In: Mayeux R, Rosen WG (eds) The dementias. Raven, New York, p 19-30

D'Amato RJ, Lipman ZP, Snyder SH (1986) Selectivity of the parkinsonian neurotoxin MPTP: toxic metabolite MPP^+ binds to neuromelanin. Science 231:987-989

D'Amato RJ, Alexander GM, Schwartzman RJ, Kitt CA, Price DL, Snyder SH (1987) Molecular mechanisms of MPTP-induced toxicity. II. Neuromelanin: a role in MPTP-induced neurotoxicity. Life Sci 40:705-712

DeLanney LE, Ricaurte GA, Irwin I, Forno LS, Langston JW (1986) Neurotoxicity of MPTP varies depending on the route of administration (abstract). Soc Neurosci 12:90

Finnegan KT, Irwin I, DeLanney LE, Ricaurte GA, Langston JW (1987) 1,2,3,6-Tetrahydro-1-methyl-4-(methylpyrrol-2-yl)pyridine: studies on the mechanism of action of 1-methyl-4-phenyl-1,2,3,6-tetrahydropyridine. J Pharmacol Exp Ther 242:1144-1151

Forno LS, DeLanney LE, Irwin I, Langston JW (1986a) Neuropathology of MPTP-treated monkeys: comparison with the neuropathology of human idiopathic Parkinson's disease. In: Markey SP, Castagnoli N Jr, Trevor AJ, Kopin IJ (eds) MPTP: a neurotoxin producing a parkinsonian syndrome. Academic, New York, p 119-140

Forno LS, Langston JW, DeLanney LE, Irwin I, Ricaurte GA (1986b) Locus ceruleus lesions and eosinophilic inclusions in MPTP-treated monkeys. Ann Neurol 20:449-455

Fries DS, de Vries J, Hazelhoff B, Horn AS (1986) Synthesis and toxicity toward nigrostriatal dopamine neurons of 1-methyl-4-phenyl-1,2,3,6-tetrahydropyridine (MPTP) analogues. J Med Chem 29 (3):424-427

Fuller RW, Hemrick-Luecke SK (1985) Effects of amfonelic acid, α-methyltyrosine, RO 4-1284 and haloperidol pretreatment on the depletion of striatal dopamine by 1-methyl-4-phenyl-1,2,3,6-tetrahydropyridine in mice. Res Commun Chem Pathol Pharmacol 48:17-25

Fuller RW, Hemrick-Luecke SK (1986)Persistent depletion of striatal dopamine in mice by *M*-hydroxy-MPTP. Res Commun Chem Pathol Pharmacol 53(2):167-172

Gupta M, Gupta BK, Thomas R, Bruemmer V, Sladek JR Jr, Felten DL (1986) Aged mice are more sensitive to 1-methyl-4-phenyl-1,2,3,6-tetrahydropyridine treatment than young adults. Neurosci Lett 70:326-331

Hallman H, Olson L, Jonsson G (1084) Neurotoxicity of the meperidine analogue *N*-methyl-4-phenyl-1,2,3,6-tetrahydropyridine on brain catecholamine neurons in the mouse. Eur J Pharmacol 97:133-136

Hallman H, Lange J, Olson L, Stromberg I, Jonsson J (1985) Neurochemical and histochemical characterization of neurotoxic effects of 1-methyl-4-phenyl-1,2,3,6-tetrahydropyridine on brain catecholamine neurons in the mouse. J Neurochem 44:117-127

Heckman RA, Best FW (1981) An investigation of the lipophilic bases of cigarette smoke condensate. Tobacco Int 33:83-89

Heikkila RE, Hess A, Duvoisin RC (1984a) Dopaminergic neurotoxicity of 1-methyl-4-phenyl-1,2,5,6-tetrahydropyridine in mice. Science 224:1451-1453

Heikkila RE, Manzino L, Cabbat FS, Duvoisin RC (1984b) Protection against the dopaminergic neurotoxicity of 1-methyl-4-phenyl-1,2,5,6-tetrahydropyridine by monoamine oxidase inhibitors. Nature 311:467-469

Herken H, Lange K, Kolbe H (1969) Brain disorders induced by pharmacological blockade of the pentose phosphate pathway. Biochem Biophys Res Commun 36(1):93-100

Huy PTB, Meulemans A, Wassef M, Manuel C, Sterkers O, Amiel C (1983) Gentamicin persistence in rat endolymph and perilymph after a two-day constant infusion. Antimicrob Agents Chemother 23:344-346

Irwin I (1986) The neurotoxin 1-methyl-4-phenyl-1,2,3,6-tetrahydropyridine (MPTP): a key to Parkinson's disease? Pharm Res 3:7-11

Irwin I, Langston JW (1985) Selective accumulation of MPP^+ in the substantia nigra: a key to neurotoxicity? Life Sci 36:207-212

Irwin I, Langston JW, DeLanney LE (1987) 4-Phenylpyridine (4PP) and MPTP: the relationship between striatal MPP^+ concentrations and neurotoxicity. Life Sci 40(8):731-740

Jacobowitz DM, Burns RS, Chiueh CC, Kopin IJ (1984) *N*-methyl-4-phenyl-1,2,3,6-tetrahydropyridine (MPTP) causes destruction of the nigrostriatal but not the mesolimbic dopamine system in the monkey. Psychopharmacol Bull 20:416-422

Jarvis MF, Wagner GC (1985) Neurochemical and functional consequences following 1-methyl-4-phenyl-1,2,5,6-tetrahydropyridine (MPTP) and methamphetamine. Life Sci 36:249-254

Javitch JA, D'Amato RJ, Strittmatter SM, Snyder SH (1985) Parkinsonism-inducing neurotoxin, *N*-methyl-4-phenyl-1,2,3,6-tetrahydropyridine: uptake of the metabolite *N*-methyl-4-phenylpyridine by dopamine neurons explains selective toxicity. Proc Natl Acad Sci USA 82:2173-3177

Jenner P, Nadia MJ, Rupniak SR, Kelly E, Kilpatrick G, Lees A, Marsden CD (1984) 1-Methyl-4-phenyl-1,2,3,6-tetrahydropyridine-induced parkinsonism in the common marmoset. Neurosci Lett 50:85-90

Johannessen JN, Savitt JM, Markey CJ, Bacon JP, Weisz A, Hanselman DS, Markey SP (1987) Molecular mechanisms of MPTP-induced toxicity. I. The development of amine substituted analogues of MPTP as unique tools for the study of MPTP toxicity and Parkinson's disease. Life Sci 40:697-704

Kalaria RN, Mitchel MJ, Harik SI (1986) Blood-brain barrier monoamine oxidase: [^{3}H]pargyline binding to cerebral microvessels (abstract). Soc Neurosci 11(2):1260

Kessler II, Diamond EL (1971) Epidemiologic studies of Parkinson's disease. III. A community-based survey. Am J Epidemiol 96:242-254

Langston JW (1985a) MPTP and Parkinson's disease. Trends Neurosci 80:79-83

Langston JW (1985b) Mechanism of MPTP toxicity: more answers, more questions. TIPS 6:375-378

Langston JW, Ballard PA (1983) Parkinson's disease in a chemist working with 1-methyl-4-phenyl-1,2,5,6-tetrahydropyridine (MPTP). N Engl J Med 309:310

Langston JW, Ballard PA (1984) Parkinsonism induced by 1-methyl-4-phenyl-1,2,3,6-tetrahydropyridine (MPTP): implications for treatment and the pathogenesis of Parkinson's disease. Can J Neurol Sci 11:160-165

Langston JW, Irwin I (1986) MPTP: current concepts and controversies. Clin Neuropharmacol 9:485-507

Langston JW, Ballard PA, Tetrud JW, Irwin I (1983) Chronic parkinsonism in humans due to a product of meperidine-analog synthesis. Science 219:979-980

Langston JW, Forno LS, Rebert CS, Irwin I (1984a) Selective nigral toxicity after systemic administration of 1-methyl-4-phenyl-1,2,5,6-tetrahydropyridine (MPTP) in the squirrel monkey. Brain Res 292:390-394

Langston JW, Irwin I, Langston EB, Forno LS (1984b) 1-Methyl-4-phenylpyridinium ion (MPP^+): identification of a metabolite of MPTP, a toxin selective to the substantia nigra. Neurosci Lett 48:87-92

Langston JW, Irwin I, Langston EB, Forno LS (1984c) Pargyline prevents MPTP-induced parkinsonism in primates. Science 225:1480-1482

Langston JW, Irwin I, Langston EB, Forno LS (1984d) The importance of the 4-5 double bond for neurotoxicity in primates of the pyridine derivative MPTP. Neurosci Lett 50:289-294

Langston JW, Irwin I, DeLanney LE (1987) The biotransformation of MPTP and disposition of MPP^+: the effects of aging. Life Sci 40(8):749-754

Langston JW, Irwin I, Ricaurte G (1987) Neurotoxins, parkinsonism and Parkinson's disease. Pharmacol Ther 32:19-49

Lent CM (1986) MPTP depletes neuronal monoamines and impairs the behavior of the medicinal leech. In: Markey SP, Castagnoli N Jr, Trevor AJ, Kopin IJ (eds) MPTP: a neurotoxin producing a parkinsonian syndrome. Academic, New York, pp 105-118

Lewin R (1986) Age factors loom in parkinsonian research. Science 234:1200-1201

Loos D, Halbhubner K, Kehr W, Herken H (1977) Action of dopamine agonists on Parkinson-like muscle rigidity induced by 6-aminonicotinamide. Neuroscience 4:667-676

Lyden A, Bondesson U, Larsson BS, Lindquist NG (1983) Melanin affinity of 1-methyl-4-phenyl-1,2,5,6-tetrahydropyridine, an inducer of chronic parkinsonism in humans. Acta Pharmacol Toxicol (Copenh) 53(5):429-432

Mann DMA, Yates PO (1983) Possible role of neuromelanin in the pathogenesis of Parkinson's disease. Mech Ageing Dev 21:193-203

Markey SP, Johannessen JN, Chiueh CC, Burns RS, Herkenham MA (1984) Intraneuronal generation of a pyridinium metabolite may cause drug-induced parkinsonism. Nature 311:464-467

Marsden CD (1961) Pigmentation in the nucleus substantiae nigrae of mammals. J Anat 95(2):256-261

Melamed E, Rosenthal J, Clobus M, Cohen O, Frucht Y, Uzzan A (1985a) Mesolimbic dopaminergic neurons are not spared by MPTP neurotoxicity in mice. Eur J Pharmacol 114:970-1000

Melamed E, Rosenthal J, Cohen O, Globus M, Uzzan A (1985b) Dopamine but not norepinephrine or serotonin uptake inhibitors protect mice against neurotoxicity of MPTP. Eur J Pharmacol 116:179-181

Mytilineou C, Cohen G (1985) 1-Methyl-4-phenyl-1,2,3,6-tetrahydropyridine destroys dopamine neurons from the neurotoxic effect of 1-methyl-4-phenylpyridinium ion. J Neurochem 45:1951-1953

Nefzger MD, Quadfasel FA, Karl VC (1968) A retrospective study of smoking and Parkinson's disease. Am J Epidemiol 88:149-158

Nicklas WJ, Vyas I, Heikkila RE (1985) Inhibition of NADH-linked oxidation in brain mitochondria by 1-methyl-4-phenylpyridine, a metabolite of the neurotoxin, 1-methyl-4-phenyl-1,2,3,6-tetrahydropyridine. Life Sci 36:2503-2508

Nicklas WJ, Youngster SK, Kindt MV, Heikkila RE (1987) MPTP, MPP^+ and mitochondrial function. Life Sci 40:721-729

Parisi JE, Burns RS (1986) The neuropathology of MPTP-induced parkinsonism in man and experimental animals. In: Markey SP, Castagnoli N Jr, Trevor AJ, Kopin IJ (eds) MPTP: a neurotoxin producing a parkinsonian syndrome. Academic, New York, p 141-148

Perry TL, Yong VW, Ito M, Jones K, Wall RA, Foulks JG, Wright JM, Kish SJ (1985) 1-Methyl-4-phenyl-1,2,3,6-tetrahydropyridine (MPTP) does not destroy nigrostriatal neurons in the scorbutic guinea pig. Life Sci 36:1233-1238

Ransom BR, Kunis DM, Irwin I, Langston JW (1987) Astrocytes convert the parkinsonism inducing neurotoxin, MPTP, to its active metabolite, MPP^+. Neurosci Lett 75:323-328

Ricaurte GA, Langston JW, DeLanney LE, Irwin I, Brooks JD (1985) Dopamine uptake blockers protect against the dopamine depleting effect of 1-methyl-4-phenyl-1,2,3,6-tetrahydropyridine (MPTP) in the mouse striatum. Neurosci Lett 59:259-264

Ricaurte GA, Langston JW, DeLanney LE, Irwin I, Peroutka SJ, Forno LS (1986) Fate of nigrostriatal neurons in young mature mice given 1-methyl-4-phenyl-1,2,3,6-tetrahydropyridine (MPTP): a neurochemical and morphological reassessment. Brain Res 376(1):117-124

Ricaurte GA, Irwin I, Forno LS, DeLanney LE, Langston EB, Langston JW (1987) Aging and MPTP-induced degeneration of dopaminergic neurons in the substantia nigra. Brain Res 403:43-51

Robinson DS, Nies A, Davis JN, Bunney WE, Davis JM, Colburn RW, Bourne HR, Shaw DM, Coppen AJ (1972) Ageing, monoamines and monoamine-oxidase levels. Lancet 1:290-291

Sahenk Z, Mendell JR (1980) Zinc pyridinethione. In: Spencer PS, Schaumburg HH (eds) Experimental and clinical neurotoxicology. Williams and Wilkins, Baltimore, pp 578-592

Sahgal A, Andrews JS, Biggins JA, Candy JM, Edwardson JA, Keith AB, Turner JD, Wright C (1984) *N*-methyl-4-phenyl-1,2,3,6-tetrahydropyridine (MPTP) affects locomotor activity without producing a nigrostriatal lesion in the rat. Neurosci Lett 48:179-184

Sayre LM, Arora PK, Iacofano LA, Harik SI (1986) Comparative toxicity of MPTP, MPP^+ and 3,3-dimethyl-$MPDP^+$ to dopaminergic neurons of the rat substantia nigra. Eur J Pharmacol 124:171-174

Schneider JS, Markham CH (1986) Neurotoxic effects of *N*-methyl-4-phenyl-1,2,3,6-tetrahydropyridine (MPTP) in the cat. Tyrosine hydroxylase immunohistochemistry. Brain Res 373:258-267

Schneider JS, Yuwiler A, Markham CH (1986) Production of a Parkinson-like syndrome in the cat with *N*-methyl-4-phenyl-1,2,3,6-tetrahydropyridine (MPTP): behavior, histology, and biochemistry. Exp Neurol 91:290-307

Snyder SH, D'Amato RJ (1986) MPTP: a neurotoxin relevant to the pathophysiology of Parkinson's disease. Neurology (NY) 36:250-258

Steranka LR, Polite LN, Perry KW, Fuller RW (1983) Dopamine depletion in rat brain by MPTP (1-methyl-4-phenyl-1,2,3,6-tetrahydropyridine). Res Commun Substances Abuse 4:315-323

Sundstrom E, Jonsson G (1985) Pharmacological interference with the neurotoxic action of 1-methyl-4-phenyl-1,2,3,6-tetrahydropyridine (MPTP) on central catecholamine neurons in the mouse. Eur J Pharmacol 110:293-299

Tetrud JW, Langston JW, Redmond DE, Roth RH, Sladek JR, Angel RW (1986) MPTP-induced tremor in human and non-human primates. Neurology (NY) 36(Suppl 1):308

Trevor AJ, Castagnoli N Jr, Caldera P, Ramsay RR, Singer TP (1987) Bioactivation of MPTP: reactive metabolites and possible biochemical sequelae. Life Sci 40:713-719

Uhl GR, Javitch JA, Snyder SH (1985) Normal MPTP binding in parkinsonian substantia nigra: evidence for extraneuronal toxin conversion in human brain. Lancet 2:956-957

Ward CD, Duvoisin RC, Ince SE, Nutt JD, Eldridge R, Calne DB (1983) Parkinson's disease in 65 pairs of twins and in a set of quadruplets. Neurology (NY) 33:815-824

Wilson JS, Wilson JA (1986) Intracellular recordings in the caudate nucleus of normal and MPTP treated dogs. In: Markey SP, Castagnoli N Jr, Trevor AJ, Kopin IJ (eds) MPTP: a neurotoxin producing a parkinsonian syndrome. Academic, Orlando, pp 695-699

Youngster SK, Duvoisin RC, Hess A, Sonsalla PK, Kindt MV, Heikkila RE (1986) 1-Methyl-4-phenyl-(2'-methylphenyl)-1,2,3,6-tetrahydropyridine [2'CH_3MPTP] is a more potent dopamine neurotoxin than MPTP in mice. Eur J Pharmacol 122:283-287

Ziering A, Berger L, Heineman SD, Lee J (1947) Piperidine derivatives. Part III. 4-Ayrlpiperidines. J Org Chem 12:894-903

Zimmerman DM, Cantrell BE, Reel JK, Hemrick-Luecke SK, Fuller RW (1986) Characterization of the neurotoxic potential of *m*-methoxy-MPTP and the use of its *n*-ethyl analogue as a means of avoiding exposure to a possible parkinsonism-causing agent. J Med Chem 29:1517-1520.

CHAPTER 8

The Relationship Between Parkinson's Disease and Other Movement Disorders

J. JANKOVIC

A. Introduction

Parkinson's disease, a specific clinical-pathologic entity, is one of the commonest causes of disability among the elderly. The prevalence figures for Parkinson's disease vary considerably among the different epidemiologic studies. A door-to-door survey in a biracial population of Copiah County, Mississippi revealed a prevalence of 347 per 100000 inhabitants (SCHOENBERG et al. 1985). The prevalence in a population over the age of 65, however, is about 1%. Although previous studies suggested a higher frequency of Parkinson's disease among males and whites, this epidemiologic survey found no substantial differences in the age-adjusted prevalence ratios by race or by sex. Over 40% of the identified cases were not diagnosed until this door-to-door survey. Thus, it is possible that for every two patients diagnosed as having Parkinson's disease there is one patient whose symptoms may not be disabling enough to warrant medical attention. Furthermore, the diagnosis is often missed in the early stages of the disease and the symptoms may be incorrectly attributed to a variety of other problems. Therefore, the variability in reported prevalence figures can be partly explained by differences in the methods of ascertainment.

Approximately 25% of Parkinson's disease patients receive an incorrect original diagnosis (KOLLER 1984). The stooped posture, small and shuffling steps, difficulty in turning, and generalized slowness are often confused with normal senescence. However, the mechanism of senile gait is probably different from the parkinsonian gait, and the former does not improve with levodopa (NEWMAN et al. 1985). The loss of postural reflexes, a frequent cause of falls among Parkinson's disease patients, may be initially attributed to postural instability associated with normal aging (WEINER et al. 1984). Painful stiffness in Parkinson's disease is usually caused by rigidity; it is often unilateral at the onset and frequently misdiagnosed as bursitis, cervical radiculopathy, or arthritis. The diagnosis may be further complicated by the presence of a variety of sensory complaints, including aches and pains, numbness, burning, or coldness. These sensory symptoms have been added to the motor symptoms as typical features of Parkinson's disease (GOETZ et al. 1986). Dysarthria, word-finding difficulties, and other speech and language problems may be incorrectly attributed to a "stroke" (ROBBINS et al. 1986). Stroke is also a common misdiagnosis when the motor symptoms are unilateral (hemiparkinsonism). The occurrence of dysphagia and weight loss often leads to unnecessary tests for occult carcinoma. These symptoms, however, may occur independently, suggesting that the motor disability interfering with eating or swallowing is not the cause of weight loss typically seen in Parkinson's disease patients (VAN DER LINDEN et al. 1985). Likewise, depression, bradyphrenia (slow thinking), and dementia may occur in combination or separately (MAYEUX 1984; CHUI et al. 1986; SANTAMARIA et al. 1986). About one-third of Parkinson's disease patients have a disabling dementia and 28%–80% of patients with clinically diagnosed Alzheimer's disease have parkinsonian symptoms (ELIZAN et al. 1986; MAYEUX et al. 1985; MOLSA et al. 1984). While Parkinson's disease and Alzheimer's disease share some clinical and pathologic features, the etiology is probably different. However, several unifying hopotheses for common pathogenesis have been proposed (APPEL 1981; CALNE et al. 1986; GAJDUSEK 1985).

The most recognizable symptom of Parkinson's disease is tremor (JANKOVIC 1987). However, tremor is present at the onset of Parkinson's disease in only 50% of patients, and 15% of Parkinson's disease patients never have tremor (MARTIN et al. 1973). While the typical tremor at rest is usually emphasized in textbook descriptions of Parkinson's disease, symptomatic postural tremor occurs at least as frequently as the classical rest tremor and it is often more disabling. This postural tremor usually has a frequency between 5.5 and 8 Hz, with the lower frequency being associated with larger amplitude (FINDLEY and GRESTY 1984). In contrast to the tremor at rest, which usually consists of a supination-pronation oscillatory movement in the hands, the postural tremor is characterized by rhythmic flexion-extension movements. Some patients also have a high frequency (9.5–11 Hz) postural tremor, but this is rarely symptomatic. The lower frequency postural tremor of Parkinson's disease is clinically identical to essential tremor. This can be a source of difficulty in differential diagnosis. When postural tremor is present for more than 5 years before the onset of other parkinsonian symptoms, and when there is a family history of tremor, the diagnosis of essential tremor should be considered. In contrast to Parkinson's tremor, essential tremor often involves the neck, producing a "negation" or "affirmation" oscillatory movement of the head, and it rarely involves the legs.

The differentiation between essential tremor and Parkinson's disease is complicated by the observation that the two disorders may coexist in the same patient (GERAGHTY et al. 1985). We found that 6.7% of all Parkinson's disease patients had preexisting essential tremor, which compares with the expected prevalence of 0.41% for essential tremor in the general population (JANKOVIC 1986). The clinical similarity between the two conditions, the overlap of some clinical manifestations, and the occasional coexistence of essential tremor and Parkinson's disease all contribute to some difficulties in the differential diagnosis. About 5%–10% of patients referred with the diagnosis of Parkinson's disease actually have essential tremor. Accurate diagnosis and differentiation of the two conditions is important because the treatments are different. While parkinsonian rest tremor responds to anticholinergic and dopaminergic medications, essential tremor tends to improve with beta-blockers, primidone, benzodiazepines, and alcohol. Essential tremor is usually a monosymptomatic condition, but it may be associated with cogwheel rigidity, dystonia, myoclonus, and Charcot-Marie-Tooth disease (Van der Linden and Jankovic 1987). In contrast to the nigrostriatal degeneration and the Lewy bodies seen in Parkinson's disease, postmortem studies of patients with essential tremor have so far revealed no discernible pathology.

In addition to essential tremor, Parkinson's disease may be associated with dystonia, myoclonus, orofacial chorea, and peripheral neuropathy. We studied four families with the combination of hereditary neuropathy, essential tremor, Parkinson's disease, and blepharospasm (Van der Linden and Jankovic 1987). The association between essential tremor and dystonia, and between Parkinson's disease and dystonia is well recognized. Therefore, it should not be surprising that Parkinson's disease (PD) and familial essential tremor (ET) may coexist. The PD-ET subset of Parkinson's disease, as well as the families with PD, ET, and hereditary neuropathy, support the hypothesis that genetic predisposition may play a role in the pathogenesis of at least some cases of Parkinson's disease.

While the pathology of Parkinson's disease seems homogeneous, the clinical manifestations are more protean. For example, some Parkinson's disease patients start with *tremor* as the predominant clinical symptom, with bradykinesia, rigidity, and postural instability occurring only in the more advanced stages. Other patients seem to have predominantly *postural instability and gait difficulty* (PIGD) with little or no tremor. An analysis of clinical findings in 334 patients with idiopathic Parkinson's disease revealed that the tremor patients tend to have an earlier onset and higher familial occurrence, whereas the PIGD patients have a greater degree of dementia, bradykinesia, functional disability, and a less favorable long-term prognosis (Zetusky et al. 1985). Therefore, Parkinson's disease, may be viewed as a syndrome with different subgroups characterized by a typical pattern of clinical features. Future research, utilizing postmortem biochemical analyses or in vivo positron emission tomographic (PET) scan studies should provide clinical-biochemical characterization of the different Parkinson's disease subsets in an attempt to answer the question whether Parkinson's disease is a single entity or different disorders with different pathogenetic mechanisms. Besides the tremor and the PIGD forms of idiopathic Parkinson's disease, there may be other subsets, including Parkinson's disease associated with *akinesia* (freezing), *dementia, depression, sensory disturbance*, or *autonomic dysfunction* (Table 1).

While recent reviews of the etiology of Parkinson's disease have favored the environmental hypothesis, in part fueled by the unfolding story of MPTP-induced parkinsonism, it is possible that genetic predisposition, based on Mendelian pattern of inheritance or somatic mutation during embryogenesis, coupled with certain triggers such as stress, infection, trauma, toxins, drugs, or environmental factors, may also contribute to the pathogenesis of Parkinson's disease (Snyder et al. 1985; Jankovic and Calne 1987). Although the twin studies found no support for the genetic predisposition in Parkinson's disease (Ward et al. 1983), some cases of Parkinson's disease may be inherited (Barbeau and Roy 1984; Alonso et al. 1986). The genetic theory for certain Parkinson's disease subsets is partly based on 15% incidence of familial cluster-

Table 1. Classification of parkinsonism

I. *Primary Parkinson's disease*
  A. Idiopathic—dominated by
    1. Tremor
    2. Postural instability and gait difficulty (PIGD)
    3. Akinesia (freezing)
    4. Dementia
    5. Depression
    6. Sensory disturbance
    7. Autonomic dysfunction
  B. Inherited—associated with essential tremor, dystonia, or peripheral neuropathy
  C. Young onset—associated with dystonia or essential tremor
  D. Juvenile Parkinson's disease

II. *Secondary parkinsonism*
  A. Drugs (dopamine blocking and depleting drugs, α-methyldopa, lithium, diazoxide, flunarizine, cinnarizine)
  B. Toxins (manganese, mercury, carbon monoxide, cyanide, carbon disulfide, methanol, ethanol, MPTP)
  C. Metabolic (parathyroid, acquired hepatocerebral degeneration, GM_1 gangliosidosis, Gaucher's disease)
  D. Postencephalitic and slow virus
  E. Vascular (multi-infarct, Binswanger's disease)
  F. Brain tumor
  G. Trauma and pugilistic encephalopathy
  H. Hydrocephalus (normal and high pressure)
  J. Syringomesencephalia

III. *Multiple system degenerations* (Parkinsonism-plus)
  A. Sporadic
    1. Progressive supranuclear palsy (ophthalmoparesis)
    2. Shy-Drager syndrome (dysautonomia)
    3. Olivopontocerebellar atrophy (ataxia)
    4. Parkinsonism-dementia-ALS complex (motor neuron disease)
    5. Striatonigral degeneration
    6. Corticodentatonigral degeneration with neuronal achromasia
    7. Alzheimer's disease
  B. Inherited
    1. Huntington's disease
    2. Wilson's disease
    3. Hallervorden-Spatz disease
    4. Familial parkinsonism-dementia syndrome
    5. Familial basal ganglia calcification
    6. Neuroacanthocytosis
    7. Spinocerebellar-nigral degeneration and Joseph disease
    8. GDH deficiency

ing, occasional concordance for Parkinson's disease in monozygotic twin pairs (JANKOVIC and RECHES 1986), higher than expected coexistence of essential tremor and Parkinson's disease (GERAGHTY et al. 1985), and the possibility of familial deficiency in hepatic cytochrome P-450 hydroxylases in some Parkinson's disease patients (BARBEAU et al. 1985). BARBEAU and ROY (1984) subdivided the familial parkinsonism into two categories: 1. Parkinson's disease associated with autosomal dominant diseases, including parkinsonism associated with essential tremor, olivopontocerebellar atrophy, depression, and peripheral neuropathy; 2. Parkinson's disease with autosomal recessive pattern, including the akinetic-rigid syndrome, juvenile parkinsonism, and glutamate dehydrogenase deficiency.

Besides the sporadic (idiopathic) Parkinson's disease and the inherited form, there is a third type of Parkinson's disease which has been referred to as juvenile Parkinson's disease (NAIDU et al. 1978; CLOUGH et al. 1981; GERSHANIK and LEIST 1987; NYGAARD and DUVOISIN 1986; QUINN et al. 1987; LIMA et al. 1987). The term "juvenile", however, is misleading since the reports include patients up to the age of 40; therefore, the term "young onset" Parkinson's disease is preferable. Approximately 5%-10% of all patients have the onset of Parkinson's symptoms before the age of 40. Many patients in this group have a family history of Parkinson's disease, but most are sporadic. In contrast to the typical Parkinson's disease, with average age of onset at 55, the young onset disease is characterized by high frequency of associated dystonia, often preceding other signs of Parkinson's disease. Another distinguishing fea-

Table 2. Parkinsonism—relative incidence (1230 patients evaluated at the Movement Disorder Clinic, Baylor College of Medicine)

Parkinsonian disorder	Number	(%)
Parkinson's disease	1045	84.9
Progressive supranuclear palsy	81	6.6
Neuroleptic-induced parkinsonism	45	3.6
Shy-Drager syndrome	22	1.8
Olivopontocerebellar degeneration	12	1.0
Vascular (multi-infarct) parkinsonism	11	
Hallervorden-Spatz disease	6	
Postencephalitic parkinsonism	5	
Striatonigral degeneration	4	
Huntington's disease	4	
Brain tumor	3	
Manganese intoxication	3	
Familial basal ganglia calcification	2	
Normal pressure hydrocephalus	2	
Aqueductal stenosis and hydrocephalus	1	
Wilson's disease	1	
Carbon monoxide	1	
Other	18	1.5

ture of the young onset Parkinson's disease is a striking response to low doses of levodopa and an early onset of levodopa-induced dyskinesias. Because of these distinguishing features, the young onset Parkinson's disease has been thought to be etiologically different from the more typical form of the disease.

While Parkinson's disease accounts for about 85% of all cases of parkinsonism (the clinical syndrome characterized by tremor at rest, rigidity, bradykinesia, and loss of postural reflexes), there are many other conditions in which parkinsonism is an important component (Table 1). In the Baylor Movement Disorder Clinic the second commonest cause of parkinsonism is progressive supranuclear palsy, accounting for about 6% of all patients referred to the Clinic for evaluation of parkinsonism (Table 2). However, in an unselected population of patients, drug-induced parkinsonism is probably the second commonest cause of Parkinson's disease symptoms. The remainder of this chapter will focus on some of the secondary causes of parkinsonism and on the multiple system degenerations (Parkinson-plus syndromes) (SCHWAB and ENGLAND 1968; FAHN 1977; JANKOVIC 1984a).

B. Secondary Parkinsonism

I. Drug-Induced Parkinsonism

Parkinsonism caused by the antipsychotic drugs has been recognized since the early 1950s. The reported incidence of drug-induced parkinsonism varies from 15.4% to 61% of the exposed patients (RAJPUT et al. 1982; KLAWANS et al. 1973). The more potent the dopamine receptor blocking drugs, the more likely they are to induce parkinsonism. Also, age seems to be a risk factor, possibly related to the normal age-related loss of dopamine neurons and of postsynaptic receptors (JANKOVIC and CALNE 1987).

The clinical manifestations of drug-induced parkinsonism are indistinguishable from idiopathic Parkinson's disease except for the presence of coexisting tardive dupkinesia (JANKOVIC and CASABONA 1987; HARDLE and LEES 1988). However, there are some useful differentiating features. For example, drug-induced parkinsonism usually produces symmetric bradykinesia and rigidity in the early stages, whereas idiopathic Parkinson's disease is often asymmetric at onset. The slow movement, lack of facial expression, absent arm swing, soft and monotonous speech, and flexed posture are often mistaken for psychomotor retardation associated with underlying depression. The characteristic pill-rolling tremor at rest seen in idiopathic Parkinson's disease is relatively uncommon in drug-induced parkinsonism (JANKOVIC 1981). The latter is often associated with an action tremor which has a faster frequency (7-8 Hz) than the typical rest tremor (4-5 Hz). The perioral tremor in drug-induced parkinsonism is sometimes reffered to as the "rabbit syndrome," but the same type of tremor can also occur in idiopathic Parkinson's disease. Signs of parkinsonism may be evident within several days after starting the neuroleptic treatment, however, the evolution is dependent on the dosage and potency of the drug (TARSY and BALDESSARINI 1986). The parkinsonian signs usually disappear within a few weeks after the offending drug is withdrawn. In some patients, particularly the elderly, the parkinsonism may persist for several months, or even years (KLAWANS et al. 1973).

Not all patients treated with the dopamine receptor blocking agents develop parkinsonism, suggesting individual susceptibility. Perhaps, the individuals who are prone to develop drug-induced parkinsonism have a subclinical Parkinson's disease and the neuroleptic drug then precipitates the parkinsonian symptoms. This notion is supported by the observation of RAJPUT et al. (1982), who found histologic evidence of idiopathic Parkinson's disease in two patients who had reversible drug-induced parkinsonism. In addition to the typical pathologic abnormalities, including Lewy bodies and depigmentation of the substantia nigra seen in both cases, there was a reduction in the levels of dopamine in the striatum of one patient. Therefore, the occasional patient in whom parkinsonism persists after withdrawal of the drug may have an underlying idiopathic Parkinson's disease. After chronic administration of phenothiazines to monkeys, BIRD and ANTON (1982) actually found increased dopamine levels in the basal ganglia after 2 months of treatment, but the levels were markedly reduced after 20 months.

It is not known whether drug-induced parkinsonism is a "predictor" of subsequent risk for the development of tardive dyskinesia. While drug-induced parkinsonism and tardive dyskinesia are presumably mediated by different mechanisms, the two movement disorders may coexist in the same individual (JANKOVIC and CASABONA 1987). This combination presents a therapeutic challenge because the treatment of one condition exacerbates the other.

There are three classes of drugs that are likely to produce parkinsonism:

1. Dopamine receptor blocking agents, including the phenothiazines, butyrophenones, and metoclopramide;
2. Dopamine-depleting agents, including reserpine and tetrabenazine;
3. Drugs that act by various known and unknown mechanisms, including α-methyldopa, which presumably blocks the dopa decarboxylase and is converted to the "false neurotransmitter" α-methylepinephrine (STRANG 1966).

In one series of 95 new cases of parkinsonism evaluated in a department of geriatric medicine, 58 were drug-induced (STEPHEN and WILLIAMSON 1984). The commonest offending drug was prochlorperazine, an antiemetic drug. In our experience, metoclopramide is one of the commonest causes of drug-induced parkinsonism (INDO and ANDO 1982; HASSAN et al. 1986). Rarely, lithium, flunarizine, cinnarizine, diazoxide, and possibly captopril may produce reversible parkinsonism (RECHES et al. 1981; SANDYK 1985; LAPORTE and CAPELLA 1986; MICHELI et al. 1987).

II. Toxin-Induced Parkinsonism

1. Manganese

After drug-induced parkinsonism, manganese is probably responsible for more cases of secondary parkinsonism than any other toxin. First noted more than 150 years ago, manganese-induced parkinsonism has been reported in almost all mining regions of South America, India, Africa, and the United States. Another possible sance of manganese intoxication is occupational exposure to pesticides containing manganese (FERRAZ et al 1988). Despite the widespread exposure only a minority of exposed averkus develops the neurologic syndrome. This suggests an individual susceptibility, perhaps related to a slow rate of elimination in some patients. More than 5 mg manganese per cubic meter in the environment is required for the symptoms of manganism to occur, and the duration of exposure is usually over 2 years (MENA et al. 1967). However, cases have been described after only 1 month's exposure.

The initial manifestations of manganese poisoning are nonspecific, consisting chiefly of asthenia, malaise, and emotional instability, including anxiety, inappropriate laughing and crying, compulsive singing and dancing, and visual or auditory hallucinations. This "manganese madness," is usually replaced by progressive bradykinesia, hypomimia, low volume and monotonous speech, and palilalia (BANTA and MARKESBERY 1977). Eventually, the patients have gait difficulty, postural instability, and dystonic postures of the arms and neck, as well as dystonic grimacing. In addition to the tremor at rest, the patients have action-postural tremor in the hands, tremor of the tongue, and tremulous handwriting. Some patients develop cerebellar and pyramidal signs, but peripheral neuropathy is rare. The diagnosis may not be easy to establish because the blood and urine manganese levels may be normal, or even low, by the time the patient is studied. The diagnosis, therefore, rests on the clinical examination and the history of occupational exposure.

In contrast to idiopathic Parkinson's disease, the brunt of the pathologic change in manganese-induced parkinsonism is in the globus pallidus and in the subthalamic nucleus rather than in the substantia nigra. In addition to the neuropathologic changes, chiefly consisting of neuronal loss and gliosis, the patients often have liver cirrhosis. Because of the behavioral and neurochemical similarities between manganese-induced parkinsonism and experimental 6-hydroxydopamine-induced parkinsonism, it has been suggested that manganese neurotoxicity is due to potentiation of dopamine auto-oxidation with concomitant production of free radicals such as supraoxide ($O_2^-\cdot$), hydrogen peroxide (H_2O_2), and hydroxyl radicals ($HO\cdot$) (DONALDSON et al. 1980; COHEN 1984). It has even been suggested that the neuroleptic-induced parkinsonism is mediated by manganese, which is increased in the striatum of monkeys exposed to chronic neuroleptic treatment (BIRD et al. 1967, 1984).

Because of the low tissue concentration of manganese, MENA et al. (1967) concluded that chelation with eidetic acid (EDTA) is not effective in manganese intoxication; penicillamine has not been studied. Symptomatic therapy with levodopa has produced variable results, and only a few patients show marked improvement. The most important step in the management of manganese intoxication is the removal of the worker from the area with the high manganese concentration.

2. Carbon Monoxide

There are two distinct parkinsonian syndromes associated with carbon monoxide poisoning. In the first syndrome, parkinsonism is associated with sei-

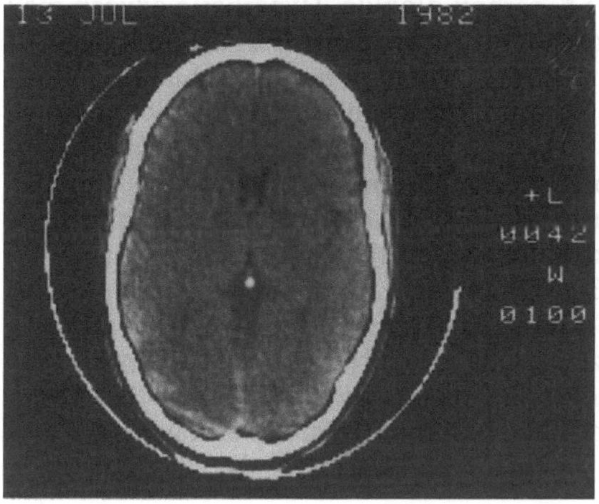

Fig. 1. CT scan of a 44-year-old man exposed to high concentration of carbon monoxide for at least 1 h, showing bilateral globus pallidus lucencies

zures, cortical blindness, various agnosias, ataxias, and cognitive impairment. Some patients have akinetic mutism, pyramidal tract signs, deafness, peripheral neuropathy, and other neurologic symptoms (Jackson and MENGES 1980; CHOI 1983). The second syndrome, much less common, consists of pure parkinsonism (KLAWANS et al. 1982). We studied a 44-year-old man who was found unconscious in a smoke-filled apartment bedroom with a fire burning in the living room. The estimated time of exposure to carbon monoxide was 1 h. When he was brought to the local emergency room he was deeply comatose and the carboxyhemoglobin level was 45%. After 12 h he awakened, but was left with a pure parkinsonian syndrome characterized by severe bradykinesia and frequent freezing, marked rigidity, and hypophonic, monotonous speech. CT scan revealed bilateral pallidal lucencies (Fig. 1). The patient failed to respond to high dosages of levodopa, bromocriptine, and other antiparkinsonian medications. An autopsy report of a 32-year-old woman with similar CT findings after carbon monoxide poisoning revealed severe necrosis and degeneration of the globus pallidus corresponding to the low density areas seen on the CT scan (SAWADA et al. 1983). In addition to the globus pallidus involvement, pathological studies have shown damage in layers II and III of the cerebral cortex, Sumner's sector of the hippocampus, the Purkinje cell layer of the cerebellum, and the cerebral white matter.

While carbon monoxide can produce an acute or subacute neurologic syndrome by producing anoxic damage in vulnerable brain regions, delayed neurologic sequelae after carbon monoxide intoxication are relatively uncommon. In one series, only 2.75% of 2360 victims of acute carbon monoxide intoxication had neurologic sequelae (CHOI 1983). In the vast majority of cases, the carbon monoxide poisoning is either rapidly fatal or causes no neurologic sequelae. About 10% of patients admitted to hospital for carbon monoxide intoxication suffer significant neurologic complications. The most important predisposing factors include age and severity of anoxia. Treatment should begin as soon as possible after removal from the carbon monoxide environment, with 95% oxygen and 5% carbon dioxide. Hyperbaric oxygen may be useful when exposure and subsequent unconsciousness are prolonged. If the patient recovers from the acute insult, it is important to schedule follow-up visits to detect delayed onset postanoxic encephalopathy or the parkinsonian syndrome.

Almost all patients with carbon monoxide-induced neurologic sequelae initially presented in coma. However, we have evaluated three young patients with chronic carbon monoxide exposure for several years who developed atypical parkinsonism in their early 30s. One man, 32 years old, had the onset of severe bradykinesia and akinesia after he had worked for 5 years in a poorly ventilated warehouse, loading frozen goods into a refrigerated truck while the engine was running to prevent melting of the goods. These, and other cases, suggest that chronic repeated exposure to carbon monoxide may rarely produce a parkinsonian syndrome.

3. Cyanide

Acute cyanide intoxication can produce neurologic symptoms similar to those seen with carbon monoxide poisoning. Anoxia from cyanide is presumably due to the inhibition of the cytochrome oxidase mitochondrial enzyme system. Chronic cyanide toxicity has been reported in silver reclamation workers, and has been suggested by epidemiologic studies to occur with chronic ingestion of naturally occurring plant cyanogenic glycosides present most notably in the tropical foodstuff cassava (Blanc et al. 1985). In the past, thiocyanates were used as antihypertensive agents, but because of the potential cyanide toxicity, the use of these drugs has diminished. Chronic cyanide exposure produces only nonspecific neurologic symptoms such as headaches, dizziness, anosmia, paresthesias, and fatigability. However, acute cyanide intoxication, seen in suicide attempts, may lead to death or a variety of neurologic sequelae such as parkinsonism. Uitti et al. (1985) reported an 18-year-old man who ingested over 1000 mg potassium cyanide in a suicide attempt and later developed a severe parkinsonian syndrome characterized chiefly by akinesia and rigidity. He died 19 months after the drug overdose and at autopsy had necrosis of the globus pallidus, putamen, and substantia nigra reticulata, whereas the zona compacta, most involved in idiopathic Parkinson's disease, was normal. In experimental animals, severe cyanide poisoning causes necrosis of not only the basal ganglia, but also the cerebellum, cerebral cortex, and central white matter (Hurst 1972).

4. Carbon Disulfide

Carbon disulfide has been used in many industries, e.g., as a solvent for rubber and in the production of viscose rayon and cellophane (Peters HL et al. 1988). Carbon disulfide is absorbed by vapor inhalation and it is retained by the brain and other tissues. During acute intoxication, the patients experience motor restlessness, delirium, mania, and violent behavior. If the patient is not removed from exposure during this toxic psychosis phase, permanent dementia may occur. In addition to the cognitive deficit, some patients have polyneuropathy, cranial nerve and brain stem signs; parkinsonism was seen in 20 % of workers in a viscose rayon plant (Lewy 1941). Pathologic changes associated with carbon disulfide poisoning are most evident in the cerebral cortex. Rhesus monkeys exposed to carbon disulfide have signs of parkinsonism, and the most consistent pathologic change is an extensive necrosis of the globus pallidus and the zona reticulata of the substantia nigra, resembling the pathologic changes of carbon monoxide and cyanide poisoning (Richter 1945). Dopamine and other monoamines are also impaired in rats exposed to carbon disulfide (Magos and Jarvis 1970).

5. Other Toxins

Most toxins that produce parkinsonism cause diffuse cerebral pathology with nonspecific involvement of the basal ganglia. In contrast, 1-methyl-4-phenyl-1,2,3,6-tetrahydropyridine (MPTP) causes relatively selective damage to the dopamine-containing cells of the substantia nigra (Allen et al. 1986; Langston and Irwin 1986; Snyder and D'Amato 1986). Because of the close resemblance to idiopathic Parkinson's disease, MPTP has received a great deal of attention from the scientific community (Ballard et al. 1985). MPTP-induced parkinsonism is reviewed in Chap. 7 and will not be discussed in detail here.

While the MPTP model provides an important tool for studying the pathogenesis and treatment of Parkinson's disease, neither human nor experimental MPTP-induced parkinsonism is identical to idiopathic Parkinson's disease. Some of the clinical, biochemical, and pathologic similarities and differences are listed in Table 3. The differences between human idiopathic Parkinson's disease and MPTP-induced parkinsonism may be less obvious after chronic administration of MPTP. For example, protracted MPTP administration in older monkeys produces not only degeneration of the neurons in the substantia nigra, but also in the locus ceruleus, which is also seen in human idiopathic Parkinson's disease (Forno et al. 1986). Furthermore, eosinophilic inclusions resembling Lewy bodies may be seen after chronic exposure to MPTP (Forno et al. 1986; Richardson 1986).

Other toxins implicated in the pathogenesis of secondary parkinsonism include mercury, methanol, and ethanol (Greenhouse 1982; Lang et al. 1982; Goetz et al. 1981; Carlen et al. 1981). Mercury and methanol produce parkinsonism by causing damage to the putamen. The growing interest in environmental causes of Parkinson's disease will undoubtedly result in more epidemiologic studies designed to uncover other environmental or occupational neurotoxins. For example, the relatively high incidence of parkinsonism among painters and other workers exposed to organic solvents seen at the Baylor Movement Disorder Clinic deserves further investigation. We recently evaluated a 44-year-old painter with a 3-year history of nonprogressive parkinsonism, high frequency postural tremor, and other atypical features starting after a period of intense exposure to epoxy paints in a poorly ventilated environment. While a case-reference study of occupational exposure to organic solvents and agricultural chemicals found no evidence of increased risk of Par-

Table 3. Idiopathic Parkinson's disease compared with human and experimental MPTP-induced parkinsonism

	Idiopathic	MPTP-induced
Onset	Insidious	Acute
Age at onset in humans (years)	51–60	22–42
Course	Progressive	Stable
Tremor at rest	+++	+
Postural tremor	++	++
Axial rigidity	+	++
Dystonic postures	+	++
Early freezing	○	++
Eyelid apraxia and blepharospasm	+	++
Hypophonia	+	++
Dementia	++	○
Dysautonomia	+	○
Diaphoresis	+	+
Progressive deterioration	+++	±
Response to levodopa	++	+++
Early onset of levodopa-related psychosis dyskinesias, and fluctuations	+	+++
Dopamine deficiency	+++	+++
Norepinephrine deficiency	++	○
Serotonin deficiency	+	○
SNc (A8, 9) degeneration	+++	+++
Ventral tegmental degeneration	++	○
Depigmentation of locus ceruleus and dorsal motor nucleus of vagus	++	+ (?)
Lewy bodies	+++	+ (?)
Prevention by antioxidants and by DA reuptake inhibitors	+ (?)	+++

kinson's disease, it is possible that chronic exposure to solvent vapors, pesticides, and other chemicals produces neurologic sequelae which may include tremor and other parkinsonian features (OLSEN and HOGSTEDT 1981; HORMES et al. 1986).

III. Metabolic Causes of Parkinsonism

1. Hypoparathyroidism

Hypoparathyroidism, often associated with basal ganglia calcification, may produce a levodopa-unresponsive parkinsonian syndrome (KLAWANS et al. 1976). The commonest cause of hypoparathyroidism is thyroidectomy, but this form of postoperative hypoparathyroidism is associated with basal ganglia calcification in only 1% of cases. In contrast, idiopathic hypoparathyroidism is associated with calcification in 21% of cases, and pseudohypoparathyroidism in 42% (MUENTER and WHISNANT 1968, EVANS and DONLEY 1988). Besides parkinsonism, other neurolcgic manifestations of postoperative hypoparathyroidism include tetany, seizures, and chorea. Parkinsonism may be delayed for several years after the surgery and may progress, despite correction of serum calcium (BERGER and ROSS 1981). Pathologic examination reveals calcium deposits, particularly in the perivascular areas of the basal ganglia. Although mineralization of the basal ganglia may be quite extensive, it is unlikely that it interferes with neurotransmission.

Besides hypoparathyroidism, there are many other causes of basal ganglia calcification, including toxoplasmosis and other infections, cranial irradiation, carbon monoxide intoxication, anoxia, vascular changes associated with aging, Kearns-Sayre syndrome, and Cockayne's syndrome (BEALL et al. 1984; KOLLER et al. 1979; MURPHY 1979; SOFFER et al. 1979). Familial calcification associated with parkinsonism will be discussed later.

2. Acquired Hepatocerebral Degeneration

The clinical syndrome of acquired hepatocerebral degeneration consists of repeated episodes of hepatic encephalopathy with subsequent insidious onset of global dementia, often associated with agnosia or lateralizing parietal findings (VICTOR et al. 1965). Many patients have dysarthria, coarse tremor, and chorea, resembling Huntington's disease. Because of the facial grimacing and dystonia, coupled with the hepatic dysfunction, the diagnosis of Wilson's disease is often entertained. In later stages, the patients develop parkinsonian tremor at rest, rigidity, bradykinesia, and postural instability. Pathological examination reveals Alzheimer type II astrocytes and areas of neuronal degeneration, particularly in the basal ganglia and in the deep layers of the frontal, parietal, and occipital cortex (VICTOR et al. 1965; FINLAYSON and SUPERVILLE 1981).

3. Other Metabolic Causes

A 51-year-old Japanese man with *type 3 GM_1 gangliosidosis* was reported by OHTA et al. (1985). He had dementia, dysarthria, gait disturbance, and limb rigidity. In contrast to other cases of GM_1 gangliosidosis, he did not have dystonia or athetosis.

We recently studied a 35-year-old woman with adult onset *Gaucher's disease* and right hemiparkinsonism. Although the association may be coincidental, the early age of onset, some atypical features, such as coarse intention tremor, and an abnormal signal intensity in the left striatum, suggest that the parkinsonism is secondary to the Gaucher's disease.

IV. Postencephalitic Parkinsonism and Slow Virus Infections

Until three decades ago, postencephalitic parkinsonism was the commonest type of secondary parkinsonism. POSKANZER and SCHWAB (1963) hypothesized that most cases of "idiopathic" parkinsonism were related to encephalitis lethargica and that Parkinson's disease will gradually disappear. However, this theory is no longer accepted and postencephalitic parkinsonism is considered a specific clinical-pathologic entity, separate from idiopathic Parkinson's disease. The epidemic of encephalitis lethargica originated in Rumania in 1915 and spread throughout Europe and other continents during the next 15 years, occurring primarily in the winter. Von Economo, an Austrian neuropathologist, provided the best description of this condition (EDITORIAL 1981; RAVENHOLT and FOEGE 1982). Encephalitis lethargica and influenza epidemics occurred in the United States at the same time, but the two conditions were quite different. Fever and other nonspecific symptoms of infection occurred in both conditions, but the encephalitis lethargica was characterized by somnolence or coma, oculomotor palsies, and focal neurologic disturbances, including hemiplegia and aphasia. Because of these differences, it is incorrect to assume that patients with a history of influenza during the 1918-1919 epidemic have postencephalitic parkinsonism.

During the early phase of encephalitis lethargica, in addition to fever, most patients had fluctuating and progressive somnolence, alteration of the sleep cycle, ocular palsies, including ptosis, impaired convergence, and pupillary involvement. Many patients experienced oculogyric crises, an involuntary conjugate dystonic deviation of the eyes, usually upward. Motor restlessness, chorea, dystonia, tics, hiccups, myoclonus, and other hyperkinesias also occurred during the illness (KRUSZ et al. 1987). Psychiatric symptoms included delirium, hypomania, catatonia, hallucinations, and personality changes. Respiratory distress was the usual cause of death during the acute illness. The overall mortality ranged between 30% and 40%, and the very young and the very old were particularly at risk of dying. The majority of the survivors had a residual deficit. Some recovered completely, but symptoms of parkinsonism may have developed several months or years later. Parkinsonism as a long-term sequela has occurred even in those individuals in whom the extrapyramidal symptoms were not present during the acute illness. About 60% of the patients who survived the epidemic developed parkinsonism within 2.5 years after the initial attack. However, the latency period may be as long as 20 years, or even longer (CALNE and LEES 1988).

Besides the early age at onset, usually in the third decade, there are other clinical features which help differentiate postencephalitic parkinsonism from the idiopathic disease. These include residual abnormal involuntary movements, e.g., dystonia, chorea, myoclonus, and facial and respiratory dyskinesias. Furthermore, the patients may have residual hyperreflexia, hemiparesis, aphasia, palilalia, oculomotor palsies, oculogyric crises, blepharospasm, psychiatric disturbances, sleep alterations, and autonomic dysfunction.

The CSF is usually normal, although during the acute stage of the illness a mild lymphocytic pleocytosis may be present. In the 13 cases reported by Von Economo, the characteristic pathologic features consisted of inflammatory changes in the tegmentum of the midbrain, pons, and the basal ganglia. In addition to perivascular cuffing there were multiple small hemorrhages present in the areas of necrosis. Of the patients who survived for many years with postencephalitic parkinsonism, the most severe site of involvement was the substantia nigra. Besides the loss of neurons and neuromelanin pigment, there is usually extensive gliosis and neurofibrillary tangles. In contrast, neurofibrillary tangles are not found in idiopathic Parkinson's disease, unless it is associated with Alzheimer's disease. Lewy bodies are rarely seen in postencephalitic parkinsonism. In addition to the substantia nigra, the corpus striatum and the globus pallidus may also be involved. The loss of postural reflexes in postencephalitic parkinsonism has been attributed to the degeneration of the globus pallidus (Purdon-Martin 1967).

Postmortem biochemical studies found marked depletion of dopamine in the basal ganglia. Interestingly, patients with postencephalitic parkinsonism are more sensitive to L-dopa than are the idiopathic cases, possibly owing to the more pronounced denervation supersensitivity of the postsynaptic striatal receptors.

Although all attempts have failed to isolate the infectious agent from patients with encephalitis lethargica, with immunostaining techniques an influenza antigen was detected in brains of patients with postencephalitic parkinsonism, but not with idiopathic Parkinson's disease (Gamboa et al. 1974). This suggests a possible antigenic relationship between the encephalitis lethargica virus and the influenza virus. Further evidence of a possible link between encephalitis lethargica and the influenza epidemic has been provided by the analysis of successive waves of influenza pneumonia showing chronologic and geologic relation between the two endemics (Ravenholt and Foege 1982). Genetic susceptibility to postencephalitic parkinsonism has been suggested by histocompatibility studies (Elizan et al. 1980).

Because of the association between encephalitis and subsequent parkinsonism, viral etiology for Parkinson's disease has been sought. However, virologic and immunologic studies have failed to provide any evidence for infectious etiology in idiopathic Parkinson's disease (Elizan et al. 1980; Martilla et al. 1981). Several sporadic cases of encephalitis lethargica have been reported after the last outbreak in 1926, and the clinical manifestations of the recent cases are similar to those seen during the 1915–1926 epidemic (Rail et al. 1981; Editorial 1981; Greenough and Davis 1983).

Besides encephalitis lethargica, presumably due to a viral pathogen, postencephalitic parkinsonism has also been attributed to other viruses, including poliomyelitis, Japanese B, Western Equine, Coxsackie, and Echoviruses (Fahn 1977; Rail et al. 1981; Howard and Lees 1987). However, these viruses usually produce only transient parkinsonism, which resolves within 3 months after the encephalitis. We reported a 36-year-old man with herpes meningoencephalitis who had persistent and disabling parkinsonism, pseudobulbar palsy, and calcification of the basal ganglia (Jankovic 1985). We also described parkinsonism and other movement disorders in patients infected with the AIDS (HIV) virus (Nath et al. 1986). This retrovirus may have a special predilection for the basal ganglia.

Slow virus diseases, including subacute sclerosing panencephalitis (SSPE) and Creutzfeldt-Jakob disease, may have parkinsonian features (Graves 1984; Fournier et al. 1985; Brown et al. 1986). We studied a 15-year-old boy with static congenital cerebral and ocular toxoplasmosis who had a 1-month course of rapidly progressive bradykinesia, rigidity, and asymmetric rest and action hand tremor (Jankovic et al. 1988). Except for cerebral calcifications, the CT and MRI scans were normal, but a biopsy of the right frontal lobe revealed the diagnosis of SSPE. A similar case of congenital toxoplasmosis and SSPE has recently been described (WEINER and DE LA MONTE 1986).

In addition to the typical triad of progressive dementia, myoclonus, and periodic EEG discharges, patients with Creutzfeldt-Jakob disease may manifest a variety of extrapyramidal findings, including parkinsonism. We recently studied a 53-year-old man who presented with rapidly progressive parkinsonism manifested chiefly by asymmetric tremor at rest, bradykinesia, rigidity, and postural instability. He later developed ataxia, myoclonus, and dementia. Creutzfeldt-Jakob disease was suspected when a repeat EEG showed diffuse slowing and periodic, 1 Hz, sharp wave complexes. Further support for the diagnosis was provided by the detection of specific proteins in the CSF (HARRINGTON et al. 1986). Besides dementia, myoclonus, and parkinsonism, many patients with Creutzfeldt-Jakob disease have ataxia, spasticity, lower motor neuron disease, and visual loss due to involvement of the occipital cortex (ROYDEN JONES et al. 1985). The spinal fluid is usually normal, but two-dimensional gel electrophoresis may detect proteins which seem specific for Creutzfeldt-Jakob disease (HARRINGTON et al. 1986).

The characteristic postmortem finding in Creutzfeldt-Jakob disease is marked cytoplasmic vacuolization of neuropil, causing a diffuse, spongiform appearance of the gray matter. This is accompanied by gliosis in later stages, but there is no evidence of inflammation (MANUELIDIS 1985). The neuropathologic examination may also show amyloid plaques, possibly composed of paracrystalline arrays of prion proteins (BOCKMAN et al. 1985; KITAMOTO et al. 1986). Prion (PrP, proteinacious infectious particle) is a protease-resistant membrane protein with molecular weight 27-30 kdalton. While the PrP 27-30 protein has not been shown to be infective, it is found in almost all brains of patients with Creutzfeldt-Jakob disease, and is considered to be a relatively specific marker for some transmissible neurologic diseases (ROBERTS et al. 1986).

Creutzfeldt-Jakob disease has been transmitted to primates and to other species. Transmission between humans has been related to corneal transplant, infected neurosurgical stereotactic apparatus, and to human growth hormone therapy (BROWN et al. 1985). Until recently the diagnosis could be confirmed only by pathological examination of brain biopsy or at autopsy, and by transmitting the disease into an animal. However, abnormal proteins, identified in the CSF with Creutzfeldt-Jakob disease may be used in the future to confirm the diagnosis quickly, thus avoiding a brain biopsy (HARRINGTON et al. 1986).

V. Vascular Parkinsonism

In the classic descriptions of parkinsonism, arteriosclerosis was often mentioned as a leading cause of this syndrome (CRITCHLEY 1929; PARKES 1974). CRITCHLEY (1929), in his original report, however, emphasized that in "arteriosclerotic parkinsonism" tremor was absent and additional signs such as dementia, incontinence, pyramidal signs, and pseudobulbar palsy were often present. These atypical features have been used to differentiate "arteriosclerotic parkinsonism" from the idiopathic variety. Thus far, however, there has been no clinical-pathologic evidence for "arteriosclerotic parkinsonism." This does not mean that vascular parkinsonism does not exist. With the advent of CT and MRI scans, multiple strokes have been recognized to produce a syndrome in which parkinsonism may be an important manifestation. Several cases of basal ganglia infarcts associated with parkinsonian features have been reported (TOLOSA and SANTAMARIA 1984; FRIEDMAN et al. 1986; MAYO et al. 1986). The patients usually have a history of sudden neurologic deficit (stroke) with subsequent onset of parkinsonism, often associated with other neurologic problems. While spontaneous improvement, unrelated to treatment, is considered important evidence for postinfarction parkinsonism, we evaluated several multi-infarct patients whose parkinsonian features did not improve. In a study of 426 patients with parkinsonism, we found 25 (5.9%) with evidence of stroke (DUBINSKY and JANKOVIC 1987). This frequency is consistent with other series which have reported 5.7%-9% stroke prevalence in several community-based surveys of Parkinson's disease patients (MARTTILA and RINNE 1976; KESSLER 1972).

Parkinsonism may be caused by multiple infarcts in the basal ganglia, but this is a rare complication of striatal infarcts or hemorrhages (BLADIN and BERKOVIC 1984; STEIN et al. 1984). None of the 11 patients with CT scan-documented striatocapsular infarcts had parkinsonian symptoms (BLADIN and BERKOVIC 1984). In contrast, parkinsonism in combination with dementia, gait disturbance, and ataxia has been observed in patients with subcortical arteriosclerotic encephalopathy (Binswanger's disease) and cerebral amyloid angiopathy (KINKEL et al. 1985; DUBINSKY and JANKOVIC 1987; THOMPSON and MARSDEN 1987). We have seen several patients, usually hypertensive, who have "lower body parkinsonism", manifested by gait disturbance and freezing without other parkinsonian features (FITZGERALD and JANKOVIC). On MRI scans they have evidence of deep white matter and periventricular ischemia (GERARD and WEISBERG 1986). Multiple, deep, small infarcts may also lead to ventricular dilatation and other features of normal pressure hydrocephalus, including the classic triad of dementia, "magnetic gait," and urinary incontinence. In addition to this triad, some patients have parkinsonism that improves after shunting (EARNEST et al. 1974).

VI. Brain Tumors

Brain tumors may produce parkinsonism by four different mechanisms. 1. Gliomas, ependymomas, craniopharyngiomas, meningiomas, and metastases (Fig. 2) may infiltrate or compress the basal ganglia and the brain stem (SCIARRA and SPROFKIN 1953). 2. Another mechanism is the involvement of the premotor frontal cortical areas which give rise to the projections to the basal ganglia. Parkinsonism has been reported to be associated with parasagittal meningiomas (NICHOLSON and TURNER 1964). 3. There may be a direct compression by the tumor or by tentorial herniation of the vessels that provide blood supply to the basal ganglia (VAN ECK 1961). 4. Finally, brain tumors may produce hydrocephalus and thus cause parkinsonism (JANKOVIC et al. 1986).

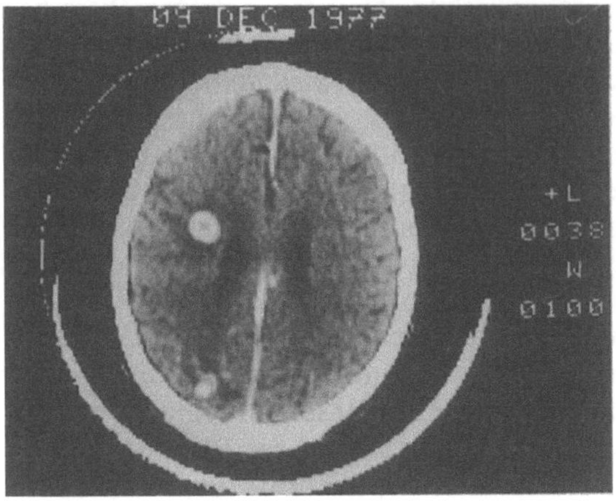

Fig. 2. CT scan of a 63-year-old man with right hemiparkinsonism secondary to left cerebral metastases of bronchogenic origin

Considering the frequent occurrence of brain tumors, it is surprising that they rarely cause parkinsonism (JANKOVIC 1984a). One possible explanation is that the brain tumor involves the corticospinal tract which is essential for the expression of tremor and rigidity. Another reason may be that the other symptoms related to the brain tumor, such as increased intracranial pressure, visual and sensory symptoms, and hemiparesis, overshadow the parkinsonian features. Postmortem biochemical analysis of a case of parkinsonism secondary to large craniopharyngioma showed a marked reduction in the striatal dopamine and norepinephrine, and in the dopamine receptors in the caudate nucleus (YEBENES et al. 1982). The postsynaptic damage may explain why patients with parkinsonism secondary to tumors are usually unresponsive to levodopa or to dopamine agonists.

VII. Trauma

Trauma, although an uncommon cause of parkinsonism, may produce this syndrome by at least four different mechanisms. Parkinsonism may occur as a sequela when the head trauma is severe enough to produce damage to the basal ganglia (GRIMBERG 1934). Subdural hematoma may compress the brain and the basal ganglia, producing atypical parkinsonism, usually associated with other neurologic findings (GLATT et al. 1983). Another cause of trauma-induced parkinsonism is the "punchdrunk" syndrome. In addition to parkinsonism, punchdrunk boxers may show signs of pyramidal, cerebellar, and cortical damage (dementia pugilistica). Postmortem examination of the brain shows diffuse cortical atrophy, severe neuronal loss in the locus ceruleus, substantia nigra, and in the striatum, as well as numerous neurofibrillary tangles distributed throughout the brain. Most of the cases of parkinsonism following head trauma occur after repeated closed head injuries, but post-traumatic parkinsonism has been reported after a single bullet injury to the substantia nigra (JELLINGER 1986a). In addition to direct damage to the upper brain stem or the striatum, parkinsonism may be caused by raised intracranial pressure and transtentorial herniation with secondary damage to the midbrain and the basal ganglia from vascular compression. It is possible that less severe trauma may be sufficient in the elderly to trigger or precipitate the subclinical form of Parkinson's disease (SCHWAB and ENGLAND 1968). In such cases, however, it is difficult to prove that the head trauma is related to the parkinsonism; furthermore, parkinsonism associated with head trauma has been seen in young adults (FAHN 1977). KLAWANS (1981) suggested that some cases with slowly progressive hemiparkinsonism represent a late sequela of a hemispherical injury in early life, causing contralateral hemiatrophy and hemiparkinsonism.

Physical stress as an aggravating or triggering factor in the development of parkinsonism is receiving more attention from those investigators interested in the etiology of Parkinson's disease. Experimental evidence supports the hypothesis that either metabolic, emotional, or physical stress precipitates parkinsonism in susceptible subjects (SNYDER et al. 1985).

In addition to head trauma, a few cases of parkinsonism have been reported following electrical injury (FARRELL and STARR 1968). Peripheral trauma may also induce tremor and other abnormal involuntary movements in the affected limb, and this can lead to a more generalized movement disorder, including parkinsonism (SCHOTT 1986; JANKOVIC and VAN DER LINDEN 1989).

VIII. Hydrocephalus

Parkinsonism has been reported in association with normal pressure hydrocephalus and with obstructive hydrocephalus (JANKOVIC et al. 1986). In normal pressure hydrocephalus, the shuffling or magnetic gait resembling the "freezing" phenomenon in Parkinson's disease, in combination with dementia and urinary incontinence, comprise the classic triad of symptoms (KNUTSSON and LYING-TUNNELL 1985; SOELBERG et al. 1986). Parkinsonism, usually without tremor, may be the sole manifestation of normal pressure hydrocephalus, occasionally seen without dementia or urinary incontinence. Like Parkinson's disease, normal pressure hydrocephalus is often ascribed to the inevitable consequences of aging or "rheumatism" (RASKER et al. 1985). CT or MRI scans show ventricular dilatation with minimal cortical atrophy, and radioactive isotope CSF flow study shows retention of the tracer in the ventricles for longer than 24 h.

Marked improvement in any of the symptoms associated with normal pressure hydrocephalus can be seen in about half of patients after shunting. Parkinsonism can also be reversed after a shunting procedure (ANTUNES et al. 1983). The cause of normal pressure hydrocephalus is often not readily apparent, and therefore the condition is labeled as idiopathic or occult. Sometimes it is a late complication of subarachnoid hemorrhage or meningitis. Another possible mechanism is softening of the periventricular white matter by multiple infarcts, resulting in ventricular dilatation (EARNEST et al. 1974).

In addition to normal pressure hydrocephalus, parkinsonism may be caused by obstructive hydrocephalus. This is usually due to a tumor or cyst in the posterior fossa compressing the aqueduct. We reported a 14-year-old girl in whom parkinsonism was the first symptom of acquired hydrocephalus caused by post-traumatic aqueductal stenosis (JANKOVIC et al. 1986). The Parkinson's disease symptoms completely resolved after successful ventricular-peritoneal shunting. Parkinsonism may not always improve with ventricular drainage, but some features, such as akinetic mutism, may respond (BERGER et al. 1985).

IX. Syringomesencephalia

Two of three reported cases of syringomesencephalia had associated parkinsonism (HARDY and STEVENSON 1957; SAMPLES et al. 1983). In the first case of a 60-year-old man with a 50-year history of uncontrollable tremor on the right side of the body, the postmortem examination showed that the substantia nigra was largely replaced by long narrow cavities lined with ependyma. The second case was a 59-year-old woman with a 27-year history of internuclear ophthalmoplegia, unsteadiness of gait, paresthesias, and rest tremor in the limbs and tongue. The diagnosis was multiple sclerosis until an autopsy was performed, which revealed a syrinx extending from the aqueduct across the mesencephalic root of the trigeminal nerve. The syrinx was confined to the lower mesencephalon and the upper pons and, in contrast to the first case, the substantia nigra was not involved. The parkinsonian features were probably due to the involvement of the nigrostriatal pathway.

C. Sporadic Multiple System Degenerations (Parkinsonism-Plus)

First coined by Oppenheimer (BANNISTER and OPPENHEIMER 1981), the term "multiple system atrophy" was used to describe a group of disorders in which there was a degeneration of selected neuronal structures or systems, such as the striatum, pigmented and pontine nuclei, inferior olives, and the cerebellar outflow system.

I. Progressive Supranuclear Palsy

Although described earlier, the syndrome of progressive supranuclear palsy (PSP) was not established as a specific clinical-pathologic entity until 1964. In that year, STEELE et al. (1964) reported the clinical features in nine patients and the neuropathologic findings in four. Since the classic report, over 400 new cases have been reported and the clinical features have been well characterized (JANKOVIC 1984 b, KRISTENSEN 1985).

While not emphasized in the report of STEELE et al. (1964), parkinsonian symptoms, particularly postural instability, have become recognized as a dominant motor disability in this disorder. Approximately 6% of patients with parkinsonism have PSP. The average age at onset in 104 patients was about 63,5 years (JANKOVIC et al. 1989), 10 years later than the average onset of Parkinson's disease. The male: female ratio was about 3:2. Impairment of gait, characterized by fluctuating unsteadiness and unexplained falls, is the usual mode of onset for this progressive disorder. Later, patients develop generalized slowness, dysarthria, and other parkinsonian symptoms, but tremor is rarely present. The classic tetrad: (a) "supranuclear ophthalmoplegia" chiefly affecting the vertical gaze; (b) axial rigidity and dystonia in extension; (c) pseudobulbar palsy with marked dysarthria; and (d) dementia, was emphasized in the original report. Although supranuclear ophthalmoparesis is the hallmark of PSP, patients seldom complain of oculomotor problems. However, they may be "sloppy eaters" because of inability to look down at the plate, and many have difficulties with focusing, owing to problems with convergence. Because of the inability to track a line of print, they have difficulties in reading. Family members often remark that the patient does not maintain normal eye contact during conversation.

Downward gaze seems to be restricted earlier and to a greater extent than upward gaze. However, in some patients in the early stages of the disease upward gaze may be more involved. Rarely, even in the more advanced stages of the disease, the extraocular movements are spared (Davis et al. 1985). The diagnosis may then be established only at autopsy. This presents a dilemma in the clinical diagnosis of PSP since without the ophthalmoparesis the diagnosis is difficult. Other ophthalmologic findings include hypometric saccades, impairment of convergence, internuclear ophthalmoplegia, ocular fixation defect with square-wave jerk, impaired optokinetic and caloric-induced nystagmus, and reversal of normal eye-closed preponderance of caloric-induced nystagmus. Eyelid abnormalities, including decreased blinking, blepharospasm, and apraxia of eyelid opening, the latter attributed to levator inhibition, may provide additional clues to the correct diagnosis (Lepore and Duvoisin 1985). In about one-third of the patients with PSP, besides the vertical gaze palsy, there may be a marked limitation of lateral gaze (Jackson et al. 1983). However, in most patients there is a marked limitation of vertical gaze which can be overcome by the oculocephalic maneuver. This supranuclear nature of the ophthalmoparesis is also evidenced by involuntary persistent ocular fixation of the subject while the body turns. The head movement may lag behind the body movement for up to 30–45 s. In the most advanced stages, in addition to the supranuclear ophthalmoparesis, there may be nuclear involvement, and neither the oculocephalic maneuver nor caloric stimulation is then sufficient to overcome the oculomotor palsy.

Other characteristic features of PSP include axial rigidity and dystonia, as well as limb dystonia (Rafal and Friedman 1987). Facial dystonia and rigidity account for the stiff, deeply furrowed face (Jankovic 1984 b). The pseudobulbar palsy is manifested by exaggerated jaw jerk, trismus, dysarthria, aphonia, palilalia, and dysphagia. Emotional incontinence and dementia usually occur later in the course of the disease. The cognitive decline in the advanced stages of PSP seems to be a consequence of associated visual and motor deficit, in contrast to the more typical dementia seen in Alzheimer's disease (Jankovic et al. 1989). The clinical course of PSP is characterized by relentless progression. In addition to frequent falls associated with the broad-based, unsteady gait, the patient develops progressive dysarthria and dysphagia, and eventually becomes bedridden and prone to aspiration and infection.

The histologic hallmarks of PSP include neuronal cell loss, gliosis, granulovacuolar degeneration, and neurofibrillary tangles. These changes are particularly noticeable in the mesencephalic tegmentum and tectum, internal segment of the globus pallidus, substantia nigra, the subthalamic, vestibular, and dentate nuclei, the superior colliculi, the periaqueductal gray matter, and the spinal cord (KATO et al. 1986; ZWEIG et al. 1987). In addition, the basal nucleus of Meynert is involved, and this may explain the cognitive deficit seen in the later stages of the disease. The cholinergic neurons in the pedunculopontine nucleus, normally spared in Alzheimer's disease, seem to be involved in PSP (ZWEIG et al. 1987).

The neurofibrillary tangles in PSP are quite specific; in addition to the paired helical filaments, the PSP neurofibrillary tangles contain straight 15-nm tubules, which actually represent an earlier stage in the development of the paired helical filament (Table 4). Using monoclonal antibodies and the peroxidase-antiperoxidase technique, DICKSON et al. (1985) found that the 15-nm straight filaments typically seen in PSP share an antigenic determinant with the Alzheimer's neurofibrillary tangles. Furthermore, this antigenic determinant is localized in the PSP perikarya and it is not found in the normal brain, aluminum-induced experimental tangles, Lewy bodies, Hirano bodies, or axonal filamentous inclusions of amyotrophic lateral sclerosis (ALS) and giant axonal neuropathy. Impaired assembly of microtubules, owing to abnormal phosphorylation of the microtubule-associated tau protein has been proposed as the mechanism of formation of the neurofibrillary tangles. However, tau antigen has been identified not only in PSP, but also in Alzheimer's disease, Down's syndrome, Parkinsonism-dementia complex of Guam, and Pick bodies in Pick's disease (IQBAL et al. 1986; POLLOCK et al. 1986; JOACHIM et al.

Table 4. Disorders with neurofibrillary tangles

A. Paired helical filaments: 20- to 24-nm tubules with 15-nm lumen, twisted every 80-120 nm; 10-nm filaments	
Degenerative	Alzheimer, Pick, ALS, Hallervorden-Spatz, Guam parkinsonism-dementia complex, PSP[a]
Viral	SSPE, herpes, rabies, Creutzfeldt-Jakob, postencephalitic parkinsonism
Other infections	Dysentery, chronic TB, scarlet fever
Toxins	Acute and chronic lead poisoning, chronic manganese poisoning
Trauma	Dementia pugilistica
Chromosomal abnormalities	Down's syndrome
Heredofamilial	Tuberous sclerosis, hereditary spastic spinal paresis, familial cerebellar ataxia
Metabolic	Storage diseases, pigment-variant lipofuscinosis
B. Proliferation of normal 10-nm filaments: motor neuron disease, vincristine neuropathy, infantile neuroaxonal dystrophy	

[a] Abnormal straight 15-nm tubules

1987). Therefore, the neurofibrillary lesions found in various neurodegenerative disorders are relatively nonspecific. Granulovacuolar degeneration in PSP consists of 2- to 5-micron cytoplasmic vacuoles with central basophilic or argyrophilic granules.

Pallidonigral-luysial atrophy is clinically similar to PSP, but the pathologic changes are confined to the pallidum, substantia nigra, and subthalamic nucleus, and some cases show massive widespread CNS accumulations of corpora amylacea (KOSAKA et al. 1981; JELLINGER 1986b). This rare disorder probably represents a variant of PSP.

In the early stages of PSP, levodopa therapy may provide improvement in the axial rigidity and in the postural instability. However, even high dosages of levodopa are usually ineffective in the later stages. Levodopa combined with dopamine agonists, including bromocriptine, pergolide, and lisuride, may be more effective than when used alone (JANKOVIC 1983). Other useful drugs in the management of PSP include amitriptyline and amantadine, but methysergide has not proven to be useful in our experience (NEWMAN 1985; JANKOVIC 1984b). Rarely, cricopharyngeal myotomy is required in the advanced stages for at least partial relief of the dysphagia.

The median duration from onset to death is about 6 years, with the range between 1 and 12 years (MAHER and LEES 1986, GOLBE et al. 1988). While the etiology of PSP has not been found, it is likely that the dopaminergic and cholinergic systems are primarily involved in the motor and cognitive deficit seen in patients with this syndrome. The pathogenesis of the neuronal degeneration is unknown, but in some patients subcortical hypoperfusion due to chronic ischemia or due to a multi-infarct state may result in a loss of dopamine receptors in the striatum, nucleus accumbens, and the substantia nigra (DUBINSKY and JANKOVIC 1987). Such dopamine receptor changes have been documented by in vivo positron tomography and by postmortem ligand binding studies (D'ANTONA et al. 1985; RUBERG et al. 1985; KISH et al. 1985). It is likely that the poor response to dopaminergic therapy in patients with PSP is due to these postsynaptic receptor changes.

II. Shy-Drager Syndrome

When parkinsonism is associated with orthostatic hypotension, impotence, sphincter disturbance, atonic bladder, and other symptoms of autonomic failure, then the diagnosis of Shy-Drager syndrome should be seriously considered. In addition to the autonomic symptoms, patients with this disorder have iris atrophy, cerebellar ataxia, corticospinal tract signs, amyotrophy, and various sleep and respiratory abnormalities (JANKOVIC 1984c). The pathologic changes are widespread and consist chiefly of neuronal loss in the intermediolateral column and pyramidal tract of the spinal cord, Onuf's nucleus in the sacral cord, anterior horn cells, inferior olives, dorsal motor nucleus of vagus, locus ceruleus, Purkinje cells of the cerebellum, oculomotor nuclei, Edinger-Westphal nucleus, the periaqueductal gray matter, and the basal ganglia, especially the striatum. There is neuronal loss and depigmentation in the substantia nigra. The hypothalamus shows no morphological abnormalities.

Peripheral nerves may be involved more frequently than previously recognized (Galassi et al. 1982; Sobue et al. 1986). The pathologic changes in Shy-Drager syndrome overlap with the olivopontocerebellar atrophy and the striatonigral degeneration. Lewy bodies have been noted rarely in the affected areas.

The nosology of Shy-Drager syndrome has been problematic since the original description by Shy and Drager (1960). One end of the spectrum of autonomic dysfunction is the pure autonomic failure (PAF) which occurs without any central or peripheral neurologic signs. This disorder has been previously called "idiopathic orthostatic hypotension," but that term has been abandoned because it ignores other autonomic symptoms besides orthostatic hypotension. At the other end of the spectrum is multiple system atrophy, a term used to describe disorders such as olivopontocerebellar atrophy, striatonigral degeneration, and Shy-Drager syndrome (Polinsky 1988). In contrast to the patients with PAF who have low supine plasma norepinephrine levels, patients with Shy-Drager syndrome have normal supine norepinephrine, and in neither group does the plasma norepinephrine level increase on standing. In the Shy-Drager syndrome, norepinephrine is readily released by tyramine, and there is no evidence of adrenoceptor supersensitivity to indirectly acting sympathomimetic drugs. In addition to the noradrenergic involvement, the low CSF homovanillic and 5-hydroxyindoleacetic acid levels suggest that the dopaminergic and serotonergic systems are also involved in this degenerative process.

The clinical investigation of patients with Shy-Drager syndrome should exclude other causes of orthostatic hypotension, such as diabetes, amyloidosis, porphyria, tabes dorsalis, adrenal insufficiency, and remote effects of cancer. Electromyographic studies may reveal evidence of denervation, but in contrast to PAF the peripheral autonomic system is spared, until the advanced stages. The reader is referred to a recent review on this subject for a description of automatic testing (Polinsky 1988).

The treatment of Shy-Drager syndrome is challenging. When parkinsonian symptoms interfere with the patient's functioning, then the usual antiparkinsonian medications must be employed. However, the dopaminergic drugs must be used with caution because they often exacerbate the orthostatic hypotension. Therefore, the usual precautions should be taken to prevent syncopal episodes. The patient should be encouraged to increase dietary salt and fluid, use the reverse Trendelenburg position during sleep, and wear tight elastic stockings. Ancillary medications include fludrocortisone, indomethacin, ibuprofen, beta-blockers, sympathomimetics, ganglionic stimulants, norepinephrine precursors, CNS stimulants, and pressor hormones. An implantable sympathetic neural prosthesis is being developed (Polinsky 1988).

III. Olivopontocerebellar Atrophies

The combination of parkinsonism associated with cerebellar signs should suggest the diagnosis of olivopontocerebellar atrophy (OPCA). Dejerine and Thomas described two patients with a nonfamilial progressive gait ataxia, bradykinesia, dysarthria, dysphagia, and incontinence, beginning in middle age (BERCIANO 1988). Postmortem examination revealed marked atrophy of the olives, pons, the middle cerebellar peduncles, and the cerebellar cortex (KOEPPEN et al. 1986). Similar cases were previously described by Menzel, but these patients had autosomal dominant inheritance, earlier onset, and choreiform involuntary movements (KONIGSMARK and WEINER 1970). Later, autosomal recessive OPCAs have been characterized clinically and pathologically. In cases of OPCA with parkinsonism, in addition to cerebellar atrophy, there is usually depigmentation and neuronal loss in the substantia nigra. Furthermore, the brain stem nuclei, the corticospinal tract, and the posterior column are also affected to a variable degree. In some cases, the anterior horn cells and peripheral nerves may also be involved. This phenotypic variability could be explained by a single gene mutation transmitted as an autosomal recessive trait and another single gene mutation transmitted as an autosomal dominant trait, with the host genes modifying the expression and penetrance of the mutant gene (HARDING 1986).

While some cases of OPCA are sporadic, most are inherited in an autosomal dominant pattern. One variety of autosomal dominant OPCA, named Joseph disease, has been described among the descendants of Azorian-Portuguese families (ROSENBERG 1984). Type I Joseph disease has its onset in the third decade of life and is characterized by progressive spasticity, rigidity, dystonia, ophthalmoparesis, and fasciculations. Type II begins in the fifth decade of life and is characterized by cerebellar, pyramidal, and parkinsonian findings. Type III begins in the sixth and seventh decades of life and is associated with progressive cerebellar ataxia and amyotrophy. Two-dimensional electrophoresis revealed 50 and 40 kdalton proteins in brains of patients with Joseph's disease, but not in control brains. These proteins seem to be most abundantly expressed in the cerebellum and the basal ganglia, but their relationship, if any, to the underlying pathogenetic mechanism is unknown.

OPCA associated with progressive ataxia, upper and lower motor neuron signs, parkinsonian symptoms, and supranuclear ophthalmoparesis has been associated with leukocyte glutamate dehydrogenase (GDH) deficiency. Since the GDH deficiency is only partial, and not all GDH-deficient individuals are clinically affected, it is likely that the GDH deficiency is not the primary biochemical defect in these patients. It is also possible that several isoenzymes of GDH exist and may be affected to a variable degree. Thus, a GDH deficiency may be seen in sporadic cases of OPCA as well as in the inherited forms. One autopsy case revealed diffuse lipofuscinosis, pancerebellar, olivary, and pontine atrophy, as well as demyelination of the posterior columns, and degeneration of the anterior horns and of the dorsal ganglion cells (CHOKROVERTY et al. 1984). Although a peripheral neuropathy is usually not seen clinically in patients with GDH deficiency, there is a reduction of myelinating fibers in the sural nerve. Furthermore, the autonomic neurons seem to be affected and account for the occasional orthostatic hypotension and other automatic symptoms.

Postmortem biochemical studies of brains of patients with OPCA revealed a reduction in aspartate and glutamate, major excitatory neurotransmitters associated with the climbing and mossy fibers respectively. In addition, some cases of OPCA show depletion of γ-aminobutyric acid (GABA). In vitro receptor autoradiography in four cases of OPCA showed increased numbers of benzodiazepine receptors in the dentate nucleus, but not in the cerebellar cortex, possibly indicating a loss of Purkinje cells or brain stem afferents (WHITEHOUSE et al. 1986).

While in some patients with OPCA the parkinsonian symptoms improve with levodopa therapy, in most cases the symptoms or the course of the disorder cannot be altered by pharmacologic manipulation (MANYAN 1986). Those few patients with OPCA who improve initially with levodopa may later develop dose-related fluctuations and dyskinesias (LANG et al. 1986).

IV. Corticobasal Degeneration

Only a few cases have been described all with a relentlessly progressive course characterized by bradykinesia, rigidity, postural tremor, focal dystonia and myoclonus, alien hand short steps, unsteadiness, limitation of convergence and upward gaze, reflex blepharospasm, dysarthria, anomia, and apraxia (REBEIZ et al. 1968; FUNKENSTEIN and MILLER 1985; GIBB et al. 1989). In addition, the patients had proprioceptive sensory loss, corticospinal tract signs, and grasp reflex. Pathologic examination showed corticobasal degeneration consisting of frontoparietal cortical atrophy, corticospinal tract degeneration, depigmentation of the substantia nigra, and in two cases there was degeneration of the dentate nucleus. The most characteristic finding was the presence of large, pale (achromatic) neurons without Nissl substance resembling Pick's cells, but without the typical argyrophilic Pick bodies.

V. Parkinsonism-Dementia-ALS Complex

Parkinsonism coupled with motor neuron disease and dementia should suggest the possibility of Shy-Drager syndrome, Creutzfeldt-Jakob disease, or the parkinsonism-dementia-ALS complex of Guam (HUDSON 1981). There are two neurologic disorders that occur with higher than expected frequency among the Chamorro people of Guam: amyotrophic lateral sclerosis (ALS) and parkinsonism-dementia complex. Except for younger age at onset and slower progression, the ALS of Guam resembles the classical ALS seen worldwide. However, the pathologic findings suggest that the ALS of Guam may be a different disorder. In addition to the neuronal loss in the anterior horn cells and degeneration of the corticospinal tract, patients with ALS of Guam have neurofibrillary tangles and abnormal helical filaments derived from normal neurotubule protein. These changes are usually not seen in idiopathic ALS or Parkinson's disease, but may be seen in a variety of neurodegenerative disorders (Table 5). The parkinsonism-dementia complex resembles idiopathic Parkinson's disease except for propulsion, festination, and tremor which are relatively mild in the first disorder. In addition to progressive dementia, there are corticospinal tract signs and supranuclear ophthalmoparesis. The age at onset is about 54 years with a male: female ratio 2.5:1. More than one-third of patients with parkinsonism-dementia complex have features of ALS, but only

5 % of the Chamorro subjects with ALS develop parkinsonism-dementia. This may be due to the fact that the ALS patients die before the onset of the parkinsonism-dementia syndrome.

The pathologic changes in brains of patients with parkinsonism-dementia complex consist of depigmentation of the substantia nigra and locus ceruleus, widespread Alzheimer neurofibrillary changes, and intracytoplasmic granular degeneration, particularly in the Sumner's sector of the Ammon's horn (Table 5). Senile plaques, Lewy bodies, or spongiform changes are not usually found in the parkinsonism-dementia cases of Guam.

The etiology of the parkinsonism-dementia complex is unknown, but an "environmental trigger" in genetically susceptible individuals seems to be the most likely mechanism (Schmitt et al. 1984). Efforts to transmit the disorder into nonhuman primates or other laboratory animals have failed. Because of

Table 5. Pathologic changes in parkinsonian disorder

Entity	Neuronal loss		Lewy bodies	Neuro-fibrillary tangles	Other sites of involvement
	Substantia nigra	Corpus striatum			
Idiopathic Parkinson's disease	+++	○	+++	○	Locus ceruleus, dorsal motor nucleus of vagus, sympathetic ganglia; cerebral cortex—variable and late, nucleus basalis in demented patients
Postencephalitic parkinsonism	+++	++	+ (?)	++	Locus ceruleus, brain stem tegmentum, dentate nuclei
Striatonigral degeneration	+++	+ + + (putamen)	○	○	Overlaps with olivopontocerebellar atrophy; putamen—brown pigmentation
Progressive supranuclear palsy	+++	+++	○	++	Globus pallidus, subthalamic and red nuclei, midbrain and pontine tegmentum, periaqueductal gray, vestibular and dentate nuclei, granulovacuolar degeneration in red nucleus and pontine nuclei
Shy-Drager syndrome	+++	+ + (putamen)	+	○	Spinal cord especially intermediolateral columns, locus ceruleus, olives, dorsal motor nucleus of vagus, Purkinje cells
Parkinsonism-dementia complex	+++	+ + + (globus pallidus)	○	++	Cerebral cortex, hippocampus, diencephalon, brain stem tegmentum, spinal cord

these and other reasons, a viral etiology of these disorders has been excluded (GARRUTO and YASE 1986). Environmental geochemical studies showed that the soil and the drinking water from Guam and the Kii peninsula of Japan, another endemic area for the parkinsonism-dementia-ALS complex, have unusually low concentrations of calcium and magnesium, and relatively high concentrations of aluminum and iron. In Guam, calcium and magnesium concentrations were lowest in those regions where the incidence rates of ALS and parkinsonism were highest. Finally, animals with experimental calcium-magnesium deficiency show degeneration of the anterior horn cells in the spinal cord.

A remarkable decline in the incidence of these disorders has been noted in the past 30 years (GARRUTO et al. 1985). This has been attributed to change in dietary habits and in the local water supply, as well as reduced dependence on locally grown foodstuffs. The epidemiologic, genetic, and environmental studies, as well as the experimental data, support the hypothesis that the parkinsonism-dementia-ALS complex results from a basic defect in mineral metabolism and secondary hyperparathyroidism due to chronic deficiency of calcium and magnesium. Because increased intestinal absorption of calcium, aluminum, and other metals may cause an impairment of slow axonal transport, it has been postulated that there is an accumulation of excess neurofilament which leads to the formation of neurofibrillary tangles (GAJDUSEK 1985; GARRUTO and YASE 1986). Recent epidemiologic studies implicated β-*N*-methylamino-L-alanine (BMAA), a component of the local cycad seed used in the production of flour, as a possible etiologic agent. When injected into monkeys, this cycad-derived BMAA produces parkinsonism and corticospinal tract signs (CALNE et al. 1986; SPENCER et al. 1987). However, this cycad hypothesis of parkinsonism-dementia-ALS complex has been recently challenged (GARRUTO et al. 1988).

Treatment of the parkinsonian component with levodopa is usually successful, but the dementia and the motor neuron disease do not improve with dopaminergic therapy. The association of parkinsonism and lower motor neuron disease in Guam and in other populations is well recognized. Much less appreciated is the rare coexistence of chronic spinal muscular atrophy (with pallidonigral degeneration) or hereditary neuropathy and parkinsonism (SERRATRICE et al. 1983). Rarely, patients with spinocerebellar degeneration may show parkinsonian signs (WEIR and FAN 1981).

VI. Striatonigral Degeneration

The term striatonigral degeneration was used by ADAMS et al. (1964) in reference to three cases described in 1961 and an additional case in 1964 of a distinct entity characterized clinically by typical and atypical parkinsonian features, but pathologically showing severe degeneration of the substantia nigra and putamen. Although clinically striatonigral degeneration may resemble idiopathic Parkinson's disease, several characteristic features have emerged among the 50 cases reported since the original report. The distinguishing features of striatonigral degeneration include lack of tremor and the presence of hyperreflexia, extensor plantar responses, and severe bulbar dysfunction, low volume and hypokinetic dysarthria, and even aphonia and stridor. Despite the putamenal involvement, chorea, dystonia, or other involuntary movements are usually not present and autonomic dysfunction and dementia are rare. Family history of a similar disorder is usually lacking and the disease is rapidly progressive, usually resulting in death within 5 years after the onset (ADAMS et al. 1964; KAN 1978).

The characteristic pathologic findings include neuronal loss and depigmentation in the substantia nigra, combined with neuronal loss and gliosis in the striatum and pallidum. The atrophy is particularly prominent in the putamen and is often accompanied by brownish discoloration due to the presence of dark-brown pigment granules in the glial and neuronal cytoplasm (Table 5). These pigmented granules apparently contain lipofuscin and iron, and small amounts of phosphorus, calcium, and sulfur. Lewy bodies and neurofibrillary tangles are conspicuously absent in this disorder. The caudate nucleus, globus pallidus, subthalamic nucleus, and the brain stem nuclei may also show neuronal loss, gliosis, and demyelination. In some cases, there is a moderate involvement of the dentate nucleus, the cerebellar cortex, and the middle cerebellar peduncle. Because of the overlap between this entity and the OPCAs, it has been suggested that striatonigral degeneration respresents a variant of a relatively common entity termed "multiple system atrophy" (TAKEI and MIRRA 1973; ROPPER 1983). Other disorders which overlap with OPCA are the Shy-Drager syndrome and PSP, and therefore, these four disorders could be designated as the parkinsonian multiple system atrophies (Fig. 3).

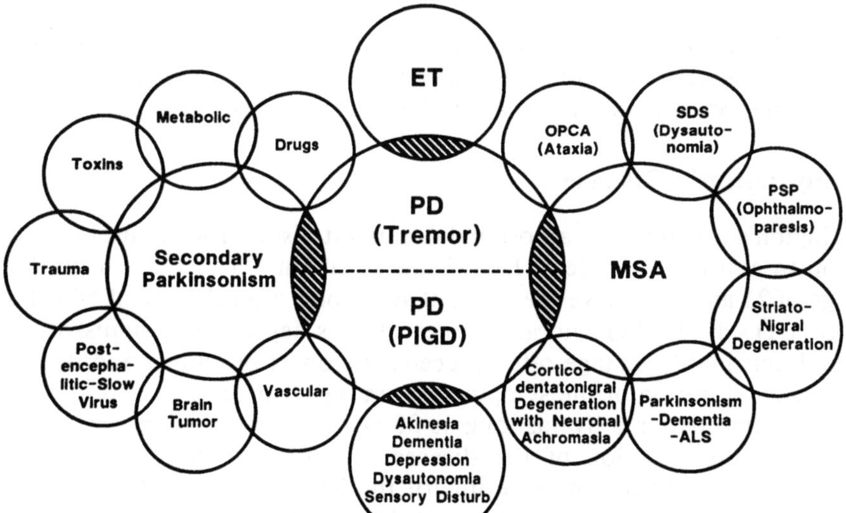

Fig. 3. Parkinson's disease, Parkinson-plus syndromes (multiple system atrophies), and secondary parkinsonian disorders showing an overlap in some clinical and pathologic features. ALS amyotrophic lateral sclerosis; ET essential tremor; MSA multiple system atrophy; OPCA olivopontocerebellar atrophy; PD Parkinson's disease; PIGD postural instability and gait difficulty; PSP progressive supranuclear palsy; SDS Shy-Drager syndrome

Until a specific biochemical or histologic marker is found, it will not be known whether these neurodegenerative disorders represent distinct entities or whether they are just variants of the same disease. Although they share some clinical and pathologic features with idiopathic Parkinson's disease, it is likely that the parkinsonian multiple system atrophies differ in etiology from idiopathic Parkinson's disease. While it is possible that examples of "pure" striatonigral degeneration, Shy-Drager syndrome, OPCA, and PSP exist, majority of cases probably represent part of a broad spectrum of multiple system degenerations which may include other sporadic or inherited disorders.

Although some cases of striatonigral degeneration may improve with anticholinergic therapy, the response is usually modest and short-lived. SHARPE et al. (1973) found very low levels of dopamine and homovanillic acid in the caudate and putamen of brains with striatonigral degeneration. Also, L-dopa decarboxylase was markedly reduced in the striatum while glutamic acid decarboxylase (GAD) was normal in the striatum, but markedly reduced in the cortex. About 25% of reported patients with striatonigral degeneration showed appreciable improvement with levodopa therapy, but only a few sustained the benefit for longer than 6 months (RAJPUT et al. 1972; SHARPE et al. 1973; KAN 1978). There has been no systematic study of dopamine agonists in striatonigral degeneration. Thalamotomy has been reported to provide moderate relief of tremor and rigidity.

D. Inherited Multiple System Degenerations

Because Parkinson's disease is usually not inherited, parkinsonian heredofamilial neurodegeneration will be discussed only briefly. The reader will be referred to recent reviews for more detailed discussion of these disorders.

I. Huntington's Disease

Huntington's disease is an autosomal dominant disorder with onset usually in the fourth and fifth decade of life. However, about 6% of cases begin before the age of 21 and 28% have the onset of symptoms after the age of 50 (MARTIN and GUSELLA 1986). Behavioral changes such as depression, impulsive behavior, and personality change often precede the onset of chorea and dementia. The average duration of the symptoms from onset to death is about 17 years, although some patients live as long as 30 years. The earlier the age of onset, the shorter the duration of the disease. Majority of children and young adults with Huntington's disease inherit the gene from their father. Juvenile Huntington's disease is characterized by bradykinesia, rigidity, dystonic movements, seizure disorder, and intellectual decline. Tremor is rare, but action tremor often accompanies the rigid state of the disease. The name "Westphal variant" has been given to the akinetic-rigid form of Huntington's disease, which accounts for up to 14% of all patients with the disease.

The defective gene, mapped to the terminal band on the short arm of chromosome 4, in some way programs premature death of neurons, particularly in the striatum (caudate). Marked reduction in caudate glucose metabolism has been demonstrated in symptomatic and presymptomatic individuals and, when combined with DNA polymorphisms, this technique may be useful in preclinical detection of Huntington's disease (HAYDEN et al. 1987). The medium-sized spiny neurons with abundant spinous processes (type I cells) usually degenerate first. Although globus pallidus, cerebral cortex, thalamus, subthalamic nucleus, brain stem, and spinal cord may show some neuronal loss, caudate atrophy is the most characteristic pathologic change in Huntington's disease. It is not known whether the rigid form of Huntington's disease preferentially affects the large aspiny cells as opposed to the medium-sized spiny neurons.

Despite the striatal atrophy, the concentration of dopamine in the Huntington's disease brain is normal. The most remarkable biochemical change involves the activity of GAD which is reduced by 85% in the striatum and by 60% in the substantia nigra, putamen, and globus pallidus. Furthermore, the GABA receptors are also markedly reduced. Selective loss of the GABA system may cause desinhibition of the nigrostriatal pathway and results in chorea. However, since GAD and GABA are normal in the other regions of the brain, and GABA agonists are ineffective in Huntington's disease, it is unlikely that the GABA system is fundamentally involved in the pathogenesis of the disease. Other substances decreased in the striatum include acetylcholine and dynorphin. In the striatum, somatostatin and neuropeptide Y are increased, but substance P is normal or slightly increased. However, substance P is markedly reduced in the medial pallidum and the pars reticulata of the substantia nigra. It has been proposed that somatostatin enhances the release of dopamine and thus contributes to the chorea seen in Huntington's disease.

Because pharmacologic manipulation of the neuropeptides is not possible, the most effective drugs in the treatment of chorea are the dopamine receptor blocking agents, including the phenothiazines and butyrophenones, and drugs that deplete dopamine such as reserpine or tetrabenazine.

Glutamate and other excitatory analogs, including kainic and ibotenic acid, produce neuronal degeneration in experimental animals, similar to Huntington's disease. Therefore it has been suggested that Huntington's disease is caused by an endogenous glutamate-like substance which may be selectively neurotoxic. Baclofen inhibits the release of excitatory neurotransmitter, including glutamine, and it is currently undergoing long-term clinical trials in Huntingtons's disease. However, until the responsible gene and its protein product are found, it is unlikely that a more specific therapy for Huntington's disease will be available. As a result of the discovery of a DNA marker linked to the Huntington's disease gene, predictive testing for this disorder is now feasible. The impact of presymptomatic testing is currently being evaluated to determine some guidelines for future genetic testing and counseling.

II. Wilson's Disease

Wilson's disease occassionally presents as "juvenile parkinsonism" associated with rigidity, drooling, bradykinesia, and loss of postural reflexes. Tremor at rest is uncommon, but coarse postural "wing-beating" tremor is very characteristic in this disease. About half of the Wilson's disease patients present with neurologic symptoms, one-third with hepatic dysfunction, 14% with psychiatric complications, 8% with osteoarticular problems, and 2% with Kayser-Fleischer ring (SCHEINBERG and STERNLIEB 1984; PATTEN 1988). Besides parkinsonism and tremor, other movement disorders typically present in Wilson's disease include dystonia, grimacing, chorea, and athetosis. Children and young women are more likely to present with liver abnormalities, and neurologic symptoms are unusual before the age of 12; they are more common in males. Pseudobulbar palsy dominates the clinical picture in most neurologically affected patients and is usually the mode of death. The diagnosis is established by the typical clinical features, Kayser-Fleischer ring, ceruloplasmin less than 30 mg/d, hepatic copper concentration more than 250 µg per gram dry weight, 24-h urine copper more than 100 µg, and low radioactive copper incorporation into the ceruloplasmin.

Penicillamine and treintine both chelate copper, and may produce a remarkable improvement in neurologic function, allowing patients to enjoy a normal life span. Wilson's disease is inherited as an autosomal recessive trait, therefore siblings have 25% risk of developing the disease and should be screened for neurologic abnormalities, liver disease, Kayser-Fleischer ring, and laboratory evidence of Wilson's disease.

III. Hallervorden-Spatz Disease

This disorder, first described in 1922, is characterized by childhood or adult onset, relentlessly progressive dementia, bradykinesia, rigidity, spasticity, dystonic posturing, choreoathetosis, and other hyperkinesias (EIDELBERG et al. 1987). Additional clinical findings include cerebellar ataxia, seizure disorder, amyotrophy, and retinitis pigmentosa. While autosomal recessive inheritance has been suggested for most cases, transmission from one generation to another has been reported rarely in atypical cases. Pathological examination reveals iron deposits in the globus pallidus and substantia nigra, accompanied by axonal swellings and neuronal degeneration in the basal ganglia, corticospinal tract, and the cerebellum.

The disorder rarely presents in the third or fourth decade of life as a familial parkinsonism-dementia complex. We reported a 68-year-old man who had a 13-year history of progressive dementia, rigidity, bradykinesia, mild tremor, stooped posture, slow and shuffling gait, dystonia, blepharospasm, apraxia of eyelid opening, aphonia, and incontinence (JANKOVIC et al. 1985). Autopsy examination of the brain confirmed the typical deposits of iron pigment in the globus pallidus, caudate, and substantia nigra, as well as axonal spheroids in the globus pallidus, substantia nigra, medulla, and the spinal cord. Neurochemical analysis of the brain revealed marked loss of dopamine in the nigrostriatal area with relative preservation of dopamine in the limbic area. Another study suggested that cysteine accumulates in the globus pallidus as a result of the enzymatic block in the metabolic pathway from cysteine to taurine (PERRY et al. 1985). The increased cysteine may chelate iron, accounting for the high concentration of iron in the selected areas of the brains of patients with this disease. The excess of cysteine and ferrous iron may result in the generation of free radicals and subsequent neuronal damage.

IV. Other Familial Parkinsonian Syndromes

Parkinsonism associated with OPCA, Joseph disease, spinocerebellar degeneration, and GDH deficiency have already been discussed. MATA et al. (1983) described three siblings with a syndrome characterized by parkinsonism, dementia, pyramidal signs, and ophthalmoparesis, beginning in the third decade. Although the disorder resembled PSP and the parkinsonism-dementia complex of Guam, because of the familial occurrence it was thought to represent a new entity.

Spinocerebellar-nigral degeneration with parkinsonism may occur as a sporadic disorder (SERRATRICE et al. 1983; WEIR and FAN 1981) or as a dominantly inherited neurodegenerative disorder (BIEMOND and SINNEGE 1955; WOODS and SCHAUMBURG 1972). Besides parkinsonism, these patients have amyotrophy, corticospinal tract signs, and cerebellar ataxia, with onset in the third decade of life. Pathological examination in a few cases revealed demyelination of the spinocerebellar tract and the posterior column, and degeneration of Clarke's column, substantia nigra, and the anterior horn cells. The cerebral cortex and nuclei are usually spared, but there is often diffuse demyelination of the cerebellar white matter.

V. Familial Basal Ganglia Calcifications

Basal ganglia calcifications, as already discussed in Sect. B.III, are usually associated with hypoparathyroidism or pseudohypoparathyroidism. When the basal ganglia calcifications occur in a family, they are often associated with chorea, dementia, and palilalia. This autosomal dominant disorder is referred to as Fahr's disease (BOLLER et al. 1977). However, some cases may have cogwheel rigidity, slow and shuffling gait, hypomimia, dysarthria, and other parkinsonian symptoms (HARATI et al. 1984; BEALL et al. 1984). We studied two brothers, one of whom had progressive parkinsonism and dementia, and the other who was asymptomatic, but both had calcifications in the basal ganglia and the dentate nuclei (Fig. 4). CT scan should be performed on affected and unaffected members of the family to establish the pattern of inheritance because the calcification may be the only manifestation of the familial disorder.

VI. Neuroacanthocytosis

Acanthocytosis has been associated with a variety of neurologic symptoms, including chorea, tics, lip biting, ataxia, amyotrophy, elevated creatinine kinase

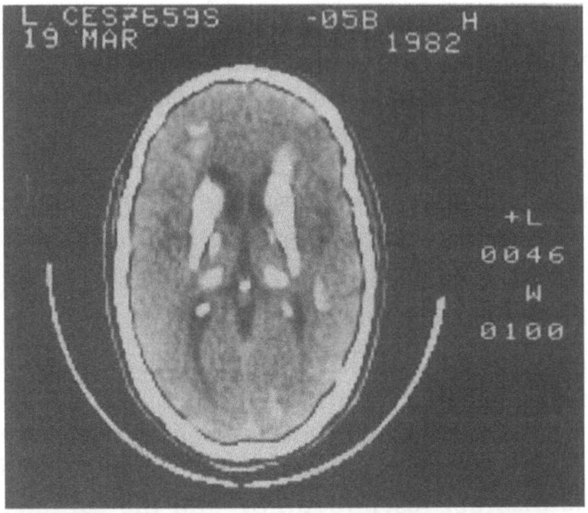

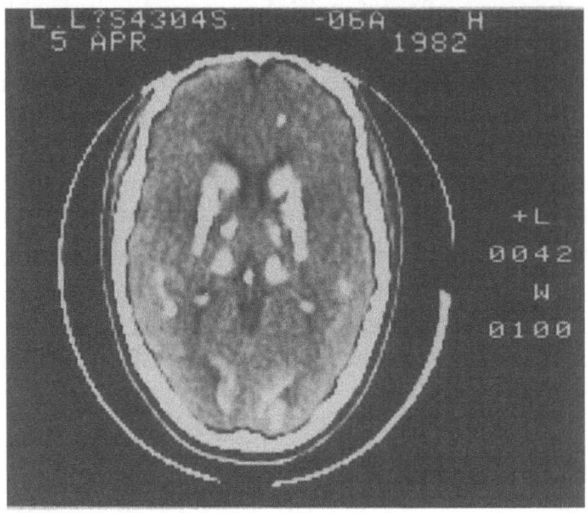

Fig. 4 a, b. CT scans of two brothers, one with parkinsonism and dementia (**a**) and the other without neurologic abnormalities (**b**), both showing dense calcifications, in the basal ganglia and the cerebellum

(CK), but normal lipoproteins. We reported two brothers of consanguineous parents who, in addition to motor and vocal tics, had parkinsonism, distal muscular atrophy, and supranuclear ophthalmoparesis associated with acanthocytosis (Spitz et al. 1985). Because of the protean neurologic manifestations associated with acanthocytosis, we suggested the term neuroacanthocytosis to describe these familial disorders (Jankovic et al. 1985). Postmortem neurochemical analysis showed depletion of dopamine, particularly in the striatum and elevation of norepinephrine in the putamen and globus pallidus (De Yebenes et al. 1988).

References

Adams RDL, van Bogaert, Van der Eecken H (1964) Striatonigral degeneration. J Neuropathol Exp Neurol 23:584-608

Allen JM, Cross AJ, Yeats JL et al. (1986) Neuropeptides and dopamine in the marmoset: effect of treatment with 1-methyl-4-phenyl-1,2,3,6-tetrohydropyridine (MPTP): an animal model for Parkinson's disease. Brain 109:143-158

Alonso MA, Otero E, D'Regules R, Figueroa M (1986) Parkinson's disease: a genetic study. Can J Neurol Sci 13:248-251

Antunes JL, Fahn S, Cote L (1983) Normal pressure hydrocephalus and Parkinson's disease. J Neurol Transm [Suppl] 19:225-231

Appel SH (1981) A unifying hypothesis for the cause of amyotrophic lateral sclerosis, parkinsonism and Alzheimer's disease. Ann Neurol 10:499-505

Ballard PA, Tetrud JW, Langston JW (1985) Permanent human parkinsonism due to 1-methyl-4-phenyl-1,2,3,6-tetrohydropyridine (MPTP): seven cases. Neurology (NY) 35:949-956

Bannister R, Oppenheimer D (1981) Parkinsonism, system degenerations and autonomic failure. In: Marsden CD, Fahn S (eds) Movement Disorders. Butterworth, London, pp 174-190

Banta RG, Markesbery WR (1977) Elevated manganese levels associated with dementia and extrapyramidal signs. Neurology (NY) 27:213-216

Barbeau A, Roy M (1984) Familial subsets in idiopathic Parkinson's disease. Can J Neurol Sci 11:144-150

Barbeau A, Cloutier T, Roy M et al. (1985) Exogenetics of Parkinson's disease: 4-hydroxylation of debrisoquine. Lancet 2:1213-1216

Beall S, Patten BM, Jankovic J (1984) Familial calcifications of the brain, iron storage, and porphyria: a new syndrome. Ann Neurol 16:137

Berciano J (1987) Olivopontocerebellar atrophies. In: Jankovic J, Tolosa E (eds) Parkinson's disease and movement disorders. Urban and Schwarzenberg, Baltimore (in press)

Berger JR, Ross DB (1981) Reversible Parkinson syndrome complicating postoperative hypoparathyroidism. Neurology (NY) 31:881-882

Berger L, Gauthier S, Le Blanc R (1985) Akinetic mutism and parkinsonism associated with obstructive hydrocephalus. Can J Neurol Sci 12:255-258

Biemond A, Sinnege JLM (1955) Tales of Friedreich with degeneration of the substantia nigra, a special type of hereditary parkinsonism. Confin Neurol 15:129-145

Bird ED, Anton AH (1982) Phenothiazine biphasic effect on dopamine concentrations in the basal ganglia of subhuman primates. Psychiatry Res 6:1-6

Bird ED, Collins GH, Dodson MH, Grant LG (1967) The effect of phenothiazine on the manganese concentration in the basal ganglia of subhuman primates. In: Barbeau A, Brunette JA (eds) Progress in neurogenetics, vol 1. Excerpta Medica, Amsterdam, pp 600-605

Bird ED, Anton AH, Bullock B (1984) The effect of manganese inhalation of basal ganglia dopamine concentrations in rhesus monkey. Neurotoxicology (Park Forest Il) 5:59-66

Bladin PF, Berkovic SF (1984) Striatocapsular infarction: large infarcts in the lenticulostriate arterial territory. Neurology (NY) 34:1423-1430

Blanc P, Hogan M, Mallin K et al. (1985). Cyanide intoxication among silver-reclaiming workers. JAMA J 253:367-371

Bockman JM, Kingsbury DT, Mc Kinley MP et al. (1985) Creutzfeldt-Jakob disease prion proteins in human brains. N Engl J Med 312:73-78

Boller F, Boller M, Gilbert J (1977) Familial idiopathic cerebral calcifications. J Neurol Neurosurg Psychiatry 40:280-285
Brown P, Gajdusek DC, Gibbs CJ, Asmer DM (1985) Potential epidemic of Creutzfeldt-Jakob disease from human growth hormone therapy. N Engl J Med 313:728-731
Brown P, Cathala F, Castaigne P, Gajdusek DC (1986) Creutzfeldt-Jakob disease: clinical analysis of a consecutive series of 230 neuropathologically verified cases. Ann Neurol 20:597-602
Calne DB, Eisen A, Mc Geer E, Spencer P (1986) Alzheimer's disease, Parkinson's disease, and motoneuron disease: abiotropic interaction between aging and environment. Lancet 2:1067-1070
Calne DB, Lees AJ (1988) Late progression of post-encephalitic Parkinson's syndrome. Can J Neurol Sci 15:135-138
Carlen PL, Lee MA, Jacob M, Livshits O (1981) Parkinsonism provoked by alcoholism. Ann Neurol 9:84-86
Choi HS (1983) Delayed neurologic sequelae in carbon monoxide intoxication. Arch Neurol 40:433-435
Chokroverty S, Khedekar R, Derby B et al. (1984) Pathology of olivopontocerebellar atrophy with glutamate dehydrogenase deficiency. Neurology (NY) 34:1451-1455
Chui HC, Mortimer JA, Slager U et al. (1986) Pathologic correlates of dementia in Parkinson's disease. Arch Neurol 43:991-995
Clough CG, Mendoza M, Yahr MD (1981) A case of sporadic juvenile Parkinson's disease. Arch Neurol 38:730-731
Cohen G (1984) Oxy-radical toxicity in catecholamine neurons. Neurotoxicology (Park Forest Il) 5:77-82
Critchley M (1929) Arteriosclerotic parkinsonism. Brain 52:23-83
D'Antona R, Baron JC, Samson Y, Serdary M et al. (1985) Subcortical dementia. Frontal cortex hypometabolism detected by positron tomography in patients with progressive supranuclear palsy. Brain 108:785-799
Davis PH, Bergeron C, Mc Lachlan DR (1985) Atypical presentation of progressive supranuclear palsy. Ann Neurol 17:337-343
Dickson DW, Kress Y, Crowe A, Yen SH (1985) Monoclonal antibodies in Alzheimer neurofibrillary tanges. 2. Demonstration of a common antigenic determinant between ANT and neurofibrillary degeneration in progressive supranuclear palsy. Am J Pathol 120:292-303
Donaldson J, La Bella FS, Gesser D (1980) Enhanced autoxidation of dopamine as a possible basis of manganese neurotoxicity. Neurotoxicology (Park Forest Il) 2:53-64
Dubinsky RM, Jankovic J (1987) Progressive supranuclear palsy and a multi-infarct state. Neurology (NY) 37:570-576
Earnest MP, Fahn S, Karp JH, Rowland LP (1974) Normal pressure hydrocephalus and hypertensive cerebrovascular disease. Arch Neurol 31:262-266
Editorial (1981) Encephalitis lethargica. Lancet 2:1396-1397 Eidelberg D, Sotrel A, Joachim C, Selkoe D, et al. (1987) Adult onset Hallervorden-Spatz disease with neurofibrillary pathology. Brain 110:993-1013
Elizan TS, Terasake PI, Yahr MD (1980) HLA-B14 antigen and postencephalitic Parkinson's disease. Arch Neurol 37:542-544
Elizan TS, Sroka H, Maker H et al. (1986) Dementia in idiopathic Parkinson's disease. Variables associated with its occurrence in 203 patients. J Neurol Transm 65:285-302

Evans BK, Donley DK (1988) Pseudohypoparathyroidism, Parkinsonism syndrome, with no basal ganglia calcification. J Neurol Neurosurg Psychiatry 51:709-713

Fahn S (1977) Secondary parkinsonism. In: Goldensohn ES, Appel SH (eds) Shy's cellular and molecular basis of neurologic disease. Lea and Febiger, Philadelphia, pp 1159-1189

Farrell DF, Starr A (1968) Delayed neurological sequelae of electrical injuries. Neurology (NY) 18:601-608

Ferraz HB, Bertolucci PHF, Pereira JS, Lima JGC, Andrade LAF (1988) Chemic exposure to the funcicide maneb may produce symptoms and signs of CNS manganese intoxication. Neurology 38:550-553

Findley LJ, Gresty MA (1984) Tremor and rhythmical involuntary movements in Parkinson's disease. In: Findley LJ, Capildeo R (eds) Movement disorders: tremor. Oxford University Press, New York, pp 295-304

Finlayson MH, Superville B (1981) Distribution of cerebral lesions in acquired hepatocerebral degeneration. Brain 104:79-95

Fitzgerald P, Jankovic J. Lower body parkinsonism: Evidence for vascular etiology. Movement Disorders (in press)

Forno LS, Langston JW, De Lanney LE et al. (1986) Locus ceruleus lesions and eosinophilic inclusions in MPTP-treated monkeys. Ann Neurol 20:449-455

Fournier JG, Tardieu M, Lebon P et al. (1985) Detection of measles virus RNA in lymphocytes from peripheral-blood and brain perivascular infiltrates of patients with subacute sclerosing panencephalitis. N Engl J Med 313:910-915

Friedman A, Kang UJ, Tatemichi TK, Burke RE (1986). A case of parkinsonism following striatal lacunar infarction. J Neurol Neurosurg Psychiatry 49:1087-1088

Funkenstein HH, Miller DC (1985) Case records of the Massachusetts General Hospital. Case 38-195. N Engl J Med 313:739-748

Gajdusek DC (1985) Hypothesis: interference with axonal transport of neurofilament as a common pathogenetic mechanism in certain diseases of the central nervous system. N Engl J Med 312:714-719

Galassi G, Nemni R, Babaldi A et al. (1982) Peripheral neuropathy in multiple system atrophy with autonomic failure. Neurology (NY) 32:1116-1121

Gamboa ET, Wolf A, Yahr MD et al. (1974) Influenza virus antigen in postencephalitic parkinsonism brain: detection by immunofluorescence. Arch Neurol 31:228-232

Garruto RM, Yase Y (1986) Neurodegenerative disorders of the western Pacific: the search for mechanisms of pathogenesis. Trends Neurosci 9:368-374

Garruto RM, Yanagihara R, Gajdusek DC (1985) Disappearance of high-incidence amyotrophic lateral sclerosis and parkinsonism-dementia on Guam. Neurology (NY) 35:193-198

Garruto RM, Yanagihara R, Gajdusek DC (1988) Cycades and amyotrophic lateral sclerosis/parkinsonism dementia. Lancet 2:1079

Geraghty JJ, Jankovic J, Zetusky WJ (1985) Association between essential tremor and Parkinson's disease. Ann Neurol 17:329-333

Gerard G, Weisberg LA (1986) MRI periventricular lesions in adults. Neurology (NY) 36:998-1001

Gershanik OS, Leist A (1987) Juvenile onset Parkinson's disease. In: Bergmann KJ, Yahr MD (eds) Parkinson's disease. Raven, New York, pp 213-216

Gibb WRG, Luthert PJ, Marsden CD (1989) Corticobasal degeneration. Brain (in press)

Glatt S, Fine S, Kaplan J (1983) Parkinsonism as a presentation of subdural hematoma. Neurology (NY) 33(Suppl 2):61

Goetz CG, Klawans HL, Cohen MM (1981) Neurotoxic agents. In: Baker AB, Joynt RJ (eds), Clinical neurology, Harper and Row, Philadelphia, chap 20, pp 1-84
Goetz CG, Tanner CM, Levy M et al. (1986) Pain in Parkinson's disease. Movement disorders 1:45-49
Golbe LI, Davis PH, Schoenberg BS, Duvoisin RC (1988) Prevalence and natural history of progressive supranuclear palsy. Neurology 38:1031-1034
Graves MC (1984) Subacute sclerosing panencephalitis. Neurol Clin 2:267-280
Greenhouse AH (1982) Heavy metals and the nervous system. Clin Neuropharmacol 5:45-92
Greenough A, Davis JA (1983) Encephalitis lethargica: mystery of the past or undiagnosed disease of the present? Lancet 1:922-923
Grimberg L (1934) Paralysis agitans and trauma. J Nerv Ment Dis 79:14-20
Harati Y, Jackson JA, Benjamin E (1984) Adult onset idiopathic familial brain calcifications. Arch Intern Med 144:2425-2427
Hardie RJ, Lees AJ (1988) Neuroleptic-induced Parkinson's syndrome: clinical features and results of treatment with levodopa. J Neurol Neurosurg Psychiatry 51:850-854
Harding AE (1986) Degenerative ataxic disorders. Trends Neurosci 9:311-313
Hardy RL, Stevenson LD (1957) Syringomesencephalia: report of a case with signs of Parkinson's disease having a syrinx of the substantia nigra. J Neuropathol Exp Neurol 16:365-369
Harrington M, Merril CR, Asher DM, Gajdusek DC (1986) Abnormal proteins in the cerebrospinal fluid of patients with Creutzfeldt-Jakob disease. N Engl J Med 315:279-283
Hassan M, Reches A, Kuhn C et al. (1986) Pharmacologic evaluation of dopaminergic receptor blockage by metoclopramide. Clin Neuropharmacol 9:71-78
Hayden MR, Hewitt J, Stoessel AJ, et al. (1987) The combined use of positron emission tomography and DNA polymorphisms for preclinical detection of Huntington's disease. Neurology (NY) 37:1441-1447
Hormes JT, Filley CM, Rosenberg NL (1986) Neurologic sequelae of chronic solvent vapor abuse. Neurology (NY) 36:698-702
Howard RS, Lees AJ (1987) Encephalitis lethargica. A report of 4 cases. Brain 110:19-34
Hudson AJ (1981) Amyotrophic lateral sclerosis and its association with dementia, parkinsonism and other neurological disorders: a review. Brain 104:217-247
Hurst EW (1972) Experimental demyelination of the central nervous system. 3. Poisoning with potassium cyanide, sodium azide, hydroxylamine, narcotics, carbon monoxide with some consideration of bilateral necrosis occurring in the basal ganglia. Aust J Exp Biol Med Sci 20:297-342
Indo T, Ando K (1982) Metoclopramide-induced parkinsonism. Clinical characteristics of ten cases. Arch Neurol 39:494-496
Iqbal K, Zaidi T, Weng Y et al. (1986) Defective brain microtubule assembly in Alzheimer's disease. Lancet 2:421-426
Jackson DL, Menges H (1980) Accidental carbon monoxide poisoning. JAMA 243:772-774
Jackson J, Jankovic J, Ford J (1983) Progressive supranuclear palsy: clinical features and response to treatment in sixteen patients. Ann Neurol 13:273-278
Jankovic J (1981) Drug-induced and other orofacial-cervical dyskinesias. Ann Intern Med 94:788-793
Jankovic J (1983) Controlled trial of pergolide mesylate in Parkinson's disease and progressive supranuclear palsy. Neurology (NY) 33:505-507.

Jankovic J (1984a) Neurological complications of cancer. In: Smith FE, Lane M (eds) Medical compiclations of malignancy. Wiley, New York, pp 107-132
Jankovic J (1984b) Progressive supranuclear palsy: clinical and pharmacologic update. Neurol Clin 2:473-486
Jankovic J (1984c) Parkinsonian disorders. In: Appel SH (ed) Current Neurology. Vol 5. Wiley, New York, pp 1-49
Jankovic J (1985) Amitriptyline in amyotrophic lateral sclerosis. N Engl J Med 313:1478-1479
Jankovic J (1986) Association between essential tremor and Parkinson's disease. A reply. Ann Neurol 19:306-307
Jankovic J (1987) Pathophysiology and clinical assessment of motor symptoms in Parkinson's disease. In: Koller WC (ed) Parkinson's disease: diagnosis and treatment. Dekker, New York, pp 99-126
Jankovic J, Armstrong D, Low NL, Goetz CG (1988) Congenital mental retardation and juvenile parkinsonism. Movement Disorders 3:352-361
Jankovic J, van der Linden C (1989) Dystonia and tremor induced by peripheral trauma: predisposing factors. J Neurol Neurosurg Psychiatry (in press)
Jankovic J, Casabona J (1987) The coexistence of tardive dyskinesia and parkinsonism. Clin Neuropharmacol 10:511-521
Jankovic J, Calne DB (1987) Parkinson's disease: etiology and treatment. In: Appel SH (ed) Current neurology, vol 7. Year Book Medical, Chicago, pp 193-234
Jankovic J, Reches A (1986) Parkinson's disease in monozygotic twins. Ann Neurol 19:405-408
Jankovic J, Killian JM, Spitz M (1985a) Neuroacanthocytosis syndrome and choreoacanthocytosis (Levine-Critchley syndrome). Neurology (NY) 35:1679
Jankovic J, Kirkpatrick JB, Blomquist K, Langlais PJ, Bird ED (1985b) Late onset Hallervorden-Spatz disease presenting as familial parkinsonism: clinical, biochemical and neuropathologic study. Neurology (NY) 35:227-234
Jankovic J, Newmark M, Peter P (1986) Parkinsonism and acquired hydrocephalus. Movement Disorders 1:59-64
Jankovic J, Friedman D, Pirozzolo FJ, McCrary JA. Progressive supranuclear palsy motor, neurobehavioral and neuro-ophthalmic findings. In: Steifler M (ed), Parkinson's Disease, Raven Press, New York, 1989 (in press)
Jellinger K (1986a). Pathology of parkinsonism. In: Fahn S, Marsden CD, Jenner P, Teychenne P (eds) Recent developments in Parkinson's disease. Raven, New York, pp 33-36
Jellinger K (1986b) Pallidal, pallidonigral and pallidoluysionigral degenerations including association with thalamic and dentate degenerations. In: Van Vinken AJ, Bruyn GW, Klawans HL (eds) Extrapyramidal disorders. Elsevier, Amsterdam, pp 445-463 (Handbook of clinical neurology, vol 49)
Joachim CL, Morris JH, Kosik KS, Selkoe DJ (1987) Tau antisera recognize neurofibrillary tangles in a range of neurodegenerative disorders. Ann Neurol 22:514-520
Kan AE (1978) Striatonigral degeneration. Pathology 10:45-52
Kato T, Hirano A, Weinberg MN, Jacobs AK (1986) Spinal cord lesions in progressive supranuclear palsy: some new observations. Acta Neuropathol (Berl) 71:11-14
Kessler II (1972) Epidemiologic studies of Parkinson's disease. III. A community-based survey. Am J Epidemiol 96:242-254
Kinkel WR, Jacobs L, Polachini I, Bates V, Heffner RR (1985) Subcortical arteriosclerotic encephalopathy (Binswanger's disease). Computer tomographic nuclear magnetic resonance and clinical correlations. Arch Neurol 42:951-959

Kish SJ, Chang LJ, Mirchandana L et al. (1985) Progressive supranuclear palsy: relationships between extrapyramidal disturbances, dementia and brain neurotransmitter markers. Ann Neurol 18:530-536

Kitamoto T, Tateishi J, Tashima T et al. (1986) Amyloid plaques in Creutzfeldt-Jakob disease stain with prion protein antibodies. Ann Neurol 20:204-208

Klawans HL (1981) Hemiparkinsonism as a late complication of hemiatrophy: a new syndrome. Neurology (NY) 31:625-628

Klawans HL, Bergen D, Bruyn GW (1973) Prolonged drug-induced parkinsonism. Confin Neurol 35:368-377

Klawans HL, Lupton M, Simon L (1976) Calcification of the basal ganglia as a cause of levodopa-resistant parkinsonism. Neurology (NY) 26:221-225

Klawans HL, Stein RW, Tanner CM, Goetz CG (1982) A pure parkinsonian syndrome following acute carbon monoxide intoxication. Arch Neurol 39:302-304

Knutsson E, Lying-Tunnell U (1985) Gait apraxia in normal-pressure hydrocephalus: patterns of movement and muscle activation. Neurology (NY) 35:155-160

Koeppen AH, Mitzen EJ, Hans MB, Barron KD (1986) Olivopontocerebellar atrophy: immunocytochemical and Golgi observations. Neurology (NY) 36:1478-1488

Koller WC (1984) The diagnosis of Parkinson's disease. Arch Intern Med 144:2146-2147

Koller WC, Cochran JW, Klawans HL (1979) Calcification of the basal ganglia: computerized tomography and clinical correlation. Neurology (NY) 29:328-333

Konigsmark BW, Weiner LP (1970) The olivopontocerebellar atrophies: a review. Medicine 49:227-241

Kosaka K, Matsushita M, Oyanagi S et al. (1981) Pallido-nigral-luysial atrophy with massive appearance of corpora amylacea in the CNS. Acta Neuropathol (Berl) 53:169-172

Kristensen MO (1985) Progressive supranuclear palsy—20 years later. Acta Neurol Scand 71:177-189

Krusz JC, Koller WC, Ziegler DK (1987) Historial review: abnormal movements associated with epidemic encephalitis lethargica. Movement Disorders 2:137-141

Lang AE, Marsden CD, Obeso JA, Parkes JD (1982) Alcohol and Parkinson disease. Ann Neurol 12:254-256

Lang AE, Birnbaum A, Blair RDG, Kierans C (1986) Levodopa dose-related fluctuations in presumed olivopontocerebellar atrophy. Movement Disorders 1:93-112

Langston JW, Irwin I (1986) MPTP: current concepts and controversies. Clin Neuropharmacol 9:485-507

LaPorte JR, Capella D (1986) Useless drugs are not placebos: lessons from flunarizine and cinnarizine. Lancet 2:853-854

Lima B, Neves G, Nora M (1987) Juvenile parkinsonism: clinical and metabolic characteristics. J Neurol Neurosurg Psychiatry 50:345-348

Lepore FE, Duvoisin RC (1985) Apraxia of eyelid opening: an involuntary levator inhibition. Neurology (NY) 35:423-427

Lewy FH (1941) Neurological, medical and biochemical signs and symptoms indicating chronic industrial carbon disulfide absorption. Ann Intern Med 15:869-878

Magos L, Jarvis JAE (1970) The effect of carbon disulfide exposure on brain catecholamines in rats. Br J Pharmacol 39:26-33

Maher ER, Lees AJ (1986) The clinical features and natural history of the Steele-Richardson-Olszewski syndrome (progressive supranuclear palsy). Neurology (NY) 36:1005-1008

Manuelidis EL (1985) Creutzfeldt-Jakob disease. J Neuropathol Exp Neurol 44:1-17

Manyam BV (1986) Recent advances in the treatment of cerebellar ataxias. Clin Neuropharmacol 9:508-516
Martin JB, Gusella JF (1986) Huntington's disease. Pathogenesis and management. N Engl J Med 315:1267-1276
Martin WE, Loewenson RB, Resch JA, Baker AB (1973) Parkinson's disease. Clinical analysis of 100 patients. Neurology (NY) 23:783-790
Marttila RJ, Rinne UK (1976) Arteriosclerosis, heredity, and some previous infections in the etiology of Parkinson's disease. Clin Neurol Neurosurg 79:46-56
Marttila RJ, Rinne UK, Halonen P et al. (1981) Herpes viruses and parkinsonism. Herpes simplex virus type 1 and 2, and cytomegalovirus antibodies in serum and CSF. Arch Neurol 38:19-21
Mata M, Dorovini-Zis K, Wilson M, Young AB (1983) New form of familial Parkinson-dementia syndrome: clinical and pathologic findings. Neurology (NY) 33:1439-1443
Mayeux R (1984) Behavioral manifestations of movement disorders. Parkinson's and Huntington's disease. In: Jankovic J (ed) Neurologic clinic, vol 2. Saunders, Philadelphia, pp 527-540
Mayeux R, Stern Y, Spanton S (1985) Heterogeneity in dementia of the Alzheimer type: evidence of subgroups. Neurology (NY) 35:453-461
Mayo J, Arias M, Leno C, Berciano J (1986) Vascular parkinsonism and periarteritis nodosa. Neurology (NY) 36:874-875
Mena I, Marin O, Fuenzalida S, Cotzias G (1967) Chronic manganese poisoning: clinical picture and manganese turnover. Neurology (NY) 17:128-136
Micheli F, Pardal MF, Gatto M, et al. (1987) Flunarizine- and cinnarizine-induced extrapyramidal reactions. Neurology (NY) 37:881-884
Molsa PK, Marttila RJ, Rinne UK (1984) Extrapyramidal signs in Alzheimer's disease. Neurology (NY) 34:1114-1116
Muenter MD, Whisnant JP (1968) Basal ganglia calcification, hypoparathyroidism and extrapyramidal motor manifestations. Neurology (NY) 18:1075-1083
Murphy MJ (1979) Clinical correlations of CT scan-detected calcifications of the basal ganglia. Ann Neurol 6:507-511
Naidu S, Wolfson LI, Sharpless NS (1978) Juvenile parkinsonism: a patient with possible primary striatal dysfunction Ann Neurol 3:453-455
Nath A, Jankovic J, Pettigrew LC (1987) Movement disorders and AIDS. Neurology (NY) 37:37-41
Newman GC (1985) Treatment of progressive supranuclear palsy with tricyclic antidepressants. Neurology (NY) 35:1189-1193
Newman RP, Le Witt PA, Jaffe M, Calne DB, Larsen TA (1985) Motor function in the normal aging population: treatment with levodopa. Neurology (NY) 35:571-573
Nicholson AN, Turner EA (1964) Parkinsonism produced by parasagittal meningiomas. J Neurosurg 21:104-113
Nygaard TG, Duvoisin RC (1986) Hereditary dystonia-parkinsonism syndrome of juvenile onset. Neurology (NY) 36:1424-1428
Ohta K, Tsuji S, Mizuno Y et al. (1985) Type 3 (adult) GM_1 gangliosidosis: case report. Neurology (NY) 35:1490-1494
Olson CG, Hogstedt C (1981) Parkinson's disease and occupational exposure to organic solvents, agricultural chemicals and mercury—a case referent study. Scand J Work Environ Health 7:252-256
Parkes JD, Marsden CD, Rees JE et al. (1974) Parkinson's disease, cerebral arteriosclerosis, and senile dementia: clinical features and response to levodopa. Q J Med 93:49-61

Patten BM (1988) Wilson's disease. In: Jankovic J, Tolosa E (eds) Parkinson's disease and movement disorders. Urban and Schwarzenberg, Baltimore, pp 179-190

Perry TL, Norman MG, Yong VW et al. (1985) Hallervorden-Spatz disease: cysteine accumulation and cysteine dioxygenase deficiency in the globus pallidus. Ann Neurol 18:482-489

Peters HL, Levin RL, Matthews CG, Chapman LJ (1988) Extrapyramidal and other neurologic manifestations associated with carbon disulfide fumigant exposure. Arch Neurol 45:537-540

Polinsky RJ (1988) Shy-Drager syndrome. In: Jankovic J, Tolosa E (eds) Parkinson's disease and movement disorders. Urban and Schwarzenberg, Baltimore, pp 153-166

Pollock NJ, Mirra SS, Binder LI et al. (1986) Filamentous aggregates in Pick's disease, progressive supranuclear palsy, and Alzheimer's disease share antigenic determinants with microtubule-associated protein, tau. Lancet 2:1211

Poskanzer DC, Schwab RS (1963) Cohort analysis of Parkinson's syndrome. Evidence for a single etiology related to subclinical infection about 1920. J Chronic Dis 16:961-972

Purdon-Martin J (1967) The basal ganglia and posture. Lippincott, Philadelphia

Quinn N, Critchley P, Marsden CD (1987) Young onset Parkinson's disease. Movement Disorders 2:73-91

Rafal RD, Friedman JH (1987) Limb dystonia in progressive supranuclear palsy. Neurology (NY) 37:1546-1549

Rail D, Scholtz C, Swash M (1981) Post-encephalitic parkinsonism: current experience. J Neurol Neurosurg Psychiatry 44:670-676

Rajput AH, Kazi KH, Rozdilsky B (1972) Striatonigral degeneration: response to levodopa therapy. J Neurol Sci 16:331-341

Rajput A, Rozdilsky B, Hornykiewicz O et al. (1982) Reversible drug-induced parkinsonism. Clinicopathologic study of two cases. Arch Neurol 39:644-646

Rasker JJ, Jansen ENG, Haan J, Oostrom J (1985) Normal-pressure hydrocephalus in rheumatic patients. A diagnostic pitfall. N Engl J Med 312:1239-1241

Ravenholt RT, Foege WH (1982) 1918 Influenza, encephalitis lethargica, parkinsonism. Lancet 2:860-864

Rebeiz JJ, Kolodny EH, Richardson EP (1968) Corticodentatonigral degeneration with neuronal achromasia. Arch Neurol 18:20-33

Reches A, Tietler J, Lavy S (1981) Parkinsonism due to lithium carbonate poisoning. Arch Neurol 38:471

Richardson EP (1986) Parkinson's disease. Ann Neurol 20:447-448

Richter R (1945) Degeneration of the basal ganglia in monkeys from chronic carbon disulfide poisoning. J Neuropathol Exp Neurol 4:324-330

Robbins JA, Logemann JA, Kirshner HS (1986) Swallowing and speech production in Parkinson's disease. Ann Neurol 19:283-287

Roberts GW, Brown P, Crow TJ et al. (1986) Prion-protein immunoreactivity in human transmittable dementias. N Engl J Med 315:1231-1233

Ropper AH (1983) Case records of the Massachusetts General Hospital. Case 23-1983. N Engl J Med 308:1406-1414

Rosenberg RN (1984) Joseph disease: an autosomal dominant motor system degeneration. Adv Neurol 41:179-193

Royden-Jones H, Hedley-Whyte T, Freidberg SR, Baker AA (1985) Ataxic Creutzfeldt-Jakob disease: diagnostic techniques and neuropathologic observations in early disease. Neurology (NY) 35:254-257

Ruberg M, Javoy-Agid F, Hirsch E et al. (1985) Dopaminergic and cholinergic lesions in progressive supranuclear palsy. Ann Neurol 18:523-529

Samples JR, Howard FM, Okazake H (1983) Syringomesencephalia. Report of a case. Arch Neurol 40:757-759

Sandyk R (1985) Parkinsonism induced by captopril. Clin Neuropharmacol 8:197-200

Santamaria J, Tolosa E, Valles A (1986) Parkinson's disease with depression: a possible subgroup of idiopathic parkinsonism. Neurology (NY) 36:1130-1133

Sawada Y, Sakamoto T, Nishide K, et al. (1983) Correlation of pathological findings with CT findings after acute carbon monoxide poisoning. N Engl J Med 308:1296

Scheinberg IH, Sternlieb I (1984) Wilson's disease. Saunders, Philadelphia, pp 1-171

Schmitt HP, Emser W, Heimes C (1984) Familial occurrence of amyotrophic lateral sclerosis, parkinsonism and dementia. Ann Neurol 16:642-648

Schoenberg BS, Anderson DW, Haerer AF (1985) Prevalence of Parkinson's disease in the biracial population of Copiah County, Mississippi. Neurology (NY) 35:841-845

Schott GD (1986) Induction of involuntary movements by peripheral trauma: an analogy with causalgia. Lancet 2:712-716

Schwab RS, England AC (1968) Parkinson syndrome due to various specific causes. In: Vinken PJ, Bruyn GW (eds) Handbook of clinical neurology, vol 6. North-Holland, Amsterdam, pp 227-247

Sciarra D, Sprofkin B (1953) Symptoms and signs referable to the basal ganglia in brain tumor. Arch Neurol Psychiatry 69:450-461

Serratrice GT, Toga M, Pellissier JF (1983) Chronic spinal muscular atrophy and pallidonigral degeneration: report of a case. Neurology (NY) 33:306-310

Sharpe JA, Rewcastle NB, Lloyd KG et al. (1973) Striatonigral degeneration: response to levodopa therapy with pathological and neurochemical correlation. J Neurol Sci 19:275-286.

Shy GM, Drager GA (1960) A neurological syndrome associated with orthostatic hypotension: a clinical-pathological study. Arch Neurol 2:511-527

Snyder AM, Stricker EM, Zigmond MJ (1985) stress-induced neurological impairments in an animal model of parkinsonism. Ann Neurol 18:544-551

Snyder SH, D'Amato RJ (1986) MPTP: a neurotoxin relevant to the pathophysiology of Parkinson's disease. Neurology (NY) 36:250-258

Sobue G, Hashizume Y, Ohya M, Takahashi A (1986) Shy-Drager syndrome: neuronal loss depends on size, function, and topography in ventral spinal outflow. Neurology (NY) 36:404-407

Soelberg, Sorensen P, Jansen EC, Gjerris F (1986) Motor disturbances in normal-pressure hydrocephalus. Special reference to stance and gait. Arch Neurol 43:34-38

Soffer D, Grotsky HW, Rapin I et al. (1979) Cockayne syndrome: unusual neuropathological findings and review of the literature. Ann Neurol 6:340-348

Spencer PS, Nunn PB, Hugon JS et al. (1987) Guam amyotrophic lateral sclerosis-parkinsonism-dementia linked to a plant excitant neurotoxin. Science 237:517-522

Spitz MC, Jankovic J, Killian JM (1985) Familial tic disorder, parkinsonism, motor neuron disease, and acanthocytosis—a new syndrome. Neurology (NY) 35:366-377

Steele JC, Richardson JC, Olszewski J (1964) Progressive supranuclear palsy: a heterogenous degeneration involving in the brain stem, basal ganglia and cerebellum

with vertical gaze and pseudobulbar palsy, nuchal dystonia and dementia. Arch Neurol 10:333-359
Stein RW, Kase CS, Hier DB et al. (1984) Caudate hemorrhage. Neurology (NY) 34:1549-1554
Stephen PJ, Williamson J (1984) Drug-induced parkinsonism in the elderly. Lancet 2:1082-1083
Strang RR (1966) Parkinsonism occurring during methyldopa therapy. Can Med Assoc J 95:928-931
Takei Y, Mirra SS (1973) Striatonigral degeneration: a form of multiple system atrophy with clinical parkinsonism. Prog Neuropathol 2:217-251
Tarsy D, Baldessarini RJ (1986) Movement disorders induced by psychotherapeutic agents. Clinical features, pathophysiology and management. In: Shah NS, Donald AG (eds) Movement Disorders. Plenum Medical, New York, pp 365-389
Thompson PD, Marsden CD (1987) Gait disorder of subcortical arteriosclerotic encephalopathy: Binswanger's disease. Movement Disorders 2:1-8
Tolosa ES, Santamaria J (1984) Parkinsonism and basal ganglia infarcts. Neurology (NY) 34:1516-1518.
Uiti RJ, Rajput AH, Ashenhurst EM, Rozdilsky B (1985) Cyanide-induced parkinsonism: a clinicopathologic report. Neurology (NY) 35:921-925
Van der Linden C, Jankovic J, Jansson B (1985) Lateral hypothalamic dysfunction in Parkinson's disease. Ann Neurol 18:137-138
Van Eck JH (1961) Parkinsonism as a misleading brain tumor syndrome. Psychiatr Neurol Neurochir 64:109-123
Victor M, Adams RD, Cole M (1965) The acquired (non-Wilsonian) type of chronic hepatocerebral degeneration. Medicine 44:345-396
Ward CD, Duvoisin RC, Ince SE et al. (1983) Parkinson's disease in 65 pairs of twins and in a set of quadruplets. Neurology (NY) 33:815-824
Weiner HL, de la Monte S (1986) CPC case 25-1986. N Engl J Med 314:1689-1700
Weiner WJ, Nora LM, Glantz RH (1984) Elderly inpatients: postural reflex impairment. Neurology (NY) 34:945-947
Weir RL, Fan KJ (1981) Spinocerebellar degeneration with parkinsonian features: a clinical and pathological report. Ann Neurol 9:87-89
Whitehouse PJ, Muramoto O, Troncoso JC, Kanazawa I (1986) Neurotransmitter receptors in olivopontocerebellar atrophy: an autoradiographic study. Neurology (NY) 36:193-197
Woods BT, Schaumburg HH (1972) Nigro-spino-dental degeneration with nuclear ophthalmoplegia. A unique and partially treatable clinicopathological entity. J Neurol Sci 17:149-156
Yebenes JG, Gervas JJ, Iglesias J et al. (1982) Biochemical findings in a case of parkinsonism secondary to brain tumor. Ann Neurol 11:313-316
Yebenes JG, Brin MF, Mena MA et al. (1988) Neurochemical findings in neuroacanthocytosis. Movement Disorders 3:300-312
Zetusky WJ, Jankovic J, Pirozzolo FJ (1985) The heterogeneity of Parkinson's disease: clinical and prognostic implications. Neurology (NY) 35:522-526
Zweig RM, Whitehouse PJ, Casanova MF et al. (1987) Loss of pedunculopontine in progressive supranuclear palsy. Ann Neurol 22:18-25

CHAPTER 9

Evaluation of Parkinson's Disease

H. TERÄVÄINEN, J. TSUI and D. B. CALNE

A. Introduction

The need for quantitation of parkinsonian symptoms in relation to drug treatment has been recognized for many years (SCHWAB and PRICHARD 1951). The advent of effective therapy has augmented the need, now that we are in an era of ever-expanding novel therapeutic agents, combined with recognition of a wider variety of deficits (such as dementia), and experience with problematic adverse reactions (such as fluctuations in response). Besides making reliable sensitive assessment more important, complications such as dementia and fluctuations in response make the process of evaluation more difficult.

The problem of assessing patients who undergo sudden changes in the severity of symptoms and signs is self-evident. The evaluation is much less complicated in mildly disabled, previously untreated patients, in whom it is possible to express the clinical deficits in a numerical form which is readily reproducible.

Consequently, different approaches have become necessary in designing studies involving subjects early in their disease, as opposed to those with more severe disability associated with fluctuations of performance. The literature concerning assessment of parkinsonian deficits is large; comprehensive reviews have been published addressing this topic (TERÄVÄINEN and CALNE 1980; MARSDEN and SCHACHTER 1981) and quantitation of neurological diseases in general (POTVIN and TOURTELOTTE 1985a,b).

Two approaches to evaluation have generally been employed: (a) *subjective* quantitation, based on the judgment of the evaluator, who assigns scores to the symptoms, signs, functional disability, and side effects, and (b) *objective* quantitation based on the use of various mechanical devices designed to measure specific aspects of motor abnormality. Significant new approaches to objective evaluation have not appeared since the topic was last reviewed (MARSDEN and SCHACHTER 1981) whereas subjective assessment is currently undergoing significant re-evaluation worldwide.

B. Subjective Assessment

Subjective assessment is designed to provide numerical quantification from the traditional clinical neurological evaluation. The usual method is assignment of scores (usually from 0 to 4: 0 = normal, 1 = mild abnormality;

2 = moderate abnormality; 3 = severe abnormality; 4 = profound disability or loss of function), defining each grade for each deficit with as much precision as is necessary to minimize interindividual variation in the interpretation of the range of values. Because clinical features fluctuate, several protocols attempt to generate scores that average deficits over a period of time, or provide separate numbers representing the best and worst phases, and the relative duration of each. Most commonly used examples, employed over a period of several years, include the Webster Scale (WEBSTER 1968), Hoehn and Yahr Staging (HOEHN and YAHR 1967), the Northwestern University Disability Scale (CANTER et al. 1961), the Columbia University Disability Scale (DUVOISIN 1970), the Schwab and England Scale for Daily Activities (SCHWAB and ENGLAND 1969), the New York University Rating Scale (ALBA et al. 1968) and the King's College Hospital Rating Scale (PARKES et al. 1970).

It is pertinent to note that many of these scales were developed prior to experiences of levodopa treatment. Some more recent gradings have been developed to deal with problems such as fluctuation in mobility and confusion (LEES et al. 1977; LIEBERMANN 1980; SCHACHTER et al. 1980; LARSEN et al. 1984). The variability of grading scales and their frequent modification by different workers has created difficulties in comparing both the patient population studied and the results obtained. Because of this problem, a meeting was held in Bermuda to develop a Unified Rating Scale for the quantitation of parkinsonian patients participating in drug studies. This protocol is a compromise and a combination of several pre-existing scales grading virtually every symptom, sign and side effect that can be expected in a large population of parkinsonian patients. It is the most comprehensive subjective scale ever introduced (FAHN and ELTON 1987).

Nevertheless, the Unified Rating Scale has its weaknesses: it is based on assessment of deficits which is subject to both intraobserver and interobserver variations. Furthermore, clinical scores are not linear and they may not correlate well with the patient's overall disability. For example, slight diminution of facial expression, moderately stooped stance and falling when posture is disturbed carry the same numerical weight. There are problems even within the same category of deficit. For example, in evaluating postural stability, a change from 0 to 2 (normal to falling when perturbed) has far greater impact for the patient than changes between grades 2 and 4. Furthermore, clinical experience indicates that some of the variables (such as posture) are relatively insensitive to treatment. This is illustrated in Table 1, which lists changes in the Columbia subscores in ten previously untreated patients given placebo, mesulergine (7 mg/day) and levodopa (750 mg/day + decarboxylase inhibitor). The mean improvement obtained with either mesulergine or levodopa was significant ($P < 0{,}001$) compared with placebo treatment. In this small population of patients, tremor, rigidity, gait and associated movement (armswing) were the most sensitive indicators of the effectiveness of treatment and less improvement occurred in various manual tests (dexterity, foot tapping, pronation-supination movements). Some variables, such as ability to maintain balance, posture and speech made a negligible contribution to the overall score, even though each is of major significance to the patient. Since the ap-

Table 1. Columbia subscores in 10 patients with Parkinson's disease treated with placebo, mesulergine (7 mg/day) and levodopa (750 mg/day + DCI)

	Patient	TRE	RIG	POST	GAIT	ARMS	BAL	CHAI	PRSU	DEXT	FOOT	HYPO	SPE
Placebo	1	0	3	1	1	4	0	1	5	1	2	1	0
	2	1	12	1	1	5	1	0	6	5	4	2	0
	3	5	11	2	1	5	0	1	6	3	6	2	0
	4	0	11	0	1	6	0	1	7	5	5	1	0
	5	6	11	1	1	2	1	0	6	5	4	1	1
	6	8	6	1	1	2	0	0	7	3	5	2	0
	7	3	14	0	1	5	0	0	4	6	4	1	0
	8	0	7	0	0	4	1	0	5	4	5	1	0
	9	0	12	1	1	7	0	1	6	7	4	0	0
	10	2	9	1	0	4	0	1	6	4	4	1	0
Mean		2.5	9.6	0.8	0.8	4.4	0.3	0.5	5.8	4.3	4.3	1.2	0.1
Mesulergine	1	0	1	1	0	2	0	1	3	2	1	0	0
	2	0	6	1	0	2	1	0	7	3	2	2	0
	3	8	5	2	1	2	0	0	6	4	4	1	0
	4	0	7	0	0	3	0	0	3	5	3	1	0
	5	2	5	1	0	0	0	0	4	3	3	1	0
	6	7	5	1	1	1	0	0	7	3	2	1	0
	7	0	6	0	0	3	0	0	4	3	2	1	0
	8	0	3	0	0	3	1	0	4	4	2	0	0
	9	1	7	1	1	4	0	1	4	2	1	1	0
	10	1	6	0	0	2	0	0	4	2	1	1	0
Mean		1.9	5.1	0.7	0.3	2.2	0.2	0.2	4.6	3.1	2.1	0.9	0
ME/PLA (%)		76	53	88	38	50	67	40	79	72	49	75	0
Levodopa	1	0	2	1	0	0	0	0	3	1	1	0	0
	2	0	6	0	0	1	1	0	5	3	1	1	0
	3	1	6	2	1	0	0	0	1	1	2	1	0
	4	0	6	1	0	0	0	1	5	3	4	0	0
	5	0	2	1	0	0	0	0	1	1	1	1	0
	6	2	2	1	0	0	0	0	6	1	2	1	0
	7	1	5	1	0	0	0	0	3	4	0	0	0
	8	0	4	0	0	1	0	0	4	4	3	1	0
	9	0	6	1	0	2	1	0	1	2	2	1	0
	10	0	0	0	0	0	0	0	3	2	3	1	0
Mean		0.4	3,9	0.8	0.1	0.4	0.2	0.1	3.2	2.2	1.9	0.7	0
LD/PLA (%)		21	40	100	13	9	67	20	55	51	44	58	0

TRE tremor; RIG rigidity; POST posture; ARMS armswing; BAL balance; CHAI arising from chair; PRSU pronation and supination; DEXT dexterity; FOOT foot tapping; HYPO hypomimia; SPE speech; PLA Placebo; ME mesulergine; LD levodopa

praisal of drug effects is facilitated by easily analysable major variables in function that are sensitive to treatment, the evaluation of motor symptoms by the Unified Rating Scale could certainly be improved by adding a separate grading for armswing during gait.

We also consider that testing of mood, memory and cognitive function is laborious and of limited value in most therapeutic investigations. Furthermore, patients with severe intellectual impairment (grade 3-4), severe depression (grade 4, perhaps even 3), persistent hallucinatons or delusions (grade 4, perhaps 3), and complete loss of motivation (grade 4) should not be included in the routine study of antiparkinson drugs. Applying the complete Bermuda battery of tests at each visit is of questionable value. Interpretation of motor scores is also difficult in the presence of fluctuations in mobility or severe dyskinesias.

The Unified Rating Scale contains several variables based on patients' information (such as sensory complaints, number of falls per day, the duration of dyskinesias) which are of special importance in patients who fluctuate. Patients selected for studies requiring self-assessment should be cooperative and attentive, and have a proper insight into their symptoms or have a reliable family member accompany them to the clinic.

There have been few studies designed to determine interobserver variability in assessing parkinsonism. In a recent report (MONTGOMERY et al. 1985), 2 evaluators rated 70 patients. Variability was low for resting tremor and gait, moderate for bradykinesia and posture and high for dyskinesia and dystonia.

Clinical scoring protocols can be evaluated by determining their ability to discriminate between variables and their ability to detect differences in value within variables. These important issues have not received the attention they deserve.

DIAMOND and MARKHAM (1983) compared four different scales in eight patients treated with pergolide. Each scale was employed by a different neurologist and the examiners were unaware that the results would be compared with other scales. Ratings from previous visits were not available for the examiner. For the sake of comparison, the disability scores were converted to percentage change relative to the initial rating before treatment. The variations were considerable. For example, at 5 months one scale (Hoehn and Yahr) showed 13% improvement and another (New York University revised) an improvment of 58%. The ICLA, and Northwestern University modified scales indicated improvement of 28% and 44% respectively. At 9 months the Hoehn and Yahr, and New York University revised scales indicated worsening of 6% and improvement of 34% respectively. The investigators pointed out that the scoring protocols have different sensitivity for different aspects of parkinsonism. FAHN and BRESSMAN (1984) have compared three evaluation systems. They concluded that the Schwab and England ADL, and Columbia scales give a reasonable assessment of functional disability, whereas the UCLA protocol gave higher relative scores.

Since the Unified Rating Scale is a combination of the pre-existing scales, its precision is no better or worse than its components. It is an elementary mistake to expect that the accuracy of assessment will increase with a new

scale. The most significant impact will be that different workers can compare their results directly, and pool their information if this is desirable. Optimal use of the Unified Rating Scale would entail the selection of subtests of components, so that frequent assessments could be undertaken, each lasting 10–15 min, which could be combined with the full Unified Rating Scale at less frequent intervals, according to the design of the study.
From these considerations we may conclude that one clinical evaluation method should not replace all others, since no single system is irrefutably better than all others. It is desirable to draw conclusions from a survey of results obtained from more than one test instrument; there will continue to be a need for several clinical scoring protocols.

C. Objective Assessment

Despite the limitations of clinical scoring systems, they still constitute the most useful method for assessing parkinsonism. Compared with various techniques for objective measurement, subjective evaluations are simple, fast and cost-effective. Most of the methods employed for objective measurement require relatively complex and expensive equipment, and they are generally restricted to sampling a small fraction of the patients' symptoms and signs.

The quest for precise objective measurement is a basic theme running through all biomedical research, and history has shown it to be ultimately worthwhile. In the context of Parkinson's disease, a distinction must be made between questions relating to pathophysiology, and the more common task of assessing novel forms of treatment. Objective measurements are particularly valuable in attempting to answer the former, but they can also have a useful role to play in the latter. We shall not review the extensive literature on comprehensive methods of objective quantifications (Teräväinen and Calne 1980; Marsden and Schachter 1981; Potvin and Tourtelotte 1985a,b); instead we shall focus attention on the practical aspects of measuring some of the commoner clinical deficits.

I. Tremor

Numerous methods of quantifying tremor have been developed over the last 30 years; the most commonly employed is monoaxial accelerometer recording, because it is relatively economical, simple and quick. Computerized triaxial accelerometers have the advantage of recording tremor in all directions, but the computer programs for the analysis of these observations are severalfold more complex and expensive. Objective tremor recording with computer analysis is better suited for studying specific neurophysiological problems rather than detecting the response to potential pharmacotherapy in patients with Parkinson's disease. Tremor is exceptionelly variable and sensitive to psychological influences. The problem is that objective techniques can only handle a limited segment of information in space and time; since parkinsonian tremor is highly variable in space and time, there is a major sampling problem in drug studies. These limitations do not apply to all types of tremor, for example the postural shake of essential tremor is more consistent, and very suitable for study by accelerometry.

II. Rigidity

Relatively few devices have been developed for the measurement of rigidity, and only the method of WEBSTER (1964, 1966) has enjoyed any widespread application (WEBSTER 1981; POTVIN and TOURTELOTTE 1985). Webster's technique took advantage of his observation that when angular position is plotted against the torque required to move the elbow passively through flexion and extension, the area of the hysteresis loop gives a more satisfactory measurement of rigidity than the torque alone.

We have employed a similar method to study rigidity at the wrist. The movement was executed by a servo-controlled torque motor, and a PDP 11 computer determined the area of the hysteresis loop. An example of a recording is shown in Fig. 1, while Fig. 2 illustrates changes in rigidity values obtained for a 55-year-old patient with Parkinson's disease experiencing marked fluctuations in mobility, while "on" and "off", at different amplitudes of movement. It may be noted that the rigidity values obtained when he was rigid were some 600% higher than when responding well to treatment. Routine testing takes some 2-3 min when one amplitude and frequency are employed. The results correlated well with subjective assessment (Columbia score). The high sensitivity of the method has as to be weighed against its major disadvantage, the substantial cost of the equipment. However, the apparatus is very versatile and has many potential applications.

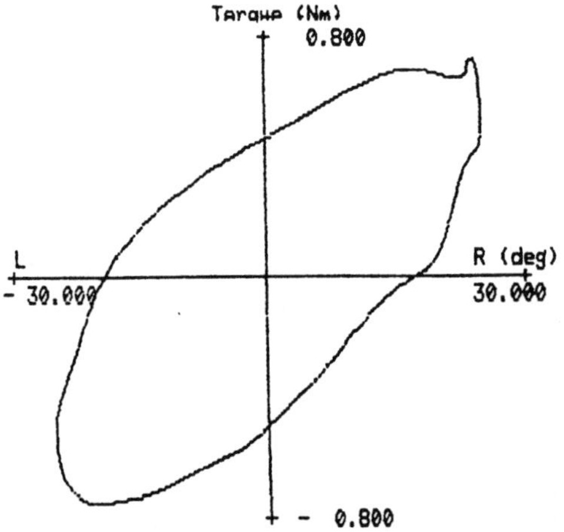

Period (s) : 0.820

Trial	Area [Nm deg.]
01	31.401
02	40.987
03	40.926
04	40.178
05	41.433

Average area : 38.985 [Nm deg]

Fig. 1. A recording of wrist rigidity in a patient with Parkinson's disease, with a calculation of the area of the hysteresis loop for 25° flexion and 25° extension. The mean rigidity score (area) over 5 trials was about 40 Nm ° when the Columbia score for rigidity was 3

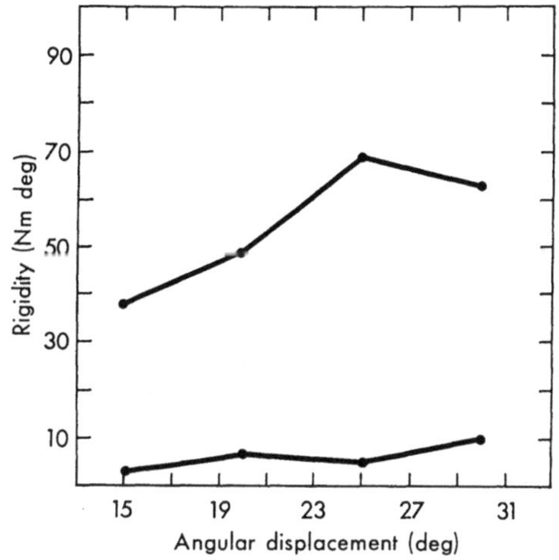

Fig. 2. Rigidity scores (Nm °) for different amplitudes of movement of the wrist obtained for a 55-year-old patient with Parkinson's disease who experienced marked fluctuations. The two series of measurements were performed when he was very rigid (*upper graph*, Columbia score for rigidity 3) and again when his mobility was optimal (*lower graph*, Columbia score 0)

III. Hypokinesia

Several components of hypokinesia can be identified: (a) the initiation of movement is delayed; (b) the speed of movement is slow; (c) accuracy of targeted complex movements is impaired; and (d) bimanual performance deteriorates. Numerous elaborate devices have been constructed to determine the reaction time and velocity of movement (movement time over a specified distance) (Teräväinen and Calne 1980; Marsden and Schachter 1981; Potvin and Tourtelotte 1985 a,b). However, few studies have employed these measurements to assess pharmacotherapy (Angel et al. 1971; F. Velasco and M. Velasco 1973). We have compared the reaction times and movement times in 72 patients with Parkinson's disease (Teräväinen and Calne 1981) and found that the movement time correlated best with clinical disability. The reaction time was less useful because the differences from normal values were relatively small.

Simple devices have been employed for the quantification of complex manipulative skills. Tests involving tapping are straightforward, but they are less satisfactory than alternating pronation-supination (Marsden and Schachter 1981).

In the Purdue pegboard test small pegs are picked up and placed in holes. This is capable of discriminating early Parkinson's disease from age-matched controls with 85 % confidence (M. Hietanen and H. Teräväinen 1987, unpublished work), so this technique may prove a useful method for providing a composite index of hypokinesia.

D. Summary

The methods employed for quantifying the neurological deficits of Parkinson's disease should be chosen according to the purpose of the evaluation. For detecting drug-induced changes in motor function, clinical scoring protocols are generally appropriate. The newest of these, the Unified Rating Scale, is the most comprehensive, but has certain limitations. For studying the pathophysiology of Parkinson's disease, objective techniques are usually the most sensitive and accurate; numbers are available for tremor, rigidity and hypokinesia; selection depends upon which particular aspect of motor function is under unvestigation.

References

Alba A, Trainor FS, Ritter W, Dacso MM (1968) A clinical disability rating for parkinsonian patients. J Chronic Dis 21:507-522

Angel RW, Alston W, Garland H (1971) L-Dopa and error correction time in Parkinson's disease. Neurology 21:1255-1260

Canter GH, de la Torre R, Mier M (1961) A method for evaluating disability in patients with Parkinson's disease. J Nerv Ment Dis 133:143-147

Diamond SG, Markham CH (1983) Evaluating the evaluations—or how to weigh the scales of parkinsonian disability. Neurology 33:1098-1099

Duvoisin RC (1970) The evaluation of extra-pyramidal disease. In: de Ajuriagerra J (ed) Monoamines, noyaux gris centraux et syndrome de Parkinson. Masson, Paris, pp 313-325

Fahn S, Bressman SB (1984) Should levodopa therapy for parkinsonism be started early or late? Evidence against early treatment. Can J Neurol Sci 11:200-205

Fahn S, and Elton RL (1987) Unified Parkinson Disease Rating Scale. In: Fahn S, Marsden CD, Calne D, Goldstein M (eds) Recent Developments in Parkinson's disease, Vol 2. Macmillan Healthcare Information, Florham Park, pp 153-163

Hietanen M, Teräväinen H, Tsui JK, Calne DB (1987) The pegboard as a measurement of parkinsonian motor deficit. Neurology 37 (Suppl 1):266.

Hoehn MM, Yahr MD (1967) Parkinsonism: onset, progression and mortality. Neurology (Minneap) 17:427-442

Larsen A, Calne S, Calne DB (1984) Assessment of Parkinson's disease. Clin Neuropharmacol 7:165-169

Lees, AJ, Shaw KM, Kohout LJ et al. (1977) Deprenyl in Parkinson's disease. Lancet 2:791-795

Lieberman A, Dziatolowski M, Gopinathan G et al. (1980) Evaluation of Parkinson's disease. In: Goldstein M, Calne, DB, Lieberman A, Thorner MD (ed) Ergot compounds and brain function: neuroendocrine and neuropsychiatric aspects. Raven, New York, pp 277-286

Marsden CD, Schachter M (1981) Assessment of extrapyramidal disorders. In: Lader MH, Richens A (eds) Methods in clinical pharmacology—central nervous system. Macmillan, London, pp 89-111

Montgomery GK, Reynolds NC, Warren RM (1985) Qualitative assessment of Parkinson's disease: study of reliability and data reduction with an abbreviated Columbia scale. Clin Neuropharmacol 8:83-92

Parkes JD, Zilkha KJ, Calver DM, Knill-Jones RP (1970) Controlled trial of amantadine hydrochloride in Parkinson's disease. Lancet I:259-262

Potvin AR, Tourtelotte WW (eds) (1985a) Quantitative examination of neurologic functions, vol I. CRC, Boca Raton
Potvin AR, Tourtelotte WW (eds) (1985b) Quantitative examination of neurologic functions, vol II. CRC, Boca Raton
Schachter M, Marsden CD, Parkes JD, Jenner P, Testa B (1980) Deprenyl in the management of response fluctuations in patients with Parkinson's disease on levodopa. J Neurol Neurosurg Psychiatry 43:1016-1021
Schwab RS, England AC (1969) Projection technique for evaluating surgery in Parkinson's disease. In: Gillingham FJ, Danoldson IML (eds) Third symposium on Parkinson's disease. Churchill Livingstone, Edinburgh, pp 152-157
Schwab RS, Prichard JS (1951) An assessment of therapy in Parkinson's disease. Arch Neurol Psychiatry 65:489-501
Teräväinen H, Calne DB (1980) Quantitative assessment of parkinsonian deficits. In: Rinne UK, Klinger M, Stamm G (eds) Parkinson's disease—current progress, problems and management. Elsevier, Amsterdam, pp 145-162
Teräväinen H, Calne DB (1981) Assessment of hypokinesia in parkinsonism. J Neural Transm 51:149-159
Velasco F, Velasco M (1973) A quantitative evaluation of the effects of L-dopa on Parkinson's disease. Neuropharmacology 12:89-99
Webster DD (1964) The dynamic quantitation of spasticity with automated integrals of passive motor resistance. Clin Pharmacol Ther 5:900-908
Webster DD (1966) Rigidity in extrapyramidal disease. J Neurosurg 24 (Suppl 2): 299-309
Webster DD (1968) Clinical analysis of the disability in Parkinson's disease. Mod Treat 5:257-282
Webster DD (1981) Assessment of deficit function in Parkinson's disease. In: Cohen BA (ed) Frontiers of engineering in health care proceedings, vol 3. IEEE Trans Biomed Eng 28:599-606

CHAPTER 10

Clinical Trials for Parkinson's Disease

J. K. Tsui, H. Teräväinen and D. B. Calne

A. Introduction

Anticholinergic agents have been used for over 100 years for treating Parkinson's disease and clinical trials have been reported since the late nineteenth century (Aquilonius 1979). In the past two decades more drug studies have been conducted in Parkinson's disease than in any other neurological disorder.

Following the discovery of the symptomatic benefits of L-dopa (Barbeau 1961, 1969; Birkmayer and Hornykiewicz 1961, 1962; Cotzias et al. 1969), numerous investigations were undertaken with this agent alone, and subsequently in combination with decarboxylase inhibitors (Barbeau et al. 1976). The introduction of dopamine agonists (Calne et al. 1974) led to a new wave of clinical trials, whereas those with anticholinergics have tapered off. Amantadine has been studied intermittently over the last decade.

In recent years, new concepts concerning the pathogenesis of Parkinson's disease have been developed. Agents such as deprenyl (a monoamine oxidase B inhibitor) (Birkmayer et al. 1977) have been investigated, and a rationale has emerged for studying antioxidants. These trends derive from the proposal that the formation of free radicals (Donaldson et al. 1980) may contribute to loss of neurons from the substantia nigra, so the underlying progress of the disease might be delayed by impeding the release of free radicals, or facilitating their removal.

Most of the therapeutic studies in Parkinson's disease have been conducted with different designs and with different statistical manipulations. There has been a lack of uniformity in the criteria for recruitment of patients, and even greater variation in the choice of techniques for monitoring the progress of the patients. Perhaps because of this diversity a persuasive consensus has evolved on the relative safety and efficacy of treatment.

B. Phases

The design chosen for investigation of a new drug will depend, in part, on the extent of experience with the compound. In phase 1 studies, first administration to human subjects is usually undertaken with single doses given to normal volunteers. In phase 2, small groups of patients are treated for a short period. Phase 3 involves more extended formal studies. Phase 4 comprises continued surveillance after the drug has been released for general use.

C. Design

Studies may be open, single-blind, or double-blind; drug dosage may be fixed or varied, and comparisons between treatment regimens may be undertaken "within patients" or "between patients". Open studies are usually employed in pilot projects before embarking on full-scale investigations. They are useful for establishing safety, determining a dose range, and studying pharmacokinetics. Single-blind studies decrease the "placebo" effect, and thus constitute a first step towards the detection of efficacy.

Double-blind studies are generally accepted as the most reliable format for evaluation of treatment, once initial information on safety has been gathered. The design depends on the circumstances. Is the study conducted to demonstrate the efficacy of a single drug or a combination? Is any special category of patient particularly important? Is the average optimal dose known? How much variation in optimal dose can be expected between patients? Can adverse reactions be predicted? Must previous therapy be continued? How long should the study run?

Provided open studies with a new compound have yielded reassuring observations on safety, and have established satisfactory dose levels, it is reasonable to proceed to a blind evaluation. For testing a single drug, comparison may be conducted either "between patients" (cross-sectional) or "within patients" (longitudinal).

In "between patient" studies one group of subjects is given the test drug and another group a placebo. The two groups are matched for age, sex, disability and disease duration. They are then assigned to the test or control group in random order. The next question is how many subjects to include. An estimate can be made provided information is available on the sensitivity of the evaluation relative to the anticipated effect, the random scatter of variables and the confidence level required from the results (Lyman 1984). Sometimes new treatment must be compared with a conventional drug because therapy cannot be withdrawn. Here the conventional drug takes the place of a placebo, and must be given in a form that patients cannot recognize.

The number of patients will be lower for "within patient" studies because each subject acts as his or her own control and thereby decreases random scatter. Sequential analysis can be performed. This technique tests each observation for the null or alternative hypothesis before going on to the next observation. Should the sample size become too large, Armitage's restriction (Armitage 1957, 1960) can be applied.

In the conduct of a "within patient" study it is important to allow sufficient "washout" time is allowed for the first agent to wear off before the effect of the next agent is measured. Each patient should receive the drug and the placebo in randomized order in case there is any time-related trend in the variables.

A double-blind "within patient" comparison with crossover is the most convenient design, although attention must be paid to eliminate bias deriving from the order in which drugs are given. The patients should be divided into two "randomized blocks" with one group receiving drug 1 before drug 2, and vice versa for the other group. If more than two drugs are to be compared separately, "between patient" studies eliminate the problem of designing complex crossovers to minimize order effects. The lack of homogeneity amongst the study groups can be limited by matching patients on relevant variables, and by increasing the sample sizes.

It may be desirable to study a combination of drugs to determine whether they are additive, synergistic or inhibitory. In a study of this type, it is necessary to allocate patients randomly into regimens of different drug combinations (cells). Comparison is made between drug 1 plus placebo, drug 2 plus placebo, and drug 1 plus drug 2 (if two drugs are studied). A double-blind procedure is undertaken, and a "between patient" study may be more practical because of order effects.

While both "within patient" and "between patient" techniques are applicable in studying the symptomatic effects of drugs, in long-term prospective studies only the latter can be conducted. All prolonged studies will face similar problems such as interindividual variation in the rate of progress of the disease, dropouts and difficulty in deciding the end point. Progress may not be linear, and the longer the study continues, the more problems will evolve in the computation of the results.

The double-blind technique is predicated upon both the patient and the evaluator being unaware of the therapeutic regimen. In general, a rigidly structured, coded dose design is less satisfactory than a flexible plan, but the latter requires an extra "unblind" physician to manage dosage. This additional physician is particularly useful when treating patients who already have problems that require fairly frequent dose adjustments before starting the study.

Should the dose be fixed or varied? The commonest single error in studies on new drug is failure to detect a therapeutic effect because dosage of the test drug was arbitrarily set too low, owing to limited experience with the drug. Dose-ranging studies are necessary in early studies with a new compound (Stoessl et al. 1985) and where prior experience with agents resembling the test drug indicate that there is wide interindividual variation in optimal dosage. In the latter case dosage must be titrated to maximum tolerated levels (often with an arbitrary upper limit based upon animal toxicity). Once the optimal (i.e. maximum tolerated) intake is reached, this should be held without change over a suitable period for the test observations to be taken for subsequent analysis. In this design, assessments prior to reaching the optimal dose are discarded. The intake of all medications other than the test drug (or placebo) should not be altered through the period of study.

To increase sample sizes, multicentre trials may be undertaken. Variability between centres then becomes a most important confounding factor: different centres having disparate patient populations, diverse interpretations of exclusion criteria and heterogeneous techniques of assessment.

D. Recruitment

The clinical features of Parkinson's disease have been well described, but there are no universally accepted diagnostic criteria. Because of the lack of objective diagnostic tests, a decision for recruitment is based on clinical findings. Two definite, cardinal features of parkinsonism (resting tremor, rigidity, bradykinesia, parkinsonian gait or posture, and mask-like facies) are usually accepted as reasonable criteria (Schoenberg et al. 1986). Exclusion of diseases such as essential tremor and an identifiable cause of parkinsonism (e.g. neuroleptics) are, of course, mandatory.

In collecting the sample, care must be taken to avoid concomitant disease, dementia or psychosis and depression. Randomization of allocation into different drug combination cells should be strictly observed. In studying patients with Parkinson's disease, the history of previous and concurrent drug therapy may be important.

E. Assessment

Clinical rating scores are the most frequently employed technique for evaluating treatment. The most widely used protocols have been the Schwab and England (England et al. 1956), Hoehn and Yahr (Hoehn et al. 1967), Webster (Webster 1968), and Columbia Scales (Duvoisin 1970). Others include the Northwestern (Canter et al. 1961), UCLA (Markham et al. 1981), New York University (Lieberman et al. 1980), King's College Hospital (Parkes et al. 1971) and UBC Scales (Larsen et al. 1984). These scoring protocols generate numerical values for the severity of clinical features, ability to perform daily activities, adverse reactions to therapy, and mental status. A more recent scale proposed by (Shoulson et al. 1987) is known as the Unified Rating Scale; it is more comprehensive than previous protocols, but takes longer to complete. This "Bermuda" scale is reviewed in more detail in Chap. 9.

All clinical evaluations, however, suffer from the limitation that the scores do not bear a linear relationship to the deficit, or the underlying pathology. In cross-sectional studies, patients are usually recruited at different stages of the disease and the end points vary widely. Changes in scores early in the disease may not be readily comparable to similar alterations in later phases of the illness. The assignment of numerical scores is also subject to "within observer" and "beween observer" variations.

Objective measurements of individual deficits have been employed to assess parkinsonism, and conform to a linear scale. Techniques are available for recording limited aspects of the disease such as tremor, rigidity, speed and accuracy of movement, gait and balance (Teräväinen and Calne 1980).

Accelerometers are employed for tremor (Larsen et al. 1983), torque motors for rigidity (Webster and Mortimer 1977), and digitized boards or pressure-sensitive pads for speed of movement and reaction time (Teräväinen et al. 1980; Evarts et al. 1981; Ward et al. 1983). Torque motors and electromyography measure the long latency stretch reflexes (Tatton et al. 1984), force-sensitive platforms record body sway (LAKKE et al. 1972), and the Purdue pegboard with bimanual execution quantifies manipulative skills (HIETANEN et al. 1987). Other techniques include video or cine analysis of limbswing (KNUTSSON 1972), measurement of saccades and smooth pursuit eye movements (TERÄVÄINEN et al. 1980), analysis of handwriting (KNOPP 1968), balance (FOLKERTS and NJIOKIKTJIEN 1971; JANSEN et al. 1982) and voice.

Any one of these can only provide information about a single deficit of the syndrome and fails to reflect the severity of parkinsonism as a whole. More than one test is required if they are to be used in therapeutic studies and the techniques tend to be slow. The Purdue pegboard (HIETANEN et al. 1987) may turn out to be a useful, simple, economical objective method; it might even provide an index of parkinsonian motor function as a whole. For a more detailed discussion of the various clinical scales and objective methods of evaluation see Chap. 9.

While the functional impact of the disease can be assessed by clinical evaluation and individual deficits can be measured by objective methods, there is no way to quantify the underlying neuropathology. The suppression of prolactin release has been employed as an index of dopaminergic integrity. The prolactin level can be elevated artifically by administering thyrotropin-releasing hormone (EISLER et al. 1981) and this increases the sensitivity of the test. Homovanillic acid concentrations in the cerebrospinal fluid have also been assayed as an index of dopaminergic function (KARTZINEL et al. 1976). Probenecid has been employed to inhibit active reabsorption of homovanillic acid in this setting (LAKKE et al. 1972).

Positron emission tomography (PET) has recently been developed to provide a semiquantitative analysis of the integrity of the nigrostriatal pathway (MARTIN et al. 1986). Further developments in this technique will involve synthesis of new ligands for dopamine in nerve endings, and for dopamine receptors (WAGNER et al. 1983). Mathematical modelling will no doubt be improved to enhance interpretation. PET may prove to be the accurate way of assessing the progress of the pathology underlying Parkinson's disease, but the equipment is complicated and expensive, so it is unlikely to be applied widely to therapeutic studies.

F. Statistical Analysis

The classical parametric methods for data analysis depend on a normal distribution of the populations under comparison. However, in Parkinson's disease, the populations of selected samples do not necessarily follow the normal Gaussian pattern. They are usually skewed and outliers are frequent. Various manipulations can be employed to smooth out the data, for example, by transformation through mathematical functions such as logarithm, sine or square root. Other more controversial techniques are also available to deal with outliers (FIENBERY 1983; HOAGLIN et al. 1983).

For data analysis in clinical studies of Parkinson's disease, the most suitable approach is usually through nonparametric techniques. These include the rank and sign tests (Wilcoxon or Mann-Whitney tests), nonparametric analysis of variance (Kruskal-Wallis test), correlation tests (Friedman or Spearman) and nonparametric regression. Contingency tests can also be employed (χ^2).

Once the null hypothesis has been formulated, an alternative hypothesis can be proposed. This can be a one-sided statement (changes in one direction only) or two-sided (changes in either direction). Correspondingly, the analysis can be one-tailed or two-tailed. While the one-tailed test is more sensitive in detecting changes in one direction, the two-tailed tests is more objective. However, the two tailed test requires a larger sample to achieve the same significance level.

G. Conclusions

Clinical studies of Parkinson's disease should generally be carried out in a double-blind, "within patient" comparison. Dosage may be fixed, but a dose-ranging design is frequently more applicable. Recruitment criteria are based on the presence of two or more definite parkinsonian features. A widely accepted clinical scoring protocol should be employed for assessment of patients; this scale should be reasonably comprehensive and detailed, but should not take more than 15 minutes. Simple objective tests, such as the Purdue pegboard, may be useful. Nonparametric statistical techniques are suitable for the analysis of results.

Clinical trials have previously been employed to test palliative therapy, but recently there has been interest in strategies treatment to slow down the underlying pathology (monoamine oxidase B inhibitors, antioxidants). Such treatment is currently being assessed in a long term, multi-centre, "between patient" study.

References

Armitage P (1957) Restricted sequential procedures. Biometrics 44:9-26

Armitage P (1960) Sequential medical trials. Blackwell, Oxford

Aquilonius S-M (1979) Cholinergic mechanisms in the CNS related to Parkinson's disease. In: Rinne UK, Klingler M, Stamm G (eds) Parkinson's disease current progress, problems and management. Elsevier, North-Holland, Amsterdam, pp 17-27

Barbeau A (1961) Biochemistry of Parkinson's disease. In: Proceedings of the seventh international congress of neurology, Rome, Sept, Societa Grafica Romana, Rome, vol 2, p 925

Barbeau A (1969) L-Dopa therapy in Parkinson's disease: a critical review of nine years' experience. Can Med. Assoc J. 101:791-800

Barbeau A Roy M (1976) Six-year results of treatment with levodopa plus benzerazide in Parkinson's disease. Neurology (NY) 267:399-404

Birkmayer W, Hornykiewicz O (1966) Der L-3,4-Dioxyphenylalanin (=DOPA)-Effekt bei der Parkinson-Akinese. Wien Klin Wochenschr 73:787-788

Birkmayer W, Hornykiewicz O (1962) Der L-Dioxyphenylalanin (= L-DOPA)-Effekt beim Parkinson-Syndrom des Menschen: zur Pathogenese und Behandlung der Parkinson-Akinese. Arch Psychiatr Nervenkr 203:560-574

Birkmayer W, Riederer P, Ambrozi L, Youdim MBH (1977) Implications of combined treatment with "Madopar" and L-deprenyl in Parkinson's disease. Lancet 1:439-443
Calne DB, Teychenne PF, Claveria LE, Eastman R, Greenacre JK, Petrie A (1974). Bromocriptine in parkinsonism. Br Med J 4:442-444
Canter GJ, deLaTorre R, Mier M (1961) A method for evaluating disability in patients with Parkinson's disease. J Nerv Ment Dis 133:143-147
Cotzias GC, Papavasiliou PS, Gellene R (1969) Modification of parkinsonism—chronic treatment with L-dopa. N Engl J Med 280:337-345
Donaldson J, LaBella FS, Gesser D (1980) Enhanced autoxidation of dopamine as a possible basis of manganese neurotoxicity. Neurotoxicology (Park Forest Il) 2:53-64
Duvoisin RC (1970) The evaluation of extra-pyramidal disease. In: de Ajuriagerra J (ed) Monoamines: noyaux gris centraux et syndrome de Parkinson. Masson, Paris, pp 313-325
Eisler T, Thorner MO, MacLeod RM, Kaiser DL, Calne DB (1981) Prolactin secretion in Parkinson's disease. Neurology (NY) 31:1356-1359
England AC, Schwab RS (1956) Post-operative evaluation of selected patients with Parkinson's disease. J Am Geriatr Soc 4:1219-1232
Evarts EV, Teräväinen H, Calne DB (1981) Reaction time in Parkinson's disease. Brain 106:167-186
Fienbery SE (1983) The analysis of cross-classified categorial data, 2nd edn. MIT Press, Cambridge
Folkerts JF, Njiokiktjien CJ (1971) The influence of L-dopa on the postural regulation of Parkinson patients. 1st Symposium international de posturographie, Madrid, 19-22 Oct
Hietanen M, Teräväinen H, Tsui JK, Calne DB (1987) The pegboard as a measurement of parkinsonian motor deficit. Neurology 37(Suppl 1):266
Hoaglin DC, Mosteller F, Tukey JW (1983) Understanding robust and exploratory data analysis. Wiley, New York
Hoehn MM, Yahr MD (1967) Parkinsonism: onset, progression and mortality. Neurology (NY) 17:427-442
Jansen EC, Larsen RE, Olesen MB (1982) Quantitative Romberg's test. Acta Neurol Scand 66:93-99
Kartzinel R, Perlow MD, Carter AC, Chase TN, Calne DB, Shoulson I (1976) Metabolic studies with bromocriptine in patients with idiopathic parkinsonism and Huntington's chorea. Trans Am Neurol Assoc 101:53-56
Knopp W (1968) Explorations in the assessment and meaning of the subclinical extrapyramidal effect of neuroleptic drugs. Pharmakopsychiatria Neuropsychopharmakol 1:54-62
Knutsson E (1972) An analysis of parkinsonian gait. Brain 95:475-486
Lakke JPWF, Korf J, Van Praag HM, Schut T (1972) Predictive value of the probenecid test for the effect of L-dopa therapy in Parkinson's disease. Nature New Biol 236:208-209
Larsen TA, LeWitt PA, Calne DB (1983) Theoretical and practical issues in assessment of deficits and therapy in parksinsonism. In: Calne DB, Horowski R, McDonald RJ, Wuttke W (eds) Lisuride and other dopamine agonists. Raven, New York, pp 363-373
Larsen TA, Calne S, Calne DB (1984) Assessment of Parkinson's disease. Clin Neuropharmacol 7(2):165-169

Lieberman A, Dziatolowski M, Gopinathan G, Kupersmith M, Neophytides A, Korein J (1980) Evaluation of Parkinson's disease. In: Goldstein M, Calne DB, Lieberman A, Thorner MO (eds) Ergot compounds and brain function: neuroendocrine and neuropsychiatric aspects. Raven, New York, pp 277-286

Lyman O (1984) An introduction to statistical methods and data analysis, 2nd ed Duxbury, Boston

Markham CH, Diamond SG (1981) Evidence to support early levodopa therapy in Parkinson's disease. Neurology (NY) 31:125-131

Martin WRW, Stoessl AJ, Adam MJ, Ammann W, Bergstrom M, Harrop R, Laihinen A, Rogers JG, Ruth TJ, Sayre CI, Pate BD, Calne DB (1986) Positron emission tomography in Parkinson's disease: glucose and DOPA metabolism. Adv Neurol 45:95-101

Parkes JD, Curzon G, Knott PJ, Tattersall R, Baxter RCH, Knill-Jones RP, Marsden CD, Vollum D (1971) Treatment of Parkinson's disease with amantadine and levodopa. Lancet 1:1083-1086

Schoenberg BS, Anderson DW, Haerer AF (1986) Racial differentials in the prevalence of Parkinson's disease. Neural 36:841-845

Shoulson I, The Parkinson Sindy Group (1987). Correlates of clinical decline in early Parkinson's disease (ABST). Neurology 37 (Suppl 1): 278

Stoessl AJ, Mak E, Calne DB (1985) (+)-4-Propyl-9-hydroxynaphthoxazine (PHNO), a new dopaminomimetic, in treatment of parkinsonism. Lancet ii: 1130-1331

Tatton WG, Eastman MJ, Bedingham W, Verrier MC, Bruce IC (1984) Defective utilization of sensory input as the basis for bradykinesia, rigidity and decreased movement repertoire in Parkinson's disease: a hypothesis. Can J Neurol Sci 11:136-143

Teräväinen H, Calne DB (1980) Quantitative assessment of Parkinsonian deficits. In: Rinne UK, Klingler M, Stamm G (eds) Parkinson's disease: current progress, problems and management. Elsevier, North-Holland, Amsterdam, pp 145-164

Teräväinen H, Evarts EV, Calne DB (1980) Studies of parkinsonian movement. 1. Programming and execution of eye movements. Acta Neurol Scand 62:137-148

Wagner HN, Burns HD, Dannals RF, Wong DF, Langstrom B, Duelfer T, Frost JJ, Ravart HT, Links JM, Rosenbloom SB, Lukas SE, Kramer AV, Kuhar MJ (1983) Imaging dopamine receptors in the human brain by positron tomography. Science 221:1264-1266

Ward CD, Sanes JN, Dambrosia JM, Calne DB (1983) Methods for evaluating treatment in Parkinson's disease. Adv Neurol 37:1-7

Webster DD (1968) Critical analysis of the disability in Parkinson's disease. Mod Treat 5:257-282

Webster DD, Mortimer JA (1977) Failure of L-dopa to relieve activated rigidity in Parkinson's disease. In: Messiha FS, Kenny AD (eds) Parkinson's disease. Neurophysiological, clinical and related aspects, Plenum, New York, pp 297-313.

CHAPTER 11

Experimental Therapeutics Directed at the Pathogenesis of Parkinson's Disease

I. SHOULSON

A. Introduction

The development of levodopa therapy for Parkinson's disease (PD) has spawned substantive clinical and scientific gains related to the experimental therapeutics of neurodegenerative disorders. Levodopa therapy has resulted in symptomatic benefit and increased life expectancy for the majority of PD patients. The rational development of levodopa therapy and related therapies to rectify the consequences of reduced dopaminergic neurotransmission in PD has also served as a prototype for the experimental therapeutics of other neurologic disorders associated with neurotransmitter deficits. In spite of these advances, the clinical and scientific limitations of replacement therapies have become increasingly evident. Symptomatic anti-PD benefits have proved transient, and clinical features intensify eventually in the setting of progressive neuronal degeneration. Disabling adverse effects, such as mental status disturbances, involuntary movements, and autonomic instability, further preclude or limit the use of dopaminergic therapies. The application of rational replacement therapy to other neurologic disorders has been constrained by the progressive and often multiple transmitter disturbances arising from neurodegeneration.

Based on our improving knowledge of the pathogenesis of nigral degeneration, preventative and protective strategies aimed at halting or slowing both neuronal loss and progress of the illness have become realistic approaches to the experimental therapeutics of PD. These strategies are focused on ameliorating the *process* underlying nigral degeneration rather than rectifying the *consequences* of nigral loss. This discussion addresses the general strategies of preventative and protective therapies for PD, the rationale for antioxidative interventions and related methodological issues, and the therapeutic prospects for neuronal grafting.

B. General Strategies for Prevention and Protection

Strategies for prevention or protection against nigral degeneration may be directed respectively at etiology or pathogenesis. The etiology of PD remains obscure; although, a variety of factors have been implicated as causes of nigral injury, including neurotoxins, infections, immunologic disturbances, genetic predisposition, aging, and the primordial population of nigral neurons (WOOTEN 1984; CALNE et al. 1986).

In certain instances, such as 1-methyl-4-phenyl-1,2,3,6-tetrahydropyridine (MPTP)-induced parkinsonism, the etiology of illness has been well established (LANGSTON et al. 1983, 1984b, BALLARD et al. 1985). Demonstration that MPTP can produce features of parkinsonism virtually indistinguishable from idiopathic PD (FORNO et al. 1986) suggests that multiple etiologies may act alone or in combination to produce what is now identified as a common illness. As additional etiologies for PD are established, rational public health measures can be undertaken to eliminate or reduce exposure to the causative factors. In MPTP-induced PD, avoiding further neurotoxic exposure is the most direct and sensible approach to prevention.

Expanded investigations are expected to help identify other causes of PD and in turn provide the basis for other *preventative* therapies. Recent epidemiologie studies suggest that exposure to certain pesticides (BARBEAU et al. 1987; BARBEAU et al. 1985b), metallic alloys (AQUILONIUS and HARTVIG 1986), and water distribution patterns (TANNER 1986) may play a role in the development of PD. Accepting the premise that multiple etiologies result in idiopathic PD, identification of manifold causes is necessary for developing focused preventative strategies.

The application of *protective* strategies to PD is not necessarily dependent on determining etiology. Substantive therapeutic interventions may be directed at the pathogenesis or process underlying disease notwithstanding etiology. De-coppering therapy for Wilson's disease is an eminent example of this approach. Although a mutant gene is the cause of this recessively inherited disorder, understanding the process whereby copper accumulation results in regional cellular pathology has led to remarkable therapeutic benefits. A variety of de-coppering agents may improve or reverse disability in patients with Wilson's disease and protect presymptomatic homozygotes from illness (SCHEINBERG and STERNLIEB 1984). A comparable understanding of the pathogenesis of PD should provide the basis for protective approaches aimed at forestalling nigral degeneration.

In the past 5 years, pharmacotherapeutic and neural grafting strategies have emerged as potential protective interventions for PD. These approaches are based generally on the notion that idiopathic PD shares a common pathogenesis, and that timely and rational interventions directed at pathogenesis will favorably alter progressive nigral depletion and advancing illness.

C. Antioxidative Pharmacotherapies

Recent animal and human studies provide several lines of evidence implicating endogenous and exogenous oxidative mechanisms in the progressive loss of dopamine-generating neurons (Fig. 1). The data supporting the role of intrinsic or *endogenous* oxidatively mediated mechanisms of nigral degeneration derive from several observations:

1. The oxidative deamination of monoamines, and particularly of nigrostriatal dopamine, results in the formation of hydrogen peroxide and other potentially toxic by-products such as hydroxyl radicals and superoxide (COHEN

1983, 1985). These in turn may produce nigral damage through generation of free radicals and related mechanisms (HALLIWELL and GUTTERIDGE 1985).

2. Nigral vulnerability to oxidative deamination may be heightened by age-related changes. As the population of nigral neurons declines in the setting of normal aging (McGEER et al. 1977), efforts of the remaining neurons to compensate with increasing dopamine output may further enhance generation of hydroxyl radicals (COHEN 1985; SLIVKA and COHEN 1985). In the aging human brain, reduced nigral neurons and increased dopamine turnover are associated with increased content of monoamine oxidase (MAO) (ROBINSON 1975; COTE and KREMZNER 1983). The enhanced oxidation of dopamine by MAO may also yield highly potent neurotoxins such as 6-hydroxydopamine and quinone metabolites (JOHNSON 1976; GRAHAM 1984).

3. Nigral vulnerability to intrinsic oxidative mechanisms may be further heightened by the specific reductions of superoxide dismutase, glutathione, peroxidase, and catalase, as demonstrated in the postmortem PD brain (AMBANI et al. 1975; ROBINSON 1975; KISH et al. 1985; PERRY and YONG 1986). These enzyme systems normally subserve scavenger functions in attenuating the potential toxicity of free radical generation.

4. Preliminary studies in the PD postmortem brain have found disproportionate increases of malondialdehyde, lending more direct support to the notion of increased lipid peroxidation in this disorder (DEXTER et al. 1986). Lipofuscin, pigmented product of lipid peroxidation, increases in neuronal content with aging and in a variety of neurodegenerative disorders, including PD. Formation of lipofuscin is thought to result from interaction of the products of polyunsaturated fatty acid peroxidation, promoted by increased generation of free radicals. Neuromelanin, which accounts for the dark pigmentation of the substantia nigra, shares structural and biochemical similarities with lipofuscin (VAN WOERT and AMBANI 1974). Investigators have long postulated that accumulation of these pigments may promote cellular dysfunction and degeneration (MANN and YATES 1947).

Extrinsic or exogenous oxidatively mediated mechanisms may also play a role in the pathogenesis of PD. This contention derives largely from the accumulating knowledge that MPTP causes parkinsonism in humans, nonhuman primates, and other vulnerable species. In humans, MPTP-induced PD has been documented in intravenous drug abusers and in individuals exposed to high ambient concentrations of this neurotoxin (LANGSTON et al. 1983; LANGSTON and BALLARD 1983; WRIGHT et al. 1983; BALLARD et al. 1985; BARBEAU et al. 1985a).

Although MPTP-induced parkinsonism is found largely in young persons, clinical features are virtually identical to idiopathic PD. Dopaminergic therapies ameliorate symptoms and produce profiles of adverse effects and motor fluctuations similar to those observed in idiopathic PD. Patients with MPTP-induced parkinsonism appear to show signs of clinical progression in spite of preventing further MPTP exposure (LANGSTON et al. 1984b, BARBEAU et al. 1985a). Some MPTP-exposed individuals who were initially asymptomatic have gradually developed early signs and symptoms of PD (TETRUD and LANG-

STON 1986). These observations indicate that parkinsonian symptoms may surface and intensify without continued MPTP exposure, and imply that self-perpetuation of PD could well be the consequence of "autotoxicity" induced by increased demands of the remaining nigral neurons. Furthermore, MPTP-exposed individuals without clinical parkinsonism show decreased uptake of labeled levodopa on positron emission tomography (PET), possibly representing a presymptomatic marker of reduced dopaminergic function (CALNE et al. 1985).

The putative role of *exogenous* oxidatively mediated mechanisms is also strengthened by accumulating experimental data:

1. The nigrotoxic action of MPTP appears to be mediated by its oxidative biotransformation to MPP^+ (CHIBA et al. 1984), a compound which is preferentially accumulated within the substantia nigra (SNYDER and D'AMATO 1986) and which may in turn exert toxicity through disruption of mitochondrial functions (VYAS et al. 1986). Oxidative biotransformation of MPTP is blocked by MAO-B inhibitors, preventing MPTP-induced neurotoxicity and the development of parkinsonism in nonhuman primates and other experimental animals (LANGSTON et al. 1984a; HEIKKILA et al. 1984; COHEN et al. 1984).

2. Neuromelanin may also provide a pathogenetic link to MPTP neurotoxicity since toxic by-products that react with glutathione are generated during melanogenesis (BANNON et al. 1984) and MPTP binds with high affinity to nigral neuromelanin (D'AMATO et al. 1986).

3. Aged experimental animals are more susceptible to the neurotoxic effects of MPTP than younger animals (FORNO et al. 1986; GUPTA et al. 1986). This observation may relate to endogenous vulnerability, including age-related increases in monoamine oxidase.

4. Recent studies suggest that pesticides (BARBEAU et al. 1987), metallic alloys (AQUILONIUS and HARTVIG 1986), and drinking water patterns (TANNER

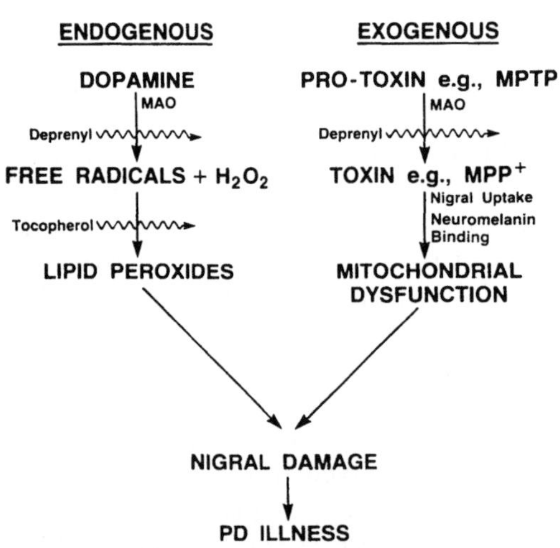

Fig. 1. Endogenous (intrinsic) and exogenous (extrinsic) mechanism proposed to account for nigral injury and the development of PD

1986) may account for the higher prevalence of PD in certain geographic regions. These epidemiologic links may be mediated by endogenous and exogenous oxidative mechanisms.

Taken together, these clinical and scientific observations provide the basis for antioxidative protective therapies in PD aimed at halting or retarding nigral degeneration and clinical decline. In particular, inhibition of oxidative deamination of dopamine and enhancement of the free radical scavenger systems represent complementary pharmacologic strategies that might be applied and examined in clinical trials of PD patients.

D. Clinical Trial of Deprenyl and Tocopherol Antioxidative Therapy of Parkinsonism (DATATOP)

A controlled clinical trial in humans represents the most direct and definitive approach for testing an oxidatively mediated hypothesis for PD. The aim of such a trial would depart from the traditional therapeutics of PD (directed at treating symptoms) and focus rather on slowing the progressive degeneration of nigral neurons and resulting clinical decline. A similar protective strategy has been proposed for Huntington's disease (SHOULSON 1983; SCHWARCZ and SHOULSON 1987).

The Parkinson Study Group (PSG), a consortium of clinical investigators and consultants from academic institutions in the United States and Canada (Table 1), was organized in 1985 to consider the rationale, drug selection, and design of an antioxidative clinical trial for PD. The PSG has proposed the clinical trial "Deprenyl and Tocopherol Antioxidative Therapy of PD (DATATOP)" based on the hypothesis that chronic deprenyl and/or α-tocopherol therapies will slow the progressive nigral loss and clinical disability of PD. The primary aim of the study is to determine whether or not chronic deprenyl and/or tocopherol treatment of early, otherwise untreated PD patients will prolong the time until levodopa is required to treat supervening disability. To this end, a proposal has been formulated for a double-blind, multi-institutional, prospective evaluation of PD patients who are in the earliest stages of illness and who are not receiving any symptomatic anti-PD therapies. Subjects are to be assigned by randomization in a 2×2 factorial design to one of four treatments: (a) deprenyl; (b) tocopherol; (c) deprenyl plus tocopherol; and (d) placebo. The following summarizes the rationale and methodology of this trial.

I. Pilot Studies and Candidate Drugs

The PSG considered therapeutic interventions for a protective clinical trial on the basis of several pilot studies, clinical experience with candidate drugs, and a variety of methodological issues. Birkmayer and colleagues have reported uncontrolled studies suggesting that the MAO-B inhibitor, deprenyl, slows the clinical decline of levodopa-treated PD patients, as well as attenuating levo-

Table 1. Parkinson Study Group (PSG)

Principal Investigator	Coprincipal Investigator	Chief Biostatisticians
*Ira Shoulson, M. D., Rochester, NY	*Stanley Fahn, M. D, New York, NY	*Charles Odoroff, Ph. D., *David Oakes, Ph. D., Rochester, NY

Clinical Investigators

James Bennett, M. D., Ph. D., Charlottesville, VA
Susan Bressman, M. D., New York, NY
Mitchell Brin, M. D., New York, NY
Roger Duvoisin, M. D., New Brunswick, NJ
J. Stephen Fink, M. D., Boston, MA
Joseph Friedman, M. D., Providence, RI
Stephen Gancher, M. D., Portland, OR
Serge Gauthier, M. D., Montreal, Quebec
Christopher Goetz, M. D., Chicago, IL
Lawrence Golbe, M. D., New Brunswick, NJ
J. David Grimes, M. D., Ottawa, Ontario
John H. Growdon, M. D., Boston, MA
Mohamed N. Hassan, M. D., Ottawa, Ontario
Margaret Hoehn, M. D., Denver, CO
Howard Hurtig, M. D., Philadelphia, PA
Joseph Jankovic, M. D., Houston, TX
Harold Klawans, M. D., Chicago, IL
William Koller, M. D., Ph. D., Kansas City, KA
Roger Kurlan, M. D., Rochester, NY
*Anthony Lang, M. D., Toronto, Ontario
*J. William Langston, M. D., San Jose, CA
*Peter LeWitt, M. D., Detroit, MI
Hamilton Moses III Jr, M. D., Baltimore, MD
John Nutt, M. D., Portland, OR
*Warren Olanow, M. D., Tampa, FL
George Paulson, M. D., Columbus, OH
*John Penney, M. D., Ann Arbor, MI
Joel Perlmutter, M. D., St. Louis, MO
Ronald Pfeiffer, M. D., Omaha, NE
Ali H. Rajput, M. D., Saskatoon, Saskatchewan
Robert Rodnitzky, M. D., Iowa City, IA
Juan Sanchez-Ramos, M. D., Ph. D., Miami, FL
Kathleen Shannon, M. D., Chicago, IL
Cliff Shults, M. D., San Diego, CA
Matthew Stern, M. D., Philadelphia, PA
*Caroline Tanner, M. D., Chicago, IL
James Tetrud, M. D., San Jose, CA
Leon Thal, M. D., San Diego, CA
William Weiner, M. D., Miami, FL
George F. Wooten, M. D., Charlottesville, VA

Consultants

Edward Bell, D.Pharm., Rochester, NY
Gerald Cohen, Ph. D., New York, NY
Harvey Cohen, Ph. D., Rochester, NY
Joseph Fleiss, Ph. D., New York, NY
W. Jackson Hall, Ph. D., Rochester, NY
Robert Herndon, M. D., Rochester, NY
Ian Irwin, Ph. D., San Jose, CA
Laurence Jacobs, M. D., Rochester, NY
C. David Marsden, M. D., London, England
Richard Mayeux, M. D., New York, NY
Arthur Moss, M. D., Rochester, NY
Robert Roth, Ph. D., New Haven, CT
Govind Vatassery, Ph. D., Minneapolis, MN

* Steering Committee

dopa-related motor fluctations (BIRKMAYER et al. 1983, 1985; RIEDERER et al. 1984). While encouraging, the effects of concurrent anti-PD therapies confound the interpretation of these uncontrolled studies. Nonetheless, the pilot studies with deprenyl have involved hundreds of PD patients and have proved reassuring regarding the safety of chronic, low dose deprenyl therapy.

Another pilot study was undertaken by Fahn and colleagves, examining the potentially protective effects of the free radical antioxidant, vitamin E (α-tocopherol), in early PD patients not yet requiring levodopa therapy (FAHN and BRESSMAN 1984). The preliminary unpublished observations from this uncontrolled study suggest that chronic, high dose vitamin E therapy may prolong the time until levodopa therapy is required to treat emerging disability. While the conclusions of this investigation are tempered by an uncontrolled design, PD patients appeared to tolerate high dose vitamin E (up to 3200 IU/day) safely for periods ranging from 31 to 110 months.

In addition to deprenyl and tocopherol, other antioxidative candidates have been considered, including pargyline, vitamin C, *N*-acetylcysteine, β-carotene, selenium, mazindol and chloroquine. However, these agents pose a variety of methodological concerns, and no pilot studies have as yet addressed their protective efficacy and long-term safety in PD patients.

II. Factorial Design

The opportunity to examine combined administration of deprenyl and tocopherol is appealing on theoretical grounds because these agents exert antioxidative effects through separate but complementary mechanisms of action. Deprenyl may confer protection if MAO-B mediates exogenous (e.g., environmental neurotoxins) or endogenous nigrotoxicity. Tocopherol may protect if free radical generation plays a role in endogenous or possibly even exogenous mechanisms of nigral damage (Fig. 1).

The ability to study concurrently two independent interventions, such as deprenyl and tocopherol, is made possible by the application of a 2×2 factorial experiment (STAMPFER et al. 1985). This design provides the means of assessing two potential antioxidative therapies with approximately the same effort required for examining a single intervention. In the context of a 2×2 factorial design for deprenyl and tocopherol, subjects are allocated by randomization to one of four treatment groups: (a) deprenyl alone; (b) tocopherol alone; (c) deprenyl plus tocopherol; and (d) placebo. As long as there are no strong adverse interactions between the two therapies, two studies can be accomplished at only a slight additional cost of sample size, and each intervention can be analyzed independently. There are no theoretical reasons or experimental evidence from human or animal studies to indicate adverse interaction between these two therapies. If a synergistic protective effect of deprenyl and tocopherol were found, this observation would be of substantive clinical and scientific importance.

III. Major Response Variable

The major response variable selected for DATATOP trial is the time interval from treatment randomization until levodopa is judged necessary to treat supervening parkinsonian disability. This time-dependent end point offers several advantages in comparison with traditional clinical measures. It is easily quantifiable to a precision of days. The statistical methodology is well developed using survivorship analyses to handle variable times of follow-up and censoring (Cox 1972; Cox and OAKES 1983). In practice, the decision of when to begin levodopa therapy depends, not on theoretical issues, but on the functional performance or capacity of the individual patient (QUINN et al. 1986). Finally, the systematic determination of when to begin levodopa therapy is exceedingly relevant to the day-to-day evaluation and care of PD patients. Demonstration that deprenyl and/or tocopherol interventions significantly forestall disability and the initiation of levodopa therapy would have a substantive impact on the treatment of PD.

Formulation of the major outcome variable was critical to the design of DATATOP. Several clinical response variables were considered, but all had drawbacks. Although tremor, rigidity, bradykinesia, and postural instability are traditional clinical measures of PD, it is not known which of these subjective measures best reflects the extent and pace of nigral degeneration. A variety of staging scales (HOEHN and YAHR 1967; SCHWAB and ENGLAND 1969) were considered as major response variables. However, the traditional clinical ratings and staging scales are nonparametric measures that are subject to floor-ceiling effects and considerable interobserver and test-retest variability. Furthermore, there are no systematically derived data from which to estimate the variance of these clinical assessment parameters for the purpose of sample size and power computations.

Ideally, one would like to examine the change of a relevant neurobiologic correlate of nigral degeneration in assessing protective effects. Although several techniques appear promising, such as changes in CSF monoamine metabolites or fluorodopa metabolism by PET imaging, the validity and applicability of these techniques have not as yet been established.

The concurrent administration of symptomatic therapies seriously confounds the assessment of protective interventions. In order to maximize the likelihood of detecting protective effects, it is necessary to evaluate early PD patients who are not receiving any symptomatic antiparkinsonian medications. Levodopa is an accepted standard antiparkinsonian therapy, and its use is clearly indicated when disabling features of PD emerge. In mildly affected PD patients, the time interval until levodopa therapy is required to treat supervening disability therefore represents a reasonable clinical outcome variable for a protective trial.

The major end point variable selected for this clinical trial is admittedly operator dependent and not strictly disease dependent. Therefore, the potential for bias and "unblinding" exists. The concern regarding bias is obviated by double-blind randomization in the setting of the 2 × 2 factorial design. As such, the clinical investigator is responsible for making a blind judgment as to if and when levodopa therapy is required to treat emerging disability. The potential for unblinding is minimized if investigators evaluate sufficient numbers of subjects. Accordingly, the randomization for DATATOP will be stratified according to the investigator who will be determining the major response variable. In practice, each participating investigator will be enrolling and evaluating a minimum of 20 subjects. Approximately 25 % of these subjects will have been randomized to one of the four experimental therapies. Since neither deprenyl nor tocopherol are known to exert symptomatic effects in non-levodopa-treated subjects, there is minimal likelihood of unblinding among the 20 subjects. The major assumption is that the investigator will determine the end point uniformly for his or her group of enrolled subjects (minimal number 20). The major response variable will therefore be analyzed among the participating investigators, each of whom will evaluate approximately equal numbers of subjects randomized to the four experimental treatments.

Since October 1985, PSG investigators have enrolled more than 300 early PD patients, not treated with levodopa, into an observational study in order to assess prospectively those clinical conditions which prompt clinicians to institute levodopa therapy (Shoulson and the Parkinson Study Group 1987). Our initial survey of investigators suggests that threatened or actual erosion of functional capacities related to employability, activities of daily living (Schwab and England 1969), domestic responsibilities, and gait are the major factors influencing the clinical judgment of when to begin levodopa therapy.

IV. Sample Size Estimates

The general design of DATATOP involves randomization of minimally disabled PD patients, otherwise untreated, into one of four treatment groups, and periodic follow-up until the primary end point is reached or until a maximum of 2 years of observation is completed. Calculation of sample size for the proposed DATATOP trial is based on several assumptions and estimates. Studies of early PD patients treated with only anticholinergic drugs and/or amantadine have found that approximately 76 % of patients required levodopa therapy to treat disability within 2 years of follow-up (Goetz et al. 1986). In patients not receiving any anti-PD medication whatsoever, it is estimated that the cumulative percentages of patients requiring levodopa therapy by 2 years will be greater, approximately 85 %. In survival terms, it is therefore expected that approximately 15 % of patients will "survive" and not require levodopa therapy by 2 years of follow-up. This estimate is currently being examined in a prospective study (Shoulson and the Parkinson Study Group 1987).

For the purpose of sample size and power computations, it is estimated that the proportion of patients requiring levodopa therapy within 2 years will be reduced by at least 10 % in the deprenyl and/or tocopherol treatment groups compared with the placebo group. Therefore, it is projected that approximately 85 % of placebo-treated patients will require levodopa therapy by

2 years (15% surviving) while approximately 75% of deprenyl-treated and 75% of tocopherol-treated patients will require levodopa therapy by 2 years (25% surviving). The sample size calculation is based further on a significance level, a, of 0.05 with the hypothesis being tested as two-sided, whether the experimental interventions are *better* or *worse* than placebo.

The power computations for sample size also assume that there is no adverse interaction between deprenyl and tocopherol; power is therefore computed separately for each treatment (deprenyl or tocopherol). Based on the foregoing assumptions and projections, the total sample size for DATATOP is estimated to be approximately 800 subjects (FLEISS 1981), with randomization of approximately 200 subjects to each of the four treatment groups. With a total sample size of 800 subjects, the likelihood (power) of detecting a significant effect of experimental drugs with respect to the major response variable is approximately 0.95.

V. Symptomatic or Protective Effects?

The proposed DATATOP study carries with it potential problems in distinguishing between truly protective effects and possible symptomatic effects of experimental interventions. In those few studies examining early, untreated PD patients, deprenyl has not produced short-term symptomatic effects (EISLER et al. 1975; CSANDA and TARCZY 1983). T.W. LANGSTON and T.W. TETRUD (unpublished work) have not observed symptomatic effects of deprenyl in their preliminary, double-blind study of 30 non-levodopa-treated subjects. S. FAHN S.B. BRESSMAN and M. BRIN (unpublished work) have not found symptomatic benefits from high dose tocopherol therapy in their early, non-levodopa-treated patients. Therefore, neither deprenyl nor tocopherol would likely produce symptomatic improvement that might confound the interpretation of protective efficacy.

Although no discernible symptomatic effects of deprenyl or tocopherol therapies have been demonstrated in those PD patients not treated concurrently with levodopa, additional procedures have been incorporated into the DATATOP proposal to help distinguish between symptomatic and protective effects. The protocol includes a 1-month wash-in period between baseline evaluation (at the time of randomization) and the first follow-up evaluation. Initial, short-term symptomatic benefits of deprenyl and tocopherol might surface between these evaluations and be detected by blind clinical and videotape assessments. A 1-month washout interval is also planned between the end point evaluation (at the time that levodopa therapy is required or at 2 years of follow-up) and the final evaluation 1 month thereafter. During this washout interval, subjects will be withdrawn from their experimental medications to help determine long-term symptomatic changes. Lumbar punctures are planned for the initial (baseline) and final evaluations (after washout) to determine CSF monoamine metabolites, particularly homovanillic acid (HVA). This will allow comparison of all untreated subjects at baseline evaluation with their untreated status at the final evaluation after washout of experimental drugs.

A secondary objective of the DATATOP trial will therefore be to examine the relationship between CSF monoamine changes and the primary response variable with respect to experimental treatments. It is hypothesized that genuinely protective interventions will attenuate the expected decline in cerebrospinal fluid HVA as well as prolonging the time until levodopa therapy is judged necessary to treat disability.

A variety of clinical measures are planned as secondary response variables, including assessment of the characteristic motor, cognitive and affective features of PD. Based on a preliminary analysis of our cohort of non-levodopa-treated PD patients, certain motor attributes, particularly posture, hand movements, ability to arise from a chair, and bradykinesia, appear to be valid indices of underlying parkinsonism and useful target signs for clinical trials designed to slow neuronal degeneration and disability (SHOULSON and the PARKINSON STUDY GROUP 1987). These or other clinical measures may ultimately prove to be sensitive correlates of the major end point variable.

VI. Recruitment

A total sample size of approximately 800 subjects is unprecedented for clinical trials in PD. To recruit the necessary number of subjects and carry out a trial of this magnitude, a multi-investigator and multi-institutional investigation is required. Recruitment of potentially eligible subjects for the proposed clinical trial is by no means trivial. The prevalence of PD is approximately 1 in 300 (CALNE 1970; MARTTILA 1983; RAJPUT et al. 1984), resulting in estimates of nearly 1 million PD patients in the United States and Canada. From this total pool, approximately 18 % or 180 000 patients are estimated to be mildly affected (HOEHN and YAHR 1967) and potentially eligible for enrollment in DATATOP. Therefore, approximately 1 in every 226 PD patients in the United States and Canada will need to be enrolled to attain the recruitment goal of 800 subjects. The cooperation of primary care physicians, referring neurologists, and the PD lay organizations will be critical in achieving the necessary sample size. The major appeal for participation in the DATATOP trial is the potential for developing therapy that may slow the progression of PD and prolong the time until levodopa is required to treat disability.

VII. Potential Pitfalls

While the potential rewards of the DATATOP trial are substantive and far reaching, there are several theoretical and practical drawbacks to such an undertaking. It may be argued that the scientific rationale for a clinical trial is not sufficiently compelling. While further preclinical testing is clearly necessary to test the oxidatively mediated hypothesis of nigral degeneration, the clinical trial approach offers a definitive and feasible strategy for testing a sensible supposition.

The cost of undertaking a clinical trial is another concern, but our analyses indicate potential cost savings. Based on unpublished pilot data generated by the PSG, it is estimated that approximately 30 % of working PD patients or nearly 61 000 persons will become unemployed annually. From data that the average United States wage earner pays approximately $ 67 per week in income tax and the average disabled person receives about $ 100 per week in social security payments (World Almanac 1986), the overall cost savings to society would exceed $ 10 000 000 annually if deprenyl and/or tocopherol prove effective in maintaining work capacity for just one additional week. This cost analysis does no take into account the potential savings of hospitalization, institutionalization, or additional medical care related to PD disability.

A rigorous trial will also avoid the enormous and unproductive effort invested in uncontrolled studies that might erroneously claim the protective efficacy of deprenyl or tocopherol. Even if the DATATOP trial fails to demonstrate therapeutic benefits, then at the very least, the natural history of PD and its CSF neurochemical correlates will have been characterized systematically. These data will in turn provide the basis for designing future trials of other promising protective interventions.

E. Animal Models

Hypothesis of oxidatively mediated nigral degeneration, especially the role of the MAO-dependent nigrotoxins, may be tested more indirectly in animals. In the MPTP animal model, several pharmacologic interventions have proved effective in preventing or limiting nigrotoxicity. MAO-B inhibitors prevent the glial oxidative conversion of MPTP to its neurotoxic derivative MPP^+ and thereby prevent nigral demage and features of PD (Langston et al. 1984a; Heikkila et al. 1984; Cohen et al. 1984). A variety of uptake blockers inhibit the entry of MPP^+ into catecholaminergic neurons and thereby attenuate neurotoxicity (Snyder and D'Amato 1986). Finally, agents such as chloroquine appear to displace the binding of MPP^+ to neuromelanin and accordingly exert some protective influence in MPTP-treated monkeys (Kitt et al. 1986).

While these strategies are encouraging, their relevance to a clinical trial to protective therapy in PD is limited primarily by the assumption that idiopathic PD is caused by an MPTP-like neurotoxin. Once such a toxin gains access to nigral neurons, there is little theoretical basis for assuming that agents which inhibit MAO-B, catecholamine uptake, or neuromelanin binding will forestall neuronal degeneration. Therefore, there is reason to test the effects of these and other treatments in animals *after* PD features have been induced by MPTP in order to assess their protective influence on the subsequent course of nigral degeneration.

No well-accepted animal models exist to study endogenous mechanisms of oxidatively mediated nigral degeneration. However, it might prove useful to test the effects of antioxidative agents in aged animals to determine if the age-related decline in nigral neurons (McGeer et al. 1977) might be attenuated by these interventions. One useful approach involves the study of agents that deplete free radical scavenger systems. One investigation has recently demonstrated that morphological changes resembling aging nigral neurons can be induced in mice treated with buthionine sulfoxine, a depletor of glutathione (McNeill et al. 1986).

F. Neural Grafting

In the past decade, the therapeutic potential for neural grafting in neurologic disease has evolved from science fiction to science. This progress reflects the recognition that neural transplants not only survive, but thrive and function under certain conditions. Neural connections and appropriate functional responses have been demonstrated following transplantation of nigral neurons into the striatum of rodents with lesioned nigrostriatal pathways (BJORKLUND and STENEVI 1979; PERLOW et al. 1979). Adrenal medullary tissue has also served as a source of striatal and intraventricular autografts in experimental models of PD, resulting in reversal of some behavioral disturbances (MORIHISA et al. 1984; FREED et al. 1985). Most recently, fetal substantia nigral cells have been transplanted successfully into the brains of MPTP-treated monkeys with resulting functional improvement and apparent morphological integration (REDMOND et al. 1986).

These animal studies have prompted serious consideration and preliminary attempts in humans to use grafted neural tissue as a therapy for PD. The initial reports of adrenal medullary autografts transplanted intracranially in Parkinson's disease have not been encouraging (BACKLUND et al. 1985a, b). However, more recent extensions of this approach purport to show benefits in Parkinson's disease patients (MADRAZO et al. 1986; DRUCKER-COLIN et al. 1987; SHOU-SHU et al. 1987).

If neural grafting does indeed produce clinical improvement in PD patients, the mechanism for improvement requires clarification. Does grafting of catecholaminergic neurons merely provide an elegant method for neurotransmitter access to striatal receptor sites? Do transplanted cells produce important trophic factors as well as dopamine that are necessary for the integrity of striatal functions? Does morphological integration actually occur? Finally, is it possible that neural implants may have reparative actions in addition to symptomatic benefits? The answers to these questions will help clarify the role of neural grafting within the context of our current conceptions of protective therapy for PD.

G. Summary

With the increasing recognition that dopamine replacement therapies confer limited symptomatic benefits for PD patients and a clearer understanding of pathogenesis, the experimental therapy of PD has taken on new directions aimed at slowing nigral degeneration and clinical decline. An oxidatively mediated hypothesis of nigral injury has been broadly formulated to account for endogenous and exogenous pathogeneses of idiopathic PD. These advances have prompted the PSG to formulate a proposal for the clinical trial DATATOP. The knowledge expected to emerge from this and similar protective strategies, as well as from animal studies examining oxidative mechanisms of nigral injury, offers promise of substantive therapeutic gains in the treatment of PD. Neural grafting also provides an innovative method for developing the protective therapy of PD.

Acknowledgments. Preparation of this manuscript was supported by grants from NIH (NS-24778) and the PEW Foundation Neuroscience Award to the University of Rochester. The planning efforts of the Parkinson Study Group (PSG) have been supported by awards from the Parkinson's Disease Foundation at Columbia University (New York, New York), the National Parkinson Foundation (Miami, Florida), the Parkinson Foundation of Canada (Toronto, Ontario), the American Parkinson's Disease Association (New York, New York), the United Parkinson Foundation (Chicago, Ilinois), and the University of Rochester. I am grateful for the advice and support of investigators and consultants in the PSG and thank Mrs. Ruth Nobel for her assistance in preparing this manuscript.

References

Ambani LM, Van Woert WH, Murphy S (1975) Brain peroxidase and catalase in Parkinson's disease. Arch Neurol 32:114–118

Aquilonius SM, Hartvig P (1986) Utilization of antiparkinson drugs in Sweden 1977–1984 (abstr). Ups J Med Sci [Suppl] 43:93

Backlund EO, Granberg PO, Hamberger B, Knutsson E, Martensson A, Sedvall G, Seiger A, Olson L (1985a) Transplantation of adrenal medullary tissue to striatum in parkinsonism: first clinical trials. J Neurosurg 62:169–173

Backlund EO, Granberg PO, Hamberger B, Sedvall G, Seiger A, Olson A (1985b) Transplantation of adrenal medullary tissue to striatum in parkinsonism. In: Bjorklund A, Stenevi U (eds) Neural grafting in the mammalian CNS. Elsevier, Amsterdam, pp 551–556

Ballard PA, Tetrud JW, Langston JW (1985) Permanent human parkinsonism due to 1-methyl-4-phenyl-1,2,3,6-tetrahydropyridine (MPTP): seven cases. Neurology (NY) 35:949–956

Bannon MJ, Goedert M, Williams B (1984) The possible relation of glutathione, melanin, and 1-methyl-4-phenyl-1,2,5,6-tetrahydropyridine (MPTP) to Parkinson's disease. Biochem Pharmacol 33:2697–2698

Barbeau A, Roy M, Langston JW (1985a) Neurological consequence of industrial exposure to 1-methyl-4-phenyl-1,2,5,6-tetrahydropyridine. Lancet I:747

Barbeau A, Roy M, Paris S, Cloutier T, Plasse L, Poirier J (1958b) Ecogentics of Parkinson's disease: 4-hydroxylation of debrisoquine. Lancet II:1213–1216

Barbeau A, Roy M, Cloutier T, Plasse L, Paris S (1987) Environmental and genetic factors in the etiology of Parkinson's disease. Adv Neurol 45:299–306

Birkmayer W, Knoll J, Riederer P, Youdim MBH (1983) Deprenyl leads to prolongation of L-dopa efficacy in Parkinson's disease. Mod Probl Pharmacopsychiatry 19:170–176

Birkmayer W, Knoll J, Riederer P, Youdim MBH, Hars V, Marton J (1985) Increased life expectancy resulting from addition of L-deprenyl to Madopar treatment in Parkinson's disease: a longterm study. J Neural Trans 64:113–127

Bjorklund A, Stenevi U (1979) Reconstruction of the nigrostriatal dopamine pathway by intracerebral nigral transplants. Brain Res 177:555–560

Calne DB (1970) Parkinsonism: physiology, pharmacology and treatment. Williams and Wilkins, Baltimore

Calne DB, Langston JW, Martin WRW et al. (1985) Positron emission tomography after MPTP: observations relating to the cause of Parkinson's disease. Nature 317:246–248

Calne DB, Eisen A, McGeer E, Spencer P (1986) Alzheimer's disease, Parkinson's disease, and motorneuron disease: abiotropic interactions between aging and environment? Lancet II:1067-1070

Chiba K, Trevor A, Castagnoli N (1984) Metabolism of the neurotoxic tertiary amine, MPTP, by brain monoamine oxidase. Biochem Biophys Res Commun 120:574-578

Cohen G (1983) The pathobiology of Parkinson's disease: biochemical aspects of dopamine neuron senescence. J Neural Transm [Suppl] 19:89-103

Cohen G (1985) Oxidative stress in the nervous system. In: Sies H (ed) Oxidative stress. Academic, London, pp 383-402

Cohen G, Pasik P, Cohen B, Leist A, Mytilineou C, Yahr MD (1984) Pargyline and deprenyl prevent the neurotoxicity of MPTP in monkeys. Eur J Pharmacol 106:209-210

Cote LJ, Kremzner LT (1983) Biochemical changes in normal aging in human brain. Adv Neurol 38:19-30

Cox DR (1972) Regression models and life tables. J R Statist Soc B 34:187-220

Cox DR, Oakes D (1983) The analysis of survial data. Chapman and Hall, New York

Csanda E, Tarczy M (1983) Clinical evaluation of deprenyl (selegiline) in the treatment of Parkinson's disease. Acta Neurol Scand [Suppl] 95:117-122

D'Amato RJ, Lipman ZP, Snyder SH (1986) Selectivity of the parkinsonian neurotoxin MPTP: toxic metabolite MPP^+ binds to neuromelanin. Science 231:987-989

Dexter D, Carter C, Agid F, Agid Y, Lees AJ, Jenner P, Marsden CD (1986) Lipid peroxidation as cause of nigral death in Parkinson's disease. Lancet II:639-640

Drucker-Colin R, Madrazo I, Diaz V, Torres C (in press) Open microsurgical autograft of adrenal medulla to caudate nucleus of patients with Parkinson's disease: report of 8 cases. In: Gash DM, Sladek JR (eds) Transplantation in the mammalian CNS: preclinical and clinical studies. Prog Brain Res 73:

Eisler T, Calne DB, Ebert MH, Kopin IJ, Ziegler MG, Levine R, Murphy DL (1975) Biochemical measurements during (−)-deprenyl treatment of parkinsonism. In: Sincer T, Van Korff R, Murphy DL (eds) Monoamine oxidase: structure, function and altered functions. Academic, New York, pp 497-505

Fahn S, Bressman SB (1984) Should levodopa therapy for parkinsonism be started early or late? Evidence against early treatment. Can J Neurol Sci 11:200-206

Fleiss JL (1981) Statistical methods for rates and proportions, 2nd edn. Wiley, New York (table 6A.3)

Forno LS, Langston JW, DeLanney Le, Irwin I, Ricaurte GA (1986) Locus ceruleus lesions and eosinophilic inclusions in MPTP-treated monkeys. Ann Neurol 20:449-455

Freed WJ, Olson L, Ko GN, Morihisa JM, Niehoff D, Stromberg I, Kuhar M Hoffer BJ, Wyatt RJ (1985) Intraventricular substantia nigra and adrenal medulla grafts: mechanisms of action and [^{3}H]spirogeridol autoradiography. In: Bjorklund A, Stenevi U (eds) Neural grafting in the mammalian CNS. Elsevier, Amsterdam, pp 471-489

Goetz CG, Shannon KM, Carroll S, Klawans HL (1986) Progression of Parkinson's disease in patients not treated with levodopa, (abstr). Ann Neurol 20:148

Graham DG (1984) Catecholamine toxicity: a proposal for the molecular pathogenesis of manganese neurotoxicity and Parkinson's disease. Neurotoxicology (Park Forest Il) 5:83-96

Gupta M, Gupta BK, Thomas R, Bruemmer V, Sladek JR, Felten DL (1986) Aged mice are more sensitive to 1-methyl-4-phenyl-1,2,5,6-tetrahydropyridine treatment than young adults. Neurosci Lett 70:326-331

Halliwell B, Gutteridge JM (1985) Oxygen radicals and the nervous system. Trends Neurosci 8:22-26

Heikkila RE, Manzino L, Cabbat RS, Duvoisin RC (1984) Protection against the dopaminergic neurotoxicity of 1-methyl-4-phenyl-1,2,5,6-tetrahydro-pyridine by monoamine oxidase inhibitors. Nature 311:467-469

Hoehn MM, Yahr MD (1967) Parkinsonism: onset, progression, and mortality. Neurology (NY) 17:427-442

Johnson G (1976) Studies of the mechanisms of 6-hydroxydopamine cytotoxicity. Med Biol 54:406-420

Kish SJ, Morito C, Hornykiewicz O (1985) Glutathione peroxidase activity in Parkinson's disease brain. Neurosci Lett 58:343-346

Kitt CA, D'Amato RJ, Alexander CM, Schwartzman RJ, Compton AL, Joh TH, Snyder SH, Price DL (1986) Effects of chloroquine on the neurotoxicity of MPTP in monkey. Soc Neurosci Abstr 12:209

Langston JW, Ballard PA (1983) Parkinson's disease in a chemist working with 1-methyl-4-phenyl-1,2,5,6-tetrahydropyridine. N Engl J Med 309:310

Langston JW, Ballard P, Tetrud JW, Irwin I (1983) Chronic parkinsonism in humans due to a product of meperidine-analog synthesis. Science 219:979-980

Langston JW, Irwin I, Langston EB (1984a) Pargyline prevents MPTP-induced parkinsonism in primates. Science 225:1480-1482

Langston JW, Langston EB, Irwin I (1984b) MPTP-induced parkinsonism in human and non-human primates—clinical and experimental aspects. Acta Neurol Scand 70(Suppl 100):49-54

Madrazo I, Drucker-Colin R, Dfaz-Simental V, Torres C (1986) Open microsurgical autograft of adrenal medulla to the right caudate nucleus in human intractable Parkinson's disease. Soc Neurosci Abstr 12:563

Mann DMA, Yates PO (1947) Lipoprotein pigments: their relationship to aging in the human nervous system. II. The melanin content of pigmented nerve cells. Brain 97:489-498

Martilla RJ (1983) Diagnosis and epidemiology of Parkinson's disease. Acta Neurol Scand [Suppl] 95:9-17

McGeer PL, McGeer EG, Suzuki JS (1977) Aging and extrapyramidal function. Arch Neurol 34:33-35

McNeill TH, Koek LL, Haycock JW, Gash DM (1986) Glutathione reduction mimics MPTP neurotoxicity and age-correlated changes in dopamine neurons in substantia nigra of the C57BL/6NNia mouse. Soc Neurosci Abstr 12:1470

Morihisa JM, Nakamura RK, Freed WJ, Mishkin M, Wyatt RJ (1984) Adrenal medulla grafts survive and exhibit catecholamine-specific fluorescence in the primate brain. Exp Neurol 84:643-653

Perlow MJ, Freed WJ, Hoffer BJ, Seiger A, Olson L, Wyatt RJ (1979) Brain grafts reduce motor abnormalities produced by destruction of nigrostriatal dopamine system. Science 204:643-647

Perry TL, Yong VW (1986) Idiopathic Parkinson's disease, progressive supranuclear palsy and glutathione metabolism in the substantia nigra of patients. Neurosci Lett 67:269-274

Quinn N, Critchley P, Parkes D, Marsden CD (1986) When should levodopa be started? Lancet II:985-986

Rajput AH, Offord KP, Beard CM, Kurland LT (1984) Epidemiology of parkinsonism: incidence, classification and mortality. Ann Neurol 16:278-282

Redmond DE, Sladek JR, Roth RH, Collier TJ, Elsworth JD, Deutch AY, Haber S (1986) Fetal neuronal grafts in monkeys given methylphenyltetrahydropyridine. Lancet I:1125-1127

Riederer P, Jellinger K, Seemann D (1984) Monoamine oxidase and parkinsonism. In: Tipton KF, Dostert P, Benedetti MS (eds) Monoamine oxidase and disease. Academic, New York, pp 403-415

Robinson DS (1975) Changes in monoamine oxidase and monoamines with human development and aging. Fed Proc 34:103-107

Scheinberg IH, Sternlieb I (1984) Wilson's disease. Saunders, Philadelphia

Schwab RS, England AC (1969) Projection technique for evaluating surgery in Parkinson's disease. In: Gillingham FJ, Donaldson MC (eds) Third symposium on Parkinson's disease. Livingston, Edinburgh, pp 152-157

Schwarcz R, Shoulson I (1987) Excitotoxins and Huntington's disease. In: Coyle JT (ed) Experimental models of dementing disorders: a synaptic neurochemical approach. Liss, New York, pp 39-68

Shoulson I (1983) Huntington's disease: anti-neurotoxic therapeutic strategies. In: Fuxe K, Roberts P, Schwarcz R (eds) Excitotoxins, (Wenner-Gren symposium series). MacMillan, New York, pp 343-353

Shoulson I, The Parkinson Study Group (1987) Correlates of clinical decline in early Parkinson's disease (abstr). Neurology 37(Suppl 1):278

Shou-Shu J, Wa-Cheng Z, Minh-Chen D, Jia-Bang S (in press) The clinical study of adrenal medullary tissue transplantation to striatum in parkinsonism. In: Gash DM, Sladek JR (eds) Transplantation in the mammalian CNS: preclinical and clinical studies. Prog Brain Res 73

Slivka A, Cohen G (1985) Hydroxyl radical attack on dopamine. J Biol Chem 260:15466-15472

Snyder SH, D'Amato RJ (1986) MPTP: a neurotoxin relevant to the pathophysiology of Parkinson's disease. Neurology 36:250-258

Stampfer MJ, Buring JE, Wilett W, Rosner B, Eberlein K, Hennekens CH (1985) The 2×2 factorial design: its application to a randomized trial of aspirin and carotene in U.S. physicians. Stat Med 4:111-116

Tanner CM (1986) Influence of environmental factors on the onset of Parkinson's disease (abstr). Neurology 36(Suppl 1):215

Tetrud J, Langston WJ (1986) Early parkinsonism in humans due to MPTP exposure. Neurology 36(Suppl 1):308

Van Woert MH, Ambani LM (1974) Biochemistry of neuromelanin. Adv Neurol 5:215-222

Vyas I, Heikkila RE, Nicklas WJ (1986) Studies on the neurotoxicity of MPTP. Inhibition of NAD-linked substrate oxidation by its metabolite, MPP^+. J Neurochem 46:1501-1507

Wooten GF (1984) Parkinsonism. In: Pearlman AL, Collins RC (eds) Pathophysiology, 3rd ed. Oxford University Press, New York, pp 365-377

World Almanac (1986) Disability insurance benefits and wage earning. Newspaper Enterprise, New York, pp 68-72, 114

Wright JM, Wall RA, Perry TL, Paty DW (1983) Chronic parkinsonism secondary to intranasal administration of a product of meperidine-analogue synthesis. N Engl J Med 310:325

CHAPTER 12

Anticholinergic Drugs and Amantadine in the Treatment of Parkinson's Disease

A. E. LANG and R. D. G. BLAIR

A. Introduction

Anticholinergics and amantadine both result in mild to moderate improvement in Parkinson's disease. These drugs still have an important role to play in the treatment of the disease despite the greater clinical effects of levodopa and dopamine agonists. They may be used either as monotherapy or as adjunctive treatment. Benefit can be obtained in all stages of the disease, but it is typically most noticeable early on when disability is less marked. Many neurologists utilize these drugs initially in hopes of delaying the need for levodopa and later to permit the use of lower doses of this agent because of increasing concern that early treatment with levodopa may create many of the long-term problems faced in the later stages of the disease. In this chapter we will review anticholinergic and amantadine treatment of Parkinson's disease, emphasizing the proposed mechanisms of action, what is known of the pharmacokinetics, the clinical effectiveness and side effects.

B. Anticholinergics

I. Introduction

Anticholinergic drugs remained the only effective medical therapy for Parkinson's disease for almost one hundred years. Charcot's pupil ORDENSTEIN (1867) first discovered their antiparkinsonian effects fortuitously when he administered tinctures of deadly nightshade *(Atropa belladonna)* and henbane *(Hyoscyamus niger)* to patients in hopes of drying their mouths in order to control the excessive salivation and drooling. Later stramonium, particularly the Bulgarian variety, was felt to be the most potent of the belladonna alkaloids and was often administered in the form of cigarettes. Wine extracts of belladonna alkaloids were found to be effective in the early 1940s (PRICE and MERRITT 1941; FABING and ZELIGS 1941). Subsequently, efforts were made to develop synthetic products with fewer disagreeable "peripheral" side effects such as blurring of vision and dry mouth. In the late 1940s and 1950s several synthetic anticholinergic and antihistaminic agents were developed with more selective central actions than the naturally occurring alkaloids. Until then hyoscine and atropine were the natural alkaloids most widely used. Since then, the commonly used anticholinergic agents have included benztropine mesylate

(Cogentin), trihexyphenidyl (benzhexol) hydrochloride (Artane, Aparkane), procyclidine hydrochloride (Kemadrin), cycrimine hydrochloride (Pagitane), biperiden hydrochloride (Akineton) and ethopropazine hydrochloride (Parsitan, Parsidol). The antihistaminic derivatives include orphenadrine hydrochloride (Disipal), chlorphenoxamine hydrochloride (Phenoxene) and diphenhydramine hydrochloride (Benadryl). These antihistaminic preparations have anticholinergic properties which are felt to be the source of their clinical effects. Antihistaminics are generally less potent than the anticholinergic agents. However, these may be better tolerated in older patients who are more prone to develop the side effects of anticholinergic drugs.

II. Mechanism of Action

The precise mechanism of action of anticholinergic drugs is not entirely clear depite a century of usage. There are good reasons to believe that the beneficial effects of anticholinergics are centrally mediated (see later in this section). Although there are many cholinergic systems in the brain, the striatum appears to have the largest concentration of acetylcholine, choline acetyltransferase and cholinesterase (KOELLE 1963). The diseased nigral dopaminergic neurones normally exert a tonic inhibitory effect on striatal cholinergic interneurones (BUNNEY 1979). This action forms the basis for the belief that anticholinergics act by normalizing the disequilibrium between striatal dopamine and acetylcholine activities which results from nigrostriatal dopamine deficiency.

There is a close correlation between the ability of anticholinergic drugs to antagonize the effects of acetylcholine on isolated guinea pig ileum (AHMED and MARSHALL 1962) and their capacity to inhibit tremorine-induced tremor in mice (EVERETT 1956). The latter is thought to be the result of a central cholinergic mechanism probably located at the brain stem level (TASKER and KERTESZ 1965; GEORGE et al. 1966). The dosage which is clinically effective in relieving parkinsonian symptoms in humans correlates well with the dose necessary to inhibit tremorine-induced tremor. Clinical efficacy of anticholinergic agents may also correlate with their ability to block EEG arousal effected by cholinergic drugs and electrical stimulation of the recticular activating system (HIMWISH 1955) which also depends on central cholinergic mechanisms.

DUVOISIN (1967) elegantly demonstrated that the cholinesterase inhibitor physostigmine, which penetrates the blood-brain barrier, consistently increased the severity of parkinsonian symptoms and that these effects could be antagonized by anticholinergic drugs such as benztropine or scopolamine. Further evidence for the importance of the central effect of anticholinergic medications was the observation by NASHOLD (1959) that acetylcholine directly injected into the globus pallidus of parkinsonian patients during stereotactic procedures resulted in increased tremor in the contralateral extremities. Injection of anticholinergic agents reduced the tremor.

Just as in the peripheral autonomic system, muscarinic and nicotinic cholinoceptive sites have been delineated in the central nervous system. There is strong evidence indicating that antimuscarinic rather than antinicotinic properties are responsible for the efficacy of anticholinergic drugs in Parkinson's disease. This is based on work by MARSHALL and SCHNEIDIN (1966) who showed that atropine had no effect on nicotine-induced tremor, but did have an antiparkinsonian action. They also showed that intravenous administration of nicotine appeared to improve rather than increase parkinsonian tremor.

Recently, different types of muscarinic receptors have been characterized by means of the novel antagonist pirenzepine. The M_1 or high affinity receptor subtype predominates in the rat forebrain, while the low affinity site, or M_2, predominates in the hindbrain (HAMMER and GIACHETTI 1982). BURKE (1986) has shown that all commonly used anticholinergics have a greater affinity for the M_1 site and the degree of selectivity of each drug for the M_1 versus M_2 receptor varies greatly. The importance of these differences to the clinical response and central side effect profile remains to be determined.

In addition to the potent anticholinergic properties, some drugs in this category, for example, benztropine, have the ability to block dopamine uptake in central dopaminergic neurones (FARNEBO et al. 1970) and in striatal synaptosomal fractions (COYLE and SNYDER 1969). It has not yet been demonstrated whether this effect on presynaptic dopamine mechanisms contributes to the antiparkinsonian action of certain synthetic anticholinergic drugs.

III. Pharmacokinetics

Most of the available anticholinergic drugs have a duration of effect of 1-6 h, with a peak in clinical activity at approximately 2-4 h. There are few studies in humans of the pharmacokinetics of this class of drugs. This may in part be the result of not having a sensitive physicochemical method for assay of serum levels. It has been shown that radioreceptor methods can be used to measure serum anticholinergic activity and BURKE and FAHN (1982) have taken advantage of this development to study the pharmacokinetics of trihexyphenidyl. They found that the half-life of trihexyphenidyl was 1.7 ± 0.2 h in normal volunteers receiving an acute dose. Dystonic patients, treated chronically with high dose trihexyphenidyl, had $t_{1/2} = 3.7 \pm 0.3$ h. Central nervous system side effects (confusion and lethargy) appeared to correlate more closely with age than with peak serum levels in the volunteers. Similar pharmacokinetic studies in patients with Parkinson's disease have not been published; however, these results suggest that a more frequent dosage schedule than the usual three times per day might increase efficacy. Side effects may supervene with more frequent administration, but smaller individual doses could be utilized and this perhaps indicates the need for availability of different dosage sizes.

IV. Clinical Effects

Until the introduction of levodopa in the 1960s, anticholinergic drugs had been the most effective treatment of Parkinson's disease for more than a century. The development of synthetic anticholinergics stimulated the beginnings of proper clinical trials in Parkinson's disease and a number of early studies repeatedly showed the antiparkinsonian effect of these preparations. For example, CORBIN (1949) and DOSHAY and CONSTABLE (1949) independently reported beneficial results with trihexyphenidyl, noting improvement in approximately three-quarters of a large number of patients. SCHWAB and LEIGH (1949) reported a mean 25% improvement in 62% of patients treated with caramiphen (Parpanit). SCHWAB and CHAFETZ (1955) found that procyclidine produced a "favourable response" in 46% of parkinsonian patients and furthermore felt that it was more effective than trihexyphenidyl.

Overall, most anticholinergic drugs produce a slight to moderate improvement in disability of 10%-25% when used alone (MARSDEN 1976). Unfortunately, the most disabling symptom of parkinsonism, akinesia, responds poorly, if at all. Prior to the advent of levodopa, many authors felt that anticholinergic preparations were more effective against rigidity than tremor (SCHWAB 1961; STRANG 1965). STRANG (1965) demonstrated that benztropine was slightly more effective against tremor than rigidity. PARKES et al. (1974) showed that benzhexol was most effective against postural deformity and rigidity. Apparently differing clinical profiles of anticholinergics encouraged the practice of choosing one drug over another, depending on the degree of tremor or rigidity which the patient demonstrated. However, there are no good control studies which provide definitive evidence for a selective clinical action of different anticholinergics and no double-blind crossover trials comparing anticholinergics in Parkinson's disease. There have been open trials, such as the one by SCHWAB and CHAFETZ (1955) who showed that procyclidine was slightly more effective than benztropine, contrary to results which would have been predicted from dopamine uptake studies.

At present, it is common practice to consider early treatment with anticholinergic drugs in patients with mild disease and in particular, when rest tremor in prominent. Most clinicians recognize that individual patients tolerate and/or benefit from one anticholinergic more than another, but this is difficult to predict in advance. Some investigators have found that benztropine is less well tolerated than other anticholinergics (HURTIG 1980). Others have casually reported better results with one particular drug than another, for example ethopropazine for tremor (FAHN 1982).

Recently, KOLLER (1986) has compared trihexyphenidyl with levodopa and amantadine for parkinsonian tremor. The former two drugs were equally effective in causing a greater than 50% reduction in tremor (not always in the same patients) while amantadine decreased tremor less than 25%. As will be discussed later, there is some evidence for an additive beneficial effect when certain anticholinergics are given in conjunction with amantadine.

Surprisingly, there is little data on the clinical and pharmacological interactions between levodopa and anticholinergics. In slowing gastric motility, anticholinergics may delay levodopa reaching its absorption sites in the small bowel, thus lowering the peak plasma level (FERMAGLICH and O'DOHERTY 1972). Over 60% of patients taking the combination of L-dopa and long-term anticholinergics are unable to tolerate anticholinergic withdrawal (HUGHES et al. 1971; HORROCKS et al. 1973) despite the mild to moderate benefit of monotherapy with these drugs.

Abrupt withdrawal of anticholinergic medications often results in a marked exacerbation of parkinsonian symptoms (GOETZ et al. 1981). This indicates the possibility that long-term anticholinergic therapy may result in acetylcholine receptor hypersensitivity which could then result in a proportionately greater clinical deterioration on medication withdrawal than the initial modest benefits would have predicted. Supporting evidence for this hypothesis is the finding of greater muscarinic acetylcholine receptor densities in the frontal cortex of patients on anticholinergic therapy (RUBERG et al. 1982). Similar changes might develop in the corpus striatum, accounting for the clinical observation described.

Long-term use of levodopa may also contribute to cholinergic hypersensitivity. YAHR et al. (1982) studied patients with fluctuations associated with long-term levodopa therapy and found that physostigmine precipitated "off" periods similar to the abrupt increase in parkinsonian symptoms documented by DUVOISIN (1967) in L-dopa naive patients. This is in contrast to the demonstrated decrease in cholinergic sensitivity which is found early in the course of levodopa therapy (WEINTRAUB and VAN WOERT 1971). Personal experience indicates that occasional patients with pronounced fluctuations on L-dopa may have their hour-to-hour status "smoothed out" with the introduction of an anticholinergic. Dystonia occurring in the "off" period may be lessened by anticholinergics and severe "off" period symptoms may be improved acutely by the administration of a parenteral preparation such as benztropine.

V. Side Effects

Anticholinergic drugs produce a number of adverse reactions related to their action in both peripheral and central cholinergic neuronal systems. Some of the peripheral effects may be of therapeutic benefit. Sialorrhea is improved significantly and urinary frequency may be better controlled with anticholinergic medication. Other patients find that dryness of the mouth, constipation, and urinary retention are limiting factors to the use of these agents. Other peripheral side effects include impotence, gingivitis occasionally resulting in tooth loss, and impairment of accommodation. Anticholinergic drugs are, of course, contraindicated in patients with narrow angle glaucoma, and blindness has been reported as a complication of this therapy (FRIEDMAN and NEUMANN 1972).

"Central" side effects are common with anticholinergic treatment. Abnormal involuntary movements (dyskinesias) may occur, particularly involving the orofacial region (FAHN and DAVID 1972), but also occasionally the limbs (BIRKET-SMITH 1974). Levodopa-induced dyskinesias may also be enhanced by this treatment (BIRKET-SMITH 1975). Changes in mental activity are the most important and serious side effects of anticholinergic medication. These represent the major limiting factor to their usage, especially in the elderly and in those with moderate to severe pre-existing cognitive impairment. The cholinergic system originating in the nucleus basalis of Meynert is known to be of importance to memory function (WHITEHOUSE et al. 1981) and this region is prominently affected in Alzheimer's disease (DAVIES and MALONEY 1976). Patients with Parkinson's disease, even in the absence of Alzheimer's changes, also have significant loss of cholinergic neurones in the nucleus basalis region (WHITEHOUSE et al. 1983). As a corollary to this, RUBERG et al. (1982) have shown a deficiency of cholinergic input to the frontal cortex in parkinsonian patients analogous to Alzheimer's disease. It is not clear to what extent these changes predispose patients to the "psychiatric" side effects of anticholinergic agents.

Anticholinergic medications have been shown to impair intermediate and delayed recall in nondemented parkinsonian patients (SYNDULKO et al. 1981; SADEH et al. 1982; KOLLER 1984). The corollary to this is the improvement in short-term and long-term memory which can be demonstrated in patients withdrawn from anticholinergic medications (LANG 1984). Toxic confusional states have been shown to be a frequent side effect of anticholinergic medications when patients have pre-existing intellectual difficulties (DESMET et al. 1982). These studies indicate that patients with pre-existing intellectual impairment should not receive anticholinergic medications. Physicians should be aware of subtle changes in mentation and memory which may be caused by this class of drugs, even in patients with normal mentation. It is not known as yet whether anticholinergic medications might contribute to the malfunction or permanent injury to cholinergic neurones in patients with early dementia.

VI. Conclusions

Anticholinergic medications are useful as initial therapy in patients with resting tremor and/or mild degrees of rigidity. This class of drugs is also effective as adjunctive therapy and commonly provides additional benefit to patients who are already taking levodopa preparations or other antiparkinsonian drugs. They also remain one of the few effective treatments for sialorrhea. Side effects are common, particularly in the elderly. When pre-existing moderate to severe cognitive impariment is present, the likelihood of mental complications is so high that anticholinergics are contraindicated in these patients. Anticholinergics are thought to improve parkinsonism by counteracting the activity of striatal cholinergic interneurones disinhibited by the depletion of their nigral dopaminergic input. Because this effect is "distal" to the postsynaptic dopamine receptor, it is possible that anticholinergics may be of benefit of those patients whose parkinsonism fails to repond to levodopa because the underlying pathology extends beyond the nigrostriatal dopamine system (e.g. the multiple system atrophies).

C. Amantadine

I. Introduction

Amantadine was first introduced in 1964 as a synthetic antiviral agent (DAVIES et al. 1964) effective in prophylaxis and symptom management of type A_2 (Asian) influenza respiratory tract infections. SCHWAB et al. (1969) reported their serendipitous discovery of amantadine's antiparkinsonian effect based on the observation of an improvement in the signs of idiopathic Parkinson's disease in a woman who was taking 200 mg daily for antiviral prophylaxis. Subsequent to this, several studies confirmed that amantadine has a consistent moderate beneficial effect in Parkinson's disease with a limited incidence of serious side effects.

II. Mechanism of Action

The mechanism of action of amantadine remains incompletely understood and somewhat controversial. Amantadine has a large ring structure and it is probably this feature, rather than the addition of side chains (as in amantadine derivatives) which determines its activity. It is generally believed that amantadine acts predominantly as an indirect dopamine agonist (BAILEY and STONE 1975; TILLEY and KRAMER 1981). Amantadine has been shown to augment the synthesis and release of dopamine (STROMBERG and SVENSSON 1971) as well as to diminish the reuptake of dopamine into the presynaptic neurone (VON VOIGTLANDER and MOORE 1971). These effects on release and reuptake may explain the improvement of parkinsonism seen with the drug. However, as pointed out by ALLEN (1983) in a review of the effects of amantadine in animal models, there are reasons to believe that the indirect dopamine agonist properties of amantadine do not satisfactorily account for its clinical efficacy. The drug has variable effects on other neurotransmitters such as noradrenaline, serotonin, GABA and cyclic nucleotides (see ALLEN 1983). Initially amantadine was found to have little or no anticholinergic action (GRELACK et al. 1970); however, contradictory evidence has been reported (NASTUCK et al. 1976). There is also clinical support for the possibility that amantadine has anticholinergic properties with the occurrence of a side effect profile which includes dry mouth, constipation, difficulty in focusing and urinary retention (see Sect. C.V). In addition, one patient has been reported with acute amantadine intoxication, which was reversed by physostigmine (CASEY 1978). Behavioural studies have also shown direct postsynaptic dopamine receptor effects, suggesting that the drug has mixed agonist and antagonist properties (STONE 1976; RANDRUP and MOGILNICKA 1976; PYCOCK et al. 1976).

ALLEN (1983) has proposed an alternative mechanism of action for amantadine which may explain the inconsistencies found in behavioural and clinical studies. He has provided preliminary data which suggest that amantadine has the ability to drive the striatal D_2 receptors towards a high affinity state. He has suggested that this effect on dopamine receptors may be secondary to amantadine's action on membrane lipids (JAIN et al. 1976) or to a direct (partial) agonist effect on the receptor which favours the D_2 high affinity state. This unique mechanism of action, if confirmed, may explain the success amantadine has maintained in remaining a part of the usual antipartkinsonian pharmacological armamentarium, while other indirect dopamine agonists (see LANG 1984) have not gained widespread usage.

III. Pharmacokinetics

Amantadine is rapidly and completely absorbed from the gastrointestinal tract. Peak plasma levels are usually reached between 2 and 6 h following a single dose of 100 mg. The biological half-life is approximately 24 h (BLEIDNER et al. 1965). At steady state, which is usually achieved in 5-7 days, the normal plasma concentration with a dosage of 200 mg/day is 0.2-0.9 µg/ml (PACIFICI et al. 1976). ING et al. (1979) have stated that plasma levels greater than 1.0 µg/ml should be viewed with concern. It is reported that 67% of amantadine is bound to plasma proteins, while 33% comprises the free drug fraction (LIU et al. 1984). This ratio remains constant over a wide range of total plasma drug levels.

In humans, there is essentially no metabolism of amantadine. Greater than 90% of the absorbed dosage is excreted unchanged by the kidneys. The renal clearance of amantadine is usually greater than the creatinine clearance (HORADAM et al. 1981), suggesting a net tubular secretion of the drug. The biological half-life may be twice as long as normal in the elderly (IEZZONI 1971) possibly secondary to compromised renal function. This emphasizes the need for caution when treating elderly parkinsonian patients with amantadine and the relative contraindication to its use in patients with a significant impairment in renal function Unlike humans, rats metabolize amantadine and excrete only 20% of the unchanged compound in the urine (BLEIDNER et al. 1965). This factor must be kept in mind when reviewing the animal pharmacology as it applies to the mechanism of action in humans.

Studying blood and CSF levels in a patient who had taken 2.8 g amantadine in a suicide attempt, FAHN et al. (1971) reported evidence of a concentration gradient between blood and CSF indicative of a blood-brain barrier for the drug. They found that, 44 h after the acute ingestion, the concentration in the CSF was 52% of that in the blood.

IV. Clinical Effects

The antipartkinsonian effects of amantadine have been demonstrated repeatedly in a number of open and double-blind studies (SCHWAB et al. 1969; PARKES et al. 1970, 1971a,b,c; APPLETON et al. 1970; BARBEAU et al. 1971; CASTAIGNE et al. 1972; CRITCHLEY 1972; HACOHOEN and GURTNER 1972; POLLOCK and JORGENSEN 1972; RAO and PEARCE 1971; SACKS et al. 1971; WALKER et al. 1972; BUTZER et al. 1975; FAHN and ISGREEN 1975). Overall, approximately two-thirds of patients respond to amantadine with an improvement in disability of 15%-25%. The benefit is noted quite early, often within the first 24-48 h. The drug is usually started in a dose of 100 mg once or twice daily and may be increased to 300-400 mg/day, depending on response and tolerance. Most studies have demonstrated an improvement in all of the major features of parkinsonism. However, some have shown little effect against tremor (SCHWAB et al. 1969; BAUER and MCHENRY 1974; KOLLER 1986). For example, KOLLER (1986) recently compared the effects of trihexyphenidyl, levodopa and amantadine on tremor. As mentioned earlier, amantadine reduced parkinsonian tremor by less than 25% while the other two drugs did so by more than 50%. There does not seem to be any correlation between the degree of disability and the occurrence of a response to amantadine. Interestingly, although they found no change in CSF homovanillic acid (HVA) or 5-hydroxyindoleacetic acid (5-HIAA), PARKES et al. (1971a) found that those patients with the lowest levels of CSF HVA tended to have the best response.

There has been considerable concern regarding the occurrence of tachyphylaxis with amantadine. However, only a few studies have shown that the initial response was short-lived (APPLETON et al. 1970; RAO and PEARCE 1971; CASTAIGNE et al. 1972; HACOHOEN and GURTNER 1972; TIMBERLAKE and VANCE 1978). The majority of long-term trials have demonstrated a sustained improvement which is clearly evident when amantadine is withdrawn or exchanged for placebo. Occasionally patients may deteriorate, even to the point of requiring hospitalization, when the drug is withdrawn. A gradual loss in clinical efficacy may simply relate to the progression of the disease. Patients seeming to lose previous benefit may improve further with an increase in the dosage up to 400 mg/day (SCHWAB et al. 1972).

Several studies have indicated a synergistic effect of amantadine given with levodopa (SACKS et al. 1971; CRITCHLEY 1972; POLLOCK and JORGENSEN 1972; WALKER et al. 1972; FAHN and ISGREEN 1975). Others have failed to show a true synergism (BARBEAU et al. 1971). It has been suggested that the effect of amantadine plus levodopa probably does not exceed the benefit obtainable from levodopa given alone in maximum dosages, but that the combination of amantadine plus levodopa may allow the use of lower doses of levodopa, potentially with fewer side effects (PARKES et al. 1971a). One study which did show a synergistic effect of amantadine combined with levodopa, also found that a certain proportion of patients who had not benefited from amantadine given alone responded later when the drug was added to ongoing L-dopa therapy (FAHN and ISGREEN 1975). In patients who have developed motoric fluctuations with chronic L-dopa therapy, amantadine may lessen the severity of these swings, improving overall disability (SHANNON et al. 1986). Although SCHWAB et al. (1972) felt that amantadine was no more effective when given with anticholinergics, PARKES et al. (1974) showed a definite additive benefit when benzhexol and amantadine were combined. The improvement in parkinsonian disability scores with benzhexol alone, amantadine alone, and benzhexol plus amantadine were 15%, 17%, and 42%, respectively.

V. Side Effects

Although a large number of side effects have been described with amantadine, its therapeutic index (risk:benefit ratio) is relatively low. As mentioned earlier, amantadine is excreted almost entirely unchanged by the kidney and therefore side effects are common in patients with impaired renal function. Guidelines have been provided for the use of amantadine in patients with renal failure (ING et al. 1979). However, in view of its limited efficacy in Parkinson's disease, it is probably better to avoid the drug entirely in these patients unless no other effective alternative is available (e.g. inability to tolerate all other medications). WILSON and RAJPUT (1983) found that the diuretic Dyazide (a combination of hydrochlorothiazide and triamterine) reduced amantadine excretion and resulted in toxicity without altering creatinine clearance. They suggested that the diuretic may have interfered with tubular secretion of amantadine. It is not clear whether this is applicable to other diuretics as well.

"Central" side effects include confusion, disorientation, depression, nervousness, insomnia, "dizziness" and lightheadedness, slurred speech, ataxia and action myoclonus. Psychiatric side effects may be pronounced, particularly in patients with compromised renal function. There may be a progression of symptoms from illusions to visual hallucinations to a paranoid hallucinatory state with delusions. Visual hallucinations include coloured lilliputian hallucinations (HARPER and KNOTHE 1973), but many other formed and unformed visual hallucinatory states also occur (POSTMA and VAN TILBURG 1975). As is the case with other antiparkinsonian agents, the psychiatric side effects are probably more common in patients with moderate to severe pre-existing cognitive disturbances. It does not seem that underlying Parkinson's disease per se predisposes to more common central side effects in the absence of age and additional cognitive disturbance factors, since these central side effects have all occurred in patients given the drug for its antiviral properties. As mentioned earlier, CASEY (1978) reported that physostigmine rapidly reversed delerium and myoclonus which occurred in a patient with amantadine intoxication.

It is commonly stated that convulsions are a side effect of amantadine and that the drug is contraindicated in patients with seizures. However, only two subjects have been reported with convulsions, one a nonparkinsonian subject receiving 800 mg/day (SCHWAB et al. 1969) and another with Parkinson's disease receiving 300 mg/day (CRITCHLEY 1972). Otherwise, there have been no reports of seizures developing in patients with Parkinson's disease treated with the usual doses of amantadine (QUINN 1984). The patient reported by FAHN et al. (1971) who took an overdose of 2.8 g amantadine developed an acute toxic psychosis with no depression in level of consciousness and no seizures. However, he was treated prophylactically with phenytoin. Another possible variable to consider is the general belief that epilepsy and Parkinson's disease rarely coexist, suggesting that Parkinson's disease may in some way protect against seizures.

Despite the proposed dopaminergic properties of amantadine, there has been only one case of facial dyskinesia developing with amantadine monotherapy (PEARCE 1971). SACKS et al. (1971) found that when amantadine was added to levodopa in patients doing poorly on this latter drug, there was an immediate exacerbation of levodopa-related side effects. On the other hand, when these patients were withdrawn from levodopa, some later showed a dramatic response to amantadine.

"Peripheral" side effects are not uncommon. Gastrointestinal intolerance is much less frequent than with levodopa (given without a decarboxylase inhibitor) or the ergot derivatives. Further support for an anticholinergic action comes from the occurrence of complications such as dry mouth, blurred vision, constipation, and urinary retention. These occur less often than with the usual anticholinergic agents. Urinary retention may result in a reduction in creatinine clearance and a further increase in amantadine toxicity.

Ankle œdema and livedo reticularis are common side effects of amantadine. They often occur together and are more frequent in women (TIMBERLAKE and VANCE 1978). The pathophysiology of these symptoms is not entirely known. PARKES et al. (1971a) found persistent moderate bilateral ankle œdema in 15 of their 66 patients, with slight swelling of the ankles present in more than 50% of the total group. They suggested that this complication was secondary to redistribution, rather than gain of body fluid, since the drug was not associated with a rise in total body water or electrolyte content. Although it has been stated that congestive heart failure occurs with amantadine, this has only been reported in four patients in the literature (QUINN 1984). Livedo reticularis refers to a purplish erythematous reticular macular rash which completely blanches with pressure. It occurs over the legs, occasionally spreading to the buttocks and less commonly occurring in the arms. It is accentuated by dependency or occlusion of venous return. The discoloration varies from almost black to barely visible, even in the same patient at different times. Livedo reticularis occurs in up to 90% of patients who are examined carefully for it (PARKES et al. 1971a). The skin biopsy is entirely normal (VOLLUN et al. 1971). It has been proposed that livedo reticularis results from the local release of vasoconstrictor catecholamines caused by amantadine. The secondary abnormal permeability of dermal blood vessels may then result in the swelling of ankles (PARKES et al. 1973). Although livedo reticularis and ankle œdema are disfiguring, they are not disabling and do not represent indications for discontinuation of the drug. Diuretic therapy may improve the swelling of the ankles; however, this approach must be used cautiously in view of its potential effect on amantadine excretion. Most acute side effects, particularly the psychiatric complications, resolve within 36-72 h of stopping the drug. Livedo reticularis and ankle œdema may take a month to subside.

Experience with amantadine in pregnancy is limited. There is one case report (NORA et al. 1975) of a woman with an extrapyramidal syndrome treated with amantadine during the first trimester of pregnancy. She gave birth to an infant with a single ventricle and pulmonary atresia. Another patient taking amantadine and levodopa/carbidopa at the time of conception who discontinued amantadine immediately when pregnancy was confirmed, gave birth to a normal child (COOK and KLAWANS 1985). Amantadine probably should be avoided in favour of a levodopa/carbidopa preparation in women of childbearing potential whose symptoms require antiparkinsonian drug therapy.

VI. Conclusions

The antiparkinsonian effects of amantadine are well established. It has mild to moderate effects on most symptoms of parkinsonism, although tremor may be less responsive. Despite the common belief to the contrary, the benefit from amantadine seems to be long-lived and there may be additive effects when it is given with levodopa and anticholinergics. Neuropsychiatric side effects are the commonest source of intolerance and these are seen more often in patients with underlying renal dysfunction or pre-existing cognitive disturbances.

The mechanism of action is controversial. It may have combined indirect dopamine agonist effects on presynaptic neurones as well as an ability to alter the affinity of postsynaptic dopamine receptors. Clearly, the former mechanism may play a less prominent role as presynaptic nigral dopaminergic neuronal degeneration progresses. The naturally occurring mechanisms which attempt to compensate for dopamine deficiency already include increased synthesis and release of dopamine as well as a reduction in reuptake. This pre-existing state probably also limits the efficacy of amantadine.

Both amantadine and the anticholinergics continue to play an important role in the symptomatic treatment of Parkinson's disease. There is increasing interest in the future possibility of directing therapy towards the prevention of progression of the disease. This focus has been renewed and directed by the discovery of the potential for the toxin 1-methyl-4-phenyl-1,2,3,6-tetrahydropyridine (MPTP) to produce an animal model of Parkinson's disease which is remarkably true to the human illness. MPP^+ is the neurotoxic oxidation product of MPTP which accumulates in nigrostriatal neurones via the dopamine neuronal uptake system. Substances which block dopamine uptake inhibit the toxic effects of MPP^+ (PILEBLAD and CARLSSON 1985) and this has recently included the anticholinergic, benztropine (BRADBURY et al. 1985) which, as mentioned earlier, has the additional property of limiting dopamine uptake. Amantadine also has this pharmacological property. If idiopathic Parkinson's disease is caused by ongoing exposure to exogenous or endogenous toxins which cause damage in a similar fashion to MPTP, it is theoretically possible that certain synthetic anticholinergics and/or amantadine could play a role in prevention of further loss of nigral neurones in addition to their symptomatic antiparkinsonian effects. Against this possibility is the lack of evidence that the introduction of either of these therapies, 40 and 20 years ago respectively, has altered the natural history of the illness. If indeed the ongoing progression of the disease does involve dopamine uptake mechanisms, then future research in drug development may provide new derivatives of these agents which have the potential of altering the natural history of the illness as well as lessening disease symptomatology.

Acknowledgment. Special thanks to Mrs. J. Lennox for assistance in typing the manuscript.

References

Ahmed A, Marshall PB (1962) Relationship between antiacetylcholine and antitremorine activity in antiparkinsonian and related drugs. Br J Pharmacol Chemother 18:247-257

Allen RM (1983) Role of amantadine in the management of neuroleptic induced extrapyramidal syndromes: overview and pharmacology. Clin Neuropharmacol 6 (Suppl 1):S64-S73

Appleton DB, Eadie MJ, Sutherland JM (1970) Amantadine hydrochloride in the treatment of parkinsonism. Med J Aust 2:626-629
Bailey EV, Stone TW (1975) The mechanism of action of amantadine in parkinsonism: a review. Arch Int Pharmacodyn Ther 216:246-262
Barbeau A, Mars H, Botez MI, Joubert M (1971) Amantadine-HCl (Symmetrel) in the management of Parkinson's disease: a double-blind cross-over study. Can Med Assoc J 105:42-46
Bauer RB, McIlenry JT (1974) Comparison of amantadine, placebo and levodopa in Parkinson's disease. Neurology (NY) 24:715-720
Birket-Smith E (1974) Abnormal involuntary movements induced by anticholinergic therapy. Acta Neurol Scand 50:801-811
Birket-Smith E (1975) Abnormal involuntary movements in relation to anticholinergics and levodopa therapy. Acta Neurol Scand 52:158-160
Bleidner WE, Harmon TB, Hewes WE et al. (1965) Absorption, distribution and excretion of amantadine hydrochloride. J Pharmacol Exp Ther 150:484-490
Bradbury AJ, Kelly ME, Costall B, Naylor RJ, Jenne P, Marsden CD (1985) Benztropine inhibits toxicity of MPTP in mice. Lancet 1:1444-1445
Bunney BS (1979) The electrophysiological pharmacology of midbrain dopaminergic systems. In: Horn AS, Korf J, Westernick BHC (eds) The neurobiology of dopamine. Academic, London, pp 417-452
Burke RE (1986) The relative selectivity of anticholinergic drugs for the M_1 and M_2 muscarinic receptor subtypes. Movement Disorders 1:135-144
Burke RE, Fahn S (1982) Pharmacokinetics of trihexyphenidyl after acute and chronic administration. Ann Neurol 12:94
Butzer NF, Silver D, Sahs AD (1975) Amantadine in Parkinson's disease. Neurology (NY) 25:603-606
Casey DE (1978) Amantadine intoxication reversed by physostigmine. Engl J Med 298:516
Castaigne P, Laplane D, Dordain S (1972) Expérimentation clinique prolongée chez 50 parkinsoniens. Nouv Presse Med 1:533-536
Cook DG, Klawans HL (1985) Levodopa during pregnancy. Clin Neuropharmacol 8:93-95
Corbin KB (1949) Trihexyphenidyl: evaluation of a new agent in treatment of parkinsonism. JAMA 141:377-382
Coyle JT, Snyder SH (1969) Antiparkinsonism drugs: inhibition of dopamine uptake in the corpus striatum as a possible mechanism of action. Science 166:899-901
Critchley E (1972) Levodopa and amantadine in the treatment of parkinsonism. Practitioner 208:499-504
Davies P, Maloney AJF (1976) Selective loss of central cholinergic neurones in Alzheimer's disease. Lancet 2:1403
Davies WL, Grunert RR, Haff RF, McGahen JW, Neumayer EM, Paulshock M, Watts JC, Wood TR, Hermann EC, Hoffmann CE (1964) Antiviral activity of 1-adamantanamine (amantadine). Science 144:862-863
DeSmet Y, Ruberg M, Serdaru M, Dubois B, Lhermitte F, Agid Y (1982) Confusion dementia and anticholinergics in Parkinson's disease. J Neurol Neurosurg Psychiatry 45:1161-1164
Doshay LJ, Constable K (1949) Artane therapy for parkinsonism: a preliminary study of results of 117 cases. JAMA 140:1317-1322
Duvoisin RC, (1967) Cholinergic-anticholinergic antagonism in parkinsonism. Arch Neurol 17:124-136
Everett GM, Blockus LE, Sheppard IM (1956) Tremor induced by tremonine and its antagonism by antiparkinsonian drugs. Science 124:79

Fabing HD, Zeligs MA (1941) Treatment of the post encephalitic parkinsonism syndrome with dessicated white wine extract of U.S.P. belladonna root. JAMA 117:332-334
Fahn S (1982) Discussion. In: The management of Parkinson's disease at different stages of the illness. Clin Neuropharm 5(1):S9
Fahn S, David E (1972) Orofacio-lingual dyskinesia due to anticholinergic medication, Trans Am Neurol Assoc 97:277-279
Fahn S, Isgreen W (1975) Long-term evaluation of amantadine and levodopa combination in parkinsonism by double blind crossover analysis. Neurology (NY) 25:695-700
Fahn S, Craddock G, Kumin G (1971) Acute toxic psychosis from suicidal overdosage of amantadine. Arch Neurol 25:45-48
Farnebo L, Fuxe K, Hamberger B, Ljungdahl H (1970) Effect of some antiparkinsonian drugs on catecholamine neurons. J Pharm Pharmacol 22:733-737
Fermaglich J, O'Doherthy S (1972) Effect of gastric motility on levodopa. Dis Nerv Syst 33:624-625
Friedman Z, Neumann E (1972) Benzhexol-induced blindness in Parkinson's disease. Br Med J 1:605
George R, Haslett WL, Jenden DJ (1966) The production of tremor by cholinergic drugs: central sites of action. Int J Neuropharmacol 5:27-34
Goetz CG, Nausieda PA, Weiner WH, Klawans HL (1981) Practical guidelines for drug holidays in parkinsonian patients. Neurology (NY) 31:641-642
Grelak RP, Clark R, Stump JM, Vernier VG (1970) Amantadine-dopamine interaction. Science 169:203-204
Hacohoen H, Gurtner B (1972) Clinical investigation into the effect of treatment with amantadine HCl in Parkinson's disease, Schweiz Med Wochenschr 102:583-586
Hammer R, Giachetti A (1982) Muscarinic receptor subtypes: M_1 and M_2 biochemical and functional characterization. Life Sci 31:2991-2998
Harper RW, Knothe BU (1973) Coloured lilliputian hallucinations with amantadine. Med J Aust 1:444
Himwish HE (1955) An analysis of the activating system including its use for screening antiparkinson drugs. Yale J Biol Med 28:308
Horadam VW, Sharp JG, Smilack JD, McAnalley BH, Garriott JC, Stephens MK, Prati RC, Brater DC (1981) Pharmacokinetics of amantadine hydrochloride in subjects with normal and impaired renal function. Ann Intern Med 94:454-458
Horrocks PM, Vicary DJ, Rees JE, Parkes JD, Marsden CD (1973) Anticholinergic withdrawal and benzhexol treatment in Parkinson's disease. J Neurol Neurosurg Psychiatry 36:936-941
Hughes RC, Polgar JG, Weightman D, Walton JN (1971) Levodopa in parkinsonism: the effects of withdrawal of anticholinergic drugs. Br Med J 2:487-491
Hurtig HI (1980) Anticholinergics for Parkinson's disease. Ann Neurol 7:495
Iezzoni DG (1971) In: Birdwood GFB, Gilder SSB, Wink CAS (eds) Parkinson's disease: a new approach to treatment. Academic, London, p 42
Ing TS, Daugirdas JT, Soung LS, Klawans HL, Mahurkar SD, Hayashi JA, Geis WP, Hano JE (1979) Toxic effects of amantadine in patients with renal failure. Can Med Assoc J 120:695-698
Jain MK, Yen-Minwu N, Morgan TK et al. (1976) Phase transition in a lipid bilayer II. Influence of adamantane derivatives. Chem Phys Lipids 17:71-78
Koelle GB (ed) (1963) Cholinesterases and anticholinesterase agents. Springer, Berlin Heidelberg New York, p 187 (Handbook of experimental pharmacology, vol 15)

Koller WC (1984) Disturbance of recent memory functions in parkinsonian patients on anticholinergic therapy. Cortex 20:307-311

Koller WC (1986) Pharmacologic treatment of parkinsonian tremor. Arch Neurol 43:126-127

Lang AE (1984) Treatment of Parkinson's disease with agents other than levodopa and dopamine agonists: controversies and new approaches. Can J Neurol Sci 11:210-220

Liu P, Cheng PJ, Ing TS, Daugirdas JT, Jeevanandhan R, Soung LS, Galinis S (1984) In vitro binding of amantadine to plasma proteins. Clin Neuropharmacol 7:159-151

Marsden CD (1976) Advances in the management of Parkinson's disease, Scott Med J 21:139-148

Marshall J, Schnieden H (1966) Effects of adrenaline, noradrenaline, atropine and nicotine on some types of human tremor. J Neurol Neurosurg Psychiatry 29:214-218

Nashold BS (1959) Cholinergic stimulation of globus pallidus in man. Proc Soc Exp Biol Med 101:68

Nastuck WL, Su PC, Doubilet P (1976) Anticholinergic and membrane activities of amantadine in neuromuscular transmission. Nature 264:76-79

Nora JJ, Nora AH, Way GL (1975) Cardiovascular maldevelopment associated with maternal exposure to amantadine. Lancet 2:607

Ordenstein L (1867) Sur la paralysie agitante et la sclérose en plaque generalisé. Martinet, Paris

Pacifici GM, Nardin M, Ferrari P, Latini R, Freschi C, Morselli PC (1976) Effect of amantadine on drug-induced parkinsonism: relationship between plasma levels and effect. Br J Clin Pharmacol 3:883-889

Parkes JD, Zilkha KJ, Calver DM, Knill-Jones RP (1970) Controlled trial of amantadine hydrochloride in Parkinson's disease. Lancet 1:259-262

Parkes JD, Curzon G, Knott PJ, Tatersall R, Baxter RCH, Knill-Jones RP, Marsden CD, Vollum D (1971 a) Treatment of Parkinson's disease with amantadine and levodopa. Lancet 1:1083-1086

Parkes JD, Knill-Jones RP, Clements PJ (1971 b) L-Dopa and amantadine hydrochloride in extrapyramidal disorders. Postgrad Med J 47:116-119

Parkes JD, Zilkha KJ, Knill-Jones RP, Clements PJ, Baxter R (1971 c) L-Dopa and amantadine hydrochloride in Parkinson's disease. Int J Clin Pharmacol Ther Toxicol 4:356-360

Parkes JD, Baxter RC, Galbraith A, Marsden CD, Rees JE (1973) Amantadine treatment in Parkinson's disease. Adv Neurol 3:105-114

Parkes JD, Baxter RC, Marsden CD, Rees JE (1974) Comparative trial of benzhexol, amantadine and levodopa in the treatment of Parkinson's disease. J Neurol Neurosurg Psychiatry 37:422-425

Pearce J (1971) Mechanism of action of amantadine. Br Med J 3:529

Pileblad E, Carlsson A (1985) Catecholamine-uptake inhibitors prevent the neurotoxicity of 1-methyl-4-phenyl-1,2,3,6-tetrahydropyridine (MPTP) in mouse brain. Neuropharmacology 24:689-592

Pollock M, Jorgenson PB (1972) Combined L-Dopa and amantadine in parkinsonism. Austr NZ J Med 3:252-255

Postma JU, Van Tilburg W (1975) Visual hallucinations and delerium during treatment with amantadine (Symmetrel). J Am Geriatr Soc 23:212-215

Price JC, Merritt HH (1941) The treatment of parkinsonism: results obtained with wine of Bulgarian belladonna and alkaloids of U.S.P. belladonna. JAMA 117-335-337

Pycock C, Milson JA, Tarsy D et al. (1976) The effects of blocking catecholamine uptake on amphetamine-induced circling behaviour in mice with unilateral destruction of striatal dopaminergic nerve terminals. J Pharm Pharmacol 28:530-532

Quinn NP (1984) Anti-parkinsonian drugs today. Drugs 28:236-262

Randrup A, Mogilnicka E (1976) Spectrum of pharmacological actions on brain dopamine. Indications for development of new psychoactive drugs. Discussion of amantadines as examples of new drugs with special actions on dopamine systems. Pol J Pharmacol Pharm 28:551-556

Rao NS, Pearce J (1971) Amantadine in parkinsonism: an extended prospective trial. Practitioner 206:241-245

Ruberg M, Ploska A, Javoy-Agid F, Agid Y (1982) Muscarinic binding and choline acetyltransferase activity in parkinsonian subjects with reference to dementia. Brain Res 232:129-139

Sacks OW, Schwartz WF, Messeloff CR (1971) Interactions of L-dopa and amantadine in parkinsonism. Clin Pharmacol Ther 12. (1):301

Sadeh M, Braham J, Modan M (1982) Effects of anticholinergic drugs on memory in Parkinson's disease. Arch Neurol 39:666-667

Schwab RS (1961) Symptomatology and medical treatment of Parkinson's disease. Int J Neurol 2:61-75

Schwab RS, England AC, Poskanzer DC, Young RR (1969) Amantadine in the treatment of Parkinson's disease. JAMA 208:1168-1170.

Schwab RS, Poskanzer DC, England AC, Young RR (1972) Amantadine in Parkinson's disease: review of more than two years' experience. JAMA 222:792-795

Schwab RS, Chafetz ME (1955) Kemadrin in the treatment of parkinsonism. Neurology (Minneap) 5:273-277

Schwab RS, Leigh D (1949) Parpanit in the treatment of Parkinson's disease. JAMA 139:629-634

Shannon KM, Goetz CG, Carroll VS, Tanner CM, Klawans HL (1986). Amantadine and motoric fluctiations in chronic Parkinson's disease. Neurology 36 (Suppl 1):182

Stone TW (1976) Responses of neurons in the cerebral cortex and caudate nucleus to amantadine, amphetamine and dopamine. Br J Pharmacol 1:101-110

Stomberg U, Svensson TM (1971) Further studies on the mode of action of amantadine. Acta Pharmacol Toxicol (Copenh) 30:161-171

Strang RR (1965) Kemadrin in the treatment of parkinsonism: a double blind and one year follow-up study. Curr Med Drugs 5:27-32

Syndulko K, Gilden ER, Hansch EC, Potvin AR, Tourtellotte WW, Potvin JH (1981) Decreased verbal memory associated with anticholinergic treatment in Parkinson's disease. Int J Neurosci 14:61-66

Tasker RR, Kertesz A (1965) The physiology of tremorine-induced tremor. J Neurosurg 22:449-456

Tilley JW, Kramer MJ (1981) Aminoadamantane derivaties. Prog Med Chem 18:1-44

Timberlake WH, Vance MA (1978) Four years treatment of patients with parkinsonism using amantadine alone or with levodopa. Ann Neurol 3:119-128

Vollum D, Parkes JD, Doyle D (1971) Livedo reticularis during amantadine treatment. Br Med J 2:627-628

Von Voigtlander PF, Moore KE (1971) Dopamine: release from the brain in vivo by amantadine. Science 174:408-410

Walker JE, Potvin A, Tourtelotte W, Albers J, Repa B, Henderson W, Snyder D (1972) Amantadine and levodopa in the treatment of Parkinson's disease. Clin Pharmacol Ther 13:28-36

Weintraub MI, Van Woert MH (1971) Reversal by levodopa of cholinergic hypersensitivity in Parkinson's disease. N Engl J Med 284:412-415

Whitehouse PJ, Price DL, Clarke AW, Coyle JT, DeLong MR (1981) Alzheimer's disease: evidence for selective cholinergic neurones in the nucleus basalis. Ann Neurol 13:243-248

Whitehouse PJ, Hedreen JC, White CL, Price DL (1983) Basal forebrain neurones in the dementia of Parkinson's disease. Ann Neurol 13:243-248

Wilson TW, Rajput AII (1983) Amantadine-Dyazide interaction. Can Med Assoc J 129:974-975

Yahr MD, Clough CG, Bergmann KJ (1982) Cholinergic and dopaminergic mechanisms in Parkinson's disease after long term levodopa administration. Lancet 2:709-710

CHAPTER 13

The Pharmacology of Levodopa in Treatment of Parkinson's Disease: An Update

P. A. LEWITT

> If it is really true that the striatal dopamine deficiency is causally connected to some of the symptoms in parkinsonism, and furthermore, if L-dopa works in parkinsonism by replenishing the missing dopamine in the striatum, then L-dopa is quite obviously the most natural substance we can have for treating what I should like to call "the striatal dopamine deficiency syndrome." However, it is quite clear to me as a pharmacologist that, whatever the mode and site of its action, L-dopa is far from being perfect as a drug.
>
> OLEH HORNYKIEWICZ (1970, p. 398)

A. Introduction

Therapy with levodopa (LD), the amino acid precursor of dopamine (DA), has been the most successful pharmacologic strategy for improving parkinsonian disability. Since its introduction in the 1960s, LD continues to impress clinician and patient alike with its almost metaphysical prowess at "awakening" paralyzed movement, unlocking rigidity, and stilling the "cruel restlessness" of shaking in parkinsonism (SACKS 1983). The degree of improvement with LD can be so dramatic that some clinicians would regard this response as a diagnostic feature of Parkinson's disease (PD). There are few other examples of symptomatic therapy for neurodegenerative disease as effectively targeted for the essential pathophysiology as is LD against PD. In the last two decades, use of LD has become so pervasive that it is virtually impossible to study the natural history of PD without the intervening factor of LD therapy.

From extensive experience with LD over the past two decades, many principles of antiparkinsonian therapeutics have emerged and undergone revision. LD can be highly efficacious as a symptomatic therapy for at least 75% of PD patients (YAHR 1975), including at least half of those with severe akinesia (BARBEAU 1976). It is unable to change the pathology or prevent further progression of the underlying neurodegenerative disorder (YAHR et al. 1972), and even after excellent initial responses many patients lose its benefits within 2 or 3 years. The reasons for the declining effectiveness may well involve more than progression in the underlying parkinsonism. As with the problems associated with long-term LD therapy (e.g., dyskinesias, dystonia, motor fluctuations, and hallucinations), the possible role of LD in the pathogenesis of declining efficacy needs further study. Because of its propensities for adverse effects, most experienced clinicians no longer regard LD therapy as a panacea, even though it is the most rational strategy for replacing the DA deficiency state of parkinsonism. With the availability of some alternatives in medication, use of LD is as often guided by attempts to avoid acute and chronic side effects as by the goal of normalizing motor deficits. This report will overview the pharmacology of LD, giving special attention to issues regarding the effectiveness of LD in clinical practice. Since the published experience is immense and has been extensively reviewed, this chapter will attempt to integrate some of the more recent clinical and basic science information, and will discuss current controversies in LD therapeutics.

B. The Efficacy of Levodopa in Parkinsonism

I. Background

It has been 20 years since the first report of striking antiparkinsonian efficacy sustained by 8 of 16 patients treated with a racemic mixture of 3,4-dihydroxyphenylalanine (dopa) (Cotzias et al. 1967). This neutral aromatic amino acid (occurring naturally as the L-isomer) came to be used as a precursor to DA because it was found to be actively transported into the brain and decarboxylated to DA (Carlsson et al. 1957). In contrast to dopa, the charged molecule DA is effectively excluded from transport (Pardridge and Oldendorf 1975; Hornykiewicz 1979). LD is derived from hydroxylation of phenylalanine (via phenylalanine hydroxylase to tyrosine) and tyrosine (via tyrosine hydroxylase). However, under natural circumstances, LD is found systemically and in the human brain only in trace amounts. As an intermediate between hydroxylation of L-tyrosine and a decarboxylation step yielding DA, LD does not accumulate because the decarboxylation step is not rate-limiting (even in the setting of severe PD) (Lloyd et al. 1975; Melamed 1986). Hence, LD undergoes rapid conversion to DA. LD can be present in the diet, though it is readily oxidized and not found in quantity sufficient for antiparkinsonian relief. While the L-isomer of tyrosine is the physiologic precursor of DA, it has not been a practical therapy for parkinsonism (Calne and Sandler 1970). Limitations to the rate of striatal uptake or conversion by tyrosine hydroxylase are such that only some mildly affected patients benefit from large oral doses (Growdon 1981). Administration of LD bypasses the rate-limiting step of DA synthesis, tyrosine hydroxylation, and permits rapid decarboxylation to produce large quantities of DA. Other physiologic roles for LD are not known, apart from its serving as an immediate precursor of DA. It lacks a nucleic acid triplet codon and is not incorporated into protein.

The earliest uses of LD in humans were to study its effects on blood pressure (Oster and Sorkin 1942) and metabolism to DA (Holtz et al. 1947). Initial trials with LD for parkinsonism were prompted by observations that DA subserved many of the motor activities mediated by the basal ganglia, where high DA concentrations were found (BERTLER and ROSENGREN 1959; CARLSSON et al. 1957; CARLSSON 1959). Identification of major striatal reductions in DA content as a characteristic feature of PD (EHRINGER and HORNYKIEWICZ 1960) set off multidisciplinary and multinational collaborations between neuroscientists and clinicians to develop the concepts of neurotransmitter replacement therapy. The earliest human experience in this regard was the finding that LD antagonized the cataleptic effects of tranquilization by reserpine in humans (DEGKWITZ et al. 1960), as it did for rodents with reserpine-induced depletion of DA (CARLSSON et al. 1957). In 1961, the rationale for replacing deficient DA in parkinsonism with LD was affirmed by acute benefits observed with relatively small doses of LD given orally (BARBEAU et al. 1962) or intravenously (BIRKMAYER and HORNYKIEWICZ 1961). Because of the minimal efficacy resulting from the low doses used, and severe adverse effects from larger quantities, many early reports doubted that there would be any potential for LD as an antiparkinsonian therapy. In fact, several investigations reported no effectiveness from trials with dopa between 1961 and the eventual recognition by 1966 of unequivocal benefits for parkinsonian disability (BARBEAU 1969).

LD finally emerged as a highly effective therapy for many parkinsonian patients from oral doses in excess of 3 g/day (up to 16 g/day) (COTZIAS et al. 1969). In the initial study by COTZIAS, 70% of patients improved and continued to do so for up to 2 years. By employing gradual LD dose buildup over several weeks to maximally tolerated doses, avoidance or reduction of the relatively common adverse effects could be achieved. Serious bone marrow depression occurring in 4 of the original 16 patients undergoing D,L-dopa therapy was attributed to the metabolically inactive D-isomer (COTZIAS et al. 1967). Development of a safe LD pharmaceutical (initially requiring special and costly methods for obtaining large quantities of the active isomer) permitted more widespread use. The next major breakthrough was the peripherally acting decarboxylase inhibitor (DCI), which has allowed the majority of patients to tolerate optimally antiparkinsonian LD doses. Current investigations with sustained-release forms of LD may soon yield a medication with improved constancy of effect.

Clinical experience and other studies pertaining to LD therapeutics have been discussed in a number of reviews and symposia over the past 18 years, among them: BARBEAU and McDOWELL (1970); BIRKMAYER and HORNYKIEWICZ (1976); BIRKMAYER and RIEDERER (1983); BOSHES (1981); BROGDEN et al. (1971); CALNE (1973); HASSLER and CHRIST (1984); LAKKE et al. (1977); MARKHAM et al. (1974); MARSDEN and PARKES (1977); McDOWELL and MARKHAM (1971); McDOWELL and BARBEAU (1974); POIRIER et al. (1979); PRESTHUIS and HOLMSEN (1974); RINNE et al. (1980b); ROSE and CAPILDEO (1981); SANDLER (1972); SHAW et al. (1980); STERN (1975); SWEET and McDOWELL (1975); SYMPOSIUM (1971, 1972); YAHR (1973, 1974a, 1975).

II. Clinical Effectiveness of Levodopa in Parkinson's Disease

LD can be highly effective at reversing virtually all of the positive (e.g., resting tremor, dystonia, rigidity, palilalia) and negative features of parkinsonism (e.g., bradykinesia, dysphagia, decreased dexterity, mask-like facies, drooling, and impaired gait, posture, and balance). Even with severe or long-standing parkinsonian disability, the reversal of incapacity can be as dramatic as for the mildly affected patient. In fact, Bernheimer et al. (1973) concluded that the more severely akinetic patients (presumably those with greater striatal DA loss) responded better to a test dose of LD than did less severely affected patients. Although LD tended to be less effective when moderate to severe cerebral atrophy was present (Schneider et al. 1979), its potential for therapeutic effectiveness was not correlated to the degree of degenerative neuronal pathology in the substantia nigra ultimately found at autopsy (Yahr et al. 1972). LD also proved to be effective in patients previously treated with other medication (such as amantadine and anticholinergics) or who underwent stereotactic thalamotomy (Kelly and Gillingham 1980).

Wider clinical experience with LD has shown that one or more features can respond earlier or show relatively greater improvement than others. However, there has been no consensus as to a particular hierarchy of symptom control resulting from increasing the dose or duration of LD therapy. Some early trials with LD (e.g., Godwin-Austin et al. 1969) reported that akinesia improved more than rigidity, and that improvement of tremor could not be demonstrated. Certain patients (including some with relatively mild disability) have tremor unresponsive to LD despite highly satisfactory improvement of other symptomatology, even though their tremor can improve with anticholinergic therapy. In many instances, however, LD can be quite efficacious at tremor suppression (Koller 1986). The benefits of LD in improving the quality of life in PD can extend to many other features of movement hard to quantify, but readily appreciated by patients, such as improved facial expressiveness, voice, writing, dressing, eating, and arising. Analysis of the physiologic basis for the effect of LD in parkinsonism has recognized improvement in motor unit frequency control and normalization of ability to activate the low threshold motor units in initiating muscle contraction (Petajan and Jarcho 1975), as well as improvement of accurate arm movement velocity (Baroni et al. 1984). However, the integrated roles of dopaminergic systems in motor function are incompletely understood in the context of the many features of motor deficit in PD that recover with LD (Marsden 1982).

When LD fails to work, the cause is often more extensive impairment to striatal systems than is generally found in idiopathic PD, in which degenerative changes for striatal DA systems are limited to the presynaptic projections from the pars compacta. Drug failure may occur if postsynaptic receptor function is disrupted (Van Woert and Weintraub 1971; Weiner et al. 1978). Parkinsonism with inadequate or absent response to LD can also portend the diagnosis of multisystem degenerations or other neurologic disorders associated with secondary parkinsonism, such as subdural hematoma, normal pressure hydrocephalus, or hypoparathyroidism (Mones 1973). Another mechanism for failure of LD responsiveness has been proposed to be the absence of adequate striatal decarboxylase activity (JANKOVIC 1981). It is also possible that, for some patients, LD dose has to be controlled within a therapeutic "window" for improvement to occur. This has been suggested by reports that increasing LD dosage produced paradoxical worsening in parkinsonian features in some PD patients with previously good responsiveness to lower doses (FAHN and BARRETT 1979).

For periods up to 3 years in most PD patients, the symptomatic effectiveness of LD can be optimal and seemingly arrests any decline in underlying symptomatology (YAHR 1975; SWEET and MCDOWELL 1975). Some patients have continued to maintain symptomatic benefit from LD since it became available more than 15 years ago. The usual situation, however, is for progression in most parkinsonian disabilities, particularly in features that from the start can be relatively unresponsive to LD (such as postural instability and start hesitation) (SWEET and MCDOWELL 1975; AMBANI and VAN WOERT 1973). The loss of clinical effectiveness is often complicated by the intervention of drug-limiting side effects and an increasing daily LD dose requirement (MENA et al. 1971). Several reports have found that, by 3-6 years of LD therapy, the disability for most patients receiving LD is essentially that of their pretreatment state.

LD can also have benefits for nonmotor features of PD, such as the hyperactivity of sebaceous secretion (KOHN et al. 1975) and impairment of external urethral spincter dysfunction (RAZ 1976). The pattern of sleep disturbance in parkinsonian patients (light and fragmented sleep architecture) has been shown to improve from LD therapy (ASKENASY and YAHR 1985), although its specific effects on rapid eye movement sleep and other sleep stages have varied among several other reports. Effects of LD on cognition (apart from adverse reactions such as somnolence, hallucination, and so on) are not commonly reported by patients. Some studies have documented that starting LD improved one or more cognitive deficits in demented or nondemented PD subjects (LORANGER et al. 1972; SWEET et al. 1976; RIKLAN et al. 1976; HALGIN et al. 1977). For patients with more advanced PD, however (whether or not in association with the development of dementia), subsequent deterioration of these initial cognitive improvements was generally found. Although their parkinsonian deficits may be quite responsive, cognitive impairment in demented PD patients does not improve, and often the use of LD is limited by psychiatric side effects (DRACHMAN and STAHL 1971). There has been no evidence that LD therapy precipitates dementia (GANDY et al. 1986).

LD does not consistently produce effects on mood for the PD patient (DAMASIO and CASTRO CALDAS 1975; MARSH and MARKHAM 1973; CELESIA and WANAMAKER 1972), who appears to be at increased risk of depression because of the disease. The reemergence of prior depressive tendencies, however, has been linked in some instances to LD therapy (MINDHAM et al. 1976). No significant improvements in cognition or behavior resulted from most trials of LD in patients with primary progressive dementia (JELLINGER et al. 1980; KRISTENSEN et al. 1977; LEWIS et al. 1978; JOHNSON et al. 1978; RENVOIZE et al. 1978).

III. Use of Levodopa in Other Forms of Parkinsonism and Other Movement Disorders

LD can improve the motor deficits of parkinsonism as the sequela of encephalitis lethargica, sometimes quite dramatically (SACKS 1983). Though this variety of parkinsonism tended to respond to lower LD doses than did idiopathic PD for symptom control (DUVOISIN et al. 1972), dose-related adverse effects as well as early loss of efficacy were more common (CALNE et al. 1969b; HUNTER et al. 1970). LD has also been effective for parkinsonism with spontaneous or familial onset in childhood (MARTIN et al. 1971), or resulting from chickenpox encephalitis (SACHDEV et al. 1977). In other instances of secondary parkinsonism, e.g., with structural lesions of nigrostriatal pathways, LD can work, provided there is some preservation of striatal dopaminergic presynaptic and postsynaptic structures. DA depletion from reserpine use can be reversed with LD (BRUNO and BRUNO 1966), but parkinsonism due to blockade of postsynaptic receptors by potent neuroleptics generally cannot be overcome (FLEMING et al. 1970).

In the wake of spectacular effects gained for idiopathic PD, LD has undergone clinical trials for a variety of disorders whose pathophysiology has also been linked to the DA system. Other movement disorders with associated parkinsonian features, such as progressive supranuclear palsy (DONALDSON 1973; KLAWANS and RINGEL 1971; JANKOVIC 1984), olivopontocerebellar degeneration (GOETZ et al. 1984; LANG et al. 1986), spinocerebellar degeneration with parkinsonism (WEIR and FAN 1981), Azorean disease (WOODS and SCHAUMBERG 1972), Shy-Drager syndrome (DELEAN and DECK 1976), pallidopyramidal syndrome (HOROWITZ and GREENBERG 1975), familial rigidity with ataxia and peripheral neuropathy (ZIEGLER et al. 1972), Wilson's disease (BARBEAU and FRIESEN 1970), striatonigral degeneration, or parkinsonism due to head trauma or hydrocephalus (LANG et al. 1982), may show modest improvement with LD. However, the extensive postsynaptic neuronal damage in these disorders can limit the therapeutic value of enhancing DA synthesis or agonism (AMINOFF et al. 1973). LD can improve bradykinesia and rigidity present in some patients with Huntington's disease (JONGEN et al. 1980). Though CSF homovanillic acid (HVA) may be low for choreic Huntington's disease patients (reflecting decrease of DA turnover), a worsening of involuntary movements, behavior, and dementia resulted from the use of LD (SISHTA and TEMPLER 1976).

Improvement in spastic and pyramidal signs in a parkinsonism patient has been reported from LD (SEYFERT and STRASCHILL 1977). In addition, other movement disorders may respond. Some patients with primary and secondary forms of dystonia (focal or generalized) have improved dramatically with LD therapy (BARRETT et al. 1970; LANG 1985; MARSDEN 1981; RAJPUT 1973; SEGAWA et al. 1976; STILL and HERBERG 1976). Progressive multisystem degenerative disorders in children can also show marked benefit in motor control in response to LD (BUGIANI and GATTI 1979). Patients with myoclonus (MINOLI and TREDICI 1974) and the "restless leg syndrome" (AKPINAR 1982) have been reported to improve with LD therapy, as has the related disorder of periodic movements in sleep (nocturnal myoclonus) (MONTPLAISIR et al. 1986). LD has also been used to suppress rubral tremor (YUILL 1980) and in monkeys has abolished tremor induced by selective lesioning of the cerebellar dentate and interpositus nuclei (GOLDBERGER and GROWDON 1971). LD treatment of patients with tardive dyskinesia (JESTE and WYATT 1982) has led to apparent recovery (ALPERT et al. 1983; SHOULSON 1983) when used with a strategy for reducing postsynaptic DA receptor supersensitivity.

Decline in motor facility with normal aging does not seem to respond measurably to LD, even though there is evidence of decline with increasing age in CNS DA. The use of typical antiparkinsonian doses of LD for 6 weeks in neurologically and cognitively normal elderly subjects did not improve movement time, reaction time, physiologic tremor, or general well-being (NEWMAN et al. 1985). However, "effortful" memory operations were enhanced (NEWMAN et al. 1984). These results imply that decline in dopaminergic functions might be one component of decline in cognitive (but not motor) features of "normal" aging.

IV. Therapeutic Principles for Levodopa

As a specific treatment for the deficiency of striatal DA, LD is a highly rational way of replacing the neurotransmitter (although controversy still exists as to its full mechanism of action, as discussed in Sect. F). The utility of LD rests in large measure on its highly favorable therapeutic index for most patients: benefit can be dramatic and sustained, and tolerance generally evolves to adverse effects developing with systemic exposure to this neurotransmitter precursor. In any one patient, however, it is difficult to anticipate the extent or duration of benefit from LD therapy. Some studies have claimed that CSF markers of DA metabolism (measured at baseline or after probenecid loading) can help in predicting the likelihood of clinical response to LD (LAKKE et al. 1971; VAN WOERT and WEINTRAUB 1971; BIANCHINE and SHAW 1976; JEQUIER and DUFRESNE 1972; GUMPERT et al. 1973; CUNHA et al. 1983). Other correlations to the clinical effect of LD have been made, using the ratio of CSF content of DA to LD as an index of DA synthesizing "capacity" (FEKETE et al. 1984). The activity of peripheral decarboxylation (when not inhibited by a peripheral DCI) may also provide some indication of central DA synthesis. In PD patients given single oral doses of LD, plasma DA levels afterwards were better correlated to clinical effect than was plasma LD, suggesting that peripheral decarboxylase activity metabolizing LD may be representative of striatal DA synthesis (ROSSOR et al. 1980). In other studies, the response to LD has not correlated to either the magnitude of DA deficiency (as reflected in CSF HVA, the major metabolite of DA) or else the severity of parkinsonian signs (BOWERS and VAN WOERT 1972; RINNE and SONNINEN 1972; CHASE and NG 1972; WEINER and KLAWANS 1973). Even when the dose of LD has correlated well to the resultant rise of CSF HVA (GODWIN-AUSTEN et al. 1971), the degree of clinical improvement has not been as closely related (RINNE et al. 1977).

Depending particularly on the goals, adverse effects, and long-term strategies of therapy, there is a wide range of what can be defined as "optimal" dose of LD. Following the initial clinical experiences with LD, clinicians were advised to use maximally tolerated LD, increasing until side effects appeared or until a satisfactory level of improvement occurred (Hunter and Shaw 1975; Barbeau 1976). Typical regimens of LD involved a starting dose of 500 mg, given in 3-4 doses daily and increasing over 4-8 weeks to a maximum of 8 g/day. Slow advancement of LD dose over weeks to months can be rewarded by a low incidence of adverse effects and recognition of the minimal dose of LD necessary for optimal benefit. The buildup period with the LD-DCI combination can be much more abbreviated than with a comparably effective amount of LD alone, which is generally 4-5 times as much (Rinne et al. 1973a; Marsden et al. 1974b; Marsden 1975). LD-DCI preparations have a markedly decreased incidence of peripheral side effects (Barbeau and Roy 1981; Rinne et al. 1975) as compared with LD regimens, which can require long periods of dose advancement for tolerance to develop.

Mild parkinsonism can generally be controlled with approximately 150-300 mg/day (with DCI) in 1-3 divided doses. In contrast with younger PD patients, elderly parkinsonian patients often achieve equivalent benefit from LD in substantially smaller amounts (Broe and Caird 1973; Sutliffe 1973). After 2 years, LD commonly needs to be to be advanced to 2-4 times as much, in order to maintain similar benefit. Dose-by-dose effects of LD are usually not recognized by patients initially, but can become measurably regular with chronic therapy.

With other antiparkinsonian drugs, such as anticholinergics and amantadine, LD may have additive or possibly synergistic effects (Parkes et al. 1971; Webster and Sawyer 1984). The same applies to bromocriptine, which as a combined regimen with LD may achieve equivalent therapeutic response and a lower incidence of dyskinesia and motor fluctuation outcomes as compared with LD regimens alone (Rinne 1985). The problems of predictable and irregular fluctuations ("wearing-off" and "on-off" responses to LD), and involuntary movements in association with chronic LD therapy, are discussed elsewhere in this volume. The presence of dementia, orthostatic hypotension, cardiac arrhythmia, active peptic ulcer disease, or concomitant monoamine oxidase inhibitor therapy (Teychenne et al. 1975) are all indications for caution and careful monitoring in the use of LD. For those adverse effects with peripheral components, including arrhythmia, hypotension, and nausea and vomiting, lessening the systemic synthesis of DA by increasing the total intake of DCI can help (Mars 1973). Peripherally active DA inhibitors such as domperidone are also useful for blocking dopaminergic side effects. Attacks of narrow angle glaucoma and elevation of intraocular pressure in other forms of glaucoma can occur from LD therapy. There has also been evidence, albeit controversial, that LD may have the potential for activating preexisting malignant melanoma (Rampen 1985). Given during gestation in animal studies, LD may initiate congenital malformations. Although it is not approved for use in pregnancy, three infants born to mothers taking antiparkinsonian doses of LD have had no complications or recognized abnormalities (Cook and Klawans 1985). Abrupt discontinuation of LD should be avoided, as it can present the hazard of severe rigidity and akinesia as well as potentially fatal hyperpyretic syndrome similar to neuroleptic malignant syndrome (Sechi et al. 1984; Friedman et al. 1985).

The relationship between medication and symptom relief can define the intervals at which some patients need LD. Particularly when regular dose-by-dose responses are experienced by a patient, careful adjustment of dose and scheduling can improve the amount of time spent each day with optimal control. Use of small doses of LD spaced as frequently as every 2 h over the course of a day can also help to "smooth out" abrupt fluctuations derived from irregularities in absorption and metabolism of the drug. Patients are often advised to take LD with meals to minimize the possibility of gastrointestinal upset. Although this practice is reasonable at the onset of LD use, many patients have no side effects from taking the drug on an empty stomach and theraby achieve more rapid and reliable absorption. Not uncommonly, smaller doses in the afternoon achieve the same result as larger doses in the morning hours, possibly as a result of LD accumulation during the course of the day. Other patients, particularly those with unpredictable fluctuations, have a less reliable response from LD later in the day, even with an increase of dose size.

Excessive self-administration of LD can occur in the setting of fluctuations, particularly if patients are not advised as to the principles governing onset and duration of LD effect. Careless adherence to optimal spacing between doses is a common cause of disabling "wearing-off" or enhanced peak-dose adverse experiences. Cases of apparent "abuse" of self-administered LD have been reported (Nausieda 1985), conceivably representing an addictive state (Priebe 1984).

C. Clinical Aspects of Levodopa Pharmacokinetics, Pharmacodynamics, and Metabolism

I. Levodopa Pharmacokinetics

The pharmacokinetics and metabolism of LD are reviewed elsewhere in this volume, as well as in other reports (Abrams et al. 1971; Bergman et al. 1974; Bianchine et al. 1971; Bianchine and Shaw 1976; Calne et al. 1969a; Goodall and Alton 1972; Hinterberger and Andrews 1972; Horne et al. 1984; Morgan et al. 1971; Peaston and Bianchine 1970; Nutt and Fellman 1984; Nutt and Woodward 1986; Sandler 1972). As a prodrug, LD is unique among therapeutic agents in utilizing GI and CNS amino acid transport mechanisms to reach its target site. For these reasons, its systemic and CNS metabolic disposition are extremely complex and cannot be fully modeled from animal studies (Rivera-Calimlim 1974) or analysis of LD blood levels. Studies have shown the systemic clearance of LD to be independent of dose, presumably following subsaturation pharmacokinetics (Sasahara et al. 1981). Absorption of LD is reflected by a prompt rise in plasma levels, which correlates fairly well in most patients with the onset of clinical effect. Before LD reaches the blood-brain barrier, a number of steps are subject to delay or competing mechanisms which can contribute to fluctuating plasma and CNS levels and, hence, variation in clinical response from oral LD. Alterations of pharmacokinetics imposed by parkinsonism or the chronic effects of the therapy itself may also be components of declining efficacy, response fluctuations, and adverse effects of LD. These changes are probably central in origin, as plasma half-life and other parameters of drug level do not seem to be different among patients with or without the problems of chronic therapy (Nutt et al. 1985; Rossor et al. 1980).

Optimal effects from LD can occur over a wide range of plasma LD concentrations (Tolosa et al. 1975). However, the usual plasma LD concentration for initiating antiparkinsonian effect ranges from 3 to 15 nmol/ml (and for most patients is 5–8 nmol/ml) (Nutt and Woodward 1986). For many patients, the LD blood level attained tends to be closely related to the extent of improvement in parkinsonian symptoms (Sweet and McDowell 1974; Yahr 1974; Rossor et al. 1980).

With the use of intravenous administration for constant delivery of LD to the striatum, clinical issues such as motor fluctuations can be differentiated into peripheral (i.e., GI absorption, liver metabolism, and distribution) and CNS components (i.e., transport across the blood-brain barrier, brain metabolism) (Nutt et al. 1984; Quinn et al. 1984). Intravenous dosing has also permitted study of the delay between a rise in plasma concentration and the onset of LD effect (Nutt and Woodward 1986). The technique of constant delivery of LD has demonstrated that the duration of LD antiparkinsonian effect is a function of plasma LD concentration minus the level needed for minimal clinical effect (Nutt and Woodward 1986). Even with maintenance of steady state plasma levels in the range 4–8 nmol/ml, deviations from optimal clinical response can occur in some patients, particularly those chronically treated (Fahn 1974; Muenter and Tyce 1971; Quinn et al. 1984; Eriksson et al. 1984; Algeri et al. 1976; Hardie et al. 1984). A number of studies have established that peripheral pharmacokinetic factors alone cannot fully explain the phenomenology of motoric fluctuations emerging during sustained LD use in PD. In support of this, LD pharmacokinetics in chronically treated patients with continuing response to LD do not differ from those of previously untreated PD subjects (Nutt and Woodward 1986).

Absorption of LD occurs primarily in the proximal small intestine (Bianchine et al. 1971; Sasahara et al. 1981). The rate of drug delivery to the duodenum from the stomach appears to be a major determinant of uptake (Evans et al. 1980, 1981). Since no significant absorption and only minimal metabolism of LD (by decarboxylation) occurs in the stomach (Rivera-Calimlim et al. 1970a, 1971; Bianchine et al. 1971), factors inhibiting release of gastric content to the small intestine can have a major influence on the rate of absorption. Meals (Morgan et al. 1971) and increasing gastric acidity (Pocelinko et al. 1972; Jenkins et al. 1973; Wade et al. 1974) can limit the onset or extent of LD effect. An attempt to improve LD uptake by alkalinization of the stomach with antacids was effective in some patients (Rivera-Calimlim et al.

1970a), but further study of this approach has not confirmed its utility (LAU et al. 1986). Drugs suppressing gastric motility can also extend the clinical benefits from a dose of LD. DA produced in the stomach from LD inhibits gastric emptying (BERKOWITZ and MCCALLUM 1980). Given with LD, the peripherally active DA receptor antagonist domperidone (LANGDON et al. 1986) enhanced peak plasma LD concentrations and bioavailability. In one study (FERMAGLICH and O'DOHERTY 1972), anticholinergics decreased gastric emptying rate, though these drugs lacked significant effect on LD pharmacokinetics in another report (MESSIHA and KNOPP 1973). Amantadine was reported to produce increases in plasma LD level in some patients in one study (BIRKMAYER et al. 1973), but not in another (PILLING et al. 1975).

Intestinal absorption of LD proceeds rapidly after delivery of LD to the proximal duodenum (BIANCHINE et al. 1971). Lesser degrees of absorption occur in the stomach (RIVERA-CALIMLIM et al. 1971) or more distally in the small intestine (SASAHARA et al. 1981). The uptake of LD occurs with the same facilitated transport mechanism that operates for other neutral L-amino acids (tryptophan, phenylalanine, tyrosine, threonine, leucine, isoleucine, methionine, histidine, valine, and cysteine (WADE et al. 1973). It is uncertain whether simple diffusion can also occur in addition to facilitated transport, since an increase in the load of LD presented to the duodenum appears to enhance absorption (MEARRICK et al. 1974).

Clinical experience showed that reducing the dietary protein load coadministered with LD increases its antiparkinsonian effectiveness (COTZIAS et al. 1973; MENA and COTZIAS 1975). In pharmacokinetic studies also, competition observed between the uptake of LD and amino acids from dietary protein (PAPAVASILIOU et al. 1974; MUENTER et al. 1977; DANIEL et al. 1976) or the LD metabolite 3-*O*-methyldopa (MUENTER et al. 1973) suggested that the saturability of carrier-mediated amino acid uptake may be a major factor limiting the absorption or CNS uptake of pharmacologic quantities of LD. Studies involving coadministration of L-methionine with LD led to deterioration in antiparkinsonian effectiveness (PEARCE and WATERBURY 1974), possibly owing to L-methionine competition with LD for uptake in the GI tract or CNS (or both). With continuous intravenous infusion of LD, however, it was possible to demonstrate deterioration of antiparkinsonian effect due to coadministration of neutral amino acids, resulting in competition for LD uptake at the blood-brain barrier (NUTT et al. 1984). Similar to the result of a high protein meal, L-phenylalanine or L-leucine loads decreased an established antiparkinsonian effect from constantly infused LD with unchanging plasma levels. For undefined reasons, some patients found that the carbohydrate intake interferes with LD effect and can seemingly precipitate "off" or other types of undermedicated states (MUENTER et al. 1977).

After long-term LD usage, some patients may be intermittently unresponsive to doses of LD (FAHN 1977). Occurring typically in the afternoon, this pattern of drug unresponsiveness seems to be unrelated to food intake. Since they have been associated with low plasma LD levels (MELAMED et al. 1986), these episodes probably reflect failures of gastrointestinal motility or duodenal absorption occurring in the context of chronic therapy. This problem could be a feature, in part, of the patient population, since episodes of delayed gastric emptying can occur in the normal elderly as well as in parkinsonian patients (Evans et al. 1981). There is also evidence for enhanced gastrointestinal absorption of LD in the elderly (Broe et al. 1981).

There are occasional patients with little or no response to conventional oral doses of LD and who have evidence of an enhanced peripheral metabolism of LD (Muenter et al. 1972; Rivera-Calimlim et al. 1977). This could conceivably be due to enhanced catabolism of LD in the gut, even if concomitant DCI is used. The intestine appears to be a catabolic "sink" for up to 10% of parenterally administered LD in rats (Tyce and Owen 1979). In humans, some metabolism of LD takes place in the gut by means other than decarboxylation (Peaston and Bianchine 1970), possibly in part by the gut flora (Nutt and Fellman 1984).

LD administered orally is almost completely absorbed from the gut. A major portion undergoes decarbocylation, either in the intestine (andersson et al. 1975) or liver. Conventional doses of peripheral DCIs such as carbidopa result in a two- to threefold increase in the bioavailability of orally administered LD into the circulation (Tissot et al. 1969; Kuruma et al. 1972; Pinder et al. 1976; Bianchine et al. 1972). During the initial systemic distribution of absorbed LD, there is rapid first-pass metabolism through the liver (Sasahara et al. 1981) and transfer to other systemic compartments. As no significant binding of LD occurs to serum proteins (Hinterberger and Andrews 1972), LD distribution is in the aqueous phase of plasma as a free drug. After systemic distribution, skeletal muscle serves as an important peripheral storage reservoir for LD. No significant metabolism is known to occur in muscle (Ordonez et al. 1974). Uptake of LD by erythrocytes occurs, and red cells possibly metabolize LD to 3-*O*-methyldopa (Ordonez et al. 1974; Reilly et al. 1980).

At the interface of the vasculature and the brain is a sodium-dependent L-stereospecific active transport system for neutral amino acids, by which LD is taken up (Pardridge 1977; Wade and Katzman 1975c). This system corresponds to a gastrointestinal transport mechanism for the same amino acids (Pardridge and Oldendorff 1975; Pardridge 1977; Wade and Katzman 1975a). The influx of LD across the blood-brain barrier can be estimated from experimentally derived rate constants and the plasma concentrations of LD and other competing amino acids. The high activity of L-aromatic amino acid decarboxylase (AAAD) in the vascular endothelium has been regarded as another physiologic barrier to the influx of LD to the brain. As LD undergoing decarboxylation to DA produces a charged species unable to be transported, this mechanism is highly effective at preventing entrance of much of the LD presented to the blood-brain barrier from entering into the brain parenchyma of the rat. This process is saturable; LD in excess enters the brain unmetabolized (Hardebo and Owman 1980). Evidence is lacking, however, that cerebral endothelial decarboxylase activity is a major factor for facilitating the influx of LD. One indication that endothelial decarboxylase activity is independent of the transport mechanism is that, in PD patients, similar clinical effects are derived from a given plasma LD level whether or not the endothelial decarboxylase activity is inhibited by a peripherally active DCI (FAHN 1974; NUTT et al. 1985). Although catechol *O*-methyltransferase (COMT) activity is present in some portions of the cerebral vasculature in animals, it does not appear to affect the rate of LD transport across the blood-brain barrier (HARDEBO and OWMAN 1980).

II. Levodopa Metabolism: 3-*O*-Methyldopa and Other Metabolites

Few routes of metabolism are known for LD other than its primary catabolism, or conversion to catecholamines (PLETSCHER et al. 1967; BIANCHINE et al. 1972). However, a significant fraction of intravenously administered labeled LD cannot be accounted for in urinary or fecal excretion (GOODALL and ALTON 1972; TYCE and OWEN 1979). *O*-Methylation via COMT is a major pathway for the peripheral metabolism of LD, particularly when inhibitors of peripheral decarboxylase activity are used concomitantly with LD to limit its systemic metabolism to catecholamines and ultimately HVA. Systemically (particularly in the liver) and in the brain, COMT generates the principal metabolite of LD, 3-*O*-methyldopa (3-*O*-MD) (SHARPLESS et al. 1972). Although *O*-methylation is directed predominantly to the 3 position, significant quantities of 4-*O*-methyldopa have also been recovered in plasma of patients given LD (ITSHIMITSU et al. 1984). 3-methoxy-4-hydroxyphenyllactic acid (formed by the actions of lactate dehydrogenase and COMT on LD; MUSKIET et al. 1978), has been found in small quantities in human CSF during LD therapy. Other metabolites derived from LD therapy, such as *m*-tyramine (detected in urine of LD-treated patients), are probably formed in the gut microbially (SANDLER et al. 1971).

3-*O*-MD does not undergo further metabolism (such as decarboxylation) to any significant degree (BARTHOLINI et al. 1972; GOODWIN et al. 1978). Shortly after intravenous dosing with [^{3}H]LD, labeled 3-*O*-MD appeared simultaneously in the striatum and cerebellum of mouse brain (HORNE et al. 1984). However, much of the 3-*O*-MD measured in the striatum after chronic systemic LD administration was of peripheral origin and then taken up into the brain (LLOYD et al. 1980). The delayed accumulation of 3-*O*-MD in the striatum after LD administration may thus be secondary to its slow rise in blood concentration, with maximal levels occuring approximately 1 h after peak LD levels are achieved from oral doses (SHARPLESS et al. 1972). The approximately 15 h half-life of 3-*O*-MD probably explains the greater accumulation and delayed clearance of this metabolite as compared with its shorter-lived parent compound (KURUMA et al. 1970; MUENTER et al. 1972). Plasma levels of 3-*O*-MD change in response to oral LD intake on a much more delayed course than LD levels. Ratios of plasma 3-*O*-MD to LD vary among LD-treated individuals, though they maintain a relatively constant value for each patient (YOKOCHI et al. 1979; RIVERA-CALIMLIM et al. 1977; FEUERSTEIN et al. 1977; BERMEJO PAREJA et al. 1985). As in the bloodstream, 3-*O*-MD accumulates to higher levels than LD does in CSF (SHARPLESS and McCANN 1971) and in the brain (RINNE et al. 1974, 1979). Some patients had little LD detectable in CSF despite relatively high levels of 3-*O*-MD (KREMZNER et al. 1974).

The use of DCIs results in enhanced systemic production of this methylated species, probably owing to elimination of the alternative pathways by which peripheral LD would be metabolized after conversion to DA (Rivera-Calimlim et al. 1977).

The metabolism and physiology of 3-*O*-MD may have major implications for LD effectiveness. High blood levels of 3-*O*-MD may be associated with a poor clinical response to LD because of competition with LD uptake into the CNS (Reches and Fahn 1982; Gervas et al. 1983). Affinity of 3-*O*-MD for facilitated neutral amino acid transport into the CNS has been shown to be greater than that for LD (Wade and Katzman 1975b). Competition by 3-*O*-MD for LD uptake into the brain has been used to explain the attenuation of DA replacement effect in nigrostriatal-lesioned rats (Reches et al. 1982; Reches and Fahn 1982), and has been linked to impairment of LD effect in parkinsonian patients. Accumulation of relatively high levels of 3-*O*-MD has been associated with suboptimal responses to LD (Muenter et al. 1972; Rivera-Calimlim et al. 1977). Similarly, coadministration of 3-*O*-MD lessens the effectiveness of LD doses (Calne et al. 1972; Muenter et al. 1972). In one study, COMT activity present in red blood cells correlated fairly well with the ratio of plasma 3-*O*-MD to LD concentrations. Those patients with increased COMT activity in red cells tended to have less favorable clinical responses to LD therapy (Reilly et al. 1980), possibly because of their greater systemic capacity to produce 3-*O*-MD. 3-*O*-MD may also have a role in the pathogenesis of LD-induced dyskinesia (Feuerstein et al. 1977; Rivera-Calimlim et al. 1977), as suggested by elevated plasma levels found in dyskinetic patients. Ogasahara et al. (1984) found no difference in CSF 3-*O*-MD content in patients sensitive to the "wearing-off" effect, as compared with patients without this problem. This study also showed that withdrawal of LD therapy for several days did not alter CSF content of 3-*O*-MD.

An attempt to block the formation of 3-*O*-MD was carried out with ascorbic acid, which serves as a weak inhibitor of COMT. This therapy had no clinical effects, though reductions occurred in both plasma 3-*O*-MD and LD levels (Reilly et al. 1983). Administration of *N*-butylgallate (an inhibitor of COMT) with LD enhanced DA effect in animal studies, but this compound proved to be toxic in humans (Ericsson 1971). Parenteral administration of another potent inhibitor of COMT, U-0521, led to dose-dependent inhibition of striatal COMT activity in rats, associated with marked decline of plasma 3-*O*-MD accumulation after LD administration (Reches and Fahn 1984). As predicted, pretreatment with intravenous U-0521 permitted, after LD, a greater rise in rat striatal levels of LD and DA. However, oral U-0521 was without peripheral or CNS effects in rodents or in a parkinsonian subject given 0.7–11 g/day on a biweekly basis (Reches and Fahn 1984).

III. Other Metabolic Effects of Levodopa

Though the product of LD decarboxylation, DA, is itself the substrate for norepinephrine (NE) synthesis, there is no consensus as to whether LD therapy has additional effects in the CNS beyond the synthesis of DA. From radiolabel studies, there is evidence in animals of some conversion of exogenous LD to NE as well as DA (EDWARDS and RIZK 1981). An increase of brain NE turnover in several sites outside the striatum following LD administration has been shown (WURTMAN and ROMERO 1972), which has been thought to be a contributing factor to hyperactive behaviors induced by LD in reserpinized mice (DOLPHIN et al. 1976). Other studies have found little or no effect of LD on NE metabolism in the brain, and postmortem studies of regional brain NE content in LD-treated patients showed no difference from controls (RINNE et al. 1974). LD administration decreases COMT activity against endogenous NE, an effect that can be reversed by coadministration of *S*-adenosylmethionine (STRAMENTINOLI et al. 1980). Hypothalamic epinephrine content was also reduced in the rodent after LD treatment, possibly from *S*-adenosylmethionine depletion resulting, in turn, from increased utilization of *O*-methylation (from intake of pharmacologic doses of LD) (WURTMAN et al. 1970a,b; WURTMAN and ORDONEZ 1978; FULLER et al. 1982). Despite increased utilization of *S*-adenosylmethionine (important as a methyl donor in neurotransmitter metabolism and protein synthesis), no deficiency of methionine resulted from sustained LD use (WURTMAN et al. 1970; WURTMAN and ORDONEZ 1978).

Sustained use of LD resulted in no change of brain L-aromatic amino acid decarboxylase (AAAD) in rats (DAIRMAN et al. 1972) or mice (TATE et al. 1971; HEFTI et al. 1981a), though there was significant increase in cats (ROBERGE and POIRIER 1973). Additional influences of LD on enzymes of monoamine neurotransmitter metabolism have been found. Peripheral measures of COMT, dopamine-β-hydroxylase, and monoamine oxidase (MAO) in patients receiving LD may be altered (MARS 1975). Mice chronically treated with LD show no change in brain COMT or MAO activity (PAPAVASILIOU et al. 1981), but rats similarly treated show a rise in activity of brain MAO-A and -B (LYLES and CALLINGHAM 1980). Although chronic LD treatment produces no change in brain tyrosine hydroxylase activity in the rat (DAIRMAN et al. 1971) and mouse (HEFTI et al. 1981a), a significant decrease in this enzyme occurred in the cat striatum and at other sites (ROBERGE and POIRIER 1973). Metabolites of LD acted as irreversible inhibitors of dihydropteridine reductase (WARING 1986), which maintains tetrahydrobiopterin in the reduced state necessary for function as the rate-limiting cofactor of tyrosine hydroxylase.

Even when derived from supraphysiologic quantities of LD used in antiparkinsonian therapeutics, DA is metabolized along its conventional catabolic pathways. The initial step for DA degradation can be either oxidative deamination (in humans, predominantly by MAO-B) or *O*-methylation (by COMT). Both reactions occur in sequence to yield, as the major product, homovanillic acid (HVA). In human brain or CSF, only small quantities are found of 3,4-dihydroxyphenylacetic acid (DOPAC), which in rodents is the predominant DA metabolite.

The formation of HVA from LD has been studied by using tracer quantities of isotope-labeled LD (PLETSCHER et al. 1967; PEASTON and BIANCHINE 1970). In PD patients, orally administered [^{14}C]LD resulted in excretion of only 0.8% as unmetabolized LD (MORGAN et al. 1971). [^{14}C]HVA appeared in lumbar CSF when sampled 7.5 h after administration of [^{14}C]LD (EXTEIN et al.

1976). From the rate of HVA accumulation as well as endogenous HVA concentration, these studies have also confirmed that the DA turnover rate is decreased in PD as compared with control subjects. A marked increase of HVA was found throughout the brain after LD administration, including sites at which DA is usually not found (Rinne et al. 1971, 1974; Davidson et al. 1971). LD and 3-O-MD (normally not present in the brain of normal subjects or untreated parkinsonian patients) were also abundant at these sites (Rinne et al. 1974). In these studies, the marked increase of DA concentration in LD-treated patients was restricted to the striatum. As was the case for DA content, the rise in HVA levels resulting from LD was not correlated with the degree of clinical response by the patients. HVA levels were proportionately more enhanced above normal values than were the increases in tissue DA concentration, a pattern that was also found in the PD brain prior to LD therapy (Hornykiewicz 1979).

In CSF, low or undetectable baseline levels of HVA are substantially elevated by LD treatment roughly in proportion to the dose given (Curzon et al. 1970). As found in several subsequent studies of CSF HVA responses to LD administration (Chase 1980), patients without clinical response to LD can have either no response or else the same magnitude of HVA elevation as in improved patients (possibly because postsynaptic systems are unable to respond to the stimulated DA synthesis). In one study, the addition of a DCI during LD therapy resulted in a major decrease of the resultant CSF HVA concentration. This study suggested that some portion of HVA in CSF is derived from metabolism of LD in the brain capillary wall (Rinne et al. 1977).

Another route for the metabolism of DA (Sharpless et al. 1981), LD (Johnson et al. 1980), and 3-O-MD (Tyce et al. 1974) is the formation of sulfate conjugates. Transamination has been described as a minor pathway of peripheral disposition of LD (Fellman et al. 1976). Tyrosine aminotransferase acts on LD to produce 3,4-dihydroxyphenylpyruvic acid. In the rat, this reaction is reversible and yields LD from the accumulated metabolite (Linden 1980). A two-step enzymatic process produces 2,4,5-trihydroxyphenylacetate, detectable in the urine of LD-treated parkinsonian patients (Wada and Fellman 1973). The latter species, because it is easily oxidized to a reactive quinone species similar to 6-hydroxydopamine, is potentially neurotoxic. However, LD also serves to inhibit the reaction generating 2,4,5-trihydroxyphenylacetate (Fellman and Roth 1971).

D. Other Effects of Levodopa

A number of other metabolic effects of LD administration have been recognized in the brain. One observation of uncertain significance is the effect of LD therapy on some CNS phospholipids. As compared with controls, PD patients have an increase of sphingomyelin and decreased phosphatidylcholine and phosphatidylethanolamine content. These differences are corrected in the LD-treated PD brain (Riekkinen et al. 1975).

LD has several influences on CNS serotonin (5-HT) metabolism. Though LD competes with the active transport into the CNS of neutral amino acids, brain tryptophan levels were actually increased in LD-treated rats (WURTMAN and ROMERO 1972). LD displaced 5-HT from the CNS (EVERTT and BORCHERDING 1970), depleting it in several regions of rat brain and spinal cord (COMMISSIONG and SEDGWICK 1979). In the cat brain, administered LD acutely caused no change in striatal 5-HT levels, but its metabolite 5-hydroxyindolcacetic acid (5-HIAA) was increased (LLOYD et al. 1980). CSF levels of 5-HIAA were diminished after LD therapy in parkinsonian patients (NG et al. 1970). However, 5-HT or 5-HIAA content in PD brain after LD treatment has been reported to be depleted (RIEDERER 1980) or else unchanged (RINNE et al. 1974). Studies of CSF 5-HIAA have suggested a correlation between LD effects on 5-HT metabolism and the occurrence of involuntary movements. As compared with nondyskinetic parkinsonian patients receiving LD, a significantly higher ratio of HVA to 5-HIAA was found in CSF of dyskinetic subjects (FRIEDMAN 1985a).

LD has influences on the disposition and metabolism of other amino acids. Acutely after LD administration, a rise in CSF content of tyrosine and phenylalanine occurs (GRUNDIG et al. 1969). Elevations of CSF leucine, valine, and arginine levels also have been reported during chronic LD therapy in humans (KREMZNER et al. 1974). Use of LD partially inhibits the incorporation into rat brain of labeled lysine (ROEL et al. 1978) and leucine (KING and BEESLEY 1978). However, overall protein synthesis is not inhibited by LD treatment in the rat, though DA has been reported to cause massive disruption of polyribosome aggregation in the brain (WEISS et al. 1972; ROEL et al. 1978). RNA content, decreased in the untreated PD brain, is found in LD-treated patients to be elevated to control levels (RINNE et al. 1974).

LD intake produces alterations in GABA metabolism (BARBEAU 1972), as judged from brain glutamate dehydrogenase activity, which, though significantly decreased in the PD brain, is elevated somewhat in response to LD therapy (RINNE et al. 1974). In the rat, administration of LD acutely elevated nigral GABA, though this effect was lost with chronic treatment. Glutamate decarboxylase activity was also decreased in the substantia nigra, though not in the striatum (LLOYD and HORNYKIEWICZ 1973, 1977). LD therapy also increased striatal acetylcholine levels in the guinea pig brain (BEANI et al. 1966). As a result of enhanced utilization of *S*-adenosylmethionine for LD metabolism, decreased levels of pyridoxyl phosphate may result from sustained LD use (WURTMAN and ROMERO 1972), though without known consequences. Studies in parkinsonian patients receiving LD have shown that chronic LD treatment resulted in increased capacity of erythrocytes to synthesize pyridoxyl-5-phosphate (MARS 1975).

Acutely, antiparkinsonian doses of LD produce no change (MELAMED et al. 1978) or else regional increases in cerebral blood flow in humans (LEENDERS et al. 1985). Most likely, the latter observation represented a primary vasodilating response. These studies showed no evidence of an associated rise in glucose utilization, nor any correlation between increased flow and antiparkinsonian effect from the administered LD. However, after LD administration to rats, regional carbohydrate metabolism was increased significantly (enhanced glucose oxidation and diminished rate of glycogen formation) in the striatum and hippocampus, but not at other brain sites (Barash et al. 1985). Metabolic studies in humans have shown no effects of LD on systemic carbohydrate metabolism (Rosati et al. 1976), though it has a kaliuretic effect (Granerus et al. 1977).

The effects of LD on peripheral muscle physiology include the depression of tension and degree of tetanic contraction fusion in the cat soleus. These effects appear to be mediated via catecholamine synthesis, as both can be antagonized by blocking decarboxylation or β_2-receptors (Bowman and Nott 1978).

As judged from endocrine responses to thyrotropin-releasing hormone, chronic LD therapy in parkinsonian patients does not lead to hypothyroidism, and only a minimal attenuation was found in the stimulation of prolactin in women (Lavin et al. 1981). Hypothalamic growth hormone-releasing factor is stimulated by LD (Chihara et al. 1986). With conventional LD therapy, plasma growth hormone was reported to be stimulated by each dose of LD on a continuing basis (Sirtori et al. 1972), though baseline levels or responses to LD stimulation are not altered by chronic therapy (Galea-Debono et al. 1977). Other effects of chronic LD therapy include a decrease in glucose tolerance and an exaggerated insulin test response (Sirtori et al. 1972).

While the existence or nature of abnormalities in EEG and evoked responses in PD has continued to be somewhat controversial (Dinner et al. 1985), some studies have shown a prominent influence of LD therapy on these measures. The conventional EEG undergoes no significant alteration when PD patients are started on LD (Yaar 1977). However, spectral analysis of the EEG power density showed significant effects from LD in 25 patients studied before and after the start of medication; there was an increase of power in all clinically relevant bands, especially in the region of the left occipital lobe (Yaar and Shapiro 1983). In one report, LD therapy in PD patients resulted acutely in increased latencies of flash and pattern reversal visual evoked responses (Cosci et al. 1984). Mintz et al. (1982) found that the amplitude of visual evoked responses was reduced over the hemisphere contralateral to parkinsonian features in hemisymptomatic patients. Sustained therapy with LD reversed this lateralized asymmetry. Human visual contrast sensitivity improved with dopaminergic therapy, potentially at the retina as well as on a central basis (Domenici et al. 1985).

E. Pharmacodynamics of Levodopa Therapy

The sustained use of LD is associated with a number of pharmacodynamic effects which may have an important bearing on its clinical use. A decrease in postsynaptic function (owing to decreased striatal DA receptors) has been implicated as a major factor to explain declining LD efficacy. Whether this occurs as a manifestation of worsening PD or as a result of therapy has been controversial. As for the lesioned nigrostriatal DA system in the rat, the dropout of nigrostriatal neurons in the PD brain is associated with the development of a supersensitive receptor state (LEE et al. 1978). In the striatum of untreated PD patients, measures of DA binding showed that this supersensitive state was not decreased as a function of either age or the duration of PD (GUTTMAN et al. 1986). However, the possible interactions between parkinsonism and dopaminergic therapy on the maintenance of receptor responsiveness have made interpretation of the currently available data a difficult matter. The effects in animals of sustained LD treatment have been used to explore the effects of therapy in both intact and dopaminergically denervated striatum. With continuous use (12 weeks) in guinea pigs, LD treatment lost the behavioral effects observed acutely (CARVEY et al. 1982). Furthermore, in association with decrease in striatal DA content, chronic LD therapy in the guinea pig resulted in a lessening of stereotypies after acute dopaminergic challenges with apomorphine (PAULSETH et al. 1984). These neurochemical-behavioral correlations are typical of many animal studies which have found evidence for a "downregulation" of dopaminergic functions after continuous LD use. No changes in the affinity of [^{3}H]spiroperidol binding sites (a correlate of postsynaptic D_2 receptors) resulted from LD treatment in some studies in rodents and in humans (PONZIO et al. 1984; GUTTMAN et al. 1986), but other investigations with sustained LD regimens in rodents have shown decreased striatal [^{3}H]spiroperidol binding (FRIEDHOFF et al. 1977; MISHRA et al. 1978; LIST and SEEMAN 1979; RECHES et al. 1984; PONZIO et al. 1984). In other reports, there was no effect (SUGA 1980; JACKSON et al. 1983) or an increase in receptor binding from sustained LD use (PYCOCK et al. 1982; WILNER et al. 1980). The differences in experimental technique, species, and data analysis may account for the disparate conclusions.

After nigrostriatal dopaminergic lesioning in rats, the resulting increased density of striatal [^{3}H] spiroperidol binding sites (indicating postsynaptic supersensitivity) was attenuated significantly by sustained LD treatment, though similar receptor changes did not occur for unlesioned controls (SCHNEIDER et al. 1984). Chronic LD regimens may (JACKSON et al. 1983) or may not (RECHES et al. 1984) have an influence on the development of presynaptic DA receptor supersensitivity (which can have autoreceptor roles in DA synthesis and release). Repeated use of LD may differentially affect other subtypes of DA receptor. In nigrostriatally lesioned rats, sustained LD treatment enhanced receptor-mediated functions at D_1 (adenylate cyclase-linked) receptors, though the number of [^{3}H]spiroperidol binding sites (primarily D_2-type DA receptors) were decreased (PARENTI et al. 1986).

The implications of LD-induced changes on receptor functions can be complicated by other less understood neurochemical or pharmacodynamic effects to sustained LD treatment. For example, short periods of LD treatment in mice can result in evidence of dopaminergic receptor supersensitivity or subsensitivity, depending on the length of delay after discontinuation of LD treatment (BAILEY et al. 1979). A 6-month period of continuous LD administration to rats (which increased [^{3}H]spiroperidol binding) was also characterized by an increase in amphetamine-induced stereotypy, decreased cataleptic potency of challenges with neuroleptic drugs, and enhanced striatal content of LD and

DA as compared with shorter periods of treatment (Pycock et al. 1982). Even in the absence of any changes in the properties of striatal DA receptor binding, repeated administration of LD to rats alters some of the dose-dependent behavioral effects of dopaminergic challenge (Hall et al. 1984). It is unlikely that changes in DA receptor function occur as direct consequences of DA metabolism, since quantitative autoradiography with [^{3}H] spiroperidol is essentially unaltered in rat striatal tissue exposed to high concentrations of LD or DA metabolites (HVA, DOPAC, 3-*O*-MD, and 3-methoxytyramine) (Wooten and Ferrari 1983).

The interpretation of DA receptor effects from chronic LD therapy in humans is further complicated by the influence of progression in the underlying PD. Some reports have concluded that an increase of striatal binding sites found in the nigrostriatal dopaminergically lesioned animal did not occur in the PD brain after chronic LD therapy (Lee et al. 1978; Rinne et al. 1980a). Other investigations showed two patterns of change in striatal DA receptor after LD treatment: either a decrease or normal numbers, but not the increase found in brains of some PD patients not treated with LD (Rinne et al. 1980c). As compared with LD-untreated patients, LD use apparently reversed the tendency for an elevated density of D_2-type receptors in the striatum as a consequence of denervation (Guttman and Seeman 1985). However, further studies indicated that the decrease of striatal DA binding sites was not correlated with duration of PD or its treatment with LD (Guttman et al. 1986).

F. Mechanism of Action of Levodopa in Parkinsonism

There is extensive evidence that replacement of DA constitutes the primary antiparkinsonian effect of LD. LD therapy resulted in nearly normal DA striatal content in the PD brain for as long as 9-h after the last dose (Lloyd et al. 1975; Rinne et al. 1979). While DA initiates a number of neurochemical and neuroendocrine effects within the CNS (such as the stimulation of growth hormone; Sirtori et al. 1972; Galea-Debono et al. 1977), there is no evidence that this hormonal effect, nor any other DA-initiated metabolic actions, have an additional role in mediating an antiparkinsonian effect.

The pharmacologic model for antiparkinsonian action calls for sufficient LD to undergo decarboxylation to replace adequate DA for receptor stimulation. Experimental animal models of neostriatal lesions have helped to explore the effects of LD in the face of extensive presynaptic destruction of DA-containing projections. Even with almost complete destruction of these pathways, systemically administered LD can restore physiologically active levels of DA in the striatum (Poirier et al. 1967). Studies in the rabbit striatum with unilateral dopaminergic denervation by 6-hydroxydopamine showed that the synthesis of DA on the side ipsilateral to the lesion was less than on the intact side, though DA production occurred rapidly and levels remained elevated for 2-h. Though the amount of DA produced on the lesioned side is less, the percentage increase in DA content was greater (Doller and Conner 1980). With a ventromedial tegmental lesion (interrupting nigrostriatal projections) in the cat brain, there was an inverse relationship between the severity of the lesion and the ability of the caudate to synthesize DA after administration of exogenous LD (LLOYD et al. 1980).

Sustained administration of LD to the rabbit resulted in the appearance of DA at extrastriatal brain sites normally with very low DA content (DOLLER and CONNER 1980). However, studies with autopsied parkinsonian brain after chronic LD treatment have shown elevated DA content only in the neostriatum (RINNE et al. 1974), though high levels of LD, HVA, and 3-*O*-MD were present throughout the brain (DAVIDSON et al. 1971; RINNE et al. 1971). LD can be symptomatically quite effective, even when resultant DA content in the striatum was substantially below normal values (YAHR et al. 1972). In one study, a favorable response to LD was associated with a greater rise of neostriatal DA than for subjects with a smaller increase (LLOYD et al. 1975). In another report, however, striatal HVA concentration and degree of clinical response were not correlated (RINNE et al. 1974).

Animal studies have also helped to confirm that the antiparkinsonian effect of LD in the striatum results from conversion to DA (LLOYD 1977; LLOYD et al. 1980; MELAMED et al. 1981; HORNYKIEWICZ 1970, 1979). When a centrally acting inhibitor of LD decarboxylation is given with LD, the antiakinetic effect of intrastriatally injected LD was abolished (MELAMED et al. 1984). Similar results have come from using the peripheral DCI benserazide (BZ) in excessively high dosage (resulting in entrance of BZ into the brain and thereby lessening the antibradykinetic response to LD; GOODALE and MOORE 1976).

However, the production of DA from LD and a concomitant anti-akinetic response does not prove that all of the antiparkinsonian effect of LD results from replacement of the neurotransmitter. It may be that the supersensitivity response of postsynaptic receptors may be in some way vital for the continuing effectiveness of LD (AGNATI and FUXE 1980). Other metabolic products of LD therapy have been proposed to interact with it or add to its effectiveness (SOURKES 1970, 1971; SANDLER 1971). The clinical experience of using LD in parkinsonism has led to questions as to whether, in addition to DA, there might be substances long-lived enough to account for the sustained effect of LD, sometimes lasting for days after discontinuation (YAHR et al. 1969). The rapid turnover of DA would make it unlikely that even sequestered DA could be solely responsible, particularly in the context of the severe nigrostriatal lesion in PD. For example, with rats ipsilaterally lesioned in the substantia nigra, the turning behavior induced by acute LD challenge persisted for much longer after the rise and return to baseline of striatal DA (SPENCER and WOOTEN 1984b). Furthemore, DA turnover resulting from LD dosing in the intact rat striatum lasted for approximately 3-h, but was attenuated significantly in magnitude and duration in the dopaminergically denervated striatum (SPENCER and WOOTEN 1984a). Similar observations have been made with positron emission tomography in parkinsonian patients and normal controls given ^{18}F-labeled LD (LEENDERS et al. 1986). The rise in striatal tracer, lasting for 240 min in controls, continued for only 100 min in the parkinsonian patients. This suggested a decreased ability to store the tracer (presumably as fluorodopamine). Studies in rabbit striatum suggested that DA derived from infusion of LD was incorporated into a rapid turnover intracellular "functional" pool, rather than a storage compartment (Doller and Conner 1980). Newly synthesized DA achieved a steady state level in the rabbit striatum after 1-h, with a tissue half-life of less than 15 min. As alternatives to kinetically distinct compartments, other mechanisms have also been suggested for the biphasic pattern of decay in DA content (Paden 1979).

In addition to the products of conventional monoamine metabolism, relatively minor metabolic products or chemical species from nonenzymatic condensation reactions with DA may result. Particularly in the setting of supraphysiologic intracellular content of LD, these by-products of therapy could conceivably accumulate and exert actions contributory to either the therapeutic or adverse effects of DA (Sourkes 1971; Sandler et al. 1973; Dougan et al. 1975; O'Leary and Baughn 1975 Coscia et al. 1977). For example, reactions between DA and aldehydes generated by alcohol dehydrogenase yield Pictet-Spengler condensation products such as salsolinol, tetrahydropapaveroline, and norlaudansoline carboxylic acids in the urine of patients treated with LD (with or without DCI) (Sourkes 1971; Sandler et al. 1973; Coscia et al. 1977). Further metabolism of tetrahydropapaveroline to compounds sharing structural and, possibly, pharmacologic similarities with the dopaminergic agonist apomorphine has been proposed (Sourkes 1971). While conventional catecholamine metabolites do not appear to affect DA receptor binding (Wooten and Ferrari 1983), studies of rodents given large loads of LD have suggested the production of novel metabolic species which can interfere with DA metabolism (El Gemayel et al. 1986). For example, tetrahydroisoquinolines derived from catecholamines may be inhibitory to tyrosine hydroxylase (Weiner and Collins 1978). 3,4-Dihydroxyphenylacetaldehyde, a relatively minor metabolite of LD decarboxylation, inhibits AAAD and could conceivably have a regulatory effect on the conversion of LD to DA [O'Leary and Baughn 1975). Although no roles for the other metabolites have been established, by-products of LD metabolism might have interactions with DA receptors or other systems that could help to explain some poorly understood clinical phenomena, such as the delay in developing optimal effect after initiation of therapy, or the gradual decline of antiparkinsonian benefit days after abrupt discontinuation of LD.

As discussed already, 3-*O*-MD is produced in substantial quantities during LD therapy and has been found throughout the brain (Davidson et al. 1971). It has been suspected of interfering with LD therapy because of its high levels and prolonged pharmacokinetic profile. However, 3-*O*-MD has no direct effects on catecholamine turnover (Gervas et al. 1983). When injected intrastriatally in rats with lesioned nigral neurons, 3-*O*-MD (like the DA metabolites DOPAC and HVA) exerts no dopaminergic effects (Melamed et al. 1984). Though 3-*O*-MD was initially suspected to be a source for regenerating LD (Bartholini et al. 1972), no evidence for this has come from animal (Reches and Fahn 1984) or human experiments with 3-*O*-MD administration (Muenter et al. 1973; Kremzner et al. 1974).

While the synthesizing capacity for nigrostriatal DA is generally not completely absent in PD, it has been uncertain whether those remaining elements of a severely degenerated dopaminergic nigrostriatal system are capable of adequate DA production to account for the clinical efficacy of LD (LLOYD et al. 1973). Initially, it was hypothesized that surviving nigrostriatal DA neurons might be able to increase their neurotransmitter output. Increased ratios of HVA to DA in the untreated PD brain supported the concept of an enhanced DA turnover in each remaining neuron (HORNYKIEWICZ 1981). Rodent models using 6-hydroxydopamine to produce partial lesions of the nigrostriatal projections (AGID et al. 1973; HEFTI et al. 1980) demonstrated increased turnover of DA among the remaining neurons. Could this compensation be present in the parkinsonian brain to permit enhanced decarboxylation of exogenous LD as well? In 6-hydroxydopamine-lesioned rat striatum, it was possible to show substantial conversion of exogenous LD to DA (MELAMED et al. 1980a, b). However, other evidence is lacking that the nigrostriatal dopaminergic neurons are by themselves capable of a major increase in DA production and release (BUNNEY et al. 1973). Intact nigrostriatal DA neurons may be incapable of a substantial increase in their net output of DA, even with alterations of neuronal impulse flow and the provision of exogenous LD (MELAMED and HEFTI 1984). Hence, it may be that the recovery of function with LD is not entirely mediated through the surviving DA neurons (HEFTI et al. 1981b). Even with complete loss of the dopaminergic system, it appears that nondopaminergic neurons with decarboxylase activity are capable of acting on LD to restore DA levels in the striatum.

Since a significant portion of decarboxylase activity resides in other than dopaminergic striatal systems, it has been proposed that decarboxylation of LD could occur with serotonin-generating neurons or in the striatal microvasculature (MELAMED et al. 1980c). However, studies with selectively lesioned striatal serotonergic neurons amid previously lesioned dopaminergic nigrostriatal pathways have shown that serotonergic neurons are not needed for generating DA from LD (MELAMED et al. 1980a). Similarly, the decarboxylation occurring in capillary walls is not likely to contribute to functional striatal DA synthesis (MELAMED et al. 1980c), as this process is outside the blood-brain barrier and subject to DCI action. Substantial decarboxylase activity in nonaminergic striatal interneurons or efferent neurons with proximity to postsynaptic DA receptors has been hypothesized as the means by which functional DA neurotransmission is restored (MELAMED et al. 1980b). Much more needs to be learned of the metabolic and regulatory properties of striatal LD metabolism, particularly if it is confirmed that the surviving DA neurons are not fully responsible for LD responsiveness.

One question relevant to the long-term problems associated with LD therapy is whether sustained use of LD affects its own utilization peripherally or in the CNS. There has been some experimental evidence in support of this from animal studies by MELAMED et al. (1983). Repeated doses of LD to intact and dopaminergically lesioned rats resulted in elevations of striatal DA that were smaller and of shorter duration at 31 days than after 10 days of treatment. However, PONZIO et al. (1984) found no change in striatal levels of DA or its metabolite 3-methoxytyramine after sustained LD treatment in the rat. In guinea pigs, continued treatment for 12 weeks with LD resulted in less accumulation of LD in plasma and a reduction in LD-induced behavioral effects from the same daily LD dose (Carvey et al. 1982). Further study is needed to learn if potentially modifiable changes in the metabolism of LD over time explain the declining efficacy or fluctuations associated with this drug.

G. Levodopa Preparations

I. Decarboxylase Inhibitors

LD is decarboxylated irreversibly by L-aromatic amino acid decarboxylase (AAAD), an enzymatic moiety found throughout the body in the vascular endothelium and also situated in gut wall, pancreas, kidney, liver, and brain. In the latter site, regional concentration is highest in the striatum (Bouchard et al. 1981). Continuing activity of AAAD is vital to the sustained effectiveness of LD: the small amount of nonenzymatic decarboxylation that occurs in the brain (Sandler 1971) is inadequate for the demands of DA synthesis. In the PD brain, the extent to which striatal AAAD activity is preserved has been difficult to study, though decarboxylase activity has been reported to be reduced to approximately 10% of normal (Marsden 1975; Lloyd et al. 1973).

Pyridoxyl-5-phosphate serves as cofactor for AAAD, and increased levels enhance enzyme activity. The clinical importance of the pyridoxine (vitamin B_6) effect is underscored by studies showing the attenuation of LD benefits by exogenously administered pyridoxine (Yahr and Duvoisin 1972; Mars 1974). Presumably, the increased levels of the vitamin "drive" the activity of peripheral AAAD, increasing the extracerebral conversion of LD to DA, and resulting in less LD delivery to the brain. AAAD activity is not thought to be a rate-limiting step for "physiologic" DA synthesis in the CNS. However, there has been evidence from parkinsonian therapeutics to suggest that supplementary pyridoxine given to patients with peripheral AAAD blockade can enhance clinical responsiveness to LD, possibly by increasing the activity of central decarboxylation (Duvoisin 1973).

Because the majority of LD undergoes peripheral decarboxylation when given as a monotherapy, dosage requirements to permit adequate quantities to enter the CNS are generally in excess of 3 days. The abundant systemic AAAD activity ensures that much of orally adminstered LD will undergo conversion to DA peripherally before reaching the brain, including sites where adverse effects are initiated. Two hydrazine derivatives highly effective at inhibition of AAAD, carbidopa (CD) (MK-486, Lodosyn) and benserazide (BZ) (Ro 4-4602), were introduced to clinical practice in the early 1970s (Pinder et al. 1976; Yahr 1973; Pletscher 1973; Marsden et al. 1974a; Papavasiliou et al. 1972). These drugs have had a major impact on LD therapy, which became much better tolerated and safer, owing to their ability to reduce or eliminate LD-induced adverse effects such as nausea, vomiting, and hypotension. Although all of these effects can also occur on a central basis, the restriction of peripheral LD conversion to DA takes away a major component of LD-initiated adverse reactions, and so has permitted the use of more efficacious and rapidly built up LD doses in clinical practice. The strategy of combining DCI with LD appears to be of greater clinical utility than that achieved from a comparable LD dose combined with a peripheral dopamine receptor antagonist, domperidone (capable of blocking most peripheral dopaminergic effects) (LANGDON et al. 1986).

LD-DCI combinations have been shown to be considerably more effective than LD alone. In one study in which the results of adding 200 mg CD daily were compared with results with an optimized amount of LD alone, a similar degree of improvement was initially achieved. However, follow-up 1 year later found the CD-LD regimen to result in 56% "good" and 13% "poor" responses, as compared with 33% "good" and 28% "poor" responses in the group receiving an LD-placebo regimen (MARSDEN et al. 1974b). While the potentiation of LD effect is approximately three- or four-fold for antiparkinsonian control, the increase of antidepressant action from LD can be substantially more when a DCI is added (DUNNER et al. 1971).

Apart from facilitating the use of LD, there has been only inconsistent evidence that therapy with DCI prolongs LD plasma half-life (DUNNER et al. 1971; DOLLER et al. 1978; RINNE et al. 1973b). Most studies have found no increase in duration of LD clinical effect (FAHN 1974), permeability into the brain (FRIIS et al. 1981), or antiparkinsonian potency. Nor is there evidence that DCI alters the essentials of LD metabolism other than by taking away the peripheral decarboxylation component of systemic metabolism. One report indicated that DCI cotherapy reduced the marked variability in achieving peak LD levels (TYCE et al. 1970). DCI therapy has been reported to increase the production of transaminated metabolites (SANDLER et al. 1974a), and decrease HVA and DOPAC production after LD treatment (BIANCHINE et al. 1972). The coadministration of BZ or CD with LD leads to decreased content of HVA in CSF, as compared with the effects of LD alone (RINNE et al. 1973a). Also, the ratio of LD to DA increases in the CSF of parkinsonian patients after DCI is added to LD therapy (PAPAVASILIOU et al. 1973). Some investigations have shown that DCI therapy improves the bioavailability of LD (TISSOT et al. 1969) by prolonging both the distribution and elimination phases of LD pharmacokinetics (NUTT et al. 1985).

As will be discussed later, most studies indicate that it is the total daily intake of the DCI rather than its ratio to LD content which influences the antiparkinsonian effectiveness of a dose of LD plus DCI. One report claims that administering the DCI prior to giving LD may enhance LD effect to a greater extent than with simultaneous administration (McLELLAN and DEAN 1982), but for most situations, coadministration of DCI with LD is adequate. Both CD and BZ are marketed as fixed-ratio products with LD (CD can be obtained as a separate product from its manufacturer in the United States, Merck Sharp & Dohme, for supplementation purposes).

CD and BZ behave as pseudoirreversible inhibitors of AAAD (DA PRADA 1984; PORTER 1973). Both are highly effective at limiting peripheral decarboxylation, but their inhibitory effects are not complete. Studies in rats with a series of irreversible AAAD inhibitors were carried out for comparison to the effects of CD, and showed increased potency in increasing the plasma half-life and bioavailability of LD (Huebert et al. 1983). These studies indicated that, even with more effective peripheral inhibition of AAAD, the practical limits to increasing LD availability to the brain are probably not complete decarboxylase inhibition, but rather the pathways for peripheral LD metabolism.

Another compound with DCI properties in humans is α-methyldopa (AMD). Though relatively weak as a DCI (Bartholini and Pletscher 1975), it is effective enough to increase central dopaminergic action with LD therapy and to reverse pyridoxine-induced antagonism of LD effect. AMD also has been suspected of having a direct or indirect effect on enhancing central LD decarboxylation (Fermaglich 1973).

Both BZ and CD are largely excluded from the brain on the basis of their lipid/water partition coefficients (Clark et al. 1973). CD and BZ (the latter metabolized in vivo to its presumed active species, trihydroxybenzylhydrazine) are potent at inhibition of enzymatic as well as nonenzymatic components of systemic decarboxylation (Vogel et al. 1972). In addition, BZ inhibits microbial metabolism of LD in the gut via aromatic dehydroxylation (Baake and Scheline 1974). CD and BZ lack the MAO inhibitory activity present in other hydrazine derivatives. Only in high dosage does the weak affinity of CD and BZ for pyridoxine-metabolizing enzymes result in significant antagonism and a pyridoxine deficiency state (Porter 1973). However, this has not been found from the use of conventional doses in patients receiving chronic therapy (Yahr and Duvoisin 1975). One advantage conferred by use of DCI and LD therapy is that there is no loss of therapeutic effectiveness if patients receive pyridoxine (Klawans et al. 1971), in contrast to the situation with LD alone (Duvoisin et al. 1969). By inhibiting kynurenine hydroxylase, both CD and BZ can potentially cause a relative deficiency of niacin (Bender et al. 1979).

CD and BZ are rapidly, though variably, absorbed from the gut: 40%–70% of an oral dose of CD (Vickers et al. 1975), and 66%–74% for BZ (Schwartz et al. 1974). For both, plasma half-lives are approximately 2–3 h, with peak levels in plasma attained within 1-h (Pinder et al. 1976). With CD and BZ, the parent drug is rapidly excreted, after which urinary metabolites appear (which are well characterized for CD, but not for BZ; Schwartz et al. 1974).

Therapy with DCI such as CD thus limits the quantity of peripherally generated DA, enhancing bioavailability by increased gut absorption and decreased first-pass metabolism in liver and kidney. The facilitative effects permit LD dose reduction in clinical use to approximately 25% of the comparably effective LD dose. With DCI, the rate and extent of LD absorption in normal subjects is the same at several sites in the upper small intestine (Gundert-Remy et al. 1983).

While some studies have concluded that as little as 75 mg/day CD in divided doses achieves the major clinical benefits of DCI (Jaffe 1973; Chase and Watanabe 1972), other reports found improvement increased from larger doses. For example, Hoehn (1980) found that increasing the mean CD dose from 60 to 120-180 mg/day led to improved control of parkinsonism. In a multicenter open-label evaluation, increasing the mean daily intake of CD from 55.8 to 141 mg/day (with no change of LD dose) resulted in small but statistically significant group improvements for some cardinal features of parkinsonism (NIBBELINK et al. 1981). Similar claims were made for further improvement from a fixed LD dose used with increase of CD from 100 to 200 mg/day (TOURTELLOTTE et al. 1980). In contrast, other studies showed no further decrease of optimal daily LD intake from a CD dose of more than 75 mg/day (JAFFE 1973; PREZIOSI et al. 1972; CHASE and WATANABE 1972), 160 mg/day (PINDER et al. 1976), 2.5 mg/per kilogram body weight per dose (WARD et al. 1984), or increasing from a mean dose of 145 to 290 mg/day (CEDARBAUM et al. 1986a).

With switch from a fixed 10-100 to a 25-100 mg CD-LD preparation, fewer gastrointestinal disturbances, dyskinesias, and psychiatric side effects were reported (HOEHN 1980; NIBBELINK et al. 1981; BERMEJO PAREJA et al. 1985). The maximally effective dose of BZ has not undergone as extensive study as that of CD, but the optimal dose is likely no greater than 125 mg/day (TISSOT et al. 1974). The results of doubling CD intake from an average of 145 to 290 mg/day include altered pharmacokinetics of oral LD such that peak plasma concentration occurred earlier for three of five subjects, and reached values approximately 44% greater (CEDARBAUM et al. 1986a). The mean increase for the area under the curve of LD concentration (a measure of LD bioavailability) was 27% with a mean daily LD dose of 1450 mg. The clinical effects of increased CD dosage included no prolongation of LD effect nor improvement in clinical fluctuations, as reported in other studies as well (PINDER et al. 1976). When studied in the context of LD infusion, oral CD decreased by one-half the rate of drug delivery (but not the plasma LD level) needed for antiparkinsonian effect (NUTT et al. 1985). In some studies, only modest improvement in clinical effect came from increase of CD dose, though it resulted in increased blood levels of LD (BERMEJO PAREJA et al. 1985). The increased CD dose also elevated plasma levels of 3-*O*-MD and tryptophan in proportion to administered dose, thereby generating substances that can compete with the uptake of LD into the brain.

To evaluate the possibility that exposure to the large doses of LD and DCI routinely used in chronic therapy might alter systemic AAAD activity, measurements of decarboxylation were carried in chronically treated and untreated parkinsonian patients, and in normal controls (WARD et al. 1984). The technique involved measurement of expired radiolabeled CO_2 derived from 1-[^{14}C] LD. No difference was found among the three groups tested, suggesting that total body decarboxylation does not change under the chronic influence of LD-CD therapy. Furthermore, this study demonstrated that 10%-15% of administered LD is decarboxylated within 3 h (either from residual AAAD or nonenzymatic decarboxylase activity when maximal systemic doses of DCI are used).

The joint location of active transport mechanisms for neutral amino acids and AAAD activity within cerebral capillary endothelium (OWMAN and ROSENGREN 1967) and choroid plexus (LINDVALL et al. 1980) has raised the possibility that inhibition of decarboxylase activity might alter LD transport across the blood-brain barrier (Hardebo and Owman 1980). Studies in rodents have estimated the enzymatic capacity of the cerebral capillary endothelium for decarboxylating LD to be approximately 3 % of the total systemic activity (Hardebo et al. 1977). Since the site of LD conversion to DA for striatal action is not the cerebral capillary endothelium (Melamed et al. 1980c), AAAD in capillary endothelium could conceivably divert some LD from transluminal transport into the striatum. However, the use of CD in doses exceeding those needed for maximal systemic decarboxylase inhibition (Pinder et al. 1976; Ward et al. 1984) did not alter the antiparkinsonian plasma levodopa concentrations (during long LD infusions) at steady state conditions (Nutt et al. 1984). These findings have confirmed that AAAD activity is not linked to the influx mechanism for LD.

Whether there is a difference between the clinical effectiveness of CD and that of BZ has been evaluated in a crossover study comparing 12-week treatment periods of CD-LD and BZ-LD, with similar dosage schedules resulting in roughly equivalent plasma LD levels (Rinne and Molsa 1979). Although there were no significant differences in the improvement of parkinsonian symptoms and overall disability, nausea and vomiting occurred more often with the CD-LD preparation (25-250 mg) than with BZ-LD (25-250 mg). Comparison between equivalent regimens of CD and BZ with LD showed evidence of greater peripheral DA formation with CD, as shown by a larger ratio of plasma DA to LD (Lieberman et al. 1978). Other evaluations of the two DCIs (Da Prada 1984; Korten et al. 1975; Greenacre et al. 1976; Pakkenberg et al. 1976; Diamond et al. 1978; Lieberman et al. 1978) found comparable beneficial effect with preparations of BZ-LD (1:4) and CD-LD (1:10), though there were some differences in adverse effect profiles. In one study with elderly parkinsonian patients, comparable doses of BZ were associated with better clinical outcomes that those achieved with CD (Admani et al. 1985). Although pharmacokinetic studies are needed for confirmation, clinical experience with parkinsonian patients who showed fluctuations suggested that combining BZ with CD resulted, in some instances, in improvements of parkinsonian control compared with CD alone (Lieberman et al. 1984).

It has been suggested that the use of DCIs may add to the risk of complications related to chronic use of LD, such as on-off syndrome and dyskinesias. One study (De Jong and Meerwaldt 1984), which compared regimes of LD approximately equivalent in clinical effect, concluded that the DCI-treated group experienced more problems associated with long-term LD therapy (such as dyskinesia and wearing-off).

II. Sustained-Release Forms

Sustained-release forms of LD with DCI have proven to be a difficult challenge for pharmaceutical development. The "brittle" parkinsonian patient has had a continuing need for a controlled drug delivery product, particularly with the difficulties in maintaining constant effect imposed by the distribution, transport, and metabolism of LD. For many patients with advanced parkinsonism and enhanced sensitivity to small changes of plasma LD concentration, the pharmacokinetic profile of gastrointestinal absorption and transfer to the CNS become critical factors for maintaining optimal effect. This has been illustrated by the improvements in fluctuations gained for patients from intravenous delivery of LD (SHOULSON et al. 1975; NUTT et al. 1984; HARDIE et al. 1984; QUINN et al. 1984). The search for preparations of LD with more uniform effect (achieved by less peak rise or drop in plasma level) began more than 10 years ago. Initially, enteric-coated forms were tried (SANDLER et al. 1974b). Although some patients improved (LAITINEN 1973; SAARINEN et al. 1978; GILLIGAN and HANCOCK 1975; GILLIGAN et al. 1979), other experience was largely unsuccessful because absorption tended to be delayed and incomplete (ECKSTEIN et al. 1973; RINNE et al. 1973b; CURZON et al. 1973; WOODS et al. 1973), as confirmed by pharmacokinetic studies (MORRIS et al. 1976). Further attempts in this direction have continued (NISHIMURA et al. 1984).

More recent developments have involved trials with four different oral controlled-release forms of CD-LD (Sinemet CR-1 through CR-4) (HUTTON et al. 1984; JUNCOS et al. 1986; GOETZ et al. 1986; CEDARBAUM et al. 1986b) and BZ-LD (Madopar HBS) (MARSDEN et al. 1987). With equivalent LD content, the CR-1 form proved to be less effective than conventional CD-LD (Sinemet 25/100). Both Sinemet CR-2 (with an erodible matrix) and CR-3 (with a polymeric nonerodible matrix) had increased mean plasma LD levels 4 and 8 h after dose, as well as more delayed and smaller peak values than the standard CD-LD preparation (HUTTON et al. 1985). Pharmacokinetic studies with Sinemet CR-3 showed more sustained plasma LD concentrations, but considerable hour-by-hour variability in levels (NUTT et al. 1986). Administered every 4 h in substitution for conventional CD-LD, Sinemet CR-3 required approximately three times as much daily LD intake. Although clinical response in the morning was improved in some patients, fluctuations or other functional disabilities were not aided by this delayed-release preparation. In a comparison of Sinemet CR-4 with conventional CD-LD, recent studies have shown clinical improvement for PD patients with LD response fluctuations, including more "on" time, fewer "off" periods, and greater plasma levels as compared with a standard CD-LD regimen (CEDARBAUM et al. 1986b; GOETZ et al. 1986).

The sustained-release preparation of BZ-LD (Madopar HBS) makes use of a gelatin capsule which is transformed in the stomach to a mucous body delivering both constituents by diffusion through a hydrated layer (ERNI and HELD 1987). Trials comparing Madopar HBS with conventional BZ-LD (Madopar) (POEWE et al. 1986a) have shown that six of ten patients had a reduction in end-of-dose and other types of motor fluctuations, although these benefits did not continue in all patients during longer follow-up. Consistent with the decreased bioavailability from this preparation, substantially more LD was needed daily for optimal clinical response. Problems with this and other preparations of sustained-release LD have included the delayed onset of effect and the lack of improvement of unpredictable fluctuations in some patients (despite the theoretical advantage of sustained drug release). The activity of gastric emptying and gastrointestinal motility in general may prove to be major determinants of efficacy with sustained-release forms, just as is the case with the current preparations (Berkowitz and McCallum 1980; Rivera-Calimlim et al. 1970a, b). It is likely that many patients would be best managed by using the sustained-release form in combination with the faster-acting conventional DCI-LD preparations.

III. Enhanced Delivery and Uptake

Since many patients with motor fluctuations are improved by intravenous delivery of LD, methods for chronic parenteral administration have been sought. Intravenous administration, although promising as a research tool, has proven to be impractical for several reasons. The daily intake needed for LD would require an extremely large volume of solution, owing to the low aqueous solubility of LD. Because it can be prepared as a more concentrated solution, LD methylester has been proposed for chronic intravenous use (Cooper et al. 1984). LD methylester is readily hydrolyzed to LD, and constant infusions give effects similar to those of LD alone (Juncos et al. 1986). However, concern has been raised as to possible toxicity from methanol generated by the hydrolysis of the methylester. Other strategies for improving delivery of LD have been devised using various transient derivatives of LD. These derivatives prevent metabolism prior to or during absorption, thus achieving better bioavailability (Bodor et al. 1977; Garzon-Arburbeh et al. 1985). Chelating LD to certain divalent metals has been one means of enhancing LD transport into the brain (Diamond et al. 1982). In animal studies, Cu^{2+} or Zn^{2+} chelated to LD increased LD transport into the brain by 100%–150% (Rajan et al. 1976).

Other routes for LD delivery have been attempted. Administration of an LD suspension rectally was ineffective in one trial (Eisler et al. 1981), but gave improvement in another (Beasley et al. 1973). Since slowed or erratic gastric emptying may account for some fluctuations in antiparkinsonian control, direct duodenal infusion of LD has been tried (Bianchine et al. 1971; Kurlan et al. 1986). This approach has given significant benefit for apparent LD nonresponders and for patients unable to improve further by adjustment of oral LD dose.

H. Therapeutic Uses for Levodopa Other Than Parkinsonian

LD therapy has been tried in other clinical settings. After severe closed head injury, improvements in level of consciousness, eye opening, and motor responsiveness have been reported in some patients from the use of LD (Van Woerkom et al. 1982). As measured by a parkinsonian rating scale, another study suggested that LD might facilitate the rehabilitation of such patients, even without the presence of overt parkinsonian features (Guidice et al. 1986). Marked improvement from experimental spinal cord injury (from ischemic insult) has occurred with the use of LD (Popovic et al. 1976).

As with other pharmacologic strategies enhancing CNS monoaminergic functions, LD significantly lessened the adverse effects of ethanol consumption on EEG changes, motor incoordination, and impairment of divided attention performance (ALKANA et al. 1977). LD has been useful for enhancing pain relief of bone metastatic lesions (NIXON 1975) and herpes zoster (KERNSBAUM and HAUCHECORNE 1981), as well as for pain of spontaneous origin from thalamic lesions (PLASCENCIA et al. 1984). Some patients with depression have improved from LD therapy (GOODWIN et al. 1970). Pathologic crying or laughter (emotional "incontinence") due to subcortical lesions (infarcts, trauma) can be controlled with LD therapy (UDAKA et al. 1984; WOLF et al. 1979). Many studies have confirmed the utility of LD in rapidly improving encephalopathic features during hepatic coma (FISCHER et al. 1976). Some clinical experience would suggest that early use of LD may help to avoid the transition of hepatic coma from a recoverable to an irreversible stage.

In congestive heart failure, oral LD therapy has improved hemodynamic performance (from the inotropic effects of DA) (BROWN et al. 1985; RAJFER et al. 1984). Since LD stimulates the release of growth hormone (GALEA-DEBONO et al. 1977), it has been effective in reversing some forms of growth retardation (HUSEMAN 1985). As a replacement therapy for DA, LD has also aided in improving alertness and motor disabilities in certain atypical forms of phenylketonuria. In the latter condition, monoamine neurotransmitter synthesis is impaired because of deficiency in tyrosine hydroxylase cofactor, tetrahydrobiopterin (MCINNES et al. 1984; MACLEOD et al. 1983). Several forms of primary dystonia may also be characterized by deficiency of this cofactor, and also by responsiveness to LD (LEWITT et al. 1986).

Treatment with LD counteracts some physiologic features of aging in certain rodent species, resulting in reinitiation of sex hormone cycles and coat growth. Constant dietary supplementation with LD in Swiss albino mice (Hale-Stoner strain) gave an approximately 50% enhancement in life span and maintenance of behavioral vigor into the older age spectrum (PAPAVASILIOU et al. 1981). In these experiments, the increase of brain MAO activity normally found with aging was attenuated, though there was no other neurochemical change. The Hale-Stoner strain does not show a decline in striatal DA levels with increasing age, as do other rodent species (and humans), and the improvement of vitality has not been validated by experiments with other rodent strains. In humans, mortality in the post-LD era for parkinsonian patients has improved greatly (JOSEPH et al. 1978; HOEHN 1983), though the basis for increased longevity is likely to be improved overall health and exercise, avoidance of falls, and so on.

J. Clinical Issues Regarding Initiation of Levodopa Therapy

I. "Early" Versus "Delayed" Use of Levodopa

The remarkable effectiveness of LD in relieving most parkinsonian symptoms (whether mild or severe) can have a seductive appeal for most clinicians and patients. It is extremely common for this drug to be started at the first sign of parkinsonism ("early" use), whether or not significant disability has developed. This practice has become an extremely important topic of controversy. There are some practical guidelines for dealing with this issue. It is clear that anticholinergics and amantadine, as alternatives to LD, can suppress much of the milder symptomatology in early PD; these drugs lack the potential for long-term complications that plague LD use. Bromocriptine and other direct-acting dopaminergic agonists are also alternatives, though they are generally less effective or tolerated than LD (Rinne 1985). Concern for withholding LD usage until development of significant disability has been founded on the relatively high incidence of declining efficacy and adverse effects from chronic therapy with it (Lesser et al. 1979). For many patients, loss of efficacy is closely linked to a decreased threshold for the occurrence of adverse effects, which then limits the dose of LD that can be used for controlling the advancing disability of the parkinsonism.

In one study, patients treated with LD for 4–8 years were significantly more impaired by parkinsonism than patients treated for 0–3 years, even when subjects were matched for total duration of their parkinsonism. Hence, deterioration of responsiveness after years of LD therapy could be in part a function of the therapy itself (Lesser et al. 1979). The risk of motor fluctuations, related to the duration of levodopa exposure, seems to be greater for the younger PD patient (Granerus 1978). If sustained LD therapy contributes to loss of receptor sensitivity, production of "on-off" and dyskinesia problems, or psychiatric disorders, then a conservative approach is warranted (initial use of alternative medications, starting LD "late," use of the smallest effective dose) (Fahn and Calne 1978). In applying this rationale, LD therapy is spared for treating more severe disability because it is considered to have a finite period for optimal utility, and a cumulative risk of adverse effects.

Follow-up studies of large groups of parkinsonian patients treated for 5 years or more have indicated that both milder and more severe stages of the disorder can maintain stable improvement from LD for up to 3 years, after which there is progressive decline in the benefit of LD (Yahr 1976). With an average daily LD dose of 3.5 G/day, one cohort of patients manifested peak improvement by 6 months, maintained significant clinical improvement for 3.5 years, and by 5 years was at the pretreatment level of disability (Rajput et al. 1984). These data suggest a similar duration for LD efficacy (regardless of PD stage) and argue for a limited term of LD effectiveness. However, holding in reserve a highly efficacious treatment may deprive a patient of months or years of optimal symptom control if, in fact, the apparent connection between the therapy and adverse outcome is not the duration of treatment. Other factors not linked to the time of starting therapy (such as quantity of LD used chronically) may also mediate the long-term decline in benefit and adverse effects. An answer to the question of when to start LD has been hindered by the difficulty of conducting controlled, randomized entry, longitudinal studies with either "early" or "delayed" use of LD. Hence, much of the available data is retrospective and subject to any number of influences, such as differences in prescribing patterns, rating techniques, or use of medication alternatives to LD.

On one hand, it has been difficult to dissociate decline in efficacy of LD from advance in the underlying parkinsonism. There has been no evidence of decline in nigrostriatal neuron counts or in enzymes metabolizing DA from the chronic use of LD in mice (Hefti et al. 1981a). No pathologic change in the substantia nigra or locus ceruleus was found in a patient treated for more than 4 years with over 2 kg of LD for incorrectly diagnosed PD (Quinn et al. 1986). There is likewise no evidence to suggest that LD hastens the progression of symptoms or changes the pathologic substrate of PD (Rinne et al. 1974; Yahr et al. 1972). Indeed, the report of Stern et al. (1972) noted that patients treated with LD for more than 2 years and subsequently withdrawn from medication were improved in motility as compared with their predrug baseline status. However, their suggestion that LD might actually have improved the underlying PD needs to be assessed in light of the methods used. Though the median drug withdrawal period was 10 days (range 3-31 days), the sustained effects of LD therapy, lasting sometimes for days after withdrawal (Muenter and Tyce 1971), may not have been absent for all of the patients at the time of their "drug-free" rating. Similar attempts to assess the status of underlying disease (Ludin and Bass-Verrey 1976) may also have had the confounding factor of long-duration LD effect.

Careful analysis of the pattern of parkinsonian signs and symptoms after 10 or more years of treatment showed most patients to have deteriorated in postural reflexes, speech, and gait, while rigidity, tremor, and handwriting were predominantly improved or unchanged (Klawans 1986). The lack of "global" deterioration in all features which are so typically those of parkinsonism may indicate that some chronic features of parkinsonism are not directly linked to the progressive loss of dopaminergic nigrostriatal projections. Subsensitive DA receptors resulting from the sustained use of LD could also be an explanation for waning clinical efficacy. This effect could well be a reversible phenomenon, since animal and human "drug holiday" studies have shown recovery of baseline responsiveness after LD withdrawal for short periods (see Sect. K). It has been argued that it may not be the duration of therapy, but, rather, progressive nigrostriatal degeneration that predisposes to dyskinesias and motor fluctuations (Muenter 1984). These problems can develop within weeks of starting LD (Lang et al. 1982; Ballard et al. 1985) to treat severe parkinsonism of abrupt onset from secondary causes. However, when LD first became available for idiopathic PD, there was little recognition, in the first months of use, of short-duration responses or marked fluctuation in control (Muenter 1984); these problems typically began on a more protracted time scale (Cotzias et al. 1969). Only rarely is dyskinesia encountered initially in LD use (LeWitt and Calne 1981). Since the first patients to receive LD included some with extremely advanced PD (both major disability and years of symptoms), the relationship between duration of LD therapy and increasing incidence of these problems has been persuasive to many investigators (Barbeau 1980; Klawans and Bergen 1975; McDowell and Sweet 1976, 1979).

A retrospective study (Markham and Diamond 1986) has claimed a direct relationship between disease (symptom) duration and disability, independent of the point at which LD was started. After the first 2 years, patients with earlier or later institution of LD started along similar rates of decline in disability ratings by the UCLA Scale. Matched for the length of their LD therapy (in excess of 6 years of LD treatment), the significant differences in disability scores persisted among groups that differed in their length of parkinsonism. This finding has implied that duration of parkinsonism (but not duration of LD use) was the major determinant of disability, not loss of LD effectiveness with increasing term of use. In this analysis, the incidence of random motor fluctuations over follow-up periods of up to 12 years did not correlate to length of disease or LD therapy, and the incidence and severity of dyskinesia likewise lacked a relationship to "early" or "late" initiation of LD (Markham and Diamond 1986). Another study (Hoehn 1983) concluded that the group of patients started with LD within 1 year of PD onset had better overall longevity and disability control than did patients waiting longer to begin LD. In a comparison between PD patients treated with LD for 15 years and a similarly disabled group followed in the pre-LD era, there was some evidence that postponing LD therapy was associated with a greater chance of drug unresponsiveness later on.

The wisdom of withholding LD therapy until significant disability intervenes is convincingly underscored from experience with some PD patients under chronic LD treatment whose underlying parkinsonism remains mild years later, but who have become disabled instead by severe dyskinesia or dystonic reactions. It is clear from other studies that, for some LD-free patients, the rate of functional decline can be so modest that activities of daily living are not compromised 5 or more years from symptom onset (Fahn and Bressman 1984). The latter study has also reassessed the data presented by Markham and Diamond (1981), concluding that the disability scale and methods of serial analysis were imprecise for comparing progression of parkinsonism between LD-treated and untreated groups.

A number of questions remain to be answered regarding how early LD can be started for optimal outcome. Several studies have shown that chronic monotherapy with bromocriptine, however unsatisfactory as a symptomatic treatment, is not associated with the frequent occurrence of motor fluctuations and dyskinesias found with comparable periods of LD therapy (Rascol et al. 1984; Stern and Lees 1983). The implication of these results is that it is not dopaminergic therapy per se, but rather the use of LD, that may be responsible for the adverse effects of long-term LD. When combined from the start with bromocriptine, LD therapy may attain a more favorable outcome in terms of efficacy and freedom from dyskinesias and fluctuations (Rinne 1985). It is not known to what extent "early" or "delayed" use of LD might influence the incidence or severity of LD-related hallucinations or dementia that develops for some parkinsonians patients. The age of the patient at the start of LD may also be a factor in the risk of adverse effects or mental deterioration with LD therapy (Pederzoli et al. 1983).

II. "Low Dose" Levodopa Regimens

Another factor having an influence on the outcome of chronic LD therapy has been the range of dose used. No guidelines have been derived from neuropharmacological research as to what constitutes a "low" or "high" dose regimen. Lacking foresight as to long-term risks associated with chronic LD therapy, many of the first clinicians to use LD gave maximally tolerated dosage (MARKS 1974). This policy undoubtedly led to the exacerbation of adverse effects and, possibly, an increased incidence of their initiation. After 5 years of LD therapy, the risk of motor oscillations has been related to greater LD intake (GRANERUS 1978). However, though greater parkinsonian symptomatology in patients requiring larger doses could also have been the basis of the increased risk. The use of low dose regimens in the elderly has a practical significance, as acute side effects that are dose related tend to affect this group to a greater extent (TURNBULL and AITKEN 1983; PEDERZOLI et al. 1983).

Attempts to restrict the amount of LD intake have been claimed to have improved outcomes in long-term follow-up (RAJPUT et al. 1984). A recent review has compared outcomes of patients so treated (with a mean LD dose of 1 g/day plus DCI; SHAW et al. 1980) with another group receiving a lower dose regimen of LD (a mean dose of 400 mg/day, though some patients received synergistic medication such as deprenyl). In this retrospective analysis (POEWE et al. 1986b), there was less peak dose dyskinesia or severe on-off fluctuations at 3 years with the lower dose regimen, though these group differences were largely gone by 6 years. From this experience, the authors concluded that, since long-term outcomes for both categories of dosage did not differ, it seemed "rational to titrate LD dosage to a patient's needs rather than pursue a rigid policy of 'low-dose' therapy on the unwarranted assumption that this will lead to better long-term control." The final word on this issue will have to come from further investigation with more rigorous prospective analysis, as in a current multicenter study in the United States comparing the extended effects of bromocriptine or CD-LD monotherapy with a combined dosage regimen.

K. Clinical Issues Regarding "Drug Holiday"

When parkinsonian disability requires the regular use of LD, interruptions in regular dosing with this lifelong therapy can be extremely difficult for patients to endure. Furthermore, the withdrawal of LD can unmask or possibly trigger severe rigidity and akinesia, which creates risks for aspiration pneumonia, deep vein thrombosis, and other serious medical consequences. These problems await PD patients dependent on LD if gastrointestinal surgery or related problems halt their regular intake of LD (EISLER et al. 1981). No parenteral alternatives exist except as research preparations of intravenous LD (NUTT et al. 1984) or the water-soluble ergot derivative lisuride (Chap. 22). Despite these hazards, temporary discontinuation of LD therapy, or "drug holiday," has been advocated for lessening the adverse effects or for improving responsiveness. Several reports (Goetz et al. 1982; Kaye and Feldman 1986) have claimed that, in the best of outcomes after drug holiday, patients can be maintained at lower daily LD dose with improved efficacy for a year or more. These studies have not identified which specific characteristics in patients tend to be associated with optimal responses to this procedure. As many patients derive no benefit from therapeutically intended drug holiday, the predictors of possible benefit need to be determined for this uncomfortable and potentially hazardous procedure.

The rationale for drug holiday came from investigations of receptor changes during chronic DA agonism. As discussed earlier, the effects of continuing LD therapy include a decrease of presynaptic (Mullen and Seeman 1979) and postsynaptic receptor density in rat striatum, both of which are reversed by a pause in LD treatment. In addition, increased stereotypy and motor hyperactivity in response to apomorphine challenge were the behavioral results of withholding LD treatment (Direnfeld et al. 1978), these may be correlates of the enhanced receptor function. Improved receptor function has been proposed as the mechanism for sustained improvement after drug holiday, although it is unclear why benefits should be sustained once dopaminergic therapy is reinstituted (resulting again in a down-regulation of receptor function within a short period of time).

The first report of this approach in patients found an enhanced response from LD doses to which patients had previously become refractory (Sweet et al. 1972), and so drug holiday came to be associated with improving DA neurotransmission or receptor function. It also permits the assessment of response characteristics to a single LD dose in the reappraisal of optimal LD dosing (Esteguy et al. 1985). One feature of LD therapy puzzling to Cotzias et al. (1967) and still inadequately understood is the apparent long-term effect of LD in PD. The rapid onset of improvement (within 2–3h) from the first few doses of LD, or even a single intravenous dose (Hornykiewicz 1974), is in keeping with the rapid turnover of LD known to occur in the CNS. However, discontinuation of LD therapy is often followed by a prolonged delay (4–14 days) in the decline of antiparkinsonian effect (Yahr et al. 1969; Muenter and Tyce 1971). It is conceivable that drug holiday, when effective, requires elimination of similar long-term LD influences. The effectiveness of drug holiday has been attributed to the elimination of accumulated levels of long-acting metabolites (such as 3-*O*-MD) or other neurotransmitter-active substances (Gervas et al. 1983). However, in patients experiencing an improvement following drug holiday, no change in the metabolism of exogenous LD (investigated repeatedly by measurement of CSF HVA) was found in one study (Ogasahara et al. 1984).

Stern et al. (1972) found that restarting the prior LD dose after a 3 to 31-day drug holiday improved the degree of motility over that immediately preceding drug withdrawal, in many instances, to a higher performance level than at any previous examination after starting LD. Further experience with drug holiday, although controversial, has in some instances found that LD-related psychiatric side effects, motor fluctuations, dyskinesia, myoclonus, and overall motor impairment could be lessened (in some instances) after temporary drug withdrawal (WEINER et al. 1980; DIRENFELD et al. 1980). After as short a drug holiday as 5–7 days, increased antiparkinsonian efficacy continued for some patients for 9 months or longer (KOLLER et al. 1981), although long-term follow-up studies with drug holiday have been limited (FRIEDMAN 1985b). GOETZ et al. (1982) have suggested guidelines for carrying out a drug holiday lasting 5 days or longer, during which time patients may be completely immobile and rigid. The methods and goals of drug holiday have attracted a great deal of controversy regarding the short- and long-term outcomes attainable. Some drug holiday protocols have become extremely doctrinaire in their methodology, although the pace of drug withdrawal or need for complete discontinuation of LD have not undergone critical review. Other evaluations of drug holiday have concluded that temporary periods of LD discontinuation lack sufficient chance of lasting improvement to be worth the considerable discomfort, dangers, and hospitalization costs (KOFMAN 1984; MAYEUX et al. 1985). In one study reporting benefit from drug holiday, the mean daily dose of LD for patients at entry was 2920 mg (KAYE and FELDMAN 1986); 1 month after drug holiday, the same patients were maintained on a mean daily LD dose of 241 mg with good effect. It seems likely that a more reasonable LD dose, rather than the period of drug withdrawal, was the basis for the claimed improvements.

Many experienced clinicians have not made regular use of drug holidays, but instead have modified the concept of drug withdrawal by using a period of partial reduction in LD dose, thereby determining if lower dose regimens have an improved therapeutic index. KOLLER (1983) has shown that alternate-day drug holiday can be as efficacious as conventional daily LD therapy, and so may permit reduction in chronic intake of LD. Another modification has given good results for lessening LD-related side effects and uses less sustained periods of drug holiday (a weekly 2-day off-drug period; GOETZ et al. 1981). Presumably, this approach can serve to lessen some of the cumulative adverse effects of LD, while maintaining sufficient antiparkinsonian effect (which can persist for 3–4 days after stopping chronic LD therapy).

References

Abrams WB, Coutinho CB, Leon AS, Spiegel HE (1971) Absorption and metabolism of levodopa. JAMA 218:1912–1914

Admani AK, Verma S, Cordingley GJ, Harris RI (1985) Patient benefits of L-dopa and a decarboxylase inhibitor in the treatment of Parkinson's disease in elderly patients. Pharmatherapeutica 4:132–140

Agid Y, Javoy F, Glowinski J (1973) Hyperactivity of remaining dopaminergic neurons after destruction of the nigrostriatal dopaminergic system in the rat. Nature 245:150–151

Agnati LF, Fuxe K (1980) On the mechanism of the antiparkinsonian action of L-dopa and bromocriptine: a theoretical and experimental analysis of dopamine receptor sub- and supersensitivity. J Neural Transm (Suppl 16):69–81

Akpinar S (1982) Treatment of restless legs syndrome with levodopa plus benserazide. Arch Neurol 39:739

Algeri S, Ruggieri S, Miranda F, Casacchia M, Morselli PL, Agnoli A (1976) Absence of relationships between L-dopa plasma levels and therapeutic effect in Parkinson's disease treated with L-dopa. Eur Neurol 14:219-228

Alkana RL, Parker ES, Cohen HB, Birch H, Noble EP (1977) Reversal of ethanol intoxication in humans: an assessment of the efficacy of L-dopa, aminophylline, and ephedrine. Psychopharmacology 55:203-212

Alpert M, Friedhoff AJ, Diamond F (1983) Use of dopamine receptor agonists to reduce dopamine receptor number as treatment for tardive dyskinesia. Adv Neurol 37:253-258

Ambani LM, Van Woert MH (1973) Start hesitation — a side effect of long term levodopa therapy. N Engl J Med 288:1113-1115

Aminoff MJ, Wilcox CS, Woakes MM, Kremer M (1973) Levodopa therapy for parkinsonism in the Shy-Drager syndrome. J Neurol Neurosurg Psychiatry 36:350-353

Andersson I, Granerus A-K, Jagenburg R, Svanborg A (1975) Intestinal decarboxylation of orally-administered L-dopa. Acta Med Scand 198:415-420

Askeny JJM, Yahr MD (1985) Reversal of sleep disturbance in Parkinson's disease by antiparkinsonian therapy: a preliminary study. Neurology 35:527-532

Baake ÒM, Scheline RR (1974) Inhibition of a minor pathway of L-dopa metabolism in the intestinal lumen using a decarboxylase inhibitor (Ro 4-4602). J Pharm Pharmacol 26:377-379

Bailey RC, Jackson DM, Bracs PU (1979) Long-term L-dopa pre-treatment of mice: central receptor subsensitivity or supersensitivity. Psychopharmacology 66:55-61

Ballard PA, Tetrud JW, Langston JW (1985) Permanent human parkinsonism due to 1-methyl-4-phenyl-1,2,3,6-tetrahydropyridine (MPTP): seven cases. Neurology 35:949-956

Barash V, Globus M, Melamed E, Weidenfeld J (1985) Effect of L-dopa on glucose oxidation and incorporation into glycogen in discrete brain regions of the rat. Brain Res 335:347-349

Barbeau A (1969) L-Dopa therapy in Parkinson's disease: a critical review of nine year's experience. Can Med Assoc J 101:791-800

Barbeau A (1972) Role of dopamine in the nervous system. Monogr Hum Genet 6:114-136

Barbeau A (1976) Six years of high-level levodopa therapy in severely akinetic parkinsonian patients. Arch Neurol 33:333-338

Barbeau A (1980) High-level levodopa therapy in severly akinetic parkinsonian patients: twelve years later. In: Rinne UK, Klinger M, Stamm G (eds) Parkinson's disease: current progress, problems and management. Elsevier/North-Holland Biomedical, New York, pp 229-239

Barbeau A, McDowell FH (eds) (1970) L-Dopa and Parkinsonism. Davis, Philadelphia

Barbeau A, Friesen H (1970) Treatment of Wilson's disease with L-dopa after failure with penicillamine. Lancet 1:1180

Barbeau A, Roy M (1976) Six-year results of treatment with levodopa plus benserazide in Parkinson's disease. Neurology 26:399-404

Barbeau A, Roy M (1981) Ten-year results of treatment with levodopa plus benserazide in Parkinson's disease. In: Rose FC, Capildeo R (eds) Research progress in Parkinson's disease. Pitman, Tunbridge, pp 241-247

Barbeau A, Sourkes TL, Murphy GF (1962) Les catecholamines dans la maladie de Parkinson. In: de Ajuriaguerra J (ed) Monoamines et systeme nerveux central. Masson, Paris, pp 247-262

Baroni A, Benvenuti F, Fantini L, Pantaleo T, Urbani F (1984) Human ballistic arm abduction movements: effects of L-dopa treatment in Parkinson's disease. Neurology 34:868-876

Barrett RE, Yahr MD, Duvoisin RC (1970) Torsion dystonia and spasmodic torticollis — results of treatment with L-dopa. Neurology 20(part 2):107-113

Bartholini G, Pletscher A (1975) Decarboxylase inhibitors. Pharmacol Ther 1:407-421

Bartholini G, Kuruma I, Pletscher A (1972) The metabolic pathways of L-3-*O*-methyldopa. J Pharmacol Exp Ther 183:65-72

Beani L, Ledda F, Bianchi C, Baldi V (1966) Reversal by 3,4-dihydroxyphenylalanine of reserpine-induced regional changes in acetylcholine content of guinea-pig brain. Biochem Pharmacol 15:779-784

Beasley BL, Hare TA, Vogel WH, Desimone S (1973) Rectal administration of L-dopa. N Engl J Med 289:919

Bender DA, Earl CJ, Lees AJ (1979) Niacin depletion in parkinsonian patients treated with L-dopa, benserazide and carbidopa. Clin Sci 56:89-93

Bergman S, Curzon G, Friedel J, Godwin-Austin RB, Marsden CD, Parkes JD (1974) The absorption and metabolism of a standard oral dose of levodopa in patients with parkinsonism. Br J Clin Pharmacol 1:417-424

Berkowitz DM, McCallum RW (1980) Interaction of levodopa and metoclopramide on gastric emptying. Clin Pharmacol Ther 27:414-420

Bermejo Pareja F, Martines-Martin P, Muradás V, de Yébenes JG (1985) Carbidopa dosage modifies L-dopa induced side-effects and blood levels of L-dopa and other amino acids in advanced parkinsonism. Acta Neurol Scand 72:506-511

Bernheimer H, Birkmayer W, Hornykiewicz O, Jellinger K, Seitelberger F (1973) Brain dopamine and the syndromes of Parkinson and Huntington. J Neurol Sci 20:415-455

Bertler A, Rosengren E (1959) Occurrence and distribution of dopamine in brain and other tissues. Experientia 15:10-11

Bianchine JR, Shaw GM (1976) Clinical pharmacokinetics of levodopa in Parkinson's disease. Clin Pharmacokinet 1:313-338

Bianchine JR, Calimlim LR, Dujovne C, Lasagna L (1971) Metabolism and absorption of L-3,4-dihydroxyphenylalanine in patients with Parkinson's disease. Ann NY Acad Sci 179:126-140

Bianchine JR, Messiha FS, Hsu TH (1972) Peripheral aromatic L-amino acid decarboxylase inhibitor in parkinsonism. Il Effect on the metabolism of L-2-^{14}C-dopa. Clin Pharmacol Ther 13:584-594

Birkmayer W, Hornykiewicz O (1961) Der L-Dioxyphenylalanin-=(L-dopa)-Effekt bei der Parkinson-Akinesie. Wien Klin Wochenschr 73:787-788

Birkmayer W, Hornykiewicz O (eds) (1976) Advances in parkinsonism. Roche, Basel

Birkmayer W, Riederer P (1983) Parkinson's disease: biochemistry, clinical pathology, and treatment. Springer Berlin Heidelberg, New York

Birkmayer W, Danielcyk W, Neumayer E, Riederer P (1973) L-Dopa level in plasma, primary condition for the kinetic effect. J Neural Transm 34:133-143

Bodor N, Sloan KB, Higuchi T, Sasahara K (1977) Improved delivery through biological membranes. 4. Prodrugs of L-dopa. J Med Chem 20:1435-1445

Boshes B (1981) Sinemet and the treatment of parkinsonism. Ann Intern Med 94:364-370

Bouchard S, Bousquet C, Roberge AC (1981) Characteristics of dihydroxyphenylalanine/5-hydroxytryptophan decarboxylase activity in brain and liver of cat. J Neurochem 37:781-787

Bowers MB, Van Woert MH (1972) The probenecid test in Parkinson's disease. Lancet 2:926-927
Bowman WC, Nott MW (1978) Peripheral muscle effects of levodopa in the anesthetized cat. Clin Exp Pharmacol Physiol 5:305-312
Broe GA, Caird FI (1973) Levodopa for parkinsonism in elderly and demented patients. Med J Aust 1:630-635
Broe GA, Evans MA, Triggs EJ (1981) Senile parkinsonism and dopa pharmacokinetics. Clin Exp Neurol 18:174-179
Brogden RN, Speight TM, Avery GS (1971) Levodopa: a review of its pharmacological properties and therapeutic uses with particular reference to parkinsonism. Drugs 2:262-400
Brown L, Lorenz B, Erdmann E (1985) The inotropic effects of dopamine and its precursor levodopa on isolated human ventricular myocardium. Klin Wochenschr 63:1117-1123
Bruno A, Bruno SC (1966) Effects of L-dopa on pharmacological parkinsonism. Acta Psychiatr Scand 42:264-271
Bugiani O, Gatti R (1979) L-Dopa in children with progressive neurological disorders. Ann Neurol 7:93
Bunney BS, Agajanian GK, Roth RH (1973) L-Dopa, amphetamine and apomorphine: comparison of effects on the firing rate of rat dopaminergic neurons. Nature 245:123-125
Calne DB (ed) (1973) Progress in the treatment of parkinsonism. Adv Neurol 3:1-326
Calne DB, Sandler M (1970) L-Dopa and parkinsonism. Nature 226:21-24
Calne DB, Karoum F, Ruthven CRJ, Sandler M (1969a) The metabolism of orally administered L-Dopa in parkinsonism. Br J Pharmacol 37:57-68
Calne DB, Stern GM, Laurence DR, Sharkey J, Armitage P (1969b) L-Dopa in postencephalitic parkinsonism. Lancet 1:744-747
Calne DB, Reid JL, Vakil SD (1972) Parkinsonism treated with 3-*O*-methyldopa. Clin Pharmacol Ther 14:386-389
Carlsson A, Linqvist M, Magnusson T (1957) 3,4-Dihydroxyphenylalanine and 5-hydroxytryptophan as reserpine antagonists. Nature 180:1200
Carlsson A (1959) The occurrence, distribution, and physiological role of catecholamines in the nervous system. Pharmacol Rev 11:490-493
Carvey P, Siegal A, Dunford J, Klawans HL (1982) Levodopa blood levels in guinea pigs. Neurology 32(2):A162
Cedarbaum JM, Kutt H, Dhar AK, Watkins S, McDowell FH (1986a) Effect of supplemental carbidopa on bioavailability of L-dopa. Clin Neuropharmacol 9:153-159
Cedarbaum JM, Breck L, Kutt H, McDowell F (1986b) Sinemet CR-4 in the treatment of response fluctuations in Parkinson's disease. Ann Neurol 20:121
Celesia GG, Wanamaker WM (1972) Psychiatric disturbances in Parkinson's disease. Dis Nerv Sys 33:577-583
Chase TN (1980) Neurochemistry of the CSF in Parkinson's disease. In: Wood JH (ed) Neurobiology of cerebrospinal fluid 1. Plenum, New York, pp 207-218
Chase TN, Ng LKY (1972) Central monoamine metabolism in Parkinson's disease. Arch Neurol 27:486-491
Chase TN, Watanabe AM (1972) Methyldopa-hydrazine as an adjunct to L-dopa therapy in parkinsonism. Neurology 22:384-392
Chihara K, Kashio Y, Kita T, Okimura Y, Kaji H, Abe H, Fujita T (1986) L-Dopa stimulates release of hypothalamic growth hormone-releasing hormone in humans. J Clin Endocrinol Metab 62:466-473

Clark WG, Oldendorf WH, Dewhurst WG (1973) Blood-brain barrier to carbidopa (MK-486) and Ro 4-4602, peripheral dopa decarboxylase inhibitors. J Pharm Pharmacol 25:416-418

Commissiong JW, Sedgwick EM (1979) Depletion of 5-HT by L-dopa in spinal cord and brainstem of rat. Life Sci 25:83-86

Cook DG, Klawans HL (1985) Levodopa during pregnancy. Clin Neuropharmacol 8:93-95

Cooper DR, Marrel C, Testa B, Van der Waterbeemd H, Quinn N, Jenner P, Marsden CD (1984) L-Dopa methylester: a candidate for chronic systemic delivery of L-dopa in Parkinson's disease. Clin Neuropharmacol 7:89-98

Cosci V, Romani A, Callieco R, Zerbi F (1984) Acute effects of levodopa plus carbidopa on evoked potentials in Parkinson's disease. Boll Soc Ital Biol Sper 60:603-609

Coscia CJ, Burke W, Jamroz G, Lasala JM, McFarlane J, Mitchell J, O'Toole MM, Wilson ML (1977) Occurrence of a new class of tetrahydroisoquinoline alkaloids in L-dopa-treated parkinsonian patients. Nature 269:617-619

Cotzias GC, Van Woert MH, Schiffer LM (1967) Aromatic amino acids and modification of parkinsonism. N Engl J Med 276:374-379

Cotzias GC, Papavasiliou PS, Gellene R (1969) Modification of parkinsonism — chronic treatment with L-dopa. N Engl J Med 280:337-345

Cotzias GC, Mena I, Papavasiliou PS (1973) Overview of present treatment of parkinsonism with levodopa. Adv Neurol 2:265-277

Cuhna L, Goncalves AF, Oliveira C, Dinis M, Amaral R (1983) Homovanillic acid in the cerebrospinal fluid of parkinsonian patients. Can J Neurol Sci 10:43-46

Curzon G, Dodwin-Austen RB, Tomlinson EB, Kantamaneni BD (1970) The cerebrospinal fluid homovanillic acid concentration in patients with parkinsonism treated with L-dopa. J Neurol Neurosurg Psychiatry 33:1-6

Curzon G, Friedel J, Grier L, Marsden CD, Parkes JD, Shipley M, Zilkha KJ (1973) Sustained-release levodopa in parkinsonism. Lancet 1:781

Dairman W, Christenson JA, Udenfriend S (1971) Decrease in liver aromatic L-amino-acid decarboxylase produced by chronic administration of L-dopa. Proc Natl Acad Sci USA 68:2117-2120

Dairman W, Christenson JG, Udenfriend S (1972) Changes in tyrosine hydroxylase and dopa decarboxylase induced by pharmacological agents. Pharmacol Rev 24:269-289

Damasio AR, Castro Caldas A (1975) Neuropsychiatric aspects. In: Stern GM (ed) The clinical uses of levodopa. University Park Press, Baltimore, pp 127-154

Daniel RM, Moorhouse SR, Pratt OE (1976) Do changes in blood levels of other aromatic amino acids influence levodopa therapy? Lancet 1:95

Da Prada M (1984) Peripheral decarboxylase inhibition: a biochemical comparison between benserazide and carbidopa. In: Birkmayer W et al. (eds) Parkinson's disease. Actual problems and management. Roche, Basel, pp 25-38

Davidson L, Lloyd K, Dankova J, Hornykiewicz O (1971) L-Dopa treatment in Parkinson's disease: effect on dopamine and related substances in discrete brain regions. Experientia 27:1048-1049

Degkwitz R, Frohwein R, Kulenkampff C, Mohs U (1960) Über die Wirkungen des L-dopa beim Menschen und deren Beeinflussung durch Reserpin, Chlorpromazin, Iproniazid und Vitamin B_6. Klin Wochenschr 38:120-123

de Jong GJ, Meerwaldt JD (1984) Response variations in the treatment of Parkinson's disease. Neurology 34:1507-1509

De Lean J, Deck JH (1976) Shy-Drager syndrome. Neuropathological correlation and response to levodopa therapy. Can J Neurol Sci 3:167-173
Diamond BI, Rajan KS, Mainer S, Borison RL (1982) Copper facilitation of L-dopa transport into brain. Neurology 32(2):A67
Diamond SG, Markham CH, Treciokas LJ (1978) A double-blind comparison of levodopa, Madopar, and Sinement in Parkinson disease. Ann Neurol 3:376-385
Dinner DS, Luders H, Hanson M, Lesser RP, Klem G (1985) Pattern evoked potentials (PEPs) in Parkinson's disease. Neurology 35:610-613
Direnfeld L, Spiro L, Marotta L, Seeman P (1978) The L-dopa on-off effect on Parkinson's disease: treatment by transient drug withdrawal and dopamine receptor sensitization. Ann Neurol 4:473-475
Direnfeld KL, Feldmann RG, Alexander MP, Kelly-Hayes M (1980) Is levodopa drug holiday useful? Neurology 30:785-788
Doller HJ, Conner JD (1980) Changes in neostriatal dopamine concentrations in response to levodopa infusions. J Neurochem 34:1264-1269
Doller HJ, Connor JD, Lock DR, Sloviter RS, Dvorchik BH, Vesell ES (1978) Levodopa pharmacokinetics. Alterations after benserazide, a decarboxylase inhibitor. Drug Metab Dispos 6:164-168
Dolphin A, Jenner P, Marsden CD (1976) Noradrenaline synthesis from L-dopa in rodents and its relationship to motor activity. Pharmacol Biochem Behav 5:431-439
Domenici L, Trimarchi C, Piccolino M, Fiorentini A, Maffei L (1985) Dopaminergic drugs improve human visual contrast sensitivity. Hum Neurobiol 4:195-197
Donaldson IM (1973) The treatment of progressive supranuclear palsy with L-dopa. Aust NZ J Med 3:413-416
Dougan D, Wade D, Mearrick P (1975) Effects of L-dopa metabolites at dopamine receptor suggests a basis for "on-off" effect in Parkinson's disease. Nature 254:70-72
Drachman DA, Stahl S (1971) Extrapyramidal dementia and levodopa. Lancet 1:809
Dunner DL, Brodie HKH, Goodwin FK (1971) Plasma dopa response to levodopa administration in man: effects of a peripheral decarboxylase inhibitor. Clin Pharmacol Ther 12:212-217
Duvoisin RC (1973) Pyridoxine as an adjunct in the treatment of parkinsonism. Adv Neurol 2:229-247
Duvoisin RC, Yahr MD, Cote LD (1969) Pyridoxine reversal of L-dopa effects in parkinsonism. Trans Am Neurol Assoc 94:81-84
Duvoisin RC, Lobo-Antunes J, Yahr MD (1972) Response of patients with postencephalitic parkinsonism to levodopa. J Neurol Neurosurg Psychiatry 35:487-495
Eckstein B, Shaw K, Stern GM (1973) Sustained-release levodopa in parkinsonism. Lancet 1:431-432
Edwards DJ, Rizk M (1981) Conversion of 3,4-dihydroxyphenylalanine and deuterated 3,4-dihydroxyphenylaline to alcoholic metabolites of catecholamines in rat brain. J Neurochem 36:1641-1647
Ehringer H, Hornykiewicz O (1960) Verteilung von Noradrenalin und Dopamin (3-hydroxytyramin) im Gehirn des Menschen und ihr Verhalten bei Erkrankungen des extrapyramidalen Systems. Klin Wochenschr 38:1236-1239 (for translation, see Marks 1974)
Eisler T, Eng M, Plotkin C, Calne DB (1981) Absorption of levodopa after rectal administration. Neurology 31:215-217
el Gemayel G, Trouvin JH, Prioux-Guyonneau M, Jacquot C, Cohen Y (1986) Re-evaluation of the L-dopa loading effect on dopamine metabolism in rat striatum. J Pharm Pharmacol 38:134-136

Ericsson AD (1971) Potentiation of the L-dopa effect in man by use of catechol-*O*-methyltransferase inhibitors. J Neurol Sci 14:193-197
Eriksson T, Magnusson T, Carlsson A, Linde A, Granerus AK (1984) "On-off" phenomenon in Parkinson's disease: correlation to the concentration of dopa in plasma. J Neural Transm 59:229-240
Erni W, Held K (1987) The hydrodynamically balanced system: a novel principle of controlled drug release. Eur Neurol 27 (Suppl 1): 21-27
Esteguy M, Bonnet AM, Kefalos J, Lhermitt F, Agid Y (1985) The L-dopa test in Parkinson's disease. Rev Neurol (Paris) 141:413-415
Evans MA, Triggs EJ, Broe GA, Saines N (1980) Systemic availability of orally administered L-dopa in the elderly parkinsonian patient. Eur J Clin Pharmacol 17:215-221
Evans MA, Broe GA, Triggs EJ, Cheung M, Creasey H, Paull PD (1981) Gastric emptying rate and the systemic availability of levodopa in the elderly parkinsonian patient. Neurology 31:1288-1294
Evertt GM, Borcherding JW (1970) Levodopa: effect on concentration of dopamine, norepinephrine, and serotonin in brains of mice. Science 168:849-850
Extein I, Van Woert MH, Roth RH, Bowers MB (1976) ^{14}C-homovanillic acid in the cerebrospinal fluid of parkinsonian patients after intravenous ^{14}C-L-dopa. Biol Psychiatry 11:227-232
Fahn S (1974) On-off phenomena with levodopa therapy in parkinsonism. Neurology 24:431-441
Fahn S (1977) Episodic failure of absorption of levodopa: a factor in the control of clinical fluctuations in the treatment of parkinsonism. Neurology 27:390
Fahn S, Barrett RE (1979) Increase of parkinsonian symptoms as a manifestation of levodopa toxicity. Adv Neurol 24:451-459
Fahn S, Bressman S (1984) Should levodopa therapy for parkinsonism be started early or late? Evidence against early treatment. Can J Neurol Sci 11[Suppl]:200-206
Fahn S, Calne DB (1978) Considerations in the management of parkinsonism. Neurology 28:5-7
Fekete M, Tárczy M, Bihari K, Katona G (1984) Dopamine/L-dopa ratio in cerebrospinal fluid of parkinsonian patients treated with L-dopa + benserazide. Psychopharmacology (Berlin) 82:93-94
Fellman JH, Roth ES (1971) Inhibition of tyrosine amino-transferase activity by L-3,4-dihydroxyphenylalanine. Biochemistry 10:408-414
Fellman JH, Roth ES, Herzia EL, Fujita TS (1976) Altered pattern of dopa metabolism. Biochem Pharmacol 25:222-223
Fermaglich J (1973) Methyldopa and L-dopa: synergistic agents in the treatment of Parkinson's disease. In: Yahr MD (ed) Current concepts in the treatment of parkinsonism. Raven, New York, pp 95-121
Fermaglich J, O'Doherty DS (1972) Effect of gastric motility on levodopa. Dis Nerv Sys 33:624-625
Feuerstein C, Tanche M, Serre F, Gavend M, Pellat J (1977) Does *O*-methyldopa play a role in levodopa-induced dyskinesias? Acta Neurol Scand 56:79-82
Fischer JE, Funovices FJ, Falcao HA, Wesdorp RI (1976) L-Dopa in hepatic coma. Ann Surg 183:386-391
Fleming P, Makar H, Hunter KR (1970) Levodopa in drug-induced extrapyramidal disorders. Lancet 2:1186
Friedhoff AJ, Bonnet K, Rosengarten H (1977) Reversal of 2 manifestations of dopamine receptor supersensitivity by administration of L-dopa.

Friedman A (1985a) Drug-induced dyskinesia during the treatment of Parkinson disease — biochemical studies, Neurol Neurochir Pol 19:401-403 (in Polish)
Friedman JH (1985b) "Drug holidays" in the treatment of Parkinson's disease. A brief review. Arch Intern Med 145:913-915
Friedman JH, Feinberg S, Feldman RG (1985) A neuroleptic malignantlike syndrome due to levodopa withdrawal. Jama 254:2792-2795
Friis ML, Paulson OB, Hertz MM, Bolwig TG (1981) Blood-brain barrier permeability of L-dopa in man. Eur J Clin Invest 11:231-234
Fuller RW, Henrick-Luecke SK, Perry KW (1982) Effects of L-dopa on epinephrine concentration in rat brain: possible inhibition of norepinephrine *N*-methyltransferase by *S*-adenosyl homocysteine. J Pharmacol Exp Ther 223:84-89
Galea-Debono A, Jenner P, Marsden CD, Parkes JD, Tarsy D, Walters J (1977) Plasma dopa levels and growth hormone response to levodopa in parkinsonism. J Neurol Neurosurg Psychiatry 40:162-167
Gandy SE, Barclay LL, Cedarbaum JM (1986) Age of onset of dementia in Parkinson's and Alzheimer's disease and the role of levodopa. Ann Neurol 20:150
Garzon-Arburbeh A, Poupaert JH, Claesen M, Dumont P (1985) A lymphotropic prodrug of L-dopa: synthesis, pharmacological properties, and pharmacokinetic behavior of 1,3-dihexadecanoyl-2-[(*S*)-2-amino-3-(3,4-dihydroxyphenyl)propanoyl] propane-1,2,3-triol. J Med Chem 29:687-691
Gervas JJ, Muradás V, Bazán E, Aguado EG, de Yébenes JG (1983) Effects of 3-*O*M-dopa on monoamine metabolism in rat brain. Neurology 33:278-282
Gilligan B, Hancock R (1975) Enteric-coated L-dopa (prodopa). A new approach to L-dopa therapy in Parkinson's disease. Med J Aust 2:824-826
Gilligan B, Wodack J, Stark R, O'Halloran M (1979) Comparison of enteric-coated levodopa with levodopa-carbidopa combination. A double-blind crossover trial. Med J Aust 2:205-207
Godwin-Austen RB, Frears CC, Tomlinson EB, Kok HWL (1969) Effects of L-dopa in Parkinson's disease. Lancet 2:165-168
Godwin-Austen RB, Kantamaneni BD, Curzon G (1971) Comparison of benefit from L-dopa in parkinsonism with increase in amine metabolites in the CSF. J Neurol Neurosurg Psychiatry 34:219-223
Goetz CG, Tanner CM, Nausieda PA (1981) Weekly drug holiday in Parkinson disease. Neurology 31:1460-1462
Goetz CG, Tanner CN, Klawans HL (1982) Drug holiday in the management of Parkinson's disease. Clin Neuropharmacol 5:351-364
Goetz CG, Tanner CM, Klawans HL (1984) The pharmacology of olivopontocerebellar atrophy. Adv Neurol 41:143-148
Goetz CG, Tanner CM, Carroll VS, Shannon K, Klawans HL (1986) Controlled-release carbidopa/levodopa combination. Neurology 36[Suppl 1]:217
Goldberger ME, Growdon JH (1971) Tremor at rest following cerebellar lesions in monkeys: effect of L-dopa administration. Brain Res 27:183-187
Goodale DB, Moore KE (1976) A comparison of the effects of decarboxylase inhibitors on L-dopa-induced circling and conversion of dopa to dopamine in the brain. Life Sci 19:701-706
Goodall MC, Alton H (1972) Metabolism of 3,4-dihydroxyphenylalanine (L-dopa) in human subjects. Biochem Pharmacol 21:2401-2408
Goodwin BL, Ruthven CR, King GS, Sandler M (1978) Metabolism of 3,4-dihydroxyphenylalanine, its metabolites, and analogues in vivo in the rat: urinary excretion pattern. Xenobiotica 8:629-651

Goodwin FK, Brodie HKH, Murphy DL, Bunney WE Jr (1970) Administration of a peripheral decarboxylase inhibitor with L-dopa to depressed patients. Lancet 1:908-911

Granerus AK (1978) Factors influencing the occurrence of "on-off" symptoms during long-term treatment with L-dopa. Acta Med Scand 203:75-85

Granerus AK, Jagenburg R, Svanborg A (1977) Kaliuretic effect of L-dopa treatment in parkinsonian patients. Acta Med Scand 201:291-297

Greenacre JK, Coxon A, Petrie A, Reid JL (1976) Comparison of levodopa with carbidopa or benserazide in parkinsonism. Lancet 2:381-384

Growdon J (1981) Tyrosine treatment in Parkinson disease: clinical effects. Neurology 31(2):134

Grundig E, Gerstenbrand F, Bruck J, Gnad H, Prosenz P, Simanyi M (1969) The effect of the administration of amino acids, especially of L-dopa and alpha-methyldopa, on the composition of cerebrospinal fluid in extrapyramidal syndromes. Dtsch Z Nervenheilk 196:235-265

Guidice MA, LeWitt PA, Berchou RC, Holland M (1986) Improvement in motor functioning with levodopa and bromocriptine following closed head injury. Neurology 36[Suppl 1]:198

Gumpert EJN, Sharp DM, Curzon G (1973) Amine metabolites in the cerebrospinal fluid in Parkinson's disease and the response to L-dopa. J Neurol Sci 19:1-12

Gundert-Remy U, Hildebrandt R, Stiehl A, Weber E, Zurcher G, Da Prada M (1983) Intestinal absorption of levodopa in man. Eur J Clin Pharmacol 25:69-72

Guttman M, Seeman P (1985) L-Dopa reverses the elevated density of D_2 dopamine receptors in Parkinson's diseased striatum. J Neural Transm 64:93-103

Guttmann M, Seeman P, Reynolds GP, Riederer P, Jellinger K, Tourtellotte WW (1986) Dopamine D_2 receptor density remains constant in treated Parkinson's disease. Ann Neurol 19:487-492

Halgin R, Riklan M, Misiak H (1977) Levodopa, parkinsonism, and recent memory. J Nerv Ment Dis 164:268-272

Hall MD, Cooper DR, Fleminger S, Rupniak NM, Jenner P, Marsden CD (1984) Behavioural and biochemical alterations in the function of dopamine receptors following repeated administration of L-dopa to rats. Neuropharmacology 23:545-553

Hardebo JE, Owman C (1980) Barrier mechanism for neurotransmitter monoamines and their precursors at the blood-brain interface. Ann Neurol 8:1-11

Hardebo JE, Edvinsson L, Owman C, Rosengren E (1977) Quantitative evaluation of the blood-brain barrier capacity to form dopamine from circulating L-dopa. Acta Physiol Scand 99:377-384

Hardie RJ, Lees AJ, Stern GM (1984) On-off fluctuations in Parkinson's disease. A clinical and neuropharmacological study. Brain 107:487-506

Hassler RG, Christ JF (eds) (1984) Parkinson-specific motor and mental disorders. Roles of the pallidum: pathophysiological, biochemical, and therapeutic Aspects. Adv Neurol 40:1-579

Hefti F, Melamend E, Wurtman RJ (1980) Partial lesions of the dopaminergic nigrostriatal system in rat brain: biochemical characterization. Brain Res 195:123-137

Hefti F, Melamed E, Bhawan J, Wurtman RJ (1981a) Longterm administration of L-dopa does not damage dopaminergic neurons in the mouse. Neurology 31:1194-1195

Hefti F, Melamed E, Wurtman RJ (1981b) The site of dopamine formation in rat striatum after L-dopa administration. J Pharmacol Exp Ther 217:189-197

Hinterberger H, Andrews CJ (1972) Catecholamine metabolism during oral administration of levodopa. Arch Neurol 26:245-252
Hoehn MM (1980) Increasing dosage of carbidopa in patients with Parkinson's disease receiving low doses of levodopa. Arch Neurol 37:146-149
Hoehn MM (1983) Comparison of the progression of Parkinson's disease before and since the levodopa era. Ann Neurol 14:135
Holtz P, Credner K, Kroneberg G (1947) Über das sympathikomimetische pressorische Prinzip des Hirn ("Urosympathin"). Arch Exp Path Pharmak 204:228-243
Horne MK, Cheng CH, Wooten GF (1984) The cerebral metabolism of L-dihydroxyphenylalanine: an autoradiographic and biochemical study. Pharmacology 28:12-26
Hornykiewicz O (1970) How does L-dopa work? In: Barbeau A, McDowell FH (eds) L-dopa and parkinsonism. Davis, Philadelphia, pp 393-399
Hornykiewicz O (1974) The mechanism of action of L-dopa in Parkinson's disease. Life Sci 15:1249-1259
Hornykiewicz O (1979) Brain dopamine in Parkinson's disease and other neurological disturbances. In: Horn AS, Korf J, Westerink BHC (eds) The neurobiology of dopamine, Academic, New York, pp 633-654
Hornykiewicz O (1981) Brain neurotransmitter changes in Parkinson's disease. In: Marsden CD, Fahn S (eds) Movement disorders. Butterworth, Boston, pp 41-58
Horowitz G, Greenberg J (1975) Pallido-pyramidal syndrome treated with levodopa. J Neurol Neurosurg Psychiatry 38:238-240
Huebert ND, Palfreyman MG, Haegele KD (1983) A comparison of the effects of reversible and irreversible inhibitors of aromatic L-amino acid decarboxylase on the half-life and other pharmacokinetic parameters of oral L-3,4-dihydroxyphenylalanine. Drug Metab Dispos 11:195-200
Hunter KR, Shaw KM (1975) Therapeutic effects. In: Stern GM (ed) The clinical uses of levodopa. University Park Press, Baltimore, pp 41-71
Hunter KR, Stern GM, Sharkey J (1970) Levodopa in postencephalitic parkinsonism. Lancet 2:1366-1367
Huseman CA (1985) Growth enhancement by dopaminergic therapy in children with intrauterine growth retardation. J Clin Endocrinol Metab 61:514-519
Hutton JT, Dippel RL, Bianchine JR, Strahlendorf HK, Meyer PG (1984) Controlled-release carbidopa/levodopa in the treatment of parkinsonism. Clin Neuropharmacol 7:135-139
Hutton JT, Albrecht JW, Roman GC, Kopetsky MT, Palet JL (1985) Prolonged bioavailability of CSR-3 and CSR-2 controlled release carbidopa/levodopa as compared with standard carbidopa/levodopa 25/100 in Parkinson's disease. Neurology 35[Suppl 1]:201
Itshimitsu T, Hirose S, Asahara K, Imaizumi M (1984) The formation of 3-hydroxy-4-methoxyphenylalanine and 3-hydroxy-4-methoxyphenethylamine in plasma during levodopa therapy in patients with Parkinson's disease. Chem Pharm Bull (Tokyo) 32:3320-3322
Jackson DM, Jenkins OF, Malor R, Christine MJ, Gregory P (1983) Chronic L-dopa treatment of rats and mice does not change the sensitivity of postsynaptic dopamine receptors. Naunyn Schmiedebergs Arch Pharmacol 324:271-274
Jaffe M (1973) Clinical studies of carbidopa and L-dopa in the treatment of Parkinson's disease. Adv Neurol 2:161-172
Jankovic J (1981) L-Dopa-resistant parkinsonism due to DOPA decarboxylase deficiency? Ann Neurol 10:64-65

Jankovic J (1984) Progressive supranuclear palsy: clinical and pharmacological update. Neurol Clin 2:473-486
Jellinger K, Flament H, Riederer P, Schmid H, Ambrozi L (1980) Levodopa in the treatment of (pre) senile dementia. Mech Ageing Dev 14:253-264
Jenkins R, Lamid S, Klawans HL (1973) Gastric acidity and levodopa in parkinsonism. JAMA 223:81
Jequier E, Dufresne JJ (1972) Biochemical investigations in patients with Parkinson's disease treated with L-dopa. Neurology 22:15-21
Jeste DV, Wyatt RJ (1982) Therapeutic strategies against tardive dyskinesia. Arch Gen Psychiatry 39:803-816
Johnson GA, Baker CA, Smith RA (1980) Radioenzymatic assay of sulfate conjugates of catecholamines and DOPA in plasma. Life Sci 26:1591-1598
Johnson K, Presly AS, Ballinger BR (1978) Levodopa in senile dementia. Br Med J [Clin Res] 1:1625
Jongen PJ, Renier WO, Gabreels FJ (1980) Seven cases of Huntington's disease in childhood and levodopa iduced improvement in the hypokinetic-rigid form. Clin Neurol Neurosurg 82:251-261
Joseph C, Chassan JB, Koch ML (1978) Levodopa in Parkinson disease: a long-term appraisal of mortality. Ann Neurol 3:116-118
Juncos JL, Mouradian MM, Fabbrini G, Seratti C, Chase TN (1986) Levodopa methyl ester treatment of Parkinson's disease. Neurology 36 [Suppl 1]:218
Kaye JA, Feldman RG (1986) The role of L-dopa holiday in the management of Parkinson's disease. Clin Neuropharmacol 9:1-13
Kelly PJ, Gillingham FJ (1980) The long-term results of stereotaxic surgery and L-dopa therapy in patients with Parkinson's disease. A 10-year follow-up study. J Neurosurg 53:332-337
Kernsbaum S, Hauchecorne J (1981) Administration of levodopa for relief of herpes zoster pain. JAMA 246:132-134
King MD, Beesley PW (1978) The effect of L-3,4-dihydroxyphenylalanine (L-dopa) on the incorporation of radioactive amino acids into rat brain protein. Biochem Soc Trans 6:1000-1002
Klawans HL (1986) Individual manifestations of Parkinson's disease after ten or more years of levodopa. Movement Disorders 1:187-192
Klawans HL, Bergen D (1975) Side effects of levodopa. In: Stern GM (ed) The clinical uses of levodopa. University Park Press, Baltimore, pp 73-106
Klawans HL, Ringel SP (1971) Observations on the efficacy of L-dopa in progressive supranuclear palsy. Eur Neurol 5:115-129
Klawans HL, Ringel SP, Shenker DM (1971) Failure of vitamin B_6 to reverse the L-dopa effect in patients on a dopa decarboxylase inhibitor. J Neurol Neurosurg Psychiatry 34:682-686
Kofman OS (1984) Are levodopa "drug holidays" justified? Can J Neurol Sci 11:206-209
Kohn SR, Pochi PE, Strauss JS, Sax DS, Feldman RG (1975) Sebaceous gland secretion in Parkinson's disease during L-dopa treatment. J Invest Dermatol 60:134-136
Koller WC (1983) Intermittent levodopa therapy in parkinsonism. Adv Neurol 37:45-50
Koller WC (1986) Pharmacologic treatment of parkinsonian tremor. Arch Neurol 43:126-127
Koller WC, Weiner WJ, Perlik S, Nausieda PA, Goetz CG, Klawans HL (1981) Complications of chronic levodopa therapy: long-term efficacy of drug holiday. Neurology 31:473-476

Korten JJ, Keyser A, Joosten EMG, Gabreëls FJM (1975) Madopar versus Sinemet. Eur Neurol 13:65-71

Kremzner LT, Berl ST, Mendoza M, Yahr MD (1974) Amino acid abnormalities in the brain and cerebrospinal fluid of parkinsonian patients. In: Yahr MD (ed) Current concepts in the treatment of parkinsonism. Raven, New York, pp 235-242

Kristensen V, Olsen M, Theilgaard A (1977) Levodopa treatment of presenile dementia. Acta Psychiatr Scand 55:41-51

Kurlan R, Rubin AJ, Miller C, Rivera-Calimlim L, Clarke A, Shoulson I (1986) Duodenal delivery of levodopa for on-off fluctuations in parkinsonism: preliminary observations. Ann Neurol 20:262-264

Kuruma I, Bartholini G, Pletscher A (1970) L-Dopa-induced accumulation of 3-*O*-methyldopa in brain and heart. Eur J Pharmacol 10:189-192

Laitinen LV (1973) Slowly absorbed L-dopa preparation in the treatment of parkinsonism. Acta Neurol Scand 49:331-338

Lakke JPWF, van Praag HM (1971) Predicting response to levodopa. Lancet 2:164-165

Lakke JPWF, Korf J, Wesseling H (eds) (1977) Parkinson's disease: concepts and prospects. Excerpta Medica, Amsterdam

Lang AE (1985) Dopamine agonists in the treatment of dystonia. Clin Neuropharmacol 8:38-57

Lang AE, Meadows JC, Parkes JD, Marsden CD (1982) Early onset of the "on-off" phenomenon in children with symptomatic parkinsonism. J Neurol Neurosurg Psychiatry 45:823-825

Lang AE, Birnbaum A, Blair RDG, Kierans C (1986) Levodopa dose-related fluctuations in presumed olivopontocerebellar atrophy. Movement Disorders 1:93-102

Langdon N, Malcolm PN, Parkes JD (1986) Comparison of levodopa with carbidopa, and levodopa with domperidone in Parkinson's disease. Clin Neuropharmacol 9:440-447

Lau E, Watermark K, Glover R, Schulzer M, Calne DB (1986) Effect of antacid on levodopa therapy. Clin Neuropharmacol 9:477-479

Lavin PJM, Gawel MJ, Das PK, Alaghband-Zadeh J, Rose FC (1981) Effect of levodopa on thyroid function and prolactin release. Arch Neurol 38:759-760

Lee T, Seeman P, Rajput A, Farley IJ, Hornykiewicz O (1978) Receptor basis for dopaminergic supersensitivity in Parkinson's disease. Nature 272:59-61

Leenders KL, Wolfson L, Gibbs JM, Wise RJ, Causon R, Jones T, Legg NJ (1985) The effects of L-dopa on regional cerebral blood flow and oxygen metabolism in patients with Parkinson's disease. Brain 108:171-191

Leenders KL, Palmer AJ, Quinn N, Clark JC, Firnau G, Garnett ES, Nahmias C, Jones T, Marsden CD (1986) Brain dopamine metabolism in patients with Parkinson's disease measured with positron emission tomography. J Neurol Neurosurg Psychiatry 49:853-860

Lesser RP, Fahn S, Snider SR, Cote LJ, Isgreen WP, Barrett RE (1979) Analysis of the clinical problems in parkinsonism and the complications of long-term levodopa therapy. Neurology 29:1253-1260

Lewis C, Ballinger BR, Presly AS (1978) Trial of levodopa in senile dementia. Br Med J [clin Res] 1:550

LeWitt PA, Calne DB (1981) Recent advances in the treatment of Parkinson's disease: the role of bromocriptine. J Neural Transm 51:175-184

LeWitt PA, Burns RS, Newman RP (1986a) Dystonia in untreated parkinsonism. Clin Neuropharmacol 9:293-297

LeWitt PA, Miller LP, Levine RA, Lovenberg W, Newman RP, Papavasiliou A, Rayes A, Eldridge R, Burns RS (1986b) Tetrahydrobiopterin in dystonia: identification of abnormal metabolism and therapeutic trials. Neurology 36:760-764

Liebermann A, Estey E, Gopinathan G, Ohashi T, Sauter A, Goldstein M (1978) Comparative effectiveness of two extracerebral DOPA decarboxylase inhibitors in Parkinson disease. Neurology 28:964-968

Liebermann AN, Goldstein M, Gopinathan G, Neophytides A, Hiesiger E, Walker R, Nelson J (1984) Combined use of benserazide and carbidopa in Parkinson's disease. Neurology 34:227-229

Linden I (1980) Effects of 3,4-dihydroxyphenylpyruvic acid on some pharmacokinetic parameters of L-dopa in the rat. J Pharm Pharmacol 32:344-348

Lindvall M, Hardebo JE, Owman C (1980) Barrier mechanism for neurotransmitter monoamines in the choroid plexus. Acta Physiol Scand 108:215-221

List SJ, Seeman P (1979) Dopamine agonists reverse the elevated [^{3}H]neuroleptic binding in neuroleptic-treated rats. Life Sci 24:1447-1452

Lloyd KG (1977) CNS compensation to dopamine loss in Parkinson's disease. Adv Exp Med Biol 90:255-266

Lloyd KG, Hornykiewicz O (1973) L-Glutamic acid decarboxylase in Parkinson's disease: effects of L-dopa therapy. Nature 243:521-522

Lloyd KG, Hornykiewicz O (1977) Effect of chronic neuroleptic or L-dopa administration on GABA levels in the rat substantia nigra. Life Sci 21:1489-1496

Lloyd KG, Davidson L, Hornykiewicz O (1973) Metabolism of levodopa in the human brain. Adv Neurol 3:173-188

Lloyd KG, Davidson L, Hornykiewicz O (1975) The neurochemistry of Parkinson's disease: effect of L-dopa therapy. J Pharmacol Exp Ther 195:453-464

Lloyd KG, Hockman CH, Davidson L, Farley IJ, Hornykiewicz O (1980) Kinetics of L-dopa metabolism in the caudate nucleus of cats with ventrotegmental lesions. J Neural Transm [Suppl] 16:33-44

Loranger AW, Goodell H, Lee JE, McDowell FH (1972) Levodopa treatment of Parkinson's syndrome. Arch Gen Psychiatry 26:163-168

Ludin HP, Bass-Verrey F (1976) Study of deterioration in long-term treatment of parkinsonism with L-dopa plus a decarboxylase inhibitor. J Neural Transm 38:249-258

Lyles Ga, Callingham BA (1980) Short- and long-term effects of L-dopa treatment upon monoamine oxidase: a comparative study in several rat tissues. Eur J Pharmacol 61:363-372

MacLeod MD, Munro JF, Ledingham JG, Farquhar JW (1983) Management of the extrapyramidal manifestations of phenylketonuria with L-dopa. Arch Dis Child 58:457-458

Markham CH, Diamond SG (1981) Evidence to support early levodopa therapy in Parkinson's disease. Neurology 31:125-131

Markham CH, Diamond SG (1986) Modification of Parkinson's disease by longterm levodopa treatment. Arch Neurol 43:405-407

Markham CH, Treciokos LJ, Diamond SG (1974) Parkinson's disease and levodopa: a five year follow-up and review. West J Med 121:188-206

Marks J (ed) (1974) The treatment of parkinsonism with L-dopa. Elsevier, New York

Mars H (1973) Modification of levodopa effect by systemic decarboxylase inhibition. Arch Neurol 28:91-95

Mars H (1974) Levodopa, carbidopa, and pyridoxine in Parkinson's disease. Arch Neurol 30:444-447

Mars H (1975) Effect of chronic levodopa treatment on pyridoxine metabolism. Neurology 25:263-266

Marsh GG, Markham CH (1973) Does levodopa alter depressiion and psychopathology in parkinsonian patients? J Neurol Neurosurg Psychiatry 36:925-935
Marsden CD (1982) The mysterious motor function of the basal ganglia: the Robert Wartenberg lecture. Neurology 32:514-539
Marsden CD (1975) Combined treatment with selective decarboxylase inhibitors. In: Stern GM (ed) The clinical uses of levodopa. University Park Press, Baltimore, pp 107-126
Marsden CD (1981) Treatment of torsion dystonia. In: Barbeau A (ed) Disorders of movement, Lippincott, Philadelphia, pp 81-104
Marsden CD, Parkes JD (1977) Success and problems of long-term levodopa therapy in Parkinson's disease. Lancet 1:345-349
Marsden CD, Parkes JD, Rees JE (1974a) Long-term treatment of Parkinson's disease with an extracerebral DOPA decarboxylase inhibitor (L-alpha-methyldopahydrazine, MK-486) and levodopa. Adv Neurol 3:79-96
Marsden CD, Parkes JD, Rees JE (1974b) A comparison of treatment of Parkinson's disease with L-dopa and with Sinemet. In: Yahr MD (ed) Current concepts in the treatment of parkinsonism. Raven, New York, pp 21-36
Marsden CD, Rinne UK, Koella W, Dubuis R (eds) (1987) International workshop on the on-off phenomenon in Parkinson's disease. Eur Neurol 27[Suppl]:1-140
Martin WE, Resch JA, Baker AB (1971) Juvenile parkinsonism. Arch Neurol 25:494-500
Mayeux R, Stern Y, Mulvey K, Cote L (1985) Reappraisal of temporary levodopa withdrawal ("drug holiday") in Parkinson's disease. N Engl J Med 313:724-728
McDowell FH, Barbeau A (eds) (1974) Second Canadian-American conference on Parkinson's disease. Adv Neurol 5
McDowell FH, Markham CH (eds) (1971) Recent advances in Parkinson's disease. Davis, Philadelphia, pp 1-511
McDowell FH, Sweet RD (1976) The "On-Off" phenomenon. In: Birkmayer W, Hornykiewicz O (eds) Advances in parkinsonism. Roche, Basel, pp 603-612
McDowell FH, Sweet R (1979) Ten year follow-up study of levodopa-treated patients with Parkinson's disease. Adv Neurol 24:475-483
McInnes RR, Kaufman S, Warsh JJ, Van Loon GR, Milstien S, Kapatos G, Soldin S, Walsh P, Macgregor D, Hanley WB (1984) Biopterin synthesis defect. Treatment with L-dopa and 5-hydroxytryptophan compared with therapy with a tetrahydropterin. J Clin Invest 73:458-469
McLellan DL, Dean BC (1982) Improved control of brittle parkinsonism by separate administration of levodopa and benserazide. Br Med J 284:1001-1002
Mearrick PT, Wade DN, Birkett DJ, Morris J (1974) Metoclopramide, gastric emptying and L-dopa absorption. Aust NZ J Med 4:144-148
Melamed E (1986) Initiation of levodopa therapy in parkinsonian patients should be delayed until the advanced stages of the disease. Arch Neurol 43:402-405
Melamed E, Hefti F (1984) Mechanism of action of short- and long-term L-dopa treatment in parkinsonism: role of the surviving nigrostriatal dopaminergic neurons. Adv Neurol 40:149-157
Melamed E, Lavy S, Cooper G, Bentin S (1978) Regional cerebral blood flow in parkinsonism. Measurement before and after L-dopa. J Neurol Sci 38:391-397
Melamed E, Hefti F, Liebman J, Schlosberg AJ, Wurtman RJ (1980a) Serotonergic neurons are not involved in action of L-dopa in Parkinson's disease. Nature 283:772-774
Melamed E, Hefti F, Wurtman RJ (1980b) Non-aminergic neurons convert exogenous L-dopa to dopamine in parkinsonism. Ann Neurol 8:558-563

Melamed E, Hefti F, Wurtman RJ (1980c) Decarboxylation of exogenous L-dopa in rat striatum after lesions of the dopaminergic nigrostriatal neurons: the role of striatal capillaries. Brain Res 198:244-248

Melamed E, Hefti F, Pettibone DJ, Liebman J, Wurtman RJ (1981) Aromatic L-amino acid decarboxylase in rat corpus striatum: implications for action of L-dopa in parkinsonism. Neurology 31:651-655

Melamed E, Globus M, Friedlander E, Rosenthal J (1983) Chronic L-dopa administration decreases striatal accumulation of dopamine from exogenous L-dopa in rats with intact nigrostriatal projections. Neurology 33:950-953

Melamed E, Hefti F, Bitton V, Globus M (1984) Suppression of L-dopa-induced circling in rats with nigral lesions by blockade of central dopa-decarboxylase: implications for the mechanism of action of L-dopa in parkinsonism. Neurology 34:1566-1570

Melamed E, Bitton V, Zelig O (1986) Episodic unresponsiveness to single doses of L-dopa in parkinsonian fluctuators. Neurology 36:100-103

Mena I, Cotzias GC (1975) Protein intake and treatment of Parkinson's disease with levodopa. N Engl J Med 292:181-184

Mena I, Court J, Cotzias GC (1971) Levodopa, involuntary movements, and fusaric acid. JAMA 218:1829-1830

Messiha FS, Knopp W (1973) Metabolic patterns and clinical response to levodopa in Parkinson's disease. Clin Pharmacol Ther 14:565-571

Mindham RHS, Marsden CD, Parkes JD (1976) Psychiatric symptoms during L-dopa therapy for Parkinson's disease and their relationship to physical disability. Psychol Med 6:23-33

Minoli G, Tredici G (1974) Levodopa in treatment of myoclonus. Lancet 2:472

Mintz M, Tomer R, Radwan H, Myslobodsky MS (1982) A comparison of levodopa treatment and task demands on visual evoked potentials in hemi-parkinsonism. Psychiatry Res 6:245-251

Mishra RK, Wong Y-W, Varmuza SL, Tuff L (1978) Chemical lesion and drug induced supersensitivity and subsensitivity of caudate dopamine receptors. Life Sci 23:443-446

Mones RJ (1973) An analysis of six patients with Parkinson's disease who have been unresponsive to L-dopa. J Neurol Neurosurg Psychiatry 36:362-367

Montplaisir J, Godbout R, Poirier G, Bedard MA (1986) Restless legs syndrome and periodic movements in sleep: physiopathology and treatment with L-dopa. Clin Neuropharmacol 9:456-463

Morgan JP, Bianchine JR, Spiegel HE, Rivera-Calimlim L, Hersey RM (1971) Metabolism of levodopa in patients with Parkinson's disease. Arch Neurol 25:39-44

Morris JGL, Parsons RL, Trounce JR, Groves MJ (1976) Plasma dopa concentrations after different preperations of levodopa in normal subjects. B J Clin Pharmacol 3:983-990

Muenter MD (1984) Should levodopa therapy be started early or late? Can J Neurol Sci 11[Suppl 1]:195-199

Muenter MD, Tyce GM (1971) L-Dopa therapy of Parkinson's disease: plasma L-dopa concentration, therapeutic response, and side-effects. Mayo Clin Proc 46:231-239

Muenter MD, Sharpless NS, Tyce GM (1972) Plasma 3-*O*-methyldopa in L-dopa therapy of Parkinson's disease. Mayo Clin Proc 47:389-395

Muenter MD, DiNapoli RP, Sharpless NS, Tyce GM (1973) 3-*O*-Methyldopa, L-dopa, and trihexiphenidyl in the treatment of Parkinson's disease. Mayo Clin Proc 48:173-183

Muenter MD, Sharpless NS, Tyce GM, Darley FL (1977) Patterns of dystonia ("I-D-I" and "D-I-D") in response to L-dopa therapy for Parkinson's disease. Mayo Clin Proc 52:163–174
Mullen P, Seeman P (1979) Pre-synaptic subsensitivity as a possible basis for sensitization by long term dopamine mimetics. Eur J Pharmacol 55:149–157
Muskiet FA, Jeuring HJ, Teelken AW, Wolthers BG, Lakke JP (1978) Identification and quantification of 3-methoxy-4-hydroxyphenyllactic acid (VLA) in cerebrospinal fluid and 3-methoxy-4-hydroxy-phenylpyruvic acid (VPA) in the urine of parkinsonian patients treated with L-dopa. J Neurochem 31:1283–1288
Nausieda PA (1985) Sinement "abusers". Clin Neuropharmacol 8:318–327
Newman RP, Weingartner H, Smallberg SA, Calne DB (1984) Effortful and automatic memory: effects of dopamine. Neurology 34:805–807
Newman RP, LeWitt PA, Jaffe M, Calne DB, Larsen TA (1985) Motor function in the normal aging population: treatment with levodopa. Neurology 35:571–573
Ng KY, Chase TN, Colburn RN, Kopin IJ (1970) L-Dopa-induced release of cerebral monoamines. Science 170:76–77
Nibbelink DW, Bauer R, Hoehn M, Muenter M, Stellar S, Berman R (1981) Sinemet 25/100. In: Rose FC, Capildeo R (eds) Research progress in Parkinson's disease. Pitman, Tunbridge, pp 226–232
Nishimura K, Sasahara K, Arai M, Nitanai T, Ikegami Y, Morioka T, Nakajima E (1984) Dosage form design for improvement of bioavailability of levodopa. VI. Formulation of effervescent enteric coated tablets. J Pharm Sci 73(7):942–946
Nixon DW (1975) Use of levodopa to relieve pain from bone metastases. N Engl J Med 292:647
Nutt JG, Fellman JH (1984) Pharmacokinetics of levodopa. Clin Neuropharmacol 7:35–49
Nutt JG, Woodward WR (1986) Levodopa pharmacokinetics and pharmacodynamics in fluctuating parkinsonian patients. Neurology 36:739–744
Nutt JG, Woodward WR, Hammerstad JP, Carter JH, Anderson JL (1984) The "on-off" phenomenon in Parkinson's disease: relation to levodopa absorption and transport. N Engl J Med 310:483–488
Nutt JG, Woodward WR, Anderson JL (1985) The effect of carbidopa on the pharmacokinetics of intravenously-administered levodopa: the mechanism of action in the treatment of parkinsonism. Ann Neurol 18:537–543
Nutt JG, Woodward WR, Carter JN (1986) Clinical and biochemical studies with controlled-release levodopa/carbidopa. Neurology 36:1206–1211
Ogasahara S, Nishikawa Y, Takahashi M, Wada K, Nakamura Y, Yorifuji S, Tarui S (1984) Dopamine metabolism in the central nervous system after discontinuation of L-dopa therapy in patients with Parkinson's disease. J Neurol Sci 66:151–163
O'Leary MH, Baughn PL (1975) New pathway for metabolism of dopa. Nature 253:52–53
Ordonez LA, Arbrus M, Boyson S, Goodman MN, Ruderman NB, Wurtman RJ (1974) Skeletal muscle resevoir for exogenous L-dopa. J Pharmacol Exp Ther 190:187–190
Oster KA, Sorkin SZ (1942) The effects of intravenous injection of L-dopa upon blood pressure. Proc Soc Exp Biol Med 51:67–71
Owman C, Rosengren E (1967) Dopamine formation in brain capillaries: an enzymatic blood-brain barrier mechanism. J Neurochem 14:547–550
Paden CM (1979) Disappearance of newly synthesized and total dopamine from the striatum of the rat after inhibition of synthesis: evidence for a homogeneous kinetic compartement. J Neurochem 33:471–479

Pakkenberg H, Birket-Smith E, Dupont E, Hansen E, Mikkelsen B, Presthus J, Rautakorpi I, Riman E, Rinne UK (1976) Parkinson's disease treated with Sinemet or Madopar. Acta Neurol Scand 53:376-385

Papavasiliou PS, Cotzias GC, Duby SE, Steck AJ, Fehling C, Bell MA (1972) Levodopa in parkinsonism: potentiation of central effects with a peripheral inhibitor. N Engl J Med 286:8-14

Papavasiliou PS, Cotzias GC, Lawrence WH (1973) Levodopa and dopamine in the cerebrospinal fluid. Neurology 23:756-759

Papavasiliou PS, Cotzias GC, Mena I (1974) Short- and long-term approaches to the "on-off" phenomenon. Adv Neurol 5:379-386

Papavasiliou PS, Miller ST, Thal LJ, Nerder LJ, Houlihan G, Rao SN, Stevens JM (1981) Age-related motor and catecholamine alterations in mice on levodopa supplemented diet. Life Sci 28:2945-2952

Pardridge WM (1977) Kinetics of competitive inhibition of neutral amino acid transport across the blood-brain barrier mechanism. J Neurochem 14:547-550

Pardridge WM, Oldendorf WH (1975) Kinetic analysis of blood-brain barrier transport of amino acids. Biochim Biophys Acta 401:128-136

Parenti M, Flauto C, Parati F, Vescovi A, Groppetti A (1986) Differential effect of repeated treatment with L-dopa on dopamine D_1 or -D_2 receptors. Neuropharmacology 25:331-334

Parkes JD, Baxter RC, Curzon G (1971) Treatment of Parkinson's disease with amantadine and levodopa. A one-year study. Lancet 1:1083-1086

Paulseth JE, Carvey PA, Klawans HL (1984) Long-term levodopa-carbidopa-induced alterations in central dopamine metabolism and turnover in guinea pigs: clinical implications. Ann Neurol 16:127-128

Pearce LA, Warterbury LD (1974) L-Methionine: a possible levodopa antagonist. Neurology 24:640-641

Peaston MJT, Bianchine JR (1970) Metabolic studies and clinical observations during L-dopa treatment of Parkinson's disease. Br Med J 1:400-403

Pederzoli M, Girotti F, Scigliano G, Aiello G, Carella F, Caraceni T (1983) L-Dopa long-term treatment in Parkinson's disease: age-related side-effects. Neurology 33:1518-1522

Petajan JH, Jarcho LW (1975) Motor unit control in Parkinson's disease and the influence of levodopa. Neurology 25:866-869

Pilling JB, Baker J, Iversen LL, Iversen SD, Robbins T (1975) Plasma concentrations of L-dopa and 3-methoxydopa and improvement in clinical ratings and motor performance in patients with parkinsonism treated with L-dopa alone or in combination with amantadine. J Neurol Neurosurg Psychiatry 38:129-135

Pinder RM, Brogden RN, Sawyer PR, Speight TM, Avery GS (1976) Levodopa and decarboxylase inhibitors: a review of their clinical pharmacology and use in the treatment of parkinsonism. Drugs 11:329-377

Plascencia RJ, Gilroy J, Cullis P (1984) Treatment of thalamic pain syndrome with levodopa. Neurology 34[Suppl 1]:137

Pletscher A (1973) Effect of inhibitors of extracerebral decarboxylase on levodopa metabolism. Adv Neurol 3:49-58

Pletscher A, Bartholine G, Tissot R (1967) Metabolic fate of L-[^{14}C]-dopa in cerebrospinal fluid and blood plasma of humans. Brain Res 4:103-106

Pocelinko R, Thomas GB, Solomon HM (1972) The effects of an antacid on the absorption and metabolism of levodopa. Clin Pharm Ther 13:149A

Poewe WH, Lees AJ, Stern GM (1986a) Treatment of motor fluctuations in Parkinson's disease with an oral sustained-release preparation of L-dopa: clinical and pharmacokinetic observations. Clin Neuropharmacol Ther 9:430-439

Poewe WH, Lees AJ, Stern GM (1986b) Low-dose L-dopa therapy in Parkinson's disease: a six-year follow-up. Neurology 36:1528-1530
Poirier LJ, Singh P, Sourkes TL (1967) Effect of amine precursors on the concentration of striatal dopamine and serotonin in cats with and without unilateral brainstem lesions. Brain Res 6:654-666
Poirier LJ, Sourkes TL, Bedard PJ (eds) (1979) The extrapyramidal system and its disorders. Adv Neurol 24:1-529
Ponzio F, Cimino M, Achilli G, Lipartiti M, Perego C, Vantini G, Algeri S (1984) In vivo and in vitro evidence of dopaminergic system down regulation induced by chronic L-dopa. Life Sci 34:2107-2116
Popovic P, Popovic V, Schaffer R (1976) Recovery from experimental paraplegia after levodopa administration. Acta Neurochir (Wien) 35:141-147
Porter CC (1973) Inhibitors of aromatic amino acid decarboxylase — their biochemistry. Adv Neurol 2:37-58
Presthuis J, Holmsen R (1974) Appraisal of long-term levodopa treatment of parkinsonism with special reference to therapy limiting factors. Acta Neurol Scand 50:774-790
Preziosi TJ, Bianchine JR, Hsu TH, Messiha FS (1972) L-Methyldopa hydrazine (MK-486) and L-dopa: a double blind study in parkinsonism. Trans Am Neurol Assoc 97:321-322
Priebe S (1984) Levodopa dependence: a case report. Pharmacopsychiatry 17:109-110
Pycock C, Dawbarn D, O'Shaughnessy C (1982) Behavioural and biochemical changes following chronic administration of L-dopa to rats. Eur J Pharmacol 79:201-215
Quinn N, Parkes JD, Marsden CD (1984) Control of on-off phenomenon by continuous intravenous infusion of levodopa. Neurology 34:1131-1136
Quinn NP, Parkes D, Janota I, Marsden CD (1986) Preservation of the substantia nigra and locus coeruleus in a patient receiving levodopa (2 kg) plus decarboxylase inhibitor over a four-year period. Movement Disorders 1:65-68
Rajan KS, Manian AA, Davis JM, Dekirmenjian H (1976) Metal chelates of L-dopa for improved replenishment of dopaminergic pools. Brain Res 107:317-331
Rajfer SI, Anton AH, Rossen JD, Goldberg LI (1984) Beneficial hemodynamic effects of oral levodopa in heart failure. Relation to the generation of dopamine. N Engl J Med 310:1357-1362
Rajput AH (1973) L-Dopa in dystonia musculorum deformans. Lancet 1:432
Rajput AH, Stern W, Laverty WH (1984) Chronic low-dose levodopa therapy in Parkinson's disease: an argument for delaying levodopa therapy. Neurology 34:991-996
Rampen FHJ (1985) Levodopa and melanoma: three cases and review of literature. J Neurol Neurosurg Psychiatry 48:585-588
Rascol A, Montastruc JL, Rascol O (1984) Should dopamine agonists be given early or late in the treatment of Parkinson's disease? Can J Neurol Sci 11:229-232
Raz S (1976) Parkinsonism and neurogenic bladder. Experimental and clinical observations. Urol Res 4:133-138
Reches A, Fahn S (1982) 3-*O*-Methyldopa blocks dopa metabolism in rat corpus striatum. Ann Neurol 12:267-271
Reches A, Fahn S (1984) Catechol-*O*-methyltransferase and Parkinson's disease. Adv Neurol 40:171-179
Reches A, Jiang D-H, Fahn S (1982a) Effect of 3,4-dihydroxy-2-methyl-propriophenone (U-0521) on catechol-*O*-methyltransferase activity and on dopa accumulation in rat blood cells and corpus striatum. Biochem Pharmacol 31:3415-3418

Reches A, Mielke LR, Fahn S (1982b) 3-*O*-Methyldopa inhibits rotations induced by levodopa in rats after unilateral destruction of the nigrostriatal pathway. Neurology 32:887–888

Reches A, Wagner HR, Jackson-Lewis V, Yablonskaya-Alter E, Fahn S (1984) Chronic levodopa or pergolide administration induces down-regulation of dopamine receptors in denervated striatum. Neurology 34:1208–1212

Reilly DK, Rivera-Calimlim L, Van Dyke D (1980) Catechol-*O*-methyltransferase activity: a determinant of levodopa response. Clin Pharmacol Ther 28:278–286

Reilly DK, Hershey L, Rivera-Calimlim L, Shoulson I (1983) On-off effects in Parkinson's disease: a controlled investigation of ascorbic acid therapy. Adv Neurol 37:51–59

Renvoize EB, Jerram T, Clough G (1978) Levodopa in senile dementia. Br Med J 2:504

Riederer P (1980) L-Dopa competes with tyrosine and tryptophan for human brain uptake. Nutr Metab 24:417–423

Riekkinen P, Rinne UK, Pelliniemi T, Sonninen V (1975) Interaction between dopamine and phospholipids. Arch Neurol 32:25–27

Riklan M, Whelihan W, Cullinan T (1976) Levodopa and psychometric test performance in parkinsonism — 5 years later. Neurology 26:173–179

Rinne UK (1985) Combined bromocriptine-levodopa therapy early in Parkinson's disease. Neurology 35:1196–1198

Rinne UK, Molsa P (1979) Levodopa with benserazide or carbidopa in Parkinson disease. Neurology 29:1584–1589

Rinne UK, Sonninen V (1972) Acid monoamine metabolites in the cerebrospinal fluid of patients with Parkinson's disease. Neurology 22:62–67

Rinne UK, Sonninen V, Hyppa M (1971) Effect of L-dopa on brain monoamines and their metabolites in Parkinson's disease. Life Sci 10:549–577

Rinne UK, Sonninen V, Siirtola T (1973a) Treatment of parkinsonian patients with levodopa and extracerebral decarboxylase inhibitor, Ro 4-4602. Adv Neurol 3:59–71

Rinne UK, Sonninen V, Siirtola T (1973b) Plasma concentration of levodopa in patients with Parkinson's disease. Eur Neurol 10:301–310

Rinne UK, Sonninen V, Riekkinen P, Laaksonen H (1974) Postmortem findings in Parkinsonian patients treated with LD: biochemical considerations. In: Yahr MD (ed) Current concepts in the treatment of parkinsonism. Raven, New York, pp 211–233

Rinne UK, Birket-Smith E, Dupont E, Hansen E, Hyppa M, Marttila R, Mikkelsen B, Pakkenberg H, Presthus J (1975) Levodopa alone and in combination with a peripheral decarboxylase inhibitor benserazide (Madopar) in the treatment of Parkinson's disease: a controlled clinical trial. J Neurol 211:1–9

Rinne UK, Sonninen V, Marttila R (1977) Brain dopamine turnover and the relief of parkinsonism. Adv Exp Med Biol 90:267–275

Rinne UK, Sonninen V, Laaksonen H (1979) Responses of brain neurochemistry to levodopa treatment in Parkinson's disease. Adv Neurol 24:259–274

Rinne UK, Sonninen V, Siirtola T, Marttila R (1980a) Long-term responses of Parkinson's disease to levodopa therapy. J Neural Transm [Suppl]:149–156

Rinne UK, Klinger M, Stamm G (eds) (1980b) Parkinson's disease: current progress, problems, and management. Elsevier/North Holland Biomedical, New York

Rinne UK, Koskinen V, Lönnberg PN (1980c) Neurotransmitter receptors in the parkinsonian brain. In: Rinne UK, Klinger M, Stamm G (eds) Parkinson's disease: current progress, problems, and management. Elsevier/North Holland Biomedical, New York, pp 93–109

Rivera-Calimlim L (1974) Absorption, metabolism, and distribution of ^{14}C-*O*-methyldopa and ^{14}C-L-dopa after oral administration to rats. Br J Pharmacol 50:259-263

Rivera-Calimlim L, Morgan JP, Dujovne CA, Bianchine JR, Lasagna L (1970a) L-Dopa absorption and metabolism by the human stomach. J Clin Invest 49:79A

Rivera-Calimlim L, Dujovne CA, Morgan JP, Lasagna L, Bianchine JR (1970b) L-Dopa failure: correction and explanation. Br Med J 4:93-94

Rivera-Calimlim L, Dujovne CA, Morgan JP, Lasagna L, Bianchine JR (1971) Absorption and metabolism of L-dopa by the human stomach. Eur J Clin Invest 1:313-320

Rivera-Calimlim L, Tandon D, Anderson F, Joynt R (1977) The clinical picture and plasma levodopa metabolite profile of parkinsonian nonresponders. Arch Neurol 34:228-232

Roberge AG, Poirier LJ (1973) Brain dopa-5-HTP decarboxylase activity after the chronic administration of L-dopa in normal and lesioned cats. Brain Res 76:401-412

Roel LE, Levine P, Rubin D, Markovitz D, Markovitz D, Munro HN, Wurtman RH (1978) Effect of L-dopa pretreatment on in vivo protein synthesis in various rat brain regions. Life Sci 22:1887-1891

Rosati G, Maioli M, Aiello I, Farris A, Agnetti V (1976) Effects of long-term L-dopa therapy on carbohydrate metabolism in patients with Parkinson's disease. Eur Neurol 14:229-239

Rose FC, Capildeo R (eds) (1981) Research progress in Parkinson's disease. Pitman, Tunbridge

Rossor MN, Watkins J, Brown MJ, Reid JL, Dollery CT (1980) Plasma levodopa, dopamine and therapeutic response following levodopa therapy of Parkinsonian patients. J Neurol Sci 46:385-392

Saarinen A, Myllylä VV, Tokola O, Hokkanen E (1978) Effect of a slow release preparation of levodopa on Parkinson's disease in combination with a peripheral decarboxylase inhibitor. Acta Neurol Scand 58:340-349

Sachdev KK, Singh N, Krishnamoorthy MS (1977) Juvenile parkinsonism treated with levodopa. Arch Neurol 34:244-245

Sacks O (1983) Awakenings (revised edition). Dutton New York, pp 1-338

Sandler M (1971) How does L-dopa work in parkinsonism? Lancet 1:784-785

Sandler M (1972) Catecholamine synthesis and metabolism in man: clinical implications (with special reference to parkinsonism). In: Blaschko H, Muscholl E (eds) Catecholamines. (Handbook of clinical pharmacology, vol 33) Springer, Berlin Heidelberg New York, pp 845-899

Sandler M, Goodwin BL, Ruthven CRJ, Calne DB (1971) Therapeutic implications in parkinsonism of *m*-tyramine formation from L-dopa in man. Nature 229:414-416

Sandler M, Bonham Carter S, Hunter KR, Stern GM (1973) Tetrahydroisoquinoline alkaloids: in vivo metabolites of L-dopa in man. Nature: 241:439-443

Sandler M, Johnson RD, Ruthven CR (1974a) Transamination is a major pathway of L-dopa metabolism following peripheral decarboxylase inhibition. Nature 247:364-366

Sandler M, Ruthven CR, Goodwin BL, Hunter KR, Stern GM (1974b) Variation of levodopa metabolism with gastrointestinal absorption site. Lancet 1:238-240

Sasahara K, Nitanai T, Habara T, Moroika T, Nakajima E (1981) Dosage form design for improvement of bioavailability of levodopa. V. Absorption and metabolism of levodopa in intestinal segments of dogs. J Pharm Sci 70:1157-1160

Schneider E, Fischer PA, Jacobi P, Becker H, Beyer M (1979) Cerebral atrophy and long-term response to levodopa in Parkinson's disease. J Neurol 222:37–43

Schneider MB, Murrin LC, Pfeiffer RF, Deupree JD (1984) Dopamine receptors: effects of chronic L-dopa and bromocriptine treatment on an animal model of Parkinson's disease. Clin Neuropharmacol 7:247–257

Schwartz DE, Jordan JC, Ziegler WH (1974) Pharmacokinetics of the decarboxylase inhibitor benserazide in man: its tissue distribution in the rat. Eur J Clin Pharmacol 7:39–45

Sechi GP, Tanda F, Mutani R (1984) Fatal hyperpyrexia after withdrawal of levodopa. Neurology 34:249–251

Segawa M, Hosaka A, Miyagawa F, Nomura Y, Imai H (1976) Hereditary progressive dystonia with marked diurnal fluctuation. Adv Neurol 14:215–233

Seyfert S, Straschill M (1977) Reversible spastic and pyramidal tract signs in a patient with Parkinson's disease. J Neurol 216:223–226

Sharpless NS, McCann DS (1971) Dopa and 3-*O*-methyldopa in cerebrospinal fluid of parkinsonian patients during treatment with oral L-dopa. Clin Chim Acta 31:155–169

Sharpless NS, Muenter MD, Tyce GM, Owen CA (1972) 3-Methoxy-4-hydroxyphenylalanine 3-methyldopa in plasma during oral L-dopa therapy of patients with Parkinson's disease. Clin Chim Acta 37:359–369

Sharpless NS, Tyce GM, Thal LJ, Waltz JM, Tabaddor K, Wolfson LI (1981) Free and conjugated dopamine in human vetricular fluid. Brain Res 217:107–118

Shaw KM, Lees AJ, Stern GM (1980) The impact of treatment with levodopa on Parkinson's disease. Q J Med 49:283–293

Shoulson I (1983) Carbidopa/levodopa therapy of coexistent drug-induced parkinsonism and tardive dyskinesia. Adv Neurol 37:259–266

Shoulson I. Glaubiger GA, Chase TN (1975) On-off response: clinical and biochemical correlations during oral and intravenous levodopa administration. Neurology 25:1144–1148

Sirtori CR, Bolme P, Azarnoff DL (1972) Metabolic response to acute and chronic L-dopa administration in patients with parkinsonism. N Engl J Med 287:729–733

Sishta SK, Templer DI (1976) Levodopa in Huntington's chorea. Can Med Assoc J 114:798–799

Sourkes TL (1970) On the mode of action of L-dopa in Parkinson's disease. Biochem Med 3:321–325

Sourkes TL (1971) Possible new metabolites mediating actions of L-dopa. Nature 229:413–414

Spencer SE, Wooten GF (1984a) Altered pharmacokinetics of L-dopa metabolism in rat striatum deprived of dopaminergic innervation. Neurology 34:1105–1108

Spencer SE, Wooten GF (1984b) Pharmacologic effects of L-dopa are nor closely linked temporally to striatal dopamine concentration. Neurology 34:1609–1611

Stern GM (ed) (1975) The clinical uses of levodopa. University Park Press, Baltimore

Stern GM, Lees AJ (1983) Sustained bromocriptine therapy in 50 previously untreated patients with Parkinson's disease. Adv Neurol 37:17–21

Stern PH, McDowell FH, Miller JM, Robinson MB (1972) Levodopa therapy effects on the natural history of parkinsonism. Arch Neurol 27:481–485

Still CN, Herberg K-P (1976) Long-term L-dopa therapy for torsion dystonia. South Med J 69:564–566

Stramentinoli G, Catto E, Algeri S (1980) Decrease of noradrenaline *O*-methylation in rat brain induced by L-dopa. Reversal effect of *S*-adenosyl-L-methionine. J Pharm Pharmacol 32:430–431

Suga M (1980) Effect of long-term L-dopa administration on the dopaminergic and cholinergic (muscarinic) receptors of striatum in 6-hydroxydopamine lesioned rats. Life Sci 27:877-882

Sutliffe RLG (1973) L-dopa therapy in elderly patients with parkinsonism. Age Ageing 2:34-38

Sweet RD, McDowell FH (1974) Plasma L-dopa concentrations and the "on-off" effect after chronic treatment of Parkinson's disease. Neurology 24:953-956

Sweet RD, McDowell FH (1975) Five years treatment of Parkinson's disease. Therapeutics, results and survival of 100 patients. Ann Intern Med 83:456-463

Sweet RD, Lee JE, Spiegel HE, McDowell F (1972) Enhanced response to low doses of levodopa after withdrawal from treatment. Neurology 22:520-525

Sweet RD, McDowell FH, Feigenson JS, Loranger AW, Godell H (1976) Mental symptoms in Parkinson's disease during chronic treatment with levodopa. Neurology 26:305-310

Symposium on levodopa in Parkinson's disease (1972) Neurology 22 (part 2):1-102

Symposium on levodopa in Parkinson's disease — clinical and pharmacological aspects (1971) Clin Pharmacol Ther 12 (part 2):317-476

Tate S, Sweet R, McDowell FH, Meistler A (1971) Decrease of the 3,4-dihydroxyphenylalanine (DOPA) decarboxylase activities in human erythrocytes and mouse tissues after administration of DOPA. Proc Natl Acad Sci USA 68:2121-2123

Teychenne P, Calne DB, Lewis PJ, Findley LJ (1975) Interactions of levodopa with inhibitors of monoamine oxidase and L-aromatic decarboxylase. Clin Pharmacol Ther 18:273-277

Tissot R, Bartholini G, Pletscher A (1969) Drug-related changes of extracerebral DOPA metabolism in man. Arch Neurol 20:187-190

Tissot R, Eisenring JJ, Constantinidis J (1974) Modes of action and optimal dosage of decarboxylase inhibitors. In: Yahr MD (ed) Current concepts in the treatment of parkinsonism. Raven, New York, pp 123-131

Tolosa ES, Martin WE, Cohen JP, Jacobson RL (1975) Patterns of clinical response and plasma dopa levels in Parkinson's disease. Neurology 25:177-183

Tourtellotte WW, Syndulko K, Potvin AR, Hirsch SB, Potvin JH (1980) Increased ratio of carbidopa to levodopa in treatment of Parkinson's disease. Arch Neurol 37:723-726

Turnbull CJ, Aitken JA (1983) Diagnosis and management of parkinsonism in the elderly. Age Ageing 12:309-316

Tyce GM, Owen CA (1979) Administration of L-3,4-dihydroxyophenylalanine to rats after complete hepatectomy. I. Metabolism in tissues. Biochem Pharmacol 28:3271-3278

Tyce GM, Muenter MD, Owen CA (1970) Dihydroxyophenalanine (DOPA) in plasma during DOPA treatment of patients with Parkinson's disease. Mayo Clin Proc 45:438-443

Tyce GM, Sharpless NS, Muenter MD (1974) Free and conjugated dopamine in plasma during levodopa therapy. Clin Pharmacol Ther 16:782-788

Udaka F, Yamao S, Nagata H, Nakamura S, Kameyama M (1984) Pathologic laughing and crying treated with levodopa. Arch Neurol 41:1095-1096

Van Woerkom TC, Minderhoud JM, Gottschal T, Nicolai G (1982) Neurotransmitters in the treatment of patients with severe head injuries. Eur Neurol 21:227-234

Van Woert MH, Weintraub MI (1971) Predicting the response to levodopa. Lancet 1:1015-1016

Vickers S, Stuart EK, Hucker HB (1975) Further studies on the metabolism of carbidopa, (−)-L-alpha-hydrazino-3,9-dihydroxy-alpha-met acid monohydrate in the human, rhesus monkey, dog and rat. J Med Chem 18:134-138

Vogel WH, Mahoney K, Hare TA (1972) Inhibition of dopa decarboxylase by Ro 4-4602, MK-485, and MK-486 in human liver homogenates. Arch Int Pharmacodyn Ther 198:85-93

Wada GM, Fellmann JH (1973) 2,3,5-Trihydroxyphenylacetic acid: a metabolite of L-3,4-dihydroxyphenylalanine. Biochemistry 12:5212-5217

Wade DN, Mearrick PT, Morris J (1973) Active transport of L-dopa in the intestine. Nature 242: 463-465

Wade DN, Mearrick PT, Birkett DJ, Morris J (1974) Variability of L-dopa absorption in man. Aust NZ J Med 4:138-143

Wade LA, Katzman R (1975a) Rat brain regional uptake and decarboxylation of L-dopa following carotid injection. Am J Physiol 228:352-359

Wade LA, Katzman R (1975b) 3-*O*-Methyldopa and inhibition of L-dopa at the blood-brain barrier. Life Sci 17:131-136

Wade LA, Katzman R (1975c) Synthetic amino acids and the nature of L-dopa transport at the blood-brain barrier. J Neurochem 25:837-842

Ward CD, Trombley IK, Calne DB, Kopin IJ (1984) L-Dopa decarboxylation in chronically treated patients. Neurology 34:198-201

Waring P (1986) The time-dependent inactivation of human brain dihydropteridine reductase by the oxidation products of L-dopa. Eur J Biochem 155:305-310

Webster DD, Sawyer GT (1984) The combined use of amantadine HCl and levodopa/carbidopa in Parkinson's disease. Curr Therap Res 35:1010-1013

Weiner CD, Collins MA (1978) Tetrahydroisoquinolines derived from catecholamines or DOPA: effects on brain tyrosine hydroxylase activity. Biochem Pharmacol 27:2699-2703

Weiner WJ, Klawans HL (1973) Failure of cerebropsinal fluid homovanillic acid to predict levodopa response in Parkinson's disease. J Neurol Neurosurg Psychiatry 36:747-752

Weiner WJ, Kramer J, Nausieda PA, Klawans HL (1978) Paradoxical response to dopaminergic agents in parkinsonism. Arch Neurol 35:453-455

Weiner WJ, Koller WC, Perlik S, Nausieda PA, Klawans HL (1980) Drug holiday and management of Parkinson's disease. Neurology 30:1257-1261

Weir RL, Fan KJ (1981) Spinocerebellar degeneration with parkinsonian features: a clinical and pathological report. Ann Neurol 9:87-89

Weiss BF, Munro HN, Ordonez LA, Wurtman RJ (1972) Dopamine: mediator of brain polysome disaggregation after L-dopa. Science 177:613-616

Wilner KD, Butler IJ, Seifert WE, Clement-Cormier YC (1980) Biochemical alterations of dopamine receptor responses following chronic L-dopa therapy. Biochem Pharmacol 29:701-706

Wolf JK, Santana HB, Thorpy M (1979) Treatment of "emotional incontinence" with levodopa, Neurology 29:1435-1436 (letter)

Woods AC, Glaubiger GA, Chase TN (1973) Sustained-release levodopa. Lancet 1:1391

Woods BI, Schaumberg HH (1972) Nigro-spino-dentatal degeneration with nuclear ophthalmoplegia: a unique and partially treatable clinicopathological entity. J Neurol Sci 17:149-166

Wooten GF, Ferrari MB (1983) Competition for the striatal dopamine receptor by metabolites of L-dopa. Ann Neurol 14:136-137

Wurtman RJ, Ordonez LA (1978) Effects of exogenous L-dopa on the metabolism of methionine and *S*-adenosylmethionine in the brain. In: Andreoli VM, Agnoli A, Fazio C (eds) Transmethylations and the central nervous system. Springer, Berlin Heidelberg New York, pp 132-143

Wurtman RJ, Romero JA (1972) Effects of levodopa on nondopaminergic brain neurons. Neurology 22[Suppl]:72-81
Wurtman RJ, Chou C, Rose C (1970a) The fate of ^{14}C-dihydroxyphenylalanine (^{14}C-dopa) in the whole mouse. J Pharmacol Ex Ther 174:351-356
Wurtman RJ, Rose CM, Matthysse S, Stephenson J, Baldessarini R (1970b) L-dihydroxyphenylalanine: effect on *S*-adenosylmethionine in brain. Science 169:395-397
Yaar I (1977) EEG power spectral changes secondary to L-dopa treatment in parkinsonian patients: a pilot study. Electroencephalogr Clin Neurophysiol 43:111-118
Yaar I, Shapiro MB (1983) A quantitative study of the electroencephalographic response to levodopa treatment in parkinsonian patients. Clin Electroencephalogr 14:82-85
Yahr MD (ed) (1973) The treatment of parkinsonism — the role of DOPA decarboxylase inhibitors. Adv Neurol 2:1-303
Yahr MD (ed) (1974a) Current concepts in the treatment of parkinsonism. Raven, New York, pp 1-257
Yahr MD (1974b) Variations in the "on-off" effect. Adv Neurol 5:397-399
Yahr MD (1975) Levodopa. Ann Intern Med 83:677-682
Yahr MD (1976) Evaluation of long-term therapy in Parkinson's disease: mortality and therapeutic efficacy. In: Birkmayer W, Hornykiewicz O (eds) Advances in parkinsonism. Roche, Basel, pp 435-443
Yahr MD, Duvoisin RG (1972) Metabolic responses to levodopa. N Engl J Med 287:1303
Yahr MD, Duvoisin RC (1975) Pyridoxine and levodopa in the treatment of parkinsonism. JAMA 220:861
Yahr MD, Duvoisin RC, Schear MJ, Barrett RE, Hoehn MM (1969) Treatment of parkinsonism with levodopa. Arch Neurol 21:343-354
Yahr MD, Wolf A, Antunes JL, Miyoshi K, Duffy P (1972) Autopsy finding in parkinsonism following treatment with levodopa. Neurology 22[Suppl]:56-71
Yokochi M, Kondo T, Hirayama K, Narabayashi H, Kuruma I (1979) Comparative studies of L-dopa alone and in combination with a peripheral dopa decarboxylase inhibitor, benserazide, on Parkinson's disease. Part II: Pharmacokinetic study [in Japanese]. Brain Nerve 31:339-348
Yuill GM (1980) Suppression of "rubral" tremor with levodopa. Br Med J [Clin Res] 281:1428
Ziegler DK, Schimke N, Kepes JJ, Rose DL, Klinkerfuss G (1972) Late onset ataxia, rigidity, and peripheral neuropathy. Arch Neurol 27:52-66

CHAPTER 14

Adverse Effects of Levodopa in Parkinson's Disease

S. FAHN

A. Historical Aspects of Adverse Effects

With the introduction of high dosage D,L-dopa therapy for Parkinson's disease by COTZIAS et al. (1967) only a few side effects were reported, namely gastrointestinal, faintness, and hematologic changes. The last appeared to be due to the presence of the dextrorotatory isomer of dopa since it has not been reported as a complication of subsequent treatment with the levorotatory form. In their next publication, using levodopa, COTZIAS et al. (1969) now listed mental effects and involuntary movements, in addition to anorexia, nausea, and vomiting. The mental effects listed were irritability, anger, hostility, paranoia, insomnia, and an awakening effect. The dyskinesias were chorea, myoclonus, hemiballism (ipsilateral to the side of a prior thalamotomy), and dystonia. These adverse effects subsided with lowering of the dosage.

Actually, Cotzias succeeded in using high dosages of levodopa to obtain benefit in parkinsonism where others had previously failed. These failures were because of adverse effects. These earlier failures (see FAHN and DUFFY 1977 for a review of early attempts at therapy) can be generally attributed to the gastrointestinal (GI) effects of dopa, occurring at subtherapeutic doses. It is generally agreed that Cotzias' success was his method of minimizing adverse effects from the high dosages of medication necessary to provide benefit. He began treatment with small doses and increased the dosage gradually until therapeutic or toxic levels were reached. The dosage was then either maintained or reduced, if necessary. If the therapeutic dosage was reached during a 3-week stay in hospital, some adverse effects became apparent later, particularly dyskinesias, attributed to increasing behavioral sensitivity to the drug (JENNER et al. 1986; KLAWANS et al. 1977), a phenomenon also seen in laboratory animals and with dopamine agonists (TRAUB et al. 1986; HASSAN et al. 1986).

In an early study of levodopa in 60 patients with parkinsonism, lasting up to 12 months, YAHR et al. (1969) reported GI side effects in 51, dyskinesias in 37, postural hypotension in 14, cardiac dysrhythmia in 12, myocardial infarction in 1, and psychic manifestations in 10. It was becoming clear that adverse toxic effects seen with short-term treatment with levodopa (less than 3 years' duration) could be divided into peripheral and central in origin. Peripheral effects are those due to levodopa's effects outside the blood-brain barrier, including the area postrema, the presumed site of GI side effects. Cardiac effects are also considered peripheral in origin. Central effects are those due to levodopa's penetration across the blood-brain barrier, and include dyskinesias and mental changes.

It was uncertain initially as to whether postural hypotension associated with levodopa treatment was a peripheral effect of the drug's action on adrenergic nerve endings or on the central nervous system (Calne et al. 1970). However, the persistence of postural hypotension in the presence of a peripheral decarboxylase inhibitor makes it clear that at least some of this problem is central in origin. The cardiac complications are the most ominous peripheral effect, but patients with a history of angina and myocardial infarction were found to tolerate levodopa with little increased hazard (Jenkins et al. 1972).

Other investigators (McDowell et al. 1970; Schwarz and Fahn 1970) published their experiences with levodopa therapy, and they agreed with those found by Cotzias' and Yahr's groups. However, dyskinesias were as common as GI effects. Moreover, the dyskinesias persisted as a problem, whereas GI effects would often abate with continuing treatment or with reduction of dosage and slower increase back to therapeutic levels. But dyskinesias became more prominent with continuing treatment, a reflection of the increasing behavioral supersensitivity phenomenon (Jenner et al. 1986; Klawans et al. 1977; Traub et al. 1986; Hassan et al. 1986). It became clear that the commonest dose-limiting adverse reaction was dyskinesia (Calne et al. 1971a); it could occur in all parts of the body, and was usually choreic in nature. Only later, with continuing therapy, would the dystonic reactions become recognized, and these become more troublesome and even painful.

With continuing use of levodopa, it was becoming clear that the drug was most effective in striatal dopamine deficiency states, such as Parkinson's disease and postencephalitic parkinsonism (Yahr 1972) — the latter is more sensitive to levodopa and a lower dosage of levodopa is necessary (Calne et al. 1969; Duvoisin et al. 1972a; Sacks 1974) — whereas it was found to be virtually ineffective in progressive supranuclear palsy and the multiple system atrophies (Yahr 1972; Jankovic 1984; Bannister and Oppenheimer 1982). In fact, levodopa would aggravate postural hypotension in those with this problem, as in the Shy-Drager syndrome. The view that levodopa (and direct-acting dopamine agonists) are effective only when dopamine receptors in the striatum are intact, and its failure in many of the "Parkinson-plus" syndromes (Fahn 1977) is that the latter have loss of these receptors as part of their pathology.

With continuing experience and longer duration treatment with levodopa, it was realized that some patients could not take levodopa because of intolerable adverse effects. Hunter et al. (1973) found that of 187 patients begun on levodopa, 22 discontinued treatment within 6 months, predominantly for central adverse effects, such as psychosis, confusion, and dyskinesias, but one-third for nausea. After 1 year of therapy, an additional eight patients stopped the medication because of dementia (three), dyskinesia (two), nausea (two), and GI bleeding (one).

The development of dyskinesias increased with increasing duration of treatment. Duvoisin (1974a) reported on 116 patients and found that an increasing percentage of them developed dyskinesias with time (Table 1). At

Table 1. Time of appearance of levodopa-induced dyskinesias in 116 patients with parkinsonism (DUVOISIN 1974a)

Length of treatment (months)	Patients with dyskinesias (%)
1	20
2	39
3	46
6	53
9	67
12	81

6 months of treatment, 53% had dyskinesias. At 12 months, the percentage with dyskinesias reached 81%.

A new adverse effect was noticed in the early 1970s — clinical fluctuations, initially referred to as the "on-off" phenomenon (SWEET and McDOWELL 1974; DUVOISIN 1974b). The initial and dramatic situation noticed by patients was that the beneficial response was suddenly terminated, like a light switch being turned off. The beneficial response could turn on again just as quickly. Variations in clinical fluctuations began to be appreciated. Whereas the sudden "on-off" effect could appear early in the course of treatment (DUVOISIN 1974b; YAHR 1974), a different type of "off" phenomenon was also being recognized, and this type would become the predominant one. This is a more gradual fading of benefit as the duration after a dose is prolonged, occurring approximately 2-3 h after taking a dose of levodopa. Patients felt as if their medication was wearing off, and hence the term "wearing-off" effect was coined for this problem (FAHN 1976). Another name widely adopted for this disorder was "end-of-dose" deterioration (MARSDEN and PARKS 1976). Both terms are synonymous. Unfortunately, the term "on-off," originally coined to refer to the more sudden "off" state, has been widely used as the generic term for any type of clinical fluctuation. It would be preferable to avoid ambiguity and use terms with specific meanings, such as "wearing-off" and "sudden off."

While central side effects became a major factor in intolerance to levodopa, GI upset was still a problem in a large minority of patients who had difficulty with the medication. The addition of a peripheral decarboxylase inhibitor markedly reduced this peripheral side effect (CALNE et al. 1971b; CHASE and WATANABE 1972; MARSDEN et al. 1973; MARKHAM et al. 1974), and other peripheral adverse effects, such as cardiac arrhythmias (MARS and KRAIL 1971). Equally important was the increased potency of levodopa when used in combination with a peripheral decarboxylase inhibitor. These factors led this combination to become the major therapeutic agent for the treatment of parkinsonism. However, with this increased central potency of levodopa comes increased frequency of central adverse effects such as involuntary movements, confusion, and psychiatric side effects.

Table 2. Limiting adverse effects of levodopa in 131 patients with Parkinson's disease (LESSER et al. 1979)

Symptom or sign	Patients (%)
Peripheral side effects	
GI symptoms	25
Sweating	7
Cardiac symptoms	1
Central side effects	
Dyskinesias	
Chorea	37
Dystonia	21
Fluctuations	
Slow wearing-off	20
Sudden on-off	18
Mental changes	
Dementia	22
Psychiatric	19
Sleepiness	14
Insomnia	7
Hypersexuality	3

One of the last reviews (LESSER et al. 1979) of adverse effects from long-term levodopa therapy alone (before widespread use in combination with a peripheral decarboxylase inhibitor) found that among 131 patients, central side effects were much commoner as major problems than were peripheral effects (Table 2). In this survey, myoclonus was not seen as a limiting adverse effect, and its appearance as a toxic phenomenon (KLAWANS et al. 1975) does not appear to be a major factor.

As levodopa continued to be used, other adverse effects were discovered. Although peak-dose dystonia appeared in combination with or in place of peak-dose chorea (Table 2), dystonia occurring during the "off" phase also made its appearance. Originally called "early morning" dystonia (MELAMED 1979) because of its timing, this problem is now recognized to appear at other times of the day in susceptible patients. Its characteristic feature is dystonic spasms when the patient is "off," and therefore can be seen with accompanying features of parkinsonism. Dystonia and chorea have also been seen at the beginning and end of dose; this form was first described by MUENTER et al. (1977) who termed it the D-I-D phenomenon, for "dystonia-improvement-dystonia." It is also known as diphasic dyskinesia (MARSDEN et al. 1982; FAHN 1982). It should be recalled that dystonia can also be a symptom of parkinsonism, e.g., posturing of the hands, the flexed posture of the neck and trunk, and the so-called striatal toe, in which the big toe is dorsiflexed in a sustained posture (DUVOISIN et al. 1972b).

Inability to respond to individual doses of levodopa was also recognized and has been found to be due to failure of gastric emptying (RIVERA-CALIMLIN et al. 1970; FAHN 1977b; MELAMED 1986). Associated with this is the recently described problem of delayed onset response with the first dose in the morning (MELAMED and BITTON 1984). This appears to be another problem of long-term levodopa therapy.

Another type of fluctuation encountered in patients with parkinsonism is "freezing," a phenomenon where the patient is transiently immobile as if stuck to the ground. This problem occurs with untreated parkinsonism, but it also appears to develop in some patients as a complication of levodopa therapy. AMBANI and VAN WOERT (1973) described one aspect of the problem and labeled it as "start hesitation." But the problem is encountered at other times when the patient is walking, such as when turning, when reaching a destination, when entering a crowded area, passing through a doorway, entering an elevator, or when trying to cross a busy street or pass through a revolving door. One may prefer to separate these phenomena with names such as "start hesitation," "turning hesitation," or "target hesitation." The label of sudden transient freezing serves as a global descriptor. It is important to determine if the freezing phenomenon occurs during an "off" period or when the patient is "on," since the pathophysiology and treatment may differ.

Finally, a nontoxic complication of levodopa therapy is the consideration of possibly aggravating melanoma, first reported by SKIBBA et al. (1972), and encountered in at least six cases (FERMAGLICH and DELANEY 1977). If indeed there is a causal relationship, this should be considered a metabolic complication of levodopa therapy because dopa is involved in the metabolism of melanomas.

B. Classification, Pathophysiology, and Treatment of Adverse Effects

There are several ways that the adverse effects of levodopa therapy could be classified, including the duration of treatment (e.g., short-term and long-term), phenomenology, site of action, pathophysiology, and treatment responses. Since the main purpose of treatment is to prolong functional ability over the long term of the illness, I will focus only on complications of long-term therapy. Table 3 classifies the adverse effects into sites of action with a subdivision based on phenomenology. Since management of adverse effects depends predominantly on their pathophysiology, I will also classify adverse effects into their mechanisms (Table 4), but I will discuss them in sequence according to the listing in Table 3. In order to do this in an integrated manner, it is important first to understand terminology relating to Table 4.

The pathophysiologic mechanisms have been divided into: (A) pharmacokinetic problems; (B) central pharmacodynamic problems; (C) those in which both of these processes seem to be playing a role; (D) peripheral pharmacodynamic problems; and (E) metabolic problems. These are listed in Table 4. By pharmacodynamic mechanisms, I refer to processes involving the dopamine

Table 3. Classification of adverse effects of long-term levodopa therapy for parkinsonism

- Peripheral side effects
  - GI symptoms: anorexia, nausea, vomiting, constipation, bloating and abdominal distension
  - Cardiac symptoms: dysrhythmia
  - Melanoma
- Central side effects
  - Dyskinesias
    - Chorea: peak-dose, diphasic
    - Dystonia: peak-dose, diphasic, "off"
    - Myoclonus
    - Simultaneous dyskinesia and parkinsonism
  - Tachykinesia with hypokinesia
    - Tachyphemia
    - Running gait
  - Fluctuations ("offs")
    - Slow "wearing-off"
    - "Sudden off"
    - "Random off"
    - Yo-yo-ing
    - Episodic failure to respond
    - "Delayed on"
    - Weak response at end of day
    - Response varies in relationship to meals
  - Freezing
    - Start hesitation
    - Target hesitation
    - Turning hesitation
    - Startle and fearfulness hesitation
    - Sudden transient freezing
  - Mental changes
    - Dementia, confusion, agitation
    - Psychiatric: hallucinosis, hallucinations, delusions, depression, mania
  - Loss of efficacy
    - Loss of efficacy in association with pyridoxine
    - Declining efficacy with continuing treatment
  - Miscellaneous
    - Altered sleep-wake cycle: insomnia, daytime drowsiness
    - Hypersexuality
    - Akathisia
    - Sweating
    - Postural hypotension
    - Respiratory distress
    - Falling
    - Pain
    - Increased parkinsonism
    - Neuroleptic malignant syndrome

receptors within the striatum and elsewhere in the central nervous system, and also actions of dopamine on peripheral organs. By pharmacokinetic mechanisms, I refer to processes involving the transport, metabolism, plasma levels, and tissue levels of levodopa and its metabolites. By metabolic problems with levodopa, I refer to the action of levodopa in metabolism, particularly in the metabolism of melanomas.

Table 4. Pathophysiologic classification of adverse effects of levodopa therapy

A. Pharmacokinetic problems
   1. "Wearing-off" motor phenomena
   2. "Wearing-off" bloating and abdominal distension
   3. "Wearing-off" depression
   4. "Wearing-off" dyspnea
   5. Peak-dose dyskinesia
   6. Episodic failure to respond to each dose of levodopa
   7. "Delayed on"
   8. Weak response at end of day
   9. Response varies in relationship to meals
   10. Loss of efficacy due to pyridoxine

B. Central pharmacodynamic problems
   1. Gastrointestinal upset
   2. Sweating
   3. Mental effects
   4. Altered sleep-wake cycle
   5. Hypersexuality
   6. Akathisia
   7. Postural hypotension
   8. Pain
   9. "Sudden off" phenomenon
   10. "Random off" phenomenon
   11. Simultaneous dyskinesia and parkinsonism
   12. Myoclonus
   13. Freezing
   14. Tachykinesia with hypokinesia
   15. Increased parkinsonism
   16. Loss of efficacy with continuing treatment
   17. Falling

C. Combined pharmacokinetic and central pharmacodynamic problems
   1. "Off" dystonia
   2. Diphasic dyskinesia
   3. Yo-yo-ing
   4. Neuroleptic malignant syndrome

D. Peripheral pharmacodynamic problems
   1. Cardiac dysrhythmia

E. Metabolic problem
   1. Aggravates melanoma

Basically, the goal of levodopa therapy is to maintain striatal levels of dopamine. This process requires levodopa to pass from the stomach to the small intestine from where it will be absorbed into the circulation; levodopa then enters the brain, and there is metabolized to dopamine in the striatum. From the plasma, levodopa can be transported into other tissues, be excreted, be metabolized via decarboxylation to form peripheral dopamine, or be metabolized via methylation to form 3-*O*-methyldopa. In brain, levodopa and dopamine can be metabolized by methylation, and dopamine can also be metabolized by deamination and oxidation (monoamine oxidase). In all situations, these processes can limit the effectiveness and duration of levodopa therapy. Adverse effects from difficulties with these processes comprise the pharmacokinetic problems. Dopamine receptors are located in several targets in the brain, particularly in the striatum, the area postrema (vomiting center), the limbic structures, some cortical regions, the retina, and in the hypothalamus. In the periphery, most effects of dopamine are best understood in regard to the cardiovascular system (GOLDBERG et al. 1974). But for purposes of understanding levodopa toxicity, it is appropriate to consider any adverse action of levodopa or dopamine outside of the blood-brain barrier as a peripheral adverse effect. Thus, dopamine's action in the area postrema, inducing nausea and vomiting, is considered a peripheral adverse effect, although the action is on a dopamine receptor located in the brain (hence listed in Table 4 as a central pharmacodynamic effect, although Table 3 states it is a peripheral side effect). Although the listing follows the structure of Table 3, I will refer to the appropriate location in Table 4 where each of the topics is also listed in order to correlate the pathophysiology with the clinical phenomenology.

I. Peripheral Adverse effects

1. Gastrointestinal Symptoms (Table 4, B. 1)

The commonest gastrointestinal complaint by patients with parkinsonism is constipation. It is never clear in most patients whether this symptom is a manifestation of Parkinson's disease itself, whether it is related to pharmacologic agents such levodopa and anticholinergics, or whether it is unrelated to either; e.g., just occurring with aging. It is probably due to a combination of all of these factors. The most accepted GI adverse effects due to levodopa are anorexia, nausea, and vomiting. These are due to activation of the dopamine receptors in the area postrema, as mentioned already. These receptors lie outside the blood-brain barrier and become exposed to peripheral dopamine. By utilizing carbidopa, thereby preventing formation of circulating dopamine (PRASAD et al. 1975), these problems can be largely eliminated.

Abdominal distension and bloating are reported by some patients receiving levodopa, but this is associated with the "off" state, and is not a toxic reaction, per se. It is not clear what the mechanism for this problem is, but it is reasonable to assume that this peripheral reaction is linked to low plasma levels of levodopa, and there is an accompanying lack of tone in the smooth muscle, leading to abdominal distension.

2. **Cardiac Dysrhythmias** (Table 4, D. 1)

Abnormalities in cardiac rhythm appear to be due to a direct effect of peripheral dopamine on cardiac monoamine receptors. Adding carbidopa can eliminate this potentially serious complication of levodopa therapy.

3. **Melanoma** (Table 4, E. 1)

Skin melanin utilizes dopa generated by tyrosinase from its precursor tyrosine. In the formation of skin melanin, dopa is converted to dopaquinone, again by the same enzyme, tyrosinase. Levodopa used in the treatment of parkinsonism could theoretically be used in the metabolism of melanomas. Because some patients on levodopa with a previous history of melanoma had a relapse and spread of this neoplasm while on this medication, it is theoretically possible that this relapse was causally related to the levodopa. Therefore, several authors have expressed caution and advised not using levodopa in patients with a prior history of melanoma (FERMAGLICH and DELANEY 1977; YAHR 1975; LIEBERMANN and SHUPACK 1974).

II. Central Adverse Effects

1. Dyskinesias

a. Chorea

α. Peak-Dose Chorea (Table 4, A. 5)

Choreic movements can occur early in the treatment with levodopa, but the incidence of these involuntary movements increases with continuing treatment (see Table 1). Peak-dose dyskinesia is due to too high a dose of levodopa and is representative of a toxic state. The plasma levels of levodopa are high (MUENTER et al. 1977), and presumably there is excess striatal dopamine. Reducing the individual dose can resolve this problem. The patient may need to take more frequent doses at this lower amount because reducing the amount of an individual dose also reduces the duration of benefit (NUTT et al. 1985). It was once believed that chorea only appears in the presence of supersensitive dopamine receptors, but if the dosage of levodopa is high enough, chorea can occur in normal individuals.

β. Diphasic Chorea (Table 4, C. 2)

Diphasic dyskinesias were first described by MUENTER et al. (1977) who labeled them as the D-I-D phenomenon. Although most of the affected individuals have dystonia as their pattern of dyskinesia, some have choreic movements, and others have a mixture of the two types. Diphasic dyskinesia is a situation where the dyskinesia develops as the plasma levels of levodopa are rising or falling, but not during the peak plasma level (MUENTER et al. 1977; LHERMITTE et al. 1978). This phenomenon is difficult to explain. It is possible that there is a differential sensitivity of at least two dopamine receptors. The more sensitive one would respond to lower levels of levodopa to induce the dyskinetic state. The other receptor would be activated at higher levels and will inhibit the dyskinesia. Treatment of the problem is difficult. Although Lhermitte et al. (1978) proposed treating this condition with higher doses of levodopa, my own experience is that higher dosages merely induce peak-dose dyskinesia and possibly other forms of central adverse effects. On the other hand, lowering the dosage is equally unsatisfactory because increasing parkinsonism ensues. Personal experience has led me to use pergolide, a direct-acting dopamine agonist with a long duration of action. When used as the major pharmacologic agent with supplementary levodopa, it is usually effective in reducing the severity of this problem (S. Fahn, unpublished work). Perhaps sustained-release levodopa will also prove beneficial.

b. Dystonia

α. Peak-Dose Dystonia (Table 4, A. 5)

In the early stages of levodopa therapy, chorea was commoner than dystonia, but with continuing treatment individual patients would develop more dystonic dyskinesias and less chorea. Many probably have a combination of chorea and dystonia. Peak-dose dystonia, like peak-dose chorea, also develops when plasma levels of levodopa are high, and subsides when the dosage is lower. Dystonia is a more serious problem than chorea because it is more disabling. In many patients, dystonia occurs at subtherapeutic doses, and lowering the dosage will render a patient with inadequate response from levodopa even more parkinsonian. There is little choice but to use smaller and more frequent doses, often coupled with dopamine agonists and other antiparkinsonian drugs, such as amantadine.

β. Diphasic Dystonia (Table 4, C. 2)

The discussion of this condition has been combined with the discussion of diphasic chorea because the D–I–D phenomenon is usually a combination of the two types of dyskinesia.

γ. "Off" Dystonia (Table 4, C. 1)

Dystonic spasms are not always a sign of levodopa overdosage. This is particularly true in many instances of painful sustained contractions. Painful dystonic cramps most often occur when the plasma level of levodopa is low, particularly in the early morning (Melamed 1979). But this type of dystonia can occur at any time the patient goes "off" (Ilson et al. 1984). In this sense, "off" dystonia is a pharmacokinetic problem. But why painful dystonic spasms should occur in addition to parkinsonism during low plasma levels of levodopa is not clear. This phenomenon may relate to some peculiarity of the dopamine receptors as well as the low plasma levels of levodopa in these patients. De Yebenes (1988) has proposed that dystonia may occur when the ratio norepinephrine/dopamine is high. However, there are only speculations about the pathophysiology of dystonia in general, and it is difficult to be certain of the explanation of either peak-dose dystonia or "off" dystonia such as early morning dystonia. Preventing "offs" is the best way to control "off" dystonia. The use of pergolide is often effective when it is the major dopaminergic agent. Perhaps sustained-release levodopa will also prove beneficial.

One should not forget that dystonia can also occur as a feature of Parkinson's disease. However, "off" dystonia does not appear to be merely a reflection of the dystonia of parkinsonism. If patients with "off" dystonia are given a drug holiday from levodopa, after a few days the painful dystonia will disappear, and the patient will be left with a baseline parkinsonian state and without painful dystonia (S. FAHN, unpublished work).

c. Myoclonus (Table 4, B. 12)

The lightning-like jerks of myoclonus can occur in untreated Parkinson's disease, but these are rarely disabling and are hardly commented on by patients. KLAWANS et al. (1975) described myoclonus occurring as a complication of long-term levodopa therapy. They reported them as single, unilateral or bilateral jerks in the extremities, most frequently during sleep. They found the serotonin antagonist methysergide to be helpful in controlling these. Such nocturnal myoclonus is infrequently encountered in my experience, and is rarely disabling. I have listed this adverse effect as being associated with a central pharmacodynamic problem (Table 4), possibly related to serotonin receptors according to the pharmacologic profile reported by KLAWANS et al. (1975).

d. Simultaneous Dyskinesias with Parkinsonism (Table 4, B. 11)

Many patients have different responses to levodopa therapy in different parts of the body. For example, the head and neck regions may be more sensitive to levodopa than are the legs. When the upper part of the body responds in this situation, the legs may remain parkinsonian and the patient may not be able to walk well. In this example, on higher dosages of levodopa, the legs improve, but now the head and neck regions are dyskinetic. This problem may be due to different sensitivities of the striatal dopamine receptors on a somatotopic basis. That is, in the case described, the head and neck areas of the striatum would have more sensitive receptors than the leg area. This problem is difficult to treat, and one can only titrate the dosage to the optimum response between the two extremes for each individual patient.

2. Tachykinesia with Hypokinesia (Table 4, B. 14)

a. Tachyphemia

A number of patients overdosed with levodopa will speak faster, running syllables together, so that it is difficult for the listener to understand what the patient is saying. At the same time as the speech is rapid (Tachyphemia), the amplitude is lower so that the voice is softer, aggravating the situation. If the patient purposely tries to enunciate each syllable distinctly, this is successful for a few words, but then the tachyphemia takes over again. With speech therapy, sometimes using a metronome for pacing, a patient can improve the pattern of speaking, but only during the treatment session. There seems to be little carry-over. Associated with tachyphemia are rapid voluntary movements in other parts of the body, displayed as such when the patient is asked to perform rapid successive movements. These are usually very fast and of small amplitude, i.e., tachykinetic and hypokinetic. Often, lowering the dosage of levodopa will allow the patient to slow down, with resulting clarity of speech. However, parkinsonian bradykinesia can become more of a problem.

b. Running Gait

This type of tachykinetic problem can also involve walking. Usually the patient moves more rapidly, but with smaller steps. Gait in this instance can be mistaken for festination, which it resembles. If postural instability is impaired, such a running gait can lead to falling. As mentioned in Sect. B.II.2.a, lowering the dosage will allow the patient to slow down.

3. Fluctuations

a. "Wearing-Off" (Table 4, A. 1-4)

The "wearing-off" phenomenon (also known as "end-of-dose" deterioration) is related to the short plasma half-life of levodopa (MUENTER and TYCE 1971). When patients have this problem, the clinical improvement from a dose of levodopa lasts only as long as the plasma level of levodopa is high. As the plasma level gradually falls, there is a gradual loss of clinical response. If the plasma level can be maintained, the clinical response can also be maintained (SHOULSON et al. 1975; HARDIE et al. 1984). Thus, benefit depends on a steady supply of levodopa in the plasma reaching the brain in a constant influx. This type of necessary bioavailability of plasma levodopa is not a factor with patients who have a smooth response to levodopa. Such individuals have the same type of plasma half-life of levodopa.

It is usually accepted that the need for constant plasma levodopa in patients with the "wearing-off" phenomenon is due to loss of storage sites of dopamine in the striatum, where dopamine is usually present in the terminals of the dopaminergic nigrostriatal fibers. In this hypothesis, severe parkinsonism should be associated with the "wearing-off" effect being present early in the course of treatment, and this was encountered in those with 1-methyl-4-phenyl-1,2,3,6-tetrahydropyridine (MPTP)-induced parkinsonism (BALLARD et al. 1985). In contrast, those with Parkinson's disease develop the "wearing-off" problem as a function of duration of treatment (MCDOWELL and SWEET 1976), which could also be a reflection of duration of illness. But this would not explain why younger patients are more prone to develop fluctuations than are older patients (PEDERZOLI et al. 1983). Nor would it explain why using low doses of levodopa instead of high doses would delay the development of this problem (POEWE et al. 1986).

Furthermore, loss of striatal storage sites of dopamine by itself does not appear to be the sole cause of this phenomenon. Treatment with direct-acting agonists does not eliminate the problem, although it does ameliorate it somewhat by making the depths of the "off" state less severe. Moreover, RINNE (1985) showed that if bromocriptine is used early in the course of treatment, the "wearing-off" phenomenon is virtually avoided. This would imply that levodopa itself or the peaks and valleys of levodopa supply to the brain may also be contribute to this problem. How levodopa or dopamine could lead to this phenomenon is unclear.

Treatment of the "wearing-off" problem consists of giving the doses of levodopa closer together or utilizing direct-acting dopamine agonists, which have a longer biologic half-life than levodopa. Using low doses of levodopa seems to delay the onset of this problem (POEWE et al. 1986), as does delaying the introduction of levodopa (BLIN et al. 1987).

Some patients with the "wearing-off" phenomenon, in addition to a motor disturbance, also notice an alteration of mood, respiratory distress, or abdominal distension when they are "off" (ILSON et al. 1983). Similar to "off" dystonia, these reactions do not appear to be a feature of the underlying parkinsonism per se, because a dopa holiday will not leave the patient with these problems. They therefore appear to be some sort of reactive pathology when the dopamine receptors are not being steadily activated, i.e., a rebound phenomenon. As with dystonic reactions in parkinsonian patients on levodopa therapy, respiratory difficulties can also appear as a peak-dose problem, particularly as a complication of dopa-induced pharyngeal dystonia (BRAUN et al. 1983).

b. "Sudden Off" (Table 4, B. 9)

As originally defined (DUVOISIN 1974b), the "on-off" phenomenon was a label for a sudden and random event in which the patient becomes parkinsonian. That is, the benefit from levodopa suddenly disappears. To avoid ambiguity, since the term "on-off" is often used to refer to any type of fluctuation in parkinsonian patients, the phenomenon of sudden "offs" is labeled here as "sudden off" in contrast to "wearing-off." In this disorder, the patient can improve just as suddenly, even without taking another dose of levodopa. Pharmacologic studies have revealed that plasma levels of levodopa are in the declining phase when the "offs" appear (FAHN 1974). It has been speculated that the "sudden off" problem is due to a sudden and transient desensitization blockade of the dopamine receptors that would be compatible with the receptor switching from a high affinity to a low affinity state (FAHN 1974). Response to apomorphine (CLOUGH et al. 1984) is compatible with the stimulation of a desensitized receptor so that it will now respond, and this result is not incompatible with the hypothesis. The "sudden off" phenomenon is a difficult problem to overcome. It is now clear that direct-acting dopamine agonists are ineffective.

c. "Random Off" (Table 4, B 10)

It was formerly believed that all "offs" occurring seemingly at random were part of the "sudden off" phenomenon. Indeed, "sudden offs" are unpredictable, but not all unpredictable "offs" occur suddenly. Some unpredictable, and seemingly random, "offs" have been found to fit a pattern in some patients. When plotted out daily, they are found to occur predominantly at the end of the dose and have been called the "complicated end-of-dose effect" (Marsden et al. 1982).

d. Yo-yo-ing (Table 4, C. 3)

Yo-yo-ing refers to that fluctuating condition in Parkinson's disease treated with levodopa in which the patient responds rapidly with a peak-dose dyskinesia, followed by a predictable "wearing-off" (Fahn 1982). The dopamine receptors are obviously intact, and they would need to be extremely supersensitive to get such dyskinesias. Hence, there appears to be a pharmacodynamic factor to account for the dyskinesias. At the same time, the short plasma half-life and the need for constant bioavailability of levodopa in the plasma accounts for the "off" states. Hence, the "offs" are the result of a pharmacokinetic factor, and this condition is listed in Table 4 as a combined pharmacokinetic and pharmacodynamic problem. Treatment is virtually impossible with regular levodopa. Perhaps sustained-release levodopa will prove to be useful. We have had success with direct-acting agonists when used as the sole or dominant form of pharmacotherapy.

e. Episodic Failure to Respond to Each Dose (Table 4, A. 6)

Failure of the patient to respond to each dose of levodopa is related to poor gastric emptying (Rivera-Calimlin et al. 1970; Fahn 1977b). This problem can be overcome by dissolving levodopa in liquid prior to ingesting it. This type of fluctuation is probably commoner than is usually recognized. In a survey of their patients, Melamed et al. (1986) found a number of patients with this difficulty. It is not clear if the delayed gastric emptying is due to levodopa therapy itself or if the disorder develops because frequent dosing will statistically result in some tablets not passing through the stomach quickly enough to reach the small intestine where absorption takes place.

f. "Delayed On" (Table 4, A. 7)

Melamed and Bitton (1984) reported that patients with fluctuations often have a problem getting an "on" with the first dose in the morning. These patients tend to have a longer delay with this dose than patients who do not experience fluctuations. The mechanism is not clear, but it may have to do with obtaining adequate plasma levels. I have noticed that many patients need a larger amount of levodopa as their first dose of the day in order to "kick in" a response to the medication. Since the first dose is often accompanied by a higher plasma level of levodopa than later doses (Fahn 1982; Shoulson et al. 1975), the problem may not be entirely pharmacokinetic. Rather, it is possible that the dopamine receptors are in a low affinity state and require more dopamine agonism to activate them.

g. Weak Levodopa Response at End of Day (Table 4, A. 8)

Another commonly heard complaint is that the response to levodopa tends to be weaker in the afternoons and evenings. It is possible that this could be due to competition of levodopa with 3-*O*-methyldopa (O-MD) which is a major peripheral metabolite of levodopa, and which is found in high levels in plasma, owing to its long half-life (MUENTER et al. 1972, 1973). O-MD competes with levodopa for penetration into brain (WADE and KATZMAN 1975) because both compounds are neutral large amino acids and use the same transport system to cross the blood-brain barrier.

h. Response Varies in Relation to Meals (Table 4, A. 9)

There are at least three variations of levodopa responsiveness in regard to meals. Because levodopa is dependent on passage through the stomach, a full meal with delayed gastric emptying will result in a delayed and weaker response to levodopa ingested after the meal. On the other hand, some patients who normally take levodopa with or after a meal will find that if they now take it before a meal, the response is much greater and they develop peak-dose dyskinesia. These two variations can easily be corrected by accommodating the timing of doses of levodopa according to the pathophysiology of their particular problem.

The third variation relates to high protein meals. Competition with other amino acids in the diet can interfere with transport of levodopa across the intestinal mucosa and across the blood-brain barrier (MUENTER et al. 1972; NUTT et al. 1984). Only rarely does this competition with other amino acids pose a serious problem for patients. Most accommodate to protein in their diet. In those rare individuals in whom any protein in any meal interferes with their response to levodopa it is necessary to plan a meal strategy. Usually having nonprotein meals at breakfast and lunch, making up for this lack by having higher protein meals at dinner, will usually be effective. If patients go "off" at night when they can best afford to be "off," they can adjust to this situation.

4. "Freezing" (Table 4, B. 13)

As pointed out in Sect. A, freezing takes many forms, and these have different names, such as listed in Table 3. However, it is not clear that any of these variants has a different pathophysiologic mechanism. The major differentiation is to distinguish between "off freezing" and "on freezing." "Off freezing" is best explained as a feature of Parkinson's disease, and its treatment is to keep the patient from getting "off." "On freezing" remains an enigma, and this problem tends to be aggravated by increasing the dosage of levodopa. It is not improved by adding direct-acting dopamine agonists. Rather, it is lessened by reducing the dosage of levodopa. Although NARABAYASHI et al. (1984) reported benefit with L-threo-DOPS, supposedly a precursor of brain norepinephrine, I have not seen any benefit with this drug in the treatment of "on freezing." Furthermore, we found no evidence that this drug increases norepinephrine in rat brain or human cerebrospinal fluid. Interestingly, in the intravenous infusions of levodopa performed by SHOULSON et al. (1975), a sudden stimulus, such as tilting the patient upright on a tilt table, resulted in sudden transient worsening of parkinsonism. This can be interpreted to indicate the induction of sudden transient freezing. It may be a useful model for future studies.

5. **Mental Changes** (Table 4, B. 3)

The central adverse effects of confusion, agitation, hallucinosis, hallucinations, delusions, depression, and mania are probably related to activation of dopamine receptors in nonstriatal regions, particularly the cortical and limbic structures. This is difficult to remedy, except by reducing the dosage of levodopa, thereby possibly lessening the beneficial response to levodopa as well. Clearly, a dopamine agonist that would be specific for striatal receptors would be welcome. Alternatively, a dopamine antagonist that would be specific for nonstriatal receptors would theoretically be effective.

6. **Loss of Efficacy**

a. Caused by Pyridoxine (Table 4, A. 10)

Excess pyridoxine can antagonize the benefit from levodopa by increasing peripheral decarboxylation of levodopa (DUVOISIN et al. 1969; LEON et al. 1971). This problem was only encountered in patients who took exogenous supplements of pyridoxine, and not from pyridoxine in regular diets. Before the introduction of a peripheral decarboxylase inhibitor, this was avoided by having patients take multivitamins without pyridoxine. Today, this potential problem is avoided by utilizing a peripheral dopa decarboxylase inhibitor, such as carbidopa. Pyridoxine is a cofactor for decarboxylase, and can thereby increase peripheral metabolism of levodopa. However, if this enzyme is inhibited by carbidopa or other peripheral inhibitors, excess pyridoxine does no harm to the peripheral pharmacokinetics of levodopa.

b. With Continuing Treatment (Table 4, B. 16)

A debated point in the treatment of parkinsonism is the cause of declining efficacy from continuing treatment with levodopa in many patients (YAHR 1976). If the postsynaptic dopamine receptors in the striatum are not lost in this disease, why should a patient get less response from medication over time? Progression of the illness with further loss of dopamine storage sites in the presynaptic terminals is the most frequently invoked explanation. However, loss of these structures does not automatically produce a loss of response to levodopa. For example, postencephalitic parkinsonism with its much greater loss of dopamine in the striatum (EHRINGER and HORNYKIEWICZ 1960; BERNHEIMER et al. 1973) has more, not less, sensitivity to levodopa (CALNE et al. 1969; DUVOISIN et al. 1972a). This observation is sufficient to argue against the concept that reduction of storage sites for dopamine is responsible for the declining efficacy of levodopa. Perhaps Parkinson's disease is associated with loss of striatal dopamine receptors as well as the presynaptic dopaminergic neuron.

Even so, there may be additional factors contributing in part to the loss of efficacy seen with continuous treatment with levodopa. Some decline may arise in part from gradual downregulation of striatal dopamine receptors (RINNE et al. 1980). Not all patients develop this problem, but it appears to be due to the receptors being constantly exposed to high levels of dopamine. Evidence to support this concept comes from the studies of levodopa drug holidays. After levodopa is eliminated for a short period, restoration of levodopa therapy usually provides enhanced temporary benefit (DIRENFELD et al. 1980; WEINER et al. 1980). Unfortunately, this enhanced sensitivity is short-lived, and the potential risks of aspiration during the drug holiday render this approach undesirable for the short-term benefit that can be obtained.

7. Miscellaneous

a. Altered Sleep-Wake Cycle (Table 4, B. 4)

The adverse effects of drowsiness during the daytime, particularly after a dose of levodopa, and insomnia at night are fairly common. They often accompany the central adverse effects of confusion. Like mental adverse effects, changes in the sleep-wake cycle most likely represent activation of nonstriatal dopamine receptors. Perhaps those located in the hypothalamus are responsible for this set of symptoms. If a patient becomes drowsy after each dose of medication, this is a sign of overdosage. Reducing the dosage is the only means of correcting this problem. If the patient is generally drowsy during the daytime and remains awake at night, this alteration of the sleep-wake cycle makes it difficult for those looking after the patient to give effective care. It is important to get the patient onto a sleep-wake schedule that fits with the rest of the household. To correct the problem it may be necessary to use a combination of approaches. Efforts must be made to keep the patient awake by physical and mental stimulation during the daytime, otherwise sleep may be difficult at night. At night, the patient should then be drowsy enough to sleep.

It may be necessary to use stimulants in the morning and sedatives at night in order to reverse the altered state of affairs. This should be done in addition to prodding the patient to remain awake during the day. Drugs such as methylphenidate and amphetamine are usually well tolerated by patients with Parkinson's disease. A10-mg dose of either of these drugs, repeated once if necessary, may be helpful. To encourage sleep at night, a hypnotic may be necessary in addition to daytime stimulants. It should be noted that strong sedatives, such as barbiturates, are poorly tolerated by patients with Parkinson's disease. Milder hypnotics, such as benzodiazepines, are usually taken without difficulty. Short-acting benzodiazepines would be preferable, but if the patient awakens too early, a longer-acting one may need to be used.

b. Hypersexuality (Table 4, B. 5)

Hypersexuality as a toxic complication of levodopa received much publicity in the lay press shortly after levodopa was introduced. It was more likely to occur in postencephalitic parkinsonism (SACKS 1974), but it is also seen as a complication of levodopa in patients with Parkinson's disease. TANNER et al. (1986) reported five male patients with disabling hypersexual behavior after years of levodopa therapy, and point out that this complication has been reported only in males.

In all reported cases, hypersexuality has been a sign of overdosage, and the adverse effect lessens on reduction of dosage. However, I have one patient with Parkinson's disease who has paradoxical hypersexuality; that is, when he is underdosed or when he is "off," he has this problem. When he is "on," the urge for sex dissipates. When the patient is in a toxic state, with choreic movements, there is no hypersexuality. Thus, he seems to be opposite to all other reported cases with this complaint.

The sexual urge is a compulsion that is overwhelming and incapacitating. Both husbands and wives bring this problem to the attention of the physician. Although the patient has compulsive thoughts about having sexual relations, and may actually attempt to carry out this activity several times a day, the desire is usually not matched by adequate performance. My patient with paradoxical hypersexuality will take showers frequently when he gets these sexual urges in an effort to take his mind off this psychic drive. Compulsive thoughts and actions caused by levodopa in postencephalitic patients account for a large number of levodopa failures in this population (SACKS 1974). It is reasonable to place hypersexuality with these other compulsive ideations. There is no effective antidote for hypersexuality other than reducing the dosage of levodopa. One could speculate that dopaminergic receptors in the limbic system, including the hypothalamus, are supersensitive in these patients and are responsible for this toxic reaction.

c. Akathisia (Table 4, B. 6)

Akathisia, meaning inability to sit still, is not a commonly recognized adverse effect of levodopa. LANG and JOHNSON (1987) asked patients with Parkinson's disease specifically for complaints of restlessness and found that 86% did have this subjective complaint. Most patients with Parkinson's disease who complain of an inner feeling of restlessness do not overtly manifest any signs such as moving about. From Lang and Johnson's study it was not clear if akathisia represented an adverse effect of levodopa or was a feature of the disease. In most of their patients it appeared only after the introduction of antiparkinsonian drugs, but a small number had this symptom early in the course of Parkinson's disease, prior to receiving any medication. It is likely that other antiparkinsonian agents may also contribute to this complaint, for I have seen it in patients with idiopathic torsion dystonia after starting anticholinergic drugs. But it is clear that levodopa can cause it since I have also seen a patient with idiopathic torsion dystonia who developed akathisia after starting levodopa as the sole pharmacologic agent.

In general, akathisia is most commonly encountered as a complication of dopamine receptor blocking drugs, predominantly the antipsychotics (FAHN 1984). With these drugs, akathisia can appear as an acute symptom following introduction of the antipsychotic (referred to as acute akathisia) and the symptom is relieved by discontinuing the offending drug. Akathisia can also appear as a late complication of these drugs and remain persistent (referred to as tardive akathisia). Tardive akathisia occurs in association with tardive dyskinesia and tardive dystonia. Unfortunately, the mechanism of akathisia induced by neuroleptics is not clearly understood. Why antidopaminergics and dopaminergics should both produce akathisia remains a mystery. It seems reasonable to conclude that levodopa-induced akathisia is a central pharmacodynamic problem. Its treatment is merely to reduce the dosage of levodopa if the symptoms are too pronounced.

d. Sweating (Table 4, B. 2)

The problem of episodic sweating remains an enigma, both in its mechanism and its treatment. These episodes, seemingly unrelated to timing of medication, cause the patient to be drenched in sweat, and could potentially cause an electrolyte imbalance. At the minimum it is an unpleasant, uncomfortable problem, and requires frequent changing of clothing. Autonomic dysfunctions do occur in Parkinson's disease (GOETZ et al. 1986a), but these do not explain the episodes of drenching sweats that can occur in some patients. It is not known whether this complication of levodopa is central or peripheral, but its occurrence in the presence of carbidopa strongly suggests that it is a central disorder. Although the receptors involved are not certain, one guess is that it may be related to dopamine receptors in the hypothalamus. Trials of propranolol have been ineffective in my experience (S. FAHN unpublished work).

e. Postural Hypotension (Table 4, B. 7)

As mentioned in the historical review (Sect. A), postural hypotension as a complication of levodopa therapy can be either a peripheral or a central adverse effect. Since this problem persists in the presence of carbidopa, a central mechanism is clearly responsible, but a peripheral action could also play a role in addition. Postural hypotension from levodopa is aggravated by other drugs taken by the patient, such as tricyclic antidepressants. The central site for producing this complication by levodopa is uncertain, but the hypothalamus should be considered since it gives rise to autonomic fibers that descend to the spinal cord. The treatment of postural hypotension can sometimes be managed by using fludrocortisone (HOEHN 1975), but often the dosage of levodopa needs to be reduced.

f. Respiratory Distress (Table 4, A. 4 and 5)

Respiratory distress such as dyspnea can occur as a symptom of Parkinson's disease in some patients. We have also encountered it during the "off" stage in a number of patients (ILSON et al. 1983). In addition, it can also occur as a complication of dystonia, usually peak-dose dystonia (BRAUN et al. 1983). Both of these situations were discussed in the discussion of the "wearing-off" phenomenon (Sect. B.II.3.a).

g. Falling (Table 4, B. 17)

Falling is a common feature of Parkinson's disease as the illness progresses and there is increasing loss of postural reflexes. Since this particular cardinal sign of Parkinson's disease is little benefited by levodopa therapy (KLAWANS 1986), this problem persists and worsens despite pharmacotherapy. Because levodopa may allow patients to be more mobile, e.g., allowing them to arise more easily from a chair and walk independently, the persistence of postural instability becomes a particular problem because it raises the hazard of increased likelihood of falling. Thus, this complication of levodopa therapy in this particular subpopulation of patients with Parkinson's disease is technically not a true adverse effect of the medication, but a complication of the improvement in mobility in a patient at risk of falling, thereby increasing that risk.

h. Pain (Table 4, B. 8)

Pain and other sensory complaints are common as a symptom of Parkinson's disease (SNIDER et al. 1976; KOLLER 1984; GOETZ et al. 1986b). Usually, these symptoms are controlled by treatment with levodopa. However, I have observed some patients with Parkinson's disease who developed sensory complaints, including pain, as a result of levodopa therapy. Often the pain is an accompaniment of "off" dystonia (MELAMED 1979; ILSON et al. 1984). In this situation, as discussed in Sect. B.II.1.b.γ, it is necessary to prevent the "off" phenomenon. A rare patient may have sensory complaints from levodopa therapy, unaccompanied by dystonia. We have treated two of these patients with electroconvulsive therapy with good results (S. FAHN unpublished work).

j. Increased Parkinsonism (Table 4, B. 15)

A few patients on high dosage levodopa may become more parkinsonian. This was initially reported unaccompanied by any other features of levodopa toxicity (FAHN and BARRETT 1979). More recently, this phenomenon has also been seen accompanied by confusion (SAGE and DUVOISIN 1986). In both situations, improvement occurs when the dosage of levodopa is reduced. In fact, the phenomenon of increased parkinsonism without other signs of toxicity probably occurs more often than is recognized. Many physicians, unaware of this complication, are inclined to increase the dosage of antiparkinsonian medication, which does not help. Rather, a reduction of dosage should be tried first to see if symptoms improve. This problem would appear to be related to reduced receptor sensitivity by high dosages of dopamine present in the striatum.

k. "Neuroleptic Malignant Syndrome" (Table 4, C. 4)

The neuroleptic malignant syndrome normally refers to that clinical condition in which a patient is taking a neuroleptic medication and suddenly develops high fever, muscle stiffness, sweating, and an alteration in mental alertness.

The mechanism for this problem has been attributed to blockade of some central dopaminergic receptors, possibly the hypothalamus.

A similar syndrome has been seen in patients with parkinsonism on suddenly discontinuing levodopa, as well as during "off" fluctuations (FRIEDMAN et al. 1985). Ordinarily, stopping levodopa suddenly presents no problem in the early years of treatment. Often there are sufficient stores of striatal dopamine for it to take several days before deterioration from the benefit of levo dopa is seen. However, with long-term treatment, particularly in patients markedly dependent on the constant bioavailability of levodopa in plasma (i.e., those with the "wearing-off" effect), sudden discontinuation of levodopa renders the patient markedly parkinsonian immediately. A few of these patients will have an associated agitation and confusion with their increased motor stiffness and bradykinesia, and develop the spectrum of the neuroleptic malignant syndrome. Thus, this poses another potential danger when suddenly discontinuing levodopa. It is assumed that the dopamine receptors are downregulated when the patient is on chronic levodopa. Suddenly discontinuing levodopa leaves the receptor without adequate stimulation, and the clinical picture of the neuroleptic malignant syndrome can develop. Treatment should be with immediate reinstitution of levodopa.

References

Ambani LM, Van Woert MH (1973) Start hesitation: side effect of long-term levodopa therapy. N Engl J Med 288:1113-1115

Ballard PA, Tetrud JW, Langston JW (1985) Permanent human parkinsonism due to 1-methyl-4-phenyl-1,2,3,6-tetrahydropyridine (MPTP): seven cases. Neurology 35:949-956

Bannister R, Oppenheimer D (1982) Parkinsonism, system degenerations and autonomic failure. In: Marsden CD, Fahn S (eds) Movement disorders. Butterworth, London, pp 174-190

Bernheimer H, Birkmayer W, Hornykiewicz O, Jellinger K, Seitelberger F (1973) Brain dopamine and the syndromes of Parkinson and Huntington. J Neurol Sci 20:415-455

Blin J, Bonnet A-M, Agid Y (1987) Does levodopa cause worsening of Parkinson's disease? In: Fahn S, Marsden CD, Calne DB, Goldstein M (eds) Recent developments in Parkinson's disease, vol 2. Macmillan, Healthcare Information, Florham Park, New Jersey, pp 165-181

Braun AR, Tanner CM, Goetz CG, Klawans HL (1983) Respiratory distress due to pharyngeal dystonia: a side effect of chronic dopamine agonism. Neurology 33 (Suppl 2):220

Calne DB, Stern GM, Laurence DR, Sharkey J, Armitage P (1969) L-Dopa in postencephalitic parkinsonism. Lancet 1:744-746

Calne DB, Brennan J, Spiers ASD, Stern GM (1970) Hypotension caused by L-dopa. Br Med J [Clin Res] 1:474-475

Calne DB, Reid JL, Vakil SD, Pallis C (1971a) Problems with L-dopa therapy. Clin Med 78:21-23

Calne DB, Reid JL, Vakil SD, Rao S, Petrie A et al. (1971b) Idiopathic parkinsonism treated with an extracerebral decarboxylase inhibitor in combination with levodopa. Br Med J [Clin Res] 3:729-732

Chase TN, Watanabe AM (1972) Methyldopahydrazine as an adjunct to L-dopa therapy in parkinsonism. Neurology 22:384-392
Clough CG, Bergmann KJ, Yahr MD (1984) Cholinergic and dopaminergic mechanisms in Parkinson's disease after long-term L-dopa administration. Adv Neurol 40:131-140
Cotzias GC, Van Woert MH, Schiffer LM (1967) Aromatic amino acids and modification of parkinsonism. N Engl J Med 276:374-379
Cotzias GC, Papavasiliou PS, Gellene R (1969) Modification of parkinsonism — chronic treatment with L-dopa. N Engl J Med 280:337-345
de Yebenes JG, Vazquez A, Martinez A, Mena MA, del Rio MN, de Felipe C, del Rio J (1988) Biochemical findings in symptomatic dystonias. Adv Neurol
Direnfeld LK, Feldman RG, Alexander MP, Kelly-Hayes M (1980) Is L-dopa drug holiday useful? Neurology 30:785-788
Duvoisin RC (1974a) Hyperkinetic reactions with L-dopa. In: Yahr MD (ed) Current concepts on the treatment of parkinsonism. Raven, New York, pp 203-210
Duvoisin RC (1974b) Variations in the "on-off" phenomenon. Adv Neurol 5:339-340
Duvoisin RC, Yahr, MD, Cote LJ (1969) Pyridoxine reversal of L-dopa effect in parkinsonism. Trans Am Neurol Assoc 94:81-84
Duvoisin RC, Antunes JL, Yahr MD (1972a) Response of patients with postencephalitic parkinsonism to levodopa. J Neurol Neurosurg Psychiatry 35:487-495
Duvoisin RC, Yahr MD, Lieberman J, Antunes J, Rhee S (1972b) The striatal foot. Trans Am Neurol Assoc 97:267
Ehringer H, Hornykiewicz O (1960) Verteilung von Noradrenalin und Dopamin (3-Hydroxytryamin) im Gehirn des Menschen und ihr Verhalten bei Erkrankungen des extrapyramidalen Systems. Klin Wochenschr 38:1238-1239
Fahn S (1974) "On-off" phenomenon with levodopa therapy in parkinsonism: clinical and pharmacologic correlations and the effect of intramuscular pyridoxine. Neurology 24:431-441
Fahn S (1976) Medial treatment of movement disorders. In: Davis FA (ed) Neurological reviews 1976. American Academy of Neurology, Minneapolis pp 72-106
Fahn S (1977a) Secondary parkinsonism. In: Goldensohn ES, Appel SH (eds) Scientific approaches to clinical neurology. Lea and Febiger, Philadelphia, pp 1159-1189
Fahn S (1977b) Episodic failure of absorption of levodopa: a factor in the control of clinical fluctuations in the treatment of parkinsonism. Neurology 27:390
Fahn S (1982) Fluctuations of disability in Parkinson's disease: pathophysiological aspects. In: Marsden CD, Fahn S (eds) Movement disorders. Butterworth, London, pp 123-145
Fahn S (1984) The tardive dyskinesias. In: Matthews WB, Glaser GH (eds) Recent advances in clinical neurology, vol 4. Churchill Livingstone, Edinburgh, pp 229-260
Fahn S, Barrett RB (1979) Increase of parkinsonian symptoms as a manifestation of levodopa toxicity. Adv Neurol 24:451-459
Fahn S, Duffy P (1977) Parkinson's disease. In: Goldensohn ES, Appel SH (eds) Scientific approaches to clinical neurology. Lea and Febiger, Philadelphia, pp 1119-1158
Fermaglich J, Delaney P (1977) Parkinson's disease, melanoma, and levodopa. J Neurol 215:221-224
Friedman JH, Feinberg SS, Feldman RG (1985) A neuroleptic malignantlike syndrome due to levodopa therapy withdrawal. JAMA 254:2792-2795
Goetz CG, Lutge W, Tanner CM (1986a) Autonomic dysfunction in Parkinson's disease. Neurology 36:73-75

Goetz CG, Tanner CM, Levy M, Wilson RS, Garron DC (1986b) Pain in Parkinson's disease. Movement Disorders 1:45-49

Goldberg LI, Tjandramaga B, Anton AH, Toda N (1974) Continuing studies of the peripheral vascular actions of dopamine. Adv Neurol 5:165-170

Hardie RJ, Lees AJ, Stern GM (1984) On-off fluctuations in Parkinson's disease. Brain 107:487-506

Hassan MN, Higgins D, Traub M, Fahn S (1986) Chronic treatment with bromocriptine induces behavioral supersensitivity in rats. Life Sci 19:513-518

Hoehn MM (1975) Levodopa-induced postural hypotension. Arch Neurol 32:50-51

Hunter KR, Laurence DR, Shaw KM, Stern GM (1973) Sustained levodopa therapy in parkinsonism. Lancet 2:929-931

Ilson J, Braun N, Fahn S (1983) Respiratory fluctuations in Parkinson's disease. Neurology (Suppl 2) 33:113

Ilson J, Fahn S, Cote L (1984) Painful dystonic spasms in Parkinson's disease. Adv Neurol 40:395-398

Jankovic J (1984) Progressive supranuclear palsy: clinical and pharmacologic update. Neurol Clin 2:473-486

Jenkins RB, Mendelson SH, Lamid S, Klawans HL (1972) Levodopa therapy of patients with parkinsonism and heart disease. Br Med J [Clin Res] 2:512-514

Jenner P, Boyce S, Marsden CD (1986) Effect of repeated L-dopa administration on striatal dopamine receptor function in the rat. In: Fahn S, Marsden CD, Jenner P, Teychenne P (eds) Recent developments in Parkinson's disease. Raven, New York, pp 189-203

Klawans HL (1986) Individual manifestations of Parkinson's disease after ten or more years of levodopa. Movement Disorders 1:187-192

Klawans HL, Goetz C, Bergen D (1975) Levodopa-induced myoclonus. Arch Neurol 32:331-334

Klawans HL, Goetz C, Nausieda PA, Weiner WJ (1977) Levodopa-induced dopamine receptor hypersensitivity. Ann Neurol 2:125-129

Koller WC (1984) Sensory symptoms in Parkinson's disease. Neurology 34:957-959

Lang AE, Johnson K (1987) Akathisia in idiopathic Parkinson's disease. Neurology 37:477-481

Leon A, Spiegel HE, Thomas G, Abrams WB (1971) Pyridoxine antagonism of levodopa in parkinsonism. JAMA 218:1924-1927

Lesser RP, Fahn S, Snider SR, Cote LJ, Isgreen WP, Barrett RE (1979) Analysis of the clinical problems in parkinsonism and the complications of long-term levodopa therapy. Neurology 29:1253-1260

Lhermitte F, Agid Y, Signoret JL (1978) Onset and end-of-dose levodopa-induced dyskinesias. Arch Neurol 35:261-262

Lieberman AN, Shupack JL (1974) Levodopa and melanoma. Neurology 24:340-343

Markham CH, Diamond SG, Treciokas LJ (1974) Carbidopa in Parkinson disease and in nausea and vomiting of levodopa. Arch Neurol 31:128-133

Mars H, Krail J (1971) L-Dopa and cardiac arrhythmias. N Engl J Med 285:1437

Marsden CD, Parkes JD (1976) "On-off" effects in patients with Parkinson's disease on chronic levodopa therapy. Lancet 1:292-295

Marsden CD, Parkes JD, Rees JE (1973) A year's comparison of treatment of patients with Parkinson's disease with levodopa combined with carbidopa versus treatment with levodopa alone. Lancet 2:1459-1462

Marsden CD, Parkes JD, Quinn N (1982) Fluctuations of disability in Parkinson's disease — clinical aspects. In: Marsden CD, Fahn S (eds) Movement disorders. Butterworth, London, pp 96-122

McDowell FH, Sweet RD (1976) The "on-off" phenomenon. In: Birkmayer W, Hornykiewicz O (eds) Advances in parkinsonism. Roche, Basel, pp 603-612
McDowell F, Lee JE, Swift T, Sweet RD, Ogsbury JS, Kessler JT (1970) Treatment of Parkinson's syndrome with dihydroxyphenylalanine (levodopa). Ann Intern Med 72:29-35
Melamed E (1979) Early-morning dystonia: a late side effect of long-term levodopa therapy in Parkinson's disease. Arch Neurol 36:308-310
Melamed E, Bitton V (1984) Delayed onset of responses to individual doses of L-dopa in parkinsonian fluctuators: an additional side effect of long-term L-dopa therapy. Neurology (Suppl 2) 34:270
Melamed E, Bitton V, Zelig O (1986) Episodic unresponsiveness to single doses of L-dopa in parkinsonian fluctuators. Neurology 36:100-103
Muenter MD, Tyce GM (1971) L-Dopa therapy of Parkinson's disease: plasma L-dopa concentration, therapeutic response, and side effects. Mayo Clin Proc 46:231-239
Muenter MD, Sharpless NS, Tyce GM (1972) Plasma 3-*O*-methyldopa in L-dopa therapy of Parkinson's disease. Mayo Clin Proc 47:389-395
Muenter MD, Di Napoli RP, Sharpless NS, Tyce GM (1973) 3-*O*-Methyldopa, L-dopa and trihexyphenidyl in the treatment of Parkinson's disease. Mayo Clin Proc 48:173-183
Muenter MD, Sharpless NS, Tyce GM, Darley FL (1977) Patterns of dystonia ("I-D-I" and "D-I-D") in response it L-dopa therapy of Parkinson's disease. Mayo Clin Proc 52:163-174
Narabayashi H, Kondo T, Nagatsu T, Hayashi A, Suzuki T (1984) D,L-Threo-3,4-dihydroxyphenylserine for freezing symptom in parkinsonism. Adv Neurol 40:497-502
Nutt JG, Woodward WR, Hammerstad JP, Carter JH, Anderson JL (1984) The "on-off" phenomenon in Parkinson's disease. N Engl J Med 310:483-488
Nutt JG, Woodward WR, Anderson JL (1985) The effect of carbidopa on the pharmacokinetics of intravenously administered levodopa: the mechanism of action in the treatment of parkinsonism. Ann Neurol 18:527-543
Pederzoli M, Girotti F, Scigliano G, Aiello G, Carella F, Caraceni T (1983) L-Dopa long-term treatment in Parkinson's disease: age-related side effects. Neurology 33:1518-1522
Poewe WH, Lees AJ, Stern GM (1986) Low-dose L-dopa therapy in Parkinson's disease: a 6-year follow-up study. Neurology 36:1528-1530
Prasad ALN, Fahn S, Isgreen WP (1975) Dopamine: a sensitive automated assay and examples of levodopa therapy. Biochem Med 13:24-32
Rinne UK (1985) Combined bromocriptine-levodopa therapy early in Parkinson's disease. Neurology 35:1196-1198
Rinne UK, Koskinen V, Lonnberg P (1980) Neurotransmitter receptors in the parkinsonian brain. In: Rinne UK, Klingler M, Stamm G (eds) Parkinson's disease: current progress, problems and management. Elsevier/North-Holland, Amsterdam, pp 93-107
Rivera-Calimlin L, Dujovne CA, Morgan JP, Lasagna L, Bianchine JR (1970) L-Dopa treatment failure: explanation and correction. Br Med J [Clin Res] 4:93-94
Sacks OW (1974) Awakenings. Doubleday, Garden City
Sage JI, Duvoisin RC (1986) Sudden onset of confusion with severe exacerbation of parkinsonism during levodopa therapy. Movement Disorders 1:267-270
Schwarz GA, Fahn S (1970) Newer medical treatments in parkinsonism. Med Clin North Am 54:773-785
Shoulson I, Glaubiger GA, Chase TN (1975) "On-off" response: clinical and biochemical correlations during oral and intravenous levodopa administration. Neurology 25:1144-1148

Skibba JL, Pickley J, Gilbert EF (1972) Multiple primary melanoma following administration of levodopa. Arch Pathol Lab Med 93:5556-5561
Snider SR, Fahn S, Isgreen WP, Cote LJ (1976) Primary sensory symptoms in parkinsonism. Neurology 26:423-429
Sweet RD, McDowell FH (1974) The "on-off" response to chronic L-dopa treatment of parkinsonism. Adv Neurol 5:331-338
Tanner CM, Goetz CG, Klawans HL (1986) Hypersexuality in Parkinson's disease. Neurology 36 (Suppl 1):183
Traub M, Wagner HR, Hassan M, Jackson-Lewis V, Fahn S (1986) Effects of chronic bromocriptine treatment on behavior and on dopamine receptor binding in the rat striatum. In: Fahn S, Marsden CD, Jenner P, Teychenne P (eds) Recent developments in Parkinson's disease, Raven, New York, pp 205-214
Wade LA, Katzman R (1975) Synthetic amino acids and the nature of L-dopa transport at the blood-brain barrier. J Neurochem 25:837-842
Weiner WJ, Koller WC, Perlik S, Nausieda PA, Klawans HL (1980) Drug holiday and management of Parkinson disease. Neurology 30:1257-1261
Yahr MD (1972) L-Dopa in neurological disease: current status. Res Publ Assoc Res Nerv Ment Dis 50:494-511
Yahr MD (1974) Variations in the "on-off" effect. Adv Neurol 5:397-399
Yahr MD (1975) Levodopa. Ann Intern Med 83:677-682
Yahr MD (1976) Evaluation of long-term therapy in Parkinson's disease: Mortality and therapeutic efficacy. In: Birkmayer W, Hornykiewicz O (eds) Advances in parkinsonism. Roche, Basel, pp 444-455
Yahr MD, Duvoisin RC, Schear MJ, Barrett RE, Hoehn MM (1969) Treatment of parkinsonism with levodopa. Arch Neurol 21:343-354

CHAPTER 15

Monoamine Oxidase Inhibitors in Parkinson's Disease

M. Sandler and V. Glover

A. Introduction

Following the seminal discovery by Ehringer and Hornykiewicz (1960) of the nigrostriatal dopamine deficiency of Parkinson's disease, one obvious therapeutic strategy was to inhibit monoamine oxidase (MAO), the major enzyme metabolizing it in human brain, in order to conserve the attenuated supply. Holzer and Hornykiewicz (1959) had previously shown that the MAO inhibitors, iproniazid and harmine, increase the dopamine content of rabbit brain. Harmine had, in fact, been used very early on for the treatment of parkinsonism (Behringer and Wilmanns 1929). Although this and related compounds are reversible MAO inhibitors (Nelson et al. 1979), the beneficial effect claimed for it may not have stemmed from this particular biochemical action; the related compound, harmalol, which was thought to have a similar therapeutic effect (Cooper and Gunn 1931), is a much less potent MAO inhibitor. Iproniazid and other nonselective irreversible MAO inhibitors were carefully considered in the early 1960s for the treatment of parkinsonism. However, their use as a single treatment did not appear to lead to any rise in brain dopamine (Bernheimer et al. 1963) and resulted in no more than modest clinical improvement (Rosen 1969). It therefore seemed more reasonable to treat affected subjects with an L-dopa-MAO inhibitor combination. Indeed, a number of groups who carried out such a trial noted beneficial results (e.g. Birkmayer and Hornykiewicz 1961, 1962; McGeer et al. 1961; Barbeau et al. 1962), although marginal benefit only was recorded by others (Hirschmann and Mayer 1964).

Both McGeer et al. (1961) and Barbeau et al. (1962) observed that the combination of drugs brought about a substantial increase in blood pressure, a finding in accord with that of Horwitz et al. (1960), who noted that the pressor effect of intravenous dopamine in humans is markedly potentiated by MAO inhibition. Since these early observations, the hypertensive action of L-dopa plus MAO inhibitor has repeatedly been noted (see Hunter et al. 1970), although no deaths have so far been recorded. Mones et al. (1970) attempted to make use of this hypertensive response in the treatment of a patient with the Shy-Drager syndrome whose existing orthostatic hypotension was exacerbated by L-dopa alone; however, the rise in blood pressure so obtained was fluctuating and uncontrolled and the treatment had to be abandoned. Because of this emergence of the equivalent of the tyramine-induced "cheese effect", the use of MAO inhibitors in Parkinson's disease fell for a time into abeyance. Even the discovery of peripheral decarboxylase inhibitors changed the situation little. It soon became obvious, when TEYCHENNE et al. (1975) used L-dopa and an inhibitor of this kind, together with the MAO inhibitor, tranylcypromine, that peripheral decarboxylase inhibition was incomplete: thus, some dopamine continued to be generated peripherally and breakthrough hypertension consequently occurred.

The stage was set for all that was to follow by the discovery that MAO exists in two forms, MAO-A and MAO-B, each with different substrate specificities and inhibitory sensitivities (Table 1). This classification was first made by JOHNSTON (1968) using a new inhibitory drug, MB 9302, later called clorgyline. He showed that clorgyline gives a double-sigmoid curve during an in vitro MAO assay procedure with tyramine as substrate, indicating that tyramine is oxidised by two forms of the enzyme, with different sensitivities to the drug. Johnston called the more clorgyline-sensitive form, MAO-A, and the less sensitive form, MAO-B. It was soon apparent that MAO-A selectively metabolises 5-hydroxytryptamine (5-HT) in vitro whereas MAO-B selectively oxidizes phenylethylamine (PEA) (YANG and NEFF 1973, 1974). These substrates are currently used widely for the assay of the two forms.

Meanwhile, KNOLL and his colleagues had been developing a new class of MAO inhibitor based on modifications of the structure of amphetamine (for a comprehensive review see KNOLL 1983). They found that one of these, E-250 [(±)-isopropylmethylpropargylamine HCl], later to be called deprenyl or, more recently, selegiline, prevents tyramine-induced release of neurotransmitter monoamines from their binding sites. It thus seemed possible that they had produced an MAO inhibitor without the "cheese effect" (KNOLL and MAGYAR 1972). Following the discovery of the two forms of MAO with clorgyline, it became apparent to these workers that (−)-deprenyl is the obverse of the clorgyline coin, a selective MAO-B inhibitor. (−)-Deprenyl can selectively inhibit MAO-B, both in vitro (EGASHIRA et al. 1976), and in vivo, (FELNER and WALDMEIER 1979) if given repeatedly in suitably low dose.

Table 1. Substrate specificity and inhibitor sensitivities of MAO-A and -B

	MAO-A	MAO-B
Selectively inhibited by	Clorgyline	(−)-Deprenyl J508 MD 780 515
Nonselectively inhibited by	Tranylcypromine Phenelzine Isocarboxazid Pargyline (more B)	
Selective substrates	5-Hydroxytryptamine	Phenylethylamine Benzylamine
Mixed substrates	Noradrenaline Dopamine Tyramine	

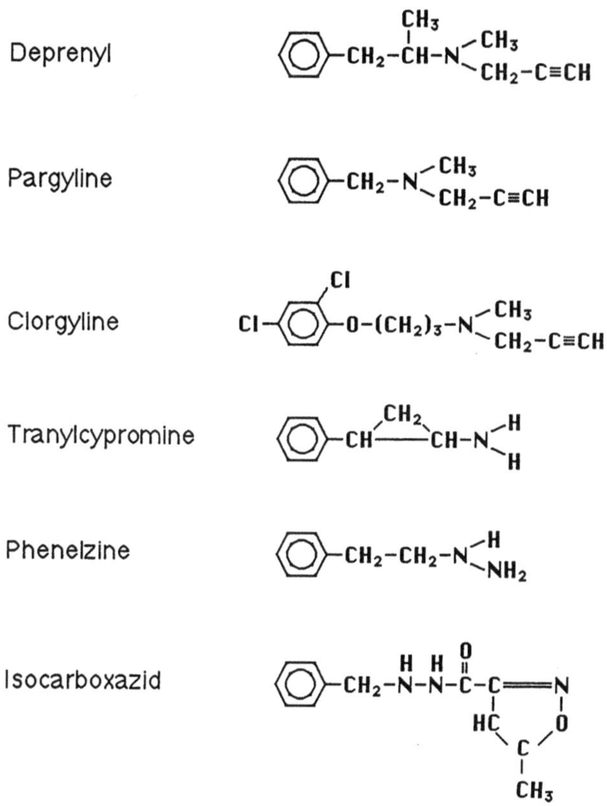

Fig. 1. Structures of some of the major MAO inhibitors

The acetylenic inhibitors of MAO, pargyline, clorgyline, (−)-deprenyl, (Fig. 1), and J508 all appear to inhibit enzyme activity by a "suicide reaction" (FOWLER et al. 1981; SINGER and SALACH 1981). An initial competitive interaction between inhibitor and enzyme is followed by formation of an irreversible adduct.

$$E + I \rightleftharpoons EI \rightarrow EI^*$$

where E = enzyme; I = inhibitor; EI = reversible enzyme-inhibitor complex; and EI* = irreversible enzyme-inhibitor adduct. It has been suggested that the MAO-B selectivity of (−)-deprenyl is determined by the initial reversible rather than the irreversible reaction (TIPTON and MANTLE 1981).

When (−)-deprenyl was first used in humans in conjunction with L-dopa in the treatment of Parkinson's disease, adverse reactions were not recorded (BIRKMAYER et al. 1975). To a greater or lesser extent, the clinical benefit conferred by L-dopa, with or without a peripheral decarboxylase inhibitor, is temporary. After a period of time which varies from individual to individual, but averages about 5 years, the drug begins to lose its effectiveness and a new series of adverse reactions supervenes, dominated by the so-called "on-off" oscillations. One subgroup of these phenomena appears to have a clear pharmacological basis, being related to fluctuations in plasma concentrations of L-dopa, "end-of-dose" and "early morning" hypokinesia. It is this small group of patients, grossly incapacitated and otherwise at the end of the line as far as treatment is concerned, for whom adjuvant therapy with the selective MAO-B inhibitor, (−)-deprenyl, may be beneficial (for reviews see SANDLER and STERN 1982 and Chap. 17). The question of whether (−)-deprenyl also helps to arrest the parkinsonian pathological process (BIRKMAYER et al. 1985) and will therefore prove to have a prophylactic role in the disease will be raised again later in this chapter.

B. The Inhibition of Dopamine Oxidation by (−)-Deprenyl

Why (−)-deprenyl should help in Parkinson's disease was, at first sight, puzzling. When the treatment was introduced by BIRKMAYER et al. (1975), there were only the results of rat experiments to provide guidance. The consensus at the time held that striatal dopamine was predominantly oxidised by MAO-A (WALDMEIER et al. 1976), whereas (−)-deprenyl was a selective inhibitor of MAO-B. Resolution of the problem came with the demonstration of a substantial species difference in the ratio of MAO-A to MAO-B in brain. Although dopamine is a substrate for both forms, the proportion oxidised by MAO-B in human brain is much greater than in rat brain because of the higher B:A ratio in the former (GLOVER et al. 1977; GARRICK and MURPHY

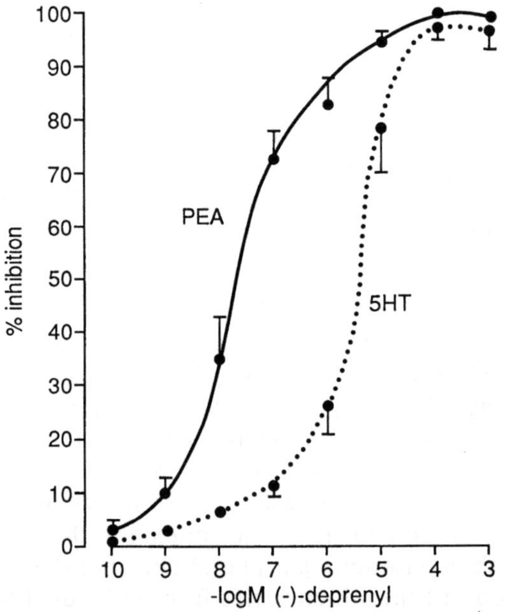

Fig. 2. Selective inhibition by (−)-deprenyl of PEA and 5-HT oxidation in human striatum. Results represent mean ± SE from separate determinations on homogenates of striatum from six individuals (GLOVER et al. 1980)

Table 2. 5-Hydroxytryptamine (5-HT), phenylethylamine (PEA) and dopamine (DA) oxidising activities in different regions of human brain, and inhibition of DA oxidation by 10^{-6} *M* (−)-deprenyl (GLOVER et al. 1980)

Region	5-HT	PEA	DA	5-HT/PEA	Inhibition of DA oxidation by deprenyl (%)
Accumbens	21.9	21.8	28.5	1.0	83
Caudate	20.2	17.2	29.4	1.2	82
Pallidus	18.0	14.1	21.8	1.3	80
Putamen	16.6	13.1	19.7	1.2	80
Hypothalamus	38.3	22.1	37.2	1.7	70
Amygdala	27.9	14.3	24.8	2.1	73
Precentral cortex	16.7	6.9	13.5	2.4	66
Occipital cortex	20.2	6.0	16.7	3.3	67
Temporal cortex	18.7	8.1	14.5	2.3	65
Frontal cortex	17.6	7.2	12.9	2.4	68
Cerebellar cortex	10.7	4.8	9.6	2.2	63

Results are the means of duplicate assays from 5 brains. Activities are expressed as nmol substrate oxidised per mg protein in 30 min.

1980). There is no difference in MAO-A or -B in parkinsonian brains compared with controls (NAGATSU et al. 1977). Within human brain, the B:A ratio varies in different regions and is highest in those rich in dopamine. Figure 2 shows that, in vitro, 10^{-6} *M* (−)-deprenyl substantially inhibits MAO-B in human brain while having little effect on MAO-A (GLOVER et al. 1980); at this concentration it shows maximal selectivity. Table 2 shows that this concentration inhibits dopamine oxidation in striatal homogenates by about 80%. K_m values for dopamine with human MAO-A and -B are very similar at 130 and 140 μM respectively (GLOVER et al. 1980), so that the proportion oxidised by the two forms should be independent of dopamine concentration. There is no evidence that any form of amine oxidase, other than MAO-A and -B, contributes to the oxidation of dopamine in human brain (GLOVER et al. 1980).

A concentration 10^{-6} *M* (−)-deprenyl is equivalent to a 10-mg dose distributed in a body water mass of 40 kg and suggests that administration of this amount might be expected to have a selective action in humans. A study of postmortem brains of patients taking 10 mg (−)-deprenyl for an average of 6 days before death (RIEDERER and YOUDIM 1986) showed MAO-B to be almost totally inhibited compared with parkinsonian controls, whilst MAO-A was much less so. Dopamine levels were substantially increased (Fig. 3), whereas 5-HT, and its deaminated metabolite, 5-hydroxyindoleacetic acid (5-HIAA), were not in general significantly changed in concentration. This finding provides evidence for the functional and selective effect of this dose of (−)-deprenyl in vivo. However, it is still possible that, over a longer time span, more gradual inhibition of MAO-A will take place, with some consequent loss in selectivity (SANDLER 1981).

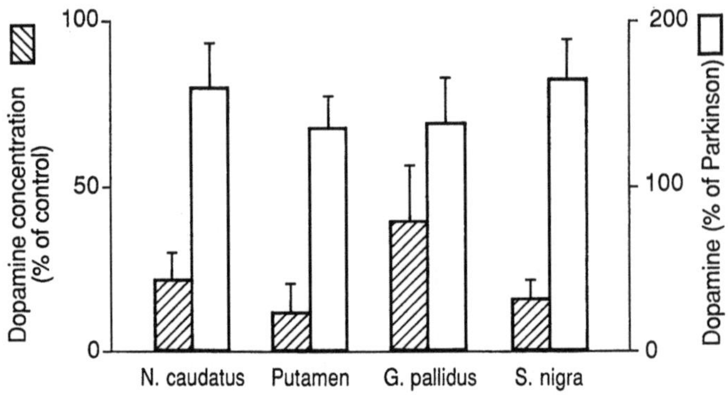

Fig. 3. Concentration of dopamine in various regions of human parkinsonian brain with and without (−)-deprenyl therapy. (−)-Deprenyl (10 mg daily dose) was given on average for a period of 6 days prior to death. The results are means ± SE of three to nine different brains. Note that dopamine concentrations in (−)-deprenyl-treated subjects are expressed as a percentage of those in parkinsonian subjects without (−)-deprenyl therapy (RIEDERER and YOUDIM 1986)

C. Cellular Localisation of MAO in the Brain

In general, histochemical studies have shown that MAO-B in brain is most highly concentrated in the 5-HT cell body regions and in glia (LEVITT et al. 1982; WESTLUND et al. 1985; GLOVER et al. 1987; KONRADI et al. 1987). Brain MAO-B levels increase in old age (FOWLER et al. 1980; MANN and STANLEY 1984), presumably owing to a rise in the relative concentration of glia. MAO-A, on the other hand, is particularly concentrated in noradrenergic neurones (WESTLUND et al. 1985; GLOVER et al. 1987; KONRADI et al. 1987).

Studies of dopamine-containing synaptosomes from either rat or human striata show them to contain MAO-A, but not -B (FAGERVALL and ROSS 1986; TIPTON et al. 1986). Thus, the high activity of MAO-B in human striatum is likely to be located in glia. Presumably, the effect of (−)-deprenyl is to preserve dopamine after it has been released and allow higher concentrations to be available for reuptake. Histochemical studies of both rat (GLOVER et al. 1987) and human (KONRADI et al. 1987) brain have failed to detect enrichment of MAO-A in the nigrostriatal tract, although activity is pronounced in the locus coeruleus. This finding suggests that intraneuronal levels of MAO-A are lower in these dopaminergic neurones than in noradrenergic ones.

D. Safety of (−)-Deprenyl

(−)-Deprenyl, used in selective inhibitory dosage, appears to be very safe. It has now been administered to thousands of parkinsonian patients without marked untoward effect or drug interaction (BIRKMAYER et al. 1982, 1985). When the selective MAO-B inhibitory dose is compared with the LD_{50}, a very

Table 3. The safety margin of deprenyl as a selective inhibitor of MAO-B (KNOLL 1983)

Species	LD_{50} (mg/kg)[a]			Highest dose (mg/kg) which blocks selectively MAO-B in brain, leaving MAO-A activity unaffected		Safety margin LD_{50}/MAO-B blocking dose[b]	
	p.o.	s.c.	i.v.	s.c.	i.v.	i.v.	s.c.
Mouse	445	205	50			410 (0.24)	
Male	(363–546)	(162–262)	(42–58)				
Female	365	190	51	0.5		378 (0.26)	
	(288–464)	(152–237)	(41–62)				
Rat	422	146	75			584 (0.17)	300 (0.33)
Male	(333–636)	(110–194)	(67–84)				
Female	303	112	70	0.25	0.25	448 (0.22)	276 (0.36)
	(228–407)	(89–142)	(62–79)				
Dog	200	95		0.25		380 (0.26)	
Cat			40		0.1		400 (0.25)

[a] Values in parentheses indicate 95% confidence limits.
[b] Values in parentheses indicate the MAO-B blocking dose expressed as a percentage of LD_{50}.

wide safety margin is apparent (Table 3). The orally active daily dose required to achieve almost complete MAO-B inhibition in the rat is about 1 mg/kg (KNOLL 1983), whilst the standard daily oral dosage regimen in Parkinson's disease, 10 mg, is 5–10 times lower. On such a dosage schedule, (−)-deprenyl shows very little potentiation of the tyramine pressor response, in humans. ELSWORTH et al. (1978) demonstrated that both normal volunteers and parkinsonian patients who had been taking 10 mg (−)-deprenyl for up to 2 months could tolerate 150–200 mg oral tyramine without any noticeable effect. This is in marked contrast to pretreatment with clorgyline, where a profound hypertensive response may be observed after less than 10 mg oral tyramine (LADER et al. 1970).

With higher dosage, (−)-deprenyl does potentiate the pressor action of tyramine. SMITH et al. (1986) found that 30 mg (−)-deprenyl significantly lowers the amount of tyramine required to produce a pressor response, and SUNDERLAND et al. (1985) similarly showed that 60 mg (−)-deprenyl results in potentiation of a similar order to tranylcypromine. It is thus only the low dose (10 or 15 mg/day) (−)-deprenyl regimen that can be considered safe, and without the "cheese effect".

It seems likely that the reduction in safety margin parallels inhibition of MAO-A. Sympathetic neurones appear to contain MAO-A (NEFF and GORIDIS 1972), the inhibition of which would allow a build-up of cytoplasmic noradrenaline. PICKAR et al. (1981), using a range of MAO inhibitors, demonstrated a correlation of 0.92 between tyramine pressor sensitivity changes and decrease in plasma 4-hydroxy-3-methoxyphenylglycol (HMPG) level. Such a reduction in plasma HMPG, an important metabolite of noradrenaline, has been employed as an indirect index of MAO-A inhibition (LIEBOWITZ et al. 1985) in the absence of any readily accessible source of the enzyme for direct assay. It remains possible that MAO inhibitors known to give rise to the cheese effect at therapeutic dosage have a second pharmacological property, the ability to augment the noradrenaline-releasing ability of tyramine (see SANDLER 1981). These two actions may, on occasion, be dissociated. The cheese effect may sometimes supervene with some drugs known not to be MAO inhibitors (e.g., SMITH and DURACK 1978; LEE et al. 1979).

E. Assessment of MAO Inhibition In Vivo

As mentioned already, a reduction in plasma HMPG concentration has been employed as an index of MAO-A inhibition. Because the small bowel mucosa exhibits high activity of this enzyme form (SQUIRES 1972; ELSWORTH et al. 1978), gut biopsy was, at one time, employed for direct measurement of MAO-A activity. Such a manoeuvre is not now considered to be ethically justifiable, however. Platelets contain MAO-B (DONNELLY and MURPHY 1977) and can readily be used to monitor in vivo inhibition of the enzyme by (−)-deprenyl and other MAO-B inhibitors. In three of four volunteers, a decline in platelet MAO-B activity could be observed 2 h after taking 1 mg (−)-deprenyl (ELSWORTH et al. 1978). We have found that an oral dose of 10 mg (−)-deprenyl usually causes 100 % inhibition of platelet MAO activity. Others have also reported that 10 mg/day for 1 week produced 99.4 %–100 % inhibition (SIMPSON et al. 1985).

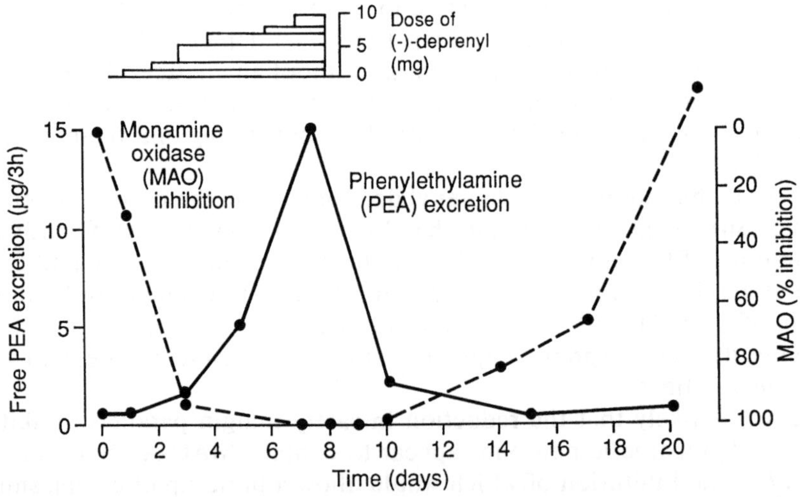

Fig. 4. Typical changes in platelet MAO activity and free urinary PEA output in a normal volunteer receiving (−)-deprenyl (ELSWORTH et al. 1978)

As (−)-deprenyl inhibits MAO-B irreversibly and the turnover of the platelet enzyme may be different from that elsewhere in the body because of rapid turnover of the platelets themselves, platelet measurement may not always provide a good index of general inhibition after the drug has been discontinued. In one positron emission tomographic (PET) scan study in the baboon, for example, brain MAO-B activity appeared to take 100 days to recover after (−)-deprenyl administration (ARNETT et al. 1986). It is also possible that certain tissues concentrate the drug to different extents. However, platelet MAO is adequate as an indicator of patient compliance in taking an irreversible MAO-B inhibitor.

Urinary PEA rises sharply when (−)-deprenyl is given (Fig. 4, ELSWORTH et al. 1978; LIEBOWITZ et al. 1985). Concentration was not altered by 1 mg of the drug, but a 10-mg dose of (−)-deprenyl brought about a 20- to 90-fold increase in output of PEA. The return to normal was much more rapid than that of platelet MAO activity. A greater than 90% inhibition of MAO was necessary for an observable increase in urinary PEA. However, such a degree of inhibition is probably not always necessary for a functional effect on the enzyme, as indicated by the linear correlation between tyramine pressor response and HMPG reduction (PICKAR et al. 1981).

Some limited studies have also been performed on CSF dopamine and dopamine metabolite levels after (−)-deprenyl administration. EISLER et al. (1981) detected an elevated dopamine concentration in the CSF of parkinsonian patients after chronic (−)-deprenyl administration. BARACZKA et al. (1983) showed that a single 10-mg dose of (−)-deprenyl did not change CSF dopamine or 3,4-dihydroxyphenylacetic acid (DOPAC) levels in nonparkinsonian patients, but that 3 days' treatment reduces the concentration of both. KOULU and LAMMINTAUSTA (1981) found that 10 mg (−)-deprenyl potentiates L-dopa induced growth hormone release, probably by increasing dopamine concentrations at receptor level.

SUNDERLAND et al. (1987) have looked at more long-term effects of (−)-deprenyl on CSF monoamine metabolites in patients with Alzheimer's disease. They found that 10 mg/day for 3–4 weeks caused a significant reduction in the dopamine and 5-HT metabolites HVA and 5-HIAA. Higher (−)-deprenyl doses of 40 mg/day also caused a significant reduction in the noradrenaline metabolite, HMPG. These results provide good evidence that the lower dose of (−)-deprenyl does have a selective effect on MAO-B, whereas the higher doses inhibit MAO-A also.

F. Distribution of (−)-Deprenyl in Brain and Body

(−)-Deprenyl is well absorbed and rapidly distributed in the body. In mice, maximal blood levels are reached 1 h after oral and 0.5 h after subcutaneous administration (MAGYAR and TOTHFALUSI 1984). These authors have also shown that, following intravenous injection of (−)-deprenyl in mice, the brain and spinal cord manifest high concentrations after 30 s. However, the radiola-

Table 4. Radioactive concentrations (dpm/100 mg) in tissues of mice after i.v. injection of [^{14}C] deprenyl[a] (MAGYAR and TOTHFALUSI 1984)

Time (min)	Blood	Brain	Liver	Kidney	Lung	Heart	Brown fat
0.5	6996	24583	8786	51248	53834	16625	101943
5	4733	9371	25278	33005	24819	15549	51714
10	4065	9537	25284	44315	20852	11626	29095
30	2340	6723	29020	29705	11495	10620	15427
60	1868	5163	18218	26609	14158	5373	6165
120	864	2685	17636	14636	16299	3232	5412

[a] Injected dose was 23 mg/kg; 5 μCi/20 g mice. Specific activity of [^{14}C] deprenyl was 10.85 μCi/mg.

belled drug is also excreted relatively rapidly from brain, with only low levels of radioactivity remaining after 2 h, compared with other tissues (Table 4).

Methods are now being developed for PET scanning with [^{11}C] (−)-deprenyl (MACGREGOR et al. 1985; FOWLER et al. 1987). Preliminary animal studies suggest that there is sufficient specific binding of (−)-deprenyl to MAO-B in the brain for the determination of levels and distribution, and studies in humans confirm these findings (FOWLER et al. 1987).

G. Metabolism of (−)-Deprenyl to Amphetamine

It has been well established that (−)-deprenyl is largely metabolised to a mixture of desmethyldeprenyl, methamphetamine and amphetamine (Fig. 5; REYNOLDS et al. 1978; SCHACHTER et al. 1980; PHILIPS 1981; KAROUM et al. 1982; MAGYAR and TOTHFALUSI 1985; LIEBOWITZ et al. 1985; YOSHIDA et al. 1986). After administration of 10 mg (−)-deprenyl, it is possible to recover 4 mg methamphetamine and rather less than 1 mg amphetamine from human urine. No unmetabolised (−)-deprenyl was detected (REYNOLDS et al. 1978). SCHACHTER et al. (1980) have shown that the metabolites generated from (−)-deprenyl also possess the (−)-configuration.

The central action of (−)-amphetamine is three or four times less than that of (+)-amphetamine (INNES and NICKERSON 1977) and it is thus unlikely that the amount of (−)-amphetamine derivatives produced in humans after a 10-mg dose of (−)-deprenyl have much pharmacological effect. Unlike (+)-amphetamine, 10 mg (−)-deprenyl does not interfere with normal sleep patterns (THORNTON et al. 1980) and is ineffective in narcolepsy (SCHACHTER et al. 1979). In recent human volunteer studies, MULLER-LIMMROTH (1985) has shown that, far from having a stimulant effect, (−)-deprenyl may, if anything, have a slight depressant action.

(+)-Amphetamine is a fairly potent selective reversible MAO-A inhibitor (MILLER et al. 1980; ROBINSON 1985) as is (+)-methamphetamine (SUZUKI et al. 1980). (−)-Amphetamine and (−)-methamphetamine are less potent (MILLER et al. 1980; ROBINSON 1985). The lack of a tyramine pressor response after administration of 10 mg (−)-deprenyl suggests that, at this dose, it or its

a
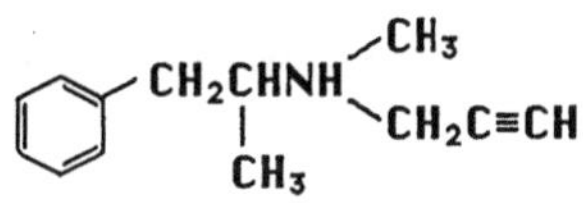

b
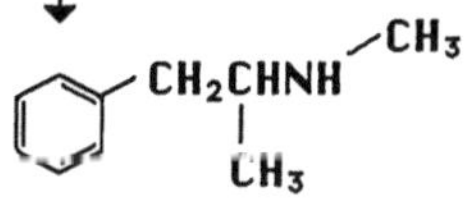

c
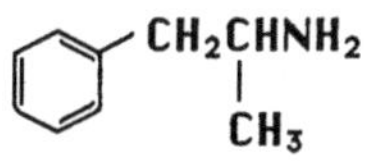

Fig. 5 a–c. The structural formulae of (**a**) deprenyl, (**b**) methamphetamine and (**c**) amphetamine

metabolites have little functional effect on MAO-A activity.

It seems unlikely that the amphetamine derivatives contribute to the therapeutic benefit achieved by administering (−)-deprenyl in Parkinson's disease. The manipulation of urinary pH in patients taking (−)-deprenyl did not result in detectable clinical change, despite substantial alterations in the excretion rate of methamphetamine and amphetamine; the substitution of (−)-deprenyl by appropriate concentrations of its amphetamine derivatives only resulted in clinical relapse (ELSWORTH et al. 1982). No amphetamine abuse behaviour pattern was observed by BIRKMAYER et al. (1982) in the many hundreds of parkinsonian patients they treated with (−)-deprenyl over the years. When patients are withdrawn rapidly from the drug, they have no craving for it, but only suffer an increase in parkinsonian disability (BIRKMAYER et al. 1982).

H. Other Actions of (−)-Deprenyl

All MAO inhibitors, including (−)-deprenyl, have long-term effects on various brain monoamine receptor sites, which are probably a consequence of the inhibition of MAO (MURPHY et al. 1984). In addition (−)-deprenyl is not a "clean drug" and, clearly, at higher doses, has many other effects apart from inhibiting MAO-B. KNOLL (1978) described its ability to inhibit amine uptake, including that of tyramine, into catecholaminergic neurones, and to inhibit the release of acetylcholine. Other more recently developed selective MAO-B inhibitors, such as J508, were devoid of these actions (KNOLL 1978). However, such effects of (−)-deprenyl were only apparent at a concentration of 5 µg/ml, approximately 25 μM, or at even higher concentrations. LAI et al. (1980) showed that it prevents noradrenaline uptake into rat synaptosomes, with an IC_{50} of 26 μM, and dopamine uptake with an IC_{50} of 330 μM. Other pharmacological effects of (−)-deprenyl have also been noted. At 10 mg/kg (approximately 50 μM), SIMPSON (1978) demonstrated sympathomimetic effects on the rat cardiovascular system. At a concentration of 30 μ*M*, FINBERG et al. (1981) observed a depression in response of the isolated vas deferens of the rat to tyramine, but a potentiation on washout of the drug. They also observed direct potentiation of noradrenaline effects with 10 μ*M* (−)-deprenyl.

ZSILLA et al. (1986) showed that repeated subcutaneous doses of (−)-deprenyl (0.25 mg/kg) to rats, which selectively inhibited MAO-B, but not MAO-A, also increased dopamine turnover rate and reduced dopamine uptake in the striatum. FOZARD et al. (1985) demonstrated that, in doses which selectively inhibit MAO-B in mice and rats, (−)-deprenyl has several other pharmacological effects, such as reversal of reserpine hypothermia, "behavioural despair", and increased blood pressure and heart rate. In doses which selectively inhibit MAO-B, another inhibitor in the literature, MDL 72 145, had none of these effects. The authors suggest that these additional effects of (−)-deprenyl may be due to its amphetamine metabolites. This is plausible, but it should be remembered that much higher doses of (−)-deprenyl (mg/kg) are needed to inhibit MAO-B selectively in rodents than in humans, perhaps because of its more rapid metabolism. Thus, the concentration of amphetamine compounds present in these models will be much greater than in humans after a selective inhibitory dose.

None of the other actions of (−)-deprenyl described here have been apparent at concentrations likely to be present after the 10-mg dose used in parkinsonian patients. At low concentration (1 μ*M*), the drug counteracts the ability of clorgyline to potentiate tyramine-induced release of noradrenaline from rat brain slices (GLOVER et al. 1983). This effect, however, is unlikely to be relevant to its primary action in Parkinson's disease, MAO-B inhibition.

I. MAO and MPTP

With the discovery that MPTP can bring about a syndrome resembling Parkinson's disease in humans and other primates (Chap. 7) and the finding that MAO-B is necessary for the conversion of MPTP to its toxic metabolite MPP^+ (HEIKKILA et al. 1984; SALACH et al. 1984; GLOVER et al. 1986a), a possible new role for MAO in the causation of Parkinson's disease, and for MAO inhibitors in its prevention, has become apparent.

Although MPTP has a different type of structure from previously known MAO substrates (GLOVER et al. 1986b), it is now firmly established (CASTA-

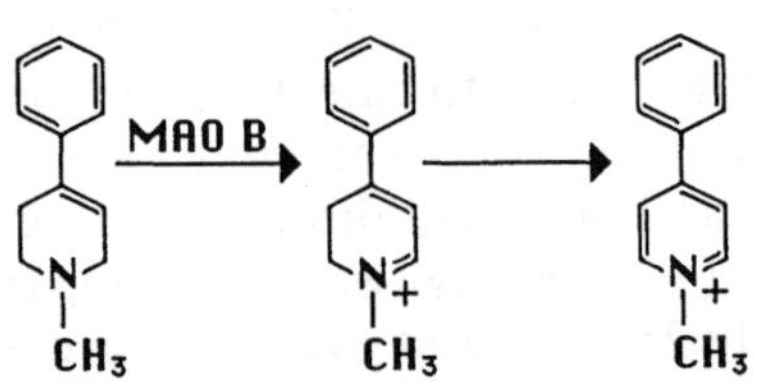

MPTP I — $MPDP^+$ II — MPP^+ III

Fig. 6. Conversion of MPTP to MPP^+

GNOLI et al. 1985) that MAO-B does catalyse the first stage of the reaction, MPTP to the intermediate MPDP, as shown in Fig. 6. Whether the conversion of MPDP to MPP^+ is also catalysed by MAO-B or whether it occurs spontaneously is still controversial. It is also well established that MAO-B is the major enzyme in the human body responsible for the initial conversion (GLOVER et al. 1986a). Prior administration of (−)-deprenyl prevents MPTP toxicity in animal models (HEIKKILA et al. 1984; COHEN et al. 1985). Dopamine uptake blockers, at least in rodents, also prevent MPTP toxicity and this may be explicable in terms of MAO-B being localised *outside* nigrostriatal neurones, probably largely in glia (see Sect. C). The MPP^+ generated outside the dopaminergic neurones may then be selectively concentrated by them, via the dopamine uptake system (JAVITCH et al. 1985). Rats are much less sensitive to the action of MPTP and this may be partially due to the fact that they have much less MAO-B surrounding their nigrostriatal tract than do primates (Fig. 7; GLOVER et al. 1987; WILLOUGHBY et al. 1988).

MPTP is a synthetic drug and presumably not itself a possible cause of idiopathic Parkinson's disease. However, there may well be related pyridines in the environment capable of acting in a similar way, perhaps more slowly, and building up their effects with time. Several analogues of MPTP are known to be substrates for MAO (GIBB et al. 1987). All are tetrahydropyridines, but some lack the methyl or phenyl substituent groups (Fig. 8). Some of these substrates are metabolised by MAO-A as well as MAO-B.

As mentioned earlier, BIRKMAYER et al. (1985), in a retrospective study, found evidence to suggest that parkinsonian patients taking (−)-deprenyl as part of their treatment regimen live longer than those who do not. This unexpected finding needs to be confirmed, of course, in a well-controlled prospective trial; if it turns out to be true, one explanation would be that (−)-deprenyl acts to block the conversion of some MPTP-like compound to an MPP^+-like neurotoxin. It is always possible, however, that the increased dopamine levels resulting from (−)-deprenyl administration (see Fig. 3) are, in themselves, protective.

K. Postscript

The rise to prominence of (−)-deprenyl has been an interesting example of scientific serendipity. The drug was initially developed as a general MAO inhibitor and found to be selective for MAO-B at a time when the two forms of MAO were just being recognised (KNOLL 1983). It was tried in Parkinson's disease and found to be effective (BIRKMAYER et al. 1975) at a time when dopamine was thought to be predominantly metabolised by MAO-A (WALDMEIER et al. 1976); only later was it shown that dopamine is largely oxidised by MAO-B in the human striatum (GLOVER et al. 1977). The finding that MPTP is converted to its neurotoxic metabolite by MAO-B (SALACH et al. 1984), an effect that can be blocked by (−)-deprenyl (HEIKKILA et al. 1984; COHEN et al. 1985), was also a surprise to scientists familiar with MAO, who at first thought

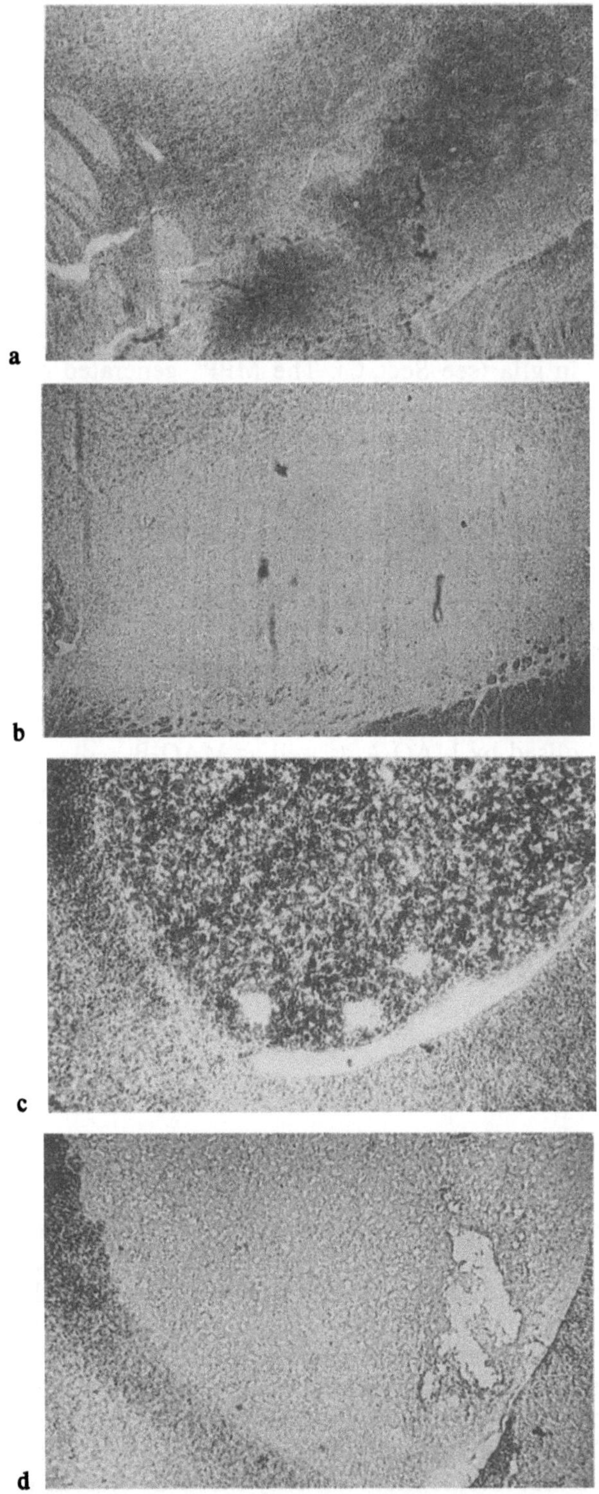

a marmoset substantia nigra, **b** marmoset substantia nigra preincubated with 10^{-6} M(−)-deprenyl, **c** marmoset caudate, **d** marmoset caudate preincubated with 10^{-6} M(−)-deprenyl,

e

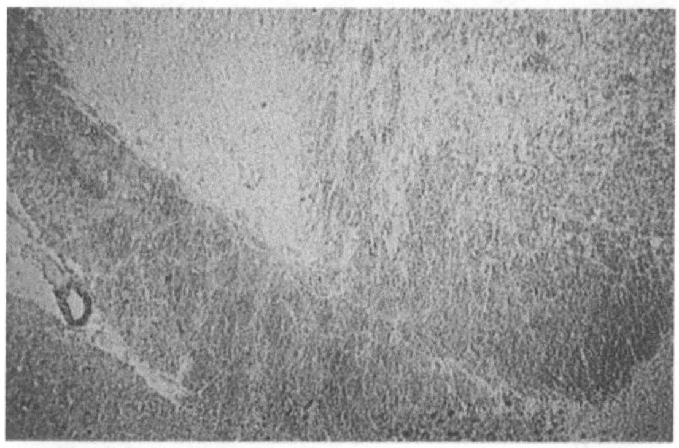
f

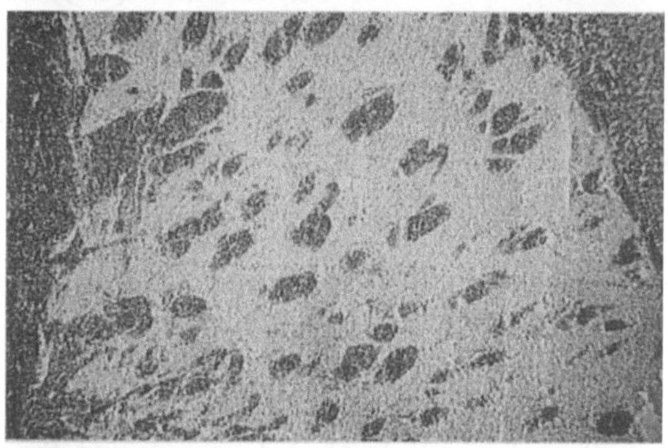
g

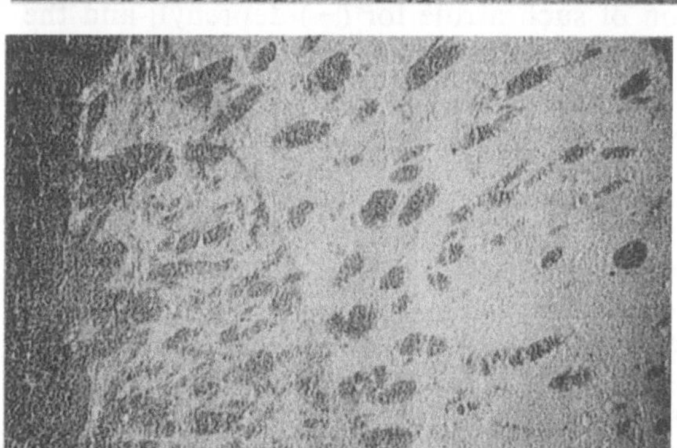
h

Fig. 7 a–h. MAO-B, stained with benzylamine, in marmoset and rat substantia nigra and striatum **e** rat substantia nigra, **f** rat substantia nigra preincubated with 10^{-6} $M(-)$-deprenyl, **g** rat striatum, **h** rat striatum preincubated with 10^{-6} $M(-)$-deprenyl

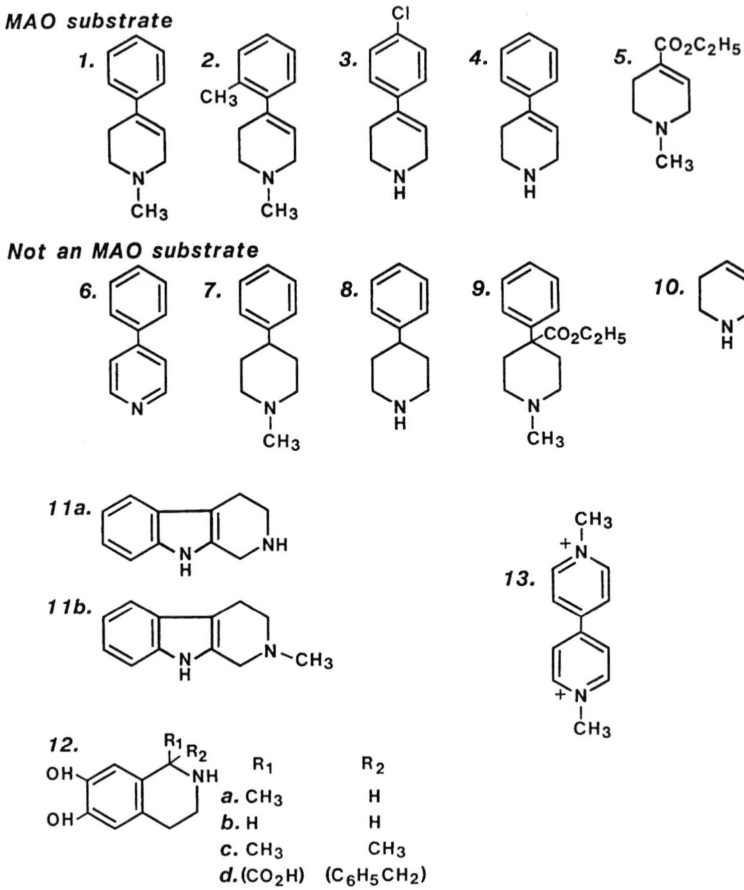

Fig. 8. Structure of some analogues of MPTP (1) showing which are, and which are not substrates for MAO (GIBB et al. 1987)

that MPTP must be a substrate for a different enzyme. At each stage there had been no prediction of such a role for (−)-deprenyl, and the MPTP story, in particular, would have seemed unlikely. Thus, the role of deprenyl has emerged purely on the basis of the acceptance of new experimental evidence, and not from any previous theoretical considerations.

References

Arnett CD, MacGregor RR, Fowler JS, Wolf AP (1986) Turnover of MAO-B in baboon brain determined in vivo by PET and [^{11}C]-*l*-deprenyl. J Nucl Med 27:982

Baraczka K, Fekete MIK, Kanyicska B (1983) Changes in dopamine and 3,4-dihydroxyphenylacetic acid (DOPAC) levels in human cerebrospinal fluid after L-dopa and deprenyl administration. J Neural Transm 58:299-304

Barbeau A, Sourkes TL, Murphy GF (1962) Les catécholamines dans la maladie de Parkinson. In: J. De Ajuriaguerra (ed) Monoamines et système nerveux central. Georg, Genève, pp 247-262

Behringer K, Wilmanns K (1929) Zur Harmin-Banisterin-Frage. Dtsch Med Wochenschr 55:2081-2086
Bernheimer H, Birkmayer W, Hornykiewicz O (1963) Zur Biochemie des Parkinson-Syndroms des Menschen. Einfluß der Monoaminoxydase-Hemmer-Therapie auf die Konzentration des Dopamins, Noradrenalins und 5-Hydroxytryptamins im Gehirn. Klin Wochenschr 41:465-469
Birkmayer W, Hornykiewicz O (1961) Der L-3,4-Dioxyphenylalanin (= DOPA)-effekt bei der Parkinsonakinese. Wien Klin Wochenschr 73:787-788
Birkmayer W, Hornykiewicz O (1962) Der L-Dioxyphenylalanin (= DOPA)-effekt beim Parkinson-Syndrom des Menschen: zur Pathogenese und Behandlung der Parkinsonakinese. Arch Psychiatr Nervenkr 203:560-574
Birkmayer W, Riederer P, Youdim MBH, Linauer W (1975) Potentiation of antiakinetic effect after L-dopa treatment by an inhibitor of MAO-B, *l*-deprenil. J Neural Transm 36:303-323
Birkmayer W, Riederer P, Youdim MBH (1982) (−)-Deprenyl in the treatment of Parkinson's disease. Clin Neuropharmacol 5:195-230
Birkmayer W, Knoll J, Riederer P, Youdim MBH, Hars V, Marton J (1985) Increased life expectancy resulting from addition of L-deprenyl to madopar treatment in Parkinson's disease: a longterm study. J Neural Transm 64:113-127
Castagnoli N, Chiba Jr K, Trevor AJ (1985) Potential bioactivation pathways for the neurotoxin 1-methyl-4-phenyl-1,2,3,6-tetrahydropyridine (MPTP). Life Sci 36:225-230
Cohen G, Pasik P, Cohen B, Leist A, Mytilineou C, Yahr MD (1985) Pargyline and deprenyl prevent the neurotoxicity of 1-methyl-4-phenyl-1,2,3,6-tetrahydropyridine (MPTP) in monkeys. Eur J Pharmacol 106:209-210
Cooper HA, Gunn JA (1931) Harmalol in the treatment of parkinsonism. Lancet 2:901-902
Donnelly CH, Murphy DL (1977) Substrate- and inhibitor-related characteristics of human platelet MAO. Biochem Pharmacol 26:853-858
Egashira T, Ekstedt B, Oreland L (1976) Inhibition by clorgyline and deprenyl of the different forms of monoamine oxidase in rat liver mitochondria. Biochem Pharmacol 25:2583-2586
Ehringer H, Hornykiewicz O (1960) Verteilung von Noradrenalin und Dopamin (3-Hydroxytyramin) im Gehirn des Menschen und ihr Verhalten bei Erkrankungen des extrapyramidalen Systems. Klin. Wochenschr 38:1236-1239
Eisler T, Teravainen H, Nelson R, Krebs H, Weise V, Lake CR, Ebert MH, Whetzel H, Murphy DL, Kopin IJ, Calne DB (1981) Deprenyl in Parkinson's disease. Neurology 31:19-23
Elsworth JD, Glover V, Reynolds GP, Sandler M, Lees AJ, Phuapradit P, Shaw KM, Stern GM, Kumar P (1978) Deprenyl administration in man: a selective monoamine oxidase B inhibitor without the "cheese effect". Psychopharmacology (Berlin) 57:33-38
Elsworth JD, Sandler M, Lees AJ, Ward C, Stern GM (1982) The contribution of amphetamine metabolites of (−)-deprenyl to its antiparkinsonian properties. J Neural Transm 54:105-110
Fagervall I, Ross SB (1986) A and B forms of monoamine oxidase within the monoaminergic neurons of the rat brain. J Neurochem 47:569-576
Felner AE, Waldmeier PC (1979) Cumulative effects of irreversible MAO inhibitors in vivo. Biochem Pharmacol 28:995-1002
Finberg JPM, Tenne M, Youdim MBH (1981) Tyramine antagonistic properties of AGN 1135, an irreversible inhibitor of monoamine oxidase type B. Br J Pharmacol 73:65-74

Fowler CJ, Wiberg A, Oreland L, Marcusson J, Winblad B (1980) The effect of age on the activity and molecular properties of human brain monoamine oxidase. J Neural Transm 49:1-20
Fowler CJ, Oreland L, Callingham BA (1981) The acetylenic monoamine oxidase inhibitors clorgyline, deprenyl, pargyline and J-508: their properties and applications. J Pharm Pharmacol 33:341-347
Fowler JS, MacGregor RR, Wolfe AP, Arnett CD, Dewey SL, Schulyer D, Christman D, Logan J, Smith M, Sachs H, Aquilonius SM, Bjurling P, Halldin C, Hartvig P, Leenders KL, Lundqvist H, Oreland L, Stalnacke C-G, Langstrom B (1987) Mapping human brain monoamine oxidase A and B with ^{11}C-labeled suicide inactivators and PET. Science 235:481-485
Fozard JR, Zreika M, Robin M, Palfreyman MG (1985) The functional consequences of inhibition of monoamine oxidase type B: comparison of the pharmacological properties of L-deprenyl and MDL 72 145. Naunyn Schmiedebergs Arch Pharmacol 331:186-193
Garrick NA, Murphy DL (1980) Species differences in the deamination of dopamine and other substrates for monoamine oxidase in brain. Psychopharmacology (Berlin) 72:27-33
Gibb C, Willoughby J, Glover V, Sandler M, Testa B, Jenner P, Marsden CD (1987) Analogues of 1-methyl-4-phenyl-1,2,3,6-tetrahydropyridine as monoamine oxidase substrates: a second ring is not necessary. Neurosci Lett 76:316-322
Glover V, Sandler M, Owen F, Riley GJ (1977) Dopamine is a monoamine oxidase B substrate in man. Nature 265:80-81
Glover V, Elsworth JD, Sandler M (1980) Dopamine oxidation and its inhibition by (−)-deprenyl in man. J Neural Transm [Suppl] 16:163-172
Glover V, Pycock CJ, Sandler M (1983) Tyramine-induced noradrenaline release from rat brain slices: prevention by (−)-deprenyl. Br J Pharmacol 80:141-148
Glover V, Gibb C, Sandler M (1986a) Monoamine oxidase B (MAO-B) is the major catalyst of 1-methyl-4-phenyl-1,2,3,6-tetrahydropyridine (MPTP) oxidation in human brain and other tissues. Neurosci Lett 64:216-220
Glover V, Gibb C, Sandler M (1986b) The role of MAO in MPTP toxicity — a review. J Neural Transm [Suppl] 20:65-76
Glover V, Willoughby J, Sandler M (1987) Histochemical localisation of MAO A and B in brain. Pharmacol Toxicol (Suppl 1) 60:22
Heikkila RE, Manzino L, Cabbat FS, Duvoisin RC (1984) Protection against the dopaminergic neurotoxicity of 1-methyl-4-phenyl-1,2,5,6-tetrahydropyridine by monoamine oxidase inhibitors. Nature 311:467-469
Hirschmann J, Mayer K (1964) Zur Beeinflussung der Akinese und anderer extrapyramidal-motorischer Störungen mit L-dopa (L-Dihydroxyphenylalanin). Dtsch Med Wochenschr 89:1877-1880
Holzer G, Hornykiewicz O (1959) Über den Dopamin-(Hydroxytyramin-)Stoffwechsel im Gehirn der Ratte. Arch Exp Pathol Pharmakol 237:27-33
Horwitz D, Goldberg LI, Sjoerdsma A (1960) Increased blood pressure responses to dopamine and norepinephrine produced by monoamine oxidase inhibitors in man. J Lab Clin Med 56:747-753
Hunter KR, Boakes AJ, Laurence DR, Stern GM (1970) Monoamine oxidase inhibitors and L-dopa Br Med J 3:388
Innes IR, Nickerson M (1977) Norepinephrine, epinephrine and the sympathomimetic amines. In: Goodman LS, A. Gilman (eds) The pharmacological basis of therapeutics. Macmillan, New York, pp 477-513
Javitch JA, d'Amato RJ, Strittmatter SM, Snyder SH (1985) Parkinsonism-inducing

neurotoxin, *N*-methyl-4-phenyl-1,2,3,6-tetrahydropyridine: uptake of the metabolite *N*-methyl-4-phenylpyridine by dopamine neurons explains selective toxicity. Proc Natl Acad Sci USA 82:2173-2177

Johnston JP (1968) Some observations upon a new inhibitor of monoamine oxidase in brain tissue. Biochem Pharmacol 17:1285-1297

Karoum F, Chuang LW, Eister P, Calne DB, Liebowitz MR, Quitkin FM, Klein DF, Wyatt RJ (1982) Metabolism of (−)-deprenyl to amphetamine and methamphetamine may be responsible for deprenyl's therapeutic benefit. A biochemical assessment. Neurology 32:503-509

Knoll J (1978) The possible mechanisms of action of (−)-deprenyl in Parkinson's disease. J Neural Transm 43:177-198

Knoll J (1983) Deprenyl (selegiline): the history of its development and pharmacological action. Acta Neurol Scand 95:57-80

Knoll J, Magyar K (1972) Some puzzling pharmacological effects of monoamine oxidase inhibitors. In: Costa E, Sandler M (eds) Monoamine oxidases — new vistas. Raven, New York, pp 393-408

Konradi C, Svoma E, Jellinger K, Riederer P, Denney RM, Arluison M, Nagatsu T (1987) Immunocytochemical differentiation of MAO-A and MAO-B in human post mortem brain. Pharmacol Toxicol 60 (Suppl 1):29

Koulu M, Lammintausta R (1981) Effects of L-deprenyl on human growth hormone secretion. J Neural Transm 51:223-231

Lader MH, Sakalis G, Tansella M (1970) Interactions between sympathomimetic amines and a new monoamine oxidase inhibitor. Psychopharmacology (Berlin) 18:118-123

Lai JCK, Leung TKC, Guest JF, Lim L, Davison AN (1980) The monoamine oxidase inhibitors clorgyline and L-deprenyl also affect the uptake of dopamine, noradrenaline and serotonin by rat brain synaptosomal preparations. Biochem Pharmacol 29:2763-2767

Lee KY, Beilin JL, Vandongen R (1979) Severe hypertension after ingestion of an appetite suppressant (phenylpropanolamine) with indomethacin. Lancet 1:1110-1111

Levitt P, Pintar JE, Breakefield XO (1982) Immunocytochemical demonstration of monoamine oxidase B in brain astrocytes and serotonergic neurons. Proc Natl Acad Sci USA 79:6385-6389

Liebowitz MR, Karoum F, Quitkin FM, Davies SO, Schwartz D, Levitt M, Linnoila M (1985) Biochemical effects of L-deprenyl in atypical depressives. Biol Psychiatr 20:558-565

MacGregor RR, Halldin C, Fowler JS, Wolf AP, Arnett CD, Langstrom B, Alexoff D (1985) Selective, irreversible in vivo binding of [^{11}C]-clorgyline and [^{11}C]-L-deprenyl in mice: potential for measurement of functional monoamine oxidase activity in brain using positron emission tomography. Biochem Pharmacol 34:3207-3210

Magyar K, Tothfalusi L (1984) Pharmacokinetic aspects of deprenyl effects. Pol J Pharmacol Pharm 36:373-384

Magyar K, Tothfalusi L (1985) Metabolism of deprenyl in rats. In: Kelemen K, Magyar K, Vizi ES (eds) Neuropharmacology. Akademiai Kiado, Budapest, pp 43-50

Mann JJ, Stanley M (1984) Postmortem monoamine oxidase enzyme kinetics in the frontal cortex of suicide victims and controls. Acta Psychiatr Scand 69:135-139

McGeer PL, Boulding JE, Gibson WC, Foulkes RG (1961) Drug-induced extrapyramidal reactions. JAMA 177:665-670

Miller HH, Shore PA, Clarke DE (1980) In vivo monoamine oxidase inhibition by *d*-amphetamine. Biochem Pharmacol 29:1347-1354

Mones RJ, Elizan TS, Siegel GJ (1970) Evaluation of L-dopa therapy in Parkinson's disease. NY State J Med 70:2309–2318
Muller-Limmroth VW (1985) Wirkung eines neuen Antiparkinsonmittels, Selegilin, auf die psychomotorische Leistungsfähigkeit des Menschen. Arzneimittelforschung 35:998–1002
Murphy DL, Garrick NA, Aulakh CS, Cohen RM (1984) New contributions from basic science to understanding the effects of monoamine oxidase inhibiting antidepressants. J Clin Psychiatr 45:37–43
Nagatsu T, Kato T, Numata Y, Ikuta K, Sano M, Nagatsu I, Kondo Y, Inagaki S, Iizuka R, Hori A, Narabayashi H (1977) Phenylethanolamine *N*-methyltransferase and other enzymes of catecholamine metabolism in human brain. Clin Chim Acta 75:221–232
Neff NH, Goridis C (1972) Neuronal monoamine oxidase: specific enzyme types and their rates of formation. In: Costa E, Sandler M (eds) Monoamine oxidase — new vistas. Raven, New York, pp 307–323
Nelson DL, Herbet A, Petillot Y, Pichat L, Glowinski J, Hamon M (1979) [^{3}H] Harmaline as a specific ligand of MAO-A — I. Properties of the active site of MAO-A from rat and bovine brains. J Neurochem 32:1817–1827
Philips SR (1981) Amphetamine, p-hydroxyamphetamine and β-phenethylamine in mouse brain and urine after (−)- and (+)-deprenyl administration. J Pharm Pharmacol 33:739–741
Pickar D, Cohen RM, Jimerson DC, Lake CR, Murphy DL (1981) Tyramine infusions and selective monoamine oxidase inhibitor treatment. Psychopharmacology (Berlin) 74:8–12
Reynolds GP, Elsworth JD, Blau K, Sandler M, Lees AJ, Stern GM (1978) Deprenyl is metabolized to methamphetamine and amphetamine in man. Br J Clin Pharmacol 6:542–544
Riederer P, Youdim MBH (1986) Monoamine oxidase activity and monoamine metabolism in brains of parkinsonian patients treated with *l*-deprenyl. J Neurochem 46:1359–1365
Robinson JB (1985) Stereoselectivity and isoenzyme selectivity of monoamine oxidase inhibitors. Biochem Pharmacol 34:4105–4108
Rosen JA (1969) The effect of a monoamine oxidase inhibitor on the bradykinesia of human parkinsonism. In: Barbeau A, Brunette JR (eds) Progress in neuro-genetics. International Congress Series no 175. Excerpta Medica, Amsterdam, pp 346–351
Salach JI, Singer TP, Castagnoli N Jr, Trevor A (1984) Oxidation of the neurotoxic amine 1-methyl-4-phenyl-1,2,3,6-tetrahydropyridine (MPTP) by monoamine oxidases A and B and suicide inactivation of the enzymes by MPTP. Biochem Biophys Res Commun 125:831–835
Sandler M (1981) Monoamine oxidase inhibitor efficacy in depression and the "cheese effect". Psychol Med 11:455–458
Sandler M, Stern GM (1982) Deprenyl in Parkinson's disease. In: Marsden CD, Fahn S (eds) Neurology 2. Movement disorders. Butterworth Scientific, London, pp 166–173
Schachter M, Price PA, Parkes JD (1979) Deprenyl in narcolepsy. Lancet 1:831–832
Schachter M, Marsden CD, Parkes JD, Jenner P, Testa B (1980) Deprenyl in the management of response fluctuations in patients with Parkinson's disease on levodopa. J Neurol Neurosurg Psychiatry 43:1016–1021
Simpson GM, Frederickson E, Palmer R, Pi E, Sloane RB, White K (1985) Platelet monoamine oxidase inhibition by deprenyl and tranylcypromine: implications for clinical use. Biol Psychiatry 20:680–684
Simpson LL (1978) Evidence that deprenyl, a type B monoamine oxidase inhibitor, is an indirectly acting sympathomimetic amine. Biochem Pharmacol 27:1591–1595

Singer TP, Salach JA (1981) Interaction of suicide inhibitors with the active site of monoamine oxidase. In: Youdim MBH, Paykel ES (eds) Monoamine oxidase inhibitors — the state of the art. Wiley, New York, pp 17-29
Smith CK, Durack DT (1978) Isoniazid and reaction to cheese. Ann Intern Med 81:793
Smith SE, Prasad A, Signy M, Sandler M (1986) Hypertensive response to oral tyramine in healthy subjects following high dose selegiline. Proceedings 3rd world conference on clinical pharmacology and therapeutics, 1986, Stockholm, p 42
Squires RF (1972) Multiple forms of monoamine oxidase in intact mitochondria as characterized by selective inhibitors and thermal stability: a comparison of eight mammalian species. In: Costa E, Sandler M (eds) Monoamine oxidases — new vistas. Raven, New York, pp 355-370
Sunderland T, Mueller EA, Cohen RM, Jimerson DC, Pickar D, Murphy DL (1985) Tyramine pressor sensitivity changes during deprenyl treatment. Psychopharmacology (Berlin) 86:432-437
Sunderland T, Tariot PN, Cohen RM, Newhouse PA, Mellow AM, Mueller EA, Murphy DL (1987) Dose-dependent effects of deprenyl on CSF monoamine metabolites in patients with Alzheimer's disease. Psychopharmacology (Berlin) 91:293-296
Suzuki O, Hattori H, Asano M, Oya M, Katsumata Y (1980) Inhibition of monoamine oxidase by *d*-methamphetamine. Biochem Pharmacol 29:2071-2073
Teychenne PF, Calne DB, Lewis PJ, Findley LJ (1975) Interactions of levodopa with inhibitors of monoamine oxidase and L-aromatic amino acid decarboxylase. Clin Pharmacol Ther 18:273-277
Thornton C, Doré CJ, Elsworth JD, Herbert M, Stern GM (1980) The effect of deprenyl, a selective monoamine oxidase B inhibitor, on sleep and mood in man. Psychopharmacology (Berlin) 70:163-166
Tipton KF, Mantle TJ (1981) The inhibition of rat liver monoamine oxidase by clorgyline and deprenyl. In: Youdim MBH, Paykel ES (eds) Monoamine oxidase inhibitors — the state of the art. Wiley, New York, pp 3-15
Tipton KF, O'Carrol AM, Sullivan JP, Hasan F (1986) Biochemical and pharmacological aspects of monoamine oxidase inhibition (abstract). British Association for Psychopharmacology Summer Meeting, Cambridge
Waldmeier PC, Delini-Stula A, Maitre L (1976) Preferential deamination of dopamine by an A type monoamine oxidase in rat brain. Naunyn-Schmiedeberg's Arch Pharmacol (Weinheim) 292:9-14
Westlund KN, Denney RM, Kochersperger LM, Rose RM, Abell CW (1985) Distinct monoamine oxidase A and B populations in primate brain. Science 230:181-183
Willoughby J, Glover V, Sandler M (1988) Histochemical localization of monoamine oxidase A and B in rat brain. J Neural Transm 74:29-42.
Yang HYT, Neff NH (1973) Beta-phenylethylamine: a specific substrate for type B monoamine oxidase of brain. J Pharmacol Exp Ther 187:365-371
Yang HYT, Neff NH (1974) The monoamine oxidases of brain: selective inhibition with drugs and the consequence for metabolism of biogenic amines. J Pharmacol Exp Ther 189:733-740
Yoshida T, Yamada Y, Yamamoto T, Kuroiwa Y (1986) Metabolism of deprenyl, a selective monoamine oxidase (MAO)-B inhibitor in rat: relationship of metabolism to MAO-B inhibitory potency. Xenobiotica 16:129-136
Zsilla G, Foldi P, Held G, Szekely AM, Knoll J (1986) The effect of repeated doses of (−)-deprenyl on the dynamics of monoaminergic transmission. Comparison with clorgyline. Pol J Pharmacol Pharm 38:57-67

CHAPTER 16

Clinical Actions of L-Deprenyl in Parkinson's Disease

M. D. YAHR and H. KAUFMANN

A. Introduction

Over the past decade a number of investigators have reported on the usefulness of L-deprenyl, a selective inhibitor of monoamine oxidase type B (MAO-B) in the treatment of Parkinson's disease. To a considerable extent these reports relate to its use as an adjunctive agent to levodopa administered when the therapeutic efficacy of the latter begins to diminish. More recently it has been suggested on theoretical grounds and in limited clinical experience that L-deprenyl may have a more valuable effect in being able to halt or retard the inexorable progressive nature of the underlying disease process of Parkinson's disease. In this chapter, the clinical experience to date of the use of L-deprenyl in regard to its pharmacologic properties, indications for use, and effectiveness in the control of parkinsonism will be discussed.

At present the treatment of Parkinson's disease is palliative and primarily directed toward control of its symptoms. To this end, it relies heavily on the concept that pharmacologic agents capable of restoring striatal dopaminergic activity, the major neurotransmitter defect in the disease, are most useful in its treatment. Of the numerous agents which have undergone clinical trial to date, levodopa, the aromatic amino acid precursor of dopamine, administered in combination with a selective extracerebral decarboxylase inhibitor, has produced the most beneficial results. Numerous reports indicating its effectiveness in the control of parkinsonian symptoms with resultant improvement of the quality of life and decreased excess mortality due to the disease are available (YAHR 1976; MARSDEN and PARKES 1977).

There are, however, a number of shortcomings to the use of levodopa. Not only does it fail to prevent the progressive disabling course of the disease, but after long-term administration the optimal therapeutic response tends to diminish, and associated side effects increase in frequency and intensity. Particularly distressing are the fluctuating responses which more than 50% of patients experience after use of levodopa for more than 5 years, as well as a continuum of increasingly severe involuntary adventitious movements (YAHR 1976; 1984; MCDOWELL et al. 1976; FAHN 1974; MARSDEN and PARKES 1977; BERGMANN et al. 1986). These limitations have prompted a search for alternate therapeutic strategies.

The central role of the enzyme monoamine oxidase (MAO) in the catabolism of dopamine and the development of agents for its inhibition (ZELLER and BARSKY 1952) has attracted particular interest. Soon after the demonstration that Parkinson's disease was associated with striatal dopamine deficiency (EHRINGER and HORNYKIEWICZ 1960), and prior to the availability of levodopa, attempts to elevate the levels of brain dopamine with MAO inhibitors were made (BERNHEIMER et al. 1962; BARBEAU et al. 1962). The results were less than promising (YAHR and DUVOISIN 1972). Only modest control of symptoms was obtained and the cumbersome dietary restrictions required to obviate the risk of hypertensive crises (BLACKWELL 1963; HORWITZ et al. 1964) made their usefulness questionable. With the introduction of levodopa, the use of MAO inhibitors was abandoned in the treatment of Parkinson's disease since they could not be given concomitantly without the risk and consequence of excessive noradrenergic activity.

The discovery in 1968 that MAO exists in at least two forms, designated type A and type B (JOHNSTON 1968), each with substrate specificity and sensitivity to inhibition by selected agents, renewed interest in their use. Clorgyline was found to inhibit the A-type enzyme selectively, with substrate preference for norepinephrine and serotonin. L-Deprenyl, (−)-isopropylmethylpropargylamine HCl, which had been developed in 1964, had the obverse effect of inhibiting the B-type enzyme (KNOLL and MAGYAR 1972; YOUDIM et al. 1972) whose substrate preference was for β-phenylethylamine and benzylamine. Dopamine, tyramine, and tryptamine were equally good substrates for both forms of the enzyme. This finding was particularly fortuitous, with considerable therapeutic potential in treating Parkinson's disease. First, the striatum of humans and nonhuman primates primarily utilizes MAO-B for the catabolism of dopamine (GLOVER et al. 1977). Second, L-deprenyl could be administered concomitantly with levodopa without the threat of inducing a hypertensive crisis. Third, dietary restrictions were not necessary since MAO in the intestinal tract is of the A variety, and would not be inhibited, allowing for tyramine to undergo oxidative deamination and avoiding the "cheese effect" (ELSWORTH et al. 1978; SANDLER et al. 1978). These pharmacologic properties of L-deprenyl offered the opportunity of safely producing a more modulated and prolonged effect of striatal dopamine derived from exogenously administered levodopa.

B. Pharmacology

L-Deprenyl is a potent, irreversible inhibitor of MAO-B. This selectivity of MAO-B is dose dependent, occurring when deprenyl is administered orally in low dosage of less than 10 mg/day. At higher dosages, it can inhibit both forms of the enzyme. It has been classified as a "suicide inhibitor" because an irreversible inhibition is produced by the action of MAO on L-deprenyl, which acts as a substrate, and it binds irreversibly and stoichiometrically to MAO at its active center (YOUDIM 1978). Deprenyl is active when given intravenously as well as orally and is not associated with the "cheese effect". MAO in the gut wall is of the A type and, hence, is not inhibited by deprenyl, thus controlling tyramine access to the circulation. Bypassing the gut with intravenous infusion of tyramine, however, does not produce a significant increase in blood pressure in animals pretreated with L-deprenyl (SANDLER et al. 1980). The fact that deprenyl inhibits the uptake of tyramine (KNOLL et al. 1968) may explain these findings.

The action of deprenyl as an antiparkinsonian agent derives from a number of pharmacologic effects (KNOLL 1978). In addition to modulating the catabolism of dopamine by its inhibition of MAO-B, it inhibits the neuronal reuptake mechanism of monoamines (KNOLL et al. 1968, 1972). Since there is a substantial conversion of L-deprenyl to amphetamine and methylamphetamine (REYNOLDS et al. 1978), which are capable of releasing dopamine from presynaptic storage sites (MOORE 1977), this effect may also play a role. An inhibitory effect of L-deprenyl on the release of acetylcholine from striatal interneurons has also been demonstrated (KNOLL 1978). All of these pharmacologic actions are most desirable in restoring normal neurotransmitter relationships in the parkinsonian state.

Human platelet MAO which is solely of the B type, provides a readily accessible way of determining pharmacologic effect (MURPHY et al. 1976), and, indirectly, bioavailability as well as compliance by patients taking deprenyl. It has been shown that complete inhibition of the platelet enzyme is achieved within 2 h by a dose of L-deprenyl as low as 2-5 mg (SANDLER et al. 1978). Patients on long-term L-deprenyl show almost complete inhibition of platelet MAO while the drug is being administered.

C. Clinical Trials

I. Combined Use of Deprenyl with Levodopa

Most clinical trials, reported to date, have assessed the therapeutic efficacy of L-deprenyl as an adjunctive agent to levodopa (Table 1). In the main, it has been administered to those on long-term levodopa therapy whose optimal control of symptoms has diminished or become complicated by fluctuating responses, i.e., end-of-dose or random "on-off" episodes. In most studies, L-deprenyl, 5 mg twice a day, has been added to an existing regimen of levodopa and a peripheral decarboxylase inhibitor (Sinemet or Madopar). Subsequently, the daily dosage of the latter has been reduced in most patients as much as 30%.

BIRKMAYER et al. (1975) first reported that the addition of L-deprenyl 10 mg/day to a combination of levodopa and benserazide resulted in improvement in akinesia and overall functional capacity as well as elimination of fluctuations. More detailed study by other investigators (LEES et al. 1977; YAHR 1978; RINNE et al. 1978; SCHACHTER et al. 1980) indicated that the most responsive fluctuations were those occurring at the end of a dosing interval, whereas the more random "off" periods were less so. Most failed to find a beneficial effect on akinetic phenomena per se, beyond that which could be obtained from an optimal dose of levodopa alone. In double-blind studies, differing results were obtained; a beneficial response was found by two groups (LEES et al. 1977; SCHACHTER et al. 1980) while a third group could not de-

Table 1. Clinical trials with L-deprenyl

Reference	number of patients	Study design	Daily dosage (mg)	Results
BIRKMAYER et al. 1975	44	Open	5-10	Significant reduction in akinesia
BIRKMAYER et al. 1977	223		5-10	35% improvement in functional capacity
LEES et al. 1977	41	DBCO	10	19 dose-related "on-off," 12 improved 8 random "on-off," 2 improved 29 early morning/nocturnal akinesi 18 improved 14 freezing, 0 improved 5 had deprenyl without L-dopa, no benefit
YAHR et al. 1978	35	Open	10	22 dose-related "on-off," 17 improved 7 random "on-off," 4 improved
YAHR et al. 1983	79	Open	10	47 dose-related "on-off," 28% increas "on" time 22 random "on-off," 17% increase "on" time
STERN et al. 1978	85	Open	10	39 dose-related "on-off," 19 improved 10 random "on-off," 1 improved
RINNE et al. 1978	47	Open	5-10	19 dose-related "on-off," 13 improved 9 random "on-off," 7 improved 12 early morning/nocturnal akinesi 5 improved 20 "freezing," 11 improved
CSANDA et al. 1978	152	DB	5-10	102 improved
CSANDA et al. 1983	53	Open	10	9 dose-related "on-off," 6 improved 18 had deprenyl without L-dopa, 12 slight benefit
SCHACHTER et al. 1980	17	DBCO	10	10 dose-related "on-off," 5 improved 7 random "on-off," 2 improved
GOLDSTEIN 1980	5	DBCO	10	2 improved
EISLER et al. 1981	9	DB	10	4 improved
GERSTENBRAND et al. 1983	48	Open	10-15	24 dose-related "on-off," 15 improved 4 random "on-off," 0 improved
PRESTHUS and HAJBA 1983	40	DB	5	18 dose-related "on-off," 9 improved 2 random "on-off," 1 improved 10 early morning akinesia, 10 improved
GIOVANNINI et al. 1985	15	Open	10	4 "on-off," 0 improved

DB double-blind; DBCO double blind crossover

monstrate such a response (EISLER et al. 1981). However, patient selection and the small numbers involved may have introduced bias in the negative study.

With few exceptions, the results of these studies support the concept that L-deprenyl enhances and prolongs the duration of action of levodopa. Not only is such potentiation evident in its beneficial effects, but in its adverse reactions as well. A substantial number of patients experience an increase in abnormal involuntary movements, others develop behavioral changes, such as confusion and hallucinations, while still others show disturbances of sleep patterns. Reducing the daily dosage of levodopa diminishes the incidence and severity of these untoward responses, but not infrequently at the expense of failing to achieve the full beneficial effects of the combined use of L-deprenyl with levodopa. It should be noted that only few side effects attributed to L-deprenyl itself have been encountered. Only isolated instances of elevated blood pressure have been reported. However, activation of preexisting gastric ulcers has occurred (YAHR 1978), presumably owing to stimulation of gastric histamine and its release. No hematopoietic, hepatic, cardiac, pulmonary, or renal disturbances have been encountered. Transitory elevations of alkaline phosphatase do occur, but are not associated with any evidence of disease.

Only a few reports are available concerning the results of long-term administration of L-deprenyl combined with levodopa. The most extensive is that of BIRKMAYER (1985) which covers a 9-year period in which patients treated with combined therapy were compared with those on levodopa alone. These investigators report that the addition of L-deprenyl not only allows patients to recoup their optimal levodopa effect, but that a significant increase in life expectancy occurs. The authors interpret these findings as being indicative of L-deprenyl's ability to retard or prevent degeneration of the nigrostriatal system (see Sect. C.II). It is indeed a most provocative conclusion. However, it must be viewed from the point of view of a retrospective study whose design was of the nonrandomized open-label and uncontrolled type. Further, it contrasts with others who have reported that initial therapeutic benefits of L-deprenyl tend to diminish after its continuous usage (YAHR et al. 1983; RINNE 1983). In the author's experience covering an 8-year period, the optimal benefits of L-deprenyl are seen in its first years of use. Indeed, the most effective response, namely the amelioration of dose-related on-off periods, tends to recur after 2 or 3 years, and there is an overall deterioration in functional capacity. Whether this represents progression of the disease process or a change in pharmacokinetics of monoamines owing to long-term inhibition of MAO cannot be determined at present. A promising approach of initiating therapy with a combination of L-deprenyl and levodopa has been suggested, but not put to the test as yet. Such would have the advantage of utilizing lower doses of levodopa, a more modulated catabolism of levodopa, and hence less potential for developing the long-term "dopa" difficulties.

II. L-Deprenyl as Monotherapy

In a few instances, investigators have administered L-deprenyl as primary therapy to naive patients in the early stage of Parkinson's disease (LEES et al. 1977; CSANDA and TARCZY 1983). Only a modest degree of control of target symptoms has been noted and most have required the use of levodopa. These data indicate that the administration of L-deprenyl alone is not effective for the control of parkinsonian symptoms. To achieve such requires the availability of an exogenously administered source of levodopa. However, the potential of L-deprenyl to halt or retard progression of the disease process was not addressed in these studies. In this regard, the following theoretical considerations are of interest.

Several lines of evidence suggest that MAO inhibitors have a "protective" effect on dopaminergic neurons. First, it has been shown that MAO inhibitors, and specifically MAO-B inhibitors, prevent the toxic effects of MPTP on dopamine neurons (HEIKKILA et al. 1984; COHEN et al. 1984). MPTP is a substrate of MAO-B and is transformed to MPP^+ which is subsequently accumulated in dopaminergic and noradrenergic neurons via the axonal membrane pump mechanism. Hence, protection by MAO-B inhibitors has been attributed to its ability to inhibit MPP^+ formation. The evolving story of MPTP's ability to produce a parkinsonian state in primates (LANGSTON et al. 1983) has led to considerable speculation that Parkinson's disease is caused by a toxin of a similar nature. If there is indeed such a toxin, either environmental or endogenous, then L-deprenyl could be used effectively to limit the development of the disease process.

An alternate proposal concerning the oxidative deamination of dopamine by MAO which generates hydrogen peroxide, either directly or through the formation of free oxy radicals, and can induce neuronal damage (COHEN 1983), is also of considerable interest. It is suggested that with destruction of some portion of the nigrostriatal system, dopamine turnover is markedly augmented in the surviving elements — nigral cells and striatal dopamine terminals. It is postulated that this leads to a flux of peroxide formation and hence accelerated neuronal damage (COHEN 1983). If this hypothesis is true, MAO inhibition may be a therapeutic strategy to prevent such from occurring and to halt the progressive neuronal loss and the emergence of further symptoms of Parkinson's disease.

D. Summary

L-Deprenyl is a potent, well-tolerated, and safe inhibitor of MAO-B. Administration in daily dosage of 10 mg produces almost complete inhibition of the enzyme. Clinical trials of the use of L-deprenyl in Parkinson's disease have shown the following. L-deprenyl as monotherapy in Parkinson's disease does not control its symptoms. In those on a therapeutic regimen containing levodopa and experiencing fluctuating responses, particularly the "wearing-off" type, the addition of L-deprenyl results in their attenuation or control. It is not fully agreed by all investigators whether such an effect is enduring or begins to wane after 2 or 3 years, nor that other symptoms of parkinsonism are improved. One investigator has reported that the combined use of these agents has resulted in an increase in life expectancy in Parkinson's disease. They have suggested that these findings indicate that L-deprenyl may be capable of preventing degeneration of the nigrostriatal system and halting progression of the Parkinson's disease process. This has raised the issue of initiating treatment with L-deprenyl during the early phases of Parkinson's disease. At present, insufficient data are available to determine the therapeutic value of such a regimen.

Acknowledgements. This work was supported in part by NIH Grant#NS-11631 and R-R 71 from the Division of Research Resources of the NIH and the International Federation of Parkinson's Societies. Requests for reprints should be addressed to: Dr. M. D. Yahr, Mount Sinai Medical Center, 1 Gustave Levy Place, New York, NY 10029, USA.

References

Barbeau A, Duchastel Y (1962) Tranylcypromine and the extrapyramidal syndrome. Can J Psychiatry 7:91-95

Bergmann KJ, Mendoza MR, Yahr MD (1986) Parkinson's disease and long term levodopa therapy. In: Yahr MD, Bergmann K (eds) Advances in neurology, vol 45. Raven, New York p 463

Bernheimer V, Birkmayer W, Horneykiewicz O (1962) Verhalten der Monoaminooxydase im Gehirn des Menschen nach Therapie mit Monoaminooxydase-Hämmern. Wien Klin Wochenschr 74:558-559

Birkmayer W, Riederer P, Youdim MBH, Linauer W (1975) The potentiation of the anti-akinetic effect after L-dopa treatment by an inhibitor of MAO-B, deprenyl. J Neural Transm 36:303-326

Birkmayer W, Riederer P, Ambrozi L, Youdim MBH (1977) Implications of combined treatment with Madopar and L-deprenyl in Parkinson's disease. Lancet 2:439-443

Birkmayer W, Knoll J, Riederer P, Youdim MBH, Hars V, Marton J (1985) Increased life expectancy resulting from addition of L-deprenyl to Madopar treatment in Parkinson's disease: a long term study. J Neural Transm 64:113-127

Blackwell B (1963) Hypertensive crisis due to monoamine-oxidase inhibitors. Lancet 2:849-851

Cohen G (1983) The pathobiology of Parkinson's disease: biochemical aspects of dopamine neuron senescence. J Neural Transm 19:89-103

Cohen G, Pasik P, Cohen B, Leist A, Mytilineou C, Yahr MD (1984) Pargyline and deprenyl prevent the neurotoxicity of 1-methyl-4-phenyl-1,2,5,6,-tetrahydropyridine in monkeys. Eur J Pharmacol 106:209-210

Csanda E, Tarczy M (1983) Clinical evaluation of deprenyl (selegiline) in the treatment of Parkinson's disease. Acta Neurol Scand [Suppl] 95:117-122

Csanda E, Antal J, Antony M, Csananaki (1978) Experiences with L-deprenyl in parkinsonism. J Neural Transm 43:263-269

Ehringer H, Horneykiewicz O (1960) Verteilung von Noradrenalin und Dopamin im Gehirn des Menschen und ihr Verhalten bei Erkrankungen des extrapyramidalen Systems. Wien Klin Wochenschr 72:1236
Eisler T, Teravainen H, Nelson R, Krebs H et al. (1981) Deprenyl in Parkinson's disease. Neurology 31:19-23
Elsworth JD, Glover V, Reynolds GP, Sandler M, Lees AJ, Phuapradit P, Shaw KM, Stern GM, Kumar P (1978) Deprenyl administration in man: a selective monoamine oxidase B inhibitor without the "cheese effect". Psychopharmacology (Berlin) 57:33-38
Fahn S (1974) On-off phenomenon with levodopa therapy in parkinsonism. Neurology 24:431-44
Gerstenbrand F, Ransmayr G, Poewe W (1983) Deprenyl (selegeline) in combination treatment of Parkinson's disease. Acta Neurol Scand [Suppl] 95:123-126
Giovannini P, Grassi MP, Scigliano G, Piccolo I, Soliveri P, Caraceni T (1985) Deprenyl in Parkinson disease: personal experience. Ital J Neurol Sci 4:207-212
Glover V, Sandler M, Owen F, Riley GJ (1977) Dopamine is a monoamine oxidase B substrate in man. Nature 265:80-81
Goldstein L (1980) The "on-off" phenomena in Parkinson's disease. Treatment and theoretical considerations. Mt Sinai J Med NY 47:80-84
Heikkila RL, Manzio L, Cabbat FS, Duvoisin RC (1984) Protection against the dopaminergic neurotoxicity of 1-methyl-4-phenyl-1,2,5,6,-tetrahydropyridine by monoamine oxidase inhibitors. Nature 311:467-469
Horwitz D, Lovenberg W, Engelman K, Sjoerdsma A (1964) Monoamine oxidase inhibitors tyramine and cheese. JAMA 188:1108-1110
Johnston JP (1968) Some observations upon a new inhibitor of monoamine oxidase in brain tissue. Biochem Pharmacol 17:1285-1297
Knoll J (1978) The possible mechanisms of action of (−)-deprenyl in Parkinson's disease. J Neural Transm 43:177-198
Knoll J, Magyar K (1972) Some puzzling effects of monoamine oxidase inhibitors. Adv Biochem Psychopharmacol 5:393-408
Knoll J, Vizi ES, Somogyi G (1968) Phenylisopropylmethylpropinyl-amine (E-250), a monoamine oxidase inhibitor antagonizing the effects of tyramine. Arzneimittelforschung 18:109
Langston JW, Ballard P, Tetrud JW, Irwin IJ (1983) Chronic parkinsonism in humans due to a product of meperidine analog synthesis. Science 219:979-980
Lees AJ, Shaw KM, Kohout LJ, Stern GM, Elsworth JD, Sandler M, Youdim MBH (1977) Deprenyl in Parkinson's disease. Lancet 2:791-796
Marsden CD, Parkes JD (1976) "On-off" effect in patients with Parkinson's disease on chronic levodopa therapy. Lancet 1:292-296
Marsden CD, Parkes JD (1977) Success and problems of chronic levodopa therapy in Parkinson's disease. Lancet 1:345-349
McDowell FH, Sweet RD (1976) The on-off phenomenon. In: Birkmayer W, Horneykiewicz O (eds) Advances in parkinsonism. Roche, Basel, pp 603-612
Moore KE (1977) The actions of amphetamine on neurotransmitters. Biol Psychiatry 12:451-462
Murphy DL, Wright C, Buchsbaum M et al. (1976) Platelet and plasma amine oxidase activity in 680 normals: sex and age differences and stability over time. Biochem Med Metab Biol 16:254-265
Presthus J, Hajba A (1983) Deprenyl (selegiline) combined with levodopa and a decarboxylase inhibitor in the treatment of Parkinson's disease. Acta Neurol Scand [Suppl] 95:127-133

Reynolds GP, Elsworth JD, Blau K, Sandler M, Lees AJ, Stern GM (1978) Deprenyl is metabolized to metamphetamine and amphetamine in man. Br J Clin Pharmacol 6:542-544

Rinne UK (1983) Deprenyl (selegiline) in the treatment of Parkinson's disease. Acta Neurol Scand [Suppl] 95:107-111

Rinne UK, Siirtola T, Sonninen V (1978) L-Deprenyl treatment of on-off phenomena in Parkinson's disease. J Neural Transm 43:253-262

Sandler M, Glover V, Ashford A, Stern GM (1978) Absence of "cheese effect" during deprenyl therapy: some recent studies. J Neural Transm 43:209-215

Sandler M, Glover V, Ashford A, Esmail A (1980) The inhibition of tyramine oxidation and the tyramine hypertensive response ("cheese effect") may be independent phenomena. J Neural Transm 48:241-247

Schacter M, Marsden CD, Parkes JD, Jenner P, Testa B (1980) Deprenyl in the management of response fluctuations in patients with Parkinson's disease on levodopa. J Neurol Neurosurg Psychiatry 43:1016-21

Stern GM, Lees AJ, Sandler M (1978) Recent observations on the clinical pharmacology of deprenyl. J Neural Transm 43:245-251

Yahr MD (1976) Evaluation of long term therapy in Parkinson's disease: mortality and therapeutic efficacy. In: Birkmayer W, Horneykiewicz O (eds) Advances in parkinsonism. Roche, Basel pp 435-444

Yahr MD (1978) Overview of present day treatment of Parkinson's disease. J Neural Transm 43:227-238

Yahr MD (1984) Limitations of long term use of antiparkinson drugs. Can J Neurol Sci 2:191-194

Yahr MD, Duvoisin R (1972) Drug therapy of parkinsonism. N Engl J Med 287:20-24

Yahr MD, Mendoza MR, Moros D, Bergmann KJ (1983) Treatment of Parkinson's disease in early and late phases. Use of pharmacological agents with special reference to deprenyl (selegiline), Acta Neurol Scand [Suppl] 95:95-102

Youdim MBH (1978) The active centers of monoamine oxidase type "A" and "B": binding with [^{14}C] clorgyline and [^{14}C] deprenyl. J Neural Transm 43:199-208

Youdim MBH, Collins GGS, Sandler M, Jones Bevan AB, Pare CMB, Nicholson WJ (1972) Human brain monoamine oxidase, multiple forms and selective inhibitors. Nature 236:225-228

Zeller AA, Barsky J (1952) In vivo inhibition of liver and brain monoamine oxidase by 1-isonicotinyl-2-isopropylhydrazine. Proc Soc Exp Biol Med 81:459

CHAPTER 17

Update on Bromocriptine in Parkinson's Disease

A. N. LIEBERMAN and M. GOLDSTEIN

A. Introduction

An important problem in the treatment of Parkinson's disease (PD) is the large number of patients who, after a good response to levodopa, fail to maintain this response and become more symptomatic. Increasing disability is usually accompanied by dyskinesias and diurnal fluctuations in performance, predominantly "wearing-off" phenomena (MARSDEN and PARKES 1977; 1976). Occasionally, patients exhibit abrupt random fluctuations: "on-off" phenomena.

The increased disability and the decreased response to levodopa may be explained as follows. In PD there is continued degeneration of the dopamine (DA) neurons in the substantia nigra and of their axonal projections in the striatum. The symptoms of PD, particularly bradykinesia, are largely due to the striatal DA deficiency and to the diminished DA neurotransmission within the striatum (BERNHEIMER et al. 1973). Levodopa acts to correct the striatal DA deficiency. Levodopa is usually combined with an extracerebral dopa decarboxylase inhibitor that facilitates levodopa's penetration into the brain (JAFFE 1973). The combination of inhibitor and levodopa permits the administration of a smaller dose of levodopa, shortens the latency from the time of initiation of levodopa to the onset of its beneficial effect, and prevents the occurrence of certain side effects such as nausea, vomiting, and orthostatic hypotension.

Two decades after its introduction, levodopa remains the cornerstone of antiparkinsonian therapy, and is still the most widely used antiparkinsonian drug. However, levodopa does not cure or arrest PD, but is rather a form of replacement therapy that can temporarily replenish the striatal DA content (HORNYKIEWICZ 1974). However, after a few years of successful levodopa treatment, there is a decline in the patient's condition with a reappearance of former symptoms or with the appearance of new symptoms (MARSDEN and PARKES 1976; 1977; LIEBERMAN et al. 1976). This deterioration may be counteracted, temporarily, by increasing the dose of levodopa. However, this increased dose may cause or amplify adverse reactions such as dyskinesias, diurnal fluctuations in performance, and mental changes.

B. Side Effects of Levodopa

Levodopa-induced dyskinesias are a common side effect of levodopa treatment, developing in up to 60% of patients. In some studies, the dyskinesias are manifested by choreoathetotic involuntary movements affecting the face, limbs, or trunk. The dyskinesias are usually dose-and-time-linked to levodopa administration and are, as a rule, maximal at the time of levodopa's peak effect. The dyskinesias may increase with time and they can become violent with marked dystonic features that may interfere with normal bodily movements. It is believed that the dyskinesias are due to the actions of DA on the supersensitive postsynaptic striatal receptors (MELAMED and HEFTI 1983; MUENTER et al. 1977). When the dyskinesias are mild, no treatment is required. When the dyskinesias are severe, they may be decreased by decreasing the total daily dose of levodopa. Unfortunately, in most patients in whom levodopa is reduced, the symptoms of PD worsen. The addition of bromocriptine (Parlodel), a DA agonist, is very helful. Bromocriptine (Fig. 1) improves bradykinesia, rigidity, and tremor but, unlike levodopa, does not cause dyskinesias. The use of bromocriptine permits a reduction in the dose of levodopa without worsening parkinsonian symptoms (LIEBERMAN et al. 1976)

Diurnal fluctuations in performance are the most incapacitating side effects associated with chronic levodopa treatment. During the first few years after starting levodopa, the beneficial effects are smooth and maintained throughout the day. Patients receive 2-4 doses of levodopa daily and cannot identify an effect of a single dose. Deterioration and reemergence of parkinsonian symptoms will occur only if treatment is stopped for at least 24 h. However, after several years of treatment, the response pattern changes and patients have to take the next dose of levodopa before the effect of the previous dose hase ended. Initially, this new situation is not too bad, and spreading the doses of levodopa more frequently throughout the day usually permits satisfactory mobility. Soon, however, the problems intensify and the patient's daily performance fluctuates between periods of mobility and immobility. The fluctuations may become more marked and complicated and occur either in a dose-related or in an unpredictable manner. The fluctuations are usually asso-

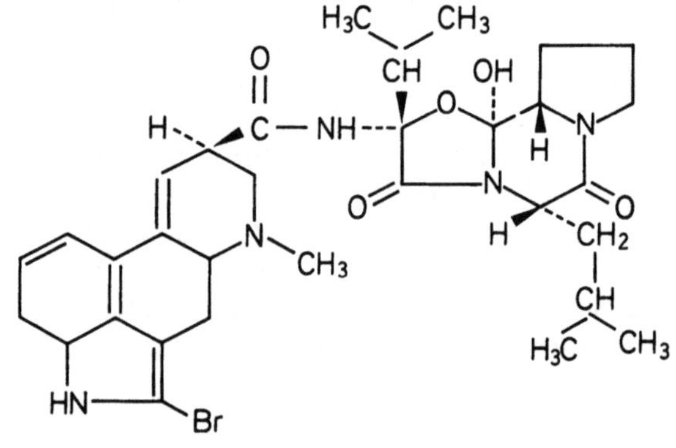

Fig. 1. Structure of bromocriptine

ciated with dyskinesias and dystonias that predominantly occur at the time of peak mobility (MELAMED and HEFTI 1983; MUENTER et al. 1977; MELAMED et al. 1980, 1983).

Levodopa-induced psychiatric reactions include confusional states, hallucinations, and psychosis. They are more frequent in demented patients. Like dyskinesias, levodopa-induced psychiatric reactions disappear when levodopa is reduced or when neuroleptic drugs such as thioridazine (Melleril) are added. However, these maneuvers usually cause worsening of the parkinsonian symptoms. Often there is no choice but to tolerate some hallucinations in order to permit mobility. The causes of these mental changes are unknown and may be related to excess DA production in the limbic structures.

C. Mechanisms for Decline in Response

Several mechanisms have been proposed to explain the declining response to levodopa and the fluctuations in performance (MELAMED and HEFTI 1983; MUENTER et al. 1977; MELAMED et al. 1980; 1983; SPENCER and WOOTEN 1984). We believe that the main reason for the declining response involves the continued degeneration of the nigral neurons whose numbers become so small that the remaining neurons cannot convert levodopa to DA in sufficient quantities to stimulate the striatal receptors (LIEBERMAN et al. 1976). As striatal DA is stored in the nerve terminals of the degenerating nigral neurons, the loss of the nigral neurons also leads to a decreased storage of DA. The decreased storage of striatal DA may be responsible, in part, for the diurnal fluctuations in performance (SPENCER and WOOTEN 1984). Another reason for the declining response to levodopa may involve a change in the sensitivity of the striatal postsynaptic DA receptors (REISINE et al. 1977; CREESE and SIBLEY 1981; DOUGAN et al. 1975). Early in PD, it is believed that the loss of the nigral neurons leads to denervation of the postsynaptic DA receptors, and that these receptors then become supersensitive. A major factor in the initial success of levodopa treatment may be the development of this postsynaptic DA receptor supersensitivity. Because they are supersensitive, the DA receptors increase their response to even the small quantities of DA that are produced by levodopa treatment. However, in time, the chronic "bombardment" of these receptors by DA may eventually downregulate the receptors and lead to a decline in the efficacy of levodopa (REISINE et al. 1977; CREESE and SIBLEY 1981; DOUGAN et al. 1975). Alternatively, overstimulation of these receptors by DA could produce a "hyperpolarization" block and, through feedback inhibition, decrease nigrostriatal DA synthesis.

Chronic levodopa treatment may also produce metabolites that could then act as false neurotransmitters and inhibit the postsynaptic receptors (DOUGAN et al. 1975). In addition, the striatal neurons that harbor the DA receptors may degenerate and cause an actual reduction in the number of receptors. New evidence indicates that problems in the absorption of orally administered levodopa may also play a role in the declining efficacy of levodopa and the associated fluctuations (NUTT et al. 1984). Several studies suggest a correlation between plasma levodopa levels and clinical responses (Muenter et al. 1977; Nutt et al. 1984). The delay in onset of a response to a single dose of levodopa may be due to slow or late absorption of the drug. In many chronically treated patients, there are episodes of total lack of response after a single dose of levodopa, associated with a complete lack of plasma elevation of levodopa (Melamed and Hefti 1983). More importantly, response fluctuations can often be abolished when the oral route is bypassed by the intravenous administration of levodopa (Quinn et al. 1984). Many factors can cause an erratic absorption of levodopa from the gut. The presence of solid food in the stomach, chronic stimulation of gastric DA receptors or anticholinergic drugs may suppress gastric motility and interfere with the dissolution of levodopa-containing tablets or capsules. Levodopa is a large, neutral, amino acid. Ingestion of heavy protein-containing meals may produce an excess of neutral amino acids in the gut which may then suppress the absorption of levodopa (Nutt et al. 1984). In addition, levodopa enters the brain by an uptake mechanism that is localized to the endothelial cells of the blood-brain barrier. This uptake mechanism is shared by the other neutral amino acids. Therefore, a heavy protein meal that produces high levels of circulating neutral amino acids may also decrease the penetration of levodopa into the brain because the neutral amino acids compete with levodopa for the common uptake mechanism. Likewise, excess formation of the major metabolite of levodopa, 3-*O*-methyldopa, may reduce the absorption of levodopa from the gut and delay its entry into the brain.

D. Treatment of Response Fluctuations: Bromocriptine

When response fluctuations, and particularly the "wearing-off" phenomena develop, levodopa may be administered in smaller doses; it may be given more frequently, and the ratio of levodopa to carbidopa may have to be changed. Indeed, many patients have to take levodopa once every 2 h. Levodopa may also have to be given during the night. The patient may require repeated readjustments of the dose and schedule of levodopa and carbidopa.

In addition, a DA agonist such as bromocriptine should be used. Bromocriptine is the most useful drug that can be added to the patient's regimen once levodopa's efficacy begins to decline (Lieberman et al. 1976; Rinne 1985; Fischer et al. 1984; Markstein and Herling 1978; Markstein 1981; Calne et al. 1978). Preferably, bromocriptine should be started early in the course of treatment before the fluctuations begin (Rinne 1985; Fischer et al. 1984). Since bromocriptine directly stimulates the DA receptors, it can bypass many of the problems that may be responsible for the failing response to levodopa, including faulty absorption from the gut, decreased entry into the brain, and erratic synthesis and storage of DA in the striatum.

In PD patients, bromocriptine can be administered in low doses (7.5-30 mg/day), in addition to levodopa. The benefits of low dose bromocriptine treatment in combination with levodopa include a reversal of the decline in function that is associated with chronic levodopa therapy, with reduction in response fluctuations and an increase in the total daily number of "on" hours.

This enables the patients to function more smoothly during the day. In some cases, bromocriptine may improve postural instability and decrease start hesitations, festinations, and short "freezing" episodes. Bromocriptine itself rarely causes dyskinesias (RINNE 1985; FISCHER et al. 1984). Therefore, the use of bromocriptine often permits a decrease in the dose of levodopa, with a reduction in levodopa-induced dyskinesias. Bromocriptine may improve early morning and "off" dystonias. In addition to its postsynaptic stimulatory effects, bromocriptine may promote the release of dopamine from the presynaptic DA terminals (MARKSTEIN and HERING 1978; MARKSTEIN 1981). This suggests that bromocriptine may be especially useful in patients with mild or moderate PD who still have a relatively large number of surviving nigrostriatal DA terminals. In these patients, when bromocriptine is used alone, it results in fewer adverse effects than levodopa, but it also results in less improvement. The best results are obtained when bromocriptine is combined with levodopa (RINNE 1985; FISCHER et al. 1984). This approach permits the use of smaller doses of levodopa, with improvement in disability equal to that achieved with higher doses of levodopa, but with fewer dyskinesias and fewer response fluctuations. The efficacy of bromocriptine may also decline during prolonged treatment. This decline in efficacy may be due to further desensitization of the postsynaptic DA receptors.

Initially, bromocriptine was used in patients with advanced PD who were no longer responding satisfactorily to levodopa (LIEBERMAN et al. 1976; CALNE et al. 1978; GLANTZ et al. 1981; KARTZINEL et al. 1976; JANSEN 1978; LEES et al. 1978; LIEBERMAN et al. 1982; RASCOL et al. 1984). In these patients, low dose bromocriptine (5-10 mg/day), when added to levodopa, had a modest effect, while high dose bromocriptine (31-100 mg/day) added to levodopa had a better effect. In our own early experience, bromocriptine was administered to 66 patients with advanced PD and increased disability despite optimal treatment with levodopa (LIEBERMAN et al. 1982). We found that 45 patients tolerated at least 25 mg/day bromocriptine. Among these 45 patients, the mean dose of bromocriptine was 47 mg, permitting a 10% reduction in the dose of levodopa. All of the patients were assessed on the New York University Disability Scale while in both their "on" and their "off" periods. (On the New York University Disability Scale a score of 0 indicates no disability and a score of 100 indicates complete disability). All of the patients also kept a 24-hour-diary that showed the number of hours they were "on". The addition of bromocriptine to levodopa resulted in a significant (36%) reduction in disability in the patients of assessed in their "on" periods (Fig. 2) and a significant (25%) reduction in disability as assessed in their "off" periods. The addition of bromocriptine also resulted in an significant (62%) increase in the number of hours the patients were "on" (Fig. 3). However, high dose bromocriptine, when added to levodopa, resulted in more adverse reactions. These included orthostatic hypotension and mental changes.

More recently, good antiparkinsonian responses have been obtained with low doses of bromocriptine, either alone or combined with levodopa, especially in patients with mild or moderate PD (RASCOL et al. 1984; CARACENI et al. 1977; FAHN et al. 1979; GRIMES and HASSAN 1983; GRON 1977; HOEHN and

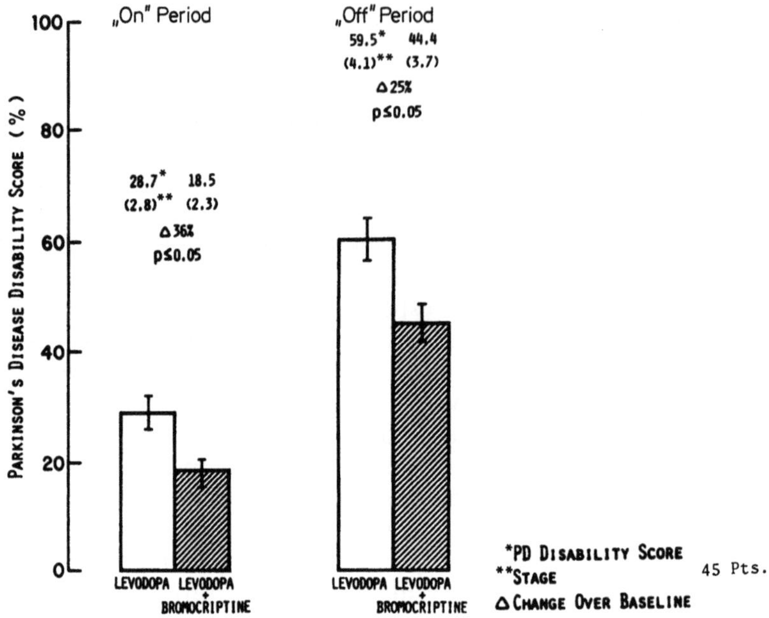

Fig. 2. Effect of levodopa, with and without bromocriptine, on Parkinson's disease disability

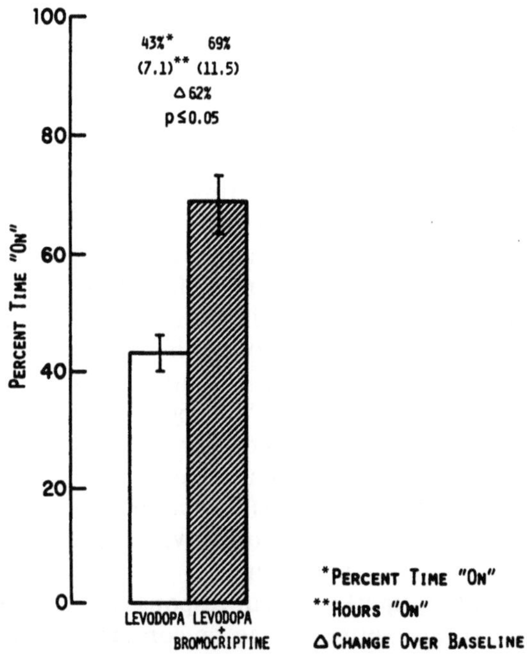

Fig. 3. Effect of levodopa, with and without bromocriptine, on time patients were "on"

ELTON 1985; PARKS et al. 1976; PFEIFER et al. 1985; TEYCHENNE et al. 1982; LARSEN et al. 1984). Adverse reactions are infrequent with low doses of bromocriptine. Initially, however, many investigators, because of their experience with high doses of bromocriptine in patients with advanced disease, were skeptical about the efficacy of the low doses of bromocriptine. Because of the questions raised about the use of bromocriptine alone or with levodopa, in a low or in a high dose regimen in patients with mild, moderate, or advanced disease, we reviewed 27 papers encompassing 790 patients. This represents most of the major studies on bromocriptine (LIEBERMAN and GOLDSTEIN 1985).

In the 27 studies, patients were selected for bromocriptine treatment for several reasons. Among patients already on levodopa, the reasons included a decreased response to levodopa, no response to levodopa, adverse effects of levodopa, or diurnal fluctuations in performance.

E. Studies of Bromocriptine Therapy

Many patients received bromocriptine as their initial treatment because of concern that chronic levodopa therapy might lead to the appearance of diurnal fluctuations in performance. The studies were divided into three groups;

1. Patients treated with bromocriptine alone, both in low doses (less than 30 mg/day) and in high doses (31-100 mg/day);
2. Patients treated with bromocriptine in low doses combined with levodopa;
3. Patients treated with bromocriptine in high doses combined with levodopa.

The first group, patients treated with bromocriptine alone, consisted mainly of patients who had never received levodopa. This group also included some patients who had been unable to tolerate levodopa (LIEBERMAN et al. 1976; RINNE 1985; KARTZINEL et al. 1976; RASCOL et al. 1984; CARACENI et al. 1977; GRIMES and HASSAN 1983; PARKES et al. 1976; TEYCHENNE et al. 1982; GERLACH 1976; GOLDWIN-AUSTEN and SMITH 1977; LEES and STERN 1981). The second group consisted mainly of patients who were on levodopa, but whose response to levodopa had decreased or who were experiencing diurnal fluctuations in performance (LIEBERMAN et al. 1982; CARACENI et al. 1977; FAHN et al. 1979; GRIMES and HASSAN 1983; GRON 1977; HOEHN and ELTON 1985; PARKES et al. 1976; PFEIFFER et al. 1985; TEYCHENNE et al. 1982). The third group consisted of patients whose response to levodopa had decreased considerably or who were experiencing marked diurnal fluctuations (CALNE et al. 1978; GLANTZ et al. 1981; KARTZINEL et al. 1976; JANSEN 1978; LEES et al. 1978; LIEBERMAN et al. 1982; RASCOL et al. 1984).

Most studies evaluated bromocriptine regardless of dose. The number of evaluable study patients listed is the number who completed the study plus the number who discontinued bromocriptine because of adverse reactions. The following parameters were analyzed for each of the groups: the mean age

Table 1. Bromocriptine in Parkinson's disease

Reference	Patients	Bromo-criptine (mg)	Age (years)	Duration of Parkinson's disease (years)
CARACENI et al. 1977	14	60	60	5.4
GERLACH 1976	20	30	65	8.0
GRIMES and HASSAN 1983	20	13		
GODWIN-AUSTEN and SMITH 1977	24	23	62	
KARTZINEL et al. 1976	12	46	57	9.1
LEES and STERN 1981	50	70	57	2.0
LIEBERMAN et al. 1976	7	70	64	9.6
PARKES et al. 1976	11	35	62	
RASCOL et al. 1984	29	56	66	5.5
RINNE 1985	24	13	64	3.8
TEYCHENNE et al. 1982	11	15	65	5.3
Mean		42	62	4.8
Total	222			

of the patients; the mean duration of PD; the number of patients who had never been treated with levodopa; the number of patients who had been treated with levodopa, but in whom levodopa was discontinued prior to or during bromocriptine treatment; the number of patients who had mild or moderate PD (stages 0, 1, 2, or 3); and the number of patients who had advanced disease (stages 4 or 5). To evaluate treatment efficacy, the number and percentage of patients in a particular study who were judged by the investigators to be improved over the baseline were noted.

With bromocriptine alone, there were 11 studies. The demographic data on these patients are summarized in Table 1. There were 222 patients of whom

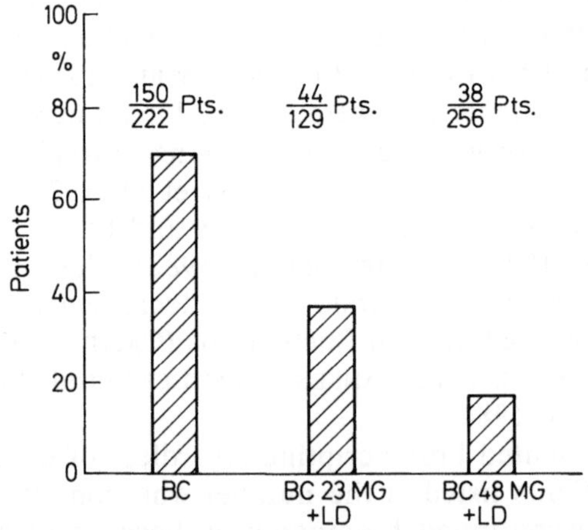

Fig. 4. Bromocriptine treatment of patients with mild to moderate Parkinson's disease

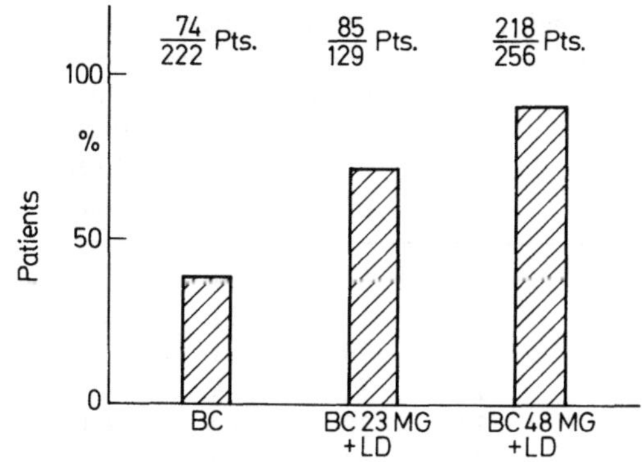

Fig. 5. Bromocriptine treatment of patients with advanced Parkinson's disease

58 % were evaluated double-blind. Among these patients, 150 (67 %) had mild or moderate PD (Fig. 4) while 72 patients (32 %) had advanced PD (Fig. 5). With low dose bromocriptine and levodopa, ther were 9 studies. The demographic data on these patients are summarized in Table 2. There were 201 patients of whom 54 % were evaluated double-blind. Among these patients, 44 patients (34 % of those on whom there were data), had mild or moderate disease; 66 % had advanced disease. With high dose bromocriptine and levodopa, there were 7 studies. The demographic data on these patients are summarized in Table 3. There were 367 patients, among whom, 38 had mild or moderate PD. This is 14 % of the patients on whom there was data on the stage of their disease (Fig. 4). It was found that 86 % of the patients had advanced PD (Fig. 5).

Among the 222 patients treated with bromocriptine alone, 134 patients (60 %) improved (Fig. 6) while 55 patients (27 %) experienced adverse effects

Table 2. Bromocriptine (low dose) + levodopa in Parkinson's disease

Reference	Patients	Bromocriptine (mg)	Age (years)	Duration of Parkinson's disease (years)	Duration of levodopa (years)	Levodopa (mg)
Carceni et al. 1977	12	25	60	5.2	3.0	700
Fahn et al. 1979	53	23	61	9.2	5.8	640
Grimes and Hassan 1983	37	24	67	9.0	5.0	
Gron 1977	15	22	64	8.5	4.0	590
Hoehn and Elton 1985	18	20				
Lieberman et al. 1982	11	26	62	8.9		
Parkes et al. 1976	20	26	62	12.9	3.9	
Pfeiffer et al. 1985	21	29	65			
Teychenne et al. 1982	14	12	70	8.5	4.5	620
Mean		23	63	8.8	4.5	
Total	201					

Table 3. Bromocriptine (high dose) + levodopa in Parkinson's disease

Reference	Patients	Bromo-criptine (mg)	Age (years)	Duration of Parkinson's disease (years)	Duration of levo-dopa (years)	Levo-dopa (mg)
Calne et al. 1978	79	53	61	10.8		700
Glantz et al. 1981	23	51	61	12.5	7.6	
Kartzinel et al. 1976	32	79	61	11.6		
Jansen 1978	12	71	59	8.8	4.6	590
Lees et al. 1978	33	30	64	10.0	10.0	
Lieberman et al. 1982	106	41	62	11.0	6.3	1150
Rascol et al. 1984	82	40	61	9.4	4.9	
Mean		48	62	10.7	6.5	
Total	367					

and discontinued treatment (Fig. 7). Among the 367 patients treated with high dose bromocriptine and levodopa, 212 patients (58%) improved (Fig. 6) while 32% experienced adverse effects and discontinued treatment (Fig. 7).

The results of treatment with low dose bromocriptine plus levodopa in patients with diurnal fluctuations in performance are as follows. In 51 of 210 patients (25%), mention is made of the presence of diurnal fluctuations. Of the 51 patient's 42 (82%) were improved during bromocriptine treatment. The results of treatment with high dose bromocriptine plus levodopa in patients with diurnal fluctuations are as follows. In 124 of 367 patients (34%), specific mention is made of the presence of diurnal fluctuations; 78 of the 124 patients (63%) improved.

The adverse effects experienced with bromocriptine consisted mainly of mental changes and were more likely to occur in patients with an underlying

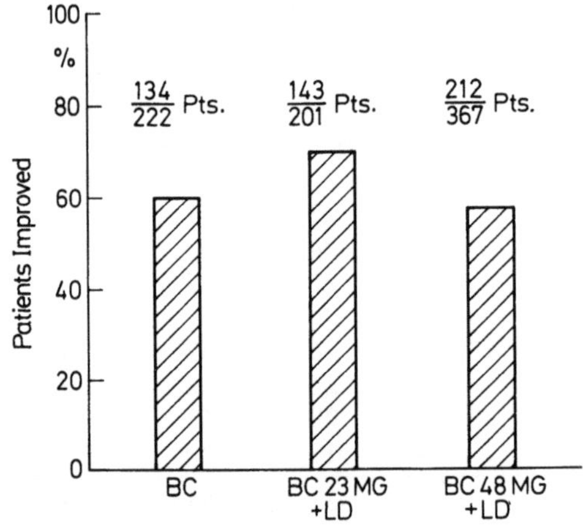

Fig. 6. Efficacy of bromocriptine treatment in Parkinson's disease

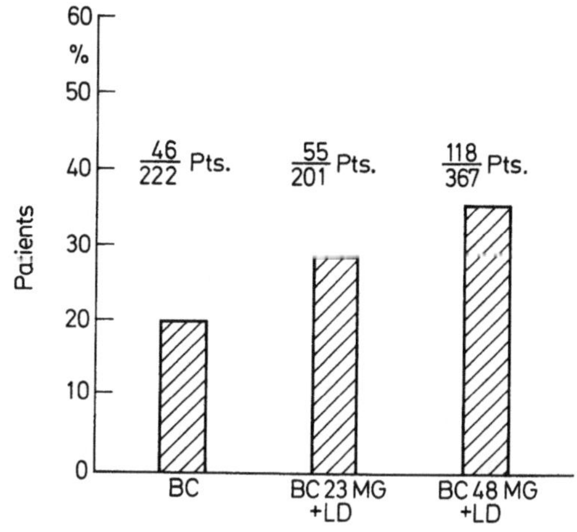

Fig. 7. Adverse effects of bromocriptine treatment in Parkinson's disease

dementia. The next most frequent adverse effects were orthostatic hypotension and nausea. A small number of patients experienced vasospastic phenomena (Raynaud's phenomenon). All adverse effects were reversible. Among 790 patients treated with bromocriptine, 217 (28 %) experienced adverse effects. The fewest adverse effects occurred on bromocriptine alone (20 %). The most adverse effects occurred with high dose bromocriptine and levodopa (32 %). This suggests that many of the adverse effects are related to the high dose of bromocriptine and the coadministration of levodopa. However, patients treated with high dose bromocriptine and levodopa also had more advanced disease. It is possible that these patients are more susceptible to adverse effects.

These studies demonstrate that the dose of bromocriptine is based on the patient's response to treatment. In analyzing the 222 patients treated with bromocriptine alone, the group could be divided into 79 patients (36 %) who were treated with low dose bromocriptine (mean dose 16 mg/day) and 143 patients (64 %) who were treated with high dose bromocriptine (mean dose 56 mg/day). Both of these subgroups consisted mainly of patients wiht mild to moderate PD. In these patients, low dose bromocriptine (16 mg/day) resulted in improvement in 58 %. Only 9 % of the patients experienced adverse effects. In these patients, high dose bromocriptine (56 mg/day) resulted in improvement in 62 % while 27 % of the patients experienced adverse effects. Thus, in patients with mild to moderate PD who have not been treated with levodopa, or in patients who cannot tolerate levodopa, bromocriptine is an effective antiparkinsonian drug.

In analyzing the 201 patients treated treated with low dose bromocriptine (mean dose 23 mg/day) combined with levodopa (see Table 2), 143 of 201 patients (71 %) improved (see Fig. 4). Although this group included a higher percentage of patients with advanced disease (66 %) than the patients treated with bromocriptine alone, more of these patients improved. This suggests that bromocriptine in low doses may have a synergistic effect when combined with levodopa. It ist possible that, if bromocriptine in low doses were combined with levodopa at an earlier time, when there were fewer patients with advanced disease, an even higher percentage of patients would improve.

In analyzing the 376 patients treated with high dose bromocriptine (mean dose 48 mg/day) combined with levodopa (see Table 3) 212 patients (58 %) improved (see Fig. 4). Although this group included the highest percentage of patients with advanced disease, nonetheless the percentage of patients improving (58 %) was comparable to the percentage of patients improving on bromocriptine alone (60 %). This again suggests that bromocriptine and levodopa may have a synergistic effect.

A reasonable question to ask is whether bromocriptine should be used as the initial treatment for PD. If it is true that the duration of the optimal response to levodopa before diurnal fluctuations in performance appears is a function of the duration of levodopa treatment (regardless of disease severity), then bromocriptine should be considered as a first treatment. Indeed, bromocriptine has been successfully used as a first treatment for PD. Patients may be started on low doses, but eventually many require higher doses, and eventually all require levodopa. Although the results of treatment with bromocriptine alone are not as good as treatment with levodopa, patients treated with bromocriptine alone do not develop diurnal fluctuations in performance or dyskinesias. The best results in terms of a good antiparkinsonian effect with few diurnal fluctuations or dyskinesias are achieved by combining bromocriptine and levodopa. This is emphasized by two recent studies comparing bromocriptine-levodopa with bromocriptine alone and levodopa alone in early PD. In one study, 76 previously untreated patients were placed on bromocriptine (MELAMED and HEFTI 1983; MUENTER et al. 1977). They were compared with 42 patients who were treated with a combination of bromocriptine and levodopa. These patients were then compared with 217 patients who had been treated with levodopa. Of the 76 patients, 21 (28 %) had a good antiparkinsonian effect and maintained it for 3 years. The mean dose of bromocriptine was 28 mg. After 3 years, the improvement declined. None of these patients developed diurnal fluctuations or dyskinesias. In 44 patients (58 %), levodopa had to be added within the first year because of an insufficient response to bromocriptine alone. Among the patients treated with bromocriptine and levodopa, the mean daily dose of bromocriptine was 16.6 mg and the mean daily dose of levodopa was 660 mg, which is lower than the dose in patients treated with levodopa alone. In this group, combined treatment resulted in an antiparkinsonian effect comparable to a higher dose of levodopa (without bromocriptine), but with fewer fluctuations and dyskinesias.

F. Pharmacology of Bromocriptine

In all of these studies, the best results regardless of disease severity were achieved with a combination of bromocriptine and levodopa. There are, thus, reasons to believe that bromocriptine and levodopa have a synergistic antiparkinsonian effect. In order to understand why this may be so it is pertinent to review the pharmacology of bromocriptine.

Bromocriptine is an ergot derivative and has some pharmacologic properties which are different from those of other ergots and of classical DA agonists such as apomorphine (MARKSTEIN and HERLING 1978; MARKSTEIN 1981). The findings that the antiparkinsonian activity of bromocriptine is diminished by pretreatment with the dopa synthesis inhibitor *a*-methyl-*para*-tyrosine suggests that a few intact nigrostriatal DA neurons are required for bromocriptine to act (MARKSTEIN and HERLING 1978; MARKSTEIN 1981; GOLDSTEIN et al. 1978; SCHRAN et al. 1980). The requirement for some presynaptic DA neurons and their nerve terminals for bromocriptine to act may explain why bromocriptine in low doses is effective in patients with mild to moderate PD where there is relative preservation of these neurons and their terminals. In patients with advanced disease with few nigral neurons, bromocriptine may act only on the postsynaptic striatal DA receptors and a higher dose may be required to stimulate these receptors. Bromocriptine stimulates the D_2 DA receptors (MARKSTEIN and HERLING 1978; MARKSTEIN 1981; GOLDSTEIN et al. 1978; SCHRAN et al. 1980; SCHWARTZ et al. 1978; FUXE et al. 1981). At these D_2 DA receptors, bromocriptine, unlike "classical" agonists such as apomorphine, has mixed agonist-antagonist properties (GOLDSTEIN et al. 1978; SCHRAN et al. 1980; SCHWARTZ et al. 1978; FUXE et al. 1981).

Binding experiments with the labeled antagonist spiperone support the concept that the D_2 and striatal DA receptors exist in both a high and a low affinity state. These two forms of the receptor correspond to discrete molecular entities (Figs. 8, 9) (SCHWARTZ et al. 1978; FUXE et al. 1981; GOLDSTEIN et al. 1985). Bromocriptine binds to the D_2 receptor, and displacement studies with radiolabeled spiperone suggest that bromocriptine, after binding to the D_2 receptor, is unable to induce a conformational change of the receptor from the low to the high affinity state.

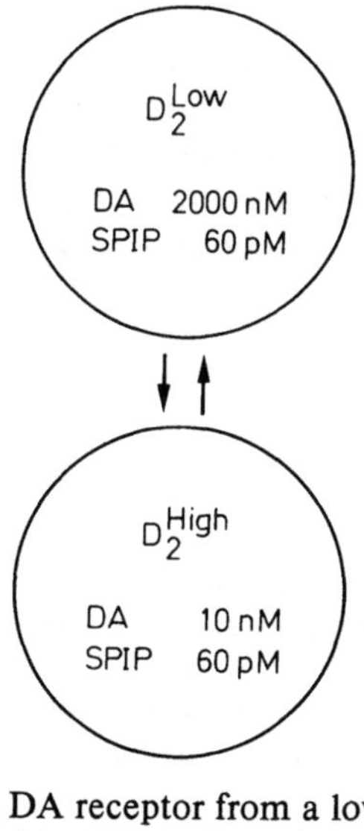

Fig. 8. Interconversion of the D_2 DA receptor from a low to a high affinity state. In the high affinity state, the concentration of DA necessary to activate the receptor is considerably lower than in the low affinity state

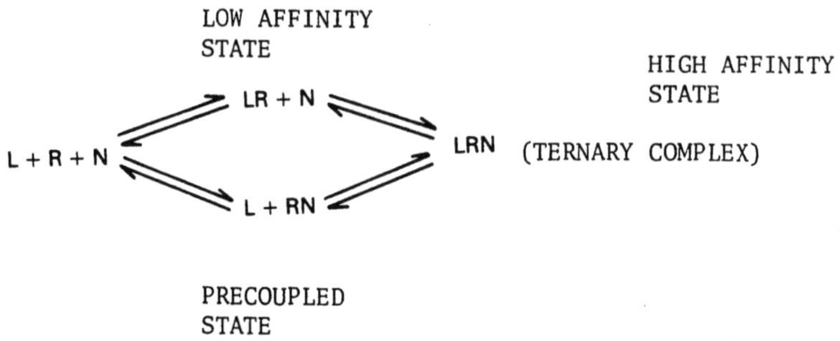

Fig. 9. Scheme for the interaction of an agonist ligand L, such as DA or bromocriptine with the D_2 striatal DA receptor and the nucleotide binding protein N that is negatively linked to the enzyme adenylate cyclase

In PD patients, who have low levels of endogenous DA, the DA formed from levodopa interacts with the D_2 striatal DA receptor (which is in a low affinity state) and a nucleotide binding protein to induce a conformational change in the receptor which leads to the conversion of the D_2 receptor to the high affinity state (called the ternary complex). It is hypothesized that formation of the ternary complex is a necessary condition for elicitation of the biologic response of the cell to receptor activation. Computer modeling predicts that the dissociation constant for the ternary complex is dependent on the intrinsic efficacy of the agonist. Thus, dopamine will convert the D_2 receptor from the low to the high affinity state. However, once this ternary complex is formed with dopamine, the complex can readily dissociate back to the low affinity state or to another precoupled state (Fig. 9). Bromocriptine will not promote conversion of the receptor to the high affinity state. However, once the high affinity state is formed, bromocriptine will displace dopamine from the ternary complex. Since the affinity of bromocriptine is greater than that of dopamine for the receptor in the high affinity state, the ternary complex with bromocriptine will dissociate more slowly than the ternary complex with dopamine, resulting in a more sustained antiparkinsonian effect.

References

Bernheimer J, Birkmayer W, Hornykiewicz O (1973) Brain dopamine and the syndromes of Parkinson and Huntington. J Neurol Sci 20:415-455

Calne DB, Williams AC, Neophytides A (1978) Long-term treatment of parkinsonism with bromocriptine. Lancet 1:735-738

Caraceni TA, Celano I, Parati E (1977) Bromocriptine alone or associated with L-dopa plus benserazide in Parkinson's disease. J Neurol Neurosurg Psychiatry 40:1142-1146

Creese I, Sibley DR (1981) Receptor adaptations to centrally acting drugs. Ann Rev Pharmacol Toxicol 21:357-391

Dougan D, Wade D, Mearrick P (1975) Effect of L-dopa metabolites at a dopamine receptor suggest a basis for "on-off" effect in Parkinson's disease. Nature 254:70-72

Fahn S, Cote LJ, Snider SR (1979) The role of bromocriptine in the treatment of parkinsonism. Neurology 29:1077-1083

Fischer PA, Przuntek H, Majer M (1984) Combined treatment of early states of Parkinson's syndrome with bromocriptine and levodopa. Dtsch Med Wochenschr 109 (34):1279-1283

Fuxe K, Agnati LF, Kohler C (1981) Characterization of normal ans supersensitive dopamine receptors: effects of ergot drugs and neuropeptides. J Neural Transm 51:3-37

Gerlach J (1976) Effect of CB 1542 (bromo-alpha ergocryptine) on paralysis agitans compared with Madopar in a double blind crossover trial. Acta Neurol Scand 53:189-219

Glantz R, Goetz CJ, Nausieda P (1981) Effect of bromocriptine on the on-off phenomena. J Neural Transm 52:41-47

Godwin-Austen RB, Smith NJ (1977) Comparison of the effects of bromocriptine and levodopa in Parkinson's disease. J Neurol Neurosurg Psychiatry 44:4479-4482

Goldstein M, Lew JY, Makamura S (1978) Dopaminephilic properties of ergot alkaloids. Fed Proc 37:2202-2205

Goldstein M, Lieberman A, Meller E (1985) A possible molecular mechanism for the antiparkinsonian action of bromocriptine in combination with levodopa. Trends Pharm Sci 6:436-437

Grimes JD, Hassan MN (1983) Bromocriptine in the long term management of advanced Parkinson's disease. Can J Neurol Sci 10:86-90

Gron U (1977) Bromocriptine versus placebo in levodopa treated patients with Parkinson's disease. Acta Neurol Scand 56:269-273

Hoehn MM, Elton RL (1985) Is low dose bromocriptine effective in parkinsonism? Neurology 35:199-206

Hornykiewicz O (1974) The mechanisms of action of L-dopa in Parkinson's disease. Life Sci 15:1249-1259

Jaffe ME (1973) Clinical studies of carbidopa and L-dopa in the treatment of Parkinson's disease. Adv Neurol 2:161-172

Jansen NH (1978) Bromocriptine in levodopa response losing parkinsonism: a double blind study. Eur Neurol 17:92-99

Kartzinel R, Teychenne P, Gillespie MM (1976) Bromocriptine and levodopa with or without carbidopa in parkinsonism. Lancet 2:272-275

Larsen TA, Newman R, Lewitt P (1984) Severty of Parkinson's disease and the dosage of bromocriptine. Neurology 34:795-797

Lees AJ, Stern GM (1981) Sustained bromocriptine therapy in previously untreated patients with Parkinson's disease. J Neurol Neurosurg Psychiatry 44:1020-1023

Lees AJ, Haddad S, Shaw KM (1978) Bromocriptine in parkinsonism a long-term study. Arch Neurol 35:503-505

Lieberman AN, Goldstein M (1985) Bromocriptine in Parkinson's disease. Pharmacol Rev 37:217-227

Lieberman AN, Kupersmith M, Estey, Goldstein M (1976) Treatment of Parkinson's disease with bromocriptine. N Engl J Med 295:1400-1404

Lieberman AN, Kupersmith M, Neophytides A (1982) Bromocriptine in Parkinson's disease: report on 106 patients treated for up to 5 years. In: Goldstein M (ed) Ergot Compounds and brain function: neuroendocrine and neuropsychiatric aspects. Raven, New York, pp 45-53

Markstein R (1981) Neurochemical effects of some ergot derivatives: a basis for their antiparkinson actions. J Neural Transm 51:39-59
Markstein R, Herrling PL (1978) The effect of bromocriptine on rat striatal adenylate cyclase and rat brain monoamine metabolism. J Neurochem 31:1163-1172
Marsden CD, Parkes JD (1976) "On-off" effects in patients with Parkinson's disease on chronic levodopa therapy. Lancet 1:292-296
Marsden CD, Parkes JD (1977) Success and problems of long-term levodopa therapy in Parkinson's disease. Lancet 1:345-349
Melamed E, Hefti F (1983) Mechanism of action of short and long term L-dopa treatment in Parkinson's disease: the role of surviving nigrostriatal dopaminergic neurons. Adv Neurol 40:149-157
Melamed E, Hefti F, Wurtman RJ (1980) Non-aminergic striatal neurons convert exogenous L-dopa to dopamine in parkinsonism. Ann Neurol 8:558-563
Melamed E, Golbus M, Friedlander E, Rosenthal J (1983) Chronic L-dopa administration decreases striatal accumulation of dopamine from exogenous L-dopa in rats with intact nigrostriatal projections Neurology 33:950-953
Muenter MD, Sharpless NS, Tyce GM (1977) Patterns of dystonia "IDI and DID" in response to L-dopa therapy for Parkinson's disease. Mayo Clin Proc 52:165-174
Nutt JG, Woodward WR, Hammerstad JP, Carter JH, Anderson AL (1984) The "on-off" phenomenon in Parkinson's disease. Relation to levodopa absorption in transport. N Engl J Med 310:488
Parkes JD, Marsden CD, Donaldson I (1976) Bromocriptine treatment in Parkinson's disease. J Neurol Neurosurg Psychiatry 39:184-193
Pfeiffer RF, Wilken K, Glaeske C (1985) Low dose bromocriptine therapy in Parkinson's disease. Arch Neurol 42:586-588
Quinn N, Parkes D, Marsden CD (1984) Control of on-off phenomenon by continuous intravenous infusion of levodopa. Neurology 34:1131-1136
Rascol A, Montastruc JL, Rascol O (1984) Should dopamine agonists be given early or late in the treatment of Parkinson's disease. J Neurol Sci 11[Suppl 1]:229-238
Reisine TD, Fields JZ, Yamamura HJ (1977) Neurotransmitter receptor alterations in Parkinson's disease. Life Sci 21:335-344
Rinne UK (1985) Combined bromocriptine-levodopa therapy early in Parkinson's disease. Neurology 35:1196-1198
Schran HF, Bhuta SI, Schwarz HJ (1980) The pharmacokinetics of bromocriptine in man. In: Goldstein M (ed) compounds and brain function: neuroendocrine and neuropsychiatric aspects. Raven, New York, pp 125-139
Schwartz R, Fuxe K, Agnati LF (1978) Effect of bromocriptine on ^{3}H-spiroperidol binding sites in rat striatum — evidence for actions of dopamine receptors not linked to adenylate cyclase. Life Sci 23:465-470
Spencer SE, Wooten GF (1984) Altered pharmacokinetics of L-dopa metabolism in rat striatum deprived of dopaminergic innervation. Neurology 34:1105-1108
Teychenne Pf, Bergsrud D, Racy A (1982) Bromocriptine: low dose therapy in parkinsonism. Neurology 32:577-583

CHAPTER 18

Pergolide in the Treatment of Parkinson's Disease

C. H. MARKHAM and S. G. DIAMOND

A. Chemistry and Pharmacology

Pergolide mesylate, a synthetic semiergoline, is a potent dopamine agonist which holds promise as a treatment for Parkinson's disease. Its chemical name is (8β)-8-[(methylthio)methyl]-6-propylergoline monomethanesulfonate; its empirical formula is $C_{20}H_{30}N_2O_3S_2$ (Fig. 1), and its molecular weight is 410.6.

Pergolide has similar actions to other ergot derivatives, in particular, lisuride and bromocriptine, but is more potent and longer acting, at least in the inhibition of prolactin secretion (KLEINBERG et al. 1980; RINNE 1981; LEMBERGER and CRABTREE 1979). It may also have a longer duration of action in the treatment of human Parkinson's disease. Pergolide is said to stimulate predominantly D_1 dopamine receptors while lisuride and bromocriptine act largely on D_2 receptors (REAVILL et al. 1981; GOLDSTEIN et al. 1980; HOROWSKI 1978). Since blockade of D_2 receptors is thought to be the mode of action of potent antipsychotic agents, it has been suggested that pergolide might have a less toxic effect on normal mentation than the D_2-stimulating bromocriptine. This view is supported by LIEBERMAN et al. (1985), although in our patients we have not noted a clear difference.

Pergolide seems to have an affinity for presynaptic and postsynaptic dopamine receptors (RABEY et al. 1981). However, its affinity for presynaptic autoreceptor sites on dopamine neurons appears to be more potent. The dose of pergolide needed to activate the presynaptic sites was less than that needed to activate the postsynaptic sites, the latter being judged by the dose of pergolide required to produce contralateral turning in rats with unilateral nigrostriatal lesions (CLEMENS and SMALSTIG 1979). Further evidence of a presynaptic ef-

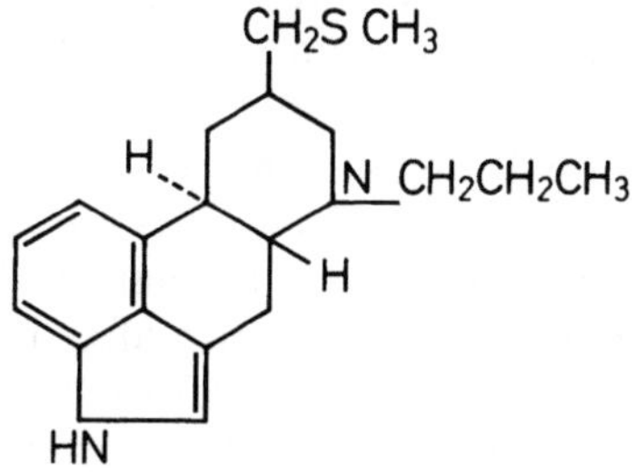

Fig. 1. Structural formula for pergolide

fect comes from the observation that pergolide blocks the enhancement of striatal dopamine turnover induced by γ-butyrolactone. The latter may produce its effect by direct action on presynaptic dopamine receptors (Roth 1979). Other work minimizes the importance of pergolide's presynaptic activity (Jiang et al. 1984). We find little firm evidence to relate either high or low dose in rats (> 0.5 mg/kg vs 0.1 mg/kg), or presumed presynaptic or postsynaptic sites of action, to pergolide's effects in humans with Parkinson's disease.

B. Clinical Studies

I. Open-Label Studies

During the years that pergolide has undergone clinical testing as an adjunct to levodopa treatment of Parkinson's disease, a substantial number of reports have been published. These open studies have reported almost unanimously favorable results in reducing disability and "off" time, and smoothing response fluctuations, although the extent and duration of the improvement varied among the reports. For the most part, these trials have focused on patients with relatively advanced disease whose response to levodopa was declining.

One of the early reports (Lees and Stern 1981) found pergolide more effective than lisuride in smoothing levodopa-induced oscillations as well as having a longer duration of action. A report on 13 patients with advanced disease (Lieberman et al. 1981) found a marked antiparkinsonian effect when pergolide was given as an adjunct to levodopa, with a significant improvement in disability and in wearing-off and on-off effects. These benefits were still apparent after 10 months. Another study found that patients taking pergolide had stable cardiac rhythms, although a few persons experienced isolated and infrequent repetitive ventricular rhythms of undetermined significance, not related to increases in premature ventricular contractions (Leibowitz et al. 1981).

The following year, a number of studies reported that pergolide had definite antiparkinsonian effects, especially beneficial in patients who were fluctuators (Ilson et al. 1982); and that it was a potent, long-lasting, useful drug when added to levodopa, having no cardiotoxicity in Parkinson's disease patients with heart disease (Tanner and Klawans 1982; Tanner et al. 1985). In one series of 56 patients with advanced disease, pergolide with levodopa decreased disability and increased "on" time. The improvement was reported to be maximal at 2 months, declining at 6 months (Lieberman et al. 1982). Another study that year found that all 23 patients enrolled in the trial improved in disability over 6 month's observation (Tanner et al. 1982); 7 patients in another center showed improvement at 1 month after initiation of pergolide, but after 6 months lost most of their initial benefits (Shoulson et al. 1982). A trial of pergolide in 26 patients with late stage Parkinson's disease reported 11 patients improved, 2 unchanged, 2 worse, and 11 unable to tolerate the medication, which was given in daily doses ranging from 0.4 to 15 mg, the latter far exceeding the typically maximum dose of 5 mg (LANG et al. 1982).

Later reports indicated that pergolide reduced "off" time in parkinsonian patients and improved dyskinesias, probably by allowing the dosage of levodopa to be reduced 48 %. Modest to no improvement in disability was noted, but fewer adverse effects were seen than on bromocriptine (ILSON et al. 1983). A report on 8 patients found that, at the end of 12 months, disability scores were improved by 30 %, mean hours "off" had declined 59 %, and choreoathetosis improved in all patients who had had this problem before taking pergolide (DIAMOND and MARKHAM 1984). Another study followed 17 patients and reported that all improved initially, but that after 2 years their disability scores were no longer significantly different from their pre-pergolide baselines (LIEBERMAN et al. 1984). Neuropsychological tests given to parkinsonian patients before and after pergolide therapy found no decline in performance after 2 years (STERN et al. 1984).

Similar results on the duration of effect were reported in a study the following year in which a 68 % improvement was recorded in 9 patients 1 month after beginning pergolide, steadily declining and disappearing by 18 months (KURLAND et al. 1985). In contrast, pergolide given to 10 patients with advanced Parkinson's disease who no longer received benefit from bromocriptine was reported still to show improvement at the end of 5 years, compared with their pre-pergolide baselines (GOETZ et al. 1985).

A study of 35 patients found that pergolide improved disability and increased "on" time for 2-50 months, but proposed that dosage as high as 8 mg/day might be necessary to maintain benefit as the disease progressed (SAGE and DUVOISIN 1985a). In yet another investigation, 18 patients who began pergolide 28 months earlier had disability scores still decreased by 42 % and "on" time still increased by 63 % (JANKOVIC 1985).

Further efforts to compare pergolide with other dopamine agonists included 56 parkinsonian patients with advanced disease who were given pergolide, and 63 who were treated with lisuride. It was found that 73 % of those on pergolide improved, compared with 59 % on lisuride. A somewhat greater number of adverse effects and lack of efficacy were observed for lisuride. Pergolide resulted in a greater increase of "on" hours, probably due to its longer half-life (LIEBERMAN et al. 1985). A double-blind comparison of pergolide and bromocriptine in 24 patients in a two-period crossover study found that both drugs improved disability; 11 patients elected to continue taking pergolide and 7 chose bromocriptine (LEWITT et al. 1983). Mesulergine and pergolide were examined in another study of 18 patients with advanced disease. These patients had taken pergolide with levodopa, and initially 12 had improved. By the end of 2 years, pergolide was discontinued in all 18 because of adverse effects or decreased efficacy, and mesulergine was started. Of the 18 patients, 12 improved, leading to the conclusion that declining response to one agonist does not preclude successful response to another of a different class, and the more they differ, the more likely the benefit (LIEBERMAN et al. 1986).

An 8- to 15-day study of pergolide used as monotherapy in 16 patients withdrawn from their customary levodopa reported that 13 of these showed similar clinical improvement on pergolide alone, given in a mean daily dose of 6.3 mg. The effect of a pergolide dosage lasted twice as long as a levodopa dosage. The remaining 3 patients required some levodopa to reduce their parkinsonian symptoms (MEAR et al. 1984).

Although open-label trials such as these are useful and necessary in testing efficacy and safety in a medication, they fail to control for the powerful placebo effect that a new treatment may have in a discouraged, chronically ill patient, hence conclusions drawn from these trials should be regarded as tentative.

II. Double-Blind Studies

In 1985, several studies reported the results of their participation in a multicenter double-blind investigation of pergolide as adjunctive treatment to levodopa in patients with Parkinson's disease. In one 6-month trial, 19 patients were reported to show improvement in disability with no change in neuropsychological tests at mean pergolide dose of 3.6 mg (HURTIG et al. 1985). Another report on 20 patients with wearing-off, on-off, or severe abnormal involuntary movements found that, during the 6-month trial, patients had more "on" time, smoother response, and decreased dystonia with pergolide and the concomitant decrease of levodopa dosage (HOEHN et al. 1985).

Two of the double-blind studies reported a substantial placebo effect. One of these examined 17 patients with advanced disease in a 6-month trial and found that all the patients on pergolide improved in disability, and about half of the placebo group also improved. The improvement in the placebo group was mostly confined to the first few weeks, whereas the pergolide group continued to improve (SAGE and DUVOISIN 1985b). Another study of 20 patients in a 6-month double-blind trial of pergolide given with levodopa found that both the pergolide and placebo groups improved significantly over the course of the study (DIAMOND et al. 1985). These double-blind studies were part of a collaborative multicenter investigation sponsored by Eli Lilly & Co. The 6-month double-blind period was followed by an open period in which patients could continue to receive the medication for as long as it was beneficial.

III. A Long-term Follow-up Study

In early 1981, a study of pergolide was undertaken at UCLA in patients with Parkinson's disease whose response to Sinemet was less than satisfactory. During the intervening years, a total of 27 patients received the medication in either open trials or a 6-month double-blind study followed by open trials, and some of these have continued on pergolide for 5 years or more.

The 27 patients consisted of 21 men and 6 women whose mean age on beginning pergolide was 61.6 years, and whose mean duration of disease was 11.3 years. On the UCLA Parkinson Disability Scale, their mean score was 134. Their mean score on the Hoehn-Yahr stage of disease was 2.7 (HOEHN and YAHR 1967).

Initially, patients were seen at weekly and then monthly intervals while their pergolide dosage was slowly titrated and their Sinemet dosage decreased as necessary. As dosages stabilized, patients were seen every 3-4 months. Electrocardiograms and lab studies were performed at each visit. At each visit, patients' disability was rated on the quantified UCLA Scale, in which 14 signs and symptoms observed by the physician and 7 acts of daily living reported by the patient were each assigned a factor weighted according to how much incapacity is produced by that symptom. Each item was then graded 0 for absent, 1 for mild, 2 for moderate, or 3 for severe; and was then multiplied by its weighting factor. The sum of these was the total disability score, the maximum possible score on this scale being 405. Also recorded at each visit was the stage of disease.

Disability scores on both scales may be seen in Fig. 2, which charts the patients' course over time. (Before beginning pergolide, the mean baseline scores were 134 on the UCLA Scale and 2.7 on the Hoehn-Yahr stage of disease.) Mean UCLA scores and Hoehn-Yahr stages improved steadily for about 1 year, and then began a gradual worsening so that at about 2 or 2½ years the group's disability was again at the baseline level. Very few patients were represented from months 36 and longer, hence these data points are subject to the vagaries of small samples, and too much weight should not be placed on them.

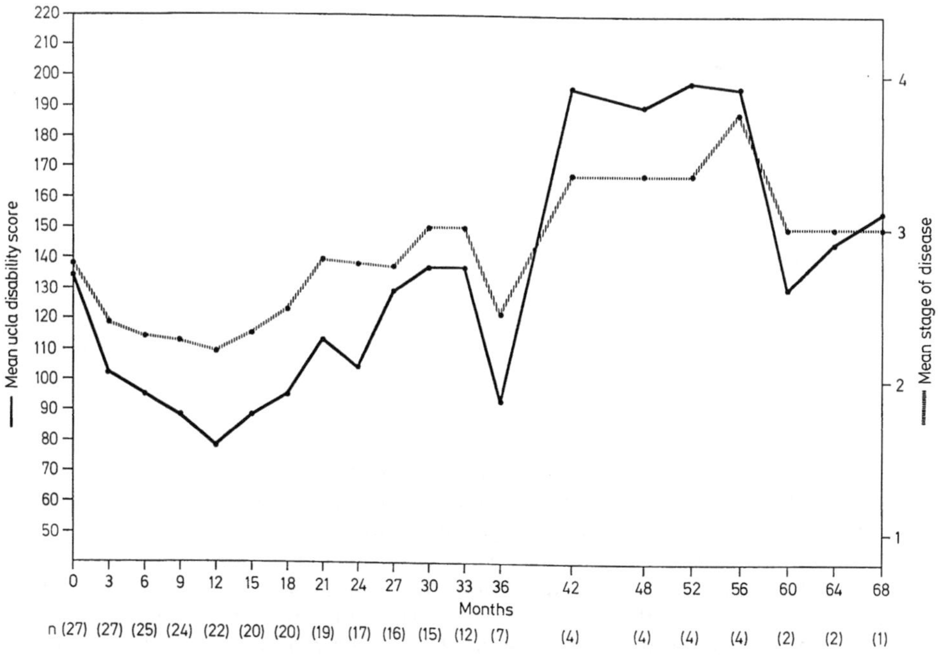

Fig. 2. Shows mean UCLA Disability Scale *(full line)* and mean Hoehn-Yahr stage of disease *(hatched line)* in patients treated with pergolide (and Sinemet) over 68 months. Note the small number of patients from 42 to 68 months

It is noteworthy that the two very different scales show such similar results. The stage of disease is a concise and useful five-point description of the broad stages through which the disease progresses. It records the major changes which mark the passage from one stage to the next, but is not designed to sense the fluctuations which occur within a given stage. As a result, it shows a somewhat blunter course, the improvement not so evident as on the UCLA Scale and, similarly, the worsening not so apparent. Nevertheless, the two scales are in agreement that on the whole the patients had a definite decrease in their disability scores lasting at least a couple of years.

Inasmuch as Parkinson's disease is a progressive disorder, it is no surprise that the patients' disability worsened over time. To say that the pre-pergolide baseline was reached once more after 2 or 2½ years in this group of patients is

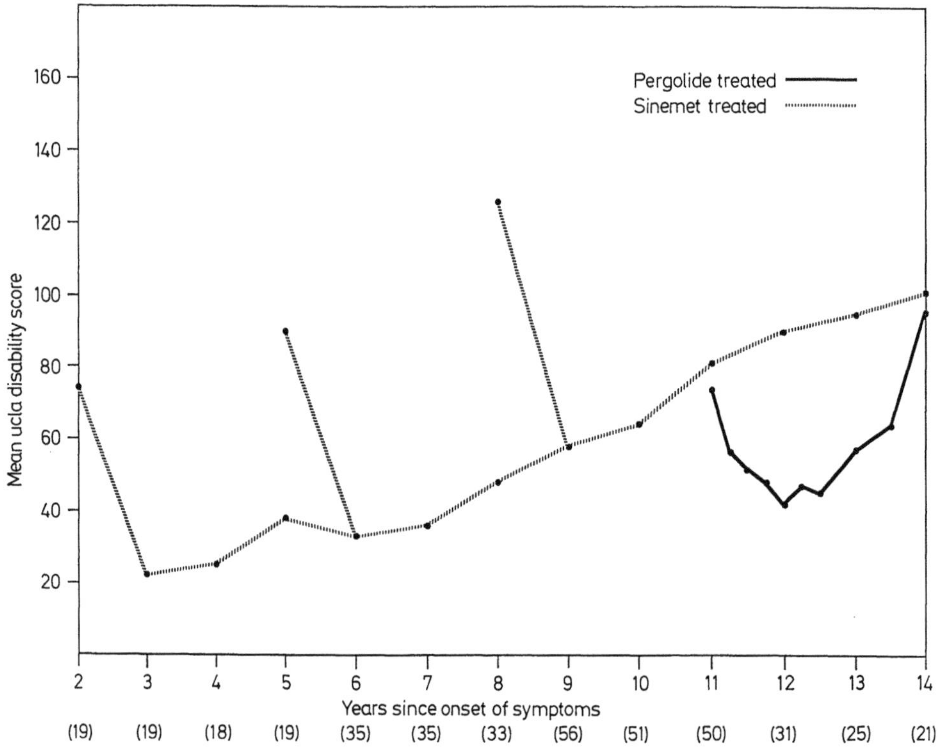

Fig. 3. The *hatched line* shows progression of symptoms in levodopa-treated patients. The *diagonal line to the left* is a group which began levodopa a mean 2 years after onset of symptoms; the *middle diagonal line*, a group beginning after a mean of 5 years; and *hatched line on the right*, a mean of 8 years. After the first year of levodopa therapy, all groups follow a common path of gradual worsening, which reaches the pre-levodopa level of disability after about 10 years of levodopa treatment (*gradually sloping hatched line*; redrawn from Markham and Diamond 1981). The *full line* shows the mean disability of the pergolide-(plus-Sinemet)-treated patients, beginning at a mean 11 years of disease symptoms and ending at 14 years. The starting and finishing disability levels fall close to the levodopa-alone curve. See text for further explanation

not to say that pergolide is not helpful for much longer periods for some individuals. The two patients who have completed more than 5 years on this medication, for example are still improved relative to their baseline scores.

To look at the effect of pergolide in a more general context, the hatched line in Fig. 3 shows the progression of symptoms in levodopa-treated patients during the 14 years following the onset of symptoms. Studies examining the longitudinal course of disability in Parkinson's disease (MARKHAM and DIAMOND 1981, 1986a, b) have found that duration of disease determines the extent of disability, and that levodopa-treated patients matched in time since onset of symptoms will tend to show remarkably similar disability scores regardless of how long they have been taking levodopa.

The pergolide patients had a mean disease duration of 11 years before beginning this medication, and their baseline disability scores appeared at the expected level on the hatched progression curve. The full line in Fig. 3 charts the course of these patients over 3 years of pergolide treatment, and during the period in which the two lines can be matched with respect to disease duration, the area between the full and the hatched lines may represent the benefit conferred when pergolide is used in conjunction with levodopa. For most patients, pergolide appears to provide significant improvement of variable interval. At this writing, it has not yet been given final approval by the US Food and Drug Administration, although the expectation is that pergolide will become available for general use in 1989, under the brand name Permax.

C. How to Administer

Pergolide has been used as monotherapy in the treatment of Parkinson's disease (MEAR et al. 1984). In 16 patients with an average Hoehn-Yahr stage of disease of 2.75, there was clear benefit for a limited period of time compared with the baseline, off-levodopa disability scale. A single dose of pergolide produced clinical improvement lasting twice as long as a single dose of levodopa plus a peripheral dopa decarboxylase inhibitor. Of these 16 patients, 3 needed a combination of levodopa and pergolide to achieve optimum effect. As a part of a double-blind study, we (DIAMOND et al. 1985) attempted to replace Sinemet with pergolide in 20 patients having a mean stage of disease of 3.27. Ultimately, all did better on the combination of the two agents. It is possible that pergolide alone may have a role in treating mild or new cases, at least for some months, as does bromocriptine (RINNE 1986).

In those cases in which pergolide is used, most will also need levodopa. When should the combination be started? Choices include early in the disease, or after some years of levodopa therapy, when the patient may have progressed to having wearing-off, on-off, and resistant fluctuations. There is little data on early, combined use of pergolide and levodopa so we may want to consider the early bromocriptine plus levodopa analogy (CALNE et al. 1974). The main reasons for starting bromocriptine and levodopa together are: (a) there may be better therapeutic response and fewer early side effects; (b) there may be advantages in having the necessarily lower oral levodopa dose to prevent dyskinesias and other adverse events; and (c) the combination may possibly reduce or lessen the late side effects such as resistant fluctuations. At the end of a 3-year trial, Rinne (1986) compared 28 patients with a mean dose of 660 mg levodopa and 16.6 mg bromocriptine with 83 patients on an average dose of 807 mg levodopa. The clinical improvement in the two groups was the same, but there were fewer freezing episodes, dystonia, and end-of-dose disturbances in the patients taking the combination therapy. One supposes that well-controlled studies with early use of pergolide plus levodopa may show similar results.

Our own experience and that of most others consists of adding pergolide to existing therapy in patients who have been on levodopa for some time and are having side effects, particularly choreoathetosis, dystonia, and wearing-off or on-off effects. Based on our open and double-blind studies, we recommend reducing the existing dose of Sinemet by about 10% for a 2-week baseline and then starting to add pergolide. The following schedule fits with our experience and is that which Eli Lilly & Co are planning to recommend.

Pergolide is typically begun with a single daily dose of 0.05 mg for 2 days. Gradually, dosage is increased by 0.1-0.15 mg/day every third day over the next 12 days. Dosage may then be increased by 0.25 mg/day every third day until optimum dosage is achieved. Levodopa should be cautiously decreased during pergolide titration. Usually, pergolide is given in divided doses, three times daily. Mean therapeutic dosage is 3.0 mg/day. Average concurrent levodopa dosage is 650 mg/day. Safety of pergolide above 5.0 mg/day has not been systematically evaluated. Pergolide (Permax) will be marketed in 0.05 mg, 0.25 mg, and 1.0 mg sizes.

As the pergolide is titrated higher, and if the levodopa has not been decreased enough, there may come a time when choreoathetoid dykinesia or occasionally an organic confusional state intervenes. Sinemet can then be decreased somewhat. In future weeks, one may try to go higher on the pergolide, and possibly lower on the Sinemet. This "juggling" of dosage has to be done cautiously, sensitively, and with full cooperation of the patient and family. Our goal has been to achieve optimum benefit of parkinsonian symptoms with minimum side effects.

Our own group of 20 patients who were taking a mean of 941 mg levodopa in Sinemet before beginning pergolide concluded a 6-month study with a mean daily dosage of 3.14 mg pergolide and 667 mg Sinemet. At the end of 48 months, 11 of those are still continuing the program with a mean of 2.64 mg/day pergolide and 634 mg/day Sinemet. These values are virtually identical to the dosages to be recommended by Eli Lilly & Co. At the end of this 4-year period, the 11 patients were continuing because they (and we) feel they are still benefiting, still have a longer interval between doses, and have somewhat fewer fluctuations than they had prior to the onset of the study.

D. Side Effects

Adverse effects following administration of pergolide resemble those seen with other antiparkinsonian medications. The most commonly experienced events are usually seen during the first days or weeks of pergolide therapy and include, in descending frequency: abnormal involuntary movements, nausea, hallucinations, orthostatic hypotension, drowsiness, confusion, lightheadedness, constipation, insomnia, depression, dizziness, and peripheral edema. Less frequently observed events are vomiting, anxiety, akinesia, dystonia, falling, tremor, nervousness, and sleep or memory disturbances. Several of these need further discussion.

Abnormal involuntary movements consist of choreoathetoid dyskinesia, peak-dose dystonia, and possibly end-of-dose dystonia, and are related to too much levodopa or pergolide. Their effect is additive. Since the intent is to increase the pergolide to 1.0–3.0 mg/day, the dyskinesia is usually dealt with by decreasing the Sinemet.

Hallucinations, confusion, and possibly lightheadedness are parts of a toxic confusional state. While pergolide can do this alone, it does appear to be more likely when pergolide is used with levodopa. It is far more likely to occur in individuals having some degree of organic dementia, the elderly, or those with cerebral atrophy. Anticholingerics or Symmetrel (amantadine hydrochloride) can also produce organic confusion. Intercurrent infection, a recent operation or other stress may also produce confusion and delirium in the elderly parkinsonian patient. Identifying the causative agent (or agents) may be difficult. If the only variable appears to be the addition of pergolide or increase in pergolide dosage, it may be judicious to reduce and maintain the dosage at the best tolerated level.

Orthostatic hypotension occurs not infrequently with the first single tablet of pergolide and not thereafter. In patients who have mild orthostatic hypotension associated with levodopa, it may be wise to have them start medication under close medical supervision. Certain individuals with Shy-Drager syndrome or a "Parkinson-plus" syndrome, may be exceedingly sensitive to pergolide or levodopa, particularly in combination.

In general, side effects can be prevented or minimized by the very slow, careful titration described in Sect. C. If side effects develop, a brief decrease in dosage or slower titration may help. Most patients develop tolerance over a reasonably short period, and should be encouraged to persevere long enough to give the medication a fair trial. Since virtually all patients receiving pergolide will be commonly taking Sinemet and possibly other antiparkinsonian drugs as well, care must be taken to avoid overdosing. Concomitant reduction in these medications as pergolide dosage is increased is usually an effective tactic in avoiding adverse reactions.

Acknowledgments. We wish to express our appreciation to Mrs. Setsuko Kashitani for secretarial work and to Leo J. Treciokas, M. D. for his care of many of the patients discussed here.

References

Calne DB, Teychenne PF, Claveria LE, Eastman R, Greenacre JK, Petrie A (1974) Bromocriptine in parkinsonism. Br Med J [Clin Res] 4:442-444

Clemens JA, Smalstig EB (1979) Effects of some new dopamine agonists on turning behavior in 6-hydroxydopamine lesioned rats. IRCS Med Sci 6:427

Diamond SG, Markham CH (1984) One-year trial of pergolide as an adjunct to Sinemet in treatment of Parkinson's disease. In: Hassler RG, Christ JF (eds) Advances in neurology. Raven, New York, pp 537-539

Diamond SG, Markham CH, Treciokas LJ (1985) Double-blind trial of pergolide for Parkinson's disease, Neurology 35:291-295

Goetz CG, Tanner CM, Glantz RH, Klawans HL (1985) Chronic agonist therapy for Parkinson's disease: a 5-year study of bromocriptine and pergolide. Neurology 35:749-751

Goldstein M, Lieberman A, Lew JY, Asano T, Rosenfeld MR, Makman MH (1980) Interaction of pergolide with central dopaminergic receptors. Proc Natl Acad Sci USA 77:3725-3728

Hoehn MM, Yahr MD (1967) Parkinsonism: onset, progression and mortality. Neurology (Minneapolis) 17:427-442

Hoehn MM, Schear MJ, Heaton A (1985) Comparison of pergolide, bromocriptine, and mesulergine in the management of complications of levodopa therapy. Neurology 35(Suppl 1):202

Horowski R (1978) Differences in the dopaminergic effects of the ergot derivatives bromocriptine, lisuride and D-LSD as compared with apomorphine. Eur J Pharmacol 51:157-166

Hurtig HI, Saykin A, Stern MB, Melvin G, Cline S, Fuller J, Gur R (1985) Pergolide improves motor function in Parkinson's disease without causing adverse mental effects: a neuropsychological study. Neurology 35(Suppl 1):202

Ilson J, Fahn S, Mayeux R, Cote LJ (1982) Pergolide treatment in parkinsonism. Neurology 32(2):181

Ilson J, Fahn S, Mayeux R, Cote LJ, Snider SR (1983) Pergolide treatment in parkinsonism. In: Fahn S, Calne DB, Shoulson I (eds) Experimental therapeutics of movement disorders, Vol 37. Raven, New York, pp 85-94

Jankovic J (1985) Long-term study of pergolide in Parkinson's disease. Neurology 35:296-299

Jiang DH, Reches A, Wagner HR, Fahn S (1984) Biochemical and behavioral evaluation of pergolide as a dopamine agonist in the rat brain. Neuropharmacology 23:295-301

Kleinberg DL, Lieberman A, Todd J, Greising J, Neophytides A, Kupersmith M (1980) Pergolide mesylate: a potent day-long inhibitor of prolactin in rhesus monkeys, and patients with Parkinson's disease. J Clin Endocrinol Metab 51:152-154

Kurland R, Miller C, Levy R, Macik B, Hamill R, Shoulson I (1985) Long-term experience with pergolide therapy of advanced parkinsonism. Neurology 35:738-742

Lang AE, Quinn N, Brincat S, Marsden CD, Parkes JD (1982) Pergolide in late-stage Parkinson disease. Ann Neurol 12:243-247

Lees AJ, Stern GM (1981) Pergolide and lisuride for levodopa-induced oscillations. Lancet 2:577

Leibowitz M, Lieberman A, Goldstein M, Neophytides A, Kupersmith M, Gopinathan G, Mehl S (1981) Cardiac effects of pergolide. Clin Pharmacol Ther 30:718-723

Lemberger L, Crabtree RE (1979) Pharmacologic effects in man of a potent, long-acting dopamine receptor agonist. Science 205:1151-1153

LeWitt PA, Ward CD, Larsen TA, Raphaelson MI, Newman RP, Foster N, Dambrosia JM, Calne DB (1983) Comparison of pergolide and bromocriptine therapy in parkinsonism. Neurology 33:1009-1014

Lieberman AN, Goldstein M, Leibowitz M, Neophytides A, Kupersmith M, Pact V, Kleinberg D (1981) Treatment of advanced Parkinson disease with pergolide. Neurology 31:675-682

Lieberman AN, Goldstein M, Gopinathan G, Leibowitz M, Neophytides A, Walker R, Hiesiger E, Nelson J (1982) Further studies with pergolide in Parkinson disease. Neurology 32:1181-1184

Lieberman AN, Goldstein M, Leibowitz M, Gopinathan G, Neophytides A, Hiesiger E, Nelson J, Walker R (1984) Long-term treatment with pergolide: decreased efficacy with time. Neurology 34:223-226

Lieberman AN, Leibowitz M, Gopinathan G, Walker R, Hiesiger E, Nelson J, Goldstein M (1985) Review: the use of pergolide and lisuride, two experimental dopamine agonists, in patients with advanced Parkinson disease. Am J Med Sci 290:102-106

Lieberman AN, Gopinathan G, Neophytides A (1986) Efficacy of pergolide and mesulergine. Eur Neurol 25:86-90

Markham CH, Diamond SG (1981) Evidence to support early levodopa therapy in Parkinson disease. Neurology 31:125-131

Markham CH, Diamond SG (1986a) Modification of Parkinson's disease by long-term levodopa. Arch Neurol 43(4):405-407

Markham CH, Diamond SG (1986b) Long-term follow up of early dopa treatment in Parkinson's disease. Ann Neurol 19:365-372

Mear JY, Barroche G, de Smet Y, Weber M, Lhermitte F, Agid Y (1984) Pergolide in the treatment of Parkinson's disease. Neurology 34:983-986

Rabey JM, Passeltiner P, Markey K, Asano T, Goldstein M (1981) Stimulation of pre- and post-synaptic dopamine receptors by an ergoline and by a partial ergoline. Brain Res 225:347-356

Reavill C, Jenner P, Marsden CD (1981) Puzzles of the mechanism of action of bromocriptine. In: Gessa EL, Corsini GU (eds) Basic pharmacology, vol 1. Apomorphine and other dopaminomimetics. Raven, New York, pp 229-239

Rinne UK (1981) Dopaminergic agonists in the treatment of Parkinson's disease. Abstracts, 12th world congress of neurology, 20-25 September, Kyoto, Japan. Excerpta Medica, Amsterdam, p 133

Rinne UK (1986) Early combination of bromocriptine and levodopa in the treatment of Parkinson's disease. In: Fahn S, Marsden CD, Jenner P, Teychenne P. (eds) Recent developments in Parkinson's disease. Raven, New York, pp 267-271

Roth RH (1979) Dopamine autoreceptors: pharmacology, function and comparison with postsynaptic dopamine receptors. Community Psychopharmacol 3:429-445

Sage JI, Duvoisin RC (1985a) Long-term efficacy of pergolide in patients with Parkinson's disease. Ann Neurol 18:137

Sage JI, Duvoisin RC (1985b) Pergolide therapy in Parkinson's disease: a double-blind, placebo-controlled study. Clin Neuropharmacol 8(3):260-265

Shoulson I, Miller C, Kurlan R, Levy R, Macik B, Hamill R (1982) Parkinsonism and on-off fluctuations: long-term effects of pergolide therapy. Ann Neurol 12:97

Stern Y, Mayeux R, Ilson J, Fahn S, Cote L (1984) Pergolide therapy for Parkinson's disease: neurobehavioral changes. Neurology 34:201-204

Tanner CM, Klawans HL (1982) Pergolide mesylate: new therapy for Parkinson disease. Ann Intern Med 96(4):522-523

Tanner CM, Goetz CG, Glantz RH, Glatt SL, Klawans HL (1982) Pergolide mesylate and idiopathic Parkinson disease. Neurology (NY) 32:1175-1179

Tanner CM, Chhablani R, Goetz CG, Klawans HL (1985) Pergolide mesylate: lack of cardiac toxicity in patients with cardiac disease. Neurology 35:918-921

CHAPTER 19

Lisuride Pharmacology and Treatment of Parkinson's Disease

G. GOPINATHAN, R. HOROWSKI and I. H. SUCHY

A. Chemistry

Lisuride hydrogen maleate [1,1 diethyl-3-(9,10-didehydro-6-methyl-8α-ergolinyl] urea is the first derivative of isolysergic acid in clinical use. It is an 8-α-aminoergoline and was synthesized by ZIKAN and SEMONSKY (1968). It is a white slightly bitter tasting substance, which is reasonably stable at room temperature if not exposed to light. Lisuride hydrogen maleate is soluble in water or saline up to 1 mg/ml. Owing to its low effective dose, it is therefore the first dopamine agonist which can also be used on a parenteral basis (DOROW et al. 1980).

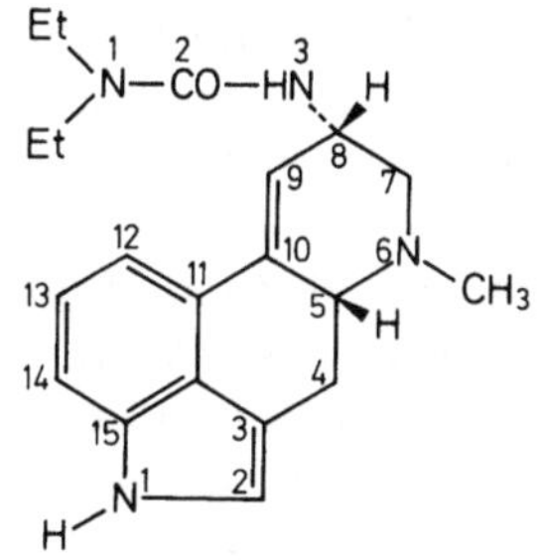

B. Toxicology

The acute LD_{50} in mice is 90 mg/kg p.o. and 14.4 mg/kg i.v. while in rats it is 10–15 mg p.o. and 1–2 mg/kg i.v. There is a large safety range if one considers that the effective oral doses in rats (e.g. reserpine antagonism and lowering of prolactin) are in the range of 0.01 mg/kg or below. In the monkey, doses up to 200 mg/kg p.o. did not kill the animals. Chronic toxicity studies in monkeys and rats did not reveal organ damage at 5 and 10 mg/kg p.o.

Carcinogenicity studies in rodents revealed a dose-dependent reduction of the incidence of pituitary and mammary tumours. More animals in the treatment groups survived the 2-year study than in the control group. There was no increased incidence of endometrial carcinomas. In a small percentage of male rats in the highest dose groups, Leydig cell hyperplasia or adenomas were observed. There was no indication for embryotoxic or teratogenic effects; only nidation and lactation in rats was impaired owing to the prolactin-lowering effect of lisuride (G. SCHUPPLER and P. GÜNZEL 1984 personal communication).

C. Pharmacokinetics

Lisuride is completely absorbed from the gut in laboratory animals as well as in young volunteers (HÜMPEL et al. 1981) and old volunteers (HÜMPEL 1983). It is in part metabolized during its first pass through the liver. The clearance rate has been found to be 0.80 ± 0.25 l/min in volunteers. In patients with Parkinson's Disease, individual clearance values derived from the estimated AUC (area under the curve) ranged from 0.39–7.13 l/min ($\bar{x}$ = 1.95 ± 1.99). There is a moderate enterohepatic circulation (BURNS and CALNE 1981; BURNS et al. 1984). Metabolism occurs in various ways via oxidative desalkylation, hydroxylation, mono-osygenation, and oxidation. No metabolite accounted for more than 10% of the high pressure liquid chromatography spectrum. It is unlikely that metabolites contribute to the pharmacological effects of lisuride. They are excreted via urine and faeces in varying proportions, mainly via urine as is the case with most ergolines (BURNS and CALNE 1981).

The proportion of lisuride that becomes systemically available varies from one individual to another more than within individuals. On the average, bioavailability has been found to be 10%–20%. The first-pass effect and the resulting bioavailability depend on individual liver plasma flow and the state of the microsomal enzyme activity in the liver. From one clinical trial a 100% increase of clearance rate during a 2 to 4 week treatment has been reported (BURNS et al. 1984). There is no indication for enzyme induction in chronic treatment (CARRUBA et al. 1985). Inhibition of liver microsomal enzyme activity by proadifen has been shown to prolong and intensify the motor action of lisuride in laboratory animals (KELLER and Da PRADA 1979).

Peak plasma levels after oral ingestion were reached at 39 ± 28 min in fasted PD patients (BURNS et al. 1984). Similar times were observed in other trials when lisuride plasma levels were determined by a specific radioimmunoassay. After the use of [^{14}C] lisuride in old volunteers it was as long as 2 h (HÜMPEL 1983). Volume of distribution has been calculated to be 2.3–2.4 l/kg which would mean that lisuride passes readily into the tissues. The time for distribution into the tissues, according to the β-phase of elimination from the plasma after i.v. injection, is 20–30 min.

Lisuride passes quickly through the blood–brain barrier into the brain where it binds to various neuronal receptors. In animal experiments the maximum level of radioactivity in the brain of the rat has been seen 15 min after i.v. injection. Brain levels of lisuride (in animals) decrease more slowly than plasma levels. From 30 min to 8 h after injection there was more radioactive material in the brain than in the plasma. Autoradiographic studies in rats with tritiated lisuride show highest levels of radioactivity 60 min after injection in caudate, putamen, parts of the thalamus and layer IV of the cortex (ratio 3–4.5:1 in corpus callosum; STUMPF et al. 1985). From experiments in rabbits it is known that [^{14}C] lisuride levels in the pituitary are ten times higher than those in putamen or other brain areas (DOROW et al. 1983). Onset of action after intravenous injection was 10 min for the motor system of PD patients (PARKES et al. 1981) and 30 min for the prolactin-lowering effect (DOROW et al. 1981).

Whole-body autoradiography has shown the labelled lisuride to accumulate in the intestine and in the kidney. This is because lisuride metabolites are excreted via the urine and via the bile. Half-life of elimination from the plasma is 1.9 ± 0.6 h in young volunteers after a test dose of 300 μg lisuride p.o. The rate of decline after i.v. injection and also after the termination of s.c. infusion is similar (KRAUSE et al. 1988). The time course of pharmacological effects does not seem to follow plasma levels (especially in the case of prolactin). Half-life of excretion of radiolabelled material via the kidney in old volunteers (in this study 90% of the material was excreted via the kidney) was 10 h (phase I) and 24 h (phase II). In chronic trials no signs of accumulation of the substance were seen (BURNS et al. 1984).

D. Biochemistry and Pharmacology

I. Receptor Binding

In the central nervous system lisuride binds to dopaminergic, serotonergic and adrenergic receptors and their subtypes, in part with outstanding affinity. The specific [^{3}H] lisuride binding to the dopaminergic striatal membranes of the rat was characterized by the maximal binding capacity $B_{max} = 490$ fmol per milligram protein and the dissociation constant $K_D = 0.5$ nM; B_{max} for [^{3}H] spiroperidol in a parallel assay was 360–390 fmol per milligram protein. (Inhibition constants were: haloperidol 7.1 nM; apomorphine 100 nM; methysergide 180 nM; dopamine 3.500 nM; FUJITA et al. 1979). The B_{max} valve of 494 fmol per milligram protein ($K_D = 0.6$ nM) for [^{3}H] lisuride in the absence of spiroperidol was reduced to 172 fmol/mg ($K_D = 0.7$ nM) in the presence of spiroperidol (BATTAGLIA and TITELER 1981). In another assay, the presence of sodium ions increased B_{max} of [^{3}H] lisuride in rat striatal homogenates from 723 to 1035 fmol/mg ($K_D = 1.8$ nM). In contrast to the findings with apomorphine, guanosine triphosphate did not influence the binding of lisuride (UZUMAKI et al. 1982).

The inhibition constant (IC_{50}) of lisuride in [^{3}H] spiroperidol binding in rat striatum was reported to be 2.1 nM (MCPHERSON and BEART 1983) and 6 nM (SUCHY et al. 1983); in [^{3}H] ADTN binding in rat caudate nucleus IC_{50} of lisuride was 3.9 nM (ROSENFELD and MAKMAN 1981). In rat pituitary tumour cells, [^{3}H] spiroperidol binding was inhibited at 0.39 ± 0.2 nM lisuride (CRONIN et al. 1981). In human putamen (crude membranes) IC_{50} was 9 nM [^{3}H] spiroperidol, $B_{max} = 17.9–27.5$ pmol/g (REYNOLDS and RIEDERER 1981). In most dopamine receptor binding studies, the affinity of lisuride was 10–20 times higher than bromocriptine and about 1000 times higher than dopamine itself.

By its action on D_2 receptors in the intermediate lobe of rat pituitaries, lisuride inhibits α-melanocyte-stimulating hormone (α-MSH) secretion at concentrations as low as 0.01 nM (Cote et al. 1983). Binding to serotonin (5-HT) receptors has been studied in the rat frontal cortex. IC_{50} of lisuride in [^{3}H] ketanserin binding was 5.0 nM, and 110 nM in [^{3}H] serotonin binding, showing a preference for 5-HT_2 receptors (SUCHY et al. 1983). In rabbit frontal cortex IC_{50} to [^{3}H] spiroperidol binding was 1.0 nM, 3.7 nM to [^{3}H] LSD binding and 14.0 nM to [^{3}H] 5-HT binding. The ability of lisuride to displace spiroperidol in the cortex is decreased in the presence of guanosine nucleotides. In rabbit caudate nucleus the IC_{50} of lisuride in [^{3}H] 5-HT binding was 7.0 nM and in [^{3}H] LSD binding 2.0 nM (ROSENFELD et al. 1980). The specific binding of lisuride in the bovine frontal cortex can be inhibited by 5-HT by 20% and by clonidine by 40% (BATTAGLIA and TITELER 1981).

Lisuride also binds to α_1- and α_2-adrenoceptors and to β-adrenoceptors. In the forebrain it displaces [^{3}H] clonidine with $IC_{50} = 8.0$ nM (SUCHY et al. 1983); [^{3}H] rauwolscine (α_2) with $IC_{50} = 0.83$ nM (MCPERSON and BEART 1983) and [^{3}H] prazosin (α_1) with $IC_{50} = 150$ nM (MCPHERSON and BEART 1983). Lisuride binds to β-adrenoceptors in the parathyroid gland (nanomolar range) and in the cerebellum (micromolar range) (COTE et al. 1983).

The interference of lisuride with the formation of cyclic adenosine monophosphate (cAMP) has remained a matter of debate until now. In the striatal dopaminergic system lisuride inhibits dopamine–(DA)-stimulated formation of cAMP ($IC_{50} = 1-1.2 \times 10^{-7}$ M; PIERI et al. 1978; SPANO et al. 1978, 1983). In homogenates of rat striatum lisuride by itself in doses up to 10^{-4} M does not stimulate adenyl cyclase. After stimulation with guanyl nucleotides lisuride (3–10 μM) induced cAMP formation, a result that was not observed with lisuride doses above 10 μM (AZUMA and OSHINO 1980). If the material is not homogenized, but cut into slices so that part of the interneuronal connections remain intact, lisuride as well as bromocriptine stimulate accumulation of cAMP (SAIANI et al. 1979). Lisuride (0.2 mg/kg) also increases cAMP formation in vivo in rat striatum from 5.9 to 9.7 pmol/mg ($P < 0.01$). After kainic acid administration, the reaction is reduced on the lesioned side by 60%–70% (SPANO et al. 1980). This finding is confirmed by FUXE et al. (1978). There is an increase of cAMP levels in the intact rabbit retina after lisuride administration (SCHORDERET 1978). In the parathyroid gland, inhibition of DA-induced stimulation of D_1 receptors by lisuride decreases adenylate cyclase activity and cAMP formation (COTE et al. 1983).

In the frontal cortex (rabbit) lisuride stimulates cAMP formation via 5-HT_1 receptors (ROSENFELD and MAKMAN 1981). In cell preparations of the intermediate lobe of the pituitary of the rat, lisuride (1 nM) inhibits the accumulation of cAMP and the release of α-MSH induced by isoproterenol while it has no effect by itself on these reactions. Similar findings have been obtained in the cerebellum (MUNEMARA et al. 1980).

From these observations lisuride seems to be a strong D_2 receptor agonist and a partial antagonist/agonist at the D_1 receptor in the striatum, an agonist at D_2 receptors of prolactin cells and α-MSH-producing cells of the pituitary, an antagonist at the D_1 receptor in the parathyroid gland, an antagonist at β-receptors in the pituitary intermediate lobe and in the cerebellum and an agonist at 5-HT_1 and 5-HT_2 receptors in the cortex. Lisuride displays its highest affinity to D_2 and 5-HT receptors.

II. Biochemistry

Neurobiochemical studies indicate that lisuride is a direct DA agonist at postsynaptic D_2 receptors and, at low doses, DA autoreceptors. In dopaminergic areas of the rat (striatum and mesolimbic forebrain) lisuride in doses of 0.03–0.1 mg/kg reduces dopa formation and the rate of tyrosine hydroxylation (KEHR 1977; RIPKA 1972), while at doses greater than 0.1 mg/kg the values return to normal (KEHR 1977). With 0.015 mg/kg lisuride, dopa accumulation produced by 700 mg/kg γ-butyrolactone can be reversed (TISSARI et al. 1983). If the activity of tyrosine hydroxylase in a homogenate from the caudate can be used as a measure of dopaminergic activity at presynaptic autoreceptors, lisuride in a dose range of 0.001–10 μM shows little inhibitory activity, in contrast to apomorphine (KEBABIAN and KEBABIAN 1978).

TISSARI and GESSA (1983) also find that apomorphine seems to have a more intense effect on tyrosine hydroxylase, inhibiting DA synthesis more completely than lisuride. In this reaction in vitro, lisuride and bromocriptine are equipotent, while in vivo bromocriptine seems to be 100-fold less potent, and has delayed onset (TISSARI and GESSA 1983). Lisuride (at 0.1 and 0.2 mg/kg i.p.) also increases acetylcholine (ACh) levels in rat striatum and rat accumbens for more than 2 h. The reaction can be blocked by pimozide, but is not prevented by α-methyl-tyrosine (α-MPT). Some 28 days after 6-hydroxydopamine (6-OHDA) lesion, when supersensitivity has developed, a dose as small as 0.025 mg/kg lisuride is enough to produce a measurable increase of ACh (LADINSKY 1977; GARATTINI et al. 1980).

In all rat brain areas investigated (striatum, mesolimbic fore brain, neocortex) lisuride in a dose of 0.3 mg/kg reduced 5-hydroxy-tryptophan (5-HTP) formation by 50% (KEHR 1977), which is in keeping with its presynaptic serotonergic stimulatory activity. In noradrenaline-rich neocortex, low doses of lisuride had no influence on dopa formation, but a 100% increase was seen after higher doses (0.3–1.0 mg/kg; KEHR 1977). This finding may be interpreted as an indicator of central adrenolytic activity of lisuride at higher doses.

III. Pharmacology

Although initially it was synthesized and characterized as a peripheral 5-HT antagonist (for review see PODVALOVA and DLABAC 1970), the most prominent characteristics of lisuride relate to its dopaminergic effects (which in some respects were comparable to apomorphine; HOROWSKI et al. 1975, HOROWSKI and WACHTEL 1976; HOROWSKI and GRÄF 1976). Indeed, rats trained to discriminate between drugs generalize the effects of lisuride to the apomorphine cue, an effect which can be inhibited by the dopamine antagonist haloperidol (WHITE and APPEL 1982).

As regards the effective doses, lisuride is in most tests of dopaminergic activity much more effective than apomorphine or other dopamine agonists, including bromocriptine, and thus has become a standard for the development of new dopaminergic drugs. It affects all dopaminergic systems, but its effects on the central modulation of the motor system are the most conspicuous in animal tests and one can relate many different effects of lisuride in animals to its DA agonist action on the nigroneostriatal DA system. These effects include the induction of stereotyped behaviour in rodents, motor activation and restoring of mobility in reserpine-pretreated mice and rats (HOROWSKI and WACHTEL 1976), but also inhibition of spontaneous motility at very low doses in intact animals, possibly by a presynaptic DA receptor activation (WACHTEL 1978). At higher doses, also in intact rats, motor stimulation is observed (FINK and MORGENSTERN 1985; CARRUBA et al. 1985).

A role of the tegmentomesolimbic DA system in these effects is suggested by the observation that injection of 7.5 μg lisuride into the nucleus accumbens in intact and DA-depleted rats also stimulated motor activity (WACHTEL et al. 1986). In rats with unilateral destruction of the nigroneostriatal pathway, contralateral turning was caused by 0.01 mg/kg lisuride i.p. (FUXE et al. 1978), 0.1 mg/kg i.p. (LOEW et al. 1978) or 0.05 mg/kg s.c. (KREJCI et al. 1985; PIERI et al. 1983).

Muscular rigidity caused by 6-aminonicotinamide or reserpine could be suppressed by 50 μg/kg lisuride to a similar extent as by 100 mg/kg L-dopa plus benserazide (25 mg/kg); this observation led Loos et al. (1977) from the group led by H. HERKEN to the prediction that lisuride should be useful in the treatment of PD. Again, these effects of lisuride could be blocked by DA antagonists (HALBHÜBNER et al. 1978) and were observed in a similar way after intrastriatal, intraventricular or intralumbar injection (BRÄUTIGAM et al. 1980, FLOSBACH et al. 1982).

The locomotor stimularoy effect of lisuride can also be obtained in animals with almost complete DA depletion caused by combined reserpine plus α-MPT pretreatment (HOROWSKI 1978). Chronic discontinuous application of lisuride (e.g. by repeated s.c. or i.p. injections) greatly enhances the stimulating effects of this drug whilst the initial motor inhibition observed in naive animals is no longer detectable (HOROWSKI and WACHTEL 1979). Also in guinea pigs, chronic s.c. injections of lisuride caused behavioural supersensitivity consisting of increased stereotyped behaviour (CARVEY and KLAWANS 1983). On the other hand, such a behavioural supersensitivity is not induced when lisuride is given over the same period of time by a continuous delivering Alzet minipump system (WACHTEL et al. 1988) thus stressing the importance of the route and mode of application.

Lisuride decreased in a dose-dependent way (0.1–0,39 mg/kg i.p.) the body temperature in mice (HOROWSKI and WACHTEL 1976) and in a similar way in rats (0.0125–0.1 mg/kg i.p.) which were in a cold environment (4 °C) (HOROWSKI and RIEDEL 1979). The hypothermic effect of lisuride was less pronounced at normal ambient temperature, but still significant (KELLER and DA PRADA 1979; CARRUBA et al. 1980). The lowering of body temperature by lisuride could be prevented by haloperidol or sulpiride pretreatment (HOROWSKI and RIEDEL 1979) and on the other hand, lisuride effectively antagonized reserpine-induced hypothermia (HOROWSKI and WACHTEL 1976).

Lisuride was also able — possibly again by acting at the level of hypothalamic DA neurons — to cause anorexia in rats at very low doses (ED_{50} = 0.06 mg/kg). This effect was prevented by pretreatment with DA antagonists, but not with the DA-depleting inhibitor of tyrosine hydroxylase, α-MPT, nor by compounds interacting with 5-HT receptors (CARRUBA et al. 1980, 1983). In dogs, lisuride has pronounced emetic effects which are thought to be due to an activation of the chemoreceptor trigger zone at the area postrema (HOROWSKI and WACHTEL 1976). This emetic effect in dogs, however, rapidly develops tolerance on repeated treatment, like the effects of lisuride on body temperature and food intake in rats, without evidence for a change in drug metabolism.

Thus, in the same rats chronic injections of lisuride cause these effects to disappear, whilst the motor stimulatory effects are clearly enhanced (HOROWSKI and WACHTEL 1979; CARRUBA et al. 1983, 1985). In the same animals again, the prolactin-lowering effect of lisuride remains unchanged; indeed, in contrast to its sometimes biphasic or state-dependent effects on more central DA systems lisuride almost invariably inhibits prolactin secretion and lowers prolactin levels in all experimental and clinical situations.

Lisuride exerts its prolactin-lowering effects by a direct action on dopaminergic receptors within the pituitary (GRÄF et al. 1976, 1977; HOROWSKI et al. 1976; HOROWSKI and GRÄF 1976). Lisuride acts on the prolactin cell itself (BURDMAN et al. 1982) and is highly effective under in vitro conditions (MAC LEOD 1977) as well as on pituitaries grafted under the renal capsule of hypophysectomized rats (GRÄF et al. 1977; HOROWSKI and GRÄF 1976). It is a potent inhibitor of both estrogen- and reserpine-induced hyperprolactinemia in rats (GRÄF et al. 1976) and inhibits prolactin secretion in dogs and monkeys.

The effects of lisuride on the 5-HT system are rather complex, and it is usually defined as a 5-HT agonist or partial agonist. However, as often in ergot pharmacology, different and even opposing effects can be observed, depending on the test model used and its actual state. In peripheral systems, 5-HT effects can be blocked by lisuride which is as active as D-lysergic acid diethylamide (LSD), but less active than methysergide (PODVALOVA and DLABAC 1972): its ID_{50} in the 5-HT-induced paw edema test is 0.05 (0.04–0.09) mg/kg i.p. and in the antagonism of 5-HT-induced papules 0.15 (0.11–0.22) mg/kg i.p., and it also inhibited 5-HT-induced contractions of the isolated rat stomach in vitro (at 1.0 μg per 20 ml).

5-HT-induced constriction of dog coronary arteries was also inhibited by lisuride ($IC_{50} = 4.2$ nM); at higher concentrations however (1000 nM), constrictor effects were also present (BRAZENOR and ANGUS 1982). 5-HT-related synergistic effects on ADP-induced platelet aggregation was inhibited in the range 1–1000 nM (GLUSA and MARKWARDT 1984) and 5-HT-induced platelet shape changes were inhibited at 8.2 (6.8–9.9) nM concentration of lisuride (GRAF and PLETSCHER 1979).

Central effects of lisuride have been interpreted as 5-HT agonism based upon effects on high limb flexor reflex (KEHR 1977) and on induction of a so-called 5-HT syndrome at high doses (SILBERGELD and HRUSKA 1979). At lower doses, however, lisuride inhibited 5-HT-induced head twitches (GERBER et al. 1985). Mounting behaviour (similar to a combined administration of the 5-HT synthesis inhibitor p-chlorophenylalanine (pCPA) and a dopaminergic drug) could be induced by high doses of lisuride (DA PRADA et al. 1977; PIERI et al.

1983; Horowski and Dorow 1981; Horowski 1983). This male-type mounting behaviour was independent of the hormonal situation of the animals (male, female, adult or juvenile, pretreated with hormones or antihormones; Horowski 1983). In castrated male rats, lisuride restored the complete pattern of male sexual behaviour in a dose-dependent way (Ahlenius et al. 1980). Neuropharmacological interaction studies revealed that these effects of lisuride were due to a combined DA agonistic effect and a functional inhibition of central 5-HT neurotransmission, maybe due to a preferential 5-HT autoreceptor stimulation by this drug (Pieri et al. 1978; Horowski 1983).

Little is known as to whether the high affinity of lisuride for α_1 and α_2 receptors leads to important pharmacological effects. Only under in vitro conditions could lisuride at a nanomolar concentration inhibit adrenaline-induced platelet aggregation (Glusa et al. 1979). There is no behavioural equivalent for a significant interaction of lisuride with β-adrenergic systems, where binding has been reported to be inhibited by higher in vitro concentrations of lisuride (Cote et al. 1979).

IV. Neurophysiology

Neurophysiological findings in rats confirm the results of behavioural pharmacology, biochemical and receptor binding studies. Firing of raphe dorsalis neurons is inhibited at doses of 1–5 μg/kg i.v. (Rogawski and Aghajanian 1979); after a cumulative i.v. dose of 15 μg/kg, inhibition is 98% of baseline activity (Walters et al. 1979). Firing of locus coeruleus neurons — depending on the predrug firing rate — may increase by up to 500% after administration of lisuride 25–50 μg/kg i.v. (Rogawski and Aghajanian 1979).

Firing of most neurons in the substantia nigra zona compacta in intact rats is inhibited by lisuride doses of 25–100 μg/kg i.v., while the firing of a few neurons is increased dose dependently (Rogawski and Aghajanian 1979). The inhibition is not complete. There is no effect on zona reticulata cells. Initial inhibitory effects were seen in another study after doses as low as 1.5 μg/kg i.v. Subsequent agonist challenges were unsuccessful, and antagonistic effects of haloperidol were weaker than in the case of apomorphine. The effects of lisuride were interpreted as possibly related to a partial agonist effect of the drug (Walters et al. 1979). The effects in the substantia nigra are reduced by previous destruction of the striatonigral feedback loop with kainic acid (Walters et al. 1978). Pallidal neurons increase their firing rate as a response to lisuride or other DA agonists. This increase is attenuated by general anaesthesia or a small, nonexcitatory dose of apomorphine or lisuride (Bergstrom et al. 1984).

Recently, G. L. Gessa's group (Mereu et al. 1986, 1987; Gessa 1988) has further investigated the neurophysiological effects of lisuride on the pars compacta cells of substantia nigra and on the ventral tegmental area (A9 and A10). Lisuride was shown to have opposite, dose-dependent effects (at 1–100 μg/kg i.v.) in animals with general anaesthesia (inhibition of firing rate) and in animals with local anaesthesia (increase). As microiontophoretic studies consistently showed an inhibitory effect, the inhibition is thought to be related to an activation of autoreceptors. These receptors seem to be blocked to subsequent application of apomorphine or haloperidol; however, 1–3 days after acute or repeated lisuride application, these receptors show a normal response.

GESSA (1988) proposes that lisuride has a higher affinity but a lower intrinsic activity than DA and thus behaves as an agonist or antagonist, depending on the state of the DA receptors (with agonist properties especially on denervated "empty" receptors). Owing to its high receptor affinity, lisuride is thought to occupy the DA receptors on the presynaptic DA neuron for long periods, making it insensitive towards DA agonist and antagonists. In line with this, lisuride was able to prevent, but not to reverse the effects of apomorphine in anaesthetized rats (MEREU et al. 1986) whilst haloperidol could prevent, but not reverse the effects of lisuride.

This concept would also explain differences between the "pure" DA agonist apomorphine and lisuride, and explain why lisuride is more potent and has longer-lasting effects in conditions of functional DA denervation. In this context, the DA receptors on prolactin cells should also be considered as "empty" receptors with no neuronal feedback loops.

From bromocriptine, lisuride not only differs in its higher receptor affinity and pharmacological activity (which could explain why DA antagonists are much less effective in inhibiting its actions (HOROWSKI and RIEDEL 1979) and why lisuride can be used for antagonizing neuroleptic overdose (H. PRZUNTEK 1988, personal communication). As only one example, lisuride had an ED_{50} of 2.6 µg/kg i.p. and bromocriptine of 220 µg/kg i.p. as antagonists of reserpine-induced rigidity in rats (LOOS et al. 1979). In addition, motor and other behavioural effects of lisuride cannot be abolished by DA depletion by reserpine and α-MPT (HOROWSKI 1978; CARRUBA et al. 1980; HARA et al. 1982) whilst the motor effects of bromocriptine seem to depend largely on the presence of endogenous DA or concomitant DA receptor activation (JOHNSON et al. 1976; HOROWSKI 1978; JACKSON 1986).

The similarity of chemical structures of LSD, which causes hallucinations in healthy volunteers in a dose range of 0.5–2.0 µg/kg, and lisuride, which in single doses of up to 600 µg p.o. and 100 µg i.v. has never caused hallucinations in healthy volunteers, has led experimental research to find the difference in hallucinogenic potential (PIERI et al. 1978). The issue has received new interest because mental side effects, including hallucinations, can occur during chronic or high dose treatment with lisuride in patients with PD (see Sect. F. I.).

In spite of superficial similarities between lisuride and LSD (both substances inhibit firing of raphe dorsalis neurons, (ROGAWSKI and AGHAJANIAN 1979; both substances increase dopa formation in corpus striatum in reserpinized rats; KEHR and SPECKENBACH 1978), various quantitative and qualitative differences have been observed. Lisuride is ten times more potent in increasing dopa formation (KEHR and SPECKENBACH 1978). Locomotion in rats is inhibited by lisuride at low doses (15–30 µg/kg), but increased at 60 µg/kg, whereas LSD inhibits locomotion at all doses (ADAMS and GEYER 1985). In the single-unit recording from A10 DA cells, lisuride suppresses firing

($ID_{50} = 0.021$ mg/kg), while LSD depresses firing in only 54% of the cells, increases firing in 23% and has no effect in the rest; LSD also reversed apomorphine-induced suppression (WHITE and WANG 1983). In reserpinized animals LSD counteracts the stimulatory effect of apomorphine, while lisuride does not (KEHR and SPECKENBACH 1978). This means that LSD acts more, or solely, at DA autoreceptors in the nigrostriatal system and shows more antagonistic effects. Therefore LSD, in contrast to lisuride, does not produce mounting behaviour, which results from presynaptic stimulation of serotonergic neurons and postsynaptic stimulation of dopaminergic neurons (KELLER and DA PRADA 1979), but even antagonizes the hypersexuality caused by lisuride (HOROWSKI 1983).

Lisuride, but not LSD, reduces 5-hydroxyindole-acetic acid (5-HIAA) levels and antagonizes acceleration of 5-HT turnover caused by 5-HT receptor blocking agents like methiothepin (PIERI et al. 1978). Lisuride increases and LSD reduces baseline firing in locus coeruleus neurons (MCCALL 1982).

In one selection of drug discrimination paradigms, lisuride substituted for LSD (HOLOHEAN et al. 1982). Animals could, however, discriminate clearly between both drugs (WHITE and APPEL 1982a) 5-HT agonists substituted only partially for lisuride (quipazine, 5-methoxy-*N,N*-dimethyltryptamine, 5-MeODMT; WHITE and APPEL 1982b) and DA agonists substituted partially for LSD. DA agonists (lergotrile, apomorphine) substituted completely for lisuride and 5-HT agonist/antagonists substitute completely for LSD (WHITE and APPEL 1982a). The lisuride cue was blocked by haloperidol but not by cyproheptadine, in contrast to the LSD cue (WHITE and APPEL 1982b).

The cat behavioural model which was once believed to be specific for hallucinogenic drugs, has turned out to be nonspecific because it failed to discriminate between the common serotonergic properties of LSD (0.08 mg/kg), lisuride (0.02 mg/kg) and methysergide (0.06 mg/kg) (MARINI et al. 1981; MARINI and SHEARD 1981).

LSD, mescaline and psilocybin facilitate the response of facial motor neurons to noradrenaline and 5-HT, while lisuride and methysergide do not show this effect (MCCALL and AGHAJANIAN 1980). Peripheral motor neurons react in a similar way to LSD which also enhances reactivity of locus coeruleus neurons to excitatory inputs (MCCALL, 1982). Parkinsonian patients develop hallucinations after treatment with all types of dopaminergic drugs. The fact that DA antagonists are useful in blocking hallucinations speaks for the involvement of the dopaminergic system in the production of hallucinations.

In a recent important and comprehensive review paper, WHITE (1986) compares the effects of LSD and lisuride on various biochemical, behavioural and electrophysiological indices of neuronal function. It is stated that in biochemical studies involving brain monoamines, differences seem to be more quantitative, but that detailed investigation of special 5-HT and DA subpopulations is necessary. Behavioural studies consistently demonstrate a greater dopaminergic activity of lisuride as compared with LSD, and these findings are supported by electrophysiological studies as well. The potentiation of excitomodulatory effects of 5-HT by LSD, but not by lisuride, is considered to be the most important difference so far which, — if confirmed for other brain areas — could explain the vivid visual hallucinations caused by LSD in humans.

Lisuride causes a reduction of blood pressure in anaesthetized cats and spontaneously hypertensive rats. This effect is explained by moderate peripheral vasodilation, a decrease of heart rate and contractility and a reduction of the ejection fraction. These effects are blocked by haloperidol (MANNESMANN et al. 1979).

In baroreceptor-denervated cats, lisuride (5–10 μg/kg i.v.) decreased mean arterial pressure, heart rate and sympathetic nervous discharge. The effects are observed with other 5-HT antagonists and have been explained by a tonic central serotonergic influence on sympathetic outflow (McCALL and HUMPHREY 1982). In spontaneously hypertensive stroke-prone rats, lisuride improved performance and prevented stroke-associated pathology (HARA et al. 1982). Glucose utilization in some rat brain areas was stimulated by lisuride, an effect which could be blocked by haloperidol and sulpiride (AZUMA et al. 1982).

With the exception of antihistamine properties of lisuride in high doses, no other general pharmacological effects were observed with lisuride in various tests. Especially, contraction or relaxation of the uterus or vasoconstricting properties were not observed (PODVALOVA and DLABAC 1972).

E. Clinical Applications

Effects and side effects of lisuride in humans are related to a great extent to its direct dopaminergic effects on D_2 receptors. This obviously applies to its therapeutic effects, alone or in combination with L-dopa, in the dopaminergic deficiency syndrome of the basal ganglia, i.e. PD, and related neurological disorders. In addition, the important role of DA in the regulation of pituitary hormones, and especially its strong inhibitory effect on prolactin secretion, is the basis for the long-lasting therapeutic use of lisuride in neuroendocrine disorders. Side effects of lisuride such as nausea, emesis and orthostatic hypotension are also due to an activation of dopaminergic systems and can be prevented or treated by DA antagonists. The effects of lisuride on cortical reflex myoclonus and its preventive action in migraine are possibly related to the additional interaction of lisuride with serotonergic mechanisms.

I. Clinical Pharmacology

In healthy volunteers, oral lisuride lowers basal and stimulated prolactin levels in a dose- and time-dependent way (HOROWSKI et al. 1977; DELITALA et al. 1979). Effective single oral doses range from 0.1 to 0.3 mg whilst similar prolactin-lowering effects can be achieved by 0.025 mg lisuride given as a parenteral bolus (DOROW et al. 1980). Patients suffering from renal insufficiency and migraine seem to be especially sensitive to the prolactin-lowering effects as well as to side effects of lisuride and therefore only low oral doses are used in these conditions (RUILOPE et al. 1985; DESAGA et al. 1985). The prolactin-lowering effect of lisuride is rather specific as no other hormones are affected and, in women, a normal cycle (BOHNET et al. 1979) and, in males, normal gonadal function is maintained (GRÄF et al. 1982).

Side effects are related to the dopaminergic effects of lisuride since they can be blocked by peripheral DA antagonists such as sulpiride or domperidone; they include nausea and emesis and, occasionally, orthostatic hypotension, dizziness and drowsiness (CANGI et al. 1985; DOROW et al. 1983); in most cases they develop tolerance in chronic treatment whilst the prolactin-lowering effect increases with time (DOROW et al. 1983). Another side effect associated with acute lisuride treatment is headache, especially in migraine sufferers; this side effect seems to be of central origin because it did not react to domperidone (CANGI et al. 1985). Also in this case, tolerance seems to develop and lisuride can be used in the prevention of migraine. In a controlled double-blind study in volunteers, less peripheral side effects were clearly observed with lisuride than with methysergide, and no hallucinogenic properties were found in contrast to the effects induced by D-LSD (VOTAVA et al. 1966). Indeed, so far no hallucinations or other psychotic side effects have been observed with lisuride in healthy volunteers — in contrast to D-LSD and to side effects of lisuride in parkinsonian patients (HOROWSKI 1986). In pharmaco-EEG studies (ITIL et al. 1974, 1975), lisuride was reported to have a "psychostimulant" type of EEG profile at oral doses ranging from 0.025 to 0.1 mg; this classification was, however, based only on electrophysiological parameters.

Based upon its dopaminergic and other properties (for reviews see HOROWSKI and MCDONALD 1983; SUCHY 1986), lisuride has also been used to improve vigilance and mood in elderly patients suffering from cerebrovascular disease (MISUREC et al. 1978; OTOMO et al. 1981; AIZAWA et al. 1980). The clinical pharmacology of oral and parenteral lisuride (as well as of its derivative terguride) has been reviewed by DOROW et al. (1983).

II. Applications in Clinical Endocrinology

As a DA agonist, lisuride has strong and long-lasting prolactin-lowering effects at low doses (0.05–0.2 mg p.o.) and thus has been used successfully in a variety of endocrine disorders such as hyperprolactinemia associated with cycle and fertility disorders in women (anovulatory cycles, luteal insufficiency, premenstrual mastodynia, amenorrhoea). These conditions are, in part, caused by prolactin-producing micro- or macroadenomas and can be associated with galactorrhoea.

In males, these tumours can cause, by increasing serum prolactin, an inhibition of testosterone synthesis with its consequences: reduction or loss of libido, potency and fertility. In both sexes, DA agonists such as lisuride not only lower prolactin and normalize gonadal and breast function, but also can cause a clear-cut reduction of tumour size (CHIODINI et al. 1978, 1981; VON WERDER et al. 1983; THORNER et al. 1984). Such a reduction in tumour size is not observed in acromegaly, where DA agonists like lisuride lower serum growth hormone levels in a proportion of patients — in contrast to the situation in healthy people where acute application of lisuride (similar to apomorphine) causes a short-lasting increase in growth hormone (VERDE et al. 1980).

Table 1. Summary of 25 studies of Lisuride for various indications in endocrine diseases

Reference	Indication	Patients sex and age	Dose (mg/day)	Duration of treatment	Effects on PRL and clinical effects	Side effects	Remarks
BESSER et al. 1983	Hyperprolactinemia, acromegaly	22 16	0.2–1.2 0.1–1.6	1–18 months	PRL ↓ in all, GH ↓ in 75%, clinical improvement	Typical of DA agonists; no adverse long-term effects	
BOHNET et al. 1979	Healthy volunteers	4 f (22–33 years)	0.4	1 month	PRL ↓	Nausea and emesis, but tolerance development	LH, FSH, E_2, progesterone unchanged
	Hyperprolactinemia	4 f (24–33 years)	0.4	1 month	PRL ↓	Nausea and emesis but tolerance development	Normalization of progesterone and cycle
BOHNET et al. 1981	Premenstrual syndrome + mastodynia	21 f (27–36 years)	0.2	1–3 months	PRL ↓, 16 improved in clinical symptoms	n.m.	
CARSTENSEN et al. 1983	Anovulation, pituitary tumour, galactorrhoea, amenorrhoea	87 f 15 f 18 f (+ 127 on bromocriptine 2.5 mg)	0.2	n.m. (several months at least)	PRL ↓, improvement of galactorrhoea, 22 pregnancies	Nausea ($n = 15$), dizziness ($n = 7$), tiredness ($n = 8$), headache ($n = 2$)	
CHIODINI et al. 1981	Macroprolactinoma	10 f, 5 m (28–58 years) (both drugs $n = 3$)	0.4–2.0	5–50 months	PRL ↓, tumour size reduction in 7, improvement in visual field	n.m.	Open trial comparison with bromocriptine ($n = 14$) 5–20 mg/day, tumour size reduction ($n = 11$)
CROSIGNANI et al. 1982	Hyperprolactinemia, 3 idiopathic, 7 microadenoma, 24 macroadenoma	34 (28 f, 6 m) (+ 123 on bromocriptine, 82 on methergoline) (16–52 years)	0.2–2.0	2–48 months	PRL ↓, biochemical and clinical improvement, potency improved in 5 m	Mild headache, nausea, vomiting, dizziness in several patients, but tolerance, 2 subjects did not tolerate owing to gastric side effects	Results of bromocriptine and lisuride superimposable, methergoline weaker, 11 of 12 patients who did not tolerate one drug did well on another

Table 1 (continued)

Reference	Indication	Patients sex and age	Dose (mg/day)	Duration of treatment	Effects on PRL and clinical effects	Side effects	Remarks
DECECCO et al. 1983	Hyperprolactinemia (17 with microadenoma)	54 f	0.2–0.8	12 weeks	PRL ↓, clinical improvement, restoration of cycles in 46, 6 pregnan cies	n.m.	
DESAGA et al. 1983	Hyperprolactinemia (after chronic renal failure)	27	0.2–2.4	2 months–2 years (mean 13 months)	PRL ↓, clinical improvement, tumour size reduction	Nausea, vertigo, orthostatic dysregulation, fatigue tolerated by patients	1 patient died from tumour haemorrhage
DESAGA et al. 1985	Chronic renal failure with hemodialysis with hyperprolactinemia	30 (21 f, 9 m)	0.075–0.15	3 months	PRL ↓	TSH and T_3, T_4, rT_3 unchanged, FT_4 normalized, by lisuride, no severe side effects observed	High sensitivity to lisuride in patients on dialysis
	or normoprolactinemia	10 m					
GIULIANI et al. 1982	Benign prostatic hypertrophy	11 m (58–604)	0.6	1 month	PRL ↓, basal and after TRH clinical improvement	n.m.	TSH unchanged, LH, FSH, testosterone unchanged
HARDT et al. 1979	Inhibition of lactation	50 f	0.6–0.9	2 weeks	PRL ↓, clinical efficacy	Nausea, drowsiness, episodic orthostatic hypotension in a total of 12, but tolerance development, exanthema in 3	
HEINLEIN et al. 1983	Hyperprolactinemia	12 f, 15 m, crossover vs bromocriptine	0.2–1.4 (mean 0.8)	3-6 months	PRL ↓ in 83 %, 10 patients could not be normalized with either lisuride or bromocriptine	Hypotension ($n = 7$), nausea ($n = 10$), vomiting ($n = 5$), dropouts ($n = 2$)	Bromocriptine: PRL ↓ in 87 %, dropout ($n = 1$), hypotension ($n = 8$), vomiting ($n = 4$), nausea ($n = 11$)
LIUZZI et al. 1985	Prolactinoma	4 f (23–45 years), parallel group with bromocriptine	0.4–0.8	61–82 months	PRL ↓, clinical improvement + tumour size reduction	n.m.	Low maintenance dose still effective (0.05 mg)

LÜDECKE et al. 1983	Prolactinoma acromegaly	27 (reported by DESAGA et al.) 22	0.1–2.4	8.8 months (mean)	Half of patients have response in GH, 7 normalized	n.m.	11 nonresponders, 1 responders
MAREK et al. 1983	Hyperprolactinemia (prolactinoma 32, idiopathic hyperprolactinemia 7)	39 (31 f, 8 m) (19–42 years)	0.075–6.0	2 months or n.m.	PRL ↓, 15 pregnancies of 16 women with desire for children, in 4 men libido, potency and testosterone normalized	Severe ($n=13$), stopped medication at higher doses ($n=3$), (nausea and vomiting, lack of appetite)	Dropouts ($n=7$)
	Acromegaly	21 (14 f, 7 m) (28–72 years)	0.075–6.0	3 months–3 years	GH ↓ below 50% ($n=7$), GH normalized ($n=2$) somatomedin improvement ($n=11$)	Weakness, tiredness, sleepiness ($n=10$); headaches, muscle cramps in a few cases	
	Cushing's disease	6 (5 f, 1 m) (25–45 years)	n.m.	1 year	improvement ($n=1$) (partially)		
RUILOPE et al. 1985	Chronic renal failure	10 m (32–48 years)	0.075	4 weeks testosterone 80% ↑)	Improvement in libido	Asthenia, nausea, anorexia	Treatment continued for 1 year
		10 m (34–48 years)	0.075	4 weeks testosterone 40% ↑)			
SCHMIDT-GOLLWITZER et al. 1977	Ablactation	18 f crossover vs, placebo and bromocriptine	0.075	10 days	PRL ↓, no clinical effect	Nausea ($n=1$), confluent exanthemas ($n=2$)	Dose-finding study, dosage too low
SCHNEIDER 1983	Infertility	78 f (28 normoprolactinemic)	0.2–?	2–6 months	PRL ↓	n.m.	43 pregnancies
SCHWIBBE et al. 1983	Premenstrual syndrome	16 f (mean 25.5 years)	0.3 (crossover vs placebo)	4 weeks	PRL ↓, improvement in symptoms of premenstrual syndrome	Nausea and vertigo	Increased power of EEG band
THORNER et al. 1984	Hyperprolactinemia	6 f	0.6	6 months	PRL ↓, bromocriptine and lisuride appear to have similar efficacy	n.m.	

Table 1 (continued)

Reference	Indication	Patients sex and age	Dose (mg/day)	Duration of treatment	Effects on PRL and clinical effects	Side effects	Remarks
VAN DAM and ROLLAND 1981	Ablactation	26 f (vs 24 f on bromocriptine, double-blind)	0.4	14 days	PRL ↓, "satisfactory inhibition of puerperal milk production"	Nausea, vomiting (1 dropout), dizziness	Bromocriptine effects on PRL slightly stronger, but more rebound and more side effects
VENTURINI et al. 1981	Ablactation	21 f (vs 21 f on bromocriptine, double-blind)	0.6	15 days	PRL ↓, inhibition of lactation and breast symptoms	Headache, nausea, vomiting, dizziness, itch, (1 dropout)	As effective and less rebound than bromocriptine (3 dropouts)
VERDE et al. 1980	Acromegaly	21 (3 m, 18 f) (27–72 years)	0.4–2.4	2–24 months	10 GH responders, PRL ↓, clinical improvement in both groups	Side effects at the beginning (nausea and orthostatic hypotension)	
	Hyperprolactinemia	25 (5 m, 20 f) (16–56 years)	0.4–2.4	3–26 months	PRL ↓		
VON WERDER et al. 1983	Prolactinoma (open crossover comparison with bromocriptine)	16	0.2–2.4	2 months	PRL ↓, clinical improvement	n.m.	Tumour shrinkage observed
WINKELMANN et al. 1983	Micro- and macroprolactinomas	34 (18 m, 16 f) (10 microadenomas, 24 macroadenomas) (17–69 years)	≦ 4.8 mg	Mean 20, maximum 38 months	PRL ↓ 100 % of microadenomas, 94 % of f and 88 % of m with macroadenomas, clinical improvement (libido, potency, menstrual cycles, ophthalmological function)	paranoid psychosis ($n = 1$), after lisuride or bromocriptine severe side effects ($n = 9$)	Open comparison with bromocriptine ($n = 61$), (micro $n = 13$, macro $n = 48$)

E_2 estradiol; FSH follicle-stimulating hormone; FT_4 (free thyroxine); GH growth hormone; LH luteinizing hormone; n.m. not mentioned; PRL↓ prolactin-lowering effect; rT_3 (reverse triodothyroxine); T_3 (triodothyroxine); T_4 (thyroxine); TRH thyrotropin-releasing hormone; TSH thyroid-stimulating hormone

Lisuride is also used for lowering prolactin levels and inhibition of lactation in women after delivery, and it can also be used to normalize prolactin and gonadal steroid synthesis in patients suffering from renal failure.

Table 1 summarizes a total of 25 published studies involving lisuride in the indications mentioned here. In these studies, 814 patients were treated with oral dosages of lisuride ranging from 0.075 to 6 mg/day over 10 days up to 50 months. In all these studies (except the first dose-finding study by SCHMIDT-GOLLWITZER et al. 1977, where the dosage was too low), lisuride was found to be an effective and well-tolerated prolactin-lowering drug similar to bromocriptine. In addition, SCHOLZ and HOROWSKI (1987) analysed 1081 pregnancies associated with the use of lisuride in female fertility disorders and found no indication for a negative influence on the development and outcome of these pregnancies.

F. Lisuride in Neurological Diseases

I. Parkinson's Disease and Related Disorders

There is a large number of studies where the dopaminergic effects of lisuride were the basis of its use in situations of disturbed dopaminergic neurotransmission and function. The use of lisuride in neurological diseases and especially in PD has been reviewed several times (McDONALD and HOROWSKI 1983; SUCHY and HOROWSKI 1986). In a recent comprehensive review, LEWITT (1986) quotes subtypes of Parkinsonism and other movement disorders which may respond to lisuride therapy:

Parkinsonism
- Idiopathic
- Neuroleptic-induced
- MPTP-induced (limited trial)
- Postencephalitic
- Shy–Drager syndrome (minimal benefits)
- Guam type (limited trial)
- Olivopontocerebellar atrophy type (limited trial)
- Progressive supranuclear palsy (minimal benefits)
- Juvenile parkinsonism (onset < age 40)
- Familial parkinsonism (1 case)

Others
- Tardive and other dyskinesias
- Focal and generalized dystonia
- Huntington's chorea
- Myoclonic disorders

LEWITT summarizes the effects of lisuride in parkinsonism as follows:

- Reverses all cardinal features of parkinsonism as monotherapy
- May improve loss of levodopa efficacy and worsening of parkinsonian stage

- May lessen severity of wearing-off and on–off phenomena, abnormal involuntary movements and dystonic features
- Available for intravenous as well as oral administration
- Possibly less potential for causing long-term problems (such as on–off effect and dyskinesia associated with chronic levodopa therapy)
- Tolerance may develop to adverse effects; antiparkinsonian efficacy is sustained

In extension of these effects, it has been proposed recently that not only patients on lisuride monotherapy, but also on a combination of L-dopa and lisuride from the beginning of the disease will experience less fluctuations in mobility than patients treated with L-dopa monotherapy (RINNE 1986; RINNE 1989, Neurology, acc. f. publ.). If confirmed by independent studies, this observation will improve not only the life expectancy (as was the case with L-dopa) but also the quality of life for patients with PD and thus will have great clinical consequences.

1. Oral Application

Lisuride's antiparkinsonian efficacy was described independently by SCHACHTER et al. (1979), LIEBERMAN et al. (1979), GOPINATHAN et al. (1980) and RUGGIERI et al. (1980). In Table 2, relevant data from the first controlled and open studies of lisuride in PD are summarized.

SCHACHTER (1980) had progressively replaced bromocriptine by lisuride on the basis of a calculated dose relation of 12–15:1 in 13 patients; 3 of them received first bromocriptine and then lisuride as monotherapy and 8 of them had a regimen combining bromocriptine or lisuride together with levodopa plus peripheral decarboxylase inhibitor (PDI). The final lisuride dosage varied interindividually between 0.6 and 4.8 mg/day. On the whole SCHACHTER found no significant difference between lisuride and bromocriptine. LIEBERMAN et al. (1979, 1981c) added 1-5 mg lisuride to an ongoing levodopa treatment, the dosage of which was reduced. So-called hours-on, the time during which patients were mobile, were increased by 130%. LIEBERMAN also replaced levodopa by lisuride in an acute study, which was possible without loss of benefit in four of five patients. GOPINATHAN et al. (1980, 1981) slowly increased doses of lisuride without changing the basic standard therapy and evaluated patients at a low dose level of 0.65 mg/day, at a high dose level of 4.65 mg/day and further during a single-blind within-patient control versus placebo. A "blinded" observer evaluated the patients throughout the trial. Clinical evaluation saw better effects with high dose lisuride, while objective measurements of reaction time and movement time noted better results during the low dose regimen. RUGGIERI (1980) and AGNOLI et al. (1983) performed an intergroup comparison with lisuride and bromocriptine as monotherapy in beginning PD (seven patients per group) and an intraindividual crossover, lisuride versus bromocriptine, in combination with levodopa plus PDI in advanced PD. All trials were performed on a single-blind basis. Dosages of lisuride and of bromocriptine were increased in a predetermined way on a 1:10 ratio. The final dose (mean effective tolerated dose) for lisuride reached in monotherapy was

Table 2. Summary of 9 studies of lisuride for Parkinson's disease

Reference	Patients			Mean age (years)	Diagnosis and severity	Duration of PD (years)	Concomitant medication (mg/day)	Lisuride (mg/day)	Duration	Design	Findings
	n	m	f								
SCHACHTER et al. 1980	10			50–71	Idiopathic III–IV	17	Levodopa + carbidopa (920)	3.5 (1.2–4.8)	≦ 9 months	Open, bromocriptine as a control	Levodopa + carbidopa reduced 30 % to 710 mg/day; PD disability scores improved 33 %; lisuride comparable with bromocriptine 15 mg; side effects: nausea, hypotension, psychiatric
GOPINATHAN et al. 1981	20			53	Idiopathic	9	Levodopa + carbidopa (850, $n = 15$)	4.65	4 weeks	Single-blind, placebo substitution	Significant improvement: total and akinesia composite scores, component scores: rigidity, posture, dexterity, face, speech; side effects: light-headedness, drowsiness, confusion; 2 patients withdrawn owing to drug-induced confusion
LEES and STERN 1981	12	9	3	70	Idiopathic	15	Levodopa + carbidopa	2.0 (0.2–6.0)		Open, historical control	Oscillations improved in 1 patient; 4 patients reported improvement in overall mobility during "on" periods; side effects: nausea, hyperkinesias, confusion
LE WITT et al. 1982	28	15	13	55	Idiopathic	10	Levodopa + carbidopa	4.5 (0.6–10)	7–10 weeks	Double-blind crossover vs bromocriptine (56.5 mg/day)	Similar to bromocriptine at optimal doses; ratio 13:1; side effects comparable: light-headedness, drowsiness, nausea, hallucinations; 15 patients continued on lisuride

Table 2 (continued)

Reference	Patients n	m	f	Mean age (years)	Diagnosis and severity	Duration of PD (years)	Concomitant medication (mg/day)	Lisuride (mg/day)	Duration	Design	Findings
Lieberman et al. 1983	63	43	20	65	Idiopathic III–V	11	Levodopa + carbidopa (1000, $n=59$)	2.6 (0.2–5.0)	1–18 months	Double-blind, placebo substitution	Significant improvement in PD disability score 34 % in the "on" period; 37 patients improved by at least one stage; patients with "on-off" phenomena: number of hours "on" increased; 96 % side effects: confusion/hallucination, dyskinesias
Agnoli et al. 1983	14	9	5	57–68	Idiopathic	1–3	None	3 (2–5.0)	12 weeks	Open, vs bromocriptine	20 % improvement in tremor 120 min after acute administration; improvement in PD total disability score, tremors and rigidity following chronic intake; prolactin levels decrease, but no correlation with clinical results
	14	9	5	53–72	Idiopathic	4–10	Levodopa + carbidopa	3 (2–5.0)	12 weeks		Improvement noted in PD total disability score and tremor more pronounced with lisuride
Riederer et al. 1983	13	1	12	72	"Benign"	12	Levodopa + benserazide (375–750)	0.4	6 months	Open, historical control	Decrease in Webster score by 17 points; 20 % reduction in levodopa + benserazide therapy; all patients continued to take lisuride

BIRKMAYER and RIEDERER 1983	12	7	5	58–78	Benign	5–10	Levodopa + benserazide (375)	0.6–1.2	7 months	Open, historical control	All benign cases: significant improvement in disability scores (21%–26%) and motor functions; daily fluctuations and long-term oscillations milder or disappeared; reduction in levodopa by 30–40%. Malignant cases: no improvement; side effects: hyperkinesia, toxic delirium, hypotension
	14	9	5	61–79	Benign	10–15	Levodopa + benserazide (375)	0.6–1.2	8 months		
	14	8	6	64–86	Benign	15–20	Levodopa + benserazide (375)	0.6–1.2	10 months		
	12	8	4	56–62	Malignant	1–5	Levodopa + benserazide (375)	0.6–1.2	12 months		
RINNE 1983	10	3	7	64	Idiopathic I–III	1–2	None	1.5 (0.6–2.4)	12 weeks	Open, vs bromocriptine, levodopa + benserazide	Significant improvement in PD total disability score and tremor; therapeutic response: mild ($n = 4$) to moderate ($n = 4$); side effects: nausea, dizziness, sedation; one dropout: severe nausea and dizziness
	10	6	4	60	Idiopathic III–IV	11	Levodopa (775)	1.5 (0.6–4.0)	12 weeks	Open, vs levodopa + benserazide	Significant improvement in PD total disability score and tremor; therapeutic response: mild ($n = 4$) to moderate ($n = 4$), beneficial effect on "on-off" phenomena; 2 patients withdrawn: hallucination/confusion and nausea/anxiety

Table 2 (continued)

Reference	Patients *n*	m	f	Mean age (years)	Diagnosis and severity	Duration of PD (years)	Concomitant medication (mg/day)	Lisuride (mg/day)	Duration	Design	Findings
ULM 1983	119	71	48	62	Idiopathic	9	Levodopa	1.0 (0.5–1.2)	≧ 6 weeks	Open, historical control	Significant improvement noted in patients with: "end-of-dose" and "on-off" response ($n = 33$), resting akinesia ($n = 24$), akinesia with falling ($n = 20$); drop-outs due to side effects: mental ($n = 12$), hypotension ($n = 7$), involuntary movements ($n = 5$) and lack of efficacy ($n = 10$)

2.3-3.8 mg; in combined treatment it was 1.6-2.8 mg. Best effects with lisuride were seen on PD tremor.

These first results were further confirmed by controlled trials: LIEBERMAN and co-workers (LIEBERMAN et al. 1980; GOPINATHAN et al. 1981; LIEBERMAN et al. 1983, 1984) treated a total of 63 patients with lisuride. They used within-patient placebo substitution, sometimes under double-blind conditions, sometimes using a "blinded" observer. LEES and STERN (1981) used a within-patient single-blind placebo control. LEWITT et al. (1982) did an intraindividual crossover, lisuride versus bromocriptine, in 30 patients, continuing basic therapy unchanged. The trial was double-blind, maintaining blindness on the physician's side by use of a "blinded" observer. RINNE's group (RINNE 1983; LAIHINEN et al. 1986) compared lisuride with bromocriptine in a double-blind randomized crossover trial, combined with standard levodopa therapy in 20 patients. Basic therapy was continued unchanged. Treatment time was 8-12 weeks per phase. RINNE (1983, 1986) started a double-blind randomized intergroup comparison in beginning PD, using lisuride in an early combination with levodopa plus PDI and comparing it either with levodopa plus PDI alone or with lisuride as monotherapy. After 3 months this trial was continued in an open manner for long-term observation (see p 495).

These reports from early studies with lisuride in PD agree in the following results. Lisuride improves all symptoms of PD: akinesia, rigidity, tremor, posture, gait, dexterity, facial expression, speech. It has been more widely used in advanced PD than in the beginning of the disease. Usually the results have been more satisfactory if lisuride was used in combination with levodopa (plus PDI) therapy rather than as monotherapy. The extent of the effect is highly dependent on the basal situation of the patient and the on-dose regimen. Oral dosage used has been between 0.4 and 5.0 mg/day. The higher doses of lisuride were mostly used in quite disabled patients or during early dose-finding studies.

LIEBERMAN and co-workers (LIEBERMAN et al. 1981) were able to replace levodopa (plus PDI) standard therapy in advanced PD by high doses of lisuride (up to 5 mg/day) as monotherapy for a short time without loss of benefit, and these findings are backed by LEIGUARDA et al. (1985) who used an average dose of 4.5 mg (2.4-6.4) lisuride per day, while others, possibly because they used smaller doses, found that it was less active as monotherapy in advanced PD (SCHACHTER et al. 1980; SCIGLIANO et al. 1985; RIEDERER et al. 1983).

In beginning PD, monotherapy with lisuride has been found to be satisfactory in a number of patients (SCIGLIANO et al. 1982; MARTIGNONI et al. 1986; SCAGLIONI et al. 1986; LAMBERTI et al. 1986). The results from these groups describe only lisuride responders. It becomes evident that lisuride as a monotherapy is not generally advisable as DA substitution therapy in beginning PD. On the other hand, in a number of patients, it may be well tolerated and absolutely sufficient for a number of months or even years in order to postpone the beginning of levodopa therapy. There is a hope that patients on long-term therapy without levodopa will experience a much lower incidence in motor fluctuations than patients on levodopa.

More often oral lisuride was used in combination with levodopa plus PDI. Usually, by such a combination levodopa intake could be reduced by 20%–50%. (ULM 1983; ULM and SUCHY 1986; BIRKMAYER and RIEDERER 1983; BRACCO et al. 1984). In other clinical trials lisuride has been added to the pre-existing therapy without changing its dose (LAIHINEN et al. 1986; GOPINATHAN et al. 1981; LEWITT et al. 1982). An improvement of bradykinesia and rigidity has been reported from all trials as measured by the Columbia University Rating Scale (CURS), the Webster Rating Scale or other scales. There has been some discussion on lisuride's effect on parkinsonian tremor: RINNE (1983), AGNOLI et al. (1983), RUGGIERI et al. (1980), SCHACHTER et al. (1980), PARKES et al. (1981) all reported that they achieved better results on tremor (clinical rating) than on the other symptoms. Others (ULM 1983; ULM and SUCHY 1986; BIRKMAYER and RIEDERER 1983; RIEDERER et al. 1985; and MARTIGNONI et al. 1986) felt that tremor was less responsive than the other characteristic symptoms (clinical rating). Presumably, this controversy is not specific for lisuride, but rather due to different patient populations.

By combining lisuride with levodopa plus PDI, fluctuations in motor performance can be reduced. This has consistently been reported by several authors who reported reduction of "off-hours" from 70% to 130% by a reduction of end-of-dose akinesia and wearing-off of levodopa effects in levodopa- (plus PDI)-treated patients (LIEBERMAN et al. 1980; LIEBERMAN et al. 1983; OBESO et al. 1985; ULM 1983; ULM and SUCHY 1986).

PARKES et al. (1981) have shown that the acute antiakinetic effect of a single i.v. dose of lisuride in patients suffering from wearing-off of levodopa efficacy lasts for about 4 h. Plasma half-life of lisuride is 2–3 h (see Sect. C). Therefore, the effect of lisuride on motor fluctuations which occurs in chronic treatment without fractioning of doses and even when levodopa dose is reduced, seems not to be due to acute direct stimulation of postsynaptic D_2 receptors only. An influence on presynaptic autoreceptors is discussed, which might in the long run normalize DA storage, release or reuptake. However, no evidence for this hypothesis is so far available.

By its dopaminergic activity, lisuride should decrease abnormal involuntary movements (AIMS) which are due to lack of dopaminergic stimulation, and increase AIMS which are due to an excess of dopaminergic activation. In most clinical trials, effects on AIMS are given as a global result. Therefore, figures are contradictory and increase of AIMS may appear as a side effect. Actually, increase of peak-dose dyskinesia is due to dopaminergic overstimulation and may be avoided by reduction either of levodopa or lisuride, or other DA agonistic drugs. In some cases, peak-dose dyskinesia has been reported to be reduced after the introduction of lisuride to the drug regimen, but this seems to be due rather to the concomitant reduction of levodopa than to a direct effect of lisuride. PARKES et al. (1981) have shown dose dependence of the acute antiparkinson effect of lisurde, and that the highest dose (0.2 mg i.v. acutely) resulted in peak-dose dyskinesias.

AGNOLI et al. 1983) reported a reduction of biphasic, "onset" and "end-of-dose" dyskinesias. RINNE (1983; personal communication 1988) pointed out that "off-phase" dystonia and "early morning" dystonia respond well to the

Table 3. Dosage, improvement of parkinsonian disability and occurrence of fluctuations in disability during long-term treatment of parkinsonian patients with levodopa, lisuride or levodopa + lisuride for 12-36 months. (RINNE 1986)

Group	Patients (*n*)	Dose (mg)	Improvement of disability (%)	Fluctuations (*n*)	Fluctuations (%)
Levodopa	30	710 ± 42	54 ± 3	11	37
Lisuride	13	1,6 ± 0,2	41 ± 5	1	8
Levodopa + lisuride	29	400 ± 20 1,1 ± 0,2	58 ± 3	4	14
Lisuride + levodopa	15	0,7 ± 0,2 640 ± 49	48 ± 4	2	13

use of oral lisuride (data on file). Such observations as well as beneficial effects of lisuride in other dystonias (QUINN et al. 1985) would support some partial agonist effect of lisuride. The effect of lisuride in combined treatment is sustained over long time periods. Eventual adjustments of dosage are due to the natural progression of the disease (CARACENI et al. 1984; SCIGLIANO et al. 1985; CAMERLINGO et al. 1986).

Quite recently, a 3-year evaluation of the same ongoing controlled prospective study by RINNE (where only additional patients had been recruited for the lisuride monotherapy group) demonstrated a significant prevention of the development of motor fluctuations by additional lisuride (combination group and secondary combination group) or lisuride monotherapy (with equal therapeutic efficacy regarding the parkinsonian symptomatology, and no clear differences in tolerance; Table 3).

When all patients had reached 3 years of treatment, the difference was even more significant (Table 4).

These results stress the importance of an early combination of lisuride with levodopa in order to avoid or postpone motor fluctuations.

CARACENI et al. (1984) also reported a comparative prospective study in 49 patients with advanced parkinsonism who had increasing problems with levodopa therapy. The study was performed on an open basis and levodopa was re-

Table 4. Prospective randomized controlled study: 3-year interim evaluation (U. RINNE 1987, personal communication)

	I Levodopa	II Levodopa + lisuride	III Lisuride monotherapy	IV Secondary combination
End-of-dose-failure	11/26	2/28	0/6	4/19
Peak-dose dyskinesia	16/28	2/28	0/6	4/19

Therapeutic effects in II slightly better than in I

duced with increasing lisuride or bromocriptine doses (up to a mean daily dose of 2.7 mg lisuride, range 1.2–4.8 mg, or 30.5 mg bromocriptine, range 20–40 mg). The levodopa dose was reduced from a mean of 722 to 476 mg/day resp. from 750 to 495 mg/day. When compared with baseline levels, lisuride caused a significant improvement in the akinesia, rigidity, tremor and total score on the Duvoisin scale, as well as an improvement in the Hoehn-Yahr stage of the patients ($P \leqq 0.05$) whilst bromocriptine had no significant effect on tremor or on the Hoehn-Yahr stage. Off-phases improved in 6 of 8 cases with lisuride, and in 5 of 9 cases with bromocriptine whilst AIMS improved in 9 of 21 cases with lisuride (5 worsened) and in 6 of 18 cases with bromocriptine (7 worsened). Bromocriptine caused more orthostatic hypotension than lisuride, but less psychiatric disturbances. In summary, lisuride was more effective, but also caused more side effects than bromocriptine.

In contrast to the studies of Rinne, who compared lisuride monotherapy with levodopa monotherapy and a combined therapy in early parkinsonism, Rabey et al. (1986) recently started a controlled, randomized, prospective study in parkinsonian patients with beginning problems of wearing-off and motor fluctuations after long-term levodopa therapy. In an elegant design, patients were treated at random either with additional lisuride, or the levodopa dose was increased. After 6 months, the group with additional lisuride treatment (0.4–5 mg/day) had a greater improvement in all parameters tested which was significant (except tremor, which had a low score in this population anyway, so that no further improvement was possible); additional lisuride, however, not only was superior to an increase in levodopa dosage in the parkinsonian symptoms, but also caused significantly more improvement in the symptoms of the levodopa long-term syndrome, i.e. dyskinesias and "off" periods. These data confirm the superiority of a combined form of treatment even in quite advanced forms of parkinsonism. This study has been followed so far for up to 4 years with similar conclusions (Rabey et al. 1987).

Adverse drug effects, which sometimes become dose-limiting and may induce interruption of therapy, include above all: nausea and vomiting; sedation, drowsiness or somnolence; and mental changes.

1. Nausea and vomiting are due to stimulation of DA receptors in the chemoreceptor trigger zone. Usually tolerance develops. More interruptions of therapy due to nausea are reported from those trials where, in the beginning, daily dosage has been increased by 0.2 mg or more, than from those where daily dose increment was 0.1 mg or less. Tolerance develops over a period of 8–12 weeks; there is almost complete cross-tolerance to other dopaminergic ergots, less to levodopa. Therefore, nausea is more severe in patients with beginning PD than in the advanced cases. Its prevalence has been reported to be as high as 66% in some trials and as low as 10% in other trials. DA antagonists, especially domperidone, may be used to suppress nausea or a slow increase in dosing is recommended.

2. Sedation, also referred to as somnolence or drowsiness, is reported frequently, especially at the beginning of therapy. It may occur together with nausea or hypotension, but it certainly occurs independently (LeWitt et al. 1982; Scigliano et al. 1982). When it occurs initially, it might be related to

DA autoreceptor stimulation which rapidly undergoes tolerance development. Somnolence has been the main symptom in two cases of attempted suicide by use of lisuride. Symptoms in these two patients, however, cleared without sequelae after about 24 h.

3. Mental changes, including heavy dreaming, pseudohallucinations, agitation, paranoic reactions and confusion occur during chronic treatment. They are in most cases reversible within 24–48 h upon dose reduction or interruption of therapy. Risk factors for the occurrence of mental changes are: advanced stage of the disease, advanced age of the patient, history of previous episodes of similar nature, symptoms of dementia (which predispose especially to the development of confusion) and concomitant treatment with amantadine and/or anticholinergics. Of great importance is the dosage of any dopaminergic therapy. It is not clear whether these side effects are more or less common with lisuride as compared with other dopaminergic therapies.

Further adverse effects are orthostatic dysregulation (rare, occurring after high dosage in sensitive persons), decrease of blood pressure (frequent, but usually minor), and slight bradycardia. Orthostatic reactions sometimes lead to interruption of therapy, though the majority of trialists state that hypotension is less of a problem with lisuride (SCIGLIANO et al. 1982; ULM and SUCHY 1986; RUGGIERI et al. 1980; LEWITT et al. 1982). Headache, light-headedness and vertigo have been reported. Sleep may be improved owing to improved nocturnal mobility, or may deteriorate by an unknown mechanism, possibly anxiety and agitation (LEIGUARDA et al. 1985).

Weight gain has been reported in a number of patients (RINNE 1986; CARACENI et al. 1984; CAMERLINGO et al. 1986); peripheral edema or fluid retention may rarely occur (LIEBERMAN et al. 1981; BRACCO et al. 1984). Allergic skin reactions have been reported to the manufacturers in a total of 14 cases treated with lisuride (H. Vetter, unpublished data at Schering AG, Berlin, 1987).

2. Parenteral Application

Lisuride has been used intravenously by bolus injection (0.025–0.2 mg acutely within 3 min): (a) to test response to DA agonists in extrapyramidal motor disorders (OBESO et al. 1983) and in acromegaly; and (b) to treat severe akinesia which has developed acutely owing to external factors like accidents, surgery, faulty drug withdrawal.

PARKES et al. (1981) and AGNOLI et al. (1983) have shown an almost immediate onset of action on akinesia and tremor, respectively, after injection of lisuride, thereby proving not only efficacy of the drug, but also rapid penetration of the blood-brain barrier and direct action at the receptor site within the basal ganglia. QUINN et al. (1983) have described a standard procedure for a lisuride test, using 0.15 mg lisuride acutely. They report on 59 tests in 46 patients with PD and other extrapyramidal disorders. They tested the responsiveness of DA receptors and suggest the use of such a test for the prediction of therapeutic results. Though parenteral lisuride seems to cause less nausea than oral lisuride if given in comparable doses (F. MUNDINGER, D. SEEMANN and G. ULM, personal communication), 0.15 mg i.v. is rather a high dosage which causes severe side effects. They may be prevented by use of domperidone.

In acute akinetic states, patients have been successfully treated by temporary use of i.v. lisuride. The applied doses vary between 3×0.025 and 6×0.05 mg/day. For continuation, 0.025 mg lisuride i.v. may be slowly substituted by 0.2 mg lisuride p.o. Side effects are similar to those with oral medication. Special attention should be paid to unexpected vomiting in akinetic patients in order to avoid aspiration. In malignant neuroleptic hyperthermia, lisuride, given at high doses parenterally, offers a promising approach (H. PRZUNTEK (1988), personal communication).

3. Subcutaneous Lisuride: Continuous Dopaminergic Stimulation (see also Chap. 22)

Subcutaneous infusion of lisuride, by means of a microinfusion pump, in PD patients with severe otherwise untreatable fluctuations of motoric performance, has been used by OBESO since 1983. The effects of lisuride infusions on the motor system are comparable to those of levodopa infusions reported by SHOULSON et al. (1975); QUINN et al. (1984) and HARDIE et al. (1984). Lisuride infusions could be used without problems in many patients over long time periods. In most patients motor fluctuations could be controlled, in some the results were extraordinary. Levodopa dosage has to be reduced by 30%–50%. AIMs are reduced too, though some choreic dyskinesias may persist.

Lisuride is mostly given at an infusion rate of about 0.1 mg/h, together with levodopa (in very severe cases) (OBESO et al. 1986a, b; LUQUIN et al. 1986) or as monotherapy (STOCCHI et al. 1986). The usual early side effects (nausea, vomiting, hypotension) can be managed by use of domperidone and do not necessitate interruption of therapy. But the avoidance of psychiatric symptoms (perceptual abnormalities and paranoic reactions) occurring with this type of high dose therapy has turned out to be a problem.

This may partly be due to the fact that those patients for whom this treatment seems appropriate are a population severely at risk (see Sect. F.I.1). They are in an advanced stage of the disease and many of them have already shown mental disturbances in the past. The most important reason may be that the lisuride dosage which becomes bioavailable per day is much higher than the dosages applied by oral treatment (SUCHY, in preparation). Methods to avoid such psychiatric symptoms include: a 12-h daytime infusion instead of 24-h infusions (RUGGIERI et al. 1986, 1987); unfortunately, this is possible only in patients who do not become too akinetic overnight. A reduction of overnight doses and a reduction of total daily doses has proved effective in the patients treated by another group (BITTKAU and PRZUNTEK 1986). The advantage of this approach was confirmed by several case reports (M. RINNE 1986 and W. H. POEWE 1987, personal communications).

Pre-lisuride levodopa dosages may play a decisive role; CRITCHLEY et al. (1986) did not find a way to avoid mental disturbances in their patients who had been formerly treated with an average levodopa dose of 1500 mg/day while those treated by OBESO had formerly been treated with an average of

850 mg/day only. CRITCHLEY found it to be impossible to reduce lisuride and levodopa far enough to avoid disturbances. Patients would become too akinetic, and preferred the former state of fluctuations. Therefore, it seems advisable not to wait for too desolate a situation, but to begin the continuous dopaminergic stimulation as soon as fluctuations develop. No tachyphylaxis has been found in an observation period of up to 2 years. Instead, it has been necessary to reduce dosage over time in most of the patients, which indicates an increasing sensitivity of receptors during continuous stimulation.

Local nonspecific reactions occur in the subcutaneous tissue at the injection site. Usually they are not troublesome and vanish after change of injection site, but, rarely, they are severe enough to induce discontinuation of therapy. As similar reactions are seen with insulin or apomorphine infusions (STIBE et al. 1988), they may be caused by mechanical or chemical irritation of the tissue. As the benefit of lisuride on the motor system is most desirable local irritations may be an acceptable risk. The results obtained with s.c. lisuride infusion have recently been summarized and discussed in a supplement volume to the *Journal of Neural Transmission* edited by J. OBESO, R. HOROWSKI and C. D. MARSDEN (1988).

II. Other Motor Disturbances

Lisuride has been reported to be effective in Meige's disease in dosages between 0.4 and 5 mg, in two of two patients by MICHELI et al. (1982), in two of ten by QUINN et al. (1985) and QUINN (1984), in five of eight by BRINKMANN et al. (1985) and in five of nine by NUTT et al. (1984). These studies were in part controlled by placebo, and there was an improvement of at least 40%. Efficacy was not altered for at least 2 months. Lisuride has also been effective in some cases of cranial dystonia, segmental dystonia and generalized dystonia, but never in writer's cramp. The effect in individual patients is not predictable, dosages needed tend to be higher (3–4 mg) than the average range used in PD, and the effect is sometimes more transient than in PD, where decrease of efficacy is usually clearly correlated with progression of the disease.

Acute lisuride injections (0.2 mg within 3 min) have been shown to reduce on-period dyskinesia in PD patients treated with high doses of levodopa and bromocriptine simultaneously (1150 ± 127 mg, and 28.5 ± 2.4 mg, respectively) though it was not specified whether the improvement was due to reduction of on-phase mobile dystonia or choreic hyperkinesias (FRATTOLA et al. 1982; ALBIZZATI et al. 1982). The same authors showed clear-cut dose-dependent reduction of choreatic movements in Huntington's chorea, with 100 and 150 μg lisuride i.v., controlled in a double-blind way with placebo (FRATTOLA et al. 1983). This effect appeared as early as 10 min, began to decrease at 40 min and could be blocked by haloperidol and sulpiride (high dosage). An effect on a presumed imbalance of D_1/D_2 activity is proposed (BOSIO et al. 1984).

In Shy-Drager syndrome (LEES and BANNISTER 1981, 1983) there is a degeneration of presynaptic neurons. Subsensitivity, but not reduction of the number of postsynaptic dopaminergic receptors has been described. Therefore a high dose of a direct DA agonist should be effective. Unfortunately, in all DA agonists of the ergot type, severe side effects have been dose-limiting. The use of lisuride has been limited by psychotic events at doses between 3 and 5 mg. (In bromocriptine, at doses of 5-10 mg, orthostatic hypotension was dose-limiting; in pergolide, at doses of 0.5-6 mg, nausea and psychiatric side effects prevented further use.) It may well be that finding out what makes Shy-Drager patients so sensitive towards dopaminergic drugs could be helpful in understanding the mechanism of psychosis that can be triggered by these compounds.

Progressive supranuclear palsy. One patient of seven at 2.5 mg (1-5 mg) lisuride showed definite improvement, but eye movements did not respond (NEOPHYTIDES et al. 1982).

Myoclonus. Lisuride reduces or blocks cortical reflex myoclonus. Though there are forms of cortical reflex myoclonus, e.g. flash-induced myoclonus, which are inhibited by dopamine agonists, the action of lisuride is presumably mediated via a serotonergic mechanism, because its effect on cortical reflex myoclonus, above all on that induced by somatosensory input, can be inhibited by methysergide and not by haloperidol and because it parallels the efficacy of 5-HTP infusions. Its clinical efficacy in this condition differs from its effects on the EEG. Lisuride has been used intravenously, 0.1 mg, in order to tests its efficacy, and orally, 3 mg/day for chronic treatment in combination with other drugs (OBESO 1984, OBESO et al. 1983 a, b, ROTHWELL et al. 1984).

III. Migraine

Lisuride is used for the preventive treatment of migraine attacks. The pharmacological basis for its use in migraine has shifted from the more peripheral anti-5-HT effect (PODVALOVA and DLABAC 1972), which should prevent pathological vasodilation of facial and skull vessels, to its central stimulatory effect on presynaptic 5-HT receptors in the raphe dorsalis which slows down central 5-HT activity (HOROWSKI 1982 a). An influence of chronic lisuride treatment on supersensitivity of migraine patients to various pharmacological and physiological stimuli has been discussed as well (HOROWSKI 1982 b).

Lisuride has been shown to reduce frequency and intensity of attacks in various double-blind trials versus placebo (SOMERVILLE and HERRMANN 1978; HERRMANN et al. 1978) or versus methysergide (HERRMANN et al. 1977). The average dosage is 0.025 mg t.i.d., effects occur after 2-3 months treatment. It is equipotent to methysergide, but less side effects have been observed.

The use of lisuride for preventive treatment in cluster headaches seems promising, but sometimes higher doses are needed (RAFFAELLI et al. 1983). The use of lisuride in migraine and its possible mechanism of action has been reviewed recently (HOROWSKI et al. 1986). In hormone-dependent and premenstrual migraine, the prolactin-lowering effect of lisuride may contribute to its therapeutic action (HOROWSKI and RUNGE 1986).

G. Conclusion

In conclusion, lisuride has been shown to be a strong DA agonist which is useful in a variety of clinical conditions. In addition, it has been a useful tool for studying CNS functions in animals and humans. Some of its effects (e.g. on dyskinesias) might be explained by a kind of partial agonistic activity whilst others may relate to its interaction with serotonergic systems. The possibility of parenteral use of lisuride has opened quite new ways of treatment in neurology, and lisuride has now become a standard against which new forms of treatment need to be tested.

References

Adams LM, Geyer MA (1985) Patterns of exploration in rats distinguishes lisuride from lysergic acid diethylamide. Pharmacol Biochem Behav 23:461-468

Agnoli A, Ruggieri S, Baldassarre M, Stocchi F, Denaro A, Falaschi P (1983) Dopaminergic ergots in Parkinsonism. In: Calne DB, Horowski R, McDonald RJ, Wuttke W (eds). Lisuride and other dopamine agonists. Raven, New York, pp 407-417

Ahlenius S, Larsson K, Svensson L (1980) Stimulating effects of lisuride on masculine sexual behaviour. Eur J Pharmacol 64:47-51

Aizawa T, Kutsuzawa T, Otomo E, Goto F, Tazaki Y, Abe H, Omae T, Kameyama M, Ito H (1980) Clinical utility of eunal in treatment of cerberovascular disorders: multi-center double-blind study in comparison with dihydroergotoxine mesylate. Clin Eval 8:577-628

Albizzati MG, Alemani A, Bassi G, Ferrarese C, Frattola L, Trabucchi M (1982) Effetti del lisuride nella patologia discinetico distonica: Considerazioni patogenetiche. In: Agnoli A, Bertolani G (eds) Morbo di Parkinson e compromissione delle attivita' nervose superiori. 8th Meeting LIMPE, Rome, 31 Oct 1981. Guanelli, Rome, pp 223-231

Azuma H, Oshino N (1980) Stimulatory action of lisuride on dopamine-sensitive adenylate cyclase in the rat striatal homogenate. Jpn J Pharmacol 30:629-639

Azuma H, Miyazawa T, Ishikawa T, Oshino N (1982) Effects of an ergot derivative lisuride, on the central nervous system. Stimulatory effect on local cerebral glucose utilization in the rat Folia Pharmacol Jpn 80:69-81

Battaglia G, Titeler M (1981) Direct binding of [^{3}H]lisuride to adrenergic and serotonergic receptors. Life Sci 29:909-916

Bergstrom DA, Bromley SD, Walters JR (1984) Dopamine agonists increase pallidal unit activity: attenuation by agonist pretreatment and anesthesia. Eur J Pharmacol 100:3-12

Besser GM, Wass JAH, Grossmann A, Moult PJA, Bouloux P (1983) Hormonal and clinical effects of dopamine agonists In: Calne DB, Horowski R, McDonald RJ, Wuttke W (eds) Lisuride and other dopamine agonists. Raven, New York, pp 239-254

Birkmayer W, Riederer P (1983) Effects of lisuride on motor function, psychomotor activity, and psychic bahaviour in Parkinson's disease. In: Calne DB, Horowski R, McDonald RJ, Wuttke W (eds) Lisuride and other dopamine agonists. Raven, New York, pp 453-461

Bittkau S, Przuntek H (1986) Psychosis and the lisuride pump. Lancet II:349 (letter)
Bohnet HG, Hanker JP, Horowski R, Wickings EJ, Schneider HPG (1979) Suppression of prolactin secretion by lisuride throughout the menstrual cycle and in hyperprolactinaemic menstrual disorders. Acta Endocrinol. 92:8-19
Bohnet HG, Hilland U, Hanker JP, Schneider HPG (1981) Hormonelle Diagnostik und Therapie der Mastodynie. Wissenschaftl Information 7:87-94
Bosio A, Bassi S, Govoni S, Spano PF, Trabucchi M, Covelli V, Frattola L (1984) Lisuride in extrapyramidal disorders: a possible mechanism of action. In: Hassler RG, Christ JF (eds) Advances in neurology vol 40. Raven, New York, pp 523-526
Bracco F, Meneghetti G, Ferla S, Giometto B, Battistin L (1984) Effetti della lisuride nel trattamento del morbo di Parkinson. Osservazioni preliminari. In: Agnoli A, Bertolani G (eds) LIMPE Atti della 10. Reunione ‚Disorderi autonomici e correlati neuroendocrini delle malattie extrapiramidali'. Guanella, Rome, pp 272-278
Bräutigam M, Flosbach CW, Herken H (1980) Depression of reserpine-induced muscular rigidity in rats after administration of lisuride into the spinal subarachnoid space. Naunyn-Schmiedeberg's Arch Pharmacol 315:177-179
Brazenor RM, Angus JA (1982) Actions of serotonin antagonists on dog coronary artery. Eur J Pharmacol 81:569-576
Brinkmann A, Schumm F, Dichgans J (1985) Therapie des Meige-Syndroms. In: Gänshirt H, Berlit P (eds) Kardiovaskuläre Erkrankungen und Nervensystem, Neurotoxikologie, Probleme des Hirntodes. Springer, Berlin Heidelberg New York, pp 662-665
Burdman JA, Calabrese MT, Harcus CT, MacLeod RM (1982) Lisuride, a dopamine agonist, inhibits DNA synthesis in the pituitary gland. Neuroendocrinology 35:282-286
Burns RS, Calne DB (1981) Treatment of parkinsonism with artificial dopaminomimetics; pharmacokinetic considerations. In: Corsini GU, Gessa GL (eds) Apomorphine and other dopaminomimetics, vol 2. Raven, New York, pp 93-106
Burns, RS, Gopinathan G, Hümpel M, Dorow R, Calne DB (1984) Disposition of oral lisuride in Parkinson's disease. Clin Pharmacol Ther 35:548-556
Camerlingo M, Bottacchi E, D'Alessandro G, Gambaro P, Crippa D, Finelli F, Mamoli A (1986) Un anno di esperienza con ergot-derivati nel trattamento di monoterapia di pazienti Parkinsoniani di prima diagnosi. In: Agnoli A, Battistin L (eds) L.I.M.P.E., Atti della 12. Riunione ‚Morbo di parkinson: le nuove terapie'. Guanella, Rome, 1986, pp 188-196
Cangi F, Fanciullacci M, Pietrini U, Boccuni M, Sicuteri F (1985) Emergence of pain and extra-pain phenomena from dopaminomimetics in migraine. In: Pfaffenrath V, Lundberg PO, Sjaastad O (eds) Updating in headache. Springer, Berlin Heidelberg New York, pp 276-280
Caraceni T, Giovannini P, Parati E, Scigliano G, Grassi MP, Carella F (1984) Bromocriptine and lisuride in Parkinson's disease. In: Hassler RG, Christ JF (eds) Advances in neurology, vol 40. Raven, New York, pp 531-535
Carruba MO, Mantegazza P (1983) Behavioural pharmacology of ergot derivatives. In: Calne DB, Horowski R, McDonald RJ, Wuttke W (eds) Lisuride and other dopamine agonists. Raven, New York, pp 65-77
Carruba MO, Ricciardi S, Negreanu J, Calogero M, Mantegazza P (1980) Effects of lisuride on body temperature of rats and rabbits: relation to microsomal biotransformation and dopaminergic receptor stimulation. Psychopharmacology 70:223-229
Carruba MO, Ferrari P, Rossi AC, Mantegazza P (1983) Effects of antihypertensive drugs on sexual behaviour of male rats. Pharmacol Res Commun 15:367-375

Carruba MO, Ricciardi S, Chiesara E, Spano PF, Mantegazza P (1985) Tolerance to some behavioural effects of lisuride, a dopamine receptor agonist, and reverse tolerance to others, after repeated administration. Neuropharmacology 24:199-206

Carstensen M, Söhnchen R, Braendle W, Bettendorf G (1983) Effects of prolactin-lowering agents in the treatment of disorders of fertility. In: Calne DB, Harowski R, McDonald RJ, Wuttke W (eds) Lisuride and other dopamine agonists. Raven, New York, pp 337-343

Carvey PM, Klawans HL (1983) Effect of chronic lisuride treatment on stereotypical and myoclonic jumping behavior in guinea pigs. In: Calne DB, Horowski R, McDonald RJ, Wuttke W (eds) Lisuride and other dopamine agonists. Raven, New York, pp 97-107

Chiodini PG, Liuzzi A, Silvestrini F, Verde G, Cozzi R, Marsili MT, Horowski R, Passerini F, Luccarelli G, Borghi PG (1978) Size reduction of a prolactin secreting adenoma during a long-term treatment with a dopamine-agonist: (lisuride). Int Symp on Pituitary Microadenomas, Milan (abstract)

Chiodini PG, Liuzzi A, Verde G, Cozzi R, Silvestrini F, Marsili MT, Horowski R, Passerini F, Luccarelli G, Borghi PG (1980) Size reduction of a prolactin secreting adenoma during long-term treatment with the dopamine agonist lisuride. Clin Endocrinol 12:47-51

Chiodini P, Liuzzi A, Cozzi R, Verde G, Oppizzi G, Dallabonzana D, Spelta B, Silvestrini F, Borghi G, Luccarelli G, Rainer E, Horowski R (1981) Size reduction of macroprolactinomas by bromocriptine or lisuride treatment. J Clin Endocrinol Metab 53:737-743

Cote T, Munemura M, Kebabian J (1979) Lisuride hydrogen maleate: an ergoline with beta-adrenergic antagonist activity. Eur J Pharmacol 59:303-306

Cote TE, Eskay RL, Frey EA, Grewe CW, Munemura M, Tsuruta K, Brown EM, Kebabian JW (1983) Actions of lisuride on adrenoceptors and dopamine receptors. In: Calne DB, Horowski R, McDonald RJ, Wuttke W (eds) Lisuride and other dopamine agonists. Raven, New York, pp 45-53

Critchley P, Grandas Perez F, Quinn N, Coleman R, Parkes D, Marsden CD (1986) Psychosis and the lisuride pump. Lancet II:349 (letter)

Cronin MJ, Valdenegro CA, Perkins SN, MacLeod RM (1981) The 7315a pituitary tumor is refractory to dopaminergic inhibition of prolactin release but contains dopamine receptors. Endocrinology 109:2160-2166

Crosignani PG, Ferrari C, Liuzzi A, Benco R, Mattei A, Rampini P, Dellabonzana D, Scarduelli C, Spelta B (1982) Treatment of hyperprolactinemic states with different drugs: a study with bromocriptine, metergoline, and lisuride. Fertil Steril 37:61-67

Da Prada M, Bonetti EP, Keller HH (1977) Induction of mounting behaviour in female and male rats by lisuride. Neurosci Lett 6:349-353

De Cecco L, Venturini PL, Ragni N, Valenzano M, Constantini S, Horowski R (1983) Dopaminergic ergots in lactation and cycle disturbances. In: Calne DB, Horowski R, McDonald RJ, Wuttke W (eds) Lisuride and other dopamine agonists. Raven, New York, pp 291-299

Delitala G, Wass JAH, Stubbs WA, Jones A, Williams S, Besser GM (1979) The effect of lisuride hydrogen maleate, an ergot derivative on anterior pituitary hormone secretion in man. Clin. Endocrinol 11:1-9

Desaga U, Lüdecke D, Kühne D (1983) Prolactin lowering effect of lisuride and reduction of tumor size in patients with prolactinomas. Periodicum Biologorum 85, Suppl 1:73-82

Desaga U, Reich-Schulze E, Gräf K-J, Dorow R, Frahm H (1985) Endocrine effects of the dopamine agonist lisuride in patients with chronic renal failure on long-term hemodialysis. In: MacLeod RM, Thorner MO, Scapagnini U (eds) Prolactin. Basic and clinical correlates. Liviana, Padova, pp 761-771

Dorow R, Gräf K-J, Nieuweboer B, Horowski R (1980) Intravenous lisuride: a new tool for testing responsiveness to dopaminergic agonists and neuroendocrine function. Acta Endocrinol (Kbh) 94, Suppl 234:9 (abstract)

Dorow R, Breitkopf M, Desaga U, Ebeling J, Zimmermann R, Horowski R (1981) Pharmacokinetics and clinical effects of lisuride. Acta Endocrinol 97, Suppl 243:413 (abstract)

Dorow R, Breitkopf M, Gräf K-J, Horowski R (1983) Neuroendocrine effects of lisuride and its 9,10-dihydrogenated analog in healthy volunteers. In: Calne DB, Horowski R, McDonald RJ, Wuttke W (eds) Lisuride and other dopamine agonists. Raven, New York, pp 161-174

Fink H, Morgenstern R (1985) Locomotor effects of lisuride: a consequence of dopaminergic and serotonergic actions. Psychopharmacology 85:464-468

Flosbach CW, Bräutigam M, Dreesen R, Herken H (1982) Action of lisuride on reserpine-induced muscular rigidity in rats after local application into the striatum, ventricular space or the spinal subarachnoid space. Arzneim Forsch Drug Res 32:488-491

Frattola L, Albizzati MG, Bassi S, Ferrarese C, Trabucchi M (1982) 'On-Off' phenomena, dyskinesias and dystonias. Acta Neurol Scand 66:227-236

Frattola L, Albizzati MG, Alemani A, Bassi S, Ferrarese C, Trabucchi M (1983) Acute treatment of Huntington's chorea with lisuride. J Neurol Sci 59:247-253

Fujita N, Saito K, Yonehara N, Watanabe Y, Yoshida H (1979) Binding of [^{3}H] lisuride hydrogen maleate to striatal membranes of rat brain. Life Sci 25:969-974

Fuxe K, Fredholm BB, Ögren S-O, Agnati LF, Hökfelt T, Gustafsson J-A (1978) Ergot drugs and central monoaminergic mechanisms: a histochemical, biochemical and behavioural analysis. Fed Proc 37:2181-2191

Garattini S, Consolo S, Ladinski H (1980) Neuronal links in the CNS: focus on dopaminergic and serotonergic regulation of striatal cholinergic neurons. Pol J Pharmacol Pharm 32:155-164

Gerber R, Barbaz BJ, Martin LL, Neale R, Williams M, Liebman JM (1985) Antagonism of L-5-hydroxytryptophan-induced head twitching in rats by lisuride: a mixed 5-hydroxytryptamine agonist-antagonist? Neurosci Lett 60:207-213

Gessa GL (1988) Agonist and antagonist actions of lisuride on dopamine neurons: electrophysiological evidence In: Obeso JA, Horowski R, Marsden CD (eds) Continuous dopaminergic stimulation in Parkinson's disease. J Neural Transm (Suppl) 27:201-210

Giuliani L, Barreca T, Giberti C, Pescatore D, Magnani G, Martorana G, Bottaro L, Damonte P, Rolandi E (1982) Clinical effects of a prolactin-inhibiting drug, lisuride, in patients with benign prostatic hypertrophy. Curr Ther Res 32:8-16

Glusa E, Markwardt F (1984) Mechanism of metallic mercury oxidation in vitro by catalase and peroxidase. Biochem Pharmacol 33:490-493

Glusa E, Markwardt F, Barthel W (1979) Studies on the inhibition of adrenaline-induced aggregation of blood platelets. Pharmacology 19:196-201

Gopinathan G, Teravainen H, Dambrosia J, Ward C, Sanes J, Stuart W, Evarts E, Calne D (1980) Studies on parkinson disease II: evaluation of lisuride as a therapeutic agent. Annu Course of the American Academy of Neurology, 32nd Meeting, New Orleans (abstract)

Gopinathan G, Teräväinen H, Dambrosia JM, Ward CD, Sanes JN, Stuart WK, Evarts EV, Calne DB (1981) Lisuride in parkinsonism. Neurology (NY) 31:371-376

Graf M, Pletscher A (1979) Shape change of blood platelets — a model for cerebral 5-hydroxytryptamine receptors? Br J Pharmacol 65:601-608

Gräf K-J, Neumann F, Horowski R (1976) Effect of the ergot derivative lisuride hydrogen maleate on serum prolactin concentrations in female rats. Endocrinology 98:598-605

Gräf K-J, Horowski R, El Etreby MF (1977) Effect of prolactin inhibitory agents on the ectopic anterior pituitary and the mammary gland in rats. Acta Endocrinol. 85:267-278

Gräf K-J, Schmidt-Gollwitzer M, Horowski R, Dorow R (1982) Effect of metoclopramide and lisuride on hypophyseal and gonadal function in men. Clin Endocrinol 17:243-251

Halbhübner K, Herken H, Loos D (1978) Experimental neuropathy with parkinson-like muscular rigidity. Arzneim Forsch Drug Res 28:1743-1752

Hara K, Ikoma Y, Oshino N (1982a) Effects of an ergot derivative, lisuride, on the central dopaminergic system — studies of behavioral pharmacology. Folia Pharmacol Jp 80:1-13

Hara K, Ikoma Y, Nakao H, Ezumi K, Oshino N, Shiota C, Sasagawa S (1982b) Central dopaminergic function in stroke-prone spontaneously hypertensive rats (SHRSP): II. Effects of chronic treatment with lisuride on the impaired swimming ability. Folia Pharmacol Jp 80:385-394

Hardie RJ, Lees AJ, Stern GM (1984) On-off fluctuations in Parkinson's disease: a clinical and neuropharmacological study. Brain 107:487-506

Hardt W, Schmidt-Gollwitzer M, Horowski R (1979) Suppression of lactation with lisuride. Gynecol Obstet Invest 10:95-105

Heinlein W, Stracke H, Schröder O, Laube H, Grothe E, Schatz H (1983) Dopaminagonistentherapie bei Hypophysentumoren. Therapiewoche 33:6123-6128

Herrmann WM, Horowski R, Dannehl K, Kramer U, Lurati K (1977) Clinical effectiveness of lisuride hydrogen maleate: a double blind trial versus methysergide. Headache 17:54-60

Herrmann WM, Kristof M, Sastre y Hernandez M (1978) Preventive treatment of migraine headache with a new isoergolenyl derivative. J Int Med Res 6:476-482

Holohean AM, White FJ, Appel JB (1982) Dopaminergic and serotonergic mediation of the discriminable effects of ergot alkaloids. Eur J Pharmacol 81:595-601

Horowski R (1978) Differences in the dopaminergic effects of the ergot derivatives bromocriptine, lisuride and *d*-LSD as compared with apomorphine. Eur J Pharmacol 51:157-166

Horowski R (1982a) Role of monoaminergic mechanisms in the mechanism of action of ergot derivatives used in migraine. In: Rose FC (ed) Advances in migraine. Raven, New York, pp 187-198

Horowski R (1982b) Some aspects of the dopaminergic action of ergot derivatives and their role in the treatment of migraine. Adv Neurol 33:325-334

Horowski R (1983) Pharmacological effects of lisuride and their potential role in further research. In: Calne DB, Horowski R, McDonald RJ, Wuttke W (eds) Lisuride and other dopamine agonists. Raven, New York, pp 127-139

Horowski R (1986) Psychiatric side-effects of high-dose lisuride therapy in parkinsonism. Lancet II:510 (letter)

Horowski R, Dorow R (1981) Influence of estraciol and other gonadal steroids on central effects of lisuride and comparable ergot derivatives. In: Wuttke W, Horowski R (eds) Gonadal steroids and brain function. Springer, Berlin Heidelberg New York, pp 169-181

Horowski R, Gräf K-J (1976) Influence of dopaminergic agonists and antagonists on serum prolactin concentrations in the rat. Neuroendocrinology 22:273-286

Horowski R, McDonald RJ (1983) Experimental and clinical aspects of ergot derivatives used in the treatment of age-related disorders. In: Agnoli A et al. (eds) Aging brain and ergot alkaloids, Raven, New York, Aging 23:283-303

Horowski R, Riedel C (1979) Hypothermic action on lisuride in rats and differences to bromocriptine in the antagonistic effect of neuroleptics. Naunyn-Schmiedeberg's Arch. Pharmacol. 306: 147-151

Horowski R, Runge I (1986) Possible role of gonadal hormones as triggering factors in migraine. Funct. Neurol. 4:405-414

Horowski R, Wachtel H (1976) Direct dopaminergic action of lisuride hydrogen maleate an ergot derivative in mice. Eur.J. Pharmacol. 36:373-383

Horowski R, Neumann F, Gräf K-J (1975) Influence of apomorphine hydrochloride, dibutyryl-apomorphine and lysenyl on plasma prolactin concentrations in the rat. J Pharm Pharmac 27:532-534

Horowski R, Wachtel H (1979) Pharmacological effects of lisuride in rodents mediated by dopaminergic receptors: mechanism of action and influence of chronic treatment with lisuride. In: Fuxe K, Calne DB (eds) Dopaminergic ergot derivatives and motor function. Pergamon, Oxford, pp 237-251

Horowski R, von Berswordt-Wallrabe R, Gräf K-J (1976) Role of serotoninergic mechanisms in the regulation of prolactin secretion and possible mechanism of action of ergot derivatives. In: Endröczi E (ed) Cellular and molecular bases of neuroendocrine processes. Akadémiai Kiadó, Budapest, pp 183-195

Horowski R, Wendt H, Gräf K-J (1977) Inhibition of basal and stimulated prolactin secretion by lisuride in humans, Acta Endocrinol. (Kbh.) 87:234-240

Horowski R, Wachtel H, Dorow R (1986) Dopamine as a deuteragonist in migraine: implications for clinical pharmacology of dopamine agonists. In: Winlow W, Markstein R (eds) The neurobiology of dopamine systems. Manchester University Press, Manchester, pp 415-426

Hümpel M (1983) Pharmacokinetics of lisuride in animals species and humans. In: Calne DB, Horowski R, McDonald RJ, Wuttke W (eds) Lisuride and other dopamine agonists. Raven, New York, pp 411-152

Hümpel M, Nieuweboer B, Hasan SH, Wendt H (1981) Radioimmunoassay of plasma lisuride in man following intravenous and oral administration of lisuride hydrogen maleate; effects on plasma prolactin level. Eur J Clin Pharmacol 20:47-51

Itil TM, Herrmann M, Akpinar S (1974) Lisuride hydrogen maleate, a new psychotropic, discovered by quantitative pharmaco-electroencephalogram. Int J Clin Pharmacol 10:143 (Abstract)

Itil TM, Akpinar S, Herrmann W (1975) Discovery of specific cns effects of lisuride hydrogen maleate — an antimigraine compound. Psychopharmacol Bull 11:67

Jackson DM, Jenkins OF (1985) Hypothesis: bromocriptine lacks intrinsic dopamine receptor stimulating properties. J Neural Transm 62:219-230

Johnson AM, Loew DM, Vigouret JM (1976) Stimulant properties of bromocriptine on central dopamine receptors in comparison to apomorphine, (+)-amphetamine and L-dopa. Br J Pharmac 56:59

Kebabian JW, Kebabian PR (1978) Lergotrile and lisuride: in vivo dopaminergic agonists which do not stimulate the presynaptic dopamine autoreceptor. Life Sci 23:2199-2204

Kehr W (1977) Effect of lisuride and other ergot derivatives on mono-aminergic mechanisms in rat brain. Eur J Pharmacol 41:261-273

Kehr W, Speckenbach W (1978) Effect of lisuride and LSD on monoamine synthesis after axotomy or reserpine treatment in rat brain. Naunyn-Schmiedeberg's Arch Pharmacol 301:163-169

Keller HH, Da Prada M (1979) Central dopamine agonistic activity and microsomal biotransformation of lisuride, lergotrile and bromocriptine. Life Sci 24:1211-1222

Krause W, Nieuweboer B, Ruggieri St, Stocchi F, Suchy I (1988) Pharmacokinetics of lisuride after subcutaneous infusion. In: Obeso JA, Horowski R, Marsden CD (eds) Continuous dopaminergic stimulation in Parkinson's disease. J Neural Trans (Suppl) 27:71-74

Krejci I, Schuh J, Pragerova H, Dlabac A (1985) Lisuride and transdihydrolisuride: differences in action on central dopaminergic function in dependence on the location and the state of receptors. Pol J Pharmacol Pharm 37:263-272

Ladinsky H (1977) Effect of lisuride on the central cholinergic system. In: Istituto di Ricerche Farmacologiche 'Mario Negri', Milan, pp 5-18

Laihinen A, Rinne UK, Sonninen V, Suchy I (1986) Lisuride in comparison to bromocriptine in the treatment of Parkinson's disease. 26th Scandinavian Congress of Neurology, Uppsala, June 11-14, 1986, Uppsala, J Med Sci Suppl 43:92

Lamberti P, Bandiera L, De Liso E, Demari M, Fiore P, Iliceto G, Margari L (1986) Rilievi clinici, neuropsicologici e neurofisiologici in Parkinsoniani trattati con lisuride. In: Agnoli A, Battistin L (eds) L.I.M.P.E., Atti della 12. Riunione Morbo di parkinson: le nuove terapie. Guanella, Rome, pp 197-206

Lees AJ, Bannister R (1981) The use of lisuride in the treatment of multiple system atrophy with autonomic failure (Shy-Drager syndrome). J Neurol Neurosurg Psychiat 44:347-351

Lees AJ, Bannister R (1983) Treatment of the Shy-Drager syndrome with dopaminergic ergots. In: Calne DB, Horowski R, McDonald RJ, Wuttke W (eds) Lisuride and other dopamine agonists. Raven, New York, pp 395-405

Lees AJ, Stern GM (1981) Pergolide and lisuride for levodopa-induced oscillations. Lancet II:577

Leiguarda R, Micheli F, Fernandez Pardal M (1985) Lisuride en la enfermedad de Parkinson. Medicina (Buenos Aires) 45:29-34

LeWitt PA (1986) Clinical and pharmacological aspects of the antiparkisonian ergolene lisuride. In: Fahn S et al. (eds) Recent developments in Parkinson's disease. Raven, New York, pp 347-354

LeWitt PA, Gopinathan G, Ward CD, Sanes JN, Dambrosia JM, Durso R, Calne DB (1982) Lisuride versus bromocriptine treatment in parkinson disease: a double-blind study. Neurology (Ny) 32:69-72

Lieberman AN, Leibowitz M, Neophytides A, Kupersmith M, Mehl S, Kleinberg D, Serby M, Goldstein M (1979) Pergolide and lisuride for Parkinson's disease. Lancet II/8152:1129-1130

Lieberman A, Neophytides A, Leibowitz M, Kupersmith M, Pact V, Walker R, Zarosin N, Goodgold A, Goldstein M (1980) The use of two new dopamine agonists: pergolide and lisuride in Parkinson's disease. In: Rinne UK, Klinger M, Stamm G (eds) Parkinson's disease - current progress, problems and management. Elsevier/North Holland Biomedical, Amsterdam, pp 335-356

Lieberman A, Goldstein M, Neophytides A, Kupersmith M, Leibowitz M, Zasorin N, Walker R, Kleinberg D (1981 a) Lisuride in parkinson disease: Efficacy of lisuride compared to levodopa Neurology (Ny) 31:961-965

Lieberman AN, Goldstein M, Leibowitz M, Neophytides A, Gopinathan G, Walker R, Pact V (1981 b) Lisuride combined with levodopa in advanced parkinson disease. Neurology (Ny) 31:1466-1469

Lieberman AN, Goldstein M, Neophytides A, Leibowitz M, Gopinathan G, Goodgold A, Pact V, Walker R (1981 c) Use of lisuride in advanced Parkinson's disease. N Y State J Medicine Nov 81:1751-1755

Lieberman AN, Goldstein M, Gopinathan G, Leibowitz M, Neophytides A, Walker R, Hiesiger E (1983 a) Further studies with lisuride in Parkinson's disease. Eur Neurol 22:119-123

Lieberman AN, Goldstein M, Gopinathan G, Neophytides A, Leibowitz M, Walker R, Hiesiger E (1983 b) Lisuride in Parkinson's disease and related disorders. In: Calne DB, Horowski R, McDonald RJ, Wuttke W (eds) Lisuride and other dopamine agonists. Raven, New York, pp 419-429

Lieberman AN, Gopinathan G, Neophytides A, Leibowitz M, Goldstein M (1984) Pergolide and lisuride in advanced Parkinson's disease. Adv Neurol 40:503-507

Liuzzi A, Dallabonzana D, Oppizzi G, Verde GG, Cozzi R, Chiodini P, Luccarelli G (1985) Low doses of dopamine agonists in the long-term treatment of macroprolactinomas. N Engl J Med 313:656-659

Loew DM, Van Deusen EB, Meier-Ruge W (1978) Effects on the central nervous system. In: Berde B, Schild HO (eds) Ergot alkaloids and related compounds. Springer, Berlin Heidelberg New York, pp 421-531

Loos D, Halbhübner K, Herken H (1977) Lisuride, a potent drug in the treatment of muscular rigidity in rats, Naunyn-Schmiedeberg's Arch Pharmacol 300:195-198

Loos D, Halbhübner K, Kehr W, Herken H (1979) Action of dopamine agonists on parkinson-like muscle rigidity induced by 6-aminonicotinamide. Neuroscience 4:667-676

Lüdecke DK, Herrmann, H-D, Hörmann C, Desaga U, Saeger W (1983) Comparison of effects of dopamine agonists and microsurgery in GH- and PRL-secreting adenomas. In: Calne DB, Horowski R, McDonald RJ, Wuttke W (eds) Lisuride and other dopamine agonists. Raven, New York, pp 271-289

Luquin MR, Obeso JA, Martinez-Lage JM, Tresguerres J, Parada J, Nieuweboer B, Dorow R, Horowski R (1986) Parenteral administration of lisuride in Parkinson's disease. In: Adv Neurol 45:561-568

MacLeod RM (1977) Influence of dopamine, serotonin, and their antagonists on prolactin secretion. Prog Reprod Biol 2:54-68

Mannesmann G, Haberey M, Müller B, Goedecke H (1979) Pharmacological characterization of the cardiovascular activity of lisuridhydrogenmaleate. Naunyn-Schmiedeberg' Arch. Pharmacol. Suppl 308:72 (abstract)

Marek J, Rezábek K, Srámková (1982) Medikamentöse Behandlung der Hyperprolaktinämie, Akromegalie und Cushing-Krankheit. Ber Ges Inn Med 13:246-248

Marini JL, Sheard MH (1981) On the specificity of a cat behavior model for the study of hallucinogens. Eur J Pharmacol 70:479-487

Marini JL, Jacobs BL, Sheard MH, Trulson ME (1981) Activity of a non-hallucionogenic ergoline derivative, lisuride, in an animal behavior model for hallucinogens. Psychopharmacology 73:328-331

Martignoni E, Suchy I, Pacchetti C, Sinforiani E, Sandrini G, Nappi G (1986) Lisuride in Parkinson's disease: naive and long-term treatment. Curr Ther Res 39:696-708

McCall RB (1982) Neurophysiological effects of hallucinogens on serotonergic neuronal systems. Neurosci Biobehav Rev 6: 509-514
McCall RB, Aghajanian GK (1980) Hallucinogens potentiate responses to serotonin and norepinephrine in the facial motor nucleus. Life Sci 26:1149-1156
McCall RB, Humphrey SJ (1982) Involvement of serotonin in the central regulation of blood pressure: Evidence for a facilitating effect on sympathetic nerve activity. J Pharm Exp Ther 222:94-102
McDonald RJ, Horowski R (1983) Lisuride in the treatment of parkinsonism. Eur Neurol 22:240-255
McPherson GA, Beart PM (1983) The selectivity of some ergot derivatives for alpha-1- and alpha-2-adrenoceptors of rat cerebral cortex. Eur J Pharmacol 91:363-369
Mereu G, Muntoni F, Collu M, Boi V, Gessa GL (1986) Delayed blockade of dopamine autoreceptors by lisuride In: Biggio S, Spano FP, Toffano G, Gessa GL (eds) Modulation of central and peripheral transmitter function. Liviana/Springer, Berlin Heidelberg New York, pp 597-601
Mereu G, Hu XT, Wang RY, Westfall TC, Gessa GL (1987) Failure of subchronic lisuride to modify A10 dopamine autoreceptors' sensitivity. Brain Res 408:210-214
Micheli F, Fernandez Pardal MM, Leiguarda RC (1982) Beneficial effects of lisuride in Meige disease. Neurology (Ny) 32:432-434
Misurek J, Morávek Z, Náhunek K (1978) Lisurid (Lysenyl SPOFA) in the treatment of organic psychosyndrome in involution, Activ Nerv Sup (Praha) 20:87-88
Munemara M, Eskay RL, Kebabian JW, Long R (1980) Release of alpha-melanocyte-stimulating hormone from dispersed cells of the intermediate lobe of the rat pituitary gland: involvement of catecholamines and adenosine 3′,5′-monophosphate, Endocrinology 106:1795-1803
Neophytides A, Lieberman AN, Goldstein M, Gopinathan G, Leibowitz M, Bock J, Walker R (1982) The use of lisuride, a potent dopamine and serotonin agonist, in the treatment of progressive supranuclear palsy. J Neurol Neurosurg Psychiat 45:261-263
Nutt JG, Hammerstad JP, Carter JH, DeGarmo PL (1984) Lisuride treatment of cranial dystonia. 36th Meeting of the American Academy of Neurology, Boston, April 1984
Obeso JA (1984) Untersuchung und Behandlung des Myoclonus. Akt. Neurol. 11:210-213
Obeso JA, Rothwell JC, Quinn NP, Lang AE, Thompson C, Marsden CD (1983) Cortical reflex myoclonus responds to intravenous lisuride, Clin Neuropharmacol 6: 231-240
Obeso JA, Luquin MR, Martinez Lage JM (1985) Lisuride oral en la enfermedad de Parkinson. Medicina Clinica (Barcelona) 85:307-312
Obeso JA, Luquin MR, Martinez-Lage JM (1986) Lisuride infusion pump: a device for the treatment of motor fluctuations in Parkinson's disease. Lancet I:467-470
Obeso JA, Horowski R, Marsden CD (eds) (1988) Continuous Dopaminergic Stimulation in Parkinson's disease. J Neural Transm (Suppl) 27
Otomo E, Hasegawa K, Kuroiwa Y et al. (1981) Effects of lisuride hydrogen maleate on EEG in patients with cerebral vascular impairments and mild senile dementia: multicenter double-blind study, Jpn J Clin Pharmacol Ther 12:377-396
Parkes JD, Schachter M, Marsden CD, Smith B, Wilson A (1981) Lisuride in parkinsonism, Ann Neurol 9:48-52
Pieri L, Keller HH, Burkard W, Da Prada M (1978) Effects of lisuride and LSD on cerebral monoamine systems and hallucinosis, Nature 272:278-280

Pieri L, Keller HH, Laurent J-P, Burkard WP, Pieri M, Bonetti EP, Da Prada M (1983) Behavioral, neurochemical, and electrophysiological effects of lisuride and LSD in animals. In: Calne DB, Horowski R, McDonald RJ, Wuttke W (eds) Lisuride and other dopamine agonists. Raven, New York, pp 89-96

Podvalovà I, Dlabac A (1972) Lysenyl, a new antiserotonin agent. Res Clin Stud Headache 3:325-334

Quinn NP (1984) Anti-parkinsonian drugs today, Drugs 28:236-262

Quinn N, Marsden CD, Schachter M, Thompson C, Lang AE, Parkes JD (1983) Intravenous lisuride in extrapyramidal disorders. In: Calne DB, Horowski R, McDonald RJ, Wuttke W (eds) Lisuride and other dopamine agonists. Raven, New York, pp 383-393

Quinn N, Parkes JD, Marsden CD (1984) Control of on/off phenomenon by continuous intravenous infusion of levodopa. Neurology 34:1131-1136

Quinn NP, Lang AE, Sheehy MP, Marsden CD (1985) Lisuride in dystonia. Neurology 35:766-769

Rabey JM, Treves T, Streifler M, Korczyn AD (1986) Comparison of efficacy of lisuride hydrogen maleate with increased doses of levodopa in parkinsonian patients. Adv Neurol 45:569-572

Rabey JM, Streifler M, Treves T, Korczyn AD (1987) Long-term lisuride in Parkinson's disease. In: Calne DB, Crippa D, Comi G (eds) Int symp parkinsonism and aging, suppl 7. Ital J Neurol Sci Suppl 5:55

Raffaelli E, Martins OJ, dos Santos P, Dágua Filho A (1983) Lisuride in cluster headache. Headache 23:117-121

Reynolds GP, Riederer P (1981) The effects of lisuride and some other dopaminergic agonists on receptor binding in human brain. J Neural Transm 51:107-111

Riederer P, Reynolds GP, Danielczyk W, Jellinger K, Seemann D (1983) Desensitization of striatal spiperone-binding sites by dopaminergic agonists in Parkinson's disease. In: Calne DB, Horowski R, McDonald RJ, Wuttke W (eds) Lisuride and other dopamine agonists. Raven, New York, pp 375-381

Rinne UK (1983) New ergot derivatives in the treatment of Parkinson's disease. In: Calne DB, Horowski R, McDonald RJ, Wuttke W (eds) Lisuride and other dopamine agonists. Raven, New York, pp 431-442

Rinne UK (1986) The importance of an early combination of a dopamine agonist and levodopa in the treatment of Parkinson's disease. In: van Manen J, Rinne UK (eds) Lisuride: a new dopamine agonist and Parkinson's disease. Excerpta Medica, Amsterdam, pp 64-71

Ripka O (1972) Effetti della soministrazione a lungo termine dell' anti-serotoninico lysenyl nelle cefalee di varia eziologia. Minerva Med 63:3266-3271

Rogawski MA, Aghajanian GK (1979) Response of central monaminergic neurons to lisuride: comparison with LSD, Life Sci 24:1289-1298

Rosenfeld MR, Makman MH (1981) The interaction of lisuride, an ergot derivative, with serotonergic and dopaminergic receptors in rabbit brain. J Pharmacol Exp Ther 216:526-531

Rosenfeld MR, Makman MH, Goldstein M (1980) Stimulation of adenylate cyclase in rat striatum by pergolide: influence of GTP. Eur J Pharmacol 68:65-68

Rothwell JC, Obeso JA, Marsden CD (1984) On the significance of giant somatosensory evoked potentials in cortical myoclonus. J Neurol Neurosurg Psychiat 47:33-42

Ruggieri S, Baldassarre M, Del Roscio S, Stocchi F, Palesse N, Martucci N, Agnoli A (1980) Confronto tra l'attivita' terapeutica della lisuride e della bromocriptina nel morbo di parkinson. In: Atti della 7. Riunione. LIMPE, Milan 24-25 Oct 1980

Ruggieri S, Stocchi F,Agnoli A (1986) Lisuride infusion pump for Parkinson's disease, Lancet II:348-349 (letter)

Ruggieri S, Stocchi F, Antonini A, Bellantuono P, Carta A, Agnoli A (1987) Lisuride infusion in chronically treated fluctuating Parkinson's disease: effects of continuous dopaminergic stimulation, New Trends Clin Neuropharmacol 1:55-60

Ruilope L, Garcia-Robles R, Paya C, de Villa LF, Miranda B, Morales JM, Parada J, Sancho J, Rodicio JL (1985) Influence of lisuride, a dopaminergic agonist, on the sexual function of male patients with chronic renal failure. Am J Kidney Dis 5:182-185

Saiani L, Trabucchi M,Tonon GC, Spano PF (1979) Bromocriptine and lisuride stimulate the accumulation of cyclic AMP in intact slices but not in homogenates of rat neostriatum. Neurosci Lett 14:31-36

Scaglioni A, Caffarra P, Passeri S, Saginario M (1986) Trattamento a lungo termine con lisuride nel morbo di Parkinson, In: Agnoli A, Battistin L (eds) LIMPE, Atti della 12. Riunione 'Morbo di parkinson: le nuove terapie'. Guanella, Rome, pp 207-217

Schachter M, Blackstock J, Dick JPR, George RJD, Marsden CD, Parkes JD (1979) Lisuride in Parkinson's disease. Lancet II/8152:1129

Schachter M, Sheehy MP, Parkes JP, Marsden CD (1980) Lisuride in the treatment of parkinsonism, Acta Neurol Scand 32:382-385

Schmidt-Gollwitzer M, Hardt W, Schmidt-Gollwitzer K, Nevinny-Stickel J (1977) Vergleichsstudie über den Einfluß verschiedener Sexual-steroide, 2-Br-Alpha-Ergokryptin und Lisurid-hydrogenmaleat auf die postpartalen Prolaktin-Serumkonzentrationen und die Laktation. Geburtsh Frauenheilk 37:500-508

Schneider WHF (1983) Lisuride and female fertility. In: Calne DB, Horowski R, McDonald RJ, Wuttke W (eds) Lisuride and other dopamine agonists. Raven, New York, pp 331-335

Scholz A, Horowski R (1987) Effects of the prolactin-lowering agent lisuride (Dopergin R) in early pregnancy. In: Teoh ES, Ratnam SS, Wong PC (eds) Ovulation and early pregnancy. Parthenon, Carnforth, pp 159-162

Schorderet M (1978) Dopamine-mimetic activity of cyclic AMP in isolated retinae of the rabbit. Gerontology 24, Suppl 1:86-93

Schwibbe M, Becker D, Wuttke W (1983) EEG and psychological effects of lisuride in women with premenstrual tension. In: Calne DB, Horowski R, McDonald RJ, Wuttke W (eds) Lisuride and other dopamine agonists. Raven, New York, pp 345-355

Scigliano G, Giovannini P, Grassi MP, Soliveri P, Parati E, Mallucci C, Caraceni T (1982) La lisuride nel trattamento del morbo di Parkinson. In: Atti della 9. Riunione LIMPE, Taormina, 29-30 Oct 1982

Scigliano G, Giovannini P, Grassi MP, Soliveri P, Piccolo I, Lamperti E, Girotti F, Caraceni T (1985) Valutazione dell'efficacia a lungo termine della lisuride nel trattamento del morbo di Parkinson. In: Agnoli A, Battistin L (eds) LIMPE, Atti della 11. Riunione 'Semeiologia e terapia: delle malattie extrapiramidali le nuove frontiere.' Guanella, Rome, pp 250-262

Shoulson I, Glaubiger GA, Chase TN (1975) On-off response: clinical and biochemical correlations during oral and intravenous levodopa administration in parkinsonian patients. Neurology 25:1144-1148

Silbergeld EK, Hruska RE (1979) Lisuride and LSD: dopaminergic and serotonergic interactions in the „serotonin syndrome". Psychopharmacology 65:233-237

Somerville BW, Herrmann WM (1978) Migraine prophylaxis with lisuride hydrogen maleate — a double blind study of lisuride versus placebo. Headache 18:75-79

Spano PF, Govoni S, Trabucchi M (1978) Studies on the pharmacological properties of dopamine receptors in various areas of the central nervous system In: Adv Biochem Psychopharmacol 19:155-165

Spano PF, Saiani L, Memo M, Trabucchi M (1980) Interaction of dopaminergic ergot derivatives with cyclic nucleotide system. In: Goldstein M et al. (eds) Ergot compounds and brain function: Neuroendocrine and neuropsychiatric aspects. Raven, New York, pp 95-102

Spano PF, Govoni S, Uzumaki H, Bosio A, Memo M, Lucchi L, Carruba M, Trabucchi M (1983) Stimulation of D_2-dopamine receptors by dopaminergic ergot alkaloids: studies on the mechanism of action. In: Calne DB, Horowski R, McDonald RJ, Wuttke W (eds) Lisuride and other dopamine agonists. Raven, New York, pp 165-117

Stibe CMH, Lees AJ, Kempster PA, Stern GM (1988) Subcutaneous apomorphine in parkinsonian on-off oscillations, Lancet I:403-406

Stocchi F, Ruggieri S, Brughitta G, Agnoli A (1986) Problems in daily motor performances in Parkinson's disease: the continuous dopaminergic stimulation. J Neural Transm 22:209-218

Stumpf WE, Detmer WM, Sar M, Horowski R, Dorow R (1985) Autoradiographic studies with [^{3}H]dopamine, and [^{3}H]domperidone in pituitary and brain. In: Macleod RM, Thorner MO, Scapagnini U (eds) Prolactin. Basic and clinical correlates, Liviana, Padova, pp 27-35

Suchy I (1986) Neurotransmitters in intellectual deterioration and dementia in presenium and senium. In: van Manen J, Rinne UK (eds) Lisuride: a new dopamine agonist and Parkinson's disease. Excerpta Medica, Amsterdam, pp 38-43

Suchy I, Horowski R (1984) Use of ergot derivative lisuride in Parkinson's disease. In: Adv Neurol 40: pp 515-521

Suchy I, Schneider HH, Riederer P, Horowski R (1983) Considerations on the clinical relevance of differences in receptor affinity of various dopaminergic ergot alkaloids. Psychopharmacol. Bull. 19:743-746

Thorner MO, Vance ML, MacLeod RM (1984) Hyperprolactinaemic infertility: some considerations on medical management. In: Rolland R (ed) Advances in fertility control and the treatment of sterility. MTP Press, Boston, pp 23-25

Tissari AH, Gessa GL (1983) Ergot-induced inhibition of dopamine synthesis in striatal synaptosomes: A D-2 DA receptor-mediated mechanism. In: Calne DB, Horowski R, McDonald RJ, Wuttke W (eds) Lisuride and other dopamine agonists. Raven, New York, pp 33-43

Tissari AH, Rossetti ZL, Meloni M, Frau MI, Gessa GI (1983) Autoreceptors mediate the inhibition of dopamine synthesis by bromocriptine and lisuride in rats. Eur J Pharmacol 91:463-468

Ulm G (1983) Experience with lisuride in the treatment of Parkinson's disease. In: Calne DB, Horowski R, McDonald RJ, Wuttke W (eds) Lisuride and other dopamine agonists. Raven, New York, pp 463-472

Ulm G, Suchy I (1986) Drug treatment of Parkinson's disease with special reference to lisuride. In: van Manen J, Rinne UK (eds) Lisuride: a new dopamine agonist and Parkinson's disease. Excerpta Medica, Amsterdam, pp 55-63

Uzumaki H, Govoni S, Memo M, Carruba MO, Trabucchi M, Spano PF (1982) Effects of GTP and sodium on rat striatal dopamine receptors labeled with lisuride. Brain Res. 248:185-187

van Dam LJ, Rolland R (1981) Lactation-inhibiting and prolactin-lowering effect of lisuride and bromocriptine: a comparative study. Eur J Obstet Gynec Reprod Biol 12:323-330

Venturini PL, Horowski R, Maganza C, Morano S, Pedretti E, Ragni N, Semino F, De Cecco L (1981) Effects of lisuride and bromocriptine on inhibition of lactation and on serum prolactin levels: comparative double-blind study. Eur J Obstet Gynec Reprod Biol 11:395-400

Verde G, Chiodini PG, Liuzzi A, Cozzi R, Favales F, Botalla L, Spelta B, Dalla Bonzana D, Rainer E, Horowski R (1980) Effectiveness of the dopamine agonist lisuride in the treatment of acromegaly and pathological hyperprolactinemic states. J Endocrinol Invest 1:105 414

Votava Z, Podvalova I, Vojtechovsky M (1966) Unterschiede in der pharmakologischen Wirkung von halluzinogenen und nicht halluzinogenen Lysergsäure-Derivaten (LSD-25, Deseril, Lysenyl, Mesenyl). Arzneim Forsch Drug Res 16:220-222

Wachtel H (1978) Inhibition of locomotor activity of rats by low doses of different dopamine (DA) receptor agonists. Naunyn-Schmiedeberg's Arch Pharmacol 302 (Suppl): R 59 (abstract)

Wachtel H, Kehr W, Schlangen M (1986) Involvement of dopamine auto- and postsynaptic receptors in locomotor effects of lisuride in rats after systemic or intracerebral administration. In: van Manen J, Rinne UK (eds) Lisuride: a new dopamine agonist and Parkinson's disease. Excerpta Medica, Amsterdam, pp 11-23

Wachtel H, Rettig KJ, Löschmann PA (1988) Effect of chronic subcutaneous minipump infusion of lisuride upon locomotor activity in rats. In: Obeso JA, Horowski R, Marsden CD (eds) Continuous dopaminergic stimulation in Parkinson's disease. J Neural Transm (Suppl) 27:177-183

Walters JR, Baring MD, Lakoski JM (1978) Effects of dopamine agonists on dopaminergic unit activity. In: Usdin E, Kopin IJ, Barchas J (eds) Catecholamines: basic and clinical effects, vol 1. Pergamon, New York, pp 637-639

Walters JR, Baring MD, Lakoski JM (1979) Effects of ergolines on dopaminergic and serotonergic single unit activity In: Fuxe K, Calne DB (eds) Dopaminergic ergot derivatives and motor function. Pergamon, Oxford, pp 207-221

von Werder K, Fahlbusch R, Rjosk H-K (1983) Dopamine agonists and prolactinomas: clinical and therapeutic aspects In: Calne DB, Horowski R, McDonald RJ, Wuttke W (eds) Lisuride and other dopamine agonists. Raven, New York, pp 255-269

White FJ (1986) Comparative effects of LSD and lisuride: clues to specific hallucinogenic drug actions, Pharmacol Biochem Behav 24:365-379

White FJ, Wang RY (1982a)Lysergic acid diethylamide (LSD) and lisuride: differentiation of their neuropharmacological actions Science 216:535-537

White FJ, Appel JB (1982b) The role of dopamine and serotonin in the discriminative stimulus effects of lisuride, J Pharmacol Exp Ther 221:421-427

White FJ, Wang RY (1983) Comparison of the effects of LSD and lisuride on A10 dopamine neurons in the rat. Neuropharmacology 22:669-676

Winkelmann W, Allolio B, Deuß U, Heesen D, Kaulen D, Wilcke O (1983) Medikamentöse Langzeittherapie bei Patienten mit Prolaktin-produzierenden Hypophysentumoren. Akt Endokr Stoffw 4:163-168

Zikán V, Semonsky M (1968) Mutterkorn-Alkaloide: 31. Mitt.: Ein Beitrag zur Herstellung von N-(D-6-Methyl-8-isoergolenyl)-*N'*,*N'*-diäthylharnstoff. Pharmazie 23:147-148

CHAPTER 20

Domperidone and Parkinson's Disease

J. D. PARKES

A. Introduction

Levodopa and dopamine agonists cause widespread stimulation of dopamine systems inside and outside the brain. The therapeutic effect of levodopa in parkinsonism is thus accompanied by cardiovascular, gastrointestinal, hormonal, respiratory, and mental problems. It has not as yet proved possible to separate the nigrostriatal from other central effects of dopamine, although peripheral effects of levodopa have been limited in two different ways.

1. The first successful strategy to achieve this was the coadministration of levodopa with inhibitors of extracerebral dopamine decarboxylase. Two such inhibitors, benserazide and carbidopa, have received extended therapeutic trials, and have been shown in particular to limit nausea and vomiting, and cardiovascular abnormalities due to levodopa. The use of decarboxylase inhibitors in combination with levodopa has, in addition, a four- to fivefold levodopa-sparing effect, but does not completely overcome all the peripheral problems connected with the use of levodopa alone. Carbidopa does not completely abolish gastric side effects induced by levodopa, which persist in 10%–15% of all subjects (RINNE and MÖLSÄ 1979). Peripheral decarboxylase inhibitors are of no value in preventing peripheral dopamine-stimulant effects of dopamine agonist drugs such as bromocriptine. Decarboxylase inhibitors per se have few if any serious unwanted effects, but both carbidopa and benserazide reduce the oxidative metabolism of tryptophan (BENDER 1980). They interfere with niacin metabolism, and may induce niacin depletion (BENDER et al. 1979), although in subjects with normal dietary intake, they do not cause pellagra. In rats, HO and SMITH (1982) have shown that decarboxylase inhibitors interfere with melatonin synthesis, possibly also the case in humans. Given without levodopa, decarboxylase inhibitors induce hyperprolactinaemia (PONTIROLLI et al. 1977).

2. Most dopamine receptor antagonists will prevent peripheral dopamine receptor stimulation, but unlike carbidopa and benserazide they enter the brain and therefore prevent the therapeutic effect of levodopa in Parkinson's disease. Domperidone holds a unique position in the management of Parkinson's disease, since it is the only available dopamine-blocking antiemetic drug with limited brain penetration that does not negate the desired central effects of dopamine receptor stimulation (CRITCHLEY et al. 1985).

B. Domperidone

Domperidone is a benzimidazolinic derivative that acts as a peripheral dopamine antagonist that does not readily cross the blood-brain barrier (COSTALL et al. 1978; LADURON and LEYSEN 1979; HEYKANTS et al. 1981; MICHIELS et al. 1981). Domperidone has been reported to produce high concentrations in the gastrointestinal tract, but low concentrations in the brain. In contrast to neuroleptic drugs and metoclopramide, it is practically devoid of central effects. Domperidone was developed, like metoclopramide hydrochloride, to control nausea and vomiting, and to increase gastrointestinal motility.

I. Animal Pharmacology

Most of the evidence that domperidone has major peripheral but few central effects derives from animal experiments. In the rat, domperidone is practically unable to reach dopamine receptors in the striatum, and only antagonises apormorphine-induced stereotypies at high dosage (LADURON and LEYSEN 1979). In vitro experiments on isolated guinea pig stomach show that domperidone inhibits the relaxation induced by exogenous dopamine (VAN NUETEN et al. 1978; VAN NUETEN and JANSSEN 1978). The wide dissociation between the peripheral and central effects of domperidone has been clearly demonstrated in animal studies in a number of ways.

1. After systemic administration of domperidone, relatively large concentrations accumulate outside the blood-brain barrier.
2. Domperidone has little effect on plasma and striatal concentration of the dopamine metabolite homovanillic acid.
3. There is a marked difference between the effect of intracerebrally and systemically administered domperidone in antagonising the behavioural effects of dopamine. Intracerebral domperidone, in contrast, is as active as haloperidol, sulpiride or fluphenazine in antagonising the behavioural effects of intracerebral dopamine or amphetamine.
4. A much lower intravenous domperidone dosage is needed to antagonise the peripheral (emetic) than central effects of apomorphine (for review see BROGDEN et al. 1982). On the other hand, domperidone antagonises peripheral effects of dopamine agonists, including apomorphine-induced increased femoral blood flow in dogs (WILLEMS et al. 1981), bromocriptine-induced increase in ventricular fibrillation threshold in dogs (FALK et al. 1981), and dopamine-induced neurogenic vasodilation in dogs (WILLEMS et al. 1981).

II. Pharmacokinetics in Humans

Domperidone is usually given *by mouth* in doses of 10–20 mg thrice daily, although it is also absorbed after intramuscular and rectal administration. The *intravenous* use of domperidone should be avoided, since there are reports of sudden death, cardiac arrest, or clinically significant arrhythmias in patients given domperidone by this route. However, most of these subjects have had malignancies, and the domperidone dose has been very high. Also, subjects have had low plasma potassium levels when recorded (Joss et al. 1982); ROUSSAK et al. 1984; GIACCONE et al. 1984; OSBORNE et al. 1985).

Following oral or intramuscular administration of domperidone, peak plasma levels are reached within 30 min. Domperidone shows extensive first-pass elimination, and bioavailability is considerably lower after oral than after parenteral administration. Rectal doses give a similar bioavailability to oral doses. Absorption of domperidone is somewhat delayed by a previous meal. Despite these indications of poor bioavailability of domperidone following oral administration, many clinical trials show the excellent therapeutic value of oral domperidone at low dose levels (e.g. 10-20 mg t.i.d.) in patients suffering from various disturbances of gastric motility and emptying (HEYKANTS et al. 1981).

III. Behavioural Effects

Domperidone is an effective drug in the control of nausea and vomiting due to a wide variety of causes. Domperidone will prevent cytotoxic, anaesthetic, and antiparkinsonian drug-induced vomiting. In Parkinson's disease, this effect has enabled subjects to tolerate higher doses of levodopa or bromocriptine than would be otherwise achieved. In animals, domperidone in dosages up to 2.5 mg/kg is not effective against copper sulphate-induced emesis (which acts centrally) but in low doses (0.026-0.056 mg/kg) will prevent levodopa-induced sickness in dogs (SHUTO et al. 1980). Although domperidone effectively alleviates gastrointestinal symptoms, and vomiting due to various diseases, including uremia, pancreatitis, hepatitis, gastroenteritis, or due to surgical procedures, it does not consistently relieve hiccup.

Studies in humans have shown that oral domperidone increases lower oesophageal sphincter pressure, increases the duration of antral and duodenal contractions, increases the gastric emptying of liquids and semisolids, such as a barium meal, and shortens the stationary phase for solids. Delayed gastric emptying following levodopa, apomorphine, and dopamine agonist ergot derivatives can be prevented by domperidone.

Domperidone in oral dosages lower than 100 mg/day is practically devoid of central activity in humans. In a review of the actions of domperidone, BROGDEN et al. (1982) stressed that no extrapyramidal reaction had been reported during controlled therapeutic trials. However, although infrequent, acute dystonia and akathisia have been attributed to domperidone.

SOL et al. (1980) described a 4-month-old infant, in good health apart from episodes of vomiting who, when given domperidone 0.33 mg/kg every 6-8 h, became stiff with retrocollis and ocular deviation. DEBONTRIDDER (1980a, b) described a 27-year-old woman, vomiting with renal infection, who developed opisthotonus and myoclonus 10 min after intravenous administration of domperidone 10 mg. This subject had a similar extrapyramidal reaction with metoclopramide. GONCE et al. (1982) reported a 14-year-old patient with myeloblastic leukaemia treated with oral domperidone 30 mg for 4 days without adverse effect, but who developed an oculogyric crisis on the 5th day, 3 h after intravenous administration of domperidone 20 mg. VAN DAELE et al. (1984) and LABATTO et al. (1982) have also described acute dystonia following administration of domperidone. LESSER and BATEMAN (1985) described akathisia in three of eight subjects with Parkinson's disease given domperidone in high oral dosage (120 mg/day) in addition to levodopa.

There are no reports of domperidone-induced parkinsonism. In a study designed to determine the potential for extrapyramidal reactions in patients treated with higher than usual doses of domperidone for 15 days, HAASE et al. (1978) showed that a daily dose of 120 mg caused no change in fine motor activity, as determined by handwriting, in 31 male subjects.

Unlike many neuroleptics, domperidone has little or no sedative effect, although a small number of parkinsonian subjects given domperidone in high oral dosage (120 mg/day) report some sedation (N. LANGDON 1985, personal communication). Disturbances of sleep-wakefulness induced by an emetic dose of apomorphine in dogs, are antagonised by pimozide 0.063 mg/kg, but not by domperidone 0.16 mg/kg (WANQUIER et al. 1980). Peripheral stimulation in vagal territory has been reported to promote sleep (KUKORELLI and JUHASZ 1977), although gastrointestinal effects of domperidone have not been associated with any major alteration in sleep or wakefulness.

IV. Effect on the Pharmacokinetic Behaviour of Levodopa

Levodopa is readily absorbed following oral administration, although there is considerable interindividual variation in the pattern of absorption in subjects with Parkinson's disease. These variations have been the subject of intensive study. WADE et al. (1974) showed that after oral administration of levodopa 1 g to parkinsonian patients, peak plasma concentrations of dopa ranged from 0.25 to 2.42 μg/ml, and occurred from 30 min to 4 h after ingestion.

Following the oral administration of levodopa, two or more peaks of plasma dopa concentration occur in approximately 40% of all subjects. The explanation of this phenomenon is uncertain. The first peak may derive partially from stomach and the second from small bowel absorption, although gastric absorption is limited and both peaks probably derive from absorption in the small bowel.

Variation in the speed and degree of levodopa absorption depends on many factors, in particular age, levodopa formulation, intake with levodopa of additional drugs such as anticholinergics with altered gastric motility, the inherent tendency for slow or fast gastric emptying, and the presence or absence of food in the stomach. Major factors accounting for slow absorption of levodopa include (a) the presence of food in the stomach (MORGAN et al. 1971); (b) high gastric acidity, which may delay stomach emptying and increase the metabolism of levodopa in the stomach; and (c) gastric stasis induced by levodopa itself. These factors may be influenced by domperidone, with consequent changes in the speed and degree of levodopa absorption.

Gastric emptying time is prolonged by dopamine and dopamine agonists, and reduced by gastrokinetic antagonists, including domperidone.

The evidence for the presence of dopamine D_2 receptors in the human gastrointestinal tract of is indirect, largely based on experiments in the guinea pig (VAN NUETEN and JANSSEN 1978; VALENZUELA and DOOLEY 1984), although there is no doubt of the major gastrokinetic effect of domperidone. SCHUURKES and VAN NUETEN (1981) suggested that dopamine stimulates gastric dopamine receptors, and levodopa may cause gastric stasis for 1-2 h.

In most normal subjects, as well as those with Parkinson's disease, stomach emptying time for liquids, as measured by impedance and dye-dilution techniques, is fairly rapid, and the addition of levodopa or domperidone in usual dosages does not greatly prolong or shorten this liquid emptying time. Results with solid foods may be different. Using the method of epigastric impedance developed by SUTTON et al. (1985) to measure gastric emptying, P. CRITCHLEY (1987, personal communication) has shown that mean stomach half-emptying times for liquid meals in untreated patients with Parkinson's disease are usually 10-13 min. However, occasionally this time is markedly prolonged following oral administration of levodopa 500 mg. In these subjects, levodopa prolongation of gastric emptying time can be prevented by oral pretreatment with domperidone 20 mg. Reduction in gastric emptying time with domperidone has two consequences in Parkinson's disease, both of which may increase levodopa bioavailability to some extent. Orally administered levodopa is extensively metabolised in the stomach prior to absorption of levodopa from the intestine (RIVIERA-CALIMLIM et al. 1970). Reduction in levodopa-induced delay in gastric emptying with domperidone may therefore result in reduced gastric levodopa wastage. In addition, rapid gastric emptying results in smooth delivery of levodopa to the small intestine, the main site of absorption.

SHINDLER et al. (1984) showed that in 15 patients with idiopathic Parkinson's disease given levodopa 500 mg orally both alone and with domperidone pretreatment, domperidone pretreatment was followed by a slight increase in the peak plasma levodopa concentrations achieved; and peak plasma levodopa levels occurred about 10 min earlier than when levodopa was given alone. These results were interpreted to indicate that domperidone prevented levodopa-induced gastric stasis, and hastened the delivery and subsequent absorption of levodopa to the small bowel.

C. Effect of Domperidone on Levodopa Response

I. Therapeutic Response to Levodopa in Parkinsonism

The therapeutic response to levodopa, bromocriptine and other dopamine agonists in parkinsonism is unaltered by low dosage domperidone (20-60 mg/day), although it may be somewhat impaired with oral dosages of domperidone above 100 mg/day. Results of reported trials are reviewed in Table 1. Some studies (e.g. AGID et al. 1979, 1981) have indicated that with bromocriptine, higher dosages and a greater therapeutic response can be achieved by combination with domperidone, owing to limitation of nausea and vomiting.

Table 1. Results of studies of domperidone in combination with levodopa, or dopamine agonists, in Parkinson's disease

Reference	No. of patients	Domperidone dose, period	Antiparkinsonian drug	Effects
BOGAERTS et al. 1979	13	30–90 mg/day p.o., 4 weeks	Levodopa 700–1500 mg/day	Levodopa-induced nausea disappeared in 11 patients, decreased in 1. Action within 1 week of starting domperidone. Domperidone better than metoclopramide. One patient dry mouth
GRIMES and MAGNER 1984	1[a]	40 mg/day	Metoclopramide, but not domperidone, induced parkinsonism in this subject. Both drugs controlled sickness	
SHINDLER et al. 1984	15	10–40 mg p.o., i.v. or i.m., single dose studies	Levodopa 500 mg	Two subjects sick on levodopa alone, none with domperidone pretreatment
LANGDON et al. 1986	20	60–120 mg/day, 4 weeks	Levodopa 1500–3000 mg/day	Slow deterioration in parkinsonism in eight subjects changed from decarboxylase inhibitors to domperidone combined with levodopa. However, levodopa dose equivalents may have been incorrect
LESSER and BATEMAN 1985	8	120 mg/day, period not stated	Levodopa dose not stated	One subject akathisia. One subject "personality change". One subject increase in parkinsonian symptoms
YAMADA and UONO 1980	6	30–60 mg, 19 months	Levodopa 2400 mg/day	No influence of domperidone on symptoms of parkinsonism

Reference	No. of patients	Domperidone dose, period	Antiparkinsonian drug	Effects
NAGAOKA 1980	148[b]	Dose variable, six subjects given > 60 mg/day	Variable dose levodopa	Improvement in levodopa-induced vomiting, nausea, upper abdominal pain, and heartburn, without alteration in therapeutic effect of levodopa
CORSINI et al. 1979	4	100 μg/kg i.m., single dose	Apomorphine 20 μg/kg	Nausea and vomiting, but not yawning with apomorphine prevented by domperidone
AGID et al. 1979	17	60 mg p.o.	Bromocriptine 148 mg/day with domperidone. Bromocriptine 92 mg/day without domperidone	Nausea, vomiting, hiccups, due to bromocriptine, prevented by domperidone. Higher bromocriptine doses tolerated with addition of domperidone
AGID et al. 1981	42	Long-term follow-up of study by AGID et al. 1979		In long term, domperidone discontinued in most subjects
POLLAK et al. 1981	6	20-100 mg p.o.	Bromocriptine 20 mg p.o.	Domperidone 20-100 mg protected against nausea and vomiting
SCHACHTER et al. 1980b	11	60 mg p.o.	Lisuride 0.6-4.8 mg/day	Not stated, but lisuride addition to levodopa resulted in vomiting in only one subject

[a] One diabetic subject with vomiting
[b] Multicentre study

The design of some other studies (e.g. NAGAOKA 1980) makes it difficult to exclude domperidone-induced limitation of antiparkinsonian effect, since antiparkinsonian drug dosage was not constant throughout the study. With oral domperidone 60-120 mg/day, LANGDON et al. (1986) reported a moderate inhibition of the antiparkinsonian effect of levodopa.

II. Dopamine Agonist-Induced Sickness (see DAVIS et al. 1986)

There is a high density of dopamine D_2 receptors in the area postrema which contains the chemotrigger receptor zone (STEFANINI and CLEMENT-CORMIER 1981; BORISON and WANG 1953). The area postrema is notable for a relative deficiency of blood-brain barrier system (WISLOCKI and LEDUC 1952). The chemotrigger receptor zone in the area postrema in the floor of the fourth ventricle is connected to although separate from, the medullary emetic centre (BORISON and WANG 1953). Dopamine, apomorphine and other dopamine agonists injected into the chemotrigger receptor zone cause vomiting. Electrical stimulation of the emetic centre causes projectile vomiting without concomitant retching (or presumably nausea), in contrast to vomiting produced in experimental animals by gastric irritants such as metal salts.

It is a common clinical observation that tolerance develops to levodopa- or bromocriptine-induced sickness after 2–6 months' continuous treatment, although tolerance is not universal. With chronic levodopa treatment, dopamine receptor supersensitivity has been observed in the chemotrigger receptor zone (SHEN et al. 1983).

Nausea and vomiting following administration of levodopa or dopamine agonist drugs may be mediated partly by stimulation of gastrointestinal afferents to the vomiting centre, as well as direct stimulation of the chemotrigger receptor zone. Domperidone may exert a useful peripheral (gastric) as well as medullary (chemotrigger receptor zone) antiemetic effect. The most appropriate dose of domperidone relative to that of dopamine agonists has not been defined clearly. Overall, oral domperidone 10–60 mg/day in three divided doses will reduce the incidence of levodopa-induced sickness from approximately 80% to 15% of all subjects.

III. Levodopa-Associated Cardiovascular Problems

1. Domperidone- and Levodopa-Induced Hypotension

The fall in systolic and diastolic blood pressure following administration of levodopa and other dopamine agonists has been attributed to at least three mechanisms, medullary stimulation, baroceptor modulation, and peripheral vasodilation, particularly in renal circulation. Following administration of levodopa or bromocriptine, in both normal and hypertensive parkinsonian subjects, there is a minor and usually asymptomatic fall in systolic and diastolic blood pressure. Centrally and peripherally acting dopamine antagonists such as sulpiride in large doses protect against or reverse hypertension following administration of levodopa or ergot derivative dopamine agonists in young healthy subjects. However, the majority of the dopamine antagonist drugs that have been investigated in elderly parkinsonian subjects give little protection against levodopa-induced hypotension.

Levodopa-induced hypotension may be slightly reduced by the combined administration of a peripheral decarboxylase inhibitor with levodopa (BARBEAU et al. 1971), although there are conflicting results (REID et al. 1976).

Domperidone does not greatly alter the severity of hypotension following bromocriptine administration (QUINN et al. 1981) and in the study of AGID et al. (1979) orthostatic hypotension with bromocriptine still occurred, despite treatment with domperidone 60 mg/day. However, POLLAK et al. (1981) suggested that domperidone limited the occurrence of orthostatic hypotension in two of three bromocriptine-treated subjects. In single-dose studies of domperidone 10–20 and 40 mg orally and intramuscularly, SHINDLER et al. (1984) did not demonstrate any protective effect of this drug on levodopa-induced hypotension.

2. Domperidone and Cardiac Dysrhythmias

A wide range of disturbances in cardiac rhythm has been recorded in parkinsonian patients receiving levodopa, but their incidence is generally low. Transient sinus tachycardia is most commonly reported. There have been occasional reports of atrial and ventricular extrasystoles. These problems do not usually require levodopa withdrawal, nor in subjects with cardiac disease do they prohibit levodopa treatment.

Many hundreds of nonparkinsonian patients have received domperidone orally without any cardiac problems and, as with the decarboxylase inhibitors, domperidone may protect against spontaneous levodopa-induced or dopamine agonist-induced cardiac dysrhythmias in Parkinson's disease. QUINN et al. (1985) showed that in three subjects with Parkinson's disease, one with cardiac disease and two without, the addition of oral domperidone 20–50 mg three times daily caused a reduction in premature ventricular, and to a lesser extent, supraventricular aberrant beats.

3. Domperidone and Cerebral Blood Flow

There is a marked increase in regional cerebral blood flow in all areas after levodopa administration. Using positron emission tomography ^{15}O steady state inhalation, LEENDERS et al. (1984) showed that this effect was prevented by domperidone in five subjects with Parkinson's disease. Results indicated that both levodopa and domperidone had a direct effect on cerebral blood vessel walls.

IV. Levodopa-Associated Respiratory Problems

Respiratory irregularities are not a prominent feature of untreated idiopathic Parkinson's disease, although sometimes conspicuous in encephalitis lethargica as well as accompanying levodopa dyskinesias. It has recently become apparent that dopamine mechanisms are important in peripheral respiratory control mechanisms, and there has been direct biochemical as well as neuropharmacological identification of dopamine D_2 receptors in the rabbit carotid body (MIR et al. 1984). There is strong evidence that this receptor is involved in dopamine-induced depression of chemosensory discharge. Dopamine is the most abundant catecholamine in the carotid body, and dopamine has an important role in the reflex control of respiration (MIR et al. 1984). Dopamine is released by hypoxia from the carotid body glomus cells which store it (HANBAUER and HELLSTROM 1978). Functionally, the chemoreceptor sensitivity to hypoxia of the carotid body is modulated by dopamine.

There have been many recent investigations of the respiratory effects of levodopa, dopamine, dopamine agonists; as well as dopamine receptor antagonists in both animals and humans (BAINBRIDGE and HEISTAD 1980; STEEN and YATES 1972; TANDON 1976; DELPIERRE et al. 1985; MCQUEEN et al. 1984; CHOW et al. 1986; ZAPATA and TORREALBA 1984). These investigations have given essentially similar results. Infused dopamine depresses ventilation in animals and humans at rest, but particularly with eucapnic hypoxia (OLSON et al. 1982). Most of these physiological studies have investigated the effect of dopamine rather than other dopamine agonists, although many different dopamine receptor blocking drugs have been investigated, all with similar stimulant respiratory properties.

In acute studies, resting ventilation is not greatly increased by dopamine antagonists such as haloperidol, which does not increase the ventilatory response to hypoxia per se, although it reverses the depressant effect of intravenously induced dopamine on the ventilatory response to eucapnic hypoxia (BAINBRIDGE and HEISTAD 1980). ZAPATA and TORREALBA (1984) have shown that in the case of domperidone, the basal frequency of spontaneous chemosensor impulses from the carotid body in the cat is immediately and sustainedly increased. With chronic treatment, the chemoreceptor responsiveness of the carotid body to hypoxia is blunted (after 8 weeks in rabbits); normal chemoreceptor responsiveness can be restored by an acute dose of domperidone.

Hypoxic drive is not usually a major respiratory stimulus to normal humans living at low altitudes, but dopamine control of carotid body hypoxic sensitivity may be important in many clinical situations including, in addition to levodopa- and dopamine agonist-induced respiratory dyskinesias, chronic tardive dyskinesia where respiratory abnormalities are often prominent; variants of oromandibular dystonia where respiration is commonly involved; and in a younger age group, the infant death syndrome, which has been associated with increased carotid body dopamine and noradrenaline content (PERIN et al. 1984). These authors showed that carotid bodies from infants dying of the sudden infant death syndrome contained significantly higher concentrations of dopamine (tenfold) and noradrenaline (threefold) than those from age-matched control infants. High levels of endogenous catecholamines found in victims of sudden infant death syndrome may compromise the normal function of the carotid body, particularly the ventilatory response to hypoxia.

OLSON and SAUNDERS (1985) showed that prochlorperazine caused a marked increase in the ventilatory response to hypoxia, mainly owing to an increase in the frequency of breathing, but also to an increase in tidal volume (Fig. 1). This effect of prochlorperazine on hypoxic responsiveness was attributed to blockade of both α-adrenergic and dopaminergic receptors that modulate carotid body discharge. Domperidone would not be expected to alter any central component in levodopa-induced respiratory dyskinesias, although

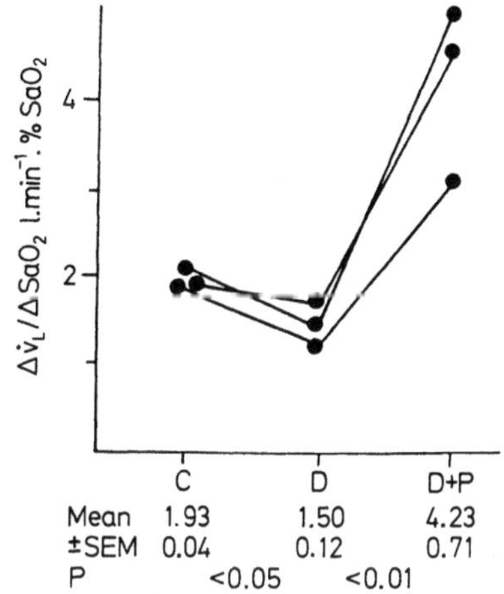

Fig. 1. Slope of ventilatory response to transient asphyxia before (C) and during (D) dopamine hydrochloride infusion and following administration of prochlorperazine in the presence of dopamine (D + P) in three subjects. (OLSON et al. 1982)

it may prevent levodopa-induced reduction in hypoxic ventilatory drive. This may be clinically relevant in parkinsonian patients with symptomatic sleep apnoea, which has a high incidence in otherwise normal elderly subjects. If the peripheral chemoreceptors do play a major role in respiratory oscillations in obstructive sleep apnoea, the use of levodopa without decarboxylase inhibitors or domperidone will potentially increase the severity of this disorder (LAHIRI et al. 1985).

D. Comparison of Domperidone with Metoclopramide and Other Neuroleptics in the Management of Parkinson's Disease

The overall incidence of acute dystonic reactions to metoclopramide is approximately 1% of the population of the United Kingdom. Reported cases of dystonia following domperidone administration are far fewer than following metoclopramide administration, and metoclopramide but not domperidone has been reported to cause dystonia in subjects with Parkinson's disease. Dystonic reactions to metoclopramide appear in general to be idiosyncratic. However, there is no doubt that they can be related to the administered dose, and there may be some relationship between the occurrence of these reactions and the peak metoclopramide plasma level attained. In the few reported examples of domperidone-induced dystonia, plasma domperidone levels have not been reported.

BATEMAN and DAVIES (1979) observed that akathisia was a very common event after dosing with intravenous metoclopramide 10 mg. However, akathisia is much less common after oral doses of metoclopramide, and relates to the peak plasma concentration of metoclopramide. This varies fivefold between different subjects given the same dose.

Metoclopramide, but not domperidone, has been reported frequently to cause parkinsonism during chronic treatment and, when given to subjects with Parkinson's disease in order to limit levodopa-induced nausea and vomiting, may increase the severity of illness. SCHACHTER et al. (1980a) showed that oral metoclopramide 60 mg caused an overall 33% reduction in the response to oral levodopa 1 g. BATEMAN et al. (1978) found that in patients on antiparkinsonian drugs, significant impairment of function occurred after the addition of metoclopramide in a dose of 60 mg/day, but not 30 mg/day.

These results all indicate that domperidone is to be preferred to metoclopramide in the control of levodopa- and dopamine agonist-associated nausea and vomiting.

E. Comparison of Domperidone with Decarboxylase Inhibitors in the Management of Parkinson's Disease

A number of studies have shown that, with both carbidopa and benserazide, combination with levodopa results in a plasma level of amino acid equivalent to that obtained with four to five times as high a dose of levodopa when given alone. Thus in acute studies, the combined administration of levodopa 200 mg with benserazide 50 mg will result in a plasma level of amino acid about equivalent to that from a single dose of levodopa 1000 mg. Also, the half-life of levodopa in plasma is sometimes prolonged when levodopa is combined with decarboxylase inhibitor. This levodopa-sparing effect is a major advantage of levodopa-decarboxylase inhibitor combinations, not shared with domperidone. However, decarboxylase inhibitors do not completely overcome all the peripheral problems connected with the use of levodopa alone. Dopa decarboxylase in brain capillaries is considered to form the major barrier to the entry of levodopa into the central nervous system. Histofluorimetric studies in animals have shown that benserazide and carbidopa specifically block dopamine formation in the capillaries, but enhance it in brain parenchyma (PLETSCHER 1973). A neglected finding is that of CONSTANTINIDIS et al. (1969) who reported that this effect was much more marked in striatal than cerebral cortex and mesencephalic capillaries.

As with domperidone, the exact dose of decarboxylase inhibitor used in combination with levodopa is critical for the successful management of Parkinson's disease. Benserazide in high dosage (300 mg/kg) inhibits cerebral decarboxylase in intact animals, although this is not observed with a similar high dose of carbidopa.

Apart from the levodopa-sparing effect, there is little to choose between decarboxylase inhibitors and domperidone in the levodopa management of parkinsonism, although with domperidone, since this is available as a single (not combined) preparation, treatment can be "tailor-made" for each individual subject. The advantages of domperidone given in combination with dopamine agonist ergot derivatives, including bromocriptine, lisuride and pergolide, are considerable. Until the ideal of a specific nigrostriatal dopamine receptor stimulant drug has been achieved, the combination of levodopa with dopa decarboxylase inhibitors or domperidone — or of bromocriptine with domperidone — forms the most effective treatment with the least side effects available for patients with Parkinson's disease.

References

Agid Y, Bonnet AM, Pollak P, Signoret JL, Lhermitte F (1979) Bromocriptine associated with a peripheral dopamine blocking agent in treatment of Parkinson's disease. Lancet 1:570-572

Agid Y, Quinn N, Lhermitte F (1981) Long-term results of treatment of Parkinson's disease with bromocriptine and domperidone. Rev Neurol 137:49-51

Bainbridge CW, Heistad DD (1980) Effect of haloperidol on ventilatory responses to dopamine in man. J Pharmacol Exp Ther 213:13-17

Barbeau A, Mars H, Gillo-Joffroy L (1971) Adverse clinical side-effects of levodopa therapy. In: Mc Dowell, Markham CH (eds) Recent advances in Parkinson's disease. Davis, Philadelphia, pp 204-237

Bateman DN, Davies DS (1979) Pharmacokinetics of metoclopramide. Lancet 1:166

Bateman DN, Kahn C, Legg NJ, Reid JL (1978) Metoclopramide in Parkinson's disease. Clin Pharmacol Ther 24:459-464

Bender DA (1980) Effect of benserazide, carbidopa and isoniazid administration on tryptophan-nicotinamide nucleotide metabolism in the rat. Biochem Pharmacol 29:707-712

Bender DA, Earl CJ, Lees AJ (1979) Niacin depletion in parkinsonian patients treated with L-dopa, benserazide and carbidopa. Clin Sci 56:89-93

Bogaerts M, Braems M, Martens C (1979) Domperidone in the prevention of L-dopa induced nausea. Postgrad Med J 55(Suppl 1):50-54

Borison HL, Wang SC (1953) Physiology and pharmacology of vomiting. Pharmacol Rev 5:193-230

Brogden RN, Carmine AA, Heel RC, Speight TM, Avery GS (1982) Domperidone: a review. Drugs 24:360-400

Chow CM, Winder C, Read DJC (1986) Influences of endogenous dopamine on carotid body discharge and ventilation. J Appl Physiol 60:370-376

Constantinidis J, Bartholini G, Geissbuhler F, Tissot R (1969) La barrière capillaire enzymatique pour la dopa au niveau de quelques noyaux du tronc cérébral du rat. Experientia 26:381-386

Corsini GU, Del Zompo M, Gessa GL, Mangoni A (1979) Therapeutic efficacy of apomorphine combined with an extracerebral inhibitor of dopamine receptors in Parkinson's disease. Lancet 1:954-956

Costall B, Fortune DH, Naylor RJ (1978) Differential activities of some benzamide derivatives on peripheral and intracerebral administration. J Pharm Pharmacol 30:796-798

Critchley P, Langdon N, Parkes JD (1985) Domperidone. Br Med J 290:788

Davis DJ, Lake-Bakaar GV, Grahme-Smith DG (1986) Nausea and vomiting: mechanisms and treatment Springer, Berlin Heidelberg New York Tokyo, pp 184

Debontridder O (1980a) Extrapyramidal reactions due to domperidone Lancet 2:802

Debontridder O (1980b) Extrapyramidal reactions due to domperidone. Lancet 2:1259

Delpierre S, Peyrot J, Guillot C, Grimaud C (1985) Ventilatory effects of domperidone, a new dopamine antagonist, in anaesthetised rabbits. Arch Int Pharmacodyn Ther 275:47-58
Falk RH, Desilva RD, Lown B (1981) Reduction in vulnerability to ventricular fibrillation by bromocriptine, a dopamine agonist. Cardiovasc Res 15:175-177
Giaccone G, Berletti O, Calciati A (1984) Two sudden deaths during prophylactic antiemetic treatment with high doses of domperidone and methylprednisolone. Lancet 2:1336-1337
Gonce M, Bury J, Burton L, Delwaide PJ (1982) Syndrome neurodysléptique induit par le dompéridone. Nouv Pres Med 11:2298
Grimes JD, Magner PO (1984) Safety of domperidone in metoclopramide-induced parkinsonism. Arch Neurol 41:363-364
Haase HJ, Kaumeier S, Schwartz H, Gundel A, Linde OK, Schnabele E, Stripf L (1978) Domperidone: a preliminary evaluation of a potential antiemetic. Nebr Symp Motiv 23:702-705
Hanbauer I, Hellstrom S (1978) The regulation of dopamine and noradrenaline in the rat carotid body and its modification by denervation and by hypoxia. J Physiol (Lond) 282:21-34
Heykants J, Knaeps A, Meuldermans W, Michiels M (1981) On the pharmacokinetics of domperidone in animals and man. 1. Plasma levels of domperidone in rats and dogs. Age-related absorption and passage through the blood-brain barrier in rats. Eur J Drug Metab Pharmacokinet 6:43-47
Ho AK, Smith JA (1982) Effects of 5-hydroxytryptamine, melatonin synthesising enzymes and serum melatonin. Biochem Pharmacol 31:2251-2255
Joss RA, Goldhirsch A, Brunner KW, Galeazzi RL (1982) Sudden death in a cancer patient on high dose domperidone. Lancet 1:1019
Kukorelli T, Juhasz G (1977) Sleep induced by intestinal stimulation in cats. Physiol Behav 19:355-358
Labatto S, Van den Broek PJ, Henny FC, Van Dijk JG (1982) Farmacotherapie van emesis ten gevolge van cytostatics. Ned Tijdschr Geneeskd 35:1608
Laduron PM, Leysen JE (1979) Domperidone, a specific in vitro dopamine antagonist, devoid of in vivo central dopaminergic activity. Biochem Pharmacol 28:2161-2165
Lahiri S, Hasiao C, Zhang R, Mokashi A, Nishino T (1985) Peripheral chemoreceptors in respiratory oscillation. Am J Physiol 249:1901-1908
Langdon N, Malcolm P, Parkes JD (1986) Comparison of levodopa with domperidone, and levodopa with carbidopa, in the treatment of Parkinson's disease. Clin Neuropharmacol (in press)
Leenders KL, Wolfson L, Jones T (1984) The effects of dopamine agonists on regional cerebral blood flow and oxygen metabolism in patients with Parkinson's disease and normal volunteers. Clin Sci 66:68-69
Lesser J, Bateman DN (1985) Domperidone. Br Med J 290:241
McQueen DS, Mir AK, Brash HM, Nahorski SR (1984) Increased sensitivity of rabbit carotid body chemoreceptors to dopamine after chronic treatment with domperidone. Eur J Pharmacol 104:39-46
Michiels M, Hendricks R, Heykants J (1981) On the pharmacokinetics od domperidone in animals and men. II. Tissue distribution, placental and milk transfer of domperidone in the Wistar rat. Eur J Drug Metab Pharmacokinet 6:47-54
Mir AK, McQueen DS, Pallot DJ, Nahorski SR (1984) Direct biochemical and neuropharmacological identification of dopamine D_2-receptors in the rabbit carotid body. Brain Res 291:273-283

Morgan JP, Bianchine JR, Spiegel HE Rivera-Calimlim L Hersey RM (1971) Metabolism of levodopa in patients with Parkinson's disease. Arch Neurol 25:39–46

Nagaoka M (1980) Effects of domperidone (KW-5338) on gastrointestinal symptoms caused by L-dopa. Multicentre study. Shinyaku Rinsho 29:2–27

Olson LG, Saunders NA (1985) Ventilatory stimulation by dopamine-receptor antagonists in the mouse. Br J Pharmacol 85:133–141

Olson LG, Hensley MJ, Saunders NA (1982) Augmentation of ventilatory response to asphyxia by prochlorperazine in humans. J Appl Physiol 53:637–643

Osbourne RJ, Slevin ML, Hunter RW, Hamer J (1985) Cardiotoxicity of intravenous domperidone. Lancet 2:385

Perin DG, Cutz E, Becker LE, Bryan AC, Madapallimatu MA, Sole M (1984) Sudden infant death syndrome: increased carotid body dopamine and noradrenaline content. Lancet 2:535–537

Pletscher A (1973) Effect of inhibitors of extracerebral decarboxylase on levodopa metabolism. Adv Neurol 3:49–55

Pollak P, Gaio JM, Hommel M, Pellat J, Chateau R (1981) Acute study of the association of bromocriptine and domperidone in parkinsonism. Therapie 36:671–676

Pontiroli AE, Castegnaro E, Vettaro MP, Viberti GC, Pozza G (1977) Stimulatory effect of the dopa-decarboxylase inhibitor Ro 4-4602 on prolactin release-inhibition by L-dopa, metergoline, methysergide and 2-Br-α-ergocryptine. Acta Endocrinol (Copenh) 84:36–44

Quinn N, Illas A, Lhermitte F, Agid Y (1981) Bromocriptine in Parkinson's disease: a study of cardiovascular effects. J Neurol Neurosurg Psychiatry 44:426–429

Quinn N, Parkes D, Jackson G, Upward J (1985) Cardiotoxicity of domperidone. Lancet 2:724

Reid JL, Greenacre JK, Teychenne PF (1976) Cardiovascular actions of L-dopa and dopaminergic agonists in parkinsonism. In: Birkmayer W, Hornykiewicz O (eds) Advances in parkinsonism. Roche, Basle, pp 566–572

Rinne UK, Mölsä P (1979) Levodopa with benserazide or carbidopa in Parkinson's disease. Neurology 29:1584–1589

Rivera-Calimlim L, Dujovne CA, Morgan JP, Lasagna L, Bianchine JR (1970) Absorption and metabolism of L-dopa by the human stomach. Br Med J 4:93–95

Roussak JB, Carey P, Parry H (1984) Cardiac arrest after treatment with intravenous domperidone. Br Med J 289:1579

Schachter M, Bedard P, Debono AG et al. (1980a) The role of D_1 and D_2 receptors. Nature 286:157–159

Schachter M, Sheehy MP, Parkes JD, Marsden CD (1980b) Lisuride in the treatment of parkinsonism. Acta Neurol Scand 62:382–385

Schuurkes JAJ, Van Nueten JM (1981) Is dopamine an inhibitory modulator of gastrointestinal motility? Scand J Gastroenterol 16(Suppl 67):33–36

Shen WW, Baig MS, Sata LS, Hofstatter L (1983) Dopamine receptor supersensitivity and the chemoreceptor trigger zone. Biol Psychiatry 18:917–921

Shindler JS, Finnerty GT, Towlson K, Dolan AL, Davies CL, Parkes JD (1984) Domperidone and levodopa in Parkinson's disease. Br J Clin Pharmacol 18:959–962

Shuto K, Shiozaki S, Kojima T, Tanaka M (1982) Antagonism of KW-5338 (domperidone) against emesis and depression of intestinal motility induced by L-dopa. J Pharmacobiodyn 3:709

Sol P, Pelet B, Guignard JP (1980) Extrapyramidal reaction due to domperidone. Lancet 2:802

Steen SN, Yates M (1972) The effects of benzquinamide and prochlorperazine, separately and combined, on the human respiratory centre. Anesthesiology 36:519–520

Stefanini E, Clement-Cormier Y (1981) Detection of dopamine receptors in the area postrema. Eur J Pharmacol 74:257-260

Sutton JA, Thompson S, Jobnack R (1985) Measurement of gastric emptying rates by radioactive isotope scanning and epigastric impedance. Lancet 1:898-900

Tandon MK (1976) Effect on respiration of diazepam, chlorpromazine and haloperidol in patients with chronic airway obstruction. Aust NZ J Med 6:561-565

Valenzuela JE, Dooley CP (1984) Dopamine antagonists in the upper gastrointestinal tract. Scand J Gastroenterol [Suppl] 19:127-134

Van Daele MC, Cyklis RD, Van de Casseye W, Verbeeck P, Wijndaele L (1984) Refusal of further cancer chemotherapy due to antiemetic drug. Lancet 1:57

Van Nueten JM, Janssen PA (1978) Is dopamine an endogenous inhibitor of gastric emptying? In: Duthie HL (ed) Gastroenterology in health and disease. MTP, Lancaster, pp 173-181

Van Nueten JM, Ennis C, Helksen L, Laduron PM, Janssen PAJ (1978) Inhibition of dopamine receptors in the stomach: an explanation of the gastrokinetic properties of domperidone. Life Sci 23:453-458

Wanquier A, Van den Broeck WAE, Niemegeers CJE (1980) On the antagonist effect of pimozide and domperidone on apomorphine-disturbed sleep-wakefulness in dogs. In: Koella WP (ed) Sleep. Karger, Basel, pp 279-282

Wade DN, Mearrick PT, Birkett DJ, Morris J (1974) Variability of L-dopa absorption in man. Aust NZ J Med 4:138-143

Willems JL, Bogaert MG, Buylaert W (1981) Preliminary observations on the interactions of domperidone with peripheral dopamine receptors. Jpn J Pharmacol 31:131-135

Wislockie GB, Leduc EH (1952) Vital staining of the hematoencephalic barrier by silver nitrate and trypan blue, and cytological comparison of the neurohypophysis, pineal body, area postrema, intercolumnar tubercle and supraoptic crest. J Comp Neurol 96:371-413

Yamada K, Uono K (1980) Safety study of long-term administration of domperidone. Yakuri Chiryo 8:269-271

Zapata P, Torrealba F (1984) Blockade of dopamine-induced chemosensory inhibition by domperidone. Neurosci Lett 51:359-364

CHAPTER 21

New Routes of Administration for Antiparkinsonian Therapy

J. A. OBESO and S. M. STAHL

A. Introduction

The combination of levodopa with a peripheral dopa decarboxylase inhibitor DCI is well established as the most useful treatment of Parkinson's disease. After serveral years of treatment a significant proportion of patients, who have initially shown an excellent response to levodopa, develop motor complications (MARSDEN and PARKES 1976). Daily fluctuations in motor performance, frequently accompanied by dyskinesias, are the commonest problem after long-term levodopa therapy. Initially, when the oscillations are predictable and related to the timing of levodopa administration, increasing the frequency of each dose and/or adding a long-acting dopaminergic agonist may temporarily provide some benefit. However, as duration of the disease, time under levodopa treatment and total daily levodopa administration (in an effort to improve mobility) increase, many patients fail to respond to individual levodopa doses and unpredictable changes in mobility occur. The treatment of complex motor fluctuations in Parkinson's disease is a major clinical challenge and a very difficult problem to resolve or avoid with conventional therapeutic strategies (FAHN 1982; QUINN 1984).

A combination of both peripheral (kinetic) and central (dynamic) factors are probably involved in the origin of the "on-off" phenomenon in patients chronically treated with levodopa. The former includes slow dissolution of the levodopa tablet in the stomach (FAHN 1977), delayed gastric emptying, interference with intestinal absorption of levodopa owing to competition with amino acids from the diet (COTZIAS et al. 1973), competition with the same neutral amino acids for levodopa transport from plasma to brain (NUTT et al. 1984), or inhibition of the active transport mechanism by 3-*O*-methyldopa accumulation when large levodopa plus DCI doses are used (GERVAS et al. 1983; Luquin et al. 1986). Central factors, such as decreased capacity to store dopamine presynaptically and desensitization of postsynaptic striatal dopaminergic receptors as a consequence of chronic dopamine stimulation, may also be contributing to the origin of motor fluctuations in Parkinson's disease (MARSDEN et al. 1982; FAHN 1982). It has also been shown that continuous administration of levodopa, levodopa methyl ester or lisuride may correct complex motor fluctuations in Parkinson's disease, suggesting a paramount role of the pharmacokinetic factor, in the "on-off" phenomenon (QUINN et al. 1984; NUTT et al. 1984; COOPER et al. 1984; OBESO et al. 1986 a, b). In this chapter, we shall examine the different techniques and drugs which may be used to apply the concept of continuous drug infusion in Parkinson's disease.

B. Delivery Methods

The enormous difficulties of getting a drug to a specific site of action within the central nervous system are well recognized. Chemical and physical barriers, such as enzymes and membranes or small blood vessels and glia cells, for example, have to be adequately overcome. The many different approaches have been reviewed elsewhere (Stahl, 1988; Stahl and Weis, 1988). Drug delivery systems currently available for the treatment of Parkinson's disease will be discussed here (Table 1).

I. Enteral Administration

Oral slow or sustained-delivery systems attempt to combat the significant disadvantages associated with conventional tablets, capsules, or liquids, which expose the body to uneven levels of drug concentrations in the blood that may produce inadequate therapy or side-effects. Oral sustained-release systems using various polymers, which slowly decay and release the drug in a programmed and sustained manner, are currently under development for the rate-controlled administration of levodopa. Madopar HBS is a hydrodynamically balanced system which is retained in the upper gastrointestinal tract, and which slowly releases the drug from a matrix. Sinemet CR-4, (200 mg levodopa and 50 mg carbidopa) is a controlled-release type of polymer which in vitro allows the drugs to be released at a rate approximating first-order kinetics. In general, either tablet (Madopar HBS or Sinement CR-4) gives higher plasma levodopa levels, but does not reduce hourly variability, and the time to peak concentration compared with Madopar or Sinemet standard tablets is prolonged (Nutt et al. 1986). The absorption of these compounds is under the influence of factors modifying gastric emptying and transit time, perhaps explaining the failure to achieve very stable plasma levels of levodopa.

Table 1. Examples of drug delivery devices that may be used for admiministration of antiparkinsonian drugs (Modified from Stahl, in press)

Implantable infusion devices
Osmotic minipumps
Polymers
Infusaid pumps (fluorocarbon propellant)
Medtronics pump (battery-driven and programmable)
Wearable infusion devices
Travenol infusion (balloon-filled reservoir)
Tecensa, Hoechst pumps (battery-driven)
Transdermal delivery systems
Ingestible infusion devices
Osmotic minipumps
Polymers

Recently duodenal delivery of levodopa has employed by KURLAN et al. (1986) in a few patients with Parkinson's disease. A very thin nasoduodenal tube was passed with the distal end localized radiographically in the proximal duodenum, Sinemet was dissolved in sterile water with a 2% concentration of ascorbic acid (KURLAN et al. 1986). In our experience the method is simple and well tolerated by the patients. Portable mechanical pumps are available so that the patients may move freely. Recently clinical trials using (+)-PHNO (4-propyl-9-hydroxynaphthoxazine) via nasogastric infusion have been started (COLEMAN et al. in press).

II. Intravenous Administration

Levodopa, levodopa methyl ester, lisuride and apomorphine are the antiparkinsonian drugs available for intravenous infusion. Pergolide is also water soluble, but has not been given intravenously in patients so far. One of the most commonly used drug delivery systems is a nonimplantable intravenous infusion device which stands at the patient's bedside, and is commonly used to control intravenous infusions of medications such as antihypertensive drugs, and agents for the treatment of shock. Levodopa, levodopa methyl ester, lisuride and apomorphine intravenous infusions have generally been carried out with the aid of this type of system.

Miniaturized wearable models, strapped to the body or carried in clothing, are also made as adaptations of the first-generation devices used at the bedside. These wearable models deliver a continuous, often adjustable, supply of drug by way of a catheter surgically implanted in a blood vessel. Lisuride and (+)-PHNO are highly soluble in normal saline and could therefore be given by means of this smaller infusion system.

III. Subcutaneous Administration

Levodopa cannot be administered by this route because of its low solubility and high acidity. Levodopa methyl ester is more soluble, but has not been given subcutaneously. Lisuride and apomorphine are readily absorbed after subcutaneous administration. One type of infusion device employed by clinical investigators is a disc-shaped implantable pump usually placed under the skin of the abdomen; it contains a compartment filled with a liquid fluorocarbon propellant that presses against a collapsible reservoir containing the desired medication. Body heat causes the propellant to vaporize and expand, compressing the reservoir and forcing the medication out through a catheter. At intervals the drug reservoir is refilled from the outside by a hypodermic injection, which also recompresses the fluorocarbon to a liquid. Other more sophisticated pumps include an implantable system which also contains a refillable drug reservoir, a motor-driven pump, and even a microelectronic control element that can be reprogrammed from outside the body, including the possibility of utilizing a telephone signal from a physician's office, in order to vary the pump's delivery rate. Many other implantable pumps exist, but are too numerous to review here.

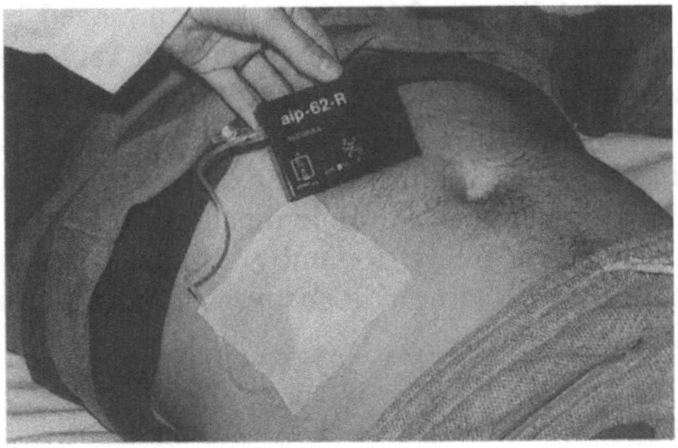

Fig. 1. Portable infusion pump for subcutaneous lisuride infusions

Wearable infusion pumps are usually attached to the abdomen (Fig. 1). The subcutaneous needle has to be replaced and changed every 2-3 days. Nodules may develop at the injection site, but infections are rare with lisuride infusions. Variables modifying skin blood flow in the abdominal wall, such as temperature, physical exercise or digestion, may change the rate of subcutaneous absorption.

IV. Transcutaneous Administration

One of the most elegant drug delivery systems is the transdermal drug delivery method. These systems are simply skin patches which use the skin as a controlled route of entry to the systemic circulation. The transdermal route is especially attractive for potent medications which require rate-controlled input, and which have a large first-pass effect. Transdermal delivery devices are thin, multilayered discs containing a reservoir of the drug, and an adhesive-coated polymer membrane that passes the drug at a controlled rate for a period of time up to 1 week. These discs are often about the size of a small coin, and are attached to various areas of the skin.

(+)-PHNO is the only dopamine agonist under investigation at present for transdermal administration. Whether or not possible modifications of skin metabolic and circulatory conditions could interfere with the therapeutic efficacy of (+)-PHNO or similar drugs is not known.

C. Clinical Studies

I. Enteral Routes

Duodenal infusion of levodopa-carbidopa via a nasoduodenal tube was successfully employed recently in three patients with "on-off" fluctuations (KURLAN et al. 1986). Levodopa was continuously infused throughout the day and evening (7 a.m. to 10 or 12 p.m.). The best effect on mobility was achieved with an infusion rate of 96, 10 and 52 mg/h in each patient. Fluctuations were attenuated, but did not disappear completely in these patients, one of whom was discharged from hospital under this form of treatment. Interestingly, intermittent duodenal administration of a levodopa bolus did not result in the same benefit in one patient.

(+)-PHNO is a new dopamine agonist, structurally different from other morphine or ergot derivatives currently used in Parkinson's disease. Both in vitro and in vivo experiments indicate that (+)-PHNO acts directly on postsynaptic dopamine receptors. (+)-PHNO is probably a selective D_2 dopamine agonist with some affinity for serotonin type 1 (5-HT_1) and α_2 receptors (MARTIN et al. 1984/1985). Two clinical trials, using (+)-PHNO by the oral route (STOESSL et al. 1985; GRANDAS et al. 1986), and experimental studies in marmosets treated with 1-methyl-4-phenyl-1,2,3,6-tetrahydropyridine (MPTP) (COLEMAN et al. in press) have confirmed that the drug is effective in acutely reversing the motor disability of Parkinson's disease. More recently (+)-PHNO was administered by continuous nasogastric infusion to six parkinsonian patients with "on-off" fluctuations. Four patients received prolonged infusion throughout day and night. In these four cases (+)-PHNO (plus a small amount of levodopa, p.o.) produced a clear-cut improvement in the total number of hours "on" during their waking time.

Several clinical trials on the value of sustained-release levodopa preparations (Madopar HBS, Sinemet CR-3,4,5) have been undertaken. The general opinion is that neither formulation provides significant benefit in the greater proportion of patients with motor fluctuations and dyskinesias (QUINN 1984). However, some patients showed improvement in early morning dystonia and night mobility (NUTT et al. 1986).

II. Intravenous Administration

1. Levodopa

The therapeutic application of infusion of levodopa associated with DCI (p.o.) in Parkinson' disease was first considered by SHOULSON et al. (1975). They showed a remarkably good response to intravenous levodopa in five patients with "wearing-off" phenomenon. However, the response was not maintained when the patients were asked to stand up and the method was abandoned in favour of new dopamine agonists. In 1979 ROSIN et al. treated a severe parkinsonian patient who had had an abdominal operation with intravenous levodopa for 7 days; the response however, was not constant during that period. Interest in the intravenous administration of levodopa reappeared recently when three independent groups showed that severe "on-off" fluctuations could be virtually abolished for many hours.

Thus, QUINN et al. (1984) reported a dramatic improvement of "on-off" swings in ten patients, who remained independent during the infusions. Two patients with diphasic dyskinesias required an extra bolus of levodopa. These findings were confirmed by HARDIE et al. (1984) and NUTT et al. (1984) who studied 12 and 6 severe patients, respectively. In addition, NUTT et al. (1984) showed that despite maintaining constant plasma levodopa levels, "off" periods could supervene following a high protein meal or the oral administration of phenylalanine, leucine or isoleucine. This was explained as competition between large neutral amino acids and levodopa for transport from plasma to brain. These two studies also suggest a certain difficulty in keeping a stable response for longer than 48 h in some patients. More recently, MARION et al. (1986) gave intravenous infusions of levodopa to three subjects on five consecutive days, each session lasting 6 h on day 1, and 12 h on days 2–5. Although two patients experienced some transient "off" episodes, they all enjoyed a better clinical control during the infusion than on chronic oral levodopa treatment. FABRINI et al. (1987) have given continuous infusions of levodopa for 10 consecutive days to patients with "wearing-off" and "on-off" phenomena, by means of a wearable "jacket-like" pump. This study confirmed the earlier impression of MARION et al. (1986), indicating that a relatively stable response can be obtained when levodopa is given continuously via the intravenous route, but particularly in patients with simple "wearing-off" fluctuations.

In summary, all studies with intravenous levodopa agree on the clear-cut therapeutic benefit of this form of treatment. Motor fluctuations in Parkinson's disease may therefore be due to a failure in delivering sufficient levodopa to the striatum once this has lost its ability to achieve a sustained response when plasma levodopa levels oscillate following oral administration. A number of problems limit the practical application of levodopa in continuous infusion. First, the high acidity and low solubility of levodopa necessitates its dilution in large volumes, which in turn causes some difficulties with the size of the pump, refilling procedure, etc. On the other hand, intravenous delivery of levodopa does not remove the competition of neutral amino acids for transport across the blood-brain barrier nor the possible inhibitory influence of 3-*O*-methyldopa accumulation following high levodopa intake (LUQUIN et al. 1986). Finally, the price of levodopa solutions for intravenous administration is probably too high to be used widely.

Levodopa methyl ester is very soluble and can have the same therapeutic effect as levodopa itself (COOPER et al. 1984; CHASE et al. in press). However, it has all the other limitations mentioned for levodopa. In addition, levodopa methyl ester is metabolized into alcoholic compounds which are potentially toxic to the organism.

2. Lisuride

The high solubility of lisuride in saline allows its administration by intravenous or subcutaneous routes in small volumes. OBESO et al. (1983, 1986 a, b) first reported the sustained antiparkinsonian effect of lisuride when given by intravenous infusion over a period of 8–72 h. However, some patients had transient "off" periods during the infusion and in about 20% of patients there was no positive effect whatsoever (LUQUIN et al. 1987). Oral levodopa treatment had to be added in all cases to maintain normal mobility throughout the infusion. All patients were pretreated with oral domperidone 10 mg t.i.d., for 3 days prior to the infusion. An intravenous bolus of 40 mg domperidone was administered 30 min before initiation of the infusion. A small number of patients experienced transient nausea, sweating and malaise. Severe hypotension occurred in two patients, requiring termination of the infusion.

3. Apomorphine

The use of apomorphine in Parkinson's disease was limited until recently by the severity and high incidence of side effects such as nausea, vomiting, hypotension, sweating and malaise. Marked elevation of blood urea, nitrogen and plasma creatine was also observed in some patients (COTZIAS et al. 1973). At present, pretreatment with domperidone (10-20 mg t.i.d.) for several days or its acute intravenous administration (20 mg) a few minutes before apomorphine produce a marked attenuation of adverse effects. A single intravenous apomorphine bolus of 1-3 mg or 15-40 g/kg is usually sufficient to obtain full motor capacity ("on") in patients with the "on-off" phenomenon.

Similar results are observed after subcutaneous administration, confirming previous findings by Cotzias and his collaborators a few years ago (COTZIAS et al. 1973; PAPAVASILIOU et al. 1974). Continuous intravenous administration of apomorphine has recently been applied to the study of parkinsonian patients with motor fluctuations (OBESO et al. 1987a). The infusion is started with a bolus of apomorphine which generally turns the patient "on", and continued with an infusion rate of about 5-10 mg/h. Intravenous domperidone (20 mg) is given 20 min before starting the infusion in a bolus, and repeated every 90-120 min according to tolerance. Peripheral side effects are not usually important, but drowsiness has been encountered in some patients and profound hypotension was seen in one case. Apomorphine infusions can keep a patient completely mobile for many hours, but complete stabilization is difficult to achieve in severe (complex) cases.

III. Subcutaneous Infusions

Lisuride has been the dopamine agonist more widely used chronically via the subcutaneous route for the treatment of Parkinson's disease (OBESO et al. 1986b). OBESO et al. (1988) recently reported their present experience with subcutaneuos lisuride infusion in 28 severe patients with long-standing Parkinson's disease, and severe motor disability associated with chronic levodopa therapy uncontrolled with standard antiparkinsonian drugs. All patients started to take oral domperidone 10 mg t.i.d. a few days before and during the initial weeks of the infusion to prevent nausea and vomiting. Oral daily levodopa treatment with levodopa plus DCI was maintained unchanged during the first days of infusion and adjusted subsequently according to the clinical response. Two different portable infusion pumps (Tecensa-62, Madrid, Spain and Hoechst Insulin Pump, Frankfurt, Federal Republic of Germany) were used for lisuride administration. The needle was changed and replaced in the abdomen every 4-5 days initially, but later most patients changed it every 2-3 days.

Four patients abandoned the trial in the first 4 weeks owing to psychiatric complications in two, inability to use the system at home in one and the developing of local nodules as well as a rejection of permanent dependence upon the pump in the fourth case. In the 24 patients treated for more than 3 months (mean 9.6 months, maximum 24 months) the average daily dose of lisuride

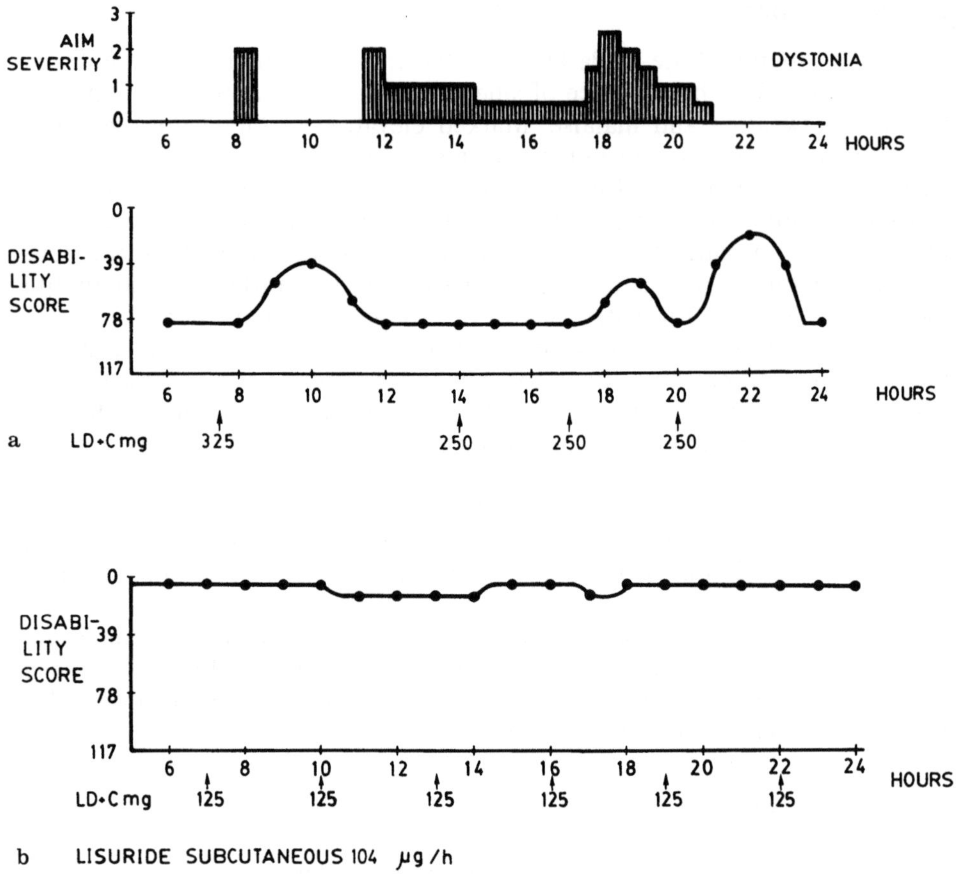

Fig. 2. a Representative evolution of motor function thoughout the day, before initiating treatment with subcutaneous lisuride. Prolonged "off" periods accompanied by painful dystonia in the limbs, in spite of levodopa treatment, were the main problem. Abnormal involuntary movements (AIM) and disability scores are plotted against time. **b** After 1 month of lisuride infusion, the patient enjoyed normal mobility. This has continued over 15 consecutive months, although reduction in the daily dose of levodopa-benserazide has been required several times in order to control dyskinetic movements

was 2.80 mg (0.5-3.5 mg), generally with a constant infusion rate (mean 117.3 µg/h). The levodopa dose was reduced from a mean of 878 to 558.1 mg/day (37% decrease).

Most patients showed considerable improvement in motor capacity after the optimum regimen of subcutaneous lisuride and oral levodopa plus DCI was established. Perhaps the most dramatic effect was that periods of total immobility disappeared. Thus, although motor capacity was not completely stable throughout the day, patients could be independent most of the time because "off" periods were considerably less severe (Fig. 2). Another notable effect of lisuride infusion was on night and early morning mobility, which were greatly improved in all patients. However, most patients still required their first levodopa dose in the morning to achieve a complete "on".

Diphasic dyskinesia and "off" period dystonia, were abolished or greatly reduced in all patients. Patients with peak-dose dyskinesias showed variable response to lisuride infusion. Subcutaneous nodules developed at the injection sites in all patients. Nodules of sufficient dimensions and inflammatory character to interfere with absorption of lisuride were present in five patients.

Nine patients showed psychiatric complications of sufficient intensity to stop infusion temporarily and responsible for permanent withdrawal in five patients. Complications took the form of visual and/or auditory hallucinations in all nine patients, an organic confusional state was observed in six and a florid paranoid state in five patients.

The positive effect of lisuride infusion on motor fluctuations has been confirmed by other investigators (Critchley et al. in press; Ruggieri et al. 1986), but psychiatric side effects have limited its practical applications in a significant proportion of severely parkinsonian patients (Critchely et al. in press).

Recently, Stibe et al. (1988) reported on the application of subcutaneous apomorphine to 18 patients with motor fluctuations for a mean period of 8 months (1-15). Most patients continued to receive levodopa (p.o) in order to obtain optimal benefit, thus confirming our experience. The response was strikingly positive in all cases with a 63% mean reduction of hours "off" per day and also less severe motor disability during the remaining "off" periods. Apomorphine infusion was not maintained over night. General tolerance was excellent. Skin nodules similar to the ones seen with lisuride infusion also developed in all cases, but psychiatric side effect were only present in 3. Stibe et al. (1988) also reported a very clear positive effect in 8 patients, with less severe motor disability, who were given repeated apomorphine injections during the day in combination with frequent oral levodopa doses.

IV. Transdermal Application

(+)-PHNO is a suitable candidate for continuous transdermal delivery, but clinical studies have not yet been initiated.

D. Conclusion

The "on-off" phenomenon is one of the major problems associated with long-term levodopa therapy. The developing of new dopamine agonists and more sophisticated delivery systems may contribute to better control of motor fluctuations. Continuous dopaminergic stimulation may become an early therapeutic tool for the prevention of motor complications associated with standard levodopa therapy (Chase et al. 1988; Horowski et al. 1988).

Acknowledgments. The authors are grateful to Mrs. Jacqueline Oehling for kindly typing and preparing the manuscript and to Mrs. Carol Elders for reviewing the text.

References

Chase TN, Mouradian MM, Fabbrini G et al. (1988) Pathogenic studies of motor fluctuations in Parkinson's disease. J Neural Transm, Suppl. 27:3-10

Coleman RJ, Grandas Perez F, Quinn N et al. (in press) 4-Propyl-9-hydroxynaphthoxazine in Parkinson's disease and MPTP models. In: Rose FC (Ed). Parkinson's disease. Libby, London.

Cooper DR, Marrel C, Testa B et al. (1984) L-Dopa methyl ester: a candidate for chronic systemic delivery of L-Dopa in Parkinson's disease. Clin Neuropharmacol 7:88-98

Cotzias GC, Mena I, Papavasilious PS (1973) Overview of present treatment of parkinsonism with levodopa. Adv Neurol 2:265-277

Critchley P, Grandas F, Quinn N et al. (1988) Continuous subcutaneous lisuride infusions in Parkinson's disease. J Neural Trans, Suppl. 27:55-60

Fahn S (1977) Episodic failure of absorption of L-dopa: a factor in the control of clinical fluctuations in the treatment of parkinsonism. Neurology 27 (Suppl 2):390

Fahn S (1982) Fluctuations of disability in Parkinson's disease: pathophysiology. In: Marsden CD, Fahn S (eds) Movement disorders. Butterworth, London, pp 123-145

Gervas JJ, Muradas V, Bazan E et al. (1983) Effects of 3-OM-dopa on monoamine metabolism in rat brain. Neurology 33:278-282

Grandas Perez F, Jenner J, Nomoto M et al, (1986) (+)-4-propyl-9-hydroxynaphthoxazine in Parkinson's disease. Lancet 1:906

Hardie RJ, Lees AJ, Stern GM (1984) On-off fluctuations in Parkinson's disease. A clinical and neuropharmacological study. Brain 107:487-506

Kurlan R, Rubin AJ, Miller CH et al. (1986) Duodenal delivery of levodopa for "on-off" fluctuations in parkinsonism: preliminary observations. Ann Neurol 20:262-265

Luquin MR, Obeso JA, Martinez Lage JM, Bossi L, Scatton B (1986) Levodopa and 3-*O*-methyldopa plasma levels in Parkinson's disease: patients with clinical fluctuations versus smooth responders. Neurology 36 (Suppl 1): 244

Luquin MR, Obeso JA, Martinez Lage JM, Tresguerres J, Parada J, Nieuweboer B, Dorow T, Horowski R (1986) Parenteral administration of lisuride in Parkinson's disease. Adv Neurol 45:561-568

Marion MH, Stocchi F, Quinn NP, Jenner P, Marsden CD (1986) Repeated levodopa infusions in fluctuating Parkinson's disease: clinical and pharmacokinetic data. Clin Neuropharmacol 9:165-181

Marsden CD, Parkes JD (1976) "On-off" effects in patients with Parkinson's disease on chronic levodopa therapy. Lancet 1:292-296

Marsden CD, Parkes JD, Quinn N (1982) Fluctuations of disability in Parkinson's disease: pathophysiological aspects. In: Marsden CD, Fahn S (eds) Movement disorders. Butterworth, London, pp 96-122

Martin GE, Williams M, Pettibone DJ, Yarbrough GG, Clineschmidt BV, Jones JH (1984) Pharmacological profile of a novel potent direct-acting dopamine agonist (+)-4-propyl-9-hydroxynapthoxazine (+)-PHNO. J Pharmacol Exp Ther 223: 298-304

Martin GE, Williams M, Pettibone DJ, Zrada MM, Lotti VJ, Taylor DA, Jones JH (1985) Selectivity of (+)-4-propyl-9-hydroxynaphthoxazine (+)-PHNO for dopamine receptors in vitro and in vivo. J Pharmacol Exp Ther 233:395-401

Nutt JG, Woodward WR, Hammerstad JP et al. (1984) The "on-off" phenomenon in Parkinson's disease. N Engl J Med 310:483-488

Nutt JG, Woodward WR, Carter JH (1986) Clinical and biochemical studies with controlled-release levodopa/carbidopa. Neurology 36:1206-1211

Obeso JA, Martinez Lage JM, Luqin MR (1983) Lisuride infusion for Parkinson's disease. Ann Neurol 14:134

Obeso JA, Luquin MR, Martinez Lage JM, (1986a) Intravenous lisuride corrects motor oscillations in Parkinson's disease. Ann Neurol 19:31-35

Obeso JA, Luquin MR, Martinez Lage JM (1986b) Lisuride infusion pump: a device for the treatment of motor fluctuations in Parkinson's disease. Lancet 1:467-470

Obeso JA, Luquin MR, Vaamonde J et al. (1987) Continuous dopaminergic stimulation for Parkinson's disease. Can J Neurol Sci 14:488-492

Obeso JA, Luquin MR, Vaamonde J et al. (1988) Subcutaneous administration of lisuride in the treatment of complex motor fluctuations in Parkinson's disease. J Neural Transm, Suppl. 27:17-26

Papavasiliou PS, Cotzias GC, Mena I (1974) (in press) Short and long-term approaches to the "on-off" phenomenon. Adv Neurol 5:379-386

Quinn NPW (1984) Anti-parkinsonian drugs today. Drugs 28:236-262

Quinn NPW, Parkes JD, Marsden CD (1984) Control of "on-off" phenomenon by continuous intravenous infusion of levodopa. Neurology 34:1131-1136

Rosin AJ, Devereux D, Calne DB (1979) Parkinsonism with "on-off" phenomena — intravenous treatment with levodopa after major abdominal surgery. Arch Neurol 36:32-34

Ruggieri S, Stocchi F, Agnoli A (1986) Lisuride infusion pump for Parkinson's disease. Lancet 2:348-349

Shoulson IRA, Glaubiger GA, Chase TN (1975) On-off response: clinical and biochemical correlations during oral and intravenous levodopa administrations in parkinsonian patients. Neurology 25:1144-1148

Stahl S (1988) Applications of new drug delivery technologies to Parkinson's disease and dopaminergic agents. J Neural Transm, Suppl. 27:123-132

Stahl SM, Wets KM (in press) Recent advances in drug delivery technology for neurology. Clin Neuropharmacol

Stibe CM, Kenpster FA, Lees AJ, Stern GM (1988) Subcutaneous apomorphine in parkinsonian on-off oscillations. Lancet I: 403-406

Stoessl AJ, Mak E, Calne DB (1985) (+)-4-prophyl-9-hydroxynaphthoxazine (PHNO), a new dopaminomimetic, in treatment of parkinsonism. Lancet 2:1330-1331

CHAPTER 22

Treatment of Parkinsonian Features in Neurological Disorders Other than Parkinson's Disease

R. F. PEPPARD and D. B. CALNE

A. Introduction

Parkinsonism is a collection of neurological features, including tremor, rigidity, akinesia and loss of postural reflexes. This syndrome is most commonly due to Parkinson's disease, but may occur with a wide variety of conditions in which parkinsonian features may be found alone or in combination with other neurological, medical and psychiatric deficits.

From a therapeutic viewpoint, the conditions causing parkinsonism can be grouped under four headings:

1. Drug-induced parkinsonism, in which management is influenced by a compromise between the desired effect of antipsychotic drug treatment and this unwanted side effect.

2. Conditions in which the primary treatment of parkinsonism is the specific management of the causative disorder rather than palliation. In this category, Wilson's disease will be discussed in detail.

3. The multiple system degenerative diseases of the nervous system in which the therapeutic approach is similar to that in Parkinson's disease, but usually only a partial or temporary response of the parkinsonism is seen and nonparkinsonian features tend to be unchanged.

4. Other types of parkinsonism.

B. Drug-Induced Parkinsonism

A necessary step in the management of drug-induced parkinsonism is an awareness of which drugs are capable of causing this syndrome. The most important group is the neuroleptic drugs (the phenothiazines, the thioxanthenes and the butyrophenones) which act as dopamine antagonists. Extrapyramidal effects are most likely to occur with fluphenazine, perphenazine, trifluoperazine, triflupromazine and haloperidol, and least likely to occur with thioridazine and molindone. Chlorpromazine, acetophenazine, chlorprothixene and thiothixene are in an intermediate category (BALDESSARINI 1980).

Metoclopramide, a centrally active D_2 receptor antagonist, is a common cause of drug-induced parkinsonism. Reserpine and its synthetic analogue tetrabenazine cause parkinsonism by depletion of presynaptic stores of dopamine.

Lithium has been reported to potentiate neuroleptic-induced parkinsonism. Lithium may inhibit striatal dopamine synthesis and decrease the sensitivity of dopamine receptors (SACHDEV 1986). Estrogens may act as dopamine antagonists and their administration to patients taking phenothiazines may precipitate parkinsonism (GRATTON 1960).

Phenytoin may interfere with dopaminergic mechanisms and has been claimed to reduce the therapeutic effect of levodopa in Parkinson's disease (MENDEZ et al. 1975). A case of parkinsonism has been attributed to phenytoin (GONI et al. 1985). In one study, 4 of 27 patients treated with diazepam in high doses (at least 100 mg/day) for schizophrenia developed parkinsonism which responded poorly to anticholinergic drugs, but resolved within 1-3 weeks of the discontinuation of diazepam (SURANYI-CADOTTE et al. 1985). However, benzodiazepines are usually well tolerated in Parkinsonism. There are single case reports of parkinsonism in association with the use of fentanyl (RIVERA et al. 1975), captopril (SANDYK 1985) and flunarizine (MEYBOOM et al. 1986).

The prevalence of parkinsonism in association with the use of antipsychotic drugs ranges from 5 to 60% (MARSDEN et al. 1975). Patients over 40 years of age are more vulnerable and some believe as many as 50% of new cases of parkinsonism in the elderly may be drug induced (STEPHEN and WILLIAMSON 1984).

Drug-induced parkinsonism may begin within several days of the initiation of neuroleptic treatment and 90% of cases appear within 3 months (TARSY 1983). CASEY et al. (1986) in a long-term follow-up study, found that drug-induced parkinsonism persisted and increased in all patients in whom neuroleptics were continued, even if there was no change in dose. Usually parkinsonism resolves within 3 weeks to 3 months after cessation of neuroleptics, but it may persist for as long as 2 years (HERSHON et al. 1972; MARSDEN and FAHN 1982; WILLIAMSON 1984). DEMARS (1966) has described worsening of parkinsonian symptoms persisting for 5 weeks following withdrawal of phenothiazines. The most important determinant in the development of drug-induced parkinsonism is individual susceptibility. ARONSON (1985) has described a case of parkinsonism persisting for as long as 1 year after a single 5-mg dose of haloperidol.

All phenothiazines and butyrophenones with parkinsonism-inducing potential produce competitive blockade of dopamine receptors in the striatum. These agents are also antagonists at central noradrenergic receptors (HORNYKIEWICZ 1975).

MILLER and HILEY (1975) have reported an inverse relationship between the antimuscarinic potencies of neuroleptic drugs and their tendency to produce parkinsonism. Thioridazine and the dibenzodiazepine, clozapine, which are associated with a low incidence of drug-induced parkinsonism, have powerful antimuscarinic activity. Haloperidol and the piperazine phenothiazines have little or no anticholinergic effect.

It has been proposed that striatal dopamine receptor blockade by neuroleptics leads, by a feedback control mechanism, to activation of nigrostriatal neurons with increased synthesis and release of dopamine in the striatum

(HORNYKIEWICZ 1975). CHASE et al. (1970) found that the use of antipsychotic drugs caused an increase in homovanillic acid (HVA) in the cerebrospinal fluid (CSF). However, this increase was not as great in those patients with extrapyramidal side effects, suggesting deficient presynaptic compensatory mechanisms for overcoming the dopamine receptor blockade. In contrast, BOWERS (1985) found that patients with a persistent elevation of HVA in the CSF had more parkinsonian side effects than patients whose CSF HVA pattern showed minimal initial elevation or an initial rise followed by a decline. Bowers suggested that hypokinetic side effects are not, in fact, due to decreased dopamine turnover and release in the presence of dopamine receptor blockade, but rather depend on postsynaptic mechanisms.

The nigrostriatal system can compensate for loss of up to 80% of striatal dopamine content before clinical parkinsonism becomes manifest (HORNYKIEWICZ 1986) so, where there is a pre-existent striatal dopamine deficiency, the additional functional deficiency induced by dopamine antagonists may produce clinical parkinsonism. The decline in nigral cell counts and in the striatal concentration of dopamine and tyrosine hydroxylase with age (MCGEER et al. 1977) may account for the increased susceptibility of the elderly to the effects of dopamine antagonists. It has been claimed that about 10% of elderly patients who recover from drug-induced parkinsonism go on to develop Parkinson's disease (STEPHEN and WILLIAMSON 1984). Reversible drug-induced parkinsonism has been reported in two patients who subsequently were found to have the pathological changes of Parkinson's disease (RAJPUT et al. 1982). However, no consistent pathology has been found in association with drug-induced parkinsonism and the factors which account for individual susceptibility remain poorly understood.

More caution in the use of neuroleptic agents may prevent some cases of drug-induced parkinsonism. Inappropriate indications for neuroleptic therapy include prochlorperazine for nonspecific dizziness and the long-term administration of oral metoclopramide for nausea (GRIMES 1981); substitution of domperidone for metoclopramide can be helpful in many cases (GRIMES and MAGNER 1984).

The anticholinergic drugs may be employed for the treatment of parkinsonism with concomitant antipsychotic drug therapy or for persistent parkinsonism after neuroleptic discontinuation. Trihexyphenidyl, procyclidine and benztropine have been extensively used in these situations. The anticholinergic antihistamines, such as diphenhydramine, have also been found to be effective (MCGEER et al. 1961; FLEMING et al. 1970). CHOUINARD et al. (1979) have reported that ethopropazine, a phenothiazine with significant anticholinergic activity, is as efficacious in drug-induced parkinsonism as procyclidine or benztropine.

Anticholinergic agents are very effective in the prevention and control of acute dystonias occurring early in the course of antipsychotic drug therapy (MANOS et al. 1986), but their use as prophylaxis of other extrapyramidal side effects remains controversial. Prophylactic anticholinergic drugs may reduce the severity, but not the incidence of drug-induced parkinsonism (RIFKIN et al. 1978).

In a double-blind study of the effects of anticholinergic drug withdrawal, Jellinek et al. (1981) reported a significant increase in motor restlessness after 2 weeks and in hypomotility and depression after 4 weeks. Adverse effects of anticholinergic cessation on behaviour and mood may suggest a continuing need for these drugs, but they can also produce euphoria and be abused (Dimascio and Demirgian 1970).

The elderly and demented are at increased risk of central side effects from anticholinergic drugs, including disorientation, impairment of recent memory and hallucinations. Peripheral anticholinergic symptoms are generally not troublesome with the small doses required to control extrapyramidal side effects of neuroleptics.

Some extrapyramidal features, especially akathisia and akinesia, may go unrecognized or may be difficult to distinguish from manifestations of underlying psychiatric illness. A therapeutic trial of antiparkinsonian therapy can help to distinguish drug-induced akinesia from psychomotor retardation related to depression or schizophrenic catatonia.

Theoretically, anticholinergic drugs may interfere with the therapeutic effects of antipsychotic drugs. Lowering of the blood levels of neuroleptics has been reported to occur during anticholinergic therapy (Donlon and Stetson 1976), but usually no loss of efficacy is seen (Davis 1985).

Anticholinergic usage has been implicated in the development and exacerbation of tardive dyskinesia (Rifkin et al. 1978). In patients with concurrent drug-induced dyskinesia and parkinsonism, paradoxical improvement in tardive dyskinesia can occur with levodopa treatment of persistent parkinsonism after stopping neuroleptics (Shoulson 1983) despite the supposed dopamine receptor hypersensitivity in the striatum.

Guidelines for the use of anticholinergic agents in association with antipsychotic therapy have been suggested (Tarsy 1983; Ayd 1974). In patients with a history of drug-induced parkinsonism, a relatively low potency drug such as thioridazine or molindone is proposed, but, if high potency antipsychotic drugs are essential, an anticholinergic agent may be administered prophylactically.

If parkinsonism develops in patients being treated for the first time with neuroleptic agents, consideration should be given to decreasing the dose of neuroleptic. If this proves ineffective, anticholinergic drugs can be introduced with the expectation of rapid improvement in parkinsonian symptoms. In patients who have been on successful prophylactic anticholinergic therapy for more than 3 months, a cautious attempt at withdrawal should be made and, if extrapyramidal side effects emerge, treatment can be reinstituted.

Amantadine has been found to be as effective as trihexyphenidyl in the control of drug-induced parkinsonism (Fann and Lake 1976) and its use may have less long-term risk of tardive dyskinesia (Davis 1985). Amantadine may also be effective in some cases of drug-induced parkinsonism which are resistant to anticholinergic agents (Gelenberg and Mandel 1977). High potency antipsychotic drugs can induce a syndrome of catatonia with associated parkinsonian rigidity which responds poorly to anticholinergics, but may be effectively treated by amantadine (Davis 1985) or lorazepam (Fricchione et al. 1983).

Levodopa therapy is effective for reserpine-induced parkinsonism, presumably acting by replacement of depleted catecholamines, but it has usually been found to be ineffective in parkinsonism due to dopamine receptor antagonists (McGeer et al. 1961; Yaryura-Tobias et al. 1970). However, intravenous levodopa has been reported to produce a transient improvement in parkinsonism associated with chlorpromazine or haloperidol (Bruno and Bruno 1966). More recently Marsden and Fahn (1982) have claimed that drug-induced parkinsonism may respond to levodopa at doses that do not provoke a psychotic disorder. In cases of parkinsonism persisting after cessation of neuroleptics, levodopa may be given with periodic attempts at withdrawal at intervals of 3 months for 2 years to assess whether remission of the parkinsonism has occurred (Williamson 1984; Marsden and Fahn 1982).

Experience with the use of dopamine agonists in drug-induced parkinsonism is limited, no doubt because of the reluctance to use theoretically mutually antagonistic drugs and the relatively high incidence of behavioural side effects with these agents (Calne 1982). Parkes et al. (1981) have reported a greater than 50% reduction in parkinsonism due to haloperidol in a patient treated with lisuride.

C. Wilson's Disease

Wilson's disease, or familial hepatolenticular degeneration, is an autosomal recessive disorder in which excess copper accumulation leads to cirrhosis of the liver, central nervous system lesions and pigmented deposits near the limbus of the cornea, known as Kayser-Fleischer rings. The main neurological pathology is in the striatum where there is atrophy, severe neuronal loss and reactive gliosis. Neuronal degeneration is also seen in the caudate nucleus, the globus pallidus, the thalami and occasionally in other brain stem nuclei, and in the cerebral cortex (Duchen and Jacobs 1984).

The neurological manifestations include: dystonia, chorea, akinesia, rigidity, tremor and dementia. Most of these features are reversible with specific treatment of the underlying disease, but recovery of long-standing symptoms may be incomplete (Purdon Martin 1968). In severe cases neurological response may not commence for some months after the initiation of treatment, but recovery may continue for up to 2 years. At the start of treatment there is frequently a marked transient exacerbation of neurological deficits lasting several weeks.

The mainstay of treatment is the lifelong administration of copper chelating agents such as D-penicillamine (Crawhall 1980). This measure should be combined with restriction of copper intake to less than 2 mg/day by the avoidance of copper-rich foods. An oral cation exchange resin, such as potassium sulfide 40 mg with each meal, may help to prevent absorption of copper from the gastrointestinal tract.

Penicillamine is a sulphydryl compound which combines with copper, mercury, lead, zinc, iron and other metals to form soluble compounds that are readily excreted by the kidneys. Penicillamine may be commenced at 125 or 250 mg/day and increased gradually over several weeks according to side effects and urinary copper excretion. The usual maintenance dose is 250 mg t.i.d. and each dose should be given on an empty stomach at least half an hour before meals (SHOULSON et al. 1983). Some patients require 2 g penicillamine daily and occasionally it is desirable to give both penicillamine and dimercaprol. Trientene is another copper chelating agent which has been employed as an effective alternative to penicillamine (WALSHE 1986); it is particularly useful in patients who develop adverse reactions to penicillamine.

Serial determinations of transaminases, serum copper, and 24-h urinary copper excretion can guide dosage adjustment of penicillamine and assess compliance. Urinary copper excretion rises when penicillamine is commenced but, as excess copper stores are eliminated, serum copper declines and the urinary copper excretion levels off at approximately 50% of the initial values.

Acute sensitivity reactions tend to occur early in the course of penicillamine therapy and may be manifested as rashes (pruritic, macular, urticarial), arthralgia, fever, eosinophilia, leukopenia and thrombocytopenia. A lupus-like syndrome, Goodpasture's syndrome, and the nephrotic syndrome occasionally develop after a few years of treatment. TU et al. (1963) have reported optic neuritis occurring in association with the use of penicillamine which responded to pyridoxine. Other serious adverse effects include myasthenia gravis, peripheral neuropathy and polymyositis (KLAASSEN 1980).

Mild sensitivity reactions may be brought under control by reduction in penicillamine dosage. Severe reactions require withdrawal of the drug, but reinstitution may be tried, starting with low doses to attempt desensitization. A course of cortisone may be useful in controlling and preventing the re-emergence of allergic reactions.

Zinc is also bound to penicillamine, but with less affinity than copper. A progressive increase in urinary zinc and decrease in serum zinc has been reported with maintenance penicillamine therapy (VAN CAILLE-BERTRAND et al. 1985). Some of the known side effects of penicillamine, including purpura, desquamative and psoriatic lesions, stomatitis and loss of taste sensation, suggest zinc deficiency. SHOULSON et al. (1983) have described a patient in whom oral zinc 200 mg/day produced a prompt response to penicillamine-induced ageusia. Zinc therapy in Wilson's disease may also have value by interfering with copper absorption from the gastrointestinal tract and increasing fecal copper excretion (HOOGENRAAD et al. 1979).

In patients with severe neurological involvement which responds only partially to penicillamine, levodopa therapy may be tried, but usually only a limited response can be achieved (GOLDSTEIN et al. 1968). SHOULSON et al. (1983) have reported a case in which trihexyphenidyl reduced dystonia and rigidity which had persisted on maintenance copper chelation therapy. Asymptomatic homozygotes for Wilson's disease should be treated to prevent the development of clinical symptoms. Most patients with Wilson's disease can be maintained in good health indefinitely on penicillamine (WALSHE and DIXON 1986).

D. Multiple System Degenerations

Parkinsonian features may be the initial or predominant manifestations of multiple system degenerations of the central nervous system, including striatonigral degeneration, olivopontocerebellar atrophy, Shy-Drager syndrome, progressive supranuclear palsy, the ALS-Parkinson-dementia complex of Guam, corticodentatonigral degeneration and Joseph disease.

If parkinsonism is the presenting feature, there may be no way to distinguish these conditions from Parkinson's disease and antiparkinson treatment should be started in the usual way. However, the response in these conditions is often unsatisfactory and subsequently atypical features appear such as conjugate gaze paresis, pyramidal or cerebellar deficits, severe dementia, apraxia or significant autonomic dysfunction unrelated to drugs. The development of such multisystem signs in the progress of a parkinsonian illness or primary failure of response to dopaminomimetic therapy justify review of the initial diagnosis of Parkinson's disease.

In the "Parkinson-plus" syndromes, the prominence of pathology in the pallidum (Young and Penney 1984) and the loss of striatal cells which are postsynaptic to the dopaminergic input (Klawans and Ringel 1971) may account for the lack of response to dopamine therapy. Nevertheless, antiparkinsonian drugs should be tried in these conditions because a useful response may occur in a minority of patients. The multiple system degenerations will now be considered individually.

I. Striatonigral Degeneration

The clinical picture of striatonigral degeneration is usually indistinguishable from Parkinson's disease. Izumi et al. (1971) reported no response to levodopa in this condition; however, Sharpe et al. (1973) and Adams (1968) have claimed transient but significant improvement in akinesia and rigidity after 1 month of levodopa treatment.

II. Olivopontocerebellar Atrophy

Goetz et al. (1984) have retrospectively analysed the treatment of 24 patients with parkinsonism and cerebellar or corticospinal dysfunction who were considered clinically to have olivopontocerebellar atrophy. Of 19 patients treated with levodopa, alone or in combination with carbidopa, 16 showed definite improvement in rigidity, akinesia and postural reflexes. Levodopa was also effective against resting tremor, but not associated intention tremor. Central dopaminergic toxicity, including dyskinesia, sleep disturbance and hallucinations occurred after an average of 2.3 years of treatment.

Prominent resting tremor failed to respond to anticholinergic therapy in the two patients in whom it was tried. Amantadine was of mild or moderate benefit in nine of the ten patients treated. Dopamine agonists, bromocriptine in five patients and lergotrile in one patient, were effective in patients who had responded to levodopa and in one patient who had not responded.

III. Shy-Drager Syndrome

In the Shy-Drager syndrome, parkinsonian features may be the first to appear or they may be preceded or accompanied by progressive autonomic failure, signs of cerebellar disease, or upper and lower motor neuron dysfunction. Profound depletion in dopamine and noradrenaline has been reported in brain regions normally rich in these catecholamines (SPOKES et al. 1979); studies of [^{3}H]spiperone binding have suggested that a loss of dopamine receptor sites occurs in the substantia nigra with reduced binding to dopamine receptors in the caudate nucleus (QUIK et al. 1979).

The combination of autonomic failure with parkinsonism produces a significant problem in management since levodopa and dopamine agonists all tend to exacerbate postural hypotension. SHARPE et al. (1972) found improvement in both postural hypotension and parkinsonism in a patient treated with levodopa and the monoamine oxidase inhibitor, tranylcypromine. However, such an approach is potentially hazardous because severe supine hypertension may be produced (CALNE and REID 1972) and, at the same time, postural hypotension may worsen (RAE-GRANT et al. 1985).

Indomethacin, which inhibits prostaglandin production, has been reported to improve postural hypotension (KOCHAR and ITSKOVITZ 1978), but central side effects such as confusion are likely to occur with its use in Shy-Drager syndrome.

The commonest approach is to administer antiparkinsonian agents cautiously with the concomitant use of independent measures to control postural hypotension, including fludrocortisone, head-up tilt of the bed at night and elastic stockings to improve venous return from the lower body (BANNISTER and OPPENHEIMER 1982).

LEES and BANNISTER (1981) have performed a controlled trial of lisuride in seven patients with Shy-Drager syndrome. Improvement in parkinsonian features was observed in only one patient and a modest reduction in orthostatic hypotension occurred in two patients. Psychiatric side effects were dose-limiting in six patients.

IV. Progressive Supranuclear Palsy

Since parkinsonism in progressive supranuclear palsy (PSP) is often not the most disabling feature, little functional improvement may occur despite some observable decreases in rigidity and akinesia with levodopa (KLAWANS and RINGEL 1971). Nonparkinsonian features such as ophthalmoplegia and nuchal dystonia only rarely respond to levodopa (MENDELL et al. 1970). Occasional improvement in the range of eye movements has been reported with the use of dopamine agonists (WILLIAMS et al. 1979; JACKSON et al. 1983). Increased confusion is likely to occur with dopaminomimetic therapy in the intellectually impaired.

JACKSON et al. (1983) found that levodopa or amantadine benefited a minority of patients with PSP, but that dopamine agonist therapy with bromocriptine or pergolide produced improvement in 9 of 14 patients. Reponse to dopamine agonists occurred despite previous failure of levodopa. On the other hand, in a placebo-controlled study of bromocriptine in six patients with PSP, WILLIAMS et al. (1979) did not find any overall beneficial response. NEOPHYTIDES et al. (1982) have reported improvement in rigidity, tremor and akinesia in one case of PSP treated with lisuride.

RAFAL and GRIMM (1981) have claimed improvement in parkinsonian and pseudobulbar signs in 10 of 12 patients treated with the antiserotonergic agent, methysergide; however, in only 5 was the response sustained for longer than 6 months and 4 of these patients were concurrently receiving levodopa. PAULSON et al. (1981) found that methysergide failed to produce a significant improvement in doses as high as 10 mg/day. Improvement in tremor, gait and mental state has been attributed to amitriptyline in one patient with PSP (KVALE 1982).

V. Senile Parkinsonism

This term has been employed to describe the simultaneous occurrence of parkinsonism and severe dementia. Akinesia and rigidity tend to be long preceded by dementia due to Alzheimer's disease, Pick's disease or a multi-infarct state; antiparkinsonian treatment is less effective and often poorly tolerated in these settings because of the frequency and severity of psychiatric side effects.

VI. The ALS-Parkinson-Dementia Complex of Guam

Conventional antiparkinsonian therapy has been used in the ALS-Parkinson-dementia complex of Guam with modest response, but no controlled observations have been published.

VII. Joseph Disease

In Joseph disease, an autosomal dominant multiple system degeneration, parkinsonian manifestations may benefit from treatment with levodopa (ROSENBERG 1984).

E. Other Types of Parkinsonism

I. Postencephalitic Parkinsonism

Sustained and functionally important improvement was described with levodopa therapy in postencephalitic parkinsonism (CALNE 1969) although some patients showed no overall change and intolerable psychiatric side effects were frequent (HUNTER et al. 1970; SACKS and KOHL 1970).

II. Parkinsonism Due to Toxins

1-Methyl-4-phenyl-1,2,3,6-tetrahydropyridine (MPTP) produces a severe selective lesion of the dopaminergic cells of the pars compacta of the substantia nigra. The resulting parkinsonism responds well to levodopa-carbidopa combinations and bromocriptine, but severe dyskinesias, response fluctuations and psychiatric side effects have been seen several months after initiating therapy (LANGSTON and BALLARD 1984). These problems are not usually encountered in Parkinson's disease until after several years of treatment.

Parkinsonism is an occasional sequel to poisoning with carbon monoxide or cyanide. Bilateral necrosis of the globus pallidus is found and usually there is no response to levodopa. However, JAECKLE and NASRALLAH (1985) have reported one patient with carbon monoxide encephalopathy in whom significant improvement in parkinsonism and depression occurred with levodopa; this patient had pallidal lesions on computerized tomographic scan.

III. Hypoparathyroidism

Parkinsonism complicating hypoparathyroidism is usually resistant to correction of hypocalcaemia and to levodopa therapy (UNCINI et al. 1985), but reversal of the extrapyramidal syndrome associated with postoperative hypoparathyroidism has been reported following repletion of serum calcium (BERGER and ROSS 1981).

IV. Huntington's Disease

Predominantly akinetic and rigid manifestations are seen most often with juvenile onset Huntington's disease and may be associated with low CSF levels of homovanillic acid. Such patients may have a beneficial response to levodopa (BARBEAU 1969; SCHENK and LEIJNSE-YBEMA 1974), but increased chorea and behavioural disturbances may develop.

F. Conclusions

We have reviewed the treatment of parkinsonism occurring in settings other than Parkinson's disease. The management is directed at removing causal factor where possible. By far the most important examples of secondary parkinsonism, from this viewpoint, are the syndromes induced by neuroleptics and copper deposition in the basal ganglia (Wilson's disease).

References

Adams RD (1968) The striatonigral degenerations. In: Vinken PJ, Bruyn GW (eds) Handbook of clinical neurology, vol 6. Elsevier/North-Holland, Amsterdam, pp 694-702

Aronson TA (1985) Persistent drug-induced parkinsonism. Biol Psychiatry 20:795-798

Ayd FJ (1974) Rational psychopharmacotherapy and the right to treatment. Ayd Medical Communications, Baltimore

Baldessarini RJ (1980) Drugs and the treatment of psychiatric disorders. In: Goodman AG, Goodman LS, Gilman A (eds) The pharmacological basis of therapeutics. MacMillan, New York, pp 391-448

Bannister R, Oppenheimer D (1982) Parkinsonism, system degenerations and autonomic failure. In: Marsden CD, Fahn S (eds) Movement disorders. Butterworth, London, pp 174-190

Barbeau A (1969) L-Dopa and juvenile Huntington's disease. Lancet 2:1066

Berger JR, Ross DB (1981) Reversible Parkinson syndrome complicating postoperative hypoparathyroidism. Neurology 31:881-882

Bowers MB (1985) Cerebrospinal fluid homovanillic acid and the hypokinetic side effects of neuroleptics. Psychopharmacology (Berlin) 85:184-186

Bruno A, Bruno SC (1966) Effects of L-dopa on pharmacological parkinsonism. Acta Psychiatr Scand 42:264-271

Calne DB (1982) Dopamine receptor agonists in the treatment of basal ganglia disorders. Semin Neurol 2:359-364

Calne DB, Reid JL (1972) Antiparkinsonian drugs: pharmacological and therapeutic aspects. Drugs 4:49-74

Calne DB, Stern GM, Laurence DR, Sharkey J, Armitage P (1969) L-Dopa in postencephalitic parkinsonism. Lancet 1:744-746

Casey DE, Provisen UJ, Meidahl B, Gerlach J (1986) Neuroleptic-induced tardive dyskinesia and parkinsonism: changes during several years of continuing treatment. Psychopharmacol Bull 22:250-253

Chase TN, Schnur JA, Gordon EK (1970) Cerebrospinal fluid monoamine catabolites in drug-induced extrapyramidal disorders. Neuropharmacology 9:265-268

Chouinard G, Annable L, Ross-Chouinard A, Kropsky ML (1979) Ethopropazine and benztropine in neuroleptic-induced parkinsonism. J Clin Psychiatry 40:147-151

Crawhall JC (1980) Penicillamine: twenty five years later. Ann Intern Med 93:367-368

Davis JM (1985) Antipsychotic drugs. In: Kaplan HI, Sadock BJ (eds) Comprehensive textbook of psychiatry, vol 4. Williams and Wilkins, Baltimore, pp 1481-1512

Demars JPCA (1966) Neuromuscular effects of long term phenothiazine medication, ECT, and leucotomy. J Nerv Ment Dis 143:73-79

DiMascio A, Demirgian E (1970) Antiparkinson drug overuse. Psychosomatics 11:569-601

Donlon PT, Stetson RL (1976) Neuroleptic-induced extrapyramidal symptoms. Dis Nerv Syst 37:629-635

Duchen LW, Jacobs JM (1984) Nutritional deficiencies and metabolic disorders. In: Adams JH, Corsellis JAN, Duchen LW (eds) Greenfield's neuropathology. Edward Arnold, London, pp 573-626

Fann WE, Lake CR (1976) Amantadine versus trihexyphenidyl in the treatment of neuroleptic-induced parkinsonism. Am J Psychiatry 133:940-943

Fleming P, Makar H, Hunter KR (1970) Levodopa in drug-induced extrapyramidal reactions. Lancet 2:1186

Fricchione GL, Cassem NH, Hooberman D, Hobson D (1983) Intravenous lorazepam in neuroleptic-induced catatonia. J Clin Psychopharmacol 3:338-342

Gelenberg AJ, Mandel MR (1977) Catatonic reactions to high potency neuroleptic drugs. Arch Gen Psychiatry 34:947-950

Goetz CG, Tanner CM, Klawans HL (1984) The pharmacology of olivopontocerebellar atrophy. In: Duvoisin RC, Plaitakis A (eds) Advances in neurology, vol 41. The olivopontocerebellar atrophies. Raven, New York, pp 143-148

Goldstein NP, Tauxe WN, McCall JT, Randall RV, Gross JB (1968) What Wilson's disease and its treatment have taught us about the metabolism of copper. Med Clin North Am 52:989-1001

Goni M, Jimenez M, Feijoo M (1985) Parkinsonism induced by phenytoin. Clin Neuropharmacol 8:383-384

Gratton L (1960) Neuroleptiques, parkinsonnisme et schizophrénie. Union Med Can 89:679-694

Grimes JD (1981) Parkinsonism and tardive dyskinesia associated with long-term metoclopramide therapy. N Engl J Med 305:1417

Grimes JD, Magner PO (1984) Safety of domperidone in metoclopramide-induced parkinsonism. Arch Neurol 41:363-364

Hershon HI, Kennedy PF, McGuire RJ (1972) Persistence of extra-pyramidal disorders and psychiatric relapse after withdrawal of long-term phenothiazine therapy. Br J Psychiatry 120:41-50

Hoogenraad TU, Koevoet R, de Ruyter Korver EGWM (1979) Oral zinc sulphate as long-term treament of Wilson's disease. Eur Neurol 18:205-211

Hornykiewicz O (1975) Parkinsonism induced by dopaminergic antagonists. In: Calne DB, Chase TN, Barbeau A (eds) Advances in neurology, vol 9. Dopaminergic mechanisms. Raven, New York

Hornykiewicz O (1986) Dopamine deficiency and dopamine substitution in Parkinson's disease. In: Winlow W, Markstein R (eds) The neurobiology of dopamine systems. Manchester University Press, Manchester

Hunter KR, Stern GM, Sharkey J (1970) Levodopa in postencephalitic parkinsonism. Lancet 2:1366-1367

Izumi K, Inoue N, Shirabe T (1971) Failed levodopa therapy in striato-nigral degeneration. Lancet 1:1355

Jackson JA, Jankovic J, Ford J (1983) Progressive supranuclear palsy: clinical features and response to treatment in 16 patients. Ann Neurol 13:273-278

Jaeckle RS, Nasrallah HA (1985) Major depression and carbon monoxide-induced parkinsonism: diagnosis, computerized axial tomography, and response to L-dopa. J Nerv Ment Dis 173:503-508

Jellinek T, Gardos G, Cole JO (1981) Adverse effects of antiparkinson drug withdrawal. Am J Psychiatry 138:1567-1571

Klaassen C (1980) Heavy metals and heavy-metal antagonists. In: Gilman AG, Goodman LS, Gilman A (eds) The pharmacological basis of therapeutics. MacMillan, New York, pp 1627-1629

Klawans HL, Ringel SP (1971) Observations on the efficacy of L-dopa in progressive supranuclear palsy. Eur Neurol 5:115-129

Kochar MS, Itskovitz HD (1978) Treatment of idiopathic orthostatic hypotension (Shy-Drager syndrome) with indomethacin. Lancet 1:1011-1014

Kvale JN (1982) Amitriptyline in the management of progressive supranuclear palsy. Arch Neurol 39:387-388

Langston JW, Ballard P (1984) Parkinsonism induced by 1-methyl-4-phenyl-1,2,3,6-tetrahydropyridine (MPTP): implications for treatment and the pathogenesis of Parkinson's disease. Can J Neurol Sci 11:160-165

Lees AJ, Bannister R (1981) The use of lisuride in the treatment of multiple system atrophy with autonomic failure (Shy-Drager syndrome). J Neurol Neurosurg Psychiatry 44:347-351

Manos, N, Lavrentiadis G, Gkiouzepas J (1986) Evaluation of the need for prophylactic antiparkinsonian medication in psychotic patients treated with neuroleptics. J Clin Psychiatry 47:114-116
Marsden CD, Fahn S (1982) Problems in Parkinson's disease. In: Marsden CD, Fahn S (eds) Movement disorders. Butterworth, London, pp 1-7
Marsden CD, Tarsy D, Baldessarini RJ (1975) Spontaneous and drug-induced movement disorders in psychotic patients. In: Benson DF, Blumer D (eds) Psychiatric aspects of neurologic disease. Grune and Stratton, Orlando
McGeer PL, Boulding JE, Gibson WC, Foulkes RG (1961) Drug-induced extrapyramidal reactions. JAMA 177:665-670
McGeer PL, McGeer ED, Suzuki JS (1977) Aging and extrapyramidal function. Arch Neurol 34:33-35
Mendell JR, Chase TN, Engel WK (1970) Modification by L-dopa of a case of progressive supranuclear palsy. Lancet 1:593-594
Mendez JS, Cotzias GC, Mena I, Papavasiliou PS (1975) Diphenylhydantoin: blocking of levodopa effects. Arch Neurol 32:44-46
Meyboom RHB, Ferrari MD, Dieleman BP (1986) Parkinsonism, tardive dyskinesia, akathisia, and depression induced by flunarizine. Lancet 2:292
Miller R, Hiley R (1975) Antimuscarinic actions of neuroleptic drugs. In: Calne DB, Chase TN, Barbeau A (eds) Advances in neurology, vol 9. Dopaminergic mechanisms. Raven, New York, pp 141-147
Neophytides A, Lieberman AN, Goldstein M, Gopinathan G, Leibowitz M, Bock J, Walker R (1982) The use of lisuride, a potent dopamine and serotonin agonist, in the treament of progressive supranuclear palsy. J Neurol Neurosurg Psychiatry 45:261-263
Parkes JD, Schachter M, Marsden CD, Smith B, Wilson A (1981) Lisuride in parkinsonism. Ann Neurol 9:48-52
Paulson GW, Lowery HW, Taylor GC (1981) Progressive supranuclear palsy: pneumoencephalography, electronystagmography and treatment with methysergide. Eur Neurol 20:13-16
Purdon Martin J (1968) Wilson's disease. In: Vinken PJ, Bruyn GW (eds) Handbook of clinical neurology, vol 6. Diseases of the basal ganglia. North-Holland, Amsterdam, pp 267-278
Quik M, Spokes EG, Mackay AVP, Bannister R (1979) Alterations in [^{3}H]spiperone binding in human caudate nucleus, substantia nigra and frontal cortex in the Shy-Drager syndrome and Parkinson's disease. J Neurol Sci 43:429-437
Rae-Grant A, Young GB, Spence JD (1985) Monoamine oxidase inhibitors and Sinemet in Shy-Drager syndrome. Neurology 35:1058-1086
Rafal RD, Grimm RJ (1981) Progressive supranuclear palsy: functional analysis of the response to methysergide and antiparkinsonian agents. Neurology 31:1507-1508
Rosenberg RN (1984) Joseph disease: an autosomal dominant motor system degeneration. In: Duvoisin R, Plaitakis A (eds) Advances in neurology, vol 41. The olivopontocerebellar atrophies. Raven, New York, pp 179-194
Rifkin A, Quitikin F, Kane J, Struve F, Klein DF (1978) Are prophylactic antiparkinson drugs necessary? A controlled study of procyclidine withdrawal. Arch Gen Psychiatry 35:483-489
Rajput AH, Rozdilsky B, Hornikiewicz O (1982) Reversible drug-induced parkinsonism: clinicopathologic study of two cases. Arch Neurol 39:644-646
Rivera VM, Keichan AH, Oliver RE (1975) Persistent parkinsonism following neuroleptanalgesia. Anaesthesiology 42:635-637
Sachdev PS (1986) Lithium potentiation of neuroleptic-related extrapyramidal side effects. Am J Psychiatry 143:942

Sacks OW, Kohl M (1970) L-Dopa and oculogyric crises. Lancet 2:215-216
Sandyk R (1985) Parkinsonism induced by captopril. Clin Neuropharmacol 8:197
Schenk G, Leijnse-Ybema HJ (1974) Huntington's chorea and levodopa. Lancet 1:364
Sharpe J, Marquez-Julio A, Ashby P (1972) Idiopathic orthostatic hypotension treated with levodopa and MAO inhibitor: a preliminary report. Can Med Assoc J 107:296-300
Sharpe JA, Rewcastle NB, Lloyd KG, Hornykiewicz O, Hill M, Tasker RR (1973) Striatonigral degeneration: response to levodopa therapy with pathological and neurochemical correlation. J Neurol Sci 19:275-286
Shoulson I (1983) Carbidopa/levdopa therapy of coexistent drug-induced parkinsonism and tardive dyskinesia. In: Fahn S, Calne DB, Shoulson I (eds) Advances in neurology, vol 37. Experimental therapeutics of movement disorders. Raven, New York
Shoulson I, Goldblatt D, Plassche W, Wilson G (1983) Some therapeutic observations in Wilson's disease. In: Fahn S, Calne DB, Shoulson I (eds) Advances in neurology, vol 37. Experimental therapeutics of movement disorders. Raven, New York
Spokes EGS, Bannister R, Oppenheimer DR (1979) Multiple system atrophy with autonomic failure. J Neurol Sci 43:59-82
Stephen PJ, Williamson J (1984) Drug-induced parkinsonism in the elderly. Lancet 2:1082-1083
Suranyi-Cadotte BE, Nestoros JN, Nair NPV, Lal S, Gauthier S (1985) Parkinsonism induced by high doses of diazepam. Biol Psychiatry 20:451-460
Tarsy D (1983) Neuroleptic-induced extrapyramidal reactions: classification, description, and diagnosis. Clin Neuropharmacol 6 (Suppl 1):9-26
Tu J, Blackwell RQ, Lee P (1963) DL-Penicillamine as a cause of optic axial neuritis. JAMA 185:83-86
Uncini A, Tartaro A, Di Stefano E, Gambi D (1985) Parkinsonism, basal ganglia calcification and epilepsy as late complications of postoperative hypoparathyroidism. J Neurol 232:109-111
Van Caille-Bertrand M, Degenhart HJ, Luijendijk I, Bouquet J, Sinaasappel M (1985) Wilson's disease: assessment of D-penicillamine treatment. Arch Dis Child 60:652-655
Walshe JM (1986) The management of pregnancy in Wilson's disease treated with trientene. Q J Med 58:81-87
Walshe JM, Dixon AK (1986) Dangers of non-compliance in Wilson's disease. Lancet 1:845-847
Williams AC, Nutt J, Lake CR, Pfeiffer R, Teychenne PE, Ebert M, Calne DB (1979) Actions of bromocriptine in the Shy-Drager and Steele-Richardson-Olszewski syndromes. In: Fuxe K, Calne DB (eds) Dopaminergic ergot derivatives and motor function. Pergamon, Oxford, pp 271-283
Williamson J (1984) Drug-induced Parkinson's disease. Br Med J 288:1457
Yaryura-Tobias JA, Wolpert A, Dana L, Merlis S (1970) Action of L-dopa in drug induced extrapyramidalism. Dis Nerv Syst 31:60-63
Young AB, Penney JB (1984) Neurochemical anatomy of movement disorders. In: Jankovic J (ed) Movement disorders. Neurol Clin 2:417-433

CHAPTER 23

Management of Psychiatric Symptoms in Parkinson's Disease

H. L. KLAWANS and C. M. TANNER

A. Introduction

Most reviews of the psychiatric problems seen during the chronic therapy of parkinsonism have focused primarily on nonaffective psychosis and related disorders. This chapter will include both these disorders and alterations in affect or mood.

B. Depression

Depression has been associated with Parkinson's disease since the original description by James Parkinson. The potential misdiagnosis of parkinsonian akinesia and facial amimia as depression has led some clinicians to caution against the diagnosis of depression on the basis of motor symptoms alone. Nonetheless, from 1922 through the present, clinical evaluations of patients with Parkinson's disease have found symptoms of depression in 20%-93% of patients interviewed (MINDHAM 1970; PATRICK and LEWY 1922).

Depression has traditionally been divided by etiology into two types: (1) reactive, in which depressive symptoms occur as a result of an identifiable cause; and (2) endogenous, in which symptoms have no identifiable contributory environmental factors. This view of depression has led to the argument that depression in Parkinson's disease is simply a reaction to chronic disability, and is not an inherent part of the disease process itself.

Undoubtedly, some parkinsonian patients with symptoms of depression are exhibiting a reaction to their chronic disability. If this were the sole etiology of depressive symptoms in Parkinson's disease, however, symptoms would be expected to be maximal at either the time of diagnosis (when the patient must adjust to the presence of a chronic disease) or at a time when significant disability occurs. Moreover, patients with Parkinson's disease and patients with other disabilities would be expected to have a similar prevalence of depression. The relationship of depression to the underlying disease process in Parkinson's disease has been addressed by several investigators.

In one group of studies, the prevalance of depressive symptoms in patients with Parkinson's disease was compared with a control group of patients with other illness. HORN (1974) compared paraplegics, parkinsonian patients, and persons with no identified illness who were similar to each other in age, sex, and socioeconomic level. He found a significant increase in depressive symptoms in the Parkinson's disease groups when compared with either paraplegics or normal subjects. Similar findings have been obtained when comparing patients with Parkinson's disease with those with other disabilities, such as hemiplegia or amputation, as well as patients with medical illnesses. These multiple observations of an increased prevalence of depression in Parkinson's disease as compared with a variety of other disorders suggest that depression may be endogenous rather than reactive in most parkinsonian patients (ROBINS 1976).

Several other investigators have explored the relationship between the onset of parkinsonian symptoms and the onset of depression. If depressive symptoms were merely a reaction to either the diagnosis or the progressive disability of Parkinson's disease, they would be expected to be maximal around the time of diagnosis. Indeed, in several studies, depressive symptoms occurred frequently during the year after the diagnosis of Parkinson's disease was made, suggesting that reaction to the diagnosis of a chronic illness played some role in depression in many patients with Parkinson's disease (PATRICK and LEWY 1922; MINDHAM 1970). Several other studies, however, using retrospective techniques, have reported depression to occur in 34%–42% of patients before Parkinson's disease was ever diagnosed (MINDHAM 1970; MJONES 1949; MAYEUX et al. 1981). These latter data support the postulate that patients with Parkinson's disease are more likely to experience depression, and that this mood abnormality may be an intrinsic part of the disease process.

There may be a neurochemical rationale for the increased frequency of depressive symptoms in Parkinson's disease. The major chemical deficit in this disease, and the one most clearly related to the motor disability, is loss of dopaminergic neurons in the substantia nigra. Pathologic alterations in Parkinson's disease include all pigmented nuclei, however, and accompanying pharmacologic deficits of all biogenic amines occur. Although dopamine is probably not an important neurotransmitter in most depressive syndromes, decreased activity of other biogenic amine neurotransmitter systems, norepinephrine and serotonin, have frequently been implicated. Both of these neurotransmitters are present in decreased concentration in patients with Parkinson's disease, and depressive symptoms in Parkinson's disease may be the result of these biochemical defects. MAYEUX et al. (1984a, 1986a) have reported a correlation between decreased metabolites of serotonin in the spinal fluid and depression in patients with Parkinson's disease. They found that CSF 5-hydroxyindoleacetic acid (5-HIAA) was lower in depressed than nondepressed parkinsonian patients and was related to both psychomotor retardation and loss of self-esteem. An open trial of L-5-hydroxytryptamine (5-HT) alleviated depression in six of seven patients, and this alleviation was associated with increased CSF 5-HIAA in three patients (MAYEUX et al. 1986b).

The treatment of depression in Parkinson's disease is currently under active clinical investigation. Levodopa and other dopaminergic agents, while of clear benefit to the primary symptoms of Parkinson's disease, do not alleviate symptoms of depression. For this reason, depression can be the major source of disability for the patient with a good therapeutic response to antiparkinsonian agents. Some early reports suggested that levodopa had a mood elevating effect in depressed parkinsonian patients (BARBEAU 1969; CELESIA and WANAMAKER 1972). This effect was often attributed to the resolution of the "reactive" depression associated with the disease, especially when levodopa actually improved the neurologic picture. In contrast, DAMASIO et al. (1971) found that levodopa had no significant antidepressant effects and might have actually reactivated depression in patients who had a prior psychiatric history. MARSH and MARKHAM (1973) found no evidence to indicate either a significant increase or decrease in the prevalence of depression in 27 patients started on chronic levodopa. MINDHAM et al. (1971) reported that almost 50% of a large group of patients developed depression during the first 6 months of levodopa therapy. Twelve of these patients had had a prior history of depression. They concluded that levodopa probably caused a reemergence of preexisting affective symptoms rather than causing a new illness. They also felt that levodopa had little, if any, antidepressant actions.

One report has suggested that levodopa may have at least contributed to the suicide of a parkinsonian patient (RAFT et al. 1972). Suicidal thoughts and attempts are not uncommon, but suicide has been rarely documented among parkinsonian patients (MAYEUX et al. 1984b). The patient who committed suicide did have a prior history of depression beginning with the onset of the parkinsonism. In spite of the neurologic improvement with levodopa, the patient's emotional state deteriorated and he eventually killed himself. It is possible that levodopa may transiently relieve depressive symptoms in some patients, particularly if the symptoms are reactive to the initial disability; on the other hand, levodopa may also reactivate depression in those patients with a premorbid history or have no demonstrable effect.

Tricyclic antidepressants have been observed to help depressive symptoms in Parkinson's disease (BROWN and WILSON 1972; CELESIA and WANAMAKER 1972). Many of these agents have prominent anticholinergic activity; their use could improve Parkinson's disease and secondarily improve the reactive component of a depressive syndrome. Studies separately assessing improvement in Parkinson's disease and mood, however, have found tricyclics to benefit depressive symptoms independently of any effects on Parkinson's disease (LAITINEN 1969). Agents reported to be of benefit include imipramine (STRANG 1965), desipramine (LAITENIN 1969), nortriptyline (ANDERSEN et al. 1980), and amitriptyline. GOETZ et al. (1984) studied bupropion, an agent believed to affect only dopamine systems by enhancing release and blocking reuptake of transmitter presynaptically. In patients receiving both bupropion and levodopa-carbidopa, depressive symptoms and parkinsonian disability were decreased, but improvement in one was independent of the other. The overall long-term efficacy of any of these agents in the depression associated with parkinsonism is unknown.

Several reports have noted improvement of depression after electroconvulsive therapy (ECT) in Parkinson's disease, often with an associated improvement of parkinsonism. LEBENSOHN and JENKINS (1975) used ECT to treat depression in two parkinsonian patients whose depression had been resistant to more conventional therapies. Improvement in bradykinesia and rigidity, as well as mood, occurred in both patients. Two other patients have been described in whom mood and parkinsonism improved following ECT (Asnis 1977; Yudofsky 1979). Mood elevation was reported to occur before neurologic improvement in both studies, and only a few treatments were required. We have observed similar responses in three patients who did not respond to or were unable to tolerate antidepressant therapy. In one case, depression recurred several years after ECT and symptoms were again ameliorated by this modality. In each case, patients demonstrated transient amelioration of all symptoms of Parkinson's disease and tolerated a decrease in daily levodopa dose. This effect was maintained for several weeks, but the antidepressant effect lasted for months or years. While unproven, it does appear that there may be a place for ECT in depressed parkinsonian patients who are otherwise unresponsive to tricyclic antidepressants or other forms of therapy.

C. Psychosis and Related Disorders

The most important of the various drug-induced changes in mental function that influence antiparkinsonian therapy are nonaffective psychiatric manifestations induced by levodopa. In their most dramatic form, these may result in frank psychosis. These levodopa-induced psychiatric disturbances may be divided into those which occur early in the course of treatment and those which present after several years of therapy.

I. Early Onset Psychosis

When levodopa-induced psychosis occurs early in the course of levodopa therapy, it usually happens within a few weeks after the levodopa regimen has started. The patient almost invariably has a past history of a severe psychiatric disorder, usually schizophrenia. This type of levodopa-induced psychosis has been reported in two circumstances. One involves the attempted use of levodopa to replace an anticholinergic agent in the treatment of drug-induced parkinsonism in schizophrenic patients receiving neuroleptic agents. The other situation involves parkinsonian patients who are not actively psychotic, but who frequently have a remote history of psychiatric disorder — often euphemistically described as a "nervous breakdown."

In either group, psychosis usually occurs early in the course of levodopa therapy, normally within the first days or weeks. The manifestations can be quite variable, often have a strong paranoid component, and frequently have features similar to previous psychiatric disturbances. Not every patient who suffers levodopa-induced psychosis has a past history of psychiatric disease, at least not one that is readily obtainable.

If the patient does have a past history of a significant schizophrenic disorder, however, it is probably best not to begin treatment of their parkinsonism with direct-acting dopamine agonists or with precursor load therapy. If their disease necessitates such therapy, however, the doses of either class of agents should be kept quite low and the goals of therapy reduced.

II. Late Onset Psychosis

The late occurring psychoses are not predictable by past history, since the vast majority of patients who have this difficulty have no previous history of psychiatric disorder. While patients with significant dementia may be more likely to have this drug-induced effect, levodopa-induced psychosis often occurs in the absence of dementia.

Late onset psychosis induced by levodopa appears after several years of therapy (MOSKOWITZ et al. 1978). It is important to note that this is not just an acute psychosis that occurs suddenly in a patient who has been receiving levodopa or a carbidopa-levodopa combination for many years. Rather, it appears to be part of a long-standing progressive process, one which is related to two other types of levodopa-induced mental alterations: sleep disturbances and hallucinations.

The dream disorders were first well described by MOSKOWITZ et al. (1978). They reported that dream disturbances occurred in at least one-third of all patients taking levodopa for 2 years or longer. They classified these dream disturbances as vivid dreams, night terrors, and nightmares (MOSKOWITZ et al. 1978; SHARF et al. 1978).

The levodopa-induced dreams were qualitatively vivid, seemingly real, temporally condensed, and internally organized and coherent. While often affectively neutral, with a frequent theme of persons and events from the dreamer's remote past, these dreams are qualitatively different from the patient's previous dream experiences, and were by far the commonest type of levodopa-induced dreams.

In contrast, night terrors were reported by other members of the family, since patients are always amnesic during these experiences. The patient typically screams, calls out, and thrashes during sleep. Patients might awaken screaming, yet forget why they screamed or woke up. Approximately 5 % of patients receiving chronic levodopa therapy had such experiences.

Classic nightmares — frightening, often paranoid, and, like the vivid dreams described, considered by patients to be distinctly different from other dreams — affected approximately 5 % of patients on chronic levodopa therapy.

Approximately one-half of the patients on levodopa for 2 years or more were found to manifest hallucinations (MOSKOWITZ et al. 1978; SWEET et al. 1976). Hallucinatory phenomena are usually stereotyped in each patient; they are often nocturnal, nonthreatening, and recurrent. The hallucinations are predominantly visual (at times with a secondary auditory component) and are superimposed on a clear sensorium. They usually conform to boundaries imposed by actual concurrent sensory input, and often concern individuals and experiences that were significant in the patient's past. At times, the hallucinations blend indistinguishably with dream phenomena possessing similar themes. At least two-thirds of the patients reporting these hallucinations also have levodopa-induced dream disturbances.

After years of these phenomena, patients who have hallucinations and dream disturbances can progress to overt psychosis. This psychosis is most commonly paranoid in nature. In fact, all of the aforementioned mental alterations may be characterized as a pure paranoid delusional system superimposed on a clear sensorium, with no other qualities of thought disorder. A rare patient will develop a full-blown schizophreniform psychosis or will progress from paranoid psychosis to a confusional toxic psychosis. This is most often preceded by a stage consisting of paranoid nonconfusional psychosis.

Moskowitz et al. (1978) felt that their data suggested a progression of medication-induced psychiatric symptomatology, from sleep disruptions, to striking dreams, to dreams and hallucinatory experiences, to a pure paranoid delusional system, and finally, to a confusional state superimposed on the rest. The actual onset of psychiatric symptoms varies from insidious to acute, and although it is occasionally triggered by an increase in levodopa dosage, it is not usually associated with any alteration in maintenance therapy.

The sleep abnormalities seen in parkinsonian patients during chronic levodopa therapy were first well described by Nausieda et al. (1982, 1984). They found that sleep disruption was a common complaint in levodopa-treated parkinsonian patients. This conclusion was based on a survey of 100 parkinsonian patients which revealed prominent sleep complaints in 74. Sleep complaints were unrelated to age and the duration of disease, but increased in prevalence with longer periods of levodopa therapy. Seven different sleep abnormalities were documented in patients receiving chronic levodopa: (1) inability to initiate sleep, (2) difficulty in maintaining sleep; (3) excessive daytime sleepiness; (4) vivid dreams or nightmares; (5) nocturnal vocalizations; (6) nocturnal myoclonus; and (7) somnambulism. Of these, the most proment complaint was that of disrupted sleep. In these patients, falling asleep is achieved without difficulty, but is followed by frequent nocturnal awakenings, with inability to get back to sleep. Total sleep time in 24 h is usually adequate because of some degree of daytime hypersomnia. These symptoms generally preceded the onset of other sleep abnormalities. The sleep abnormalities tended to increase in severity with continued treatment, and insomnia tended to be followed by daytime somnolence, altered dream events, and episodic nocturnal vocalization and myoclonus. While dyskinetic side effects and on-off syndrome were encountered in patients with and without sleep complaints, 98% of patients experiencing psychiatric side effects also reported sleep disruption. Nausieda et al. (1982) felt that the sleep-related symptoms constituted an early stage of levodopa-induced dopaminergic psychiatric toxicity in the parkinsonian population and suggested that serotonergic mechanisms were important in the pathophysiology of this symptom complex (Nausieda et al. 1982).

Some practical observations on the treatment of levodopa-induced sleep disruption should be mentioned. Patients with recent onset of sleep fragmentation and altered dreaming often benefit from an alteration in levodopa dosage and avoidance of administration later than 6 or 7 p.m. Moving the last dose to an earlier time of evening frequently improves sleep complaints for a number of months with a no significant increase in functional disability from parkinsonian symptoms. In many patients with isolated complaints of sleep fragmentation, amitriptyline (25 mg at bedtime) is effective at consolidating nocturnal sleep, and the response is not restricted to patients with obvious depression. Usually, the improvement in sleep is quite prompt and is noticed in 1-2 days. As noted previously, patients with altered dream content, nocturnal vocalizations, or myoclonus are often adversely affected by this form of treatment, and we no longer advise using amitriptyline in patients with these complaints. The use of benzodiazepines or antihistaminic anticholinergics to improve sleep has been of little or no efficacy in our experience and seems to potentiate nocturnal episodes of confusion in many patients. While slow wave sleep nightmares have been reported to respond to low doses of diazepam, parkinsonian patients with this complaint do not appear to have the same response (NAUSIEDA et al. 1982). In situations in which sleep disruption becomes a major problem in therapy, drug holidays are usually helpful in reestablishing more restorative, consolidated sleep. Monitoring of patients' sleep-wake behavior and reinstituting dopaminergic medications when nocturnal sleep becomes consolidated have proven to be useful means of deciding when medications should be restarted.

A serotonergic basis for levodopa-induced psychosis has been raised by other investigators on the basis of postmortem neurochemical studies of patients with this side effect, which demonstrated significantly lowered levels of serotonin in the brain stem (BIRKMAYER et al. 1974; BIRKMAYER and RIEDERER 1975). It had been suggested that L-tryptophan might improve psychiatric symptoms in patients receiving levodopa, though the results of such treatment were not conclusive (MILLER and NIEBURG 1974; BEASLEY et al. 1980). In the study by NAUSIEDA et al. (1982), hallucinosis appeared to correlate well with the appearance of myoclonic activity on polysomnograms. The myoclonic disorder in levodopa-treated patients has been found to respond to the administration of methysergide (KLAWANS et al. 1975) and the authors noted exacerbation of myoclonic activity in parkinsonian patients treated with amitriptyline, a central serotonin reuptake inhibitor. Myoclonus has also been reported in patients with Down's syndrome receiving L-5-HT (COLEMAN 1971). These observations raised the possibility that overactivity in a central serotonergic system mediates myoclonus in some settings, and may underlie myoclonus in parkinsonian patients receiving levodopa chronically.

Data indicate that the acute administration of levodopa reduces brain stem serotonin levels by competing with tryptophan for uptake at the blood-brain barrier (SCHANBERG 1963), inhibiting tryptophan hydroxylase (BARTHOLINI et al. 1970), and replacing serotonin in presynaptic storage sites (NG et al. 1970). In humans, chronic levodopa treatment reduces plasma levels of tryptophan (LEHMANN 1973), which would reduce central serotonin levels (MOIR and ECCLESTON 1968). Clinically, insomnia that accompanies early levodopa therapy responds to the administration of amitriptyline, while the same treatment in patients with complaints of nightmares or myoclonic movements paradoxically worsens the complaints and may induce overt hallucinosis or waking myoclonic movements (NAUSIEDA et al. 1982). It may be that this divergent response to amitriptyline (a central serotonin reuptake blocker) is related to a changing response to serotonergic potentiation over time. Such an explanation may also clarify the high frequency of sleep complaints and previously noted high frequency of psychiatric toxicity reactions in patients receiving bromocriptine for parkinsonism (PARKES et al. 1976; LIEBERMAN et al. 1976), since bromocriptine is thought to be a central serotonergic agonist as well as a direct dopaminergic agonist (CORREDI et al. 1975).

In a logical extension to this theory, NAUSIEDA et al. (1983) attempted to use the central serotonergic antagonists methysergide and cyprohepine in the treatment of levodopa-induced psychiatric reactions. They found that methysergide reduced hallucinations in a significant percentage of patients with levodopa-induced psychiatric side effects. In general, their results were similar to those reported previously for L-tryptophan treatment in that patients with recent onset of mental symptoms were more likely to respond to therapy (BIRKMAYER and NEUMAYER 1972; COPPEN and METCALFE 1972; MILLER and NIEBERG 1974; SWEET et al. 1976). Since L-tryptophan and methysergide are felt to exert opposing effects on central serotonergic activity, this finding would appear to be inexplicable and contradictory. This contradiction may be more apparent than real. While considerable data exist to support the belief that the acute or short-term administration of levodopa results in a decrease in central serotonin levels, the consequences of this alteration in central serotonin levels have not specifically been studied. The finding of elevated 5-HIAA levels in patients with low central levels of serotonin suggests that an acceleration of serotonin synthesis may be partially compensatory (BIRKMAYER et al. 1974). Another mechanism may involve postsynaptic sensitization of serotonin receptors, an event that is reported to follow chronic blockade by methysergide (KLAWANS et al. 1977). The appearance in patients with levodopa-induced thought disorders of a myoclonic movement disorder that is responsive to methysergide suggests overactivity in a central serotonergic system following chronic levodopa administration, and supports such a hypothesis. The role of either serotonin precursors or serotonin antagonists in the management of these disorders remains unproven.

D. Drug-Induced Hallucinatory Syndromes

Levodopa is not the only agent that can cause psychiatric problems in parkinsonian patients. It is well known that both aminergic and anticholinergic drugs can induce hallucinations in nonparkinsonian patients, and parkinsonian patients are chronically exposed to both classes of drug as part of their therapy; separate types of (nonpsychiatric) hallucinatory syndromes occur: (1) the toxic delirium of anticholinergic overdosage, and (2) hallucinations in the presence of a clear sensorium.

I. Anticholinergic Delirium

The toxic delirium produced by anticholinergic drugs is a symptom of overdosage and does not necessarily relate to duration of therapy. This differs from hallucinations produced by levodopa, which appear to be the result of chronic exposure (MOSKOWITZ et al. 1978; NAUSIEDA et al. 1982). Different patients have different sensitivities to the central effects of anticholinergic agents, and elderly persons and persons with preexisting cognitive deficits are more likely to have symptoms of toxicity at "therapeutic" doses.

The hallucinations accompanying this syndrome are typically vague and disturbing. Visual hallucinations predominate, and these are often poorly formed, described as dark spots or crawling insects. Auditory or olfactory hallucinations are uncommon, but a tactile component of the hallucination may also occur. Patients with this syndrome have a toxic delirium, with prominent memory loss, disorientation, and confusion. Speech and behavior may be highly inappropriate; fearfulness, agitation, and combativeness are common. Characteristic autonomic signs of anticholinergic toxicity include tachycardia; intestinal ileus; urinary retention; mydriasis; anhidrosis with temperature elevation; flushed, dry skin; and decreased salivation. In severe cases, complete cardiovascular collapse may follow. In most cases, discontinuation of anticholinergics and supportive care will avoid this outcome. While some investigators have reported good response to the cautious administration of intravenous physostigmine, a centrally active cholinesterase inhibitor, in anticholinergic overdosage (Burks et al. 1974; Duvoisin and Katz 1968; Granacher and Baldessarini 1975, 1976; Holinger and Klawans 1976), this therapy should be reserved for the most severely affected patients, since it has its own potential morbidity, and should be given only in an intensive care unit.

II. Hallucinations with a Clear Sensorium

Goetz et al. (1982) evaluated the clinical pharmacology of these hallucinations in 20 hallucinating patients with Parkinson's disease. They found that identical hallucinations occurred after the addition or increase of either a dopaminergic or an anticholinergic drug. Moreover, hallucinations resolved after withdrawal of either pharmacologic class of agent. They suggested that a common pathophysiologic mechanism is triggered by either of these neurotransmitter changes.

Tanner et al. (1983) evaluated the relationship of cumulative drug dose, type of antiparkinsonian therapy, degrees of disability, and presence of dementia or depression to the development of hallucinations in patients with idiopathic Parkinson's disease receiving treatment with levodopa alone or with other drugs. Of 775 parkinsonian outpatients, 257 (33%) experienced hallucinations at some time during their treatment. Purely visual phenomena or visual phenomena plus a second sensory sensation comprised 97% of the hallucinations. The commonest visual phenomena were people and animals. Hallucinations were nonthreatening in 97%, and occurred in a clear sensorium, without an associated confusional state, in 86%. Hallucinations were much more likely to occur after many years of treatment, with only 13% occurring within 2 years of beginning levodopa, and 30% within 5 years. When 90 hallucinators were compared with 90 age and sex-matched nonhallucinating parkinsonian patients, no significant difference in age, sex, or Hoehn-Yahr stage was observed between hallucinators and controls. Neither cumulative does nor duration of levodopa, amantadine, or dopaminergic agonists distinguished hallucinators from nonhallucinators, but the mean cumulative dose of anticholinergic drugs was greater in hallucinators (6.5 mg years benztropine equivalent vs 4.7 mg years benztropine equivalent in nonhallucinators; $P < 0.02$). The prevalence of dyskinesias and of depression were similar in hallucinators and nonhallucinators. In contrast, 17% of hallucinators had clinically significant dementia at the onset of their first hallucination, and 27% were demented 2 years after their first hallucination, while no nonhallucinators developed dementia.

Wilson et al. (1986) performed psychometric tests in three groups of Parkinson's disease patients: those who were currently hallucinating; those who had hallucinated in the past, but were no longer hallucinating; and those who had never hallucinated. Patients with significant dementia were excluded. Patients who were currently hallucinating had significantly poorer cognition than either nonhallucinators or former hallucinators, while former hallucinators were not different from those who had never hallucinated. These results demonstrate that hallucinations and the cognitive disability that accompanies them can be transient and reversible. Hallucinations by themselves do not imply a progressive dementing process.

Some patients with hallucination have this symptom as an isolated phenomenon for years, but such hallucinations are due to a drug-induced alteration in central nervous system function. However, in another group of patients, the hallucinations may be followed by more disabling psychiatric symptoms. For this reason, an attempt to treat these phenomena is usually indicated. The simplest treatment is reduction of daily drug dosage. In many patients, this can be accomplished without a significant reduction in antiparkinsonian efficacy. If the patient is receiving amantadine, the onset of a hallucinatory syndrome should prompt a check of renal function. Amantadine is not metabolized, and its clearance is purely renal. Patients with even mild degrees of renal impairment may develop high blood levels of amantadine and subsequent hallucinations. If the hallucinating patient is receiving an anticholinergic agent, tapering this drug over several days may allow resolution of symptoms.

Similarly, modest reduction in the total daily dose of levodopa or a direct-acting dopamine agonist will control hallucinations in many patients. Bromocriptine is a dopamine receptor agonist that also stimulates serotonin receptors. Bromocriptine is associated with an increased incidence of hallucination, possibly resulting from its serotonergic activity. This drug's dosage should be reduced if patients taking it develop hallucinations.

E. Treatment of Psychosis

In general, patients who experience confusion or paranoia are older and more likely to have an underlying dementia or a prior history of psychosis. However, other patients without previous psychiatric disease, increased age, or dementia may follow the typical progression noted by Moskowitz, and proceed from nonthreatening to threatening hallucinations after months or years of chronic levodopa therapy. Paranoia is often associated with an hallucinatory syndrome (threatening visual hallucinations), but in some patients fixed paranoid delusions may occur as isolated phenomena, with a clear sensorium, and without associated hallucinations.

Treatment of these drug-induced symptoms is mandatory. If symptoms are severe, the patient should be hospitalized immediately. Attempts at reduction of drug dosage should then be instituted. Neuroleptic agents are commonly used to treat similar psychiatric disorders in nonparkinsonian patients. These drugs act to block central dopamine receptors, however, and can produce parkinsonism in normal patients. For this reason, phenothiazine compounds (including prochlorperazine, chlorpromazine, and fluphenazine) and butyrophenones such as haloperidol should be avoided when possible in psychotic parkinsonian patients to avoid aggravation of Parkinson's disease.

If the patient does not improve by any of these manipulations, levodopa should be stopped. Two types of drug holidays have been tried to prevent progression or to treat such levodopa-induced problems. The first is the formal drug holiday, in which a patient is admitted to hospital and levodopa is withdrawn for at least 5 days or until all mental changes clear. After the holiday, the medications are reintroduced slowly. Follow-up studies indicate that after the holiday, patients can be maintained on lower doses of medication and tend to show enhanced antiparkinsonian efficacy and reduced side effects for up to 1 year. The patient can be restarted on half the previous levodopa dosage. If this is tolerated well, the dosage can be slowly increased as necessary to two-thirds or three-quarters of the previous dosage (Weiner et al. 1980; Direnfeld et al. 1980; Koller et al. 1981).

During the hospitalization, patients usually show a precipitous decline in the absence of medication so that physical therapy, respiratory therapy, and nursing care are of paramount importance. Because this therapeutic venture is not without risk, it is reserved for patients with frank psychosis, or when all other therapeutic manipulations have failed.

A variant of the formal drug holiday is the weekly 2-day holiday, in which levodopa is withdrawn at home for 2 days each week. This can be done safely in most patients, particularly if amantadine is continued during the 2 "off" days, and will usually either reverse the already present psychiatric changes or at least prevent progression. In our experience, no patients on the 2-day holiday experienced any further progression of side effects, whereas controls matched for age, severity and duration of disease, and duration of levodopa therapy exhibited progression of the side effects over time in 7-day therapy (Goetz et al. 1981). While the efficacy of any type of drug holiday in parkinsonism in controversial, the role in levodopa-direct-acting agonist-induced psychosis seems well established (Calne and Rinne 1986).

In a few unfortunate patients, disabling psychotic symptoms recur after drug holiday. In these patients, symptoms of Parkinson's disease should be controlled with the smallest possible number of drugs, and anticholinergic drugs and bromocriptine (which is associated with an increased incidence of drug-induced hallucinations) should be avoided. Similarly, if patients have concomitant renal disease, amantadine should be avoided. If disabling hallucinations are still present, treatment with one of the neuroleptic agents that is not associated with a high incidence of drug-induced parkinsonism may be cautiously attemtpted. In the United States, the only such agent available is thioridazine, a phenothiazine with high anticholinergic activity.

F. Aberrant Sexual Behavior

In the first years of levodopa therapy in Parkinson's disease, much attention was given to reports of increased sexual function after the initiation of levodopa therapy. Reports of changes in sexual function were attributed to 1%-24% of all those interviewed (Quinn et al. 1983; Bowers et al. 1971; Brown et al. 1978). In most cases, the nature of the change reported was characterized as "increased sexual interest" or "increased libido." Critics of the claims for an aphrodisiac effect of levodopa argued that any change in sexual interest or performance could be explained on the basis of improved motor function with consequent enhanced self-esteem in these patients. Attempts to use levodopa in cases of psychogenic impotence in nonparkinsonian men were unsuccessful, and interest in the effects of levodopa on sexuality waned.

Quinn et al. (1983) and Tanner et al. (1986) have described small groups of patients receiving chronic antiparkinsonian therapy who developed disabling hypersexuality. Hypersexuality was usually associated with other symptoms of levodopa-related psychiatric side effects, including sleep disorders, hallucinations and delusions with sexual content, and mild confusion. Reduction or withdrawal of levodopa resulted in resolution of symptoms. The majority of the cases reported have been men, suggesting that this may reflect a gender-related difference in susceptibility to this side effect.

Acknowledgments. This work was supported in part by grants from the United Parkinson Foundation, Chicago, Illinois, and the Boothroyd Foundation, Chicago, Illinois.

References

Andersen J, Aabro E, Gulmann N, Hjelmsted A, Pedersen HE (1980) Antidepressive treatment of Parkinson's disease. Acta Neurol Scand 62:210-219

Asnis G (1977) Parkinson's disease, depression and ECT: review and case study. Am J Psychiatry 134:191-195

Barbeau A (1969) L-Dopa therapy in Parkinson's disease: a critical review of nine years experience. J Can Med Assoc 101:791-800

Bartholini G, Key KF, Pletscher A (1970) Enhancement of tyrosine transamination in vivo by catecholamines. Experientia 26:280-283

Beasley B, Nutt J, Chase T (1980) Treatment with tryptophan of levodopa associated psychiatric disturbances. Arch Neurol 37:155-156

Birkmayer W, Neumayer E (1972) Die Behandlung der Dopa-Psychosen mit L-tryptophan. Nervenarzt 43:76-78

Birkmayer W, Riederer P (1975) Responsibility of extrastriatal areas for the appearance of psychotic symptoms. J Neural Transm 35:175-182

Birkmayer W, Damelczyk W, Neumayer E, Riederer P (1974) Nucleus ruber and L-dopa psychosis: biochemical and postmortem findings. J Neural Transm 35:93-116

Bowers MB, Van Woert M, Davis L (1971) Sexual behavior during *l*-dopa treatment for parkinsonism. Am J Psychiatry 127:1691-1693

Brown E, Brown GM, Kafman O, Quaningten B (1978) Sexual function and affect in parkinsonian men treated with *l*-dopa. Am J Psychiatry 135:1552-1555

Brown GL, Wilson WP (1972) Parkinsonism and depression. South Med J 65:540-545

Burks JS, Walker JE, Rumack BH, Ott JE (1974) Tricyclic antidepressant poisoning: reversal of coma, choreathetosis, and myoclonus by physostigmine. JAMA 230:1405-1407

Calne DB, Rinne UK (1986) Controversies in the management of Parkinson's disease. Movement disorders 1:159-162

Celesia GG, Wanamaker WM (1972) Psychiatric disturbances in Parkinson's disease. Dis Nerv Syst 33:577-583

Coleman M (1971) Infantile spasms associated with 5-hydroxytryptophan administration in patients with Down's syndrome. Neurology 21:911-919

Coppen A, Metcalfe M (1972) Levodopa and L-tryptophan therapy in parkinsonism. Lancet 1:654-657

Corredi H, Farnebo LO, Fuxe K, Hamberger B (1975) Effect of ergot drugs on central 5-hydroxytryptamine neurons. Eur J Pharmacol 30:172-181

Damasio AR, Lobo-Antunes J, Macedo C (1971) Psychiatric aspects in parkinsonism treated with L-dopa. J Neurol Neurosurg Psychiatry 34:502-507

Direnfeld LK, Feldman RG, Alexander MP, Kelly-Hayes M (1980) Is levodopa drug holiday useful? Neurology 30:785-788

Duvoisin RC, Katz R (1968) Reversal of central anticholinergic syndrome in man by physostigmine. JAMA 206:1963-1965

Goetz CG, Tanner CM, Nausieda PA (1981) Weekly drug holiday in Parkinson disease. Neurology 31:1460-1462

Goetz CG, Tanner CM, Klawans HL (1982) Pharmacology of hallucinations induced by long term drug therapy. Am J Psychiatry 139:494-498

Goetz CG, Tanner CM, Klawans HL (1984) Bupropion in Parkinson's disease. Neurology 34:1092-1094

Granacher RP, Baldessarini RJ (1975) Physostigmine: its use in acute anticholinergic syndrome with antidepressant antiparkinson drugs. Arch Gen Psychiatry 23:375-380

Granacher RP, Baldessarini RJ (1976) The usefulness of physostigmine in neurology and psychiatry. In: Klawans HL (ed) Clinical neuropharmacology, vol 1. Raven, New York, pp 63-79

Hollinger PC, Klawans HL (1976) Reversal of tricyclic overdosage-induced central anticholinergic syndrome by physostigmine. Am J Psych 133:1018-1023

Horn S (1974) Some psychological factors in parkinsonism. J Neurol Neurosurg Psychiatry 37:27-31

Klawans HL, Goetz C, Bergen D (1975) Levodopa-induced myoclonus. Arch Neurol 32:331-334

Klawans HL, D'Amico D, Nausieda PA, Weiner WJ (1977) The specificity of neuroleptic and methysergide-induced behavorial supersensitivity. Psychopharmacology (Berlin) 55:49-52

Laitinen L (1969) Desipramine in treatment of Parkinson's disease. Acta Neurol Scand 45:109-113

Lebensohn Z, Jenkins RB (1975) Improvement of parkinsonism in depressed patients treated with ECT. Am J Psychiatry 132:283-285

Lehman J (1973) Tryptophan malabsorption in levodopa treated parkinsonian patients. Acta Med Scand 194:181-189
Lieberman A, Kupersmith M, Estey E, Goldstein M (1976) Treatment of Parkinson's disease with bromocriptine. N Engl J Med 295:1400-1435
Marsh GG, Markham CH (1973) Does levodopa alter depression and psychopathology in parkinsonism patients? J Neurol Neurosurg Psychiatry 36:925-935
Mayeux R, Stern Y, Roson J, Leventhal J (1981) Depression, intellectual impairment, and Parkinson's disease. Neurology 31:645-650
Mayeux R, Stern Y, Cote L, Williams JB (1984a) Altered serotonin metabolism in depressed patients with Parkinson's disease. Neurology 34:642-645
Mayeux R, Williams JBW, Stern Y, Cote L (1984b) Depression in Parkinson's disease. Adv Neurol 40:241-250
Mayeux R, Stern Y, Sano M, Williams JBW, Cote L (1986a) The relationship of serotonin to depression in Parkinson's disease. Ann Neurol 18:149
Mayeux R, Stern Y, Williams JBW, Cote L, Frantz A, Dyrenfurth T (1986b) Clinical and biochemical features of depression in Parkinson's disease. Am J Psychiatry 143:756-760
Miller E, Nieburg H (1974) L-Tryptophan in the treatment of levodopa-induced psychiatric disorders. Dis Nerv Syst 35:20-23
Mindham RHS (1970) Psychiatric syndromes in parkinsonism. J Neurol Neurosurg Psychiatry 30:88-191
Mindham RHS, Marsden CD, Parkes JD (1971) Psychiatric symptoms during *l*-dopa therapy for Parkinson's disease and their relationship to physical disability. Psychol Med 6:23-33
Mjones H (1949) Paralysis agitans. Acta Psychiatr Scand [Suppl] 54:1-195
Moir ATB, Eccleston D (1968) The effects of precursor loading in the cerebral metabolism of 5-hydroxyindoles. J Neurochem 15:1093-1108
Moskovitz C, Moses H, Klawans HL (1978) Levodopa-induced psychosis: a kindling phenomena. Am J Psychiatry 135:669-675
Nausieda PA, Weiner WJ, Kaplan LR, Weber S, Klawans HL (1982) Sleep disruption in the course of chronic levodopa therapy. Clin Neuropharmacol 5:183-194
Nausieda PA, Tanner C, Klawans HL (1983) Serotonin blocking agents in the treatment of levodopa induced psychiatric toxicity. Adv Neurol 37:23-31
Nausieda PA, Glantz R, Weber S, Baum R, Klawans HL (1984) Psychiatric complications of levodopa therapy of Parkinson's disease. Adv Neurol 40:271-277
Ng KY, Chase TN, Colburn RW, Kopin IJ (1970) L-Dopa induced release of cerebral monoamines. Science 1970:76-78
Parkes J, Debono A, Marsden C (1976) Bromocriptine in parkinsonism: long term treatment dose response, and comparison with levodopa. J Neurol Neurosurg Psychiatry 39:1101-1107
Patrick HT, Levy DM (1922) Parkinson's disease: a clinical study of 146 cases. Arch Neurol 7:711-720
Quinn NP, Toone B, Lang AE, Marsden CD, Parkes JD (1983) Dopa dose-dependent sexual deviation. Br J Psychiatry 142:296-298
Robins AH (1976) Depression in patients with parkinsonism. Br J Psychiatry 128:141-145
Schanberg SM (1963) A study of the transport of 5-hydroxytryptophan and 5-hydroxytryptamine (serotonin) into brain. J Pharmacol Exp Ther 139:191-199
Sharf B, Moskovitz C, Lupton MD, Klawans HL (1978) L-Dopa-induced sleep disorders. J Neural Transm 43:143-151

Strang RR (1965) Imipramine in treatment of parkinsonism: a double-blind placebo study. Br J Med 2:33-34
Sweet RD, McDowell FH, Feigenson (1976) Mental symptoms in Parkinson's disease during chronic treatment with levodopa. Neurology 26:305-310
Tanner CM, Vogel C, Goetz CG, Klawans HL (1983) Hallucinations in Parkinson disease: a population study (abstract). Ann Neurol 14:136
Tanner CM, Goetz CG, Klawans HL (1986) Hypersexuality in Parkinson's disease (abstract). Neurology 36 (Suppl 1):183
Weiner WJ, Perlik S, Koller WC, Nausieda PA, Klawans HL (1980) The role of drug holiday in the management of Parkinson's disease. Neurology 30:1257-1261
Wilson RW, Tanner CM, Weingarten R, Goetz CG (1986) Hallucinosis in Parkinson's disease (abstract) Neurology 36 (Suppl 1):216
Yudofsky SC (1979) Parkinson's disease, depression, and electroconvulsive therapy: a clinical and neurobiologic synthesis. Compr Psychiatry 20:579-581

CHAPTER 24

Intracranial Grafts for the Treatment of Parkinson's Disease

A. FINE and H. A. ROBERTSON

A. Introduction

Current treatment for Parkinson's disease is palliative. The development of unacceptable side effects invariably limits drug therapy. Ideally, we seek to prevent the disease, and, in those patients already afflicted, to halt its progression and restore normal function. Progress in preventing or arresting the disease is considered elsewhere in this volume. Here, we evaluate one experimental strategy for restoring normal movement: replacing the dopaminergic nigrostriatal and mesolimbic projections that have degenerated in Parkinson's disease by transplantation of dopaminergic cells to the parkinsonian brain.

Parkinson's disease is well suited to the application of neural grafts. It results from the degeneration of a discrete population of neurochemically homogeneous cells. There is an age-dependent reduction in brain levels of dopamine and in numbers of dopaminergic cells, but clinical symptoms of Parkinson's disease appear only when dopamine depletion exceeds about 80% (RIEDERER and WUKETICH 1976) and cell loss exceeds about 50% (MCGEER et al. 1977). We might expect partial recovery, therefore, following even the partial replacement of this one population of dopaminergic cells. If the grafted dopaminergic neurons were placed in the substantia nigra, their axons would then have to grow forward and innervate their targets in the basal ganglia. Unfortunately, nerve fibres do not readily grow over such long distances in the adult brain, and dopaminergic grafts to the degenerated substantia nigra are ineffective in experimental parkinsonism (DUNNETT et al. 1983). However, it is known that systemic administration of bromocriptine and related dopaminergic agonists can ameliorate experimental parkinsonism in rats, as may infusion of dopamine directly into the striatum (STRÖMBERG et al. 1985). Thus, dopamine may function in the extrapyramidal system, at least in part, in a humoral manner, without requiring specific, point-to-point connections. This may account for the ability of dopaminergic tissue to ameliorate parkinsonian symptoms when grafted adjacent to the striatum, within the lateral ventricles (PERLOW et al. 1979).

Consideration of the site for transplants must also include discussion of the role of the dopamine cells in the substantia nigra. Current indications are that placing dopaminergic grafts in the substantia nigra is without effect (DUNNETT et al. 1983). However, there is convincing evidence that the dopamine neurons in the substantia nigra pars compacta release dopamine not only from their axon terminals in the basal ganglia, but also from their dendrites within the substantia nigra pars reticulata (CHERAMY et al. 1981). This intranigral dopamine release may play an important role in the flow of information through the basal ganglia (CHERAMY et al. 1981; GAUCHY et al. 1987; TRUGMAN and WOOTEN 1987; ROBERTSON and ROBERTSON 1987a, b). It is possible, therefore, that better effects might be had by transplanting dopamine-secreting neurons into both the basal ganglia and the substantia nigra.

Autografts of adrenal medullary tissue for the treatment of Parkinson's disease are in clinical trial in a number of centres around the world; work on foetal neuronal grafts has just begun. Definitive results of these trials are not yet available. Thus we can only review in a critical fashion the available data, calling attention to the problems that remain. One of these problems is to identify the best source of tissue for transplantation. A second problem is to identify the optimal placement for the grafts. Most grafts of adrenal tissue have been placed in the caudate nucleus. However, there are indications that the putamen is more important than the caudate nucleus for regulation of movement (BERNHEIMER et al. 1973; GOLDMAN and NAUTA 1977; SELEMAN and GOLDMAN-RAKIC 1985; MITCHELL et al. 1986). The extensive studies on rats do not address this problem, as the rat has a fused caudate-putamen. Furthermore, the effects of dopaminergic grafts may not all be attributable to the release of dopamine. Other transmitters or as yet unidentified trophic factors released by the grafts may aid in restoring defective neurotransmission or may promote regeneration of damaged nerve fibres. A number of collections or reviews dealing with various aspects of neural transmission have been published (BJÖRKLUND and STENEVI 1985; FINE 1986; AZMITIA and BJÖRKLUND 1987).

B. Sources of Dopamine-Secreting Cells

It has long been known that grafts of skin or tumour tissue can survive for prolonged periods in the brain, even in an animal which quickly rejects similar grafts to the skin. The specialized blood vessels that constitute the "blood-brain barrier" and the relative lack of lymph nodes and lymphatic vessels all limit access of the immune system to antigens within the brain. Foetal neurons are essentially devoid of the major histocompatibility antigens that distinguish one individual's cells from another's. Together, these factors may account for the prolonged and apparently unlimited survival of foetal neurons transplanted to the brain of unrelated conspecifics. The antigens that distinguish cells from different species, however, are present on fetal neurons, and xenografts (grafts from another species) are generally rejected unless the recipient is immunosuppressed.

The most biologically appropriate source of dopaminergic cells for transplantation is the foetal ventral mesencephalon, wherein the dopaminergic cells of the substantia nigra and the ventral tegmental area arise. Several techniques have been used for neuronal transplantation. Solid pieces of embryonic brain tissue have been introduced directly into the host brain, into a prepared cavity or into the cerebral ventricles. Alternatively, the cells in the tissue pieces can be dissociated and then injected as a suspension directly into the brain parenchyma. This latter technique has many advantages and, in our view, is the method of choice for studies of functional restitution or for therapeutic use.

The adrenal medulla may be a useful alternative source of dopaminergic cells for transplantation. The normal adrenal medulla synthesizes dopamine in larve amounts as a precursor to adrenaline. Adrenal chromaffin cells are of neuroectodermal origin and, in the presence of glucocorticoids from the adrenal cortex, express an endocrine phenotype which includes production of the enzyme phenylethanolamine-*N*-methyl transferase (PNMT) that converts noradrenaline to adrenaline. In the absence of a source of glucocorticoids and in the presence of nerve growth factor, they develop axons, cease production of PNMT, and therefore secrete dopamine and noradrenaline, but not adrenaline (UNSICKER et al. 1980). Adrenal medullary allografts were introduced by FREED et al. (1981) and shown to reverse the effects of 6-hydroxydopamine (6-OHDA) lesions in the rat. It was thought that similar autografts in Parkinson's disease patients would avoid possible problems of rejection as well as ethical objections anticipated in the use of foetal tissue. Several assumptions implicit here may be false. First, the phenotypic plasticity of adrenal medullary cells may diminish with age (TISCHLER and GREENE 1980); since Parkinson's disease is in general a disease of the elderly, it is not clear that adrenal autografts from these patients will be useful. Second, it is possible that the disease process will have damaged the adrenal medulla cells as well. For example, experimental Parkinson's disease induced in monkeys by the drug MPTP (1-methyl-4-phenyl-1,2,3,6-tetrahydropyridine) affects the adrenal glands as well as the brain (FINE et al. 1985).

C. Neural and Adrenal Grafts in Animal Models of Parkinson's Disease

Animal models of Parkinson's disease generally involve destruction of the dopaminergic projections from the substantia nigra and ventral tegmental area to the striatum and/or limbic forebrain. The two models most widely used are the rat rotational model (UNGERSTEDT 1971) and the MPTP-treated primate (BURNS et al. 1983). Rotation in the rat provides a quantitative measure, but the neuroanatomy of the rodent striatum is significantly different from that of other mammals, including primates. Thus, for example, the caudate nucleus and putamen form a single continuous structure in rodents, so that a comparison of the effects of caudate grafts with putamen grafts is not possible. On the other hand, whereas brain structure in monkeys closely resembles that in the human, quantitative assessment of parkinsonism and recovery from parkinsonism is difficult. Both these models have important roles in developing transplantation techniques.

I. Studies in the Rat

The rat rotational model takes advantage of the bilateral nature of the nigrostriatal system. The neurotoxin is injected into the substantia nigra, the ascending dopaminergic fibre pathway, or the striatum on one side of the brain, selectively destroying the dopaminergic cells, fibres or terminals. As a result, postsynaptic neurons in the regions to which the nigra projects (the caudate nucleus-putamen) then become supersensitive to dopamine. When such an animal is challenged with a dopamine agonist such as apomorphine or bromocriptine, motor output is preferentially affected on the supersensitive side of the brain, leading to asymmetric movement (rotation). The extent of the dopaminergic lesion and the efficacy of a transplant in restoring function can then be assessed quantitatively by simply measuring rotation. Unilaterally lesioned rats also exhibit asymmetric body posture, contralateral sensory inattention and bradykinesia. Bilateral 6-OHDA lesions produce a marked behavioural loss of responsiveness which includes akinesia, bilateral sensory inattention and a hunched posture. Furthermore, bilaterally lesioned animals may not eat, drink or groom themselves.

The first demonstrations of functional effects of intracerebral neural grafts were made in 1979 (Perlow et al. 1979; Björklund and Stenevi 1979) and this finding has now been repeated in dozens of laboratories. Appropriately placed grafts of dopaminergic tissue from rat foetal ventral mesencephalon have been found to ameliorate all aspects of this dopamine deficiency syndrome with the exception of the adipsia and aphagia (Björklund et al. 1987).

This is in many ways an unexpected result. After all, the dopamine neurons are being grafted ectopically, not in the substantia nigra, but in the striatum. The mechanisms by which nigral grafts may exert their functional effects have been recently reviewed (Björklund et al. 1987). Because both adrenal medullary grafts (Freed et al. 1981) and nigral grafts (Perlow et al. 1979) placed in a cortical cavity above the denervated striatum can restore impaired behaviours, it is likely that diffuse, humoral release of dopamine from the grafts accounts for at least some of their effects. Moreover, diffusion of dopamine into the denervated striatum from a dialysis probe over a 2-week period leads to a reduction in apomorphine-induced turning (Strömberg et al. 1985), presumably reflecting reduced supersensitivity of the denervated dopamine receptors. Other graft-mediated effects appear to depend on growth of dopamine fibres into the host striatum, where transmitter can be released in close proximity to deafferentated cells. Since effects on motor asymmetry, sensory deficits or akinesia are seen with localized grafts only in certain regions of the striatum (Björklund et al. 1980, 1983 and Dunnett et al. 1981; Björklund et al. 1987), it is possible that reciprocal connections contribute to the functional effects of these grafts.

Such connections have been observed in electron micrographs. Freund et al. (1985) were able to show that grafts of dopaminergic neurons form abundant symmetric synapses with neurons of the host striatum. Up to 90% of these are on dendritic shafts or spines, with the remainder on the cell bodies. Synapses are formed with two principal types of neurons: the medium spiny cells which are presumably descending projection neurons, constituting about 95% of the striatal neurons; and the giant neurons that constitute 1% of the neurons in the striatum and are probably cholinergic interneurons (BRODAL 1983). The synapses on the dendritic spines appear normal while those on the giant cell bodies form dense pericellular "baskets" around the cells. This sort of arrangement appears to be abnormal, and unique to graft-derived inputs. FREUND et al. (1985) have suggested that the grafts function by providing powerful inhibition of the cholinergic interneurons in the host striatum. In a further study using electron microscopic (EM) immunocytochemistry, BOLAM et al. (1987) implanted grafts into a cortical cavity and studied the reinnervation of host striatum. Five distinct types of boutons contacting the tyrosine hydroxylase(TH)-positive graft neurons were distinguished based on their immunocytochemistry, morphology or membrane specialization. One was itself TH-positive; this type of connection was not seen in the contralateral substantia nigra nor in substantia nigra of a control rat. Three of the bouton types had ultrastructural properties which suggested either GABA-containing or substance P-containing neurons, resembling normal striatonigral connections.

In addition to structural evidence for the functioning of the grafted dopaminergic neurons, there is also some electrophysiological evidence. STRÖMBERG et al. (1985) measured sensitivity to locally applied phencyclidine (an indirect dopamine agonist) and spontaneous neuronal discharge rate, in dopaminergic-deafferentated striatum with foetal nigral grafts. In this study, "distal neurons" (those >2.0 mm from the nigral graft) fired at an average rate of 13.4 Hz and were relatively insensitive to the effects of locally iontophoresed phencyclidine, whereas "proximal neurons" (<1.0 mm from the graft) fired at a lower rate (4.9 Hz) and were sensitive to the effects of phencyclidine. These findings support the presence of functional synaptic contacts as suggested by EM immunocytochemistry.

Recent experiments have suggested another important mechanism by which adrenal medullary grafts might produce recovery from loss of dopamine function in the striatum. BOHN et al. (1987) transplanted adrenal medullary tissue unilaterally into mice previously treated with MPTP (a treatment producing almost total loss of TH-immunoreactive fibres in the striatum). Dense TH-immunoreactive fibres were seen in the grafted striatum, while only sparse fibres were seen on the ungrafted side. Surprisingly, the fibres appeared to derive from host, not graft, neurons. This would suggest that, at least in mice, grafted adrenal medullary tissue can exert neurotrophic effects on the host brain, enhancing recovery and growth into the striatum of the host's own damaged dopaminergic neurons. However, these results must not be overextended. While the effects of MPTP in humans resemble very closely the pathology of Parkinson's disease, in mice fibres are damaged, but cell bodies survive (LANGSTON et al. 1983; HEIKKILA et al. 1984). Thus, before it can be concluded that this is a likely mechanism in humans, further studies must be carried out in primates.

While loss of dopaminergic innervation of the striatum is associated principally with postural abnormalities, abnormalities of exploratory behaviour and locomotion appear to be associated more with dopamine depletion in the nucleus accumbens septi. Thus, grafting of foetal dopaminergic neurons to the nucleus accumbens restored amphetamine- and apomorphine-induced locomotion to normal (Nadaud et al. 1984; Brundin et al. 1987). These results support the notion that postural and locomotor deficits are caused by dopamine depletion of different parts of the striatal complex and that grafts may need to be placed in multiple sites to produce complete recovery.

Cross-species neural grafts may survive in some cases (Low et al. 1985) and can be maintained indefinitely with immunosuppression (Brundin et al. 1985). Kamo et al. (1986) showed that grafts of foetal human sympathetic neurons (from 8- to 11-week foetal paravertebral sympathetic ganglion cultured for between 3 weeks and 3 months) were able to decrease amphetamine-induced turning in rats by 70%–92%. These effects correlated well with the numbers of TH-positive cells in the grafts, and the effects persisted in four of six animals for 4½ months. This study was carried out without immunosuppression, which may reflect shedding of antigens in tissue culture. Tissue culture of neurons before transplantation may offer other possible advantages, for example selection for a particular type of neuron (e.g. dopaminergic).

Brundin et al. (1986) transplanted human foetal ventral mesencephalon to 6-OHDA-lesioned, immunosuppressed rats and were able to determine that, at least in rats, only 9-week gestation material successfully reversed amphetamine-induced turning. By contrast, grafts from 11- to 19-week-old donors contained few or no dopamine neurons and were without effect on amphetamine-induced turning. Similar results were obtained by Strömberg et al. (1986), although these workers also reported some success with material from 12-week foetuses.

II. Studies in Primates

Despite the fact that a substantial number of human adrenal medulla tissue autografts have been carried out in patients with Parkinson's disease (at least 80 such procedures have been performed worldwide as of January, 1988), there has at the time of writing been only one published report of adrenal medulla autografts in primate brain (Morihisa et al. 1987). In this brief report, seven mature male rhesus monkeys *(Macaca mulatta)* were lesioned unilaterally with 6-OHDA. After 2 months, two animals received foetal nigral grafts and five received adrenal medulla autografts. After 3 months, the animals were killed and histology carried out. While the nigral grafts did not survive, the adrenal grafts did. However, the number of cells surviving in the adrenal grafts was very low and no behavioural effects were observed.

The situation with respect to foetal neuronal grafts is only marginally better: we are at this time aware of only very few published reports concerning foetal neuronal transplants in monkeys (Redmond et al. 1986; Bakay et al. 1987; Sladek et al. 1987; Fine et al. 1988). In the first of these studies (Redmond et al. 1986), five adult male African green monkeys *(Cercopithecus aethiops sabaeus)*, given 0.25–0.40 mg/kg MPTP, i.m., four or five times over 5 days, were compared with six untreated control monkeys in biochemical studies. Three monkeys treated with 0.4 mg/kg MPTP (given six times over a

5-day period) served as hosts for foetal neural transplants; six monkeys treated with 0.25-0.4 mg/kg MPTP over 5 days were used for behavioural or anatomical experimentation. For the transplants, 1-mm^3 blocks of foetal ventral mesencephalon tissue were transplanted stereotactically into the three monkeys 3-4 weeks after MPTP treatment. In two monkeys the grafts were made to the caudate nucleus, while in the third "control" monkey earlier-gestational foetal ventral mesencephalon tissue was transplanted to the cingulate cortex, hypothalamic tissue was transplanted into the left striatum and pontine tissue was transplanted to the right striatum.

The two monkeys with caudate nucleus grafts showed improvement in a subjective parkinsonian rating as early as 2 days after surgery, whereas the "control" monkey with grafts from various places implanted in various places appeared to improve briefly after surgery before declining to the level of MPTP-only monkeys. Improvement in the two monkeys with "appropriate" grafts could conceivably have reflected graft-derived reinnervation of the caudate, but it is highly unlikely that this could account for the extremely rapid onset of improvements. Alternatively, the grafts or some aspect of the surgical procedure may have influenced surviving elements of the hosts' dopaminergic system, perhaps by expressing or inducing some trophic factor, as suggested by the same group on the basis of observations in rats (Bohn et al. 1987). In addition, it is important to note that in these studies insufficient time was allowed between MPTP administration and transplantation to rule out spontaneous recovery. By comparing the data in Redmond et al. (1986) with that in Sladek et al. (1987), it is possible to see that the monkey that improved the most also started with fewer behavioural deficits (cf. Fig. 1 in Redmond et al. 1986 with Fig. 10 in Sladek et al. 1987; monkeys A, B and C in the former paper appear to be the same as monkeys S092, S054 and S114 in the latter). Among the more disturbing aspects of these reports (Redmond et al. 1986; Sladek et al. 1987) is the fact that most of the TH-positive cells depicted in the composite drawings of the grafted monkeys' striata almost certainly correspond to intrinsic TH-positive cells seen in the striata of ungrafted, normal Old-World monkeys (Dubach et al. 1987).

Bakay et al. (1987) studied 8 rhesus monkeys that had received MPTP (at least 0.33 mg $kg^{-1}day^{-1}$ i.v. for 5 days as described by Burns et al. (1983). The behaviour of the MPTP-treated monkeys was monitored for at least 2 months after the final dose of MPTP. Nontransplanted animals with parkinsonian symptoms served as controls to ensure that spontaneous recovery did not occur. If animals showed recovery, more MPTP was administered and followed by a further 2 months of behavioural observation. As severe Parkinson-like symptoms developed, the animals were maintained on L-dopa and carbidopa, but were withdrawn for behavioural observation. Three animals received grafts of foetal ventral mesencephalon tissue. After the transplantation procedure, Bakay et al. (1987) concluded, on the basis of video-recorded home cage activity, that all three monkeys exhibited some degree of functional recovery. Unfortunately, the number of animals was too small to permit statistical evaluation. After 2 months, when the manuscript was written, there was some indication that the behavioural effect was decreasing. However, in two of these animals, CSF levels of dopamine and L-dopa were substantially increased after transplantation.

FINE et al. (1988) rendered ten common marmosets *(Callithrix jacchus jacchus)* parkinsonian with intraperitoneal injections of MPTP. These animals as well as eight untreated controls were observed over periods of 3-5 months before transplantation. This important procedure was required because MPTP-treated animals recovered partially over 4-10 weeks, but remained unchanged thereafter. Once animals stabilized, they received either control grafts (from the striatal eminence) or dopaminergic grafts (from the ventral mesencephalon) from unrelated foetuses of embryonic age 60-80 days (marmoset gestation is approximately 150 days). In eight cases, transplants were made with cell suspensions. In two cases, undissociated tissue fragments were used instead. In this study the putamen was chosen as the site for transplantation on the basis of indications that the putamen plays a more important role than the caudate in the motor impairments of Parkinson's disease. Four animals received unilateral dopaminergic grafts, two received bilateral dopaminergic grafts and two received bilateral control grafts. Transplanted and untransplanted MPTP-treated marmosets were observed for a further 4-7 months before being killed. In addition to spontaneous locomotor activity, the motor behaviour of the experimental animals and eight normal controls was observed following administration of D-amphetamine (0.5 mg/kg i.m.) or L-dopa (12.5 mg/kg i.p. in combination with 12.5 mg/kg of the peripheral dopa decarboxylase inhibitor benserazide). At postmortem, brains were examined by either conventional means (cresyl violet) or immunocytochemical procedures for TH, Met-enkephalin, substance P, neuropeptide Y, glutamic acid decarboxylase and serotonin.

Dopaminergic grafts produced a marked improvement, after 1-2 months, in spontaneous locomotion and in the response of the marmosets to D-amphetamine, indicating that the grafts were functional. In contrast, MPTP-treated animals with control nondopaminergic grafts or no grafts, but equivalent initial drug-induced hypoactivity, showed little or no increase in spontaneous locomotion over the subsequent 6-7 months. The improvement in locomotor activity associated with the dopaminergic grafts partially subsided over the second half of the survival period. However, postmortem histology demonstrated the continued viability of the grafts, so that this effect may have been due to habituation to the test procedure rather than loss of graft function.

D. Dopaminergic Grafts in Humans

The first transplantation of dopaminergic cells to the brain of a patient with Parkinson's disease was carried out in 1981 in Sweden. By the time this chapter was being written, at least 80 similar operations had been performed in Sweden, Mexico, China and the United States. Initial reports of success in the operations led the editors of the *Lancet* (2 May 1985) to conclude there were grounds for "cautious optimism" for the successful application of this new treatment.

The initial trial on two patients was reported by BACKLUND et al. (1985). In this first trial, medullary tissue from the patient's own adrenal gland was transplanted to a site deep in the caudate nucleus. While no significant deleterious effects were observed, the motor symptoms of Parkinson's disease were affected only to a slight degree and for only about 10 days. In a subsequent study, two patients (aged 46 and 63) received unilateral adrenal autografts to the putamen (LINDVALL et al. 1987). The first patient in this study, a 46-year-old male, showed a transient 2-day improvement in motor performance in the limbs contralateral to the transplantation site. This patient also had significantly longer periods of normal function for about 2 months. The second patient, a 63-year-old male, reported minor improvement in balance and gait, lasting for about 2 months. Positron emission tomography showed no alterations in D_2 dopamine receptor density in the putamen. Electrophysiological studies were consistent with increased catecholamine secretion in the basal ganglia following transplantation in both patients. The authors concluded that transplantation of adrenal tissue to the putamen had transient beneficial effects in patients with severe Parkinson's disease. No deleterious effects were observed in these patients over the 14 months between transplantation and the submission of the manuscript.

The most dramatic effects following autografts of adrenal medullary tissue have been claimed by MADRAZO et al. (1987). Two young patients (35- and 39-year-old males) received autografts of adrenal medullary tissue into a cavity, approximately $3 \times 3 \times 3$ mm, in the head of one caudate nucleus. Fragments of one adrenal medulla were anchored in the cavity by three or four standard stainless steel miniature staples. The graft was thus placed so that the grafted tissue was in contact both with the caudate nucleus and with the CSF. Improvements in motor impairments were seen bilaterally, nearly immediately after recovery from surgery. At 5 months after surgery, one patient was able to walk and eat without help and "did not require drugs". Rigidity and akinesia were "practically absent on both sides", while tremor was dramatically reduced on the right (transplanted side) and slightly reduced on the left side; 10 months after transplantation, the patient was "playing soccer with his five-year-old son". In the second patient, rigidity and tremor were also abolished.

This dramatic report has aroused great interest. However, there are a number of difficulties in interpreting its significance. First, the authors did not evaluate their patients by widely accepted, standardized methods before and after surgery. The patients were uncharacteristically young for Parkinson's disease. In the first patient, despite the fact that the graft was placed in the right caudate, the greater improvement was in right-side tremor. While MADRAZO et al. had operated on 11 patients by the time their article was written, only 2 were described in their article. Furthermore, 2 of the 11 patients had died, but pathological observations were not presented. Thus, we do not know if the grafts survived in those cases, or if observed recovery was due to causes unrelated to the graft. Unfortunately, CSF levels of catecholamines were not determined. Clearly, questions concerning this first study remain.

Nevertheless, the dramatic claims have led to a veritable epidemic of subsequent adrenal medullary autograft operations for the surgical treatment of Parkinson's disease. A voluntary registry of these procedures has been established at Emory University, Atlanta, Georgia (R.A.E. Bakay 1988, personal communication). Encouraging results have apparently been obtained in several centres in the United States when this operation was performed on young, less severely affected Parkinson's disease patients. Improvements have been less often obtained with the older and more severely affected, and at least four such patients in the United States, as well as one of the two in Mexico previously mentioned, have died as a result of the surgery.

As this chapter was being prepared, a letter from Madrazo et al. (1988) appeared describing the results of transplantation of "substantia nigra" and "adrenal medulla" from a spontaneously aborted 13-week foetus to the caudate nucleus in two patients with Parkinson's disease. The two patients were evaluated by standardized procedures (video, computed tomographic scanning, electromyography evoked potentials, neuropsychological testing and the Unified Parkinsonism Rating Scale, UPRS). One patient, a 50-year-old man with Parkinson's disease of 9 years duration, had a UPRS score of 59 while on 1000 mg of L-dopa plus carbidopa, presumably per day. After transplantation of foetal substantia nigra by an approach similar to that of Madrazo et al. (1987), this patient's UPRS score improved to 45. The second patient was a 35-year-old woman with a 5-year history of Parkinson's disease and a UPRS score of 71 while on 750 mg L-dopa plus carbidopa. After surgery and transplantation of foetal adrenal medullary tissue, this patient's UPRS score improved to 35. (Separating medulla from cortex of a 13-week foetal adrenal rudiment is a remarkable feat.) These two patients have been maintained on oral cyclosporin and prednisone, although in our opinion the survival of foetal neuronal grafts in unrelated monkeys renders immunosuppression an unjustified risk to the patients. A second unsatisfactory aspect of the procedure described by Madrazo et al. (1988) in their brief letter was the use of tissue from a spontaneously aborted foetus. While this may, to some, circumvent perceived ethical problems, it introduces a serious biological problems: spontaneously aborted foetuses have generally been dead for some time, and are unlikely to contain viable neurons. (The particular spontaneous abortion described by Madrazo et al. apparently was due to cervicouterine incompetence, in which case an otherwise normal foetus may have been ejected. Criteria used to establish foetal death were not discussed in their letter.) Appropriate regions of the foetal brain, containing viable neurons, can with practice often be identified among the foetal fragments collected during abortion by suction curettage (Brundin et al. 1986). We believe that, when the abortion is performed for reasons entirely unrelated to any subsequent transplantation procedure, these foetal fragments provide an ethically acceptable and biologically superior source of tissue for grafting (Fine, 1988).

E. Conclusions

It appears that the motor impairments of Parkinson's disease patients can be improved by adrenal or foetal neuronal (ventral mesencephalic) transplants, but the effectiveness of these procedures has not yet been conclusively demonstrated. The improvements obtained to date do not match the impressive results achieved in animal models of the disorder. The adrenalectomy and open neurosurgical procedures most widely used are associated with high risk of severe complications in older, more profoundly afflicted patients. Direct comparisons of adrenal and foetal neuronal grafts in parkinsonian primates are under way, and preliminary results indicate the superiority of foetal neuronal grafts (K. Bankiewicz, NIH, 1988 personal communication). Better results may therefore be anticipated with the advent of foetal neuronal transplantation, provided that social objections to this procedure can be overcome. Crucial issues, including identification of the optimal donor age, methods of tissue collection and implantation, graft placement, and role of trophic factors, still need to be resolved. For this purpose, considerable additional animal experimentation, particularly in primates, will be essential. However, it is our conclusion that use of these experimental techniques on appropriately selected patients, in the context of carefully designed clinical trials, may now be justified.

References

Azmitia EC, Björklund A (1987) Cell and tissue transplantation into the adult brain. Ann NY Acad Sci 495:1-813

Backlund E-O, Granberg P-O, Hamberger B, Knutsson E, Martensson A, Sedvall G, Seiger Å, Olsen L (1985) Transplantation of adrenal medullary tissue to striatum in parkinsonism. First clinical trials. J Neurosurg 62:169-173

Bakay RAE, Barrow DL, Fiandaca MS, Iuvone PM, Schiff A, Collins DC (1987) Biochemical and behavioural correction of MPTP Parkinson-like syndrome by fetal cell transplantation. Ann NY Acad Sci 495:623-638

Bernheimer H, Birkmayer W, Hornykiewicz O, Jellinger K, Seitelberger F (1973) Brain dopamine and the symptoms of Parkinson and Huntington. J Neurol Sci 20:415-455

Björklund A, Stenevi U (1979) Reconstruction of the nigrostriatal dopamine pathway by intracerebral nigral transplants. Brain Res 177:555-560

Björklund A, Dunnett SB, Stenevi U, Lewis ME, Iversen SD (1980) Reinnervation of the denervated striatum by substantia nigra transplants: functional consequences as revealed by pharmacological and sensorimotor testing. Brain Res 199:307-333.

Björklund A, Stenevi U, Schmidt RH, Dunnett SB, Gage FH (1983) Intracerebral grafting of neuronal cell suspensions. I. Survival and growth of nigral cell suspensions implanted in different brain sites. Acta Physiol Scand [suppl] 522:9-18.

Björklund A, Stenevi U (1985) Neural grafting in the mammalian CNS. Elsevier, Amsterdam

Björklund A, Lindvall O, Isacson O, Brundin P, Wictorin K, Strecker RE, Clarke DJ, Dunnett SJ (1987) Mechanisms of action of intracerebral neural implants: studies on nigral and striatal grafts to the lesioned striatum. Trends Neurosci 10:509-516

Bohn MC, Cupit L, Marciano F, Gash DM (1987) Adrenal medulla grafts enhance recovery of striatal dopaminergic fibers. Science 237:913-916

Bolam JP, Freund TF, Björklund A, Dunnett SB, Smith AD (1987) Synaptic input and local output of dopaminergic neurons in grafts that functionally reinnervate the host striatum. Exp Brain Res 68:131-146

Brodal A (1981) Neurological anatomy, 3rd edn. Oxford University Press, Oxford

Brundin P, Nilsson OG, Gage FH, Prochiantz A, Björklund A (1985) Cyclosporin A increases survival of cross-species intrastriatal grafts of embryonic dopamine-containing neurons. Exp Brain Res 60:204-208

Brundin P, Nilsson OG, Strecker RE, Lindvall O, Astedt B, Björklund A (1986) Behavioural effects of human fetal dopamine neurons grafted in a rat model of Parkinson's disease. Exp Brain Res 65:235-240

Brundin P, Strecker RE, Loudos E, Björklund A (1987) Dopamine neurons grafted unilaterally to the nucleus accumbens affect drug-induced circling and locomotion. Exp Brain Res 69:183-194

Burns RS, Chieueh CC, Markey S, Ebert MH, Jacobowitz D, Kopin IJ (1983) A primate model of Parkinson's disease: selective destruction of substantia nigra, pars compacta dopaminergic neurons by *N*-methyl-4-phenyl-1,2,3,6-tetrahydropyridine. Proc Nat Acad Sci USA 80:4546-4550

Cheramy A, Leviel V, Glowinski J (1981) Dendritic release of dopamine in the substantia nigra. Nature 289:537-542

Dubach M, Schmidt R, Kunkel D, Bowden DM, Martin R, German DC (1987) Primate neostriatal neurons containing tyrosine hydroxylase: immunohistochemical evidence. Neurosci Lett 75:205-210

Dunnett SB, Björklund A, Stenevi U, Iversen SD (1981) Grafts of embryonic substantia nigra reinnervating the ventrolateral striatum ameliorate sensorimotor impairments and akinesia in rats with 6-OHDA lesions of the nigrostriatal pathway. Brain Res 229:209-217

Dunnett SB, Björklund A, Schmidt RH, Stenevi U, Iversen SD (1983) Intracerebral grafting of neuronal cell suspensions IV. Behavioural recovery in rats with unilateral 6-OHDA lesions following implantation of nigral cell suspensions in different forebrain sites. Acta Physiol Scand [Suppl] 522:29-37

Fine A (1986) Transplantation in the central nervous system. Sci Am 255:52-58B

Fine A, Reynolds GP, Nakijima N, Jenner P, Marsden CD (1985) Acute administration of 1-methyl-4-phenyl-1,2,3,6-tetrahydropyridine affects the adrenal glands as well as the brain in the marmoset. Neurosci Lett 58:123-126

Fine A, Oertel WH, Hunt SP, Nomoto M, Chong PN, Jenner P, Ryatt J, Bond A, Marsden CD, Annett L, Dunnett SB, Temlett J (1988) Transplantation of embryonic marmoset dopaminergic neurons to the corpus striatum of MPTP-induced parkinsonian monkeys. Prog Brain Res (in press)

Fine A (1988) The ethics of fetal tissue transplantation. Hastings Centre Report 18:5-8

Freed WJ, Morihisa JM, Spoor E, Hoffer BJ, Olsen L, Seiger Å, Wyatt RJ (1981) Transplanted adrenal chromaffin cells in rat brain reduce lesioninduced rotational behaviour. Nature 292:351-352

Freed WJ, Hoffer BJ, Olsen L, Wyatt RJ (1984) Transplantation of catecholamine-containing tissues to restore the functional capacity of the damaged nigrostriatal system. In: Sladek JR, Gash DM (eds) Neural transplants. Plenum, New York, pp 373-406

Freund TF, Bolam JP, Björklund A, Stenevi U, Dunnett SB, Powell JF, Smith AD (1985) Efferent synaptic connections of grafted dopaminergic neurons reinnervating the host neostriatum: a tyrosine hydroxylase immunocytochemical study. J Neurosci 5:603-616

Gauchy C, Kemel ML, Romo R, Glowiniski J, Besson MJ (1987) The role of dopamine released from distal and proximal dendrites of nigrostriatal dopaminergic neurons in the control of GABA transmission in the thalamic nucleus ventralis medialis in the cat. Neuroscience 22:935-946

Goldman PS, Nauta WJH (1977) An intricately patterned prefrontal projection in the rhesus monkey. J Comp Neurol 171:369-386

Heikkila RE, Cabbat FS, Manzino L, Duvoisin RC (1984) Effects of 1-methyl-4-phenyl-1,2,5,6-tetrahydropyridine on neostriatal dopamine in mice. Neuropharmocology 23:711-713

Jaeger CB (1985) Cytoarchitectonics of substantia nigra grafts: a light and electron microscopic study of immunocytochemically identified dopaminergic neurons and fibrous astrocytes. J Comp Neurol 231:121-135

Kamo H, Kim SU, McGeer PL, Shin DH (1986) Functional recovery in a rat model of Parkinson's disease following transplantation of cultured sympathetic neurons. Brain Res 397:372-376.

Langston JW, Ballard P, Tetrud JW, Irwin I (1983) Chronic parkinsonism in humans due to a product of meperidine-analog synthesis. Science 219:979-980

Lindvall O, Backlund E-O, Farde L, Sedvall G, Freedman R, Hoffer B, Nobin A, Seiger A, Olson L (1987) Transplantation in Parkinson's disease: two cases of adrenal medullary grafts to the putamen. Ann. Neurol. 22:457-468

Low WC, Daniloff JK, Bodony RP, Wells J (1985) Cross-species transplants of cholinergic neurons and the recovery of function. In: Neural grafting in the mammalian CNS, Björklund A, Stenevi U (eds) Elsevier, Amsterdam, pp 575-584

Madrazo I, Drucker-Colin R, Diaz V, Martinez-Mata J, Torres C, Becerril JJ (1987) Open microsurgical autograft of adrenal medulla to the right caudate nucleus in two patients with intractable Parkinson's disease. N Engl J Med 316:831-834

Madrazo I, Leon V, Torres C, del Carman Aguilera M, Varela G, Alvarez F, Fraga A, Drucker-Colin R, Drucker-Colin R, Ostrrosky F, Skurovich M, Franco R (1988) Transplantation of fetal substantia nigra and adrenal medulla to the caudate nucleus in two patients with Parkinson's disease. N Eng J Med 318:51

McGeer PL, McGeer EG, Suzuki JS (1977) Aging and extrapyramidal function. Arch Neurol 34:33-35

Mitchell IJ, Cross AJ, Sambrook MA, Crossman AR (1986) Neural mechanism mediating MPTP-induced parkinsonism in the monkey: relative contributions of the striatopallidal and striatonigral pathways as suggested by 2-deoxyglucose uptake. Neurosci Lett 63:61-65

Morihisha JM, Nakamura RK, Freed WJ, Mishkin M, Wyatt RJ (1987) Transplantation techniques and the survival of adrenal medulla autografts in the primate brain. Ann NY Acad Sci 495:599-605

Nadaud D, Herman HP, Simon H, Le Moel M (1984) Functional recovery following transplantation of ventral mesencephalic cells in rats subjected to 6-OHDA lesion of the mesolimbic dopaminergic neurons. Brain Res 304:(1984)137-141

Olsen L, Backlund E-O, Freed W, Herrera-Marschitz M, Hoffer B, Seiger Å, Stromberg I (1985) Transplantation of monoamine-producing cell systems in oculo and intracranially: experiments in search of a treatment for Parkinson's disease. Ann NY Acad Sci 457:105-126

Perlow MJ, Freed WJ, Hoffer BJ, Seiger Å, Olsen L, Wyatt RJ (1979) Brain grafts reduce motor abnormalities produced by destruction of the nigrostriatal dopamine system. Science 204:643-647

Redmond DE, Roth RH, Elsworth JD, Sladek JR Jr, Collier TJ, Deutch AY, Haber S (1986) Fetal neural grafts in monkeys given methylphenyltetrahydropyridine, Lancet I:1125-1127

Riederer P, Wuketich St (1976) Time course of nigrostriatal degeneration in Parkinson's disease. J Neural Transm 38:277-301

Robertson GS, Robertson HA (1987a) D_1 and D_2 dopamine agonist synergism: separate sites of action. Trends Pharmacol Sci 8:295-299

Robertson HA, Robertson GS (1987b) Combined L-dopa and bromocriptine therapy for Parkinson's disease: a proposed mechanism of action. Clin Neuropharmacol 10:384-387

Schmidt RA, Björklund A, Stenevi U (1981) Intracerebral grafting of dissociated CNS tissue suspensions: a new approach for neuronal transplantations to deep brain sites. Brain Res 218:347-356

Seleman LD, Goldman-Rakic PS (1985) Longitudinal topography and interdigitation of corticostriatal projections in the rhesus monkey. J Neurosci 5:776-794

Sladek JR Jr, Collier TJ, Haber SN, Deutch AY, Elsworth JD, Roth RH, Redmond DE Jr (1987) Reversal of parkinsonism by fetal nerve cell transplants in primate brain. Ann NY Acad Sci 495:641-657

Spencer SE, Wooten GF (1984) Pharmacological effects of L-dopa are not closely linked temporally to striatal dopamine concentration. Neurology 34:1609-1611

Strömberg I, Johnson S, Hoffer B, Olson L (1985) Reinnervation of dopamine-denervated striatum by substantia nigra transplants: immunohistochemical and electrophysiological correlates. Neuroscience 14:981-990

Strömberg I, Bygdeman M, Goldstein M, Seiger Å, Olsen L (1986) Human fetal substantia nigra grafted to the dopamine-denervated striatum of immunosuppressed rats: evidence for functional reinnervation. Neurosci Lett 71:271-276

Tischler AS, Greene LA (1980) Phenotypic plasticity of pheochromocytoma and normal adrenal medullary cells. Adv Biochem Pharmacol 25:61-68

Trugman JM, Wooten GF (1987) Selective D_1 and D_2 dopamine agonists differentially alter basal ganglia glucose utilization in rats with unilateral 6-hydroxydopamine substantia nigra lesions J Neurosci 7:2927-2935

Ungerstedt U (1971) Postsynaptic supersensitivity after 6-hydroxy-dopamine induced degeneration of the nigro-straital dopamine system. Acta Physiol Scand [Suppl] 367:69-93

Unsicker K, Rieffert B, Ziegler W (1980) Effects of cell culture conditions, nerve growth factor dexamethasone and cyclic AMP on adrenal chromaffin cells in vitro. Adv Biochem Pharmacol 25:51-59

Unsicker K, Tschechne B, Tschechne D (1981) Differentiation and transdifferentiation of adrenal chromaffin cells of the guinea pig. 1. Transplants to the anterior chamber of the eye. Cell Tiss Res 215:341-361.

Subject Index

MIX
Papier aus verantwortungsvollen Quellen
Paper from responsible sources
FSC® C105338

If you have any concerns about our products,
you can contact us on
ProductSafety@springernature.com

In case Publisher is established outside the EU,
the EU authorized representative is:
Springer Nature Customer Service Center GmbH
Europaplatz 3, 69115 Heidelberg, Germany

Printed by Libri Plureos GmbH
in Hamburg, Germany